Student Solutions Manual

College Algebra

TWELFTH EDITION

R. David Gustafson
Rock Valley College

Jeff Hughes
Hinds Community College

Prepared by

Michael G. Welden
Mt. San Jacinto College

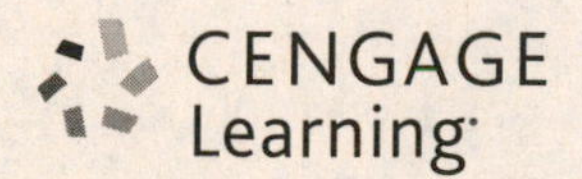

Australia • Brazil • Mexico • Singapore • United Kingdom • United States

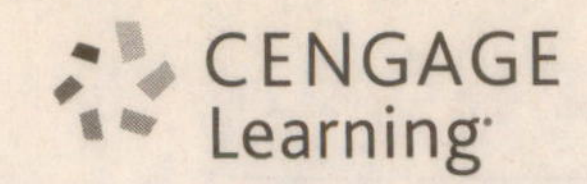

ISBN: 978-1-305-87874-7

Cengage Learning
20 Channel Center Street
Boston, MA 02210
USA

Cengage Learning is a leading provider of customized learning solutions with office locations around the globe, including Singapore, the United Kingdom, Australia, Mexico, Brazil, and Japan. Locate your local office at: **www.cengage.com/global**.

Cengage Learning products are represented in Canada by Nelson Education, Ltd.

To learn more about Cengage Learning Solutions, visit **www.cengage.com**.

Purchase any of our products at your local college store or at our preferred online store **www.cengagebrain.com**.

Printed in the United States of America
Print Number: 01 Print Year: 2015

Preface

This manual contains detailed solutions to all of the odd exercises of the text *College Algebra*, twelfth edition, by R. David Gustafson and Jeff Hughes. It also contains solutions to all chapter review, chapter test, and cumulative review exercises found in the text.

Many of the exercises in the text may be solved using more than one method, but it is not feasible to list all possible solutions in this manual. Also, some of the exercises may have been solved in this manual using a method that differs slightly from that presented in the text. There are a few exercises in the text whose solutions may vary from person to person. Some of these solutions may not have been included in this manual. For the solution to an exercise like this, the notation "answers may vary" has been included.

Please remember that only reading a solution does not teach you how to solve a problem. To repeat a commonly used phrase, mathematics is not a spectator sport. You MUST make an honest attempt to solve each exercise in the text without using this manual first. This manual should be viewed more or less as a last resort. Above all, DO NOT simply copy the solution from this manual onto your own paper. Doing so will not help you learn how to do the exercise, nor will it help you to do better on quizzes or tests.

I would like to thank Paul McCombs from Rock Valley College and Samantha Lugtu of Cengage Learning for their help and support. This solutions manual was prepared using EXP 5.1.

This book is dedicated to John, who helps me to realize that mathematics cannot describe everything in life.

May your study of this material be successful and rewarding.

Michael G. Welden

Contents

Exercises 0.1 (page 13)

1. set **3.** union **5.** decimal **7.** 2

9. composite **11.** decimals **13.** negative **15.** $x + (y + z)$

17. $5m + 5 \cdot 2$ **19.** interval **21.** two **23.** positive

25. Every natural number is a whole number, so **N** $\subset$ **W**. **TRUE**

27. The rational number $\frac{1}{2}$ is **not** a natural number, so **Q** $\not\subset$ **N**. **FALSE**

29. Every whole number is an integer, so **W** $\subset$ **Z**. **TRUE**

31. $A \cup B =$ {a, b, c, d, e, f, g}

33. $A \cap C =$ {a, c, e}

35. $\frac{9}{16} = 0.5625$; terminates

37. $\frac{3}{11} = 0.272727...$; repeats

39. natural: 1, 2, 6, 7

41. integers: $-5, -4, 0, 1, 2, 6, 7$

43. irrational: $\sqrt{2}$

45. composite: 6

47. odd: $-5, 1, 7$

49. 2 3 4

51. 11 13 17 19

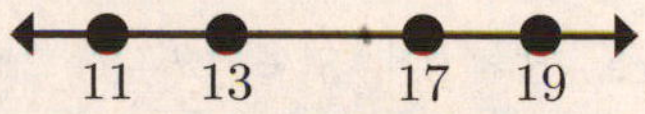

53. -4 -3 -2 -1

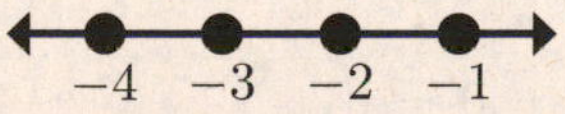

55. -5 -3 -1 1 3

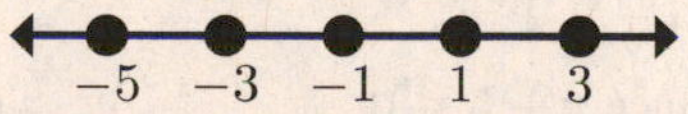

57. $x > 2 \rightarrow (2, \infty)$

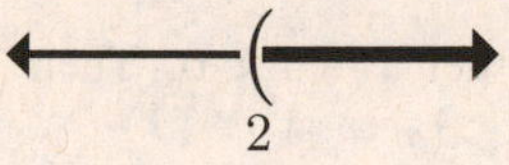

2

59. $0 < x < 5 \rightarrow (0, 5)$

0 5

61. $x > -4 \rightarrow (-4, \infty)$

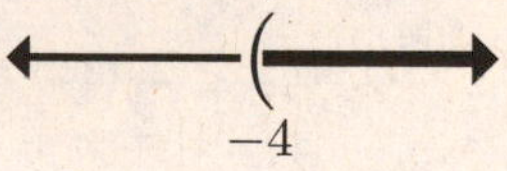

-4

63. $-2 \le x < 2 \rightarrow [-2, 2)$

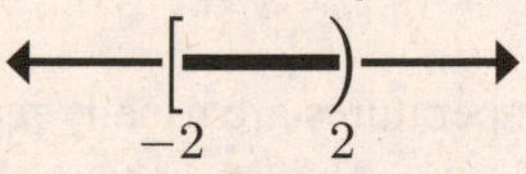

-2 2

65. $x \le 5 \rightarrow (-\infty, 5]$

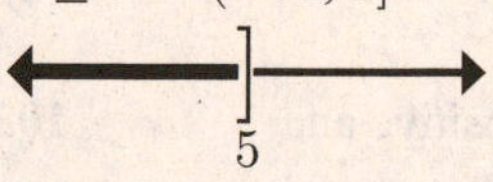

5

67. $-5 < x \le 0 \rightarrow (-5, 0]$

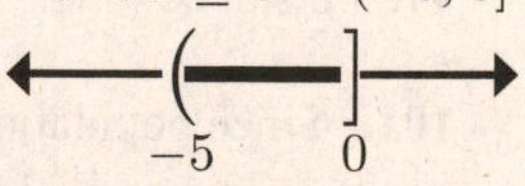

-5 0

69. $-2 \le x \le 3 \rightarrow [-2, 3]$

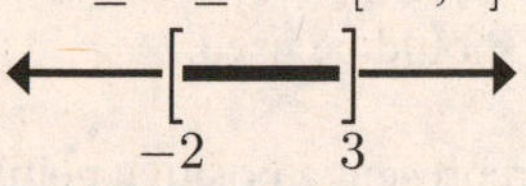

-2 3

71. $6 \ge x \ge 2 \rightarrow 2 \le x \le 6 \rightarrow [2, 6]$

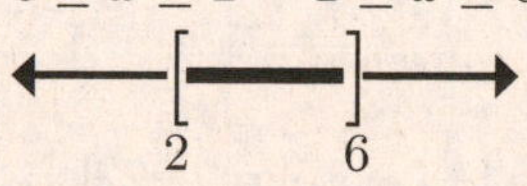

2 6

73. $x > -5$ and $x < 4 \rightarrow (-5, \infty) \cap (-\infty, 4)$

$(-5, \infty)$ 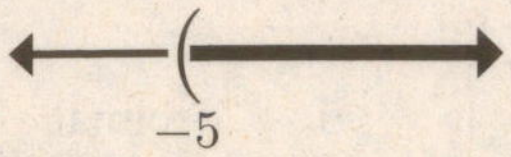

$(-\infty, 4)$ 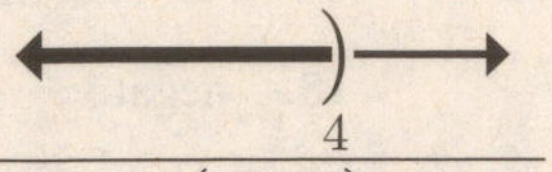

$(-5, \infty) \cap (-\infty, 4)$ 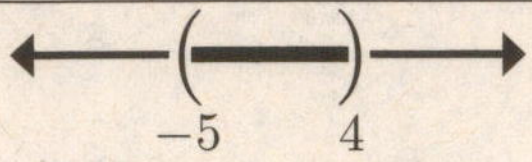

75. $x \geq -8$ and $x \leq -3 \rightarrow [-8, \infty) \cap (-\infty, -3]$

$[-8, \infty)$ 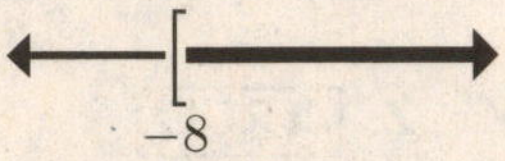

$(-\infty, -3]$ 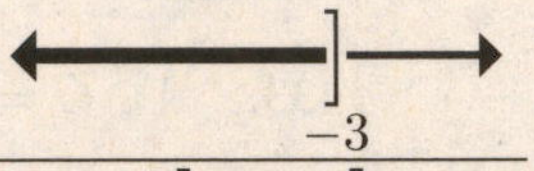

$[-8, \infty) \cap (-\infty, -3]$ 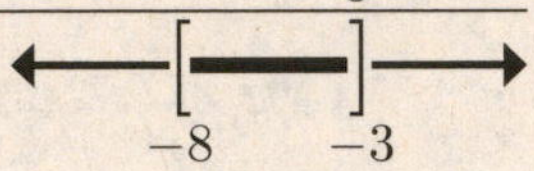

77. $x < -2$ or $x > 2 \rightarrow (-\infty, -2) \cup (2, \infty)$

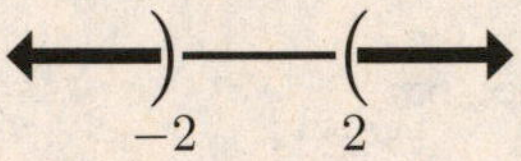

79. $x \leq -1$ or $x \geq 3 \rightarrow (-\infty, -1] \cup [3, \infty)$

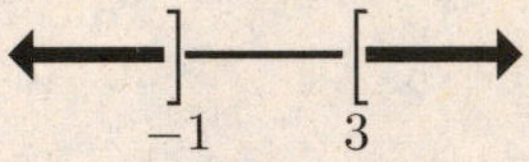

81. Since $13 \geq 0$, $|13| = 13$.

83. Since $0 \geq 0$, $|0| = 0$.

85. Since $-8 < 0$, $|-8| = -(-8) = 8$.
$-|-8| = -(8) = -8$

87. Since $32 \geq 0$, $|32| = 32$.
$-|32| = -(32) = -32$

89. Since $\pi - 5 < 0$,
$|\pi - 5| = -(\pi - 5) = -\pi + 5 = 5 - \pi$.

91. $|\pi - \pi| = |0| = 0$

93. If $x \geq 2$, then $x + 1 \geq 0$. Then
$|x + 1| = x + 1$.

95. If $x < 0$, then $x - 4 < 0$. Then
$|x - 4| = -(x - 4)$.

97. distance $= |8 - 3| = |5| = 5$

99. distance $= |-3 - (-8)| = |5| = 5$

101. Since population must be positive and never has a fractional part, the set of **natural numbers** should be used.

103. Since temperatures are usually reported without fractional parts and may be either positive or negative (or zero), the set of **integers** should be used.

105. change $= |-7 - 15| = |-22| = 22$
The change is 22° F.

107. $-x$ will represent a positive number if x itself is negative. For instance, if $x = -3$, then $-x = -(-3) = 3$, which is a positive number.

109. The statement is always true.

111. The statement is not always true. (For example, let $a = 5$ and $b = -2$.)

113. False. There are 5 integers: $-2, -1, 0, 1,$ and 2.

115. False. ∞ is not a number at all.

117. True. (You cannot find an element in the 1st set that is not in the 2nd set.)

119. False. There are eight subsets.

Exercises 0.2 (page 25)

1. factor

3. $3, 2x$

5. scientific, integer

7. $x^m x^n = x^{m+n}$

9. $(xy)^n = x^n y^n$

11. $x^0 = 1$

13. $13^2 = 13 \cdot 13 = 169$

15. $-5^2 = -1 \cdot 5 \cdot 5 = -25$

17. $4x^3 = 4 \cdot x \cdot x \cdot x$

19. $(-5x)^4 = (-5x)(-5x)(-5x)(-5x)$

21. $-8x^4 = -8 \cdot x \cdot x \cdot x \cdot x$

23. $7xxx = 7x^3$

25. $(-x)(-x) = (-1)(-1)x^2 = x^2$

27. $(3t)(3t)(-3t) = (3)(3)(-3)t^3 = -27t^3$

29. $xxxyy = x^3y^2$

31. $2.2^3 = 10.648$

33. $-0.5^4 = -0.0625$

35. $x^2x^3 = x^{2+3} = x^5$

37. $(z^2)^3 = z^{2 \cdot 3} = z^6$

39. $(y^5y^2)^3 = (y^7)^3 = y^{21}$

41. $(z^2)^3(z^4)^5 = z^6z^{20} = z^{26}$

43. $(a^2)^3(a^4)^2 = a^6a^8 = a^{14}$

45. $(3x)^3 = 3^3x^3 = 27x^3$

47. $(x^2y)^3 = (x^2)^3y^3 = x^6y^3$

49. $\left(\frac{a^2}{b}\right)^3 = \frac{(a^2)^3}{b^3} = \frac{a^6}{b^3}$

51. $(-x)^0 = 1$

53. $(4x)^0 = 1$

55. $z^{-4} = \frac{1}{z^4}$

57. $y^{-2}y^{-3} = y^{-5} = \frac{1}{y^5}$

59. $(x^3x^{-4})^{-2} = (x^{-1})^{-2} = x^2$

61. $\frac{x^7}{x^3} = x^{7-3} = x^4$

63. $\frac{a^{21}}{a^{17}} = a^{21-17} = a^4$

65. $\frac{(x^2)^2}{x^2x} = \frac{x^4}{x^3} = x^{4-3} = x^1 = x$

67. $\left(\frac{m^3}{n^2}\right)^3 = \frac{(m^3)^3}{(n^2)^3} = \frac{m^9}{n^6}$

69. $\frac{(a^3)^{-2}}{aa^2} = \frac{a^{-6}}{a^3} = a^{-6-3} = a^{-9} = \frac{1}{a^9}$

71. $\left(\frac{a^{-3}}{b^{-1}}\right)^{-4} = \frac{(a^{-3})^{-4}}{(b^{-1})^{-4}} = \frac{a^{12}}{b^4}$

EXERCISES 0.2

73. $\left(\dfrac{r^4 r^{-6}}{r^3 r^{-3}}\right)^2 = \left(\dfrac{r^{-2}}{r^0}\right)^2 = \left(r^{-2}\right)^2 = r^{-4}$

$= \dfrac{1}{r^4}$

75. $\left(\dfrac{x^5 y^{-2}}{x^{-3} y^2}\right)^4 = \left(\dfrac{x^5 x^3}{y^2 y^2}\right)^4 = \left(\dfrac{x^8}{y^4}\right)^4 = \dfrac{x^{32}}{y^{16}}$

77. $\left(\dfrac{5x^{-3}y^{-2}}{3x^2y^{-3}}\right)^{-2} = \left(\dfrac{3x^2y^{-3}}{5x^{-3}y^{-2}}\right)^2 = \left(\dfrac{3x^2x^3y^2}{5y^3}\right)^2 = \left(\dfrac{3x^5}{5y}\right)^2 = \dfrac{9x^{10}}{25y^2}$

79. $\left(\dfrac{3x^5y^{-3}}{6x^{-5}y^3}\right)^{-2} = \left(\dfrac{6x^{-5}y^3}{3x^5y^{-3}}\right)^2 = \left(\dfrac{2y^3y^3}{1x^5x^5}\right)^2 = \left(\dfrac{2y^6}{x^{10}}\right)^2 = \dfrac{4y^{12}}{x^{20}}$

81. $\dfrac{\left(8^{-2}z^{-3}y\right)^{-1}}{\left(5y^2z^{-2}\right)^3\left(5yz^{-2}\right)^{-1}} = \dfrac{8^2z^3y^{-1}}{5^3y^6z^{-6}\cdot 5^{-1}y^{-1}z^2} = \dfrac{64z^3y^{-1}}{5^2y^5z^{-4}} = \dfrac{64z^3z^4}{25y^5y^1} = \dfrac{64z^7}{25y^6}$

83. $-\dfrac{5[6^2+(9-5)]}{4(2-3)^2} = -\dfrac{5[36+4]}{4(-1)^2} = -\dfrac{5[40]}{4(1)} = -\dfrac{200}{4} = -50$

85. $x^2 = (-2)^2 = 4$

87. $x^3 = (-2)^3 = -8$

89. $(-xz)^3 = [-1\cdot(-2)\cdot 3]^3 = 6^3 = 216$

91. $\dfrac{-\left(x^2z^3\right)}{z^2-y^2} = \dfrac{-\left[(-2)^2\cdot 3^3\right]}{3^2-0^2} = \dfrac{-[4\cdot 27]}{9-0} = \dfrac{-108}{9} = -12$

93. $5x^2 - 3y^3z = 5(-2)^2 - 3(0)^3(3) = 5(4) - 3(0)(3) = 20 - 0 = 20$

95. $\dfrac{-3x^{-3}z^{-2}}{6x^2z^{-3}} = \dfrac{-1z^3}{2x^2x^3z^2} = \dfrac{-z}{2x^5} = \dfrac{-3}{2(-2)^5} = \dfrac{-3}{2(-32)} = \dfrac{-3}{-64} = \dfrac{3}{64}$

97. $372{,}000 = 3.72\times 10^5$

99. $-177{,}000{,}000 = -1.77\times 10^8$

101. $0.007 = 7\times 10^{-3}$

103. $-0.000000693 = -6.93\times 10^{-7}$

105. $1{,}000{,}000{,}000{,}000 = 1\times 10^{12}$

107. $9.37\times 10^5 = 937{,}000$

109. $2.21\times 10^{-5} = 0.0000221$

111. $0.00032\times 10^4 = 3.2$

113. $-3.2\times 10^{-3} = -0.0032$

115. $\dfrac{(65{,}000)(45{,}000)}{250{,}000} = \dfrac{\left(6.5\times 10^4\right)\left(4.5\times 10^4\right)}{2.5\times 10^5} = \dfrac{(6.5)(4.5)}{2.5}\times 10^{4+4-5} = 11.7\times 10^3$

$= 1.17\times 10^1\times 10^3$

$= 1.17\times 10^4$

EXERCISES 0.2

117. $\dfrac{(0.00000035)(170{,}000)}{0.00000085} = \dfrac{(3.5 \times 10^{-7})(1.7 \times 10^{5})}{8.5 \times 10^{-7}} = \dfrac{(3.5)(1.7)}{8.5} \times 10^{(-7)+5-(-7)}$
$= 0.7 \times 10^{5}$
$= 7 \times 10^{-1} \times 10^{5} = 7 \times 10^{4}$

119. $\dfrac{(45{,}000{,}000{,}000)(212{,}000)}{0.00018} = \dfrac{(4.5 \times 10^{10})(2.12 \times 10^{5})}{1.8 \times 10^{-4}} = \dfrac{(4.5)(2.12)}{1.8} \times 10^{10+5-(-4)}$
$= 5.3 \times 10^{19}$

121. 3.31×10^{4} cm/sec $= \dfrac{3.31 \times 10^{4} \text{ cm}}{1 \text{ sec}} \cdot \dfrac{1 \text{ m}}{100 \text{ cm}} \cdot \dfrac{60 \text{ sec}}{1 \text{ min}} = \dfrac{(3.31 \times 10^{4})(6 \times 10^{1})}{1 \times 10^{2}}$ m/min
$= \dfrac{(3.31)(6)}{1} \times 10^{4+1-2}$ m/min
$= 19.86 \times 10^{3}$ m/min
$= 1.986 \times 10^{4}$ m/min

123. mass $= 1{,}000{,}000{,}000(0.00000000000000000000000167248 \text{ g})$
$= (1 \times 10^{9})(1.67248 \times 10^{-24} \text{ g}) = 1.67248 \times 10^{-15}$ g

125. $10 \cdot 10 \cdot 10 \cdot 26 \cdot 26 \cdot 26 = 10^{3} \cdot 26^{3}$; $10^{3} \cdot 26^{3} = 17{,}576{,}000 = 1.7576 \times 10^{7}$

127. polar radius $= 6.356750 \times 10^{3}$ km
equatorial radius $= 6.378135 \times 10^{3}$ km

129. $x^{n}x^{2} = x^{n+2}$

131. $\dfrac{x^{m}x^{2}}{x^{3}} = \dfrac{x^{m+2}}{x^{3}} = x^{m+2-3} = x^{m-1}$

133. $x^{m+1}x^{3} = x^{m+1+3} = x^{m+4}$

135. In the expression $-x^{4}$, the base of the exponent is x, while in the expression $(-x)^{4}$, the base of the exponent is $-x$.

137. **Answers will vary.**

139. $x^{11} \cdot x^{11} = x^{11+11} = x^{22}$

141. $\dfrac{y^{50}}{y^{10}} = y^{50-10} = y^{40}$

143. False. 0^{0} is undefined.

145. False. $x^{-n} = \dfrac{1}{x^{n}}$

147. True. $\left(2^{-1} = \frac{1}{2}, 2^{-2} = \frac{1}{4}\right)$

149. $110 \times 365 \times 31{,}500{,}000 = 1.1 \times 10^{2} \times 3.65 \times 10^{2} \times 3.15 \times 10^{7}$
$= 12.64725 \times 10^{11}$
$= 1.264725 \times 10^{1} \times 10^{11} = 1.264725 \times 10^{12}$

Exercises 0.3 (page 39)

1. 0

3. not

5. $a^{1/n}$

7. $\sqrt[n]{ab}$

9. $\neq$

11. $9^{1/2} = (3^2)^{1/2} = 3$

13. $\left(\frac{1}{25}\right)^{1/2} = \left[\left(\frac{1}{5}\right)^2\right]^{1/2} = \frac{1}{5}$

15. $-81^{1/4} = -(3^4)^{1/4} = -3$

17. $(10{,}000)^{1/4} = (10^4)^{1/4} = 10$

19. $\left(-\frac{27}{8}\right)^{1/3} = \left[\left(-\frac{3}{2}\right)^3\right]^{1/3} = -\frac{3}{2}$

21. $(-64)^{1/2} \Rightarrow$ not a real number

23. $(16a^2)^{1/2} = \left[(4a)^2\right]^{1/2} = 4|a|$

25. $(16a^4)^{1/4} = \left[(2a)^4\right]^{1/4} = 2|a|$

27. $(-32a^5)^{1/5} = \left[(-2a)^5\right]^{1/5} = -2a$

29. $(-216b^6)^{1/3} = \left[(-6b^2)^3\right]^{1/3} = -6b^2$

31. $\left(\frac{16a^4}{25b^2}\right)^{1/2} = \left[\left(\frac{4a^2}{5b}\right)^2\right]^{1/2} = \left|\frac{4a^2}{5b}\right| = \frac{4a^2}{5|b|}$

33. $\left(-\frac{1000x^6}{27y^3}\right)^{1/3} = \left[\left(-\frac{10x^2}{3y}\right)^3\right]^{1/3}$
$= -\frac{10x^2}{3y}$

35. $4^{3/2} = (4^{1/2})^3 = 2^3 = 8$

37. $-16^{3/2} = -(16^{1/2})^3 = -(4)^3 = -64$

39. $-1000^{2/3} = -(1000^{1/3})^2 = -(10)^2$
$= -100$

41. $64^{-1/2} = \frac{1}{64^{1/2}} = \frac{1}{8}$

43. $64^{-3/2} = \frac{1}{64^{3/2}} = \frac{1}{(64^{1/2})^3} = \frac{1}{8^3} = \frac{1}{512}$

45. $-9^{-3/2} = -\frac{1}{9^{3/2}} = -\frac{1}{(9^{1/2})^3} = -\frac{1}{3^3}$
$= -\frac{1}{27}$

47. $\left(\frac{4}{9}\right)^{5/2} = \left[\left(\frac{4}{9}\right)^{1/2}\right]^5 = \left(\frac{2}{3}\right)^5 = \frac{32}{243}$

49. $\left(-\frac{27}{64}\right)^{-2/3} = \left(-\frac{64}{27}\right)^{2/3} = \left[\left(-\frac{64}{27}\right)^{1/3}\right]^2 = \left(-\frac{4}{3}\right)^2 = \frac{16}{9}$

51. $(100s^4)^{1/2} = 100^{1/2}(s^4)^{1/2} = 10s^2$

EXERCISES 0.3

53. $\left(32y^{10}z^5\right)^{-1/5} = \dfrac{1}{\left(32y^{10}z^5\right)^{1/5}} = \dfrac{1}{32^{1/5}\left(y^{10}\right)^{1/5}\left(z^5\right)^{1/5}} = \dfrac{1}{2y^2z}$

55. $\left(x^{10}y^5\right)^{3/5} = x^{30/5}y^{15/5} = x^6y^3$

57. $\left(r^8s^{16}\right)^{-3/4} = r^{-24/4}s^{-48/4} = r^{-6}s^{-12} = \dfrac{1}{r^6s^{12}}$

59. $\left(-\dfrac{8a^6}{125b^9}\right)^{2/3} = \dfrac{(-8)^{2/3}a^{12/3}}{125^{2/3}b^{18/3}} = \dfrac{(-2)^2a^4}{5^2b^6} = \dfrac{4a^4}{25b^6}$

61. $\left(\dfrac{27r^6}{1000s^{12}}\right)^{-2/3} = \left(\dfrac{1000s^{12}}{27r^6}\right)^{2/3} = \dfrac{1000^{2/3}s^{24/3}}{27^{2/3}r^{12/3}} = \dfrac{10^2s^8}{3^2r^4} = \dfrac{100s^8}{9r^4}$

63. $\dfrac{a^{2/5}a^{4/5}}{a^{1/5}} = \dfrac{a^{6/5}}{a^{1/5}} = a^{5/5} = a$

65. $\sqrt{49} = \sqrt{7^2} = 7$

67. $\sqrt[3]{125} = \sqrt[3]{5^3} = 5$

69. $\sqrt[3]{-125} = \sqrt[3]{(-5)^3} = -5$

71. $\sqrt[5]{-\dfrac{32}{100{,}000}} = \sqrt[5]{\left(-\dfrac{2}{10}\right)^5} = -\dfrac{2}{10} = -\dfrac{1}{5}$

73. $\sqrt{36x^2} = \sqrt{(6x)^2} = |6x| = 6|x|$

75. $\sqrt{9y^4} = \sqrt{(3y^2)^2} = |3y^2| = 3y^2$

77. $\sqrt[3]{8y^3} = \sqrt[3]{(2y)^3} = 2y$

79. $\sqrt[4]{\dfrac{x^4y^8}{z^{12}}} = \sqrt[4]{\left(\dfrac{xy^2}{z^3}\right)^4} = \left|\dfrac{xy^2}{z^3}\right| = \dfrac{|x|y^2}{|z^3|}$

81. $\sqrt{8} - \sqrt{2} = \sqrt{4}\sqrt{2} - \sqrt{2} = 2\sqrt{2} - \sqrt{2} = \sqrt{2}$

83. $\sqrt{200x^2} + \sqrt{98x^2} = \sqrt{100x^2}\sqrt{2} + \sqrt{49x^2}\sqrt{2} = 10x\sqrt{2} + 7x\sqrt{2} = 17x\sqrt{2}$

85. $2\sqrt{48y^5} - 3y\sqrt{12y^3} = 2\sqrt{16y^4}\sqrt{3y} - 3y\sqrt{4y^2}\sqrt{3y} = 2\left(4y^2\right)\sqrt{3y} - 3y(2y)\sqrt{3y}$
$= 8y^2\sqrt{3y} - 6y^2\sqrt{3y} = 2y^2\sqrt{3y}$

87. $2\sqrt[3]{81} + 3\sqrt[3]{24} = 2\sqrt[3]{27}\sqrt[3]{3} + 3\sqrt[3]{8}\sqrt[3]{3} = 2(3)\sqrt[3]{3} + 3(2)\sqrt[3]{3} = 6\sqrt[3]{3} + 6\sqrt[3]{3} = 12\sqrt[3]{3}$

89. $\sqrt[4]{768z^5} + \sqrt[4]{48z^5} = \sqrt[4]{256z^4}\sqrt[4]{3z} + \sqrt[4]{16z^4}\sqrt[4]{3z} = 4z\sqrt[4]{3z} + 2z\sqrt[4]{3z} = 6z\sqrt[4]{3z}$

91. $\sqrt{8x^2y} - x\sqrt{2y} + \sqrt{50x^2y} = \sqrt{4x^2}\sqrt{2y} - x\sqrt{2y} + \sqrt{25x^2}\sqrt{2y}$
$= 2x\sqrt{2y} - x\sqrt{2y} + 5x\sqrt{2y} = 6x\sqrt{2y}$

93. $\sqrt[3]{16xy^4} + y\sqrt[3]{2xy} - \sqrt[3]{54xy^4} = \sqrt[3]{8y^3}\sqrt[3]{2xy} + y\sqrt[3]{2xy} - \sqrt[3]{27y^3}\sqrt[3]{2xy}$
$= 2y\sqrt[3]{2xy} + y\sqrt[3]{2xy} - 3y\sqrt[3]{2xy} = 0$

EXERCISES 0.3

95. $\dfrac{3}{\sqrt{3}} = \dfrac{3}{\sqrt{3}} \cdot \dfrac{\sqrt{3}}{\sqrt{3}} = \dfrac{3\sqrt{3}}{3} = \sqrt{3}$

97. $\dfrac{2}{\sqrt{x}} = \dfrac{2}{\sqrt{x}} \cdot \dfrac{\sqrt{x}}{\sqrt{x}} = \dfrac{2\sqrt{x}}{x}$

99. $\dfrac{2}{\sqrt[3]{2}} = \dfrac{2}{\sqrt[3]{2}} \cdot \dfrac{\sqrt[3]{4}}{\sqrt[3]{4}} = \dfrac{2\sqrt[3]{4}}{\sqrt[3]{8}} = \dfrac{2\sqrt[3]{4}}{2} = \sqrt[3]{4}$

101. $\dfrac{5a}{\sqrt[3]{25a}} = \dfrac{5a}{\sqrt[3]{25a}} \cdot \dfrac{\sqrt[3]{5a^2}}{\sqrt[3]{5a^2}} = \dfrac{5a\sqrt[3]{5a^2}}{\sqrt[3]{125a^3}} = \dfrac{5a\sqrt[3]{5a^2}}{5a} = \sqrt[3]{5a^2}$

103. $\dfrac{2b}{\sqrt[4]{3a^2}} = \dfrac{2b}{\sqrt[4]{3a^2}} \cdot \dfrac{\sqrt[4]{27a^2}}{\sqrt[4]{27a^2}} = \dfrac{2b\sqrt[4]{27a^2}}{\sqrt[4]{81a^4}} = \dfrac{2b\sqrt[4]{27a^2}}{3a}$

105. $\sqrt[3]{\dfrac{2u^4}{9v}} = \dfrac{\sqrt[3]{2u^4}}{\sqrt[3]{9v}} = \dfrac{\sqrt[3]{u^3}\sqrt[3]{2u}}{\sqrt[3]{9v}} \cdot \dfrac{\sqrt[3]{3v^2}}{\sqrt[3]{3v^2}} = \dfrac{u\sqrt[3]{6uv^2}}{\sqrt[3]{27v^3}} = \dfrac{u\sqrt[3]{6uv^2}}{3v}$

107. $\dfrac{\sqrt{5}}{10} = \dfrac{\sqrt{5}}{10} \cdot \dfrac{\sqrt{5}}{\sqrt{5}} = \dfrac{5}{10\sqrt{5}} = \dfrac{1}{2\sqrt{5}}$

109. $\dfrac{\sqrt[3]{9}}{3} = \dfrac{\sqrt[3]{9}}{3} \cdot \dfrac{\sqrt[3]{3}}{\sqrt[3]{3}} = \dfrac{\sqrt[3]{27}}{3\sqrt[3]{3}} = \dfrac{3}{3\sqrt[3]{3}} = \dfrac{1}{\sqrt[3]{3}}$

111. $\dfrac{\sqrt[5]{16b^3}}{64a} = \dfrac{\sqrt[5]{16b^3}}{64a} \cdot \dfrac{\sqrt[5]{2b^2}}{\sqrt[5]{2b^2}} = \dfrac{\sqrt[5]{32b^5}}{64a\sqrt[5]{2b^2}} = \dfrac{2b}{64a\sqrt[5]{2b^2}} = \dfrac{b}{32a\sqrt[5]{2b^2}}$

113. $\sqrt{\dfrac{1}{3}} - \sqrt{\dfrac{1}{27}} = \dfrac{\sqrt{1}}{\sqrt{3}} - \dfrac{\sqrt{1}}{\sqrt{27}} = \dfrac{1}{\sqrt{3}} \cdot \dfrac{\sqrt{3}}{\sqrt{3}} - \dfrac{1}{\sqrt{27}} \cdot \dfrac{\sqrt{3}}{\sqrt{3}} = \dfrac{\sqrt{3}}{3} - \dfrac{\sqrt{3}}{\sqrt{81}} = \dfrac{\sqrt{3}}{3} - \dfrac{\sqrt{3}}{9}$

$$= \dfrac{3\sqrt{3}}{9} - \dfrac{\sqrt{3}}{9} = \dfrac{2\sqrt{3}}{9}$$

115. $\sqrt{\dfrac{x}{8}} - \sqrt{\dfrac{x}{2}} + \sqrt{\dfrac{x}{32}} = \dfrac{\sqrt{x}}{\sqrt{8}} - \dfrac{\sqrt{x}}{\sqrt{2}} + \dfrac{\sqrt{x}}{\sqrt{32}} = \dfrac{\sqrt{x}}{\sqrt{8}} \cdot \dfrac{\sqrt{2}}{\sqrt{2}} - \dfrac{\sqrt{x}}{\sqrt{2}} \cdot \dfrac{\sqrt{2}}{\sqrt{2}} + \dfrac{\sqrt{x}}{\sqrt{32}} \cdot \dfrac{\sqrt{2}}{\sqrt{2}}$

$$= \dfrac{\sqrt{2x}}{\sqrt{16}} - \dfrac{\sqrt{2x}}{\sqrt{4}} + \dfrac{\sqrt{2x}}{\sqrt{64}}$$

$$= \dfrac{\sqrt{2x}}{4} - \dfrac{\sqrt{2x}}{2} + \dfrac{\sqrt{2x}}{8}$$

$$= \dfrac{2\sqrt{2x}}{8} - \dfrac{4\sqrt{2x}}{8} + \dfrac{\sqrt{2x}}{8} = -\dfrac{\sqrt{2x}}{8}$$

117. $\sqrt[4]{9} = 9^{1/4} = \left(3^2\right)^{1/4} = 3^{2/4} = 3^{1/2} = \sqrt{3}$

119. $\sqrt[10]{16x^6} = \left(16x^6\right)^{1/10} = \left(2^4x^6\right)^{1/10} = 2^{4/10}x^{6/10} = 2^{2/5}x^{3/5} = \left(2^2x^3\right)^{1/5} = \sqrt[5]{4x^3}$

121. $s = \sqrt{120} = \sqrt{4}\sqrt{30} = 2\sqrt{30}$ inches

123. $\sqrt{2}\sqrt[3]{2} = 2^{1/2} \cdot 2^{1/3} = 2^{3/6} \cdot 2^{2/6} = \sqrt[6]{2^3}\sqrt[6]{2^2} = \sqrt[6]{8}\sqrt[6]{4} = \sqrt[6]{32}$

125. $\dfrac{\sqrt[4]{3}}{\sqrt{2}} = \dfrac{3^{1/4}}{2^{1/2}} = \dfrac{3^{1/4}}{2^{2/4}} = \dfrac{\sqrt[4]{3}}{\sqrt[4]{2^2}} = \dfrac{\sqrt[4]{3}}{\sqrt[4]{4}} = \dfrac{\sqrt[4]{3}}{\sqrt[4]{4}} \cdot \dfrac{\sqrt[4]{4}}{\sqrt[4]{4}} = \dfrac{\sqrt[4]{12}}{\sqrt[4]{16}} = \dfrac{\sqrt[4]{12}}{2}$

127. If $a^{1/n} = x$, then $x^n = a$. However, if n is even, x^n cannot be negative.

129. To rationalize a denominator means to write an equivalent fraction with a denominator equal to a rational number.

131. $\left(\dfrac{x}{y}\right)^{-m/n} = \dfrac{x^{-m/n}}{y^{-m/n}} = \dfrac{x^{-m/n}}{y^{-m/n}} \cdot \dfrac{x^{m/n}y^{m/n}}{x^{m/n}y^{m/n}} = \dfrac{y^{m/n}}{x^{m/n}} = \dfrac{(y^m)^{1/n}}{(x^m)^{1/n}} = \left(\dfrac{y^m}{x^m}\right)^{1/n} = \sqrt[n]{\dfrac{y^m}{x^m}}$

133. $(-16)^{1/4}$ is undefined. g

135. $0^{-111/19} = 0$. d

137. $\sqrt[87]{-1} = -1$. h

139. $\dfrac{1}{\sqrt[87]{x}} = \dfrac{1}{\sqrt[87]{x}} \cdot \dfrac{\sqrt[87]{x^{86}}}{\sqrt[87]{x^{86}}} = \dfrac{\sqrt[87]{x^{86}}}{x}$. e

Exercises 0.4 (page 52)

1. monomial, variables

3. trinomial

5. one

7. like

9. coefficients, variables

11. yes, trinomial, 2nd degree

13. no

15. yes, binomial, 3rd degree

17. yes, monomial, 0th degree

19. yes, monomial, no degree

21. $(x^3 - 3x^2) + (5x^3 - 8x) = x^3 - 3x^2 + 5x^3 - 8x = x^3 + 5x^3 - 3x^2 - 8x = 6x^3 - 3x^2 - 8x$

23.
$$\begin{aligned}(y^5 + 2y^3 + 7) - (y^5 - 2y^3 - 7) &= y^5 + 2y^3 + 7 - y^5 + 2y^3 + 7 \\ &= y^5 - y^5 + 2y^3 + 2y^3 + 7 + 7 = 4y^3 + 14\end{aligned}$$

25.
$$\begin{aligned}2(x^2 + 3x - 1) - 3(x^2 + 2x - 4) + 4 &= 2(x^2) + 2(3x) + 2(-1) - 3(x^2) - 3(2x) - 3(-4) + 4 \\ &= 2x^2 + 6x - 2 - 3x^2 - 6x + 12 + 4 \\ &= 2x^2 - 3x^2 + 6x - 6x - 2 + 12 + 4 = -x^2 + 14\end{aligned}$$

27.
$$\begin{aligned}&8(t^2 - 2t + 5) + 4(t^2 - 3t + 2) - 6(2t^2 - 8) \\ &\quad = 8(t^2) + 8(-2t) + 8(5) + 4(t^2) + 4(-3t) + 4(2) - 6(2t^2) - 6(-8) \\ &\quad = 8t^2 - 16t + 40 + 4t^2 - 12t + 8 - 12t^2 + 48 \\ &\quad = 8t^2 + 4t^2 - 12t^2 - 16t - 12t + 40 + 8 + 48 = -28t + 96\end{aligned}$$

29.
$$\begin{aligned}y(y^2 - 1) - y^2(y + 2) - y(2y - 2) &= y(y^2) + y(-1) - y^2(y) - y^2(2) - y(2y) - y(-2) \\ &= y^3 - y - y^3 - 2y^2 - 2y^2 + 2y \\ &= y^3 - y^3 - 2y^2 - 2y^2 - y + 2y = -4y^2 + y\end{aligned}$$

EXERCISES 0.4

31. $xy(x-4y) - y(x^2+3xy) + xy(2x+3y)$
$$= xy(x) + xy(-4y) - y(x^2) - y(3xy) + xy(2x) + xy(3y)$$
$$= x^2y - 4xy^2 - x^2y - 3xy^2 + 2x^2y + 3xy^2$$
$$= x^2y - x^2y + 2x^2y - 4xy^2 - 3xy^2 + 3xy^2 = 2x^2y - 4xy^2$$

33. $2x^2y^3(4xy^4) = 2(4)x^2xy^3y^4 = 8x^3y^7$

35. $-3m^2n(2mn^2)\left(-\frac{mn}{12}\right) = (-3)(2)\left(-\frac{1}{12}\right)m^2mmnn^2n = \frac{6}{12}m^4n^4 = \frac{m^4n^4}{2}$

37. $-4rs(r^2+s^2) = -4rs(r^2) - 4rs(s^2) = -4r^3s - 4rs^3$

39. $6ab^2c(2ac+3bc^2-4ab^2c) = 6ab^2c(2ac) + 6ab^2c(3bc^2) + 6ab^2c(-4ab^2c)$
$$= 12a^2b^2c^2 + 18ab^3c^3 - 24a^2b^4c^2$$

41. $(a+2)(a+2) = a^2 + 2a + 2a + 4$
$$= a^2 + 4a + 4$$

43. $(a-6)^2 = (a-6)(a-6)$
$$= a^2 - 6a - 6a + 36$$
$$= a^2 - 12a + 36$$

45. $(x+4)(x-4) = x^2 - 4x + 4x - 16$
$$= x^2 - 16$$

47. $(x-3)(x+5) = x^2 + 5x - 3x - 15$
$$= x^2 + 2x - 15$$

49. $(u+2)(3u-2) = 3u^2 - 2u + 6u - 4$
$$= 3u^2 + 4u - 4$$

51. $(5x-1)(2x+3) = 10x^2 + 15x - 2x - 3$
$$= 10x^2 + 13x - 3$$

53. $(3a-2b)^2 = (3a-2b)(3a-2b) = 9a^2 - 6ab - 6ab + 4b^2 = 9a^2 - 12ab + 4b^2$

55. $(3m+4n)(3m-4n) = 9m^2 - 12mn + 12mn - 16n^2 = 9m^2 - 16n^2$

57. $(2y-4x)(3y-2x) = 6y^2 - 4xy - 12xy + 8x^2 = 6y^2 - 16xy + 8x^2$

59. $(9x-y)(x^2-3y) = 9x^3 - 27xy - x^2y + 3y^2 = 9x^3 - x^2y - 27xy + 3y^2$

61. $(5z+2t)(z^2-t) = 5z^3 - 5tz + 2tz^2 - 2t^2 = 5z^3 + 2tz^2 - 5tz - 2t^2$

63. $\left(\sqrt{5}+3x\right)\left(2-\sqrt{5}x\right) = 2\sqrt{5} - 5x + 6x - 3\sqrt{5}x^2 = -3\sqrt{5}\,x^2 + x + 2\sqrt{5}$

65. $(3x-1)^3 = (3x-1)(3x-1)(3x-1)$
$$= (9x^2 - 3x - 3x + 1)(3x-1)$$
$$= (9x^2 - 6x + 1)(3x-1)$$
$$= 9x^2(3x) + 9x^2(-1) - 6x(3x) - 6x(-1) + 1(3x) + 1(-1)$$
$$= 27x^3 - 9x^2 - 18x^2 + 6x + 3x - 1 = 27x^3 - 27x^2 + 9x - 1$$

EXERCISES 0.4

67. $(3x+1)(2x^2+4x-3) = 3x(2x^2)+3x(4x)+3x(-3)+1(2x^2)+1(4x)+1(-3)$
$= 6x^3+12x^2-9x+2x^2+4x-3 = 6x^3+14x^2-5x-3$

69. $(3x+2y)(2x^2-3xy+4y^2)$
$= 3x(2x^2)+3x(-3xy)+3x(4y^2)+2y(2x^2)+2y(-3xy)+2y(4y^2)$
$= 6x^3-9x^2y+12xy^2+4x^2y-6xy^2+8y^3 = 6x^3-5x^2y+6xy^2+8y^3$

71. $2y^n(3y^n+y^{-n}) = 2y^n(3y^n)+2y^n(y^{-n}) = 6y^{n+n}+2y^{n+(-n)} = 6y^{2n}+2y^0 = 6y^{2n}+2$

73. $-5x^{2n}y^n(2x^{2n}y^{-n}+3x^{-2n}y^n) = -5x^{2n}y^n(2x^{2n}y^{-n})-5x^{2n}y^n(3x^{-2n}y^n)$
$= -10x^{2n+2n}y^{n+(-n)}-15x^{2n+(-2n)}y^{n+n}$
$= -10x^{4n}y^0-15x^0y^{2n} = -10x^{4n}-15y^{2n}$

75. $(x^n+3)(x^n-4) = x^nx^n-4x^n+3x^n-12 = x^{2n}-x^n-12$

77. $(2r^n-7)(3r^n-2) = 2r^n(3r^n)-2r^n(2)-7(3r^n)+14$
$= 6r^{2n}-4r^n-21r^n+14 = 6r^{2n}-25r^n+14$

79. $x^{1/2}(x^{1/2}y+xy^{1/2}) = x^{1/2}x^{1/2}y+x^{1/2}xy^{1/2} = x^{2/2}y+x^{3/2}y^{1/2} = xy+x^{3/2}y^{1/2}$

81. $(a^{1/2}+b^{1/2})(a^{1/2}-b^{1/2}) = a^{1/2}a^{1/2}-a^{1/2}b^{1/2}+a^{1/2}b^{1/2}-b^{1/2}b^{1/2}$
$= a^{2/2}-b^{2/2} = a-b$

83. $$\frac{2}{\sqrt{3}-1} = \frac{2}{\sqrt{3}-1}\cdot\frac{\sqrt{3}+1}{\sqrt{3}+1} = \frac{2\left(\sqrt{3}+1\right)}{\left(\sqrt{3}\right)^2-1^2} = \frac{2\left(\sqrt{3}+1\right)}{3-1} = \frac{2\left(\sqrt{3}+1\right)}{2} = \sqrt{3}+1$$

85. $$\frac{3x}{\sqrt{7}+2} = \frac{3x}{\sqrt{7}+2}\cdot\frac{\sqrt{7}-2}{\sqrt{7}-2} = \frac{3x\left(\sqrt{7}-2\right)}{\left(\sqrt{7}\right)^2-2^2} = \frac{3x\left(\sqrt{7}-2\right)}{7-4} = \frac{3x\left(\sqrt{7}-2\right)}{3} = x\left(\sqrt{7}-2\right)$$

87. $$\frac{x}{x-\sqrt{3}} = \frac{x}{x-\sqrt{3}}\cdot\frac{x+\sqrt{3}}{x+\sqrt{3}} = \frac{x\left(x+\sqrt{3}\right)}{x^2-\left(\sqrt{3}\right)^2} = \frac{x\left(x+\sqrt{3}\right)}{x^2-3}$$

89. $$\frac{y+\sqrt{2}}{y-\sqrt{2}} = \frac{y+\sqrt{2}}{y-\sqrt{2}}\cdot\frac{y+\sqrt{2}}{y+\sqrt{2}} = \frac{\left(y+\sqrt{2}\right)\left(y+\sqrt{2}\right)}{y^2-\left(\sqrt{2}\right)^2} = \frac{y^2+2y\sqrt{2}+2}{y^2-2}$$

EXERCISES 0.4

91. $$\frac{\sqrt{2}-\sqrt{3}}{1-\sqrt{3}}=\frac{\sqrt{2}-\sqrt{3}}{1-\sqrt{3}}\cdot\frac{1+\sqrt{3}}{1+\sqrt{3}}=\frac{\sqrt{2}+\sqrt{6}-\sqrt{3}-\left(\sqrt{3}\right)^2}{1^2-\left(\sqrt{3}\right)^2}=\frac{\sqrt{2}+\sqrt{6}-\sqrt{3}-3}{1-3}$$
$$=\frac{\sqrt{2}+\sqrt{6}-\sqrt{3}-3}{-2}$$
$$=\frac{-\left(\sqrt{2}+\sqrt{6}-\sqrt{3}-3\right)}{2}$$
$$=\frac{\sqrt{3}+3-\sqrt{2}-\sqrt{6}}{2}$$

93. $$\frac{\sqrt{x}-\sqrt{y}}{\sqrt{x}+\sqrt{y}}=\frac{\sqrt{x}-\sqrt{y}}{\sqrt{x}+\sqrt{y}}\cdot\frac{\sqrt{x}-\sqrt{y}}{\sqrt{x}-\sqrt{y}}=\frac{\sqrt{x^2}-\sqrt{xy}-\sqrt{xy}+\sqrt{y^2}}{\left(\sqrt{x}\right)^2-\left(\sqrt{y}\right)^2}=\frac{x-2\sqrt{xy}+y}{x-y}$$

95. $$\frac{\sqrt{2}+1}{2}=\frac{\sqrt{2}+1}{2}\cdot\frac{\sqrt{2}-1}{\sqrt{2}-1}=\frac{\left(\sqrt{2}\right)^2-1^2}{2\left(\sqrt{2}-1\right)}=\frac{2-1}{2\left(\sqrt{2}-1\right)}=\frac{1}{2\left(\sqrt{2}-1\right)}$$

97. $$\frac{y-\sqrt{3}}{y+\sqrt{3}}=\frac{y-\sqrt{3}}{y+\sqrt{3}}\cdot\frac{y+\sqrt{3}}{y+\sqrt{3}}=\frac{y^2-\left(\sqrt{3}\right)^2}{y^2+y\sqrt{3}+y\sqrt{3}+\sqrt{9}}=\frac{y^2-3}{y^2+2y\sqrt{3}+3}$$

99. $$\frac{\sqrt{x+3}-\sqrt{x}}{3}=\frac{\sqrt{x+3}-\sqrt{x}}{3}\cdot\frac{\sqrt{x+3}+\sqrt{x}}{\sqrt{x+3}+\sqrt{x}}=\frac{\left(\sqrt{x+3}\right)^2-\left(\sqrt{x}\right)^2}{3\left(\sqrt{x+3}+\sqrt{x}\right)}$$
$$=\frac{x+3-x}{3\left(\sqrt{x+3}+\sqrt{x}\right)}$$
$$=\frac{3}{3\left(\sqrt{x+3}+\sqrt{x}\right)}=\frac{1}{\sqrt{x+3}+\sqrt{x}}$$

101. $$\frac{36a^2b^3}{18ab^6}=2a^{2-1}b^{3-6}=2a^1b^{-3}=\frac{2a}{b^3}$$

103. $$\frac{16x^6y^4z^9}{-24x^9y^6z^0}=-\frac{2}{3}x^{6-9}y^{4-6}z^{9-0}=-\frac{2}{3}x^{-3}y^{-2}z^9=-\frac{2z^9}{3x^3y^2}$$

105. $$\frac{5x^3y^2+15x^3y^4}{10x^2y^3}=\frac{5x^3y^2}{10x^2y^3}+\frac{15x^3y^4}{10x^2y^3}=\frac{x}{2y}+\frac{3xy}{2}$$

107. $$\frac{24x^5y^7-36x^2y^5+12xy}{60x^5y^4}=\frac{24x^5y^7}{60x^5y^4}-\frac{36x^2y^5}{60x^5y^4}+\frac{12xy}{60x^5y^4}=\frac{2y^3}{5}-\frac{3y}{5x^3}+\frac{1}{5x^4y^3}$$

EXERCISES 0.4

109.
$$\begin{array}{r|r} & 3x + 2 \\ \hline x+3 & 3x^2 + 11x + 6 \\ & 3x^2 + 9x \\ \hline & 2x + 6 \\ & 2x + 6 \\ \hline & 0 \end{array}$$

111.
$$\begin{array}{r|r} & x - 7 + \frac{2}{2x-5} \\ \hline 2x-5 & 2x^2 - 19x + 37 \\ & 2x^2 - 5x \\ \hline & -14x + 37 \\ & -14x + 35 \\ \hline & 2 \end{array}$$

113.
$$\begin{array}{r|r} & 2x^2 + 2x + 2 + \frac{3}{x-1} \\ \hline x-1 & 2x^3 + 0x^2 + 0x + 1 \\ & 2x^3 - 2x^2 \\ \hline & 2x^2 + 0x + 1 \\ & 2x^2 - 2x \\ \hline & 2x + 1 \\ & 2x - 2 \\ \hline & 3 \end{array}$$

115.
$$\begin{array}{r|r} & x - 3 \\ \hline x^2+x-1 & x^3 - 2x^2 - 4x + 3 \\ & x^3 + x^2 - x \\ \hline & -3x^2 - 3x + 3 \\ & -3x^2 - 3x + 3 \\ \hline & 0 \end{array}$$

117.
$$\begin{array}{r|r} & x^2 - 2 + \frac{-x^2+5}{x^3-2} \\ \hline x^3-2 & x^5 + 0x^4 - 2x^3 - 3x^2 + 0x + 9 \\ & x^5 \qquad - 2x^2 \\ \hline & -2x^3 - x^2 + 0x + 9 \\ & -2x^3 \qquad + 4 \\ \hline & -x^2 \qquad + 5 \end{array}$$

119.
$$\begin{array}{r|r} & x^4 + 2x^3 + 4x^2 + 8x + 16 \\ \hline x-2 & x^5 + 0x^4 + 0x^3 + 0x^2 + 0x - 32 \\ & x^5 - 2x^4 \\ \hline & 2x^4 + 0x^3 \\ & 2x^4 - 4x^3 \\ \hline & 4x^3 + 0x^2 \\ & 4x^3 - 8x^2 \\ \hline & 8x^2 + 0x \\ & 8x^2 - 16x \\ \hline & 16x - 32 \\ & 16x - 32 \\ \hline & 0 \end{array}$$

121.
$$\begin{array}{r|r} & 6x^2 + x - 12 \\ \hline 6x^2+11x-10 & 36x^4 + 72x^3 - 121x^2 - 142x + 120 \\ & 36x^4 + 66x^3 - 60x^2 \\ \hline & 6x^3 - 61x^2 - 142x \\ & 6x^3 + 11x^2 - 10x \\ \hline & -72x^2 - 132x + 120 \\ & -72x^2 - 132x + 120 \\ \hline & 0 \end{array}$$

123. Area $=$ length $\cdot$ width $= (x+5)(x-2)\text{ ft}^2 = (x^2 - 2x + 5x - 10)\text{ ft}^2 = (x^2 + 3x - 10)\text{ ft}^2$

125.
$$\begin{aligned} \text{Volume} = l \cdot w \cdot h = (12-2x)(12-2x)x\text{ in.}^3 &= \left(144 - 48x + 4x^2\right)x\text{ in.}^3 \\ &= \left(144x - 48x^2 + 4x^3\right)\text{ in.}^3 \\ &= \left(4x^3 - 48x^2 + 144x\right)\text{ in.}^3 \end{aligned}$$

127.
$$\begin{aligned} (a+b+c)^2 = (a+b+c)(a+b+c) &= a(a+b+c) + b(a+b+c) + c(a+b+c) \\ &= a^2 + ab + ac + ab + b^2 + bc + ac + bc + c^2 \\ &= a^2 + b^2 + c^2 + 2ab + 2bc + 2ac \end{aligned}$$

129. Answers may vary.

131. Check the formula with $a = 1$ and $b = 2$.

133. False. Some polynomials are trinomials.

135. True. $(12x - 5y)^2 = (12x - 5y)(12x - 5y) = 144x^2 - 60xy - 60xy + 25y^2$
$= 144x^2 - 120xy + 25y^2$

137. False. $\left(x^{1/3} - 6\right)\left(4x^{1/3} + 7\right) = 4x^{2/3} + 7x^{1/3} - 24x^{1/3} - 42 = 4x^{2/3} - 17x^{1/3} - 42$

139. Profit = Revenue − Cost = $\left(x^2 + 200x\right) - (-200x + 500) = x^2 + 200x + 200x - 500$
$= x^2 + 400x - 500$

Exercises 0.5 (page 63)

1. factor

3. $ax + bx = x(a + b)$

5. $x^2 + 2xy + y^2 = (x + y)(x + y) = (x + y)^2$

7. $x^3 + y^3 = (x + y)(x^2 - xy + y^2)$

9. $3x - 6 = 3(x - 2)$

11. $8x^2 + 4x^3 = 4x^2(2 + x)$

13. $7x^2y^2 + 14x^3y^2 = 7x^2y^2(1 + 2x)$

15. $a(x + y) + b(x + y) = (x + y)(a + b)$

17. $4a + b - 12a^2 - 3ab = 4a + b - 3a(4a + b) = 1(4a + b) - 3a(4a + b) = (4a + b)(1 - 3a)$

19. $4x^2 - 9 = (2x)^2 - 3^2 = (2x + 3)(2x - 3)$

21. $4 - 9r^2 = 2^2 - (3r)^2 = (2 + 3r)(2 - 3r)$

23. $81x^4 - 1 = (9x^2)^2 - 1^2 = (9x^2 + 1)(9x^2 - 1) = (9x^2 + 1)(3x + 1)(3x - 1)$

25. $(x + z)^2 - 25 = (x + z)^2 - 5^2$
$= (x + z + 5)(x + z - 5)$

27. $x^2 + 8x + 16 = (x + 4)(x + 4) = (x + 4)^2$

29. $b^2 - 10b + 25 = (b - 5)(b - 5) = (b - 5)^2$

31. $m^2 + 4mn + 4n^2 = (m + 2n)(m + 2n)$
$= (m + 2n)^2$

33. $12x^2 - xy - 6y^2 = (4x - 3y)(3x + 2y)$

35. $x^2 + 10x + 21$: $a = 1, b = 10, c = 21$
key number $= ac = 1(21) = 21$
$x^2 + 10x + 21 = x^2 + 7x + 3x + 21$
$= x(x + 7) + 3(x + 7)$
$= (x + 7)(x + 3)$

37. $x^2 - 4x - 12$: $a = 1, b = -4, c = -12$
key number $= ac = 1(-12) = -12$
$x^2 - 4x - 12 = x^2 - 6x + 2x - 12$
$= x(x - 6) + 2(x - 6)$
$= (x - 6)(x + 2)$

39. $6p^2 + 7p - 3$: $a = 6, b = 7, c = -3$
key number $= ac = 6(-3) = -18$
$6p^2 + 7p - 3 = 6p^2 + 9p - 2p - 3$
$= 3p(2p + 3) - (2p + 3)$
$= (2p + 3)(3p - 1)$

EXERCISES 0.5

41. $t^3 + 343 = t^3 + 7^3 = (t+7)[t^2 - (t)(7) + 7^2] = (t+7)(t^2 - 7t + 49)$

43. $125y^3 + 216z^3 = (5y)^3 + (6z)^3 = (5y + 6z)\left[(5y)^2 - (5y)(6z) + (6z)^2\right]$
$= (5y + 6z)\left(25y^2 - 30yz + 36z^2\right)$

45. $8z^3 - 27 = (2z)^3 - 3^3 = (2z - 3)\left[(2z)^2 + (2z)(3) + 3^2\right] = (2z - 3)(4z^2 + 6z + 9)$

47. $343y^3 - z^3 = (7y)^3 - z^3 = (7y - z)\left[(7y)^2 + (7y)(z) + z^2\right] = (7y - z)(49y^2 + 7yz + z^2)$

49. $3a^2bc + 6ab^2c + 9abc^2 = 3abc(a + 2b + 3c)$

51. $3x^3 + 3x^2 - x - 1 = 3x^2(x+1) - 1(x+1) = (x+1)(3x^2 - 1)$

53. $2txy + 2ctx - 3ty - 3ct = t(2xy + 2cx - 3y - 3c) = t[2x(y+c) - 3(y+c)] = t(y+c)(2x - 3)$

55. $ax + bx + ay + by + az + bz = x(a+b) + y(a+b) + z(a+b) = (a+b)(x+y+z)$

57. $x^2 - (y-z)^2 = [x + (y-z)][x - (y-z)] = (x + y - z)(x - y + z)$

59. $(x-y)^2 - (x+y)^2 = [(x-y) + (x+y)][(x-y) - (x+y)] = (x - y + x + y)(x - y - x - y)$
$= (2x)(-2y) = -4xy$

61. $x^4 - y^4 = (x^2)^2 - (y^2)^2 = (x^2 + y^2)(x^2 - y^2) = (x^2 + y^2)(x+y)(x-y)$

63. $3x^2 - 12 = 3(x^2 - 4) = 3(x+2)(x-2)$

65. $18xy^2 - 8x = 2x\left(9y^2 - 4\right)$
$= 2x(3y + 2)(3y - 2)$

67. $x^2 - 2x + 15 \Rightarrow$ prime

69. $-15 + 2a + 24a^2 = 24a^2 + 2a - 15$
$= (6a + 5)(4a - 3)$

71. $6x^2 + 29xy + 35y^2 = (3x + 7y)(2x + 5y)$

73. $12p^2 - 58pq - 70q^2 = 2(6p^2 - 29pq - 35q^2) = 2(6p - 35q)(p + q)$

75. $-6m^2 + 47mn - 35n^2 = -(6m^2 - 47mn + 35n^2) = -(6m - 5n)(m - 7n)$

77. $-6x^3 + 23x^2 + 35x = -x(6x^2 - 23x - 35) = -x(6x + 7)(x - 5)$

79. $6x^4 - 11x^3 - 35x^2 = x^2(6x^2 - 11x - 35) = x^2(2x - 7)(3x + 5)$

81. $x^4 + 2x^2 - 15 = (x^2 + 5)(x^2 - 3)$

83. $a^{2n} - 2a^n - 3 = (a^n - 3)(a^n + 1)$

85. $6x^{2n} - 7x^n + 2 = (3x^n - 2)(2x^n - 1)$

87. $4x^{2n} - 9y^{2n} = (2x^n)^2 - (3y^n)^2$
$= (2x^n + 3y^n)(2x^n - 3y^n)$

EXERCISES 0.5

89. $10y^{2n} - 11y^n - 6 = (5y^n + 2)(2y^n - 3)$

91. $2x^3 + 2000 = 2(x^3 + 1000) = 2(x^3 + 10^3) = 2(x + 10)(x^2 - 10x + 100)$

93. $$\begin{aligned}(x+y)^3 - 64 = (x+y)^3 - 4^3 &= [(x+y) - 4]\left[(x+y)^2 + 4(x+y) + 4^2\right] \\ &= (x + y - 4)\left(x^2 + 2xy + y^2 + 4x + 4y + 16\right)\end{aligned}$$

95. $$\begin{aligned}64a^6 - y^6 = \left(8a^3\right)^2 - \left(y^3\right)^2 &= \left(8a^3 + y^3\right)\left(8a^3 - y^3\right) \\ &= (2a + y)\left(4a^2 - 2ay + y^2\right)(2a - y)\left(4a^2 + 2ay + y^2\right) \\ &= (2a + y)(2a - y)\left(4a^2 - 2ay + y^2\right)\left(4a^2 + 2ay + y^2\right)\end{aligned}$$

97. $a^3 - b^3 + a - b = (a - b)(a^2 + ab + b^2) + (a - b)1 = (a - b)(a^2 + ab + b^2 + 1)$

99. $$\begin{aligned}64x^6 + y^6 = \left(4x^2\right)^3 + \left(y^2\right)^3 &= \left(4x^2 + y^2\right)\left(\left(4x^2\right)^2 - 4x^2y^2 + \left(y^2\right)^2\right) \\ &= \left(4x^2 + y^2\right)\left(16x^4 - 4x^2y^2 + y^4\right)\end{aligned}$$

101. $$\begin{aligned}x^2 - 6x + 9 - 144y^2 = (x-3)(x-3) - 144y^2 &= (x-3)^2 - (12y)^2 \\ &= (x - 3 + 12y)(x - 3 - 12y)\end{aligned}$$

103. $(a+b)^2 - 3(a+b) - 10 = [(a+b) - 5][(a+b) + 2] = (a + b - 5)(a + b + 2)$

105. $x^6 + 7x^3 - 8 = (x^3 + 8)(x^3 - 1) = (x + 2)(x^2 - 2x + 4)(x - 1)(x^2 + x + 1)$

107. $$\begin{aligned}x^4 + x^2 + 1 &= x^4 + 2x^2 + 1 - x^2 \\ &= \left(x^2 + 1\right)\left(x^2 + 1\right) - x^2 \\ &= \left(x^2 + 1\right)^2 - x^2 \\ &= \left(x^2 + 1 + x\right)\left(x^2 + 1 - x\right) \\ &= \left(x^2 + x + 1\right)\left(x^2 - x + 1\right)\end{aligned}$$

109. $$\begin{aligned}x^4 + 7x^2 + 16 &= x^4 + 8x^2 + 16 - x^2 \\ &= \left(x^2 + 4\right)\left(x^2 + 4\right) - x^2 \\ &= \left(x^2 + 4\right)^2 - x^2 \\ &= \left(x^2 + 4 + x\right)\left(x^2 + 4 - x\right) \\ &= \left(x^2 + x + 4\right)\left(x^2 - x + 4\right)\end{aligned}$$

111. $$\begin{aligned}4a^4 + 1 + 3a^2 = 4a^4 + 4a^2 + 1 - a^2 = \left(2a^2 + 1\right)\left(2a^2 + 1\right) - a^2 &= \left(2a^2 + 1\right)^2 - a^2 \\ &= \left(2a^2 + 1 + a\right)\left(2a^2 + 1 - a\right) \\ &= \left(2a^2 + a + 1\right)\left(2a^2 - a + 1\right)\end{aligned}$$

113. $$\begin{aligned}V &= \frac{4}{3}\pi r_1^3 - \frac{4}{3}\pi r_2^3 \\ &= \tfrac{4}{3}\pi(r_1^3 - r_2^3) \\ &= \tfrac{4}{3}\pi(r_1 - r_2)(r_1^2 + r_1r_2 + r_2^2)\end{aligned}$$

115-117. Answers may vary.

119. $3x + 2 = 2\left(\frac{3x}{2} + \frac{2}{2}\right) = 2\left(\frac{3}{2}x + 1\right)$

121. $$\begin{aligned}x^2 + 2x + 4 &= 2\left(\frac{x^2}{2} + \frac{2x}{2} + \frac{4}{2}\right) \\ &= 2\left(\tfrac{1}{2}x^2 + x + 2\right)\end{aligned}$$

123. $a+b=a\left(\frac{a}{a}+\frac{b}{a}\right)=a\left(1+\frac{b}{a}\right)$

125. $x+x^{1/2}=x^{1/2}\left(x^{1-1/2}+x^{1/2-1/2}\right)$
$=x^{1/2}\left(x^{1/2}+1\right)$

127. $2x+\sqrt{2}y=\sqrt{2}\left(\dfrac{2x}{\sqrt{2}}+\dfrac{\sqrt{2}y}{\sqrt{2}}\right)$
$=\sqrt{2}\left(\sqrt{2}x+y\right)$

129. $ab^{3/2}-a^{3/2}b=ab\left(\dfrac{ab^{3/2}}{ab}-\dfrac{a^{3/2}b}{ab}\right)$
$=ab\left(b^{1/2}-a^{1/2}\right)$

131. $x^2+x-6+xy-2y=(x+3)(x-2)+y(x-2)=(x-2)(x+3+y)$

133. $a^4+2a^3+a^2+a+1=a^2\left(a^2+2a+1\right)+a+1=a^2(a+1)(a+1)+1(a+1)$
$=(a+1)\left[a^2(a+1)+1\right]$
$=(a+1)\left(a^3+a^2+1\right)$

135. True.

137. False. $p^3q^3r^3+64=(pqr)^3+4^3=(pqr+4)(p^2q^2r^2-4pqr+16)$

139. True.

Exercises 0.6 (page 73)

1. numerator

3. $ad=bc$

5. $\frac{ac}{bd}$

7. $\frac{a+c}{b}$

9. $\dfrac{8x}{3y}\stackrel{?}{=}\dfrac{16x}{6y}$
$8x\cdot 6y\stackrel{?}{=}3y\cdot 16x$
$48xy=48xy$
EQUAL

11. $\dfrac{25xyz}{12ab^2c}\stackrel{?}{=}\dfrac{50a^2bc}{24xyz}$
$25xyz\cdot 24xyz\stackrel{?}{=}12ab^2c\cdot 50a^2bc$
$600x^2y^2z^2\neq 600a^3b^3c^2$
NOT EQUAL

13. $\dfrac{7a^2b}{21ab^2}=\dfrac{a\cdot 7ab}{3b\cdot 7ab}=\dfrac{a}{3b}\cdot\dfrac{7ab}{7ab}=\dfrac{a}{3b}$

15. $\dfrac{4x}{7}\cdot\dfrac{2}{5a}=\dfrac{4x\cdot 2}{7\cdot 5a}=\dfrac{8x}{35a}$

17. $\dfrac{8m}{5n}\div\dfrac{3m}{10n}=\dfrac{8m}{5n}\cdot\dfrac{10n}{3m}=\dfrac{80mn}{15mn}=\dfrac{16}{3}$

19. $\dfrac{3z}{5c}+\dfrac{2z}{5c}=\dfrac{3z+2z}{5c}=\dfrac{5z}{5c}=\dfrac{z}{c}$

21. $\dfrac{15x^2y}{7a^2b^3}-\dfrac{x^2y}{7a^2b^3}=\dfrac{14x^2y}{7a^2b^3}=\dfrac{2x^2y}{a^2b^3}$

23. $\dfrac{2x-4}{x^2-4}=\dfrac{2(x-2)}{(x+2)(x-2)}=\dfrac{2}{x+2}$

25. $\dfrac{4-x^2}{x^2-5x+6}=\dfrac{(2+x)(2-x)}{(x-3)(x-2)}=-\dfrac{x+2}{x-3}$

27. $\dfrac{6x^3+x^2-12x}{4x^3+4x^2-3x}=\dfrac{x(6x^2+x-12)}{x(4x^2+4x-3)}=\dfrac{x(2x+3)(3x-4)}{x(2x+3)(2x-1)}=\dfrac{3x-4}{2x-1}$

EXERCISES 0.6

29. $$\frac{x^3-8}{x^2+ax-2x-2a}=\frac{x^3-2^3}{x(x+a)-2(x+a)}=\frac{(x-2)(x^2+2x+4)}{(x+a)(x-2)}=\frac{x^2+2x+4}{x+a}$$

31. $$\frac{x^2-1}{x}\cdot\frac{x^2}{x^2+2x+1}=\frac{(x+1)(x-1)}{x}\cdot\frac{x^2}{(x+1)(x+1)}=\frac{x(x-1)}{x+1}$$

33. $$\frac{3x^2+7x+2}{x^2+2x}\cdot\frac{x^2-x}{3x^2+x}=\frac{(3x+1)(x+2)}{x(x+2)}\cdot\frac{x(x-1)}{x(3x+1)}=\frac{x-1}{x}$$

35. $$\frac{x^2+x}{x-1}\cdot\frac{x^2-1}{x+2}=\frac{x(x+1)}{x-1}\cdot\frac{(x+1)(x-1)}{x+2}=\frac{x(x+1)^2}{x+2}$$

37. $$\frac{2x^2+32}{8}\div\frac{x^2+16}{2}=\frac{2x^2+32}{8}\cdot\frac{2}{x^2+16}=\frac{2(x^2+16)}{8}\cdot\frac{2}{x^2+16}=\frac{1}{2}$$

39. $$\frac{z^2+z-20}{z^2-4}\div\frac{z^2-25}{z-5}=\frac{z^2+z-20}{z^2-4}\cdot\frac{z-5}{z^2-25}=\frac{(z+5)(z-4)}{(z+2)(z-2)}\cdot\frac{z-5}{(z+5)(z-5)}$$
$$=\frac{z-4}{(z+2)(z-2)}$$

41. $$\frac{3x^2+5x-2}{x^3+2x^2}\div\frac{6x^2+13x-5}{2x^3+5x^2}=\frac{3x^2+5x-2}{x^3+2x^2}\cdot\frac{2x^3+5x^2}{6x^2+13x-5}$$
$$=\frac{(3x-1)(x+2)}{x^2(x+2)}\cdot\frac{x^2(2x+5)}{(3x-1)(2x+5)}=1$$

43. $$\frac{x^2+7x+12}{x^3-x^2-6x}\cdot\frac{x^2-3x-10}{x^2+2x-3}\cdot\frac{x^3-4x^2+3x}{x^2-x-20}$$
$$=\frac{(x+3)(x+4)}{x(x-3)(x+2)}\cdot\frac{(x-5)(x+2)}{(x+3)(x-1)}\cdot\frac{x(x-3)(x-1)}{(x-5)(x+4)}=1$$

45. $$\frac{x^2-2x-3}{21x^2-50x-16}\cdot\frac{3x-8}{x-3}\div\frac{x^2+6x+5}{7x^2-33x-10}=\frac{x^2-2x-3}{21x^2-50x-16}\cdot\frac{3x-8}{x-3}\cdot\frac{7x^2-33x-10}{x^2+6x+5}$$
$$=\frac{(x-3)(x+1)}{(7x+2)(3x-8)}\cdot\frac{3x-8}{x-3}\cdot\frac{(7x+2)(x-5)}{(x+5)(x+1)}$$
$$=\frac{x-5}{x+5}$$

47. $$\frac{3}{x+3}+\frac{x+2}{x+3}=\frac{3+x+2}{x+3}=\frac{x+5}{x+3}$$

49. $$\frac{4x}{x-1}-\frac{4}{x-1}=\frac{4x-4}{x-1}=\frac{4(x-1)}{x-1}=4$$

51. $$\frac{2}{5-x}+\frac{1}{x-5}=\frac{-2}{x-5}+\frac{1}{x-5}=\frac{-1}{x-5}$$

EXERCISES 0.6

53. $$\frac{3}{x+1}+\frac{2}{x-1}=\frac{3(x-1)}{(x+1)(x-1)}+\frac{2(x+1)}{(x-1)(x+1)}=\frac{3x-3}{(x+1)(x-1)}+\frac{2x+2}{(x-1)(x+1)}=\frac{5x-1}{(x+1)(x-1)}$$

55. $$\frac{a+3}{a^2+7a+12}+\frac{a}{a^2-16}=\frac{a+3}{(a+3)(a+4)}+\frac{a}{(a+4)(a-4)}=\frac{1}{a+4}+\frac{a}{(a+4)(a-4)}=\frac{1(a-4)}{(a+4)(a-4)}+\frac{a}{(a+4)(a-4)}=\frac{a-4}{(a+4)(a-4)}+\frac{a}{(a+4)(a-4)}=\frac{2a-4}{(a+4)(a-4)}=\frac{2(a-2)}{(a+4)(a-4)}$$

57. $$\frac{x}{x^2-4}-\frac{1}{x+2}=\frac{x}{(x+2)(x-2)}-\frac{1}{x+2}=\frac{x}{(x+2)(x-2)}-\frac{1(x-2)}{(x+2)(x-2)}=\frac{x}{(x+2)(x-2)}-\frac{x-2}{(x+2)(x-2)}=\frac{2}{(x+2)(x-2)}$$

59. $$\frac{3x-2}{x^2+2x+1}-\frac{x}{x^2-1}=\frac{3x-2}{(x+1)(x+1)}-\frac{x}{(x+1)(x-1)}=\frac{(3x-2)(x-1)}{(x+1)(x+1)(x-1)}-\frac{x(x+1)}{(x+1)(x-1)(x+1)}=\frac{3x^2-5x+2}{(x+1)(x+1)(x-1)}-\frac{x^2+x}{(x+1)(x+1)(x-1)}=\frac{2x^2-6x+2}{(x+1)(x+1)(x-1)}=\frac{2(x^2-3x+1)}{(x+1)^2(x-1)}$$

61. $$\frac{2}{y^2-1}+3+\frac{1}{y+1}=\frac{2}{(y+1)(y-1)}+\frac{3}{1}+\frac{1}{y+1}=\frac{2}{(y+1)(y-1)}+\frac{3(y+1)(y-1)}{1(y+1)(y-1)}+\frac{1(y-1)}{(y+1)(y-1)}=\frac{2}{(y+1)(y-1)}+\frac{3y^2-3}{(y+1)(y-1)}+\frac{y-1}{(y+1)(y-1)}=\frac{3y^2+y-2}{(y+1)(y-1)}=\frac{(3y-2)(y+1)}{(y+1)(y-1)}=\frac{3y-2}{y-1}$$

EXERCISES 0.6

63. $$\frac{1}{x-2}+\frac{3}{x+2}-\frac{3x-2}{x^2-4}=\frac{1}{x-2}+\frac{3}{x+2}-\frac{3x-2}{(x+2)(x-2)}$$
$$=\frac{1(x+2)}{(x-2)(x+2)}+\frac{3(x-2)}{(x+2)(x-2)}-\frac{3x-2}{(x+2)(x-2)}$$
$$=\frac{x+2}{(x+2)(x-2)}+\frac{3x-6}{(x+2)(x-2)}-\frac{3x-2}{(x+2)(x-2)}$$
$$=\frac{x-2}{(x+2)(x-2)}=\frac{1}{x+2}$$

65. $$\left(\frac{1}{x-2}+\frac{1}{x-3}\right)\cdot\frac{x-3}{2x}=\left(\frac{1(x-3)}{(x-2)(x-3)}+\frac{1(x-2)}{(x-3)(x-2)}\right)\cdot\frac{x-3}{2x}$$
$$=\left(\frac{x-3}{(x-2)(x-3)}+\frac{x-2}{(x-2)(x-3)}\right)\cdot\frac{x-3}{2x}$$
$$=\frac{2x-5}{(x-2)(x-3)}\cdot\frac{x-3}{2x}=\frac{2x-5}{2x(x-2)}$$

67. $$\frac{3x}{x-4}-\frac{x}{x+4}-\frac{3x+1}{16-x^2}=\frac{3x}{x-4}-\frac{x}{x+4}-\frac{3x+1}{(4+x)(4-x)}$$
$$=\frac{3x}{x-4}-\frac{x}{x+4}+\frac{3x+1}{(x+4)(x-4)}$$
$$=\frac{3x(x+4)}{(x-4)(x+4)}-\frac{x(x-4)}{(x+4)(x-4)}+\frac{3x+1}{(x+4)(x-4)}$$
$$=\frac{3x^2+12x}{(x+4)(x-4)}-\frac{x^2-4x}{(x+4)(x-4)}+\frac{3x+1}{(x+4)(x-4)}$$
$$=\frac{2x^2+19x+1}{(x+4)(x-4)}$$

69. $$\frac{1}{x^2+3x+2}-\frac{2}{x^2+4x+3}+\frac{1}{x^2+5x+6}$$
$$=\frac{1}{(x+2)(x+1)}-\frac{2}{(x+3)(x+1)}+\frac{1}{(x+2)(x+3)}$$
$$=\frac{1(x+3)}{(x+2)(x+1)(x+3)}-\frac{2(x+2)}{(x+3)(x+1)(x+2)}+\frac{1(x+1)}{(x+2)(x+3)(x+1)}$$
$$=\frac{x+3}{(x+2)(x+1)(x+3)}-\frac{2x+4}{(x+2)(x+1)(x+3)}+\frac{x+1}{(x+2)(x+1)(x+3)}$$
$$=\frac{x+3-2x-4+x+1}{(x+2)(x+1)(x+3)}=\frac{0}{(x+2)(x+1)(x+3)}=0$$

EXERCISES 0.6

71. $\frac{3x-2}{x^2+x-20}-\frac{4x^2+2}{x^2-25}+\frac{3x^2-25}{x^2-16}$

$$=\frac{3x-2}{(x+5)(x-4)}-\frac{4x^2+2}{(x+5)(x-5)}+\frac{3x^2-25}{(x+4)(x-4)}$$

$$=\frac{(3x-2)(x-5)(x+4)}{(x+5)(x-4)(x-5)(x+4)}-\frac{(4x^2+2)(x-4)(x+4)}{(x+5)(x-5)(x-4)(x+4)}\cdots+\frac{(3x^2-25)(x+5)(x-5)}{(x+4)(x-4)(x+5)(x-5)}$$

$$=\frac{3x^3-5x^2-58x+40}{(x+5)(x-4)(x-5)(x+4)}-\frac{4x^4-62x^2-32}{(x+5)(x-4)(x-5)(x+4)}\cdots+\frac{3x^4-100x^2+625}{(x+5)(x-4)(x-5)(x+4)}$$

$$=\frac{3x^3-5x^2-58x+40}{(x+5)(x-4)(x-5)(x+4)}-\frac{4x^4-62x^2-32}{(x+5)(x-4)(x-5)(x+4)}\cdots+\frac{3x^4-100x^2+625}{(x+5)(x-4)(x-5)(x+4)}$$

$$=\frac{3x^3-5x^2-58x+40-4x^4+62x^2+32+3x^4-100x^2+625}{(x+5)(x-4)(x-5)(x+4)}$$

$$=\frac{-x^4+3x^3-43x^2-58x+697}{(x+5)(x-4)(x-5)(x+4)}$$

73. $\frac{\frac{3a}{b}}{\frac{6ac}{b^2}}=\frac{3a}{b}\div\frac{6ac}{b^2}=\frac{3a}{b}\cdot\frac{b^2}{6ac}=\frac{b}{2c}$

75. $\frac{3a^2b}{\frac{ab}{27}}=\frac{3a^2b}{1}\div\frac{ab}{27}=\frac{3a^2b}{1}\cdot\frac{27}{ab}=81a$

77. $\frac{\frac{x-y}{ab}}{\frac{y-x}{ab}}=\frac{x-y}{ab}\div\frac{y-x}{ab}=\frac{x-y}{ab}\cdot\frac{ab}{y-x}=-1$

79. $\frac{\frac{1}{x}+\frac{1}{y}}{xy}=\frac{xy\left(\frac{1}{x}+\frac{1}{y}\right)}{xy(xy)}=\frac{xy\left(\frac{1}{x}\right)+xy\left(\frac{1}{y}\right)}{x^2y^2}=\frac{y+x}{x^2y^2}$

81. $\frac{\frac{1}{x}+\frac{1}{y}}{\frac{1}{x}-\frac{1}{y}}=\frac{xy\left(\frac{1}{x}+\frac{1}{y}\right)}{xy\left(\frac{1}{x}-\frac{1}{y}\right)}=\frac{xy\left(\frac{1}{x}\right)+xy\left(\frac{1}{y}\right)}{xy\left(\frac{1}{x}\right)-xy\left(\frac{1}{y}\right)}=\frac{y+x}{y-x}$

83. $\frac{\frac{3a}{b}-\frac{4a^2}{x}}{\frac{1}{b}+\frac{1}{ax}}=\frac{abx\left(\frac{3a}{b}-\frac{4a^2}{x}\right)}{abx\left(\frac{1}{b}+\frac{1}{ax}\right)}=\frac{abx\left(\frac{3a}{b}\right)-abx\left(\frac{4a^2}{x}\right)}{abx\left(\frac{1}{b}\right)+abx\left(\frac{1}{ax}\right)}=\frac{3a^2x-4a^3b}{ax+b}=\frac{a^2(3x-4ab)}{ax+b}$

85. $\frac{x+1-\frac{6}{x}}{x+5+\frac{6}{x}}=\frac{x\left(x+1-\frac{6}{x}\right)}{x\left(x+5+\frac{6}{x}\right)}=\frac{x(x)+x(1)-x\left(\frac{6}{x}\right)}{x(x)+x(5)+x\left(\frac{6}{x}\right)}=\frac{x^2+x-6}{x^2+5x+6}=\frac{(x+3)(x-2)}{(x+2)(x+3)}=\frac{x-2}{x+2}$

87. $\frac{3xy}{1-\frac{1}{xy}}=\frac{xy(3xy)}{xy\left(1-\frac{1}{xy}\right)}=\frac{3x^2y^2}{xy(1)-xy\left(\frac{1}{xy}\right)}=\frac{3x^2y^2}{xy-1}$

EXERCISES 0.6

89. $\dfrac{3x}{x+\frac{1}{x}} = \dfrac{x(3x)}{x\left(x+\frac{1}{x}\right)} = \dfrac{3x^2}{x^2+1}$

91. $$\dfrac{\frac{x}{x+2}-\frac{2}{x-1}}{\frac{3}{x+2}+\frac{x}{x-1}} = \dfrac{(x+2)(x-1)\left(\frac{x}{x+2}-\frac{2}{x-1}\right)}{(x+2)(x-1)\left(\frac{3}{x+2}+\frac{x}{x-1}\right)} = \dfrac{(x+2)(x-1)\left(\frac{x}{x+2}\right)-(x+2)(x-1)\left(\frac{2}{x-1}\right)}{(x+2)(x-1)\left(\frac{3}{x+2}\right)+(x+2)(x-1)\left(\frac{x}{x-1}\right)}$$
$$= \dfrac{(x-1)(x)-(x+2)(2)}{(x-1)(3)+(x+2)(x)}$$
$$= \dfrac{x^2-x-2x-4}{3x-3+x^2+2x} = \dfrac{x^2-3x-4}{x^2+5x-3}$$

93. $\dfrac{1}{1+x^{-1}} = \dfrac{1}{1+\frac{1}{x}} = \dfrac{x(1)}{x\left(1+\frac{1}{x}\right)} = \dfrac{x}{x+1}$

95. $$\dfrac{3(x+2)^{-1}+2(x-1)^{-1}}{(x+2)^{-1}} = \dfrac{\frac{3}{x+2}+\frac{2}{x-1}}{\frac{1}{x+2}} = \dfrac{(x+2)(x-1)\left(\frac{3}{x+2}+\frac{2}{x-1}\right)}{(x+2)(x-1)\left(\frac{1}{x+2}\right)}$$
$$= \dfrac{(x+2)(x-1)\left(\frac{3}{x+2}\right)+(x+2)(x-1)\left(\frac{2}{x-1}\right)}{x-1}$$
$$= \dfrac{(x-1)(3)+(x+2)(2)}{x-1}$$
$$= \dfrac{3x-3+2x+4}{x-1} = \dfrac{5x+1}{x-1}$$

97. $\dfrac{1}{\frac{1}{k_1}+\frac{1}{k_2}} = \dfrac{k_1k_2(1)}{k_1k_2\left(\frac{1}{k_1}+\frac{1}{k_2}\right)} = \dfrac{k_1k_2}{k_1k_2\left(\frac{1}{k_1}\right)+k_1k_2\left(\frac{1}{k_2}\right)} = \dfrac{k_1k_2}{k_2+k_1}$

99-103. Answers may vary.

105. $\frac{a}{b}+\frac{c}{d} = \frac{a}{b}\cdot\frac{d}{d}+\frac{c}{d}\cdot\frac{b}{b} = \frac{ad}{bd}+\frac{bc}{bd} = \frac{ad+bc}{bd}$.

107. $\dfrac{x}{1+\frac{1}{3x^{-1}}} = \dfrac{x}{1+\frac{1}{\frac{3}{x}}} = \dfrac{x}{1+\frac{x(1)}{x\left(\frac{3}{x}\right)}} = \dfrac{x}{1+\frac{x}{3}} = \dfrac{3x}{3\left(1+\frac{x}{3}\right)} = \dfrac{3x}{3+x}$

109. $\dfrac{1}{1+\frac{1}{1+\frac{1}{x}}} = \dfrac{1}{1+\frac{x(1)}{x\left(1+\frac{1}{x}\right)}} = \dfrac{1}{1+\frac{x}{x+1}} = \dfrac{(x+1)1}{(x+1)\left(1+\frac{x}{x+1}\right)} = \dfrac{x+1}{x+1+x} = \dfrac{x+1}{2x+1}$

111. False. The denominator can never equal 0.

113. False. $\frac{x+7}{x+7} = 1$ for all values of x except $x = -7$.

115. $-\dfrac{(x-y)^3}{(y-x)^3} = -\dfrac{(x-y)^3}{[-(x-y)]^3} = -\dfrac{(x-y)^3}{(-1)^3(x-y)^3} = -\dfrac{1}{-1} = 1.$ True.

EXERCISES 0.6

117. False. $\frac{5}{x}+\frac{5}{y}=\frac{5}{x}\cdot\frac{y}{y}+\frac{5}{y}\cdot\frac{x}{x}=\frac{5y+5x}{xy}$.

119. The domain is the set of all real numbers except $x=6$.

Chapter 0 Review (page 76)

1. natural: 3, 6, 8

2. whole: 0, 3, 6, 8

3. integers: $-6, -3, 0, 3, 6, 8$

4. rational: $-6, -3, 0, \frac{1}{2}, 3, 6, 8$

5. irrational: π, $\sqrt{5}$

6. real: $-6, -3, 0, \frac{1}{2}, 3, \pi, \sqrt{5}, 6, 8$

7. prime: 3

8. composite: 6, 8

9. even integers: $-6, 0, 6, 8$

10. odd integers: $-3, 3$

11. Associative Property of Addition

12. Commutative Property of Addition

13. Associative Property of Multiplication

14. Distributive Property

15. Commutative Property of Multiplication

16. Commutative Property of Addition

17. Double Negative Rule

18. 11 13 17 19

19. 6 8 10 12 14

20. $-3 < x \le 5$

-3 5

21. $x \ge 0$ or $x < -1$

-1 0

22. $(-2, 4]$

-2 4

23. $(-\infty, 2) \cap (-5, \infty)$

$(-\infty, 2)$

2

$(-5, \infty)$

-5

$(-\infty, 2) \cap (-5, \infty)$

-5 2

24. $(-\infty, -4) \cup [6, \infty)$

$-4 \qquad 6$

25. Since $6 \geq 0$, $|6| = 6$.

26. Since $-25 < 0$, $|-25| = -(-25) = 25$.

27. Since $1 - \sqrt{2} < 0$,
$\left|1 - \sqrt{2}\right| = -\left(1 - \sqrt{2}\right) = \sqrt{2} - 1.$

28. Since $\sqrt{3} - 1 \geq 0$,
$\left|\sqrt{3} - 1\right| = \sqrt{3} - 1.$

29. distance $= |7 - (-5)| = |12| = 12$

30. $-5a^3 = -5aaa$

31. $(-5a)^2 = (-5a)(-5a)$

32. $3ttt = 3t^3$

33. $(-2b)(3b) = (-2)(3)bb = -6b^2$

34. $n^2n^4 = n^{2+4} = n^6$

35. $(p^3)^2 = p^{3\cdot 2} = p^6$

36. $(x^3y^2)^4 = (x^3)^4(y^2)^4 = x^{12}y^8$

37. $\left(\dfrac{a^4}{b^2}\right)^3 = \dfrac{(a^4)^3}{(b^2)^3} = \dfrac{a^{12}}{b^6}$

38. $\left(m^{-3}n^0\right)^2 = \left(m^{-3} \cdot 1\right)^2 = m^{-6} = \dfrac{1}{m^6}$

39. $\left(\dfrac{p^{-2}q^2}{2}\right)^3 = \left(\dfrac{q^2}{2p^2}\right)^3 = \dfrac{(q^2)^3}{(2p^2)^3} = \dfrac{q^6}{8p^6}$

40. $\dfrac{a^5}{a^8} = a^{5-8} = a^{-3} = \dfrac{1}{a^3}$

41. $\left(\dfrac{a^2}{b^3}\right)^{-2} = \left(\dfrac{b^3}{a^2}\right)^2 = \dfrac{b^6}{a^4}$

42. $\left(\dfrac{3x^2y^{-2}}{x^2y^2}\right)^{-2} = \left(\dfrac{x^2y^2}{3x^2y^{-2}}\right)^2 = \left(\dfrac{x^2y^2y^2}{3x^2}\right)^2 = \left(\dfrac{y^4}{3}\right)^2 = \dfrac{y^8}{9}$

43. $\left(\dfrac{a^{-3}b^2}{ab^{-3}}\right)^{-2} = \left(\dfrac{ab^{-3}}{a^{-3}b^2}\right)^2 = \left(\dfrac{aa^3}{b^2b^3}\right)^2 = \left(\dfrac{a^4}{b^5}\right)^2 = \dfrac{a^8}{b^{10}}$

44. $\left(\dfrac{-3x^3y}{xy^3}\right)^{-2} = \left(\dfrac{xy^3}{-3x^3y}\right)^2 = \left(\dfrac{y^2}{-3x^2}\right)^2 = \dfrac{y^4}{9x^4}$

45. $\left(-\dfrac{2m^{-2}n^0}{4m^2n^{-1}}\right)^{-3} = \left(-\dfrac{4m^2n^{-1}}{2m^{-2}n^0}\right)^3 = \left(-\dfrac{2m^2m^2}{n^1n^0}\right)^3 = \left(-\dfrac{2m^4}{n}\right)^3 = -\dfrac{8m^{12}}{n^3}$

46. $-x^2 - xy^2 = -(-3)^2 - (-3)(3)^2 = -(+9) - (-3)(9) = -9 - (-27) = -9 + 27 = 18$

47. $6750 = 6.750 \times 10^3$

48. $0.00023 = 2.3 \times 10^{-4}$

49. $4.8 \times 10^2 = 480$

50. $0.25 \times 10^{-3} = 0.00025$

51. $\dfrac{(45{,}000)(350{,}000)}{0.000105} = \dfrac{(4.5 \times 10^4)(3.5 \times 10^5)}{1.05 \times 10^{-4}} = \dfrac{4.5 \times 3.5 \times 10^4 \times 10^5}{1.05 \times 10^{-4}} = \dfrac{15.75 \times 10^9}{1.05 \times 10^{-4}}$
$= 15 \times 10^{13}$
$= 1.5 \times 10^{14}$

52. $121^{1/2} = (11^2)^{1/2} = 11$

53. $\left(\dfrac{27}{125}\right)^{1/3} = \left[\left(\dfrac{3}{5}\right)^3\right]^{1/3} = \dfrac{3}{5}$

54. $(32x^5)^{1/5} = 32^{1/5}(x^5)^{1/5} = 2x$

55. $(81a^4)^{1/4} = 81^{1/4}(a^4)^{1/4} = 3|a|$

56. $(-1000x^6)^{1/3} = (-1000)^{1/3}(x^6)^{1/3}$
$= -10x^2$

57. $(-25x^2)^{1/2} = (-25)^{1/2}(x^2)^{1/2}$
$\Rightarrow$ not a real number

58. $(x^{12}y^2)^{1/2} = (x^{12})^{1/2}(y^2)^{1/2} = x^6|y|$

59. $\left(\dfrac{x^{12}}{y^4}\right)^{-1/2} = \left(\dfrac{y^4}{x^{12}}\right)^{1/2} = \dfrac{y^2}{x^6}$

60. $\left(\dfrac{-c^{2/3}c^{5/3}}{c^{-2/3}}\right)^{1/3} = \left(\dfrac{-c^{7/3}}{c^{-2/3}}\right)^{1/3} = (-c^{9/3})^{1/3} = (-c^3)^{1/3} = -c$

61. $\left(\dfrac{a^{-1/4}a^{3/4}}{a^{9/2}}\right)^{-1/2} = \left(\dfrac{a^{9/2}}{a^{-1/4}a^{3/4}}\right)^{1/2} = \left(\dfrac{a^{9/2}}{a^{2/4}}\right)^{1/2} = \left(\dfrac{a^{9/2}}{a^{1/2}}\right)^{1/2} = (a^{8/2})^{1/2} = (a^4)^{1/2} = a^2$

62. $64^{2/3} = (64^{1/3})^2 = 4^2 = 16$

63. $32^{-3/5} = \dfrac{1}{32^{3/5}} = \dfrac{1}{(32^{1/5})^3} = \dfrac{1}{2^3} = \dfrac{1}{8}$

64. $\left(\dfrac{16}{81}\right)^{3/4} = \dfrac{16^{3/4}}{81^{3/4}} = \dfrac{(16^{1/4})^3}{(81^{1/4})^3} = \dfrac{2^3}{3^3} = \dfrac{8}{27}$

65. $\left(\dfrac{32}{243}\right)^{2/5} = \dfrac{32^{2/5}}{243^{2/5}} = \dfrac{(32^{1/5})^2}{(243^{1/5})^2} = \dfrac{2^2}{3^2} = \dfrac{4}{9}$

66. $\left(\dfrac{8}{27}\right)^{-2/3} = \left(\dfrac{27}{8}\right)^{2/3} = \dfrac{27^{2/3}}{8^{2/3}} = \dfrac{(27^{1/3})^2}{(8^{1/3})^2} = \dfrac{3^2}{2^2} = \dfrac{9}{4}$

67. $\left(\dfrac{16}{625}\right)^{-3/4} = \left(\dfrac{625}{16}\right)^{3/4} = \dfrac{625^{3/4}}{16^{3/4}} = \dfrac{(625^{1/4})^3}{(16^{1/4})^3} = \dfrac{5^3}{2^3} = \dfrac{125}{8}$

68. $(-216x^3)^{2/3} = (-216)^{2/3}(x^3)^{2/3} = 36x^2$

69. $\dfrac{p^{a/2}p^{a/3}}{p^{a/6}} = \dfrac{p^{3a/6}p^{2a/6}}{p^{a/6}} = \dfrac{p^{5a/6}}{p^{a/6}} = p^{4a/6} = p^{2a/3}$

70. $\sqrt{36} = 6$

71. $-\sqrt{49} = -7$

72. $\sqrt{\dfrac{9}{25}} = \dfrac{\sqrt{9}}{\sqrt{25}} = \dfrac{3}{5}$

73. $\sqrt[3]{\dfrac{27}{125}}=\dfrac{\sqrt[3]{27}}{\sqrt[3]{125}}=\dfrac{3}{5}$

74. $\sqrt{x^2y^4}=\sqrt{x^2}\sqrt{y^4}$
$=|x|y^2$

75. $\sqrt[3]{x^3}=x$

76. $\sqrt[4]{\dfrac{m^8n^4}{p^{16}}}=\dfrac{\sqrt[4]{m^8}\sqrt[4]{n^4}}{\sqrt[4]{p^{16}}}=\dfrac{m^2|n|}{p^4}$

77. $\sqrt[5]{\dfrac{a^{15}b^{10}}{c^5}}=\dfrac{\sqrt[5]{a^{15}}\sqrt[5]{b^{10}}}{\sqrt[5]{c^5}}=\dfrac{a^3b^2}{c}$

78. $\sqrt{50}+\sqrt{8}=\sqrt{25}\sqrt{2}+\sqrt{4}\sqrt{2}=5\sqrt{2}+2\sqrt{2}=7\sqrt{2}$

79. $\sqrt{12}+\sqrt{3}-\sqrt{27}=\sqrt{4}\sqrt{3}+\sqrt{3}-\sqrt{9}\sqrt{3}=2\sqrt{3}+\sqrt{3}-3\sqrt{3}=3\sqrt{3}-3\sqrt{3}=0$

80. $\sqrt[3]{24x^4}-\sqrt[3]{3x^4}=\sqrt[3]{8x^3}\sqrt[3]{3x}-\sqrt[3]{x^3}\sqrt[3]{3x}=2x\sqrt[3]{3x}-x\sqrt[3]{3x}=x\sqrt[3]{3x}$

81. $\dfrac{\sqrt{7}}{\sqrt{5}}=\dfrac{\sqrt{7}}{\sqrt{5}}\cdot\dfrac{\sqrt{5}}{\sqrt{5}}=\dfrac{\sqrt{35}}{5}$

82. $\dfrac{8}{\sqrt{8}}=\dfrac{8}{\sqrt{8}}\cdot\dfrac{\sqrt{2}}{\sqrt{2}}=\dfrac{8\sqrt{2}}{\sqrt{16}}=\dfrac{8\sqrt{2}}{4}$
$=2\sqrt{2}$

83. $\dfrac{1}{\sqrt[3]{2}}=\dfrac{1}{\sqrt[3]{2}}\cdot\dfrac{\sqrt[3]{4}}{\sqrt[3]{4}}=\dfrac{\sqrt[3]{4}}{\sqrt[3]{8}}=\dfrac{\sqrt[3]{4}}{2}$

84. $\dfrac{2}{\sqrt[3]{25}}=\dfrac{2}{\sqrt[3]{25}}\cdot\dfrac{\sqrt[3]{5}}{\sqrt[3]{5}}=\dfrac{2\sqrt[3]{5}}{\sqrt[3]{125}}=\dfrac{2\sqrt[3]{5}}{5}$

85. $\dfrac{\sqrt{2}}{5}=\dfrac{\sqrt{2}}{5}\cdot\dfrac{\sqrt{2}}{\sqrt{2}}=\dfrac{2}{5\sqrt{2}}$

86. $\dfrac{\sqrt{5}}{5}=\dfrac{\sqrt{5}}{5}\cdot\dfrac{\sqrt{5}}{\sqrt{5}}=\dfrac{5}{5\sqrt{5}}=\dfrac{1}{\sqrt{5}}$

87. $\dfrac{\sqrt{2x}}{3}=\dfrac{\sqrt{2x}}{3}\cdot\dfrac{\sqrt{2x}}{\sqrt{2x}}=\dfrac{2x}{3\sqrt{2x}}$

88. $\dfrac{3\sqrt[3]{7x}}{2}=\dfrac{3\sqrt[3]{7x}}{2}\cdot\dfrac{\sqrt[3]{49x^2}}{\sqrt[3]{49x^2}}=\dfrac{3\sqrt[3]{343x^3}}{2\sqrt[3]{49x^2}}$
$=\dfrac{21x}{2\sqrt[3]{49x^2}}$

89. 3rd degree, binomial

90. 2nd degree, trinomial

91. 2nd degree, monomial

92. 4th degree, trinomial

93. $2(x+3)+3(x-4)=2x+6+3x-12=5x-6$

94. $3x^2(x-1)-2x(x+3)-x^2(x+2)=3x^3-3x^2-2x^2-6x-x^3-2x^2=2x^3-7x^2-6x$

95. $(3x+2)(3x+2)=9x^2+6x+6x+4=9x^2+12x+4$

96. $(3x+y)(2x-3y)=6x^2-9xy+2xy-3y^2=6x^2-7xy-3y^2$

97. $(4a+2b)(2a-3b)=8a^2-12ab+4ab-6b^2=8a^2-8ab-6b^2$

98. $(z+3)(3z^2+z-1)=3z^3+z^2-z+9z^2+3z-3=3z^3+10z^2+2z-3$

99. $(a^n+2)(a^n-1)=a^{2n}-a^n+2a^n-2=a^{2n}+a^n-2$

100. $\left(\sqrt{2}+x\right)^2=\left(\sqrt{2}+x\right)\left(\sqrt{2}+x\right)=\left(\sqrt{2}\right)^2+x\sqrt{2}+x\sqrt{2}+x^2=2+2x\sqrt{2}+x^2$

101. $\left(\sqrt{2}+1\right)\left(\sqrt{3}+1\right)=\sqrt{6}+\sqrt{2}+\sqrt{3}+1$

102. $\left(\sqrt[3]{3}-2\right)\left(\sqrt[3]{9}+2\sqrt[3]{3}+4\right)=\sqrt[3]{27}+2\sqrt[3]{9}+4\sqrt[3]{3}-2\sqrt[3]{9}-4\sqrt[3]{3}-8=3-8=-5$

103. $\dfrac{2}{\sqrt{3}-1}=\dfrac{2}{\sqrt{3}-1}\cdot\dfrac{\sqrt{3}+1}{\sqrt{3}+1}=\dfrac{2\left(\sqrt{3}+1\right)}{\left(\sqrt{3}\right)^2-1^2}=\dfrac{2\left(\sqrt{3}+1\right)}{3-1}=\dfrac{2\left(\sqrt{3}+1\right)}{2}=\sqrt{3}+1$

104.
$$\begin{aligned}\frac{-2}{\sqrt{3}-\sqrt{2}}=\frac{-2}{\sqrt{3}-\sqrt{2}}\cdot\frac{\sqrt{3}+\sqrt{2}}{\sqrt{3}+\sqrt{2}}=\frac{-2\left(\sqrt{3}+\sqrt{2}\right)}{\left(\sqrt{3}\right)^2-\left(\sqrt{2}\right)^2}&=\frac{-2\left(\sqrt{3}+\sqrt{2}\right)}{3-2}\\&=\frac{-2\left(\sqrt{3}+\sqrt{2}\right)}{1}\\&=-2\left(\sqrt{3}+\sqrt{2}\right)\end{aligned}$$

105. $\dfrac{2x}{\sqrt{x}-2}=\dfrac{2x}{\sqrt{x}-2}\cdot\dfrac{\sqrt{x}+2}{\sqrt{x}+2}=\dfrac{2x(\sqrt{x}+2)}{\left(\sqrt{x}\right)^2-2^2}=\dfrac{2x(\sqrt{x}+2)}{x-4}$

106. $\dfrac{\sqrt{x}-\sqrt{y}}{\sqrt{x}+\sqrt{y}}=\dfrac{\sqrt{x}-\sqrt{y}}{\sqrt{x}+\sqrt{y}}\cdot\dfrac{\sqrt{x}-\sqrt{y}}{\sqrt{x}-\sqrt{y}}=\dfrac{\sqrt{x^2}-\sqrt{xy}-\sqrt{xy}+y}{\left(\sqrt{x}\right)^2-\left(\sqrt{y}\right)^2}=\dfrac{x-2\sqrt{xy}+y}{x-y}$

107. $\dfrac{\sqrt{x}+2}{5}=\dfrac{\sqrt{x}+2}{5}\cdot\dfrac{\sqrt{x}-2}{\sqrt{x}-2}=\dfrac{\left(\sqrt{x}\right)^2-2^2}{5(\sqrt{x}-2)}=\dfrac{x-4}{5(\sqrt{x}-2)}$

108. $\dfrac{1-\sqrt{a}}{a}=\dfrac{1-\sqrt{a}}{a}\cdot\dfrac{1+\sqrt{a}}{1+\sqrt{a}}=\dfrac{1^2-\left(\sqrt{a}\right)^2}{a(1+\sqrt{a})}=\dfrac{1-a}{a(1+\sqrt{a})}$

109. $\dfrac{3x^2y^2}{6x^3y}=\dfrac{y}{2x}$

110.
$$\begin{aligned}\frac{4a^2b^3+6ab^4}{2b^2}&=\frac{4a^2b^3}{2b^2}+\frac{6ab^4}{2b^2}\\&=2a^2b+3ab^2\end{aligned}$$

111.
$$\begin{array}{r} x^2 + 2x + 1 \\ 2x+3 \overline{)\, 2x^3 + 7x^2 + 8x + 3} \\ \underline{2x^3 + 3x^2} \qquad\qquad \\ 4x^2 + 8x \qquad \\ \underline{4x^2 + 6x} \qquad \\ 2x + 3 \\ \underline{2x + 3} \\ 0 \end{array}$$

112.
$$\begin{array}{r} x^3 + 2x - 3 + \frac{-6}{x^2-1} \\ x^2 - 1 \overline{)\, x^5 + 0x^4 + x^3 - 3x^2 - 2x - 3} \\ \underline{x^5 \qquad\quad - x^3} \qquad\qquad\qquad \\ 2x^3 - 3x^2 - 2x \qquad \\ \underline{2x^3 \qquad\quad - 2x} \qquad \\ -3x^2 \qquad - 3 \\ \underline{-3x^2 \qquad + 3} \\ -6 \end{array}$$

113. $3t^3 - 3t = 3t(t^2 - 1) = 3t(t+1)(t-1)$

114. $5r^3 - 5 = 5(r^3 - 1) = 5(r^3 - 1^3) = 5(r-1)(r^2 + r + 1)$

115. $6x^2 + 7x - 24 = (3x + 8)(2x - 3)$

116. $3a^2 + ax - 3a - x = a(3a + x) - 1(3a + x) = (3a + x)(a - 1)$

117. $8x^3 - 125 = (2x)^3 - 5^3 = (2x - 5)\left[(2x)^2 + (2x)(5) + 5^2\right] = (2x - 5)(4x^2 + 10x + 25)$

118. $6x^2 - 20x - 16 = 2(3x^2 - 10x - 8) = 2(3x + 2)(x - 4)$

119. $x^2 + 6x + 9 - t^2 = (x+3)(x+3) - t^2 = (x+3)^2 - t^2 = (x + 3 + t)(x + 3 - t)$

120. $3x^2 - 1 + 5x = 3x^2 + 5x - 1 \Rightarrow$ prime

121. $8z^3 + 343 = (2z)^3 + 7^3 = (2z + 7)\left[(2z)^2 - (2z)(7) + 7^2\right] = (2z + 7)(4z^2 - 14z + 49)$

122. $1 + 14b + 49b^2 = 49b^2 + 14b + 1 = (7b + 1)(7b + 1) = (7b + 1)^2$

123. $121z^2 + 4 - 44z = 121z^2 - 44z + 4 = (11z - 2)(11z - 2) = (11z - 2)^2$

124. $64y^3 - 1000 = 8(8y^3 - 125) = 8\left[(2y)^3 - 5^3\right] = 8(2y - 5)(4y^2 + 10y + 25)$

125. $2xy - 4zx - wy + 2zw = 2x(y - 2z) - w(y - 2z) = (y - 2z)(2x - w)$

126.
$$\begin{aligned} x^8 + x^4 + 1 = x^8 + 2x^4 + 1 - x^4 &= (x^4 + 1)(x^4 + 1) - x^4 \\ &= (x^4 + 1)^2 - (x^2)^2 \\ &= (x^4 + 1 + x^2)(x^4 + 1 - x^2) \\ &= (x^4 + 2x^2 + 1 - x^2)(x^4 - x^2 + 1) \\ &= \left[(x^2 + 1)(x^2 + 1) - x^2\right](x^4 - x^2 + 1) \\ &= (x^2 + 1 + x)(x^2 + 1 - x)(x^4 - x^2 + 1) \end{aligned}$$

127. $\dfrac{2 - x}{x^2 - 4x + 4} = \dfrac{-(x - 2)}{(x - 2)(x - 2)} = \dfrac{-1}{x - 2}$

128. $\dfrac{a^2 - 9}{a^2 - 6a + 9} = \dfrac{(a + 3)(a - 3)}{(a - 3)(a - 3)} = \dfrac{a + 3}{a - 3}$

129. $$\frac{x^2-4x+4}{x+2}\cdot\frac{x^2+5x+6}{x-2}=\frac{(x-2)(x-2)}{x+2}\cdot\frac{(x+2)(x+3)}{x-2}=(x-2)(x+3)$$

130. $$\frac{2y^2-11y+15}{y^2-6y+8}\cdot\frac{y^2-2y-8}{y^2-y-6}=\frac{(2y-5)(y-3)}{(y-4)(y-2)}\cdot\frac{(y-4)(y+2)}{(y-3)(y+2)}=\frac{2y-5}{y-2}$$

131. $$\frac{2t^2+t-3}{3t^2-7t+4}\div\frac{10t+15}{3t^2-t-4}=\frac{2t^2+t-3}{3t^2-7t+4}\cdot\frac{3t^2-t-4}{10t+15}$$
$$=\frac{(2t+3)(t-1)}{(3t-4)(t-1)}\cdot\frac{(3t-4)(t+1)}{5(2t+3)}=\frac{t+1}{5}$$

132. $$\frac{p^2+7p+12}{p^3+8p^2+4p}\div\frac{p^2-9}{p^2}=\frac{p^2+7p+12}{p^3+8p^2+4p}\cdot\frac{p^2}{p^2-9}=\frac{(p+3)(p+4)}{p(p^2+8p+4)}\cdot\frac{p^2}{(p+3)(p-3)}$$
$$=\frac{p(p+4)}{(p^2+8p+4)(p-3)}$$

133. $$\frac{x^2+x-6}{x^2-x-6}\cdot\frac{x^2-x-6}{x^2+x-2}\div\frac{x^2-4}{x^2-5x+6}$$
$$=\frac{x^2+x-6}{x^2-x-6}\cdot\frac{x^2-x-6}{x^2+x-2}\cdot\frac{x^2-5x+6}{x^2-4}$$
$$=\frac{(x+3)(x-2)}{(x-3)(x+2)}\cdot\frac{(x-3)(x+2)}{(x+2)(x-1)}\cdot\frac{(x-2)(x-3)}{(x+2)(x-2)}=\frac{(x+3)(x-2)(x-3)}{(x+2)^2(x-1)}$$

134. $$\left(\frac{2x+6}{x+5}\div\frac{2x^2-2x-4}{x^2-25}\right)\frac{x^2-x-2}{x^2-2x-15}=\frac{2x+6}{x+5}\cdot\frac{x^2-25}{2(x^2-x-2)}\cdot\frac{x^2-x-2}{x^2-2x-15}$$
$$=\frac{2(x+3)}{x+5}\cdot\frac{(x+5)(x-5)}{2(x-2)(x+1)}\cdot\frac{(x-2)(x+1)}{(x-5)(x+3)}=1$$

135. $$\frac{2}{x-4}+\frac{3x}{x+5}=\frac{2(x+5)}{(x-4)(x+5)}+\frac{3x(x-4)}{(x+5)(x-4)}=\frac{2x+10}{(x-4)(x+5)}+\frac{3x^2-12x}{(x-4)(x+5)}$$
$$=\frac{3x^2-10x+10}{(x-4)(x+5)}$$

136. $$\frac{5x}{x-2}-\frac{3x+7}{x+2}+\frac{2x+1}{x+2}=\frac{5x}{x-2}+\frac{-x-6}{x+2}=\frac{5x(x+2)}{(x-2)(x+2)}+\frac{(-x-6)(x-2)}{(x+2)(x-2)}$$
$$=\frac{5x^2+10x}{(x-2)(x+2)}+\frac{-x^2-4x+12}{(x-2)(x+2)}$$
$$=\frac{4x^2+6x+12}{(x-2)(x+2)}=\frac{2(2x^2+3x+6)}{(x-2)(x+2)}$$

137. $\dfrac{x}{x-1}+\dfrac{x}{x-2}+\dfrac{x}{x-3}$

$$= \frac{x(x-2)(x-3)}{(x-1)(x-2)(x-3)} + \frac{x(x-1)(x-3)}{(x-2)(x-1)(x-3)} + \frac{x(x-1)(x-2)}{(x-3)(x-1)(x-2)}$$
$$= \frac{x^3-5x^2+6x}{(x-1)(x-2)(x-3)} + \frac{x^3-4x^2+3x}{(x-1)(x-2)(x-3)} + \frac{x^3-3x^2+2x}{(x-1)(x-2)(x-3)}$$
$$= \frac{3x^3-12x^2+11x}{(x-1)(x-2)(x-3)} = \frac{x(3x^2-12x+11)}{(x-1)(x-2)(x-3)}$$

138. $\dfrac{x}{x+1}-\dfrac{3x+7}{x+2}+\dfrac{2x+1}{x+2} = \dfrac{x}{x+1}+\dfrac{-3x-7}{x+2}+\dfrac{2x+1}{x+2}$

$$= \frac{x}{x+1}+\frac{-x-6}{x+2}$$
$$= \frac{x(x+2)}{(x+1)(x+2)} + \frac{(-x-6)(x+1)}{(x+2)(x+1)}$$
$$= \frac{x^2+2x}{(x+1)(x+2)} + \frac{-x^2-7x-6}{(x+1)(x+2)} = \frac{-5x-6}{(x+1)(x+2)}$$

139. $\dfrac{3(x+1)}{x}-\dfrac{5(x^2+3)}{x^2}+\dfrac{x}{x+1} = \dfrac{3x(x+1)(x+1)}{x^2(x+1)} - \dfrac{5(x^2+3)(x+1)}{x^2(x+1)} + \dfrac{x(x^2)}{x^2(x+1)}$

$$= \frac{3x^3+6x^2+3x}{x^2(x+1)} - \frac{5x^3+5x^2+15x+15}{x^2(x+1)} + \frac{x^3}{x^2(x+1)}$$
$$= \frac{-x^3+x^2-12x-15}{x^2(x+1)}$$

140. $\dfrac{3x}{x+1}+\dfrac{x^2+4x+3}{x^2+3x+2}-\dfrac{x^2+x-6}{x^2-4} = \dfrac{3x}{x+1}+\dfrac{(x+3)(x+1)}{(x+1)(x+2)}-\dfrac{(x+3)(x-2)}{(x+2)(x-2)}$

$$= \frac{3x}{x+1}+\frac{x+3}{x+2}-\frac{x+3}{x+2} = \frac{3x}{x+1}$$

141. $\dfrac{\frac{5x}{2}}{\frac{3x^2}{8}} = \dfrac{5x}{2} \div \dfrac{3x^2}{8} = \dfrac{5x}{2}\cdot\dfrac{8}{3x^2} = \dfrac{20}{3x}$

142. $\dfrac{\frac{3x}{y}}{\frac{6x}{y^2}} = \dfrac{3x}{y} \div \dfrac{6x}{y^2} = \dfrac{3x}{y}\cdot\dfrac{y^2}{6x} = \dfrac{y}{2}$

143. $\dfrac{\frac{1}{x}+\frac{1}{y}}{x-y} = \dfrac{xy\left(\frac{1}{x}+\frac{1}{y}\right)}{xy(x-y)} = \dfrac{xy\left(\frac{1}{x}\right)+xy\left(\frac{1}{y}\right)}{xy(x-y)} = \dfrac{y+x}{xy(x-y)}$

144. $\dfrac{x^{-1}+y^{-1}}{y^{-1}-x^{-1}} = \dfrac{\frac{1}{x}+\frac{1}{y}}{\frac{1}{y}-\frac{1}{x}} = \dfrac{xy\left(\frac{1}{x}+\frac{1}{y}\right)}{xy\left(\frac{1}{y}-\frac{1}{x}\right)} = \dfrac{xy\left(\frac{1}{x}\right)+xy\left(\frac{1}{y}\right)}{xy\left(\frac{1}{y}\right)-xy\left(\frac{1}{x}\right)} = \dfrac{y+x}{x-y}$

Chapter 0 Test (page 84)

1. odd integers: $-7, 1, 3$

2. prime numbers: 3

3. Commutative Property of Addition

4. Distributive Property

5. $-4 < x \leq 2 \Rightarrow$ 

6. $(-\infty, -3) \cup [6, \infty) \Rightarrow$ 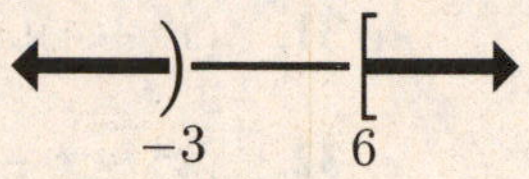

7. Since $-17 < 0$, $|-17| = -(-17) = 17$

8. If $x < 0$, then $x - 7 < 0$. Then $|x - 7| = -(x - 7)$.

9. distance $= |12 - (-4)| = |16| = 16$

10. distance $= |-12 - (-20)| = |8| = 8$

11. $x^4x^5x^2 = x^{4+5+2} = x^{11}$

12. $\dfrac{r^2r^3s}{r^4s^2} = \dfrac{r^5s}{r^4s^2} = \dfrac{r}{s}$

13. $\dfrac{(a^{-1}a^2)^{-2}}{a^{-3}} = \dfrac{(a^1)^{-2}}{a^{-3}} = \dfrac{a^{-2}}{a^{-3}} = a$

14. $\left(\dfrac{x^0x^2}{x^{-2}}\right)^6 = \left(\dfrac{x^2}{x^{-2}}\right)^6 = (x^4)^6 = x^{24}$

15. $450{,}000 = 4.5 \times 10^5$

16. $0.000345 = 3.45 \times 10^{-4}$

17. $3.7 \times 10^3 = 3{,}700$

18. $1.2 \times 10^{-3} = 0.0012$

19. $(25a^4)^{1/2} = 25^{1/2}(a^4)^{1/2} = 5a^2$

20. $\left(\dfrac{36}{81}\right)^{3/2} = \dfrac{36^{3/2}}{81^{3/2}} = \dfrac{(36^{1/2})^3}{(81^{1/2})^3} = \dfrac{216}{729} = \dfrac{8}{27}$

21. $\left(\dfrac{8t^6}{27s^9}\right)^{-2/3} = \left(\dfrac{27s^9}{8t^6}\right)^{2/3} = \dfrac{27^{2/3}(s^9)^{2/3}}{8^{2/3}(t^6)^{2/3}} = \dfrac{(27^{1/3})^2s^6}{(8^{1/3})^2t^4} = \dfrac{9s^6}{4t^4}$

22. $\sqrt[3]{27a^6} = \sqrt[3]{27}\sqrt[3]{a^6} = 3a^2$

23. $\sqrt{12} + \sqrt{27} = \sqrt{4}\sqrt{3} + \sqrt{9}\sqrt{3}$
$= 2\sqrt{3} + 3\sqrt{3} = 5\sqrt{3}$

24. $2\sqrt[3]{3x^4} - 3x\sqrt[3]{24x} = 2\sqrt[3]{x^3}\sqrt[3]{3x} - 3x\sqrt[3]{8}\sqrt[3]{3x} = 2x\sqrt[3]{3x} - 3x(2)\sqrt[3]{3x} = 2x\sqrt[3]{3x} - 6x\sqrt[3]{3x}$
$= -4x\sqrt[3]{3x}$

25. $\dfrac{x}{\sqrt{x} - 2} = \dfrac{x}{\sqrt{x} - 2} \cdot \dfrac{\sqrt{x} + 2}{\sqrt{x} + 2} = \dfrac{x(\sqrt{x} + 2)}{(\sqrt{x})^2 - 2^2} = \dfrac{x(\sqrt{x} + 2)}{x - 4}$

26. $\dfrac{\sqrt{x} - \sqrt{y}}{\sqrt{x} + \sqrt{y}} = \dfrac{\sqrt{x} - \sqrt{y}}{\sqrt{x} + \sqrt{y}} \cdot \dfrac{\sqrt{x} + \sqrt{y}}{\sqrt{x} + \sqrt{y}} = \dfrac{(\sqrt{x})^2 - (\sqrt{y})^2}{\sqrt{x^2} + \sqrt{xy} + \sqrt{xy} + \sqrt{y^2}} = \dfrac{x - y}{x + 2\sqrt{xy} + y}$

CHAPTER 0 TEST

27. $$\left(a^2+3\right)-\left(2a^2-4\right)=a^2+3-2a^2+4 \\ =-a^2+7$$

28. $(3a^3b^2)(-2a^3b^4)=-6a^6b^6$

29. $$(3x-4)(2x+7)=6x^2+21x-8x-28 \\ =6x^2+13x-28$$

30. $$(a^n+2)(a^n-3)=a^{2n}-3a^n+2a^n-6 \\ =a^{2n}-a^n-6$$

31. $(x^2+4)(x^2-4)=x^4-4x^2+4x^2-16=x^4-16$

32. $(x^2-x+2)(2x-3)=2x^3-3x^2-2x^2+3x+4x-6=2x^3-5x^2+7x-6$

33.
$$\begin{array}{r|l} & 6x+\ \ 19+\frac{34}{x-3} \\ \hline x-3 & 6x^2+\ \ x-\ \ 23 \\ & \underline{6x^2-18x} \\ & \quad 19x-\ \ 23 \\ & \quad \underline{19x-\ \ 57} \\ & \quad\quad\quad 34 \end{array}$$

34.
$$\begin{array}{r|l} & x^2+\ \ 2x+\ \ 1 \\ \hline 2x-1 & 2x^3+3x^2+0x-1 \\ & \underline{2x^3-\ \ x^2} \\ & \quad 4x^2+0x \\ & \quad \underline{4x^2-2x} \\ & \quad\quad 2x-1 \\ & \quad\quad \underline{2x-1} \\ & \quad\quad\quad 0 \end{array}$$

35. $3x+6y=3(x+2y)$

36. $x^2-100=x^2-10^2=(x+10)(x-10)$

37. $$45x^2-20y^2=5\left(9x^2-4y^2\right) \\ =5(3x+2y)(3x-2y)$$

38. $10t^2-19tw+6w^2=(5t-2w)(2t-3w)$

39. $64m^3+125n^3=(4m+5n)(16m^2-20mn+25n^2)$

40. $3a^3-648=3(a^3-216)=3(a-6)(a^2+6a+36)$

41. $$x^4-x^2-12=\left(x^2-4\right)\left(x^2+3\right) \\ =(x+2)(x-2)\left(x^2+3\right)$$

42. $6x^4+11x^2-10=(3x^2-2)(2x^2+5)$

43. $\dfrac{44p^3q^6}{33p^4q^2}=\dfrac{4q^4}{3p}$

44. $\dfrac{49-x^2}{x^2+14x+49}=\dfrac{-(x+7)(x-7)}{(x+7)(x+7)}=-\dfrac{x-7}{x+7}$

45. $\dfrac{x}{x+2}+\dfrac{2}{x+2}=\dfrac{x+2}{x+2}=1$

46. $\dfrac{x}{x+1}-\dfrac{x}{x-1}=\dfrac{x(x-1)}{(x+1)(x-1)}-\dfrac{x(x+1)}{(x+1)(x-1)}=\dfrac{x^2-x-x^2-x}{(x+1)(x-1)}=\dfrac{-2x}{(x+1)(x-1)}$

47. $\dfrac{x^2+x-20}{x^2-16}\cdot\dfrac{x^2-25}{x-5}=\dfrac{(x+5)(x-4)}{(x+4)(x-4)}\cdot\dfrac{(x+5)(x-5)}{x-5}=\dfrac{(x+5)^2}{x+4}$

48. $\dfrac{x+2}{x^2+2x+1}\div\dfrac{x^2-4}{x+1}=\dfrac{x+2}{(x+1)(x+1)}\cdot\dfrac{x+1}{(x+2)(x-2)}=\dfrac{1}{(x+1)(x-2)}$

49. $\frac{\frac{1}{a}+\frac{1}{b}}{\frac{1}{b}} = \frac{ab\left(\frac{1}{a}+\frac{1}{b}\right)}{ab\left(\frac{1}{b}\right)} = \frac{ab\left(\frac{1}{a}\right)+ab\left(\frac{1}{b}\right)}{a} = \frac{b+a}{a}$

50. $\frac{x^{-1}}{x^{-1}+y^{-1}} = \frac{\frac{1}{x}}{\frac{1}{x}+\frac{1}{y}} = \frac{xy\left(\frac{1}{x}\right)}{xy\left(\frac{1}{x}+\frac{1}{y}\right)} = \frac{y}{xy\left(\frac{1}{x}\right)+xy\left(\frac{1}{y}\right)} = \frac{y}{y+x}$

Exercises 1.1 (page 95)

1. root, solution **3.** no **5.** linear **7.** one

9. $2x+5=-17$; no restrictions

11. $\frac{1}{x}=12;\ x\neq 0$

13.
$$\frac{8}{x-6}=\frac{5}{x+2}$$
$$x-6\neq 0 \quad x+2\neq 0$$
$$x\neq 6 \quad x\neq -2$$
$$x\neq 6,\ x\neq -2$$

15.
$$\frac{1}{x-3}=\frac{5x}{x^2-16}$$
$$\frac{1}{x-3}=\frac{5x}{(x+4)(x-4)}$$
$$x-3\neq 0 \quad x+4\neq 0 \quad x-4\neq 0$$
$$x\neq 3 \quad x\neq -4 \quad x\neq 4$$
$$x\neq 3,\ x\neq -4,\ x\neq 4$$

17.
$$2x+5=15$$
$$2x+5-5=15-5$$
$$2x=10$$
$$\frac{2x}{2}=\frac{10}{2}$$
$$x=5$$
conditional equation

19.
$$2(n+2)-5=2n$$
$$2n+4-5=2n$$
$$2n-1=2n$$
$$2n-2n-1=2n-2n$$
$$-1\neq 0$$
no solution; contradiction

21.
$$\frac{x+7}{2}=7$$
$$2\cdot\frac{x+7}{2}=2(7)$$
$$x+7=14$$
$$x+7-7=14-7$$
$$x=7$$
conditional equation

23.
$$2(a+1)=3(a-2)-a$$
$$2a+2=3a-6-a$$
$$2a+2=2a-6$$
$$2a-2a+2=2a-2a-6$$
$$2\neq -6$$
no solution; contradiction

25.
$$\begin{aligned} 3(x-3) &= \frac{6x-18}{2} \\ 3x-9 &= \frac{6x-18}{2} \\ 2(3x-9) &= 2\cdot\frac{6x-18}{2} \\ 6x-18 &= 6x-18 \end{aligned}$$
all real numbers; identity

27.
$$\begin{aligned} \frac{3}{b-3} &= 1 \\ (b-3)\cdot\frac{3}{b-3} &= (b-3)(1) \\ 3 &= b-3 \\ 3+3 &= b-3+3 \\ 6 &= b \end{aligned}$$
conditional equation

29.
$$\begin{aligned} 2x^2+5x-3 &= (2x-1)(x+3) \\ 2x^2+5x-3 &= 2x^2+5x-3 \end{aligned}$$
all real numbers; identity

31.
$$\begin{aligned} 2x+7 &= 10-x \\ 3x+7 &= 10 \\ 3x &= 3 \\ x &= 1 \end{aligned}$$

33.
$$\begin{aligned} 5(x-2) &= 2(x+4) \\ 5x-10 &= 2x+8 \\ 3x-10 &= 8 \\ 3x &= 18 \\ x &= 6 \end{aligned}$$

35.
$$\begin{aligned} 7(2x+5)-6(x+8) &= 7 \\ 14x+35-6x-48 &= 7 \\ 8x-13 &= 7 \\ 8x &= 20 \\ x &= \frac{20}{8} = \frac{5}{2} \end{aligned}$$

37.
$$\begin{aligned} \frac{5}{3}z-8 &= 7 \\ \frac{5}{3}z &= 15 \\ 3\cdot\frac{5}{3}z &= 3(15) \\ 5z &= 45 \\ z &= 9 \end{aligned}$$

39.
$$\begin{aligned} \frac{z}{5}+2 &= 4 \\ \frac{z}{5} &= 2 \\ 5\cdot\frac{z}{5} &= 5(2) \\ z &= 10 \end{aligned}$$

41.
$$\begin{aligned} \frac{3x-2}{3} &= 2x+\frac{7}{3} \\ 3\cdot\frac{3x-2}{3} &= 3\left(2x+\frac{7}{3}\right) \\ 3x-2 &= 6x+7 \\ -3x-2 &= 7 \\ -3x &= 9 \\ x &= -3 \end{aligned}$$

43.
$$\begin{aligned} \frac{3x+1}{20} &= \frac{1}{2} \\ 20\cdot\frac{3x+1}{20} &= 20\cdot\frac{1}{2} \\ 3x+1 &= 10 \\ 3x &= 9 \\ x &= 3 \end{aligned}$$

45.
$$\frac{3+x}{3}+\frac{x+7}{2}=4x+1$$
$$6\left(\frac{3+x}{3}+\frac{x+7}{2}\right)=6(4x+1)$$
$$2(3+x)+3(x+7)=24x+6$$
$$6+2x+3x+21=24x+6$$
$$5x+27=24x+6$$
$$-19x+27=6$$
$$-19x=-21$$
$$x=\frac{21}{19}$$

47.
$$\frac{3}{2}(3x-2)-10x-4=0$$
$$2\left[\frac{3}{2}(3x-2)-10x-4\right]=2(0)$$
$$3(3x-2)-20x-8=0$$
$$9x-6-20x-8=0$$
$$-11x-14=0$$
$$-11x=14$$
$$x=-\frac{14}{11}$$

49.
$$\frac{(y+2)^2}{3}=y+2+\frac{y^2}{3}$$
$$3\left[\frac{(y+2)^2}{3}\right]=3\left(y+2+\frac{y^2}{3}\right)$$
$$(y+2)^2=3y+6+y^2$$
$$y^2+4y+4=y^2+3y+6$$
$$4y+4=3y+6$$
$$y+4=6$$
$$y=2$$

51.
$$x(x+2)=(x+1)^2-1$$
$$x^2+2x=(x+1)(x+1)-1$$
$$x^2+2x=x^2+2x+1-1$$
$$x^2+2x=x^2+2x$$
$0=0 \Rightarrow$ all real numbers

53.
$$\frac{4}{x}-\frac{2}{5}=\frac{6}{x}$$
$$5x\left(\frac{4}{x}-\frac{2}{5}\right)=5x\cdot\frac{6}{x}$$
$$20-2x=30$$
$$-2x=10$$
$$x=-5$$

55.
$$\frac{2}{x+1}+\frac{1}{3}=\frac{1}{x+1}$$
$$3(x+1)\left(\frac{2}{x+1}+\frac{1}{3}\right)=3(x+1)\cdot\frac{1}{x+1}$$
$$6+1(x+1)=3(1)$$
$$6+x+1=3$$
$$x+7=3$$
$$x=-4$$

57.
$$\frac{9t+6}{t(t+3)}=\frac{7}{t+3}$$
$$t(t+3)\left[\frac{9t+6}{t(t+3)}\right]=t(t+3)\cdot\frac{7}{t+3}$$
$$9t+6=7t$$
$$2t+6=0$$
$$2t=-6$$
$$t=-3$$
The answer does not check. $\Rightarrow$ no solution

59.
$$\frac{2}{(a-7)(a+2)} = \frac{4}{(a+3)(a+2)}$$
$$(a-7)(a+2)(a+3)\cdot\frac{2}{(a-7)(a+2)} = (a-7)(a+2)(a+3)\cdot\frac{4}{(a+3)(a+2)}$$
$$2(a+3) = 4(a-7)$$
$$2a+6 = 4a-28$$
$$-2a = -34$$
$$a = 17$$

61.
$$\frac{2x+3}{x^2+5x+6} + \frac{3x-2}{x^2+x-6} = \frac{5x-2}{x^2-4}$$
$$\frac{2x+3}{(x+3)(x+2)} + \frac{3x-2}{(x+3)(x-2)} = \frac{5x-2}{(x+2)(x-2)}$$
$$(x-2)(2x+3)+(x+2)(3x-2) = (x+3)(5x-2) \quad \{\text{multiply by common denominator}\}$$
$$2x^2-x-6+3x^2+4x-4 = 5x^2+13x-6$$
$$5x^2+3x-10 = 5x^2+13x-6$$
$$3x-10 = 13x-6$$
$$-10x = 4$$
$$x = -\frac{4}{10} = -\frac{2}{5}$$

63.
$$\frac{3x+5}{x^3+8} + \frac{3}{x^2-4} = \frac{2(3x-2)}{(x-2)(x^2-2x+4)}$$
$$\frac{3x+5}{(x+2)(x^2-2x+4)} + \frac{3}{(x+2)(x-2)} = \frac{2(3x-2)}{(x-2)(x^2-2x+4)}$$
$$(x-2)(3x+5)+\left(x^2-2x+4\right)(3) = 2(x+2)(3x-2) \quad \{\text{multiply by common denominator}\}$$
$$3x^2-x-10+3x^2-6x+12 = 6x^2+8x-8$$
$$6x^2-7x+2 = 6x^2+8x-8$$
$$-15x = -10$$
$$x = \frac{-10}{-15} = \frac{2}{3}$$

65.
$$\frac{1}{11-n} - \frac{2(3n-1)}{-7n^2+74n+33} = \frac{1}{7n+3}$$
$$\frac{-1}{n-11} + \frac{2(3n-1)}{7n^2-74n-33} = \frac{1}{7n+3}$$
$$\frac{-1}{n-11} + \frac{6n-2}{(7n+3)(n-11)} = \frac{1}{7n+3}$$
$$-(7n+3)+6n-2 = (n-11)1 \quad \{\text{multiply by common denominator}\}$$
$$-7n-3+6n-2 = n-11$$
$$-n-5 = n-11$$
$$-2n = -6$$
$$n = 3$$

67.
$$\begin{aligned}\frac{5}{y+4}+\frac{2}{y+2}&=\frac{6}{y+2}-\frac{1}{y^2+6y+8}\\ \frac{5}{y+4}&=\frac{4}{y+2}-\frac{1}{(y+2)(y+4)}\\ 5(y+2)&=4(y+4)-1 \quad \{\text{multiply by common denominator}\}\\ 5y+10&=4y+16-1\\ 5y+10&=4y+15\\ y&=5\end{aligned}$$

69.
$$\begin{aligned}\frac{3y}{6-3y}+\frac{2y}{2y+4}&=\frac{8}{4-y^2}\\ \frac{3y}{3(2-y)}+\frac{2y}{2(y+2)}&=\frac{8}{(2+y)(2-y)}\\ \frac{y}{2-y}+\frac{y}{2+y}&=\frac{8}{(2+y)(2-y)}\\ y(2+y)+y(2-y)&=8 \quad \{\text{multiply by common denominator}\}\\ 2y+y^2+2y-y^2&=8\\ 4y&=8\\ y&=2 \Rightarrow \text{The solution does not check, so the equation has no solution.}\end{aligned}$$

71.
$$\begin{aligned}\frac{a}{a+2}-1&=-\frac{3a+2}{a^2+4a+4}\\ \frac{a}{a+2}-\frac{1}{1}&=-\frac{3a+2}{(a+2)(a+2)}\\ (a+2)(a+2)\left[\frac{a}{a+2}-1\right]&=(a+2)(a+2)\cdot\left[-\frac{3a+2}{(a+2)(a+2)}\right]\\ a(a+2)-(a+2)(a+2)&=-(3a+2)\\ a^2+2a-\left(a^2+4a+4\right)&=-3a-2\\ a^2+2a-a^2-4a-4&=-3a-2\\ -2a-4&=-3a-2\\ a&=2\end{aligned}$$

73.
$$\begin{aligned}f&=ma\\ \frac{f}{a}&=\frac{ma}{a}\\ \frac{f}{a}&=m\end{aligned}$$

75.
$$\begin{aligned}P&=2l+2w\\ P-2l&=2w\\ \frac{P-2l}{2}&=\frac{2w}{2}\\ \frac{P-2l}{2}&=w\end{aligned}$$

77.
$$\begin{aligned}V&=\frac{1}{3}\pi r^2h\\ 3V&=3\cdot\frac{1}{3}\pi r^2h\\ 3V&=\pi r^2h\\ \frac{3V}{\pi h}&=\frac{\pi r^2h}{\pi h}\\ \frac{3V}{\pi h}&=r^2\end{aligned}$$

79.
$$\begin{aligned} P_n &= L + \frac{si}{f} \\ P_n - L &= \frac{si}{f} \\ f(P_n - L) &= f \cdot \frac{si}{f} \\ f(P_n - L) &= si \\ \frac{f(P_n - L)}{i} &= \frac{si}{i} \\ \frac{f(P_n - L)}{i} &= s \end{aligned}$$

81.
$$\begin{aligned} F &= \frac{mMg}{r^2} \\ Fr^2 &= \frac{mMg}{r^2} \cdot r^2 \\ Fr^2 &= mMg \\ \frac{Fr^2}{Mg} &= \frac{mMg}{Mg} \\ \frac{Fr^2}{Mg} &= m \end{aligned}$$

83.
$$\begin{aligned} \frac{x}{a} + \frac{y}{b} &= 1 \\ \frac{y}{b} &= 1 - \frac{x}{a} \\ b \cdot \frac{y}{b} &= b\left(1 - \frac{x}{a}\right) \\ y &= b\left(1 - \frac{x}{a}\right) \end{aligned}$$

85.
$$\begin{aligned} \frac{1}{r} &= \frac{1}{r_1} + \frac{1}{r_2} \\ rr_1r_2 \cdot \frac{1}{r} &= rr_1r_2\left(\frac{1}{r_1} + \frac{1}{r_2}\right) \\ r_1r_2 &= rr_2 + rr_1 \\ r_1r_2 &= r(r_2 + r_1) \\ \frac{r_1r_2}{r_2 + r_1} &= \frac{r(r_2 + r_1)}{r_2 + r_1} \\ \frac{r_1r_2}{r_2 + r_1} &= r \end{aligned}$$

87.
$$\begin{aligned} l &= a + (n-1)d \\ l &= a + nd - d \\ l - a + d &= nd \\ \frac{l - a + d}{d} &= \frac{nd}{d} \\ \frac{l - a + d}{d} &= n \end{aligned}$$

89.
$$\begin{aligned} a &= (n-2)\frac{180}{n} \\ an &= (n-2)\frac{180}{n} \cdot n \\ an &= (n-2)180 \\ an &= 180n - 360 \\ 360 &= 180n - an \\ 360 &= n(180 - a) \Rightarrow n = \frac{360}{180 - a} \end{aligned}$$

91.
$$\begin{aligned} R &= \frac{1}{\frac{1}{r_1} + \frac{1}{r_2} + \frac{1}{r_3}} \\ R &= \frac{r_1r_2r_3(1)}{r_1r_2r_3\left(\frac{1}{r_1} + \frac{1}{r_2} + \frac{1}{r_3}\right)} \\ R &= \frac{r_1r_2r_3}{r_2r_3 + r_1r_3 + r_1r_2} \\ R(r_2r_3 + r_1r_3 + r_1r_2) &= r_1r_2r_3 \\ Rr_2r_3 + Rr_1r_3 + Rr_1r_2 &= r_1r_2r_3 \\ Rr_1r_3 + Rr_1r_2 - r_1r_2r_3 &= -Rr_2r_3 \\ r_1(Rr_3 + Rr_2 - r_2r_3) &= -Rr_2r_3 \\ r_1 &= \frac{-Rr_2r_3}{Rr_3 + Rr_2 - r_2r_3} \end{aligned}$$

93. Answers may vary.

95. Answers may vary.

97.
$$\begin{aligned} 4x + 5(x-3) &= 9x - 15 \\ 4x + 5x - 15 &= 9x - 15 \\ 9x - 15 &= 9x - 15 \end{aligned}$$
False. The equation is an identity.

99.
$$\begin{aligned} -2x - 8 &= -(2x+8) \\ -2x - 8 &= -2x - 8 \end{aligned}$$
True. The equation is an identity, so all real numbers are solutions.

101.
$$\begin{aligned} \frac{1}{\left(\frac{1}{x-3}\right)} + \frac{1}{\left(\frac{1}{x-4}\right)} &= 7 \\ (x-3) + (x-4) &= 7 \\ 2x - 7 &= 7 \\ 2x &= 14 \\ x &= 7 \quad \text{True.} \end{aligned}$$

Exercises 1.2 (page 104)

1. add

3. amount

5. rate, time

7. Let x = the score on the final exam. Since the final is weighted as two test grades, it counts as two test grades.
$$\begin{aligned} \frac{\boxed{\text{Sum of scores}}}{7} &= 80 \\ \frac{60 + 78 + 80 + 90 + 88 + 2x}{7} &= 80 \\ \frac{2x + 396}{7} &= 80 \\ 2x + 396 &= 560 \\ 2x &= 164 \\ x &= 82 \end{aligned}$$
His score on the final exam needs to be 82.

9. Let x = the score on the first exam. Then $x + 5$ = the score on the midterm, and $x + 13$ = the score on the final.
$$\begin{aligned} \frac{\boxed{\text{Sum of scores}}}{3} &= 90 \\ \frac{x + x + 5 + x + 13}{3} &= 90 \\ \frac{3x + 18}{3} &= 90 \\ 3x + 18 &= 270 \\ 3x &= 252 \\ x &= 84 \end{aligned}$$
His score on the first exam was 84.

11. Let x = the program development score.
$$\begin{aligned} \frac{\boxed{\text{Sum of scores}}}{4} &= 86 \\ \frac{82 + 90 + x + 78}{4} &= 86 \\ \frac{x + 250}{4} &= 86 \\ x + 250 &= 344 \\ x &= 94 \end{aligned}$$
The program development score was 94.

13. Let x = the number of locks replaced.
$$\begin{aligned} 40 + 28 \cdot \boxed{\text{Number of locks}} &= 236 \\ 40 + 28x &= 236 \\ 28x &= 196 \\ x &= 7 \end{aligned}$$
7 locks can be changed for \$236.

EXERCISES 1.2

15. Let $x =$ the width.
Then $x + 26 =$ the height.

$$\boxed{\text{Perimeter}} = 92$$
$$2x + 2(x + 26) = 92$$
$$2x + 2x + 52 = 92$$
$$4x = 40$$
$$x = 10$$

The dimensions are 10 ft by 36 ft.

17.
$$\boxed{\text{Perimeter}} = 14$$
$$x + (x + 2) + x + (x + 2) = 14$$
$$4x + 4 = 14$$
$$4x = 10$$
$$x = \frac{5}{2} = 2\frac{1}{2}$$

The width is $2\frac{1}{2}$ feet.

19. Let $x =$ the width of the border.

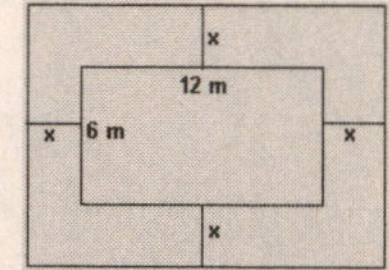

$$\boxed{\text{Perimeter of fence}} = 116$$
$$2(2x + 6) + 2(2x + 12) = 116$$
$$4x + 12 + 4x + 24 = 116$$
$$8x + 36 = 116$$
$$8x = 80$$
$$x = 10$$

The border has a width of 10 meters.

21.
$$\boxed{\text{Total Area}} = 2 \cdot \boxed{\text{Triangular Area}}$$
$$20x + \frac{1}{2}(16)(20) = 2 \cdot \frac{1}{2}(16)(20)$$
$$20x + 160 = 320$$
$$20x = 160$$
$$x = 8$$

The dimensions are 8 feet by 20 feet.

23.
$$\boxed{\begin{array}{c}\text{New}\\\text{Area}\end{array}} = \boxed{\begin{array}{c}\text{Old}\\\text{Area}\end{array}} + 0.50 \cdot \boxed{\begin{array}{c}\text{Old}\\\text{Area}\end{array}}$$
$$12(x + 10) + 12x = 12(x + 10) + 0.50 \cdot 12(x + 10)$$
$$12x + 120 + 12x = 12x + 120 + 6x + 60$$
$$24x + 120 = 18x + 180$$
$$6x = 60$$
$$x = 10 \Rightarrow \text{The length of the living room is } x + 10 = 20 \text{ feet.}$$

25. Let $x =$ the amount invested at 4%. Then $16000 - x =$ the amount invested at 6%.

$$\boxed{\text{Interest at 4\%}} + \boxed{\text{Interest at 6\%}} = \boxed{\text{Total interest}}$$
$$0.04x + 0.06(16000 - x) = 815$$
$$0.04x + 960 - 0.06x = 815$$
$$-0.02x = -145$$
$$x = 7250$$

\$7250 was invested at 4% and \$8750 was invested at 6%.

EXERCISES 1.2

27. Let $x =$ the amount invested at 7%.

$\boxed{\text{Interest at 7\%}} + \boxed{\text{Interest at 9\%}} = \boxed{\text{Total interest}}$

$$0.07x + 0.09(20000) = 5000$$
$$0.07x + 1800 = 5000$$
$$0.07x = 3200$$
$$x \approx 45714.29$$

She needs to invest \$45,714.29 at 7% to reach her goal.

29. Let $x =$ the amount invested at each rate.

$\boxed{\text{Interest at 6\%}} + \boxed{\text{Interest at 7\%}} + \boxed{\text{Interest at 8\%}} = \boxed{\text{Total interest}}$

$$0.06x + 0.07x + 0.08x = 2037$$
$$0.21x = 2037$$
$$x = 9700$$

\$9,700 was invested at each rate, for a total investment of \$29,100.

31. Let $x =$ the number of full-price tickets sold. Then $585 - x =$ the number of student tickets sold.

$2.50 \cdot \boxed{\text{\# of full-price}} + 1.75 \cdot \boxed{\text{\# of student}} = 1217.25$

$$2.50x + 1.75(585 - x) = 1217.25$$
$$2.50x + 1023.75 - 1.75x = 1217.25$$
$$0.75x = 193.50$$
$$x = 258 \Rightarrow \text{There were 327 student tickets sold.}$$

33. Let $p =$ the original price.

$\boxed{\text{Original price}} - \boxed{\text{Discount}} = \boxed{\text{New price}}$

$$p - 0.08p = 413.08$$
$$0.92p = 413.08$$
$$p = 449$$

The original price was \$449.

35. Let $w =$ the wholesale cost.

$\boxed{\text{Wholesale cost}} + \boxed{\text{Markup}} = \boxed{\text{Selling price}}$

$$w + 0.70w = 365.50$$
$$1.70w = 365.50$$
$$w = 215$$

The wholesale cost is \$215.

37. Let $x =$ # of plates for equal costs.

$\boxed{\text{Cost of 1st machine}} = \boxed{\text{Cost of 2nd machine}}$

$$600 + 3x = 800 + 2x$$
$$x = 200$$

The break point is 200 plates.

39. Let $x =$ # of computers to break even.

$\boxed{\text{Income}} = \boxed{\text{Expenses}}$

$$1275x = 8925 + 850x$$
$$425x = 8925$$
$$x = 21$$

21 computers need to be sold to break even.

EXERCISES 1.2

41. Let $x =$ days for both working together.

$$\boxed{\text{Man in 1 day}} + \boxed{\text{Roofer in 1 day}} = \boxed{\text{Total in 1 day}}$$

$$\frac{1}{7} + \frac{1}{4} = \frac{1}{x}$$

$$28x\left(\frac{1}{7} + \frac{1}{4}\right) = 28x\left(\frac{1}{x}\right)$$

$$4x + 7x = 28$$

$$11x = 28$$

$$x = \frac{28}{11} = 2\frac{6}{11}$$

They can roof the house in $2\frac{6}{11}$ days.

43. Let $x =$ hours for both working together.

$$\boxed{\text{Woman in 1 hour}} + \boxed{\text{Man in 1 hour}} = \boxed{\text{Total in 1 hour}}$$

$$\frac{1}{2} + \frac{1}{4} = \frac{1}{x}$$

$$4x\left(\frac{1}{2} + \frac{1}{4}\right) = 4x\left(\frac{1}{x}\right)$$

$$2x + x = 4$$

$$3x = 4$$

$$x = \frac{4}{3} = 1\frac{1}{3}$$

They can mow the lawn in $1\frac{1}{3}$ hours.

45. Let $x =$ hours for pool to fill with drain open.

$$\boxed{\text{Pipe in 1 hour}} - \boxed{\text{Drain in 1 hour}} = \boxed{\text{Total in 1 hour}}$$

$$\frac{1}{10} - \frac{1}{19} = \frac{1}{x}$$

$$190x\left(\frac{1}{10} - \frac{1}{19}\right) = 190x\left(\frac{1}{x}\right)$$

$$19x - 10x = 190$$

$$9x = 190$$

$$x = \frac{190}{9} = 21\frac{1}{9}$$

The pool can be filled in $21\frac{1}{9}$ hours.

47. Let $x =$ the ounces of water added.

$$\boxed{\text{Oz of alc. at start}} + \boxed{\text{Oz of alc. added}} = \boxed{\text{Oz of alc. at end}}$$

$$0.15(20) + 0(x) = 0.10(20 + x)$$

$$3 = 2 + 0.1x$$

$$1 = 0.1x$$

$$\frac{1}{0.1} = x$$

$$10 = x$$

10 oz of water should be added.

49. Let $x =$ the liters of liquid replaced with pure antifreeze.

$$\boxed{\text{Liters of a.f. at start}} - \boxed{\text{Liters of a.f. removed}} + \boxed{\text{Liters of a.f. replaced}} = \boxed{\text{Liters of a.f. at end}}$$

$$0.40(6) - 0.40x + x = 0.50(6)$$

$$2.4 + 0.6x = 3$$

$$0.6x = 0.6$$

$x = 1 \Rightarrow$ 1 liter should be replaced with pure antifreeze.

51. Let $x =$ the liters of pure alcohol added.

$$\boxed{\text{Liters of alcohol at start}} + \boxed{\text{Liters of alcohol added}} = \boxed{\text{Liters of alcohol at end}}$$

$$0.20(1) + x = 0.25(1 + x)$$

$$0.20 + x = 0.25 + 0.25x$$

$$0.75x = 0.05$$

$x = \frac{0.05}{0.75} = \frac{1}{15} \Rightarrow \frac{1}{15}$ of a liter of pure alcohol should be added.

EXERCISES 1.2

53. Let $x =$ the gallons of pure chlorine added.

[Gallons of chlorine at start] + [Gallons of chlorine added] = [Gallons of chlorine at end]

$$\begin{aligned} 0(15000) + x &= 0.0003(15000 + x) \\ x &= 4.5 + 0.0003x \\ 0.9997x &= 4.5 \\ x &\approx 4.5 \Rightarrow \text{About 4.5 gallons of pure chlorine should be added.} \end{aligned}$$

55. Let $x =$ the liters of water evaporated.

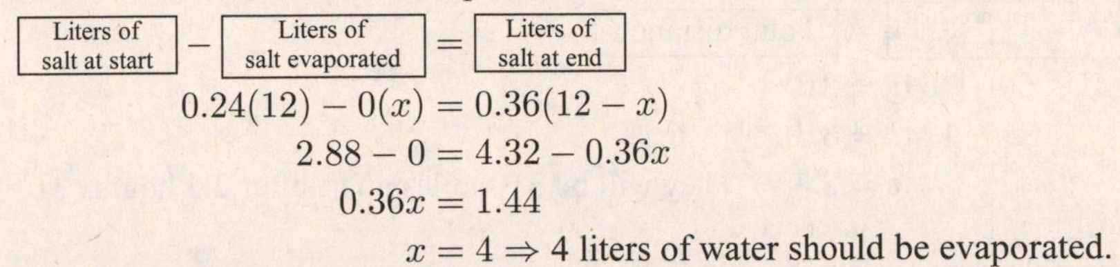

[Liters of salt at start] − [Liters of salt evaporated] = [Liters of salt at end]

$$\begin{aligned} 0.24(12) - 0(x) &= 0.36(12 - x) \\ 2.88 - 0 &= 4.32 - 0.36x \\ 0.36x &= 1.44 \\ x &= 4 \Rightarrow \text{4 liters of water should be evaporated.} \end{aligned}$$

57. Let $x =$ the pounds of extra-lean hamburger used.

[Pounds of fat in hamburger] + [Pounds of fat in lean hamburger] = [Pounds of fat in mixture]

$$\begin{aligned} 0.15(30) + 0.07(x) &= 0.10(30 + x) \\ 4.5 + 0.07x &= 3 + 0.1x \\ 1.5 &= 0.03x \\ 50 &= x \end{aligned}$$

50 pounds of the extra-lean hamburger should be used.

59. Let $x =$ the gallons of 5% solution used.

[Gallons of alc. in 5% solution] + [Gallons of alc. in 1% solution] = [Gallons of alc. in 2% solution]

$$\begin{aligned} 0.05(x) + 0.01(90) &= 0.02(x + 90) \\ 0.05x + 0.9 &= 0.02x + 1.8 \\ 0.03x &= 0.9 \\ x &= 30 \end{aligned}$$

30 gallons of the 5% solution should be used.

61. Since the mixture is to be 25% barley, there will be $0.25(2400) = 600$ pounds of barley used. Thus, the other 1800 pounds will be either oats or soybean meal. Let $x =$ the number of pounds of oats used. Then $1800 - x =$ the number of pounds of meal used.

[Pounds of protein from barley] + [Pounds of protein from oats] + [Pounds of protein from soybean meal] = [Total pounds of protein]

$$\begin{aligned} 0.117(600) + 0.118x + 0.445(1800 - x) &= 0.14(2400) \\ 70.2 + 0.118x + 801 - 0.445x &= 336 \\ 871.2 - 0.327x &= 336 \\ -0.327x &= -535.2 \\ x &\approx 1637 \end{aligned}$$

The farmer should use 600 pounds of barley, 1,637 pounds of oats and 163 pounds of soybean meal.

63. Let $r =$ his first rate. Then $r + 26 =$ his return rate.

$\boxed{\text{Distance to city}} = \boxed{\text{Return distance}}$

$$5r = 3(r + 26)$$
$$5r = 3r + 78$$
$$2r = 78$$
$$r = 39 \Rightarrow \text{He drove 39 mph going and 65 mph returning.}$$

65. Let $t =$ the time the cars travel.

$\boxed{\text{Distance 1st car travels}} + \boxed{\text{Distance 2nd car travels}} = \boxed{\text{Total distance}}$

$$60t + 64t = 310$$
$$124t = 310$$
$$t = 2.5 \Rightarrow \text{They will be 310 miles apart after 2.5 hours.}$$

67. Let $t =$ the time the runners run.

$\boxed{\text{Distance 1st runs}} + \boxed{\text{Distance 2nd runs}} = \boxed{\text{Distance between them (in miles)}}$

$$8t + 10t = \frac{440}{1760}$$
$$18t = \frac{1}{4}$$
$$t = \frac{1}{72} \text{ hour} = \frac{1}{72}(60 \text{ minutes}) = \frac{5}{6} \text{ minute} = 50 \text{ seconds}$$

They will meet after 50 seconds.

69. Let $r =$ the speed of the boat in still water.
Then the speed of the boat is $r + 2$ downstream and $r - 2$ upstream.

$\boxed{\text{Time upstream}} = \boxed{\text{Time downstream}}$ {Note: Time = Distance ÷ Rate}

$$\frac{5}{r-2} = \frac{7}{r+2}$$
$$(r+2)(r-2)\frac{5}{r-2} = (r+2)(r-2)\frac{7}{r+2}$$
$$5(r+2) = 7(r-2)$$
$$5r + 10 = 7r - 14$$
$$24 = 2r$$
$$12 = r \Rightarrow \text{The speed of the boat is 12 mph.}$$

71.
$$V = \pi r^2 h$$
$$712.51 = \pi(4.5)^2 d$$
$$\frac{712.51}{\pi(4.5)^2} = d$$
$$11.2 \approx d$$

The hole is about 11.2 millimeters deep.

73-75. Answers may vary.

Exercises 1.3 (page 118)

1. imaginary

3. imaginary

5. $2 - 5i$

7. real

9. $\sqrt{-144} = \sqrt{-1}\sqrt{144} = 12i$

11. $-\sqrt{-128} = -\sqrt{-1}\sqrt{64}\sqrt{2} = -8i\sqrt{2}$

13. $-2\sqrt{-24} = -2\sqrt{-1}\sqrt{24} = -2i \cdot 2\sqrt{6}$
$= -4i\sqrt{6}$

15. $\sqrt{-\frac{50}{9}} = \sqrt{-1} \cdot \frac{\sqrt{50}}{\sqrt{9}} = i \cdot \frac{5\sqrt{2}}{3}$
$= \frac{5\sqrt{2}}{3}i$

17. $-7\sqrt{-\frac{3}{8}} = -7\sqrt{-1} \cdot \frac{\sqrt{3}}{\sqrt{8}} = -7i \cdot \frac{\sqrt{3}}{\sqrt{8}} \cdot \frac{\sqrt{2}}{\sqrt{2}} = -7i \cdot \frac{\sqrt{6}}{\sqrt{16}} = -\frac{7\sqrt{6}}{4}i$

19. Equate real parts: $\boxed{x = 3}$ Equate imaginary parts: $x + y = 8$
$3 + y = 8$; $\boxed{y = 5}$

21. Equate real parts: $3x = 2$
$\boxed{x = \frac{2}{3}}$
Equate imaginary parts: $-2y = x + y$
$-3y = x$
$y = -\frac{1}{3}x$
$y = -\frac{1}{3} \cdot \frac{2}{3}$; $\boxed{y = -\frac{2}{9}}$

23. $(2 - 7i) + (3 + i) = 2 - 7i + 3 + i$
$= 5 - 6i$

25. $(5 - 6i) - (7 + 4i) = 5 - 6i - 7 - 4i$
$= -2 - 10i$

27. $(14i + 2) + \left(2 - \sqrt{-16}\right) = (14i + 2) + (2 - 4i) = 14i + 2 + 2 - 4i = 4 + 10i$

29. $\left(3 + \sqrt{-4}\right) - \left(2 + \sqrt{-9}\right) = (3 + 2i) - (2 + 3i) = 3 + 2i - 2 - 3i = 1 - i$

31. $(4 + 7i) + (8 - 2i) - (5 + 4i) = 4 + 7i + 8 - 2i - 5 - 4i = 7 + i$

33. $\left(3 + \sqrt{-16}\right) - \left(4 - \sqrt{-36}\right) + \left(5 - \sqrt{-144}\right) = (3 + 4i) - (4 - 6i) + (5 - 12i)$
$= 3 + 4i - 4 + 6i + 5 - 12i = 4 - 2i$

35. $-5(3 + 5i) = -15 - 25i$

37. $7i(4 - 8i) = 28i - 56i^2 = 28i - 56(-1) = 28i + 56 = 56 + 28i$

39. $(2 + 3i)(3 + 5i) = 6 + 19i + 15(-1) = 6 + 19i - 15 = -9 + 19i$

41. $(2 + 3i)^2 = (2 + 3i)(2 + 3i) = 4 + 12i + 9i^2 = 4 + 12i + 9(-1) = 4 + 12i - 9 = -5 + 12i$

EXERCISES 1.3

43. $\left(11+\sqrt{-25}\right)\left(2-\sqrt{-36}\right) = (11+5i)(2-6i) = 22-56i-30i^2 = 22-56i-30(-1)$
$= 22-56i+30 = 52-56i$

45. $\left(\sqrt{-16}+3\right)\left(2+\sqrt{-9}\right) = (4i+3)(2+3i) = 6+17i+12i^2 = 6+17i+12(-1)$
$= 6+17i-12 = -6+17i$

47. $\dfrac{1}{-i} = \dfrac{1}{-i}\cdot\dfrac{i}{i} = \dfrac{i}{-i^2} = \dfrac{i}{1} = 0+i$

49. $\dfrac{-4}{3i} = \dfrac{-4}{3i}\cdot\dfrac{i}{i} = \dfrac{-4i}{3i^2} = \dfrac{-4i}{-3} = 0+\dfrac{4}{3}i$

51. $\dfrac{1}{2+i} = \dfrac{1(2-i)}{(2+i)(2-i)} = \dfrac{2-i}{2^2-i^2} = \dfrac{2-i}{4-(-1)} = \dfrac{2-i}{5} = \dfrac{2}{5}-\dfrac{1}{5}i$

53. $\dfrac{2i}{7+i} = \dfrac{2i(7-i)}{(7+i)(7-i)} = \dfrac{14i-2i^2}{7^2-i^2} = \dfrac{14i-2(-1)}{49-(-1)} = \dfrac{14i+2}{50} = \dfrac{7i+1}{25} = \dfrac{1}{25}+\dfrac{7}{25}i$

55. $\dfrac{2+i}{3-i} = \dfrac{(2+i)(3+i)}{(3-i)(3+i)} = \dfrac{6+5i+i^2}{9-i^2} = \dfrac{5+5i}{10} = \dfrac{5}{10}+\dfrac{5}{10}i = \dfrac{1}{2}+\dfrac{1}{2}i$

57. $\dfrac{4-5i}{2+3i} = \dfrac{(4-5i)(2-3i)}{(2+3i)(2-3i)} = \dfrac{8-22i+15i^2}{4-9i^2} = \dfrac{-7-22i}{13} = -\dfrac{7}{13}-\dfrac{22}{13}i$

59. $\dfrac{5-\sqrt{-16}}{-8+\sqrt{-4}} = \dfrac{5-4i}{-8+2i} = \dfrac{(5-4i)(-8-2i)}{(-8+2i)(-8-2i)} = \dfrac{-40+22i+8i^2}{64-4i^2} = \dfrac{-48+22i}{68} = -\dfrac{48}{68}+\dfrac{22}{68}i$
$= -\dfrac{12}{17}+\dfrac{11}{34}i$

61. $\dfrac{2+i\sqrt{3}}{3+i} = \dfrac{\left(2+i\sqrt{3}\right)(3-i)}{(3+i)(3-i)} = \dfrac{6-2i+3i\sqrt{3}-i^2\sqrt{3}}{9-i^2} = \dfrac{6+\sqrt{3}+\left(3\sqrt{3}-2\right)i}{10}$
$= \dfrac{6+\sqrt{3}}{10}+\dfrac{3\sqrt{3}-2}{10}i$

63. $i^9 = i^8 i = \left(i^4\right)^2 i = 1^2 i = i$

65. $i^{38} = i^{36}i^2 = \left(i^4\right)^9 i^2 = 1^9 i^2 = i^2 = -1$

67. $i^{87} = i^{84}i^3 = \left(i^4\right)^{21} i^3 = 1^{21}i^3 = i^3 = -i$

69. $i^{100} = \left(i^4\right)^{25} = 1^{25} = 1$

71. $i^{-6} = \dfrac{1}{i^6} = \dfrac{1\cdot i^2}{i^6\cdot i^2} = \dfrac{i^2}{i^8} = \dfrac{i^2}{1} = i^2 = -1$

73. $i^{-10} = \dfrac{1}{i^{10}} = \dfrac{1\cdot i^2}{i^{10}\cdot i^2} = \dfrac{i^2}{i^{12}} = \dfrac{i^2}{1} = i^2 = -1$

75. $\dfrac{1}{i^3} = \dfrac{1\cdot i}{i^3\cdot i} = \dfrac{i}{i^4} = \dfrac{i}{1} = i = i$

77. $\dfrac{-4}{i^{10}} = \dfrac{-4\cdot i^2}{i^{10}\cdot i^2} = \dfrac{-4i^2}{i^{12}} = \dfrac{-4(-1)}{1} = 4$

79. $|3+4i| = \sqrt{3^2+4^2} = \sqrt{9+16}$
$= \sqrt{25} = 5$

81. $|2+3i| = \sqrt{2^2+3^2} = \sqrt{4+9} = \sqrt{13}$

EXERCISES 1.3

83. $\left|-7+\sqrt{-49}\right| = |-7+7i| = \sqrt{(-7)^2+7^2} = \sqrt{49+49} = \sqrt{98} = 7\sqrt{2}$

85. $\left|\frac{1}{2}+\frac{1}{2}i\right| = \sqrt{\left(\frac{1}{2}\right)^2+\left(\frac{1}{2}\right)^2} = \sqrt{\frac{1}{4}+\frac{1}{4}} = \sqrt{\frac{1}{2}} = \frac{\sqrt{2}}{2}$

87. $|-6i| = |0-6i| = \sqrt{0^2+(-6)^2} = \sqrt{0+36} = \sqrt{36} = 6$

89. $\left|\frac{2}{1+i}\right| = \left|\frac{2(1-i)}{(1+i)(1-i)}\right| = \left|\frac{2(1-i)}{1-i^2}\right| = \left|\frac{2(1-i)}{2}\right| = |1-i| = \sqrt{1^2+(-1)^2} = \sqrt{2}$

91. $\left|\frac{-3i}{2+i}\right| = \left|\frac{-3i(2-i)}{(2+i)(2-i)}\right| = \left|\frac{-3i(2-i)}{4-i^2}\right| = \left|\frac{-3i(2-i)}{5}\right| = \left|\frac{-6i+3i^2}{5}\right|$

$$= \left|-\frac{3}{5}-\frac{6}{5}i\right|$$

$$= \sqrt{\left(-\frac{3}{5}\right)^2+\left(-\frac{6}{5}\right)^2}$$

$$= \sqrt{\frac{9}{25}+\frac{36}{25}}$$

$$= \sqrt{\frac{45}{25}} = \frac{\sqrt{45}}{5} = \frac{3\sqrt{5}}{5}$$

93. $\left|\frac{i+2}{i-2}\right| = \left|\frac{(i+2)(i+2)}{(i-2)(i+2)}\right| = \left|\frac{i^2+4i+4}{i^2-4}\right| = \left|\frac{3+4i}{5}\right| = \left|\frac{3}{5}+\frac{4}{5}i\right| = \sqrt{\left(\frac{3}{5}\right)^2+\left(\frac{4}{5}\right)^2}$

$$= \sqrt{\frac{9}{25}+\frac{16}{25}}$$

$$= \sqrt{\frac{25}{25}} = \sqrt{1} = 1$$

95. $x^2+4 = x^2-(-4) = x^2-(2i)^2 = (x+2i)(x-2i)$

97. $25p^2+36q^2 = (5p)^2-(-36q^2) = (5p)^2-(6qi)^2 = (5p+6qi)(5p-6qi)$

99. $2y^2+8z^2 = 2(y^2+4z^2) = 2[y^2-(-4z^2)] = 2\left[y^2-(2zi)^2\right] = 2(y+2zi)(y-2zi)$

101. $50m^2+2n^2 = 2(25m^2+n^2) = 2\left[(5m)^2-(-n^2)\right] = 2\left[(5m)^2-(ni)^2\right] = 2(5m+ni)(5m-ni)$

103. $V = IR = (3-2i)(3+6i) = 9+18i-6i-12i^2 = 9+12i+12 = 21+12i$

105. $V = IZ = (0.5+2.0i)(0.4-3.0i) = 0.2-1.5i+0.8i-6i^2 = 0.2-0.7i+6 = 6.2-0.7i$

107. $(a+bi)+(c+di) = a+bi+c+di$
$= a+c+bi+di$
$= (a+c)+(b+d)i$

$(c+di)+(a+bi) = c+di+a+bi$
$= c+a+di+bi$
$= (c+a)+(d+b)i$
$= (a+c)+(b+d)i$

109. $[(a+bi)+(c+di)]+(e+fi) = a+bi+c+di+e+fi$
$= a+c+e+bi+di+fi$
$= (a+c+e)+(b+d+f)i$
$(a+bi)+[(c+di)+(e+fi)] = a+bi+c+di+e+fi$
$= a+c+e+bi+di+fi$
$= (a+c+e)+(b+d+f)i$

111. **Answers will vary.**

113. False. $\sqrt{-300} = \sqrt{-1}\sqrt{100}\sqrt{3} = 10i\sqrt{3}$

115. True. $\frac{\pi}{i} = \frac{\pi}{i}\cdot\frac{i}{i} = \frac{\pi i}{i^2} = -\pi i$

117. True. $4444i^{4444} = 4444(i^4)^{1111} = 4444\cdot 1^{1111} = 4444$

119. True. $(5-6i)(5+6i)$ is a real number and $(2-i)(2+i)$ is a real number, so their product is too.

Exercises 1.4 (page 131)

1. $ax^2+bx+c=0$

3. $\sqrt{c}, -\sqrt{c}$

5. rational numbers

7.
$$\begin{aligned} x^2-x-6&=0\\ (x+2)(x-3)&=0 \end{aligned}$$
$x+2=0$ **or** $x-3=0$
$x=-2$ $\quad x=3$

9.
$$\begin{aligned} x^2-144&=0\\ (x+12)(x-12)&=0 \end{aligned}$$
$x+12=0$ **or** $x-12=0$
$x=-12$ $\quad x=12$

11.
$$\begin{aligned} 2x^2+x-10&=0\\ (2x+5)(x-2)&=0 \end{aligned}$$
$2x+5=0$ **or** $x-2=0$
$2x=-5$ $\quad x=2$
$x=-\frac{5}{2}$ $\quad x=2$

13.
$$\begin{aligned} 5x^2-13x+6&=0\\ (5x-3)(x-2)&=0 \end{aligned}$$
$5x-3=0$ **or** $x-2=0$
$5x=3$ $\quad x=2$
$x=\frac{3}{5}$ $\quad x=2$

15.
$$\begin{aligned} 15x^2+16x&=15\\ 15x^2+16x-15&=0\\ (3x+5)(5x-3)&=0 \end{aligned}$$
$3x+5=0$ **or** $5x-3=0$
$3x=-5$ $\quad 5x=3$
$x=-\frac{5}{3}$ $\quad x=\frac{3}{5}$

17.
$$\begin{aligned} 12x^2+9&=24x\\ 12x^2-24x+9&=0\\ 3(4x^2-8x+3)&=0\\ (2x-1)(2x-3)&=0 \end{aligned}$$
$2x-1=0$ **or** $2x-3=0$
$2x=1$ $\quad 2x=3$
$x=\frac{1}{2}$ $\quad x=\frac{3}{2}$

EXERCISES 1.4

19.
$$x^2 = 9$$
$$x = \sqrt{9} \quad \textbf{or} \quad x = -\sqrt{9}$$
$$x = 3 \qquad\qquad x = -3$$

21.
$$x^2 = -169$$
$$x = \sqrt{-169} \quad \textbf{or} \quad x = -\sqrt{-169}$$
$$x = 13i \qquad\qquad x = -13i$$

23.
$$y^2 - 50 = 0$$
$$y^2 = 50$$
$$y = \sqrt{50} \quad \textbf{or} \quad y = -\sqrt{50}$$
$$y = 5\sqrt{2} \qquad\qquad y = -5\sqrt{2}$$

25.
$$y^2 + 54 = 0$$
$$y^2 = -54$$
$$y = \sqrt{-54} \quad \textbf{or} \quad y = -\sqrt{-54}$$
$$y = i\sqrt{9}\sqrt{6} \qquad\qquad y = -i\sqrt{9}\sqrt{6}$$
$$y = 3i\sqrt{6} \qquad\qquad y = -3i\sqrt{6}$$

27.
$$2x^2 = 40$$
$$x^2 = 20$$
$$x = \sqrt{20} \quad \textbf{or} \quad x = -\sqrt{20}$$
$$x = 2\sqrt{5} \qquad\qquad x = -2\sqrt{5}$$

29.
$$2x^2 = -90$$
$$x^2 = -45$$
$$x = \sqrt{-45} \quad \textbf{or} \quad x = -\sqrt{-45}$$
$$x = i\sqrt{9}\sqrt{5} \qquad\qquad x = -i\sqrt{9}\sqrt{5}$$
$$x = 3i\sqrt{5} \qquad\qquad x = -3i\sqrt{5}$$

31.
$$4x^2 = 7$$
$$x^2 = \tfrac{7}{4}$$
$$x = \sqrt{\tfrac{7}{4}} \quad \textbf{or} \quad x = -\sqrt{\tfrac{7}{4}}$$
$$x = \tfrac{\sqrt{7}}{2} \qquad\qquad x = -\tfrac{\sqrt{7}}{2}$$

33.
$$9x^2 = -7$$
$$x^2 = -\frac{7}{9}$$
$$x = \sqrt{-\frac{7}{9}} \quad \textbf{or} \quad x = -\sqrt{-\frac{7}{9}}$$
$$x = i\frac{\sqrt{7}}{\sqrt{9}} \qquad\qquad x = -i\frac{\sqrt{7}}{\sqrt{9}}$$
$$x = \frac{\sqrt{7}}{3}i \qquad\qquad x = -\frac{\sqrt{7}}{3}i$$

35.
$$2x^2 - 13 = 0$$
$$2x^2 = 13$$
$$x^2 = \tfrac{13}{2}$$
$$x = \sqrt{\tfrac{13}{2}} \quad \textbf{or} \quad x = -\sqrt{\tfrac{13}{2}}$$
$$x = \tfrac{\sqrt{13}}{\sqrt{2}} \cdot \tfrac{\sqrt{2}}{\sqrt{2}} \qquad\qquad x = -\tfrac{\sqrt{13}}{\sqrt{2}} \cdot \tfrac{\sqrt{2}}{\sqrt{2}}$$
$$x = \tfrac{\sqrt{26}}{2} \qquad\qquad x = -\tfrac{\sqrt{26}}{2}$$

37.
$$2x^2 + 15 = 0$$
$$2x^2 = -15$$
$$x^2 = -\frac{15}{2}$$
$$x = \sqrt{-\frac{15}{2}} \quad \textbf{or} \quad x = -\sqrt{-\frac{15}{2}}$$
$$x = i\frac{\sqrt{15}}{\sqrt{2}} \cdot \frac{\sqrt{2}}{\sqrt{2}} \qquad\qquad x = -i\frac{\sqrt{15}}{\sqrt{2}} \cdot \frac{\sqrt{2}}{\sqrt{2}}$$
$$x = \frac{\sqrt{30}}{2}i \qquad\qquad x = -\frac{\sqrt{30}}{2}i$$

39.
$$\begin{aligned}(x-1)^2-8&=0\\(x-1)^2&=8\end{aligned}$$
$$\begin{aligned}x-1&=\sqrt{8} &\quad \textbf{or} \quad x-1&=-\sqrt{8}\\x-1&=\sqrt{4}\sqrt{2} & x-1&=-\sqrt{4}\sqrt{2}\\x&=1+2\sqrt{2} & x&=1-2\sqrt{2}\end{aligned}$$

41.
$$\begin{aligned}(x+1)^2+12&=0\\(x+1)^2&=-12\end{aligned}$$
$$\begin{aligned}x+1&=\sqrt{-12} &\quad \textbf{or} \quad x+1&=-\sqrt{-12}\\x&=-1+i\sqrt{4}\sqrt{3} & x&=-1-i\sqrt{4}\sqrt{3}\\x&=-1+2i\sqrt{3} & x&=-1-2i\sqrt{3}\end{aligned}$$

43.
$$(2x+1)^2=27$$
$$\begin{aligned}2x+1&=\sqrt{27} &\quad \textbf{or} \quad 2x+1&=-\sqrt{27}\\2x+1&=3\sqrt{3} & 2x+1&=-3\sqrt{3}\\2x&=-1+3\sqrt{3} & 2x&=-1-3\sqrt{3}\\x&=\tfrac{-1+3\sqrt{3}}{2} & x&=\tfrac{-1-3\sqrt{3}}{2}\end{aligned}$$

45.
$$(5x+1)^2=-8$$
$$\begin{aligned}5x+1&=\sqrt{-8} &\quad \textbf{or} \quad 5x+1&=-\sqrt{-8}\\5x&=-1+i\sqrt{4}\sqrt{2} & 5x&=-1-i\sqrt{4}\sqrt{2}\\x&=-\frac{1}{5}+\frac{2i\sqrt{2}}{5} & x&=-\frac{1}{5}-\frac{2i\sqrt{2}}{5}\end{aligned}$$

47. $$\begin{aligned}x^2+6x+\left[\tfrac{1}{2}(6)\right]^2&=x^2+6x+3^2\\&=x^2+6x+9\end{aligned}$$

49. $$\begin{aligned}x^2-4x+\left[\tfrac{1}{2}(-4)\right]^2&=x^2-4x+(-2)^2\\&=x^2-4x+4\end{aligned}$$

51. $$a^2+5a+\left[\tfrac{1}{2}(5)\right]^2=a^2+5a+\left(\frac{5}{2}\right)^2=a^2+5a+\frac{25}{4}$$

53. $$r^2-11r+\left[\tfrac{1}{2}(-11)\right]^2=r^2-11r+\left(\frac{-11}{2}\right)^2=r^2-11r+\frac{121}{4}$$

55. $$y^2+\frac{3}{4}y+\left[\frac{1}{2}\left(\frac{3}{4}\right)\right]^2=y^2+\frac{3}{4}y+\left(\frac{3}{8}\right)^2=y^2+\frac{3}{4}y+\frac{9}{64}$$

57. $$q^2-\frac{1}{5}q+\left[\frac{1}{2}\left(-\frac{1}{5}\right)\right]^2=q^2-\frac{1}{5}q+\left(\frac{-1}{10}\right)^2=q^2-\frac{1}{5}q+\frac{1}{100}$$

EXERCISES 1.4

59.

$$x^2 + 12x = -8$$
$$x^2 + 12x + 36 = -8 + 36$$
$$(x+6)^2 = 28$$

$x + 6 = \sqrt{28}$ **or** $x + 6 = -\sqrt{28}$

$x + 6 = 2\sqrt{7}$ $\quad$ $x + 6 = -2\sqrt{7}$

$x = -6 + 2\sqrt{7}$ $\quad$ $x = -6 - 2\sqrt{7}$

61.

$$x^2 - 10x + 37 = 0$$
$$x^2 - 10x = -37$$
$$x^2 - 10x + 25 = -37 + 25$$
$$(x-5)^2 = -12$$

$x - 5 = \sqrt{-12}$ **or** $x - 5 = -\sqrt{-12}$

$x = 5 + i\sqrt{4}\sqrt{3}$ $\quad$ $x = 5 - i\sqrt{4}\sqrt{3}$

$x = 5 + 2i\sqrt{3}$ $\quad$ $x = 5 - 2i\sqrt{3}$

63.

$$x^2 + 5 = -5x$$
$$x^2 + 5x = -5$$
$$x^2 + 5x + \frac{25}{4} = -5 + \frac{25}{4}$$
$$\left(x + \frac{5}{2}\right)^2 = \frac{5}{4}$$

$x + \frac{5}{2} = \sqrt{\frac{5}{4}}$ **or** $x + \frac{5}{2} = -\sqrt{\frac{5}{4}}$

$x + \frac{5}{2} = \frac{\sqrt{5}}{2}$ $\quad$ $x + \frac{5}{2} = -\frac{\sqrt{5}}{2}$

$x = \frac{-5 + \sqrt{5}}{2}$ $\quad$ $x = \frac{-5 - \sqrt{5}}{2}$

65.

$$y^2 + 11y = -49$$
$$y^2 + 11y + \frac{121}{4} = -\frac{196}{4} + \frac{121}{4}$$
$$\left(y + \frac{11}{2}\right)^2 = -\frac{75}{4}$$

$y + \frac{11}{2} = \sqrt{-\frac{75}{4}}$ **or** $y + \frac{11}{2} = -\sqrt{-\frac{75}{4}}$

$y = -\frac{11}{2} + i\frac{\sqrt{75}}{\sqrt{4}}$ $\quad$ $y = -\frac{11}{2} - i\frac{\sqrt{75}}{\sqrt{4}}$

$y = -\frac{11}{2} + \frac{5\sqrt{3}}{2}i$ $\quad$ $y = -\frac{11}{2} - \frac{5\sqrt{3}}{2}i$

67.

$$\begin{aligned} 2x^2 - 20x &= -49 \\ x^2 - 10x &= -\frac{49}{2} \\ x^2 - 10x + 25 &= -\frac{49}{2} + \frac{50}{2} \\ (x-5)^2 &= \frac{1}{2} \end{aligned}$$

$$x - 5 = \sqrt{\frac{1}{2}} \quad \textbf{or} \quad x - 5 = -\sqrt{\frac{1}{2}}$$

$$x - 5 = \frac{1}{\sqrt{2}} \qquad x - 5 = -\frac{1}{\sqrt{2}}$$

$$x - \frac{10}{2} = \frac{\sqrt{2}}{2} \qquad x - \frac{10}{2} = -\frac{\sqrt{2}}{2}$$

$$x = \frac{10+\sqrt{2}}{2} \qquad x = \frac{10-\sqrt{2}}{2}$$

69.

$$\begin{aligned} 3x^2 &= 1 - 4x \\ 3x^2 + 4x &= 1 \\ x^2 + \frac{4}{3}x &= \frac{1}{3} \\ x^2 + \frac{4}{3}x + \frac{4}{9} &= \frac{1}{3} + \frac{4}{9} \\ \left(x + \frac{2}{3}\right)^2 &= \frac{7}{9} \end{aligned}$$

$$x + \frac{2}{3} = \sqrt{\frac{7}{9}} \quad \textbf{or} \quad x + \frac{2}{3} = -\sqrt{\frac{7}{9}}$$

$$x + \frac{2}{3} = \frac{\sqrt{7}}{3} \qquad x + \frac{2}{3} = -\frac{\sqrt{7}}{3}$$

$$x = \frac{-2+\sqrt{7}}{3} \qquad x = \frac{-2-\sqrt{7}}{3}$$

71.

$$\begin{aligned} 2x^2 &= 3x + 1 \\ 2x^2 - 3x &= 1 \\ x^2 - \frac{3}{2}x &= \frac{1}{2} \\ x^2 - \frac{3}{2}x + \frac{9}{16} &= \frac{1}{2} + \frac{9}{16} \\ \left(x - \frac{3}{4}\right)^2 &= \frac{17}{16} \end{aligned}$$

continued on next page...

71. continued

$$\left(x-\frac{3}{4}\right)^2=\frac{17}{16}$$

$$x-\frac{3}{4}=\sqrt{\frac{17}{16}} \quad \textbf{or} \quad x-\frac{3}{4}=-\sqrt{\frac{17}{16}}$$

$$x-\frac{3}{4}=\frac{\sqrt{17}}{4} \qquad x-\frac{3}{4}=-\frac{\sqrt{17}}{4}$$

$$x=\frac{3+\sqrt{17}}{4} \qquad x=\frac{3-\sqrt{17}}{4}$$

73. $9x^2=18x-14 \Rightarrow 9x^2-18x+14=0 \Rightarrow a=9, b=-18, c=14$

$$x=\frac{-b\pm\sqrt{b^2-4ac}}{2a}=\frac{-(-18)\pm\sqrt{(-18)^2-4(9)(14)}}{2(9)}=\frac{18\pm\sqrt{324-504}}{18}$$

$$=\frac{18\pm\sqrt{-180}}{18}=\frac{18\pm 6i\sqrt{5}}{18}=\frac{6\left(3\pm i\sqrt{5}\right)}{18}=\frac{3\pm i\sqrt{5}}{3}=1\pm\frac{\sqrt{5}}{3}i$$

75. $2x^2=14x-30 \Rightarrow 2x^2-14x+30=0 \Rightarrow a=2, b=-14, c=30$

$$x=\frac{-b\pm\sqrt{b^2-4ac}}{2a}=\frac{-(-14)\pm\sqrt{(-14)^2-4(2)(30)}}{2(2)}=\frac{14\pm\sqrt{196-240}}{4}$$

$$=\frac{14\pm\sqrt{-44}}{4}=\frac{14\pm 2i\sqrt{11}}{4}=\frac{2\left(7\pm i\sqrt{11}\right)}{4}=\frac{7\pm i\sqrt{11}}{2}=\frac{7}{2}\pm\frac{\sqrt{11}}{2}i$$

77. $3x^2=-5x-1 \Rightarrow 3x^2+5x+1=0 \Rightarrow a=3, b=5, c=1$

$$x=\frac{-b\pm\sqrt{b^2-4ac}}{2a}=\frac{-(5)\pm\sqrt{(5)^2-4(3)(1)}}{2(3)}=\frac{-5\pm\sqrt{25-12}}{6}=\frac{-5\pm\sqrt{13}}{6}$$

79. $x^2+1=-7x \Rightarrow x^2+7x+1=0 \Rightarrow a=1, b=7, c=1$

$$x=\frac{-b\pm\sqrt{b^2-4ac}}{2a}=\frac{-(7)\pm\sqrt{(7)^2-4(1)(1)}}{2(1)}=\frac{-7\pm\sqrt{49-4}}{2}=\frac{-7\pm\sqrt{45}}{2}$$

$$=\frac{-7\pm 3\sqrt{5}}{2}$$

81. $3x^2+6x=-1 \Rightarrow 3x^2+6x+1=0 \Rightarrow a=3, b=6, c=1$

$$x=\frac{-b\pm\sqrt{b^2-4ac}}{2a}=\frac{-(6)\pm\sqrt{(6)^2-4(3)(1)}}{2(3)}=\frac{-6\pm\sqrt{36-12}}{6}=\frac{-6\pm\sqrt{24}}{6}$$

$$x=\frac{-6\pm\sqrt{24}}{6}=\frac{-6\pm 2\sqrt{6}}{6}=\frac{2\left(-3\pm\sqrt{6}\right)}{6}=\frac{-3\pm\sqrt{6}}{3}$$

EXERCISES 1.4

83. $7x^2 = 2x + 2 \Rightarrow 7x^2 - 2x - 2 = 0 \Rightarrow a = 7, b = -2, c = -2$

$$x = \frac{-b \pm \sqrt{b^2 - 4ac}}{2a} = \frac{-(-2) \pm \sqrt{(-2)^2 - 4(7)(-2)}}{2(7)} = \frac{2 \pm \sqrt{4 + 56}}{14} = \frac{2 \pm \sqrt{60}}{14}$$

$$x = \frac{2 \pm \sqrt{60}}{14} = \frac{2 \pm 2\sqrt{15}}{14} = \frac{1 \pm \sqrt{15}}{7}$$

85. $x^2 + 2x + 2 = 0 \Rightarrow a = 1, b = 2, c = 2$

$$x = \frac{-b \pm \sqrt{b^2 - 4ac}}{2a} = \frac{-2 \pm \sqrt{2^2 - 4(1)(2)}}{2(1)} = \frac{-2 \pm \sqrt{4 - 8}}{2} = \frac{-2 \pm \sqrt{-4}}{2} = \frac{-2 \pm 2i}{2}$$
$$= -1 \pm i$$

87. $y^2 + 4y + 5 = 0 \Rightarrow a = 1, b = 4, c = 5$

$$y = \frac{-b \pm \sqrt{b^2 - 4ac}}{2a} = \frac{-4 \pm \sqrt{4^2 - 4(1)(5)}}{2(1)} = \frac{-4 \pm \sqrt{16 - 20}}{2} = \frac{-4 \pm \sqrt{-4}}{2} = \frac{-4 \pm 2i}{2}$$
$$= -2 \pm i$$

89. $x^2 - 2x = -5 \Rightarrow x^2 - 2x + 5 = 0 \Rightarrow a = 1, b = -2, c = 5$

$$x = \frac{-b \pm \sqrt{b^2 - 4ac}}{2a} = \frac{-(-2) \pm \sqrt{(-2)^2 - 4(1)(5)}}{2(1)} = \frac{2 \pm \sqrt{4 - 20}}{2} = \frac{2 \pm \sqrt{-16}}{2} = \frac{2 \pm 4i}{2}$$
$$= 1 \pm 2i$$

91. $x^2 - \frac{2}{3}x = -\frac{2}{9} \Rightarrow x^2 - \frac{2}{3}x + \frac{2}{9} = 0 \Rightarrow 9x^2 - 6x + 2 = 0 \Rightarrow a = 9, b = -6, c = 2$

$$x = \frac{-b \pm \sqrt{b^2 - 4ac}}{2a} = \frac{-(-6) \pm \sqrt{(-6)^2 - 4(9)(2)}}{2(9)} = \frac{6 \pm \sqrt{36 - 72}}{18} = \frac{6 \pm \sqrt{-36}}{18}$$
$$= \tfrac{6 \pm 6i}{18} = \tfrac{1}{3} \pm \tfrac{1}{3}i$$

93.

$$h = \tfrac{1}{2}gt^2$$
$$2h = gt^2$$
$$\tfrac{2h}{g} = t^2$$
$$\pm\sqrt{\frac{2h}{g}} = t$$
$$\pm\frac{\sqrt{2h}}{\sqrt{g}} \cdot \frac{\sqrt{g}}{\sqrt{g}} = t$$
$$\pm\frac{\sqrt{2hg}}{g} = t$$

95.

$$h = 64t - 16t^2$$
$$16t^2 - 64t + h = 0;\ a = 16, b = -64, c = h$$
$$t = \frac{-b \pm \sqrt{b^2 - 4ac}}{2a}$$
$$= \frac{-(-64) \pm \sqrt{(-64)^2 - 4(16)h}}{2(16)}$$
$$= \frac{64 \pm \sqrt{4096 - 64h}}{32}$$
$$= \frac{64 \pm \sqrt{64(64 - h)}}{32}$$
$$= \frac{64 \pm 8\sqrt{64 - h}}{32} = \frac{8 \pm \sqrt{64 - h}}{4}$$

EXERCISES 1.4

97.
$$\begin{aligned}\frac{x^2}{a^2}+\frac{y^2}{b^2}&=1\\ \frac{y^2}{b^2}&=1-\frac{x^2}{a^2}\\ \frac{y^2}{b^2}&=\frac{a^2-x^2}{a^2}\\ y^2&=\frac{b^2(a^2-x^2)}{a^2}\\ y&=\pm\sqrt{\frac{b^2(a^2-x^2)}{a^2}}\\ y&=\pm\frac{b\sqrt{a^2-x^2}}{a}\end{aligned}$$

99.
$$\begin{aligned}\frac{x^2}{a^2}-\frac{y^2}{b^2}&=1\\ a^2b^2\left(\frac{x^2}{a^2}-\frac{y^2}{b^2}\right)&=a^2b^2(1)\\ b^2x^2-a^2y^2&=a^2b^2\\ b^2x^2&=a^2b^2+a^2y^2\\ b^2x^2&=a^2\left(b^2+y^2\right)\\ \frac{b^2x^2}{b^2+y^2}&=a^2\\ \pm\sqrt{\frac{b^2x^2}{b^2+y^2}}&=a\\ \pm\frac{bx\sqrt{b^2+y^2}}{b^2+y^2}&=a\end{aligned}$$

101. $x^2+xy-y^2=0 \Rightarrow a=1, b=y, c=-y^2$

$$x=\frac{-b\pm\sqrt{b^2-4ac}}{2a}=\frac{-(y)\pm\sqrt{(y)^2-4(1)(-y^2)}}{2(1)}=\frac{-y\pm\sqrt{y^2+4y^2}}{2}=\frac{-y\pm\sqrt{5y^2}}{2}=\frac{-y\pm y\sqrt{5}}{2}$$

103. $x^2+6x+9=0 \Rightarrow a=1, b=6, c=9$

$b^2-4ac=6^2-4(1)(9)=36-36=0$

one repeated rational number

105. $3x^2-2x+5=0 \Rightarrow a=3, b=-2, c=5$

$$\begin{aligned}b^2-4ac&=(-2)^2-4(3)(5)\\ &=4-60=-56\end{aligned}$$

two different nonreal complex numbers

107. $10x^2+29x=21 \Rightarrow 10x^2+29x-21=0$

$a=10, b=29, c=-21$

$$\begin{aligned}b^2-4ac&=(29)^2-4(10)(-21)\\ &=841+840=1681\end{aligned}$$

two different rational numbers

109. $x^2-5x+2=0 \Rightarrow a=1, b=-5, c=2$

$b^2-4ac=(-5)^2-4(1)(2)=25-8=17$

two different irrational numbers

111.
$$\begin{aligned}x^2+kx+3k-5&=0\\ a=1, b=k, c&=3k-5\end{aligned}$$

Set the discriminant equal to 0:

$$\begin{aligned}b^2-4ac&=0\\ k^2-4(1)(3k-5)&=0\\ k^2-4(3k-5)&=0\\ k^2-12k+20&=0\\ (k-2)(k-10)&=0\\ k=2 \text{ or } k&=10\end{aligned}$$

113.
$$\begin{aligned}x+1&=\frac{12}{x}\\ x(x+1)&=x\left(\frac{12}{x}\right)\\ x^2+x&=12\\ x^2+x-12&=0\\ (x+4)(x-3)&=0\end{aligned}$$

$x+4=0$ **or** $x-3=0$

$x=-4 \qquad x=3$

EXERCISES 1.4

115.
$$\begin{aligned} 8x - \frac{3}{x} &= 10 \\ x\left(8x - \frac{3}{x}\right) &= x(10) \\ 8x^2 - 3 &= 10x \\ 8x^2 - 10x - 3 &= 0 \\ (4x+1)(2x-3) &= 0 \end{aligned}$$
$4x + 1 = 0$ **or** $2x - 3 = 0$
$4x = -1$ $\quad 2x = 3$
$x = -\frac{1}{4}$ $\quad x = \frac{3}{2}$

117.
$$\begin{aligned} \frac{5}{x} &= \frac{4}{x^2} - 6 \\ x^2\left(\frac{5}{x}\right) &= x^2\left(\frac{4}{x^2} - 6\right) \\ 5x &= 4 - 6x^2 \\ 6x^2 + 5x - 4 &= 0 \\ (3x+4)(2x-1) &= 0 \end{aligned}$$
$3x + 4 = 0$ **or** $2x - 1 = 0$
$3x = -4$ $\quad 2x = 1$
$x = -\frac{4}{3}$ $\quad x = \frac{1}{2}$

119.
$$\begin{aligned} x\left(30 - \frac{13}{x}\right) &= \frac{10}{x} \\ 30x - 13 &= \frac{10}{x} \\ x(30x - 13) &= x\left(\frac{10}{x}\right) \\ 30x^2 - 13x &= 10 \\ 30x^2 - 13x - 10 &= 0 \\ (5x+2)(6x-5) &= 0 \end{aligned}$$
$5x + 2 = 0$ **or** $6x - 5 = 0$
$5x = -2$ $\quad 6x = 5$
$x = -\frac{2}{5}$ $\quad x = \frac{5}{6}$

121.
$$\begin{aligned} \frac{1}{x} + \frac{3}{x+2} &= 2 \\ x(x+2)\left(\frac{1}{x} + \frac{3}{x+2}\right) &= x(x+2)(2) \\ 1(x+2) + 3x &= 2x(x+2) \\ x + 2 + 3x &= 2x^2 + 4x \\ 0 &= 2x^2 - 2 \\ 0 &= 2(x+1)(x-1) \end{aligned}$$
$x + 1 = 0$ **or** $x - 1 = 0$
$x = -1$ $\quad x = 1$

123.
$$\begin{aligned} \frac{1}{x+1} + \frac{5}{2x-4} &= 1 \\ (x+1)(2x-4)\left(\frac{1}{x+1} + \frac{5}{2x-4}\right) &= (x+1)(2x-4)1 \\ 1(2x-4) + 5(x+1) &= (x+1)(2x-4) \\ 2x - 4 + 5x + 5 &= 2x^2 - 2x - 4 \\ 0 &= 2x^2 - 9x - 5 \\ 0 &= (2x+1)(x-5) \end{aligned}$$
$2x + 1 = 0$ **or** $x - 5 = 0$
$2x = -1$ $\quad x = 5$
$x = -\frac{1}{2}$ $\quad x = 5$

EXERCISES 1.4

125.
$$x+1+\frac{x+2}{x-1}=\frac{3}{x-1}$$
$$(x-1)\left(\frac{x+1}{1}+\frac{x+2}{x-1}\right)=(x-1)\frac{3}{x-1}$$
$$(x-1)(x+1)+x+2=3$$
$$x^2-1+x+2=3$$
$$x^2+x-2=0$$
$$(x+2)(x-1)=0$$
$$x+2=0 \quad \textbf{or} \quad x-1=0$$
$$x=-2 \qquad\qquad x=1$$
Since $x=1$ does not check, the only solution is $x=-2$.

127.
$$\frac{4+a}{2a}=\frac{a-2}{3}$$
$$6a\left(\frac{4+a}{2a}\right)=6a\left(\frac{a-2}{3}\right)$$
$$12+3a=2a^2-4a$$
$$0=2a^2-7a-12$$
$$a=2,\ b=-7,\ c=-12$$
$$a=\frac{-b\pm\sqrt{b^2-4ac}}{2a}=\frac{-(-7)\pm\sqrt{(-7)^2-4(2)(-12)}}{2(2)}=\frac{7\pm\sqrt{145}}{4}$$

129.
$$x+\frac{36}{x}=0$$
$$x\left(x+\frac{36}{x}\right)=x\cdot 0$$
$$x^2+36=0$$
$$x^2=-36$$
$$x=\pm\sqrt{-36}$$
$$x=\pm 6i$$

131. Answers may vary.

133. If r_1 and r_2 are the roots of $ax^2+bx+c=0$, then their values are
$$r_1=\frac{-b+\sqrt{b^2-4ac}}{2a} \text{ and } r_2=\frac{-b-\sqrt{b^2-4ac}}{2a}.$$
$$r_1+r_2=\frac{-b+\sqrt{b^2-4ac}}{2a}+\frac{-b-\sqrt{b^2-4ac}}{2a}=\frac{-2b}{2a}=-\frac{b}{a}$$

135. Rewrite the equation as $16t^2 - v_0 t + h = 0$ and solve for t using the quadratic formula.
$a = 16, b = -v_0, c = h$

$$t = \frac{-b \pm \sqrt{b^2 - 4ac}}{2a} = \frac{-(-v_0) \pm \sqrt{(-v_0)^2 - 4(16)(h)}}{2(16)} = \frac{v_0 \pm \sqrt{v_0^2 - 64h}}{32}$$

Since t_1 and t_2 are the solutions to the equation, we have

$t_1 = \dfrac{v_0 - \sqrt{v_0^2 - 64h}}{32}$ and $t_2 = \dfrac{v_0 + \sqrt{v_0^2 - 64h}}{32}$. Calculate $16t_1t_2$:

$$16t_1t_2 = 16 \cdot \frac{v_0 - \sqrt{v_0^2 - 64h}}{32} \cdot \frac{v_0 + \sqrt{v_0^2 - 64h}}{32} = 16 \cdot \frac{v_0^2 - (v_0^2 - 64h)}{1024}$$
$$= \frac{16 \cdot 64h}{1024} = \frac{1024h}{1024} = h$$

Thus, $h = 16t_1t_2$.

137. b

139. c

141. $1492x^2 + 1984x - 1776 = 0$
$a = 1492, b = 1984, c = -1776$

$$\begin{aligned} b^2 - 4ac &= (1984)^2 - 4(1492)(-1776) \\ &= 3{,}936{,}256 + 10{,}599{,}168 \\ &= 14{,}535{,}424 \end{aligned}$$

The solutions are real numbers. True.

Exercises 1.5 (page 140)

1. $A = lw$

3. Let $w =$ the width of the rectangle. Then $w + 4 =$ the length.

$\boxed{\text{Width}} \cdot \boxed{\text{Length}} = \boxed{\text{Area}}$

$$\begin{aligned} w(w+4) &= 32 \\ w^2 + 4w &= 32 \\ w^2 + 4w - 32 &= 0 \\ (w+8)(w-4) &= 0 \end{aligned}$$

$w + 8 = 0$ **or** $w - 4 = 0$
$w = -8$ $\quad$ $w = 4$

Since the width cannot be negative, the only reasonable solution is $w = 4$. The dimensions are 4 feet by 8 feet.

5. Let $w =$ the width of the screen. Then $w + 88 =$ the length.

$\boxed{\text{Width}} \cdot \boxed{\text{Length}} = \boxed{\text{Area}}$

$$\begin{aligned} w(w+88) &= 11520 \\ w^2 + 88w &= 11520 \end{aligned}$$

$w^2 + 88w - 11520 = 0 \Rightarrow a = 1, b = 88, c = -11520$

$$w = \frac{-b \pm \sqrt{b^2 - 4ac}}{2a} = \frac{-(88) \pm \sqrt{(88)^2 - 4(1)(-11520)}}{2(1)} = \frac{-88 \pm \sqrt{53824}}{2}$$

$w = 72$ or $w = -160$; Since the screen cannot have a negative width, the solution is $w = 72$ and the dimensions of the screen are 160 ft by 72 ft.

EXERCISES 1.5

7. Let s = the side of the second square.
Then $s - 4$ = the side of the first square.

$$\boxed{\text{Area of first}} + \boxed{\text{Area of second}} = 106$$
$$(s-4)^2 + s^2 = 106$$
$$s^2 - 8s + 16 + s^2 = 106$$
$$2s^2 - 8s - 90 = 0$$
$$2(s^2 - 4s - 45) = 0$$
$$2(s+5)(s-9) = 0$$
$$s + 5 = 0 \quad \textbf{or} \quad s - 9 = 0$$
$$s = -5 \qquad\qquad s = 9$$

Since the side cannot be negative, the only reasonable solution is $s = 9$.
The larger square has a side of length 9 cm.

9. Let the dimensions be x and $1.9x$.

$$\boxed{\text{Width}} \cdot \boxed{\text{Length}} = \boxed{\text{Area}}$$
$$x(1.9x) = 100$$
$$1.9x^2 = 100$$
$$x^2 = \frac{100}{1.9}$$
$$x = \pm\sqrt{\frac{100}{1.9}}$$
$$x \approx \pm 7.25$$
$$1.9x \approx 1.9(7.25) \approx 13.75$$

Since the dimensions cannot be negative, the only reasonable solution $7\frac{1}{4}$ ft by $13\frac{3}{4}$ ft.

11. The floor area of the box is a square with a side of length $12 - 2x$.

$$\boxed{\text{Floor area}} = 64$$
$$(12 - 2x)^2 = 64$$
$$144 - 48x + 4x^2 = 64$$
$$4x^2 - 48x + 80 = 0$$
$$4(x^2 - 12x + 20) = 0$$
$$4(x-2)(x-10) = 0$$
$$x - 2 = 0 \quad \textbf{or} \quad x - 10 = 0$$
$$x = 2 \qquad\qquad x = 10$$

The solution $x = 10$ does not make sense in the problem, so the depth is 2 inches.

13. Let h = the height of the triangle.
Then $\frac{1}{3}h$ = the base of the triangle.

$$\frac{1}{2} \cdot \boxed{\text{Base}} \cdot \boxed{\text{Height}} = \boxed{\text{Area}}$$
$$\frac{1}{2} \cdot \frac{1}{3}h \cdot h = 24$$
$$\frac{1}{6}h^2 = 24$$
$$h^2 = 144$$
$$h = \sqrt{144} \quad \textbf{or} \quad h = -\sqrt{144}$$
$$h = 12 \qquad\qquad h = -12$$

Since the height cannot be negative, the only reasonable solution is $h = 12$.
The base has a length of 4 meters.

15. Let the legs have lengths x and $x - 14$.

$$x^2 + (x-14)^2 = 26^2$$
$$x^2 + x^2 - 28x + 196 = 676$$
$$2x^2 - 28x - 480 = 0$$
$$2(x^2 - 14x - 240) = 0$$
$$2(x-24)(x+10) = 0$$
$$x - 24 = 0 \quad \textbf{or} \quad x + 10 = 0$$
$$x = 24 \qquad\qquad x = -10$$

Since lengths are positive, the answer is $x = 24$, and the legs have length 10 meters and 24 meters.

EXERCISES 1.5

17. Let $x =$ the height of the screen. Then $x + 17.5 =$ the width. Use the Pythagorean Theorem:

$$\text{height}^2 + \text{width}^2 = \text{diagonal}^2$$
$$x^2 + (x + 17.5)^2 = 46^2$$
$$x^2 + x^2 + 35x + 306.25 = 2116$$
$$2x^2 + 35x - 1809.75 = 0$$
$$a = 2, b = 35, c = -1809.75$$

$$x = \frac{-b \pm \sqrt{b^2 - 4ac}}{2a} = \frac{-35 \pm \sqrt{35^2 - 4(2)(-1809.75)}}{2(2)} = \frac{-35 \pm \sqrt{15703}}{4} \approx \frac{-35 \pm 125.312}{4}$$

The only positive solution is 22.6. The dimensions are 22.6 inches by 40.1 inches.

19. Let $r =$ the cyclist's rate from DeKalb to Rockford. Then his return rate is $r - 10$.

	Rate	Time	Dist.
First trip	r	$\frac{40}{r}$	40
Return trip	$r - 10$	$\frac{40}{r-10}$	40

$$\boxed{\text{Return time}} = \boxed{\text{First time}} + 2$$
$$\frac{40}{r - 10} = \frac{40}{r} + 2$$
$$r(r - 10)\frac{40}{r - 10} = r(r - 10)\left(\frac{40}{r} + 2\right)$$
$$40r = 40(r - 10) + 2r(r - 10)$$
$$40r = 40r - 400 + 2r^2 - 20r$$
$$0 = 2r^2 - 20r - 400$$
$$0 = 2(r - 20)(r + 10)$$
$$r - 20 = 0 \quad \textbf{or} \quad r + 10 = 0$$
$$r = 20 \qquad\qquad r = -10$$

Since $r = -10$ does not make sense, the solution is $r = 20$. The cyclist rides 20 mph going and 10 mph returning.

21. Let $r =$ the slower rate. Then the faster rate is $r + 10$.

	Rate	Time	Dist.
Slower trip	r	$\frac{420}{r}$	420
Faster trip	$r + 10$	$\frac{420}{r+10}$	420

$$\boxed{\text{Faster time}} = \boxed{\text{Slower time}} - 1$$
$$\frac{420}{r + 10} = \frac{420}{r} - 1$$
$$r(r + 10)\frac{420}{r + 10} = r(r + 10)\left(\frac{420}{r} - 1\right)$$
$$420r = 420(r + 10) - r(r + 10)$$
$$420r = 420r + 4200 - r^2 - 10r$$
$$r^2 + 10r - 4200 = 0$$
$$(r - 60)(r + 70) = 0$$
$$r - 60 = 0 \quad \textbf{or} \quad r + 70 = 0$$
$$r = 60 \qquad\qquad r = -70$$

Since $r = -70$ does not make sense, the solution is $r = 60$. The slower speed results in a trip of length 7 hours.

EXERCISES 1.5

23. Set $h = 0$:

$$h = -16t^2 + 400t$$
$$0 = -16t^2 + 400t$$
$$16t^2 - 400t = 0$$
$$16t(t - 25) = 0$$
$$16t = 0 \quad \textbf{or} \quad t - 25 = 0$$
$$t = 0 \qquad\qquad t = 25$$

$t = 0$ represents when the projectile was fired, so it returns to earth after 25 seconds.

25. Set $s = 1454$:

$$s = 16t^2$$
$$1454 = 16t^2$$
$$\frac{1454}{16} = t^2$$
$$t = \sqrt{\frac{1454}{16}} \quad \textbf{or} \quad t = -\sqrt{\frac{1454}{16}}$$
$$t \approx 9.5 \qquad\qquad t \approx -9.5$$

$t = -9.5$ does not make sense, so it takes it about 9.5 seconds to hit the ground.

27. Set $h = 5$ and $s = 48$.

$$h = s - 16t^2$$
$$5 = 48 - 16t^2$$
$$16t^2 = 43$$
$$t = \sqrt{\frac{43}{16}} \quad \textbf{or} \quad t = -\sqrt{\frac{43}{16}}$$
$$t \approx 1.6 \qquad\qquad t \approx -1.6$$

$t = -1.6$ does not make sense, so she has about 1.6 seconds to get out of the way.

29. Let $x =$ the number of nickel increases. The new fare $= 25 + 5x$ (in cents), while the number of passengers $= 3000 - 80x$.

$$\boxed{\text{Number of passengers}} \cdot \boxed{\text{Fare}} = \boxed{\text{Revenue}}$$
$$(3000 - 80x)(25 + 5x) = 99400$$
$$75000 + 13000x - 400x^2 = 99400$$
$$400x^2 - 13000x + 24400 = 0$$
$$200(2x^2 - 65x + 122) = 0$$
$$200(2x - 61)(x - 2) = 0$$
$$2x - 61 = 0 \quad \textbf{or} \quad x - 2 = 0$$
$$x = 30.5 \qquad\qquad x = 2$$

Since you cannot have half of a nickel increase, $x = 30.5$ does not make sense. Thus, there should be 2 nickel increases, for a fare increase of 10 cents.

31. Let $x =$ the number of \$0.50 decreases. The new price $= 15 - 0.5x$, while the number attending $= 1200 + 40x$.

$$\boxed{\text{Number attending}} \cdot \boxed{\text{Price}} = \boxed{\text{Revenue}}$$
$$(1200 + 40x)(15 - 0.5x) = 17280$$
$$18000 - 20x^2 = 17280$$
$$20x^2 = 720$$
$$x^2 = 36$$
$$x = \sqrt{36} \quad \textbf{or} \quad x = -\sqrt{36}$$
$$x = 6 \qquad\qquad x = -6$$

$x = -6$ does not make sense. Thus, there should be six 50-cent decreases, for a ticket price of \$12 and an attendance of 1440 people.

EXERCISES 1.5

33. Let x = Chloe's principal.

	I	P	r
Chloe	240	x	$\frac{240}{x}$
Morgan	280	$x+1000$	$\frac{280}{x+1000}$

$$\boxed{\text{Morgan's rate}} = \boxed{\text{Chloe's rate}} - 0.01$$

$$\frac{280}{x+1000} = \frac{240}{x} - 0.01$$

$$x(x+1000)\frac{280}{x+1000} = x(x+1000)\left(\frac{240}{x} - 0.01\right)$$

$$280x = 240(x+1000) - 0.01x(x+1000)$$

$$0.01x^2 + 50x - 240{,}000 = 0$$

$$x^2 + 5000x - 24{,}000{,}000 = 0$$

$$(x-3000)(x+8000) = 0$$

$$x - 3000 = 0 \quad \textbf{or} \quad x + 8000 = 0$$

$$x = 3000 \qquad x = -8000 \Rightarrow$$

$x = -8000$ does not make sense. The principal amounts were \$3000 and \$4000. The interest rates were 8% for Chloe and 7% for Morgan.

35. Let x = the total number of professors.

$$\boxed{\text{New share with lower number}} = \boxed{\text{Original share}} + 10$$

$$\frac{150}{x-4} = \frac{150}{x} + 10$$

$$x(x-4)\frac{150}{x-4} = x(x-4)\left(\frac{150}{x} + 10\right)$$

$$150x = 150(x-4) + 10x(x-4)$$

$$150x = 150x - 600 + 10x^2 - 40x$$

$$0 = 10x^2 - 40x - 600$$

$$0 = x^2 - 4x - 60$$

$$0 = (x-10)(x+6)$$

$$x - 10 = 0 \quad \textbf{or} \quad x + 6 = 0$$

$$x = 10 \qquad x = -6$$

$x = -6$ does not make sense, so there are 10 professors in the department.

37. Let x = time for the second pipe to fill tank.

$$\boxed{\text{First in 1 hour}} + \boxed{\text{Second in 1 hour}} = \boxed{\text{Total in 1 hour}}$$

$$\frac{1}{4} + \frac{1}{x} = \frac{1}{x-2}$$

$$4x(x-2)\left(\frac{1}{4} + \frac{1}{x}\right) = 4x(x-2)\cdot\frac{1}{x-2}$$

$$x(x-2) + 4(x-2) = 4x$$

$$x^2 - 2x + 4x - 8 = 4x$$

$$x^2 - 2x - 8 = 0$$

$$(x-4)(x+2) = 0$$

$$x - 4 = 0 \quad \textbf{or} \quad x + 2 = 0$$

$$x = 4 \qquad x = -2$$

Since $x = -2$ does not make sense, the solution is $x = 4$. It takes the second pipe 4 hours to fill the tank alone.

39. Let x = time for the Steven to mow lawn.

$$\boxed{\text{Steven in 1 hour}} + \boxed{\text{Kristy in 1 hour}} = \boxed{\text{Total in 1 hour}}$$

$$\frac{1}{x} + \frac{1}{x-1} = \frac{1}{5}$$

$$5x(x-1)\left(\frac{1}{x} + \frac{1}{x-1}\right) = 5x(x-1)\cdot\frac{1}{5}$$

$$5(x-1) + 5x = x(x-1)$$

$$5x - 5 + 5x = x^2 - x$$

$$0 = x^2 - 11x + 5$$

$a = 1, b = -11, c = 5$

$$x = \frac{-b \pm \sqrt{b^2 - 4ac}}{2a}$$

$$= \frac{-(-11) \pm \sqrt{(-11)^2 - 4(1)(5)}}{2(1)}$$

$$= \frac{11 \pm \sqrt{101}}{2} \approx 10.5 \text{ or } 0.5$$

$x = 0.5$ does not make sense, so Kristy could mow the lawn in about 9.5 hours alone.

41. The number of trees is the length of the row divided by the space between the trees, plus 1.

Let $x =$ the original spacing.

$$\boxed{\text{Original number}} - 44 = \boxed{\text{New number}}$$

$$\frac{1320}{x} + 1 - 44 = \frac{1320}{x+1} + 1$$

$$\frac{1320}{x} - 44 = \frac{1320}{x+1}$$

$$x(x+1)\left(\frac{1320}{x} - 44\right) = x(x+1)\frac{1320}{x+1}$$

$$1320(x+1) - 44x(x+1) = 1320x$$

$$1320(x+1) - 44x(x+1) = 1320x$$

$$44x^2 + 44x - 1320 = 0$$

$$x^2 + x - 30 = 0$$

$$(x-5)(x+6) = 0$$

$$x - 5 = 0 \quad \textbf{or} \quad x + 6 = 0$$

$$x = 5 \qquad\qquad x = -6$$

$x = -6$ does not make sense. The original spacing was 5 feet, resulting in 265 trees, so the new spacing will require 221 trees.

43. Let $h = 15$:

$$\frac{l}{h} = \frac{h}{l-h}$$

$$\frac{l}{15} = \frac{15}{l-15}$$

$$15(l-15)\cdot\frac{l}{15} = 15(l-15)\cdot\frac{15}{l-15}$$

$$l(l-15) = 15^2$$

$$l^2 - 15l - 225 = 0$$

$a = 1, b = -15, c = -225$

$$l = \frac{-b \pm \sqrt{b^2 - 4ac}}{2a}$$

$$= \frac{-(-15) \pm \sqrt{(-15)^2 - 4(1)(-225)}}{2(1)}$$

$$= \frac{15 \pm \sqrt{1125}}{2} \approx \frac{15 \pm 33.541}{2}$$

The only positive solution is $l = 24.3$ ft.

45.

$$V = \pi r^2 h$$

$$47.75 = \pi r^2 (5.25)$$

$$\frac{47.75}{5.25\pi} = r^2$$

$$\pm\sqrt{\frac{47.75}{5.25\pi}} = r$$

$$\pm 1.70 \approx r$$

The radius is about 1.70 in.

47-49. Answers may vary.

Exercises 1.6 (page 151)

1. equal

3. extraneous

5.

$$x^3 + 9x^2 + 20x = 0$$

$$x(x^2 + 9x + 20) = 0$$

$$x(x+5)(x+4) = 0$$

$$x = 0 \quad \textbf{or} \quad x + 5 = 0 \quad \textbf{or} \quad x + 4 = 0$$

$$x = 0 \qquad x = -5 \qquad x = -4$$

7.

$$6a^3 - 5a^2 - 4a = 0$$

$$a(6a^2 - 5a - 4) = 0$$

$$a(2a+1)(3a-4) = 0$$

$$a = 0 \quad \textbf{or} \quad 2a + 1 = 0 \quad \textbf{or} \quad 3a - 4 = 0$$

$$a = 0 \qquad 2a = -1 \qquad 3a = 4$$

$$a = 0 \qquad a = -\tfrac{1}{2} \qquad a = \tfrac{4}{3}$$

EXERCISES 1.6

9.

$$y^4 - 26y^2 + 25 = 0$$
$$(y^2 - 25)(y^2 - 1) = 0$$
$$y^2 - 25 = 0 \quad \textbf{or} \quad y^2 - 1 = 0$$
$$y^2 = 25 \qquad y^2 = 1$$
$$y = \pm 5 \qquad y = \pm 1$$

11.

$$2y^4 - 46y^2 = -180$$
$$2(y^4 - 23y^2 + 90) = 0$$
$$2(y^2 - 18)(y^2 - 5) = 0$$
$$y^2 - 18 = 0 \quad \textbf{or} \quad y^2 - 5 = 0$$
$$y^2 = 18 \qquad y^2 = 5$$
$$y = \pm\sqrt{18} \qquad y = \pm\sqrt{5}$$
$$y = \pm 3\sqrt{2} \qquad y = \pm\sqrt{5}$$

13.

$$x^4 = 8x^2 + 9$$
$$x^4 - 8x^2 - 9 = 0$$
$$(x^2 - 9)(x^2 + 1) = 0$$
$$x^2 - 9 = 0 \quad \textbf{or} \quad x^2 + 1 = 0$$
$$x^2 = 9 \qquad x^2 = -1$$
$$x = \pm\sqrt{9} \qquad x = \pm\sqrt{-1}$$
$$x = \pm 3 \qquad x = \pm i$$

15.

$$4y^4 + 7y^2 - 36 = 0$$
$$(4y^2 - 9)(y^2 + 4) = 0$$
$$4y^2 - 9 = 0 \quad \textbf{or} \quad y^2 + 4 = 0$$
$$y^2 = \tfrac{9}{4} \qquad y^2 = -4$$
$$y = \pm\sqrt{\tfrac{9}{4}} \qquad y = \pm\sqrt{-4}$$
$$y = \pm\tfrac{3}{2} \qquad y = \pm 2i$$

17.

$$x^4 - 37x^2 + 36 = 0$$
$$(x^2 - 36)(x^2 - 1) = 0$$
$$x^2 - 36 = 0 \quad \textbf{or} \quad x^2 - 1 = 0$$
$$x^2 = 36 \qquad x^2 = 1$$
$$x = \pm 6 \qquad x = \pm 1$$

19.

$$2m^{2/3} + 3m^{1/3} - 2 = 0$$
$$(2m^{1/3} - 1)(m^{1/3} + 2) = 0$$
$$2m^{1/3} - 1 = 0 \quad \textbf{or} \quad m^{1/3} + 2 = 0$$
$$m^{1/3} = \tfrac{1}{2} \qquad m^{1/3} = -2$$
$$(m^{1/3})^3 = (\tfrac{1}{2})^3 \qquad (m^{1/3})^3 = (-2)^3$$
$$m = \tfrac{1}{8} \qquad m = -8$$

Both answers check.

21.

$$x - 13x^{1/2} + 12 = 0$$
$$(x^{1/2} - 12)(x^{1/2} - 1) = 0$$
$$x^{1/2} - 12 = 0 \quad \textbf{or} \quad x^{1/2} - 1 = 0$$
$$x^{1/2} = 12 \qquad x^{1/2} = 1$$
$$(x^{1/2})^2 = (12)^2 \qquad (x^{1/2})^2 = (1)^2$$
$$x = 144 \qquad x = 1$$

Both answers check.

23.

$$6p + p^{1/2} = 1$$
$$6p + p^{1/2} - 1 = 0$$
$$(2p^{1/2} + 1)(3p^{1/2} - 1) = 0$$
$$2p^{1/2} + 1 = 0 \quad \textbf{or} \quad 3p^{1/2} - 1 = 0$$
$$p^{1/2} = -\tfrac{1}{2} \qquad p^{1/2} = \tfrac{1}{3}$$
$$(p^{1/2})^2 = (-\tfrac{1}{2})^2 \qquad (p^{1/2})^2 = (\tfrac{1}{3})^2$$
$$p = \tfrac{1}{4} \qquad p = \tfrac{1}{9}$$

$p = \frac{1}{4}$ does not check and is extraneous.

EXERCISES 1.6

25.

$$2t^{1/3} + 3t^{1/6} - 2 = 0$$
$$\left(2t^{1/6} - 1\right)\left(t^{1/6} + 2\right) = 0$$

$$2t^{1/6} - 1 = 0 \quad \textbf{or} \quad t^{1/6} + 2 = 0$$
$$t^{1/6} = \tfrac{1}{2} \qquad t^{1/6} = -2$$
$$\left(t^{1/6}\right)^6 = \left(\tfrac{1}{2}\right)^6 \qquad \left(t^{1/6}\right)^6 = (-2)^6$$
$$t = \tfrac{1}{64} \qquad t = 64$$

$t = 64$ does not check and is extraneous.

27.

$$x^{-2} - 10x^{-1} + 16 = 0$$
$$\left(x^{-1} - 8\right)\left(x^{-1} - 2\right) = 0$$

$$x^{-1} - 8 = 0 \quad \textbf{or} \quad x^{-1} - 2 = 0$$
$$x^{-1} = 8 \qquad x^{-1} = 2$$
$$\left(x^{-1}\right)^{-1} = (8)^{-1} \qquad \left(x^{-1}\right)^{-1} = (2)^{-1}$$
$$x = \tfrac{1}{8} \qquad x = \tfrac{1}{2}$$

Both answers check.

29.

$$z^{3/2} - z^{1/2} = 0$$
$$z^{1/2}\left(z^{2/2} - 1\right) = 0$$
$$z^{1/2}(z - 1) = 0$$
$$z^{1/2} = 0 \quad \textbf{or} \quad z - 1 = 0$$
$$\left(z^{1/2}\right)^2 = 0^2 \qquad z = 1$$
$$z = 0 \qquad z = 1$$

Both answers check.

31.

$$\sqrt{x-2} - 3 = 2$$
$$\sqrt{x-2} = 5$$
$$\left(\sqrt{x-2}\right)^2 = 5^2$$
$$x - 2 = 25$$
$$x = 27$$

The solution checks.

33.

$$3\sqrt{x+1} = \sqrt{6}$$
$$\left(3\sqrt{x+1}\right)^2 = \left(\sqrt{6}\right)^2$$
$$9(x+1) = 6$$
$$9x + 9 = 6$$
$$9x = -3$$
$$x = -\tfrac{1}{3}$$

The solution checks.

35.

$$\sqrt{5a-2} = \sqrt{a+6}$$
$$\left(\sqrt{5a-2}\right)^2 = \left(\sqrt{a+6}\right)^2$$
$$5a - 2 = a + 6$$
$$4a = 8$$
$$a = 2$$

The solution checks.

37.

$$2\sqrt{x^2+3} = \sqrt{-16x-3}$$
$$\left(2\sqrt{x^2+3}\right)^2 = \left(\sqrt{-16x-3}\right)^2$$
$$4\left(x^2+3\right) = -16x - 3$$
$$4x^2 + 12 = -16x - 3$$
$$4x^2 + 16x + 15 = 0$$
$$(2x+3)(2x+5) = 0$$
$$2x + 3 = 0 \quad \textbf{or} \quad 2x + 5 = 0$$
$$x = -\tfrac{3}{2} \qquad x = -\tfrac{5}{2}$$

Both solutions check.

39.

$$\sqrt{x^2+21} = x + 3$$
$$\left(\sqrt{x^2+21}\right)^2 = (x+3)^2$$
$$x^2 + 21 = x^2 + 6x + 9$$
$$21 = 6x + 9$$
$$12 = 6x$$
$$2 = x$$

The solution checks.

EXERCISES 1.6

41.

$$\sqrt{x+37} = x-5$$
$$\left(\sqrt{x+37}\right)^2 = (x-5)^2$$
$$x+37 = x^2-10x+25$$
$$0 = x^2-11x-12$$
$$0 = (x-12)(x+1)$$
$$x-12=0 \quad \textbf{or} \quad x+1=0$$
$$x=12 \qquad x=-1$$

$x=-1$ does not check and is extraneous.

43.

$$\sqrt{3z+1} = z-1$$
$$\left(\sqrt{3z+1}\right)^2 = (z-1)^2$$
$$3z+1 = z^2-2z+1$$
$$0 = z^2-5z$$
$$0 = z(z-5)$$
$$z=0 \quad \textbf{or} \quad z-5=0$$
$$z=0 \qquad z=5$$

$z=0$ does not check and is extraneous.

45.

$$x-\sqrt{7x-12}=0$$
$$x=\sqrt{7x-12}$$
$$x^2=\left(\sqrt{7x-12}\right)^2$$
$$x^2=7x-12$$
$$x^2-7x+12=0$$
$$(x-4)(x-3)=0$$
$$x-4=0 \quad \textbf{or} \quad x-3=0$$
$$x=4 \qquad x=3$$

Both solutions check.

47.

$$x+4=\sqrt{\frac{6x+6}{5}}+3$$
$$x+1=\sqrt{\frac{6x+6}{5}}$$
$$(x+1)^2=\left(\sqrt{\frac{6x+6}{5}}\right)^2$$
$$x^2+2x+1=\frac{6x+6}{5}$$
$$5x^2+10x+5=6x+6$$
$$5x^2+4x-1=0$$
$$(5x-1)(x+1)=0$$
$$5x-1=0 \quad \textbf{or} \quad x+1=0$$
$$x=\tfrac{1}{5} \qquad x=-1$$

Both solutions check.

49.

$$\sqrt{\frac{x^2-1}{x-2}}=2\sqrt{2}$$
$$\left(\sqrt{\frac{x^2-1}{x-2}}\right)^2=\left(2\sqrt{2}\right)^2$$
$$\frac{x^2-1}{x-2}=8$$
$$x^2-1=8x-16$$
$$x^2-8x+15=0$$
$$(x-3)(x-5)=0$$
$$x-3=0 \quad \textbf{or} \quad x-5=0$$
$$x=3 \qquad x=5$$

Both solutions check.

51.

$$\sqrt{2p+1}-1=\sqrt{p}$$
$$\left(\sqrt{2p+1}-1\right)^2=\left(\sqrt{p}\right)^2$$
$$2p+1-2\sqrt{2p+1}+1=p$$
$$p+2=2\sqrt{2p+1}$$
$$(p+2)^2=\left(2\sqrt{2p+1}\right)^2$$
$$p^2+4p+4=4(2p+1)$$
$$p^2+4p+4=8p+4$$
$$p^2-4p=0$$
$$p(p-4)=0$$
$$p=0 \quad \textbf{or} \quad p-4=0$$
$$p=0 \qquad p=4$$

Both solutions check.

EXERCISES 1.6

53.

$$\sqrt{x+3} = \sqrt{2x+8} - 1$$
$$\left(\sqrt{x+3}\right)^2 = \left(\sqrt{2x+8} - 1\right)^2$$
$$x+3 = 2x+8-2\sqrt{2x+8}+1$$
$$2\sqrt{2x+8} = x+6$$
$$\left(2\sqrt{2x+8}\right)^2 = (x+6)^2$$
$$4(2x+8) = x^2+12x+36$$
$$8x+32 = x^2+12x+36$$
$$0 = x^2+4x+4$$
$$0 = (x+2)(x+2)$$
$$x+2=0 \quad \textbf{or} \quad x+2=0$$
$$x=-2 \qquad\qquad x=-2$$

The solution checks.

55.

$$\sqrt{y+8} - \sqrt{y-4} = -2$$
$$\sqrt{y+8} = \sqrt{y-4} - 2$$
$$\left(\sqrt{y+8}\right)^2 = \left(\sqrt{y-4} - 2\right)^2$$
$$y+8 = y-4-4\sqrt{y-4}+4$$
$$4\sqrt{y-4} = -8$$
$$\left(4\sqrt{y-4}\right)^2 = (-8)^2$$
$$16(y-4) = 64$$
$$16y-64 = 64$$
$$16y = 128$$
$$y = 8$$

The solution does not check. $\Rightarrow$ No solution.

57.

$$\sqrt{2b+3} - \sqrt{b+1} = \sqrt{b-2}$$
$$\left(\sqrt{2b+3} - \sqrt{b+1}\right)^2 = \left(\sqrt{b-2}\right)^2$$
$$2b+3-2\sqrt{(2b+3)(b+1)}+b+1 = b-2$$
$$3b+4-2\sqrt{2b^2+5b+3} = b-2$$
$$2b+6 = 2\sqrt{2b^2+5b+3}$$
$$(2b+6)^2 = \left(2\sqrt{2b^2+5b+3}\right)^2$$
$$4b^2+24b+36 = 4\left(2b^2+5b+3\right)$$
$$4b^2+24b+36 = 8b^2+20b+12$$
$$0 = 4b^2-4b-24$$
$$0 = 4(b-3)(b+2)$$
$$b-3=0 \quad \textbf{or} \quad b+2=0$$
$$b=3 \qquad\qquad b=-2$$

$b=-2$ does not check, so it is an extraneous solution.

59.

$$\sqrt{\sqrt{b}+\sqrt{b+8}} = 2$$
$$\left(\sqrt{\sqrt{b}+\sqrt{b+8}}\right)^2 = 2^2$$
$$\sqrt{b}+\sqrt{b+8} = 4$$
$$\sqrt{b+8} = 4-\sqrt{b}$$
$$\left(\sqrt{b+8}\right)^2 = \left(4-\sqrt{b}\right)^2$$
$$b+8 = 16-8\sqrt{b}+b$$
$$8\sqrt{b} = 8$$
$$\sqrt{b} = 1$$
$$\left(\sqrt{b}\right)^2 = 1^2$$

$b=1 \Rightarrow$ The solution checks.

EXERCISES 1.6

61.
$$\begin{aligned} \sqrt[3]{7x+1} &= 4 \\ \left(\sqrt[3]{7x+1}\right)^3 &= 4^3 \\ 7x+1 &= 64 \\ 7x &= 63 \\ x &= 9 \end{aligned}$$
The solution checks.

63.
$$\begin{aligned} \sqrt[3]{x^3+7} &= x+1 \\ \left(\sqrt[3]{x^3+7}\right)^3 &= (x+1)^3 \\ x^3+7 &= x^3+3x^2+3x+1 \\ 0 &= 3x^2+3x-6 \\ 0 &= 3(x+2)(x-1) \end{aligned}$$
$x+2=0$ **or** $x-1=0$

$x=-2$ $\qquad x=1$

Both solutions check.

65.
$$\begin{aligned} \sqrt[3]{8x^3+61} &= 2x+1 \\ \left(\sqrt[3]{8x^3+61}\right)^3 &= (2x+1)^3 \\ 8x^3+61 &= 8x^3+12x^2+6x+1 \\ 0 &= 12x^2+6x-60 \\ 0 &= 6(2x+5)(x-2) \end{aligned}$$
$2x+5=0$ **or** $x-2=0$

$x=-\frac{5}{2}$ $\qquad x=2$

Both solutions check.

67.
$$\begin{aligned} \sqrt[4]{30t+25} &= 5 \\ \left(\sqrt[4]{30t+25}\right)^4 &= 5^4 \\ 30t+25 &= 625 \\ 30t &= 600 \\ t &= 20 \end{aligned}$$
The solution checks.

69.
$$\begin{aligned} \sqrt[5]{2x-11} &= \sqrt[5]{14} \\ \left(\sqrt[5]{2x-11}\right)^5 &= \left(\sqrt[5]{14}\right)^5 \\ 2x-11 &= 14 \\ 2x &= 25 \\ x &= \tfrac{25}{2} \end{aligned}$$
The solution checks.

71.
$$\begin{aligned} t &= \sqrt{\frac{d}{16}} \\ 5 &= \sqrt{\frac{d}{16}} \\ 5^2 &= \left(\sqrt{\frac{d}{16}}\right)^2 \\ 25 &= \frac{d}{16} \end{aligned}$$
$400 = d \Rightarrow$ The bridge is 400 feet high.

73.
$$\begin{aligned} l &= \sqrt{f^2+h^2} \\ 10 &= \sqrt{f^2+6^2} \\ 10^2 &= \left(\sqrt{f^2+36}\right)^2 \\ 100 &= f^2+36 \\ 64 &= f^2 \\ \pm 8 &= f \end{aligned}$$
He should nail the brace to the floor 8 feet from the wall.

75.
$$r = \sqrt[n]{\frac{A}{P}} - 1$$
$$0.065 = \sqrt[4]{\frac{4000}{P}} - 1$$
$$1.065 = \sqrt[4]{\frac{4000}{P}}$$
$$(1.065)^4 = \left(\sqrt[4]{\frac{4000}{P}}\right)^4$$
$$(1.065)^4 = \left(\sqrt[4]{\frac{4000}{P}}\right)^4$$
$$1.286466 \approx \frac{4000}{P}$$
$$1.286466P \approx 4000$$
$$P \approx 3109$$
The original price was about \$3109.

77-81. Answers may vary.

83. False. The equation is not quadratic in form. $x^4 + 6x^2 + 5 = 0$ can be solved by factoring.

85. True.

87. False.
$$\sqrt{\sqrt{\sqrt{x}}} = 2$$
$$\sqrt{\sqrt{x}} = 4$$
$$\sqrt{x} = 16$$
$$x = 256$$

89. False. Isolate one of the radicals first, then square both sides of the equation.

Exercises 1.7 (page 166)

1. right

3. $a < c$

5. $b - c$

7. $>$

9. linear

11. equivalent

13.
$$3x + 2 < 5$$
$$3x < 3$$
$$x < 1 \Rightarrow (-\infty, 1)$$

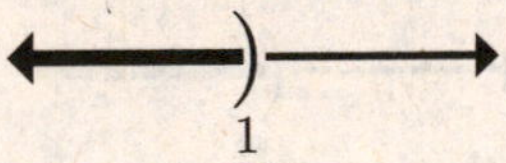

15.
$$3x + 2 \geq 5$$
$$3x \geq 3$$
$$x \geq 1 \Rightarrow [1, \infty)$$

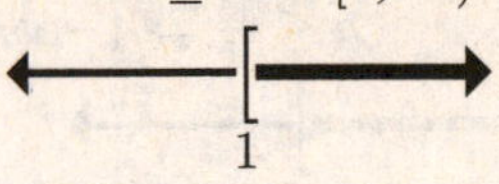

17.
$$-5x + 3 > -2$$
$$-5x > -5$$
$$x < 1 \Rightarrow (-\infty, 1)$$

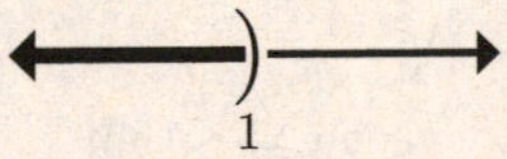

19.
$$-5x + 3 \leq -2$$
$$-5x \leq -5$$
$$x \geq 1 \Rightarrow [1, \infty)$$

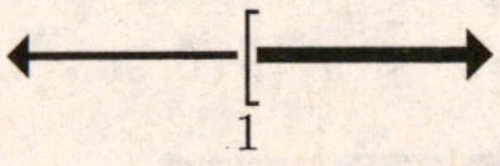

EXERCISES 1.7

21.
$$2(x-3) \le -2(x-3)$$
$$2x-6 \le -2x+6$$
$$4x \le 12$$
$$x \le 3 \Rightarrow (-\infty, 3]$$

3

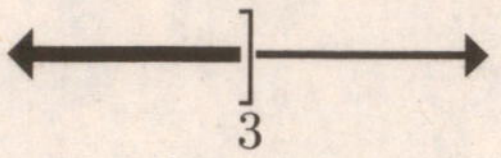

23.
$$\frac{3}{5}x+4 > 2$$
$$5\left(\frac{3}{5}x+4\right) > 5(2)$$
$$3x+20 > 10$$
$$3x > -10$$
$$x > -\tfrac{10}{3} \Rightarrow \left(-\tfrac{10}{3}, \infty\right)$$

$-\frac{10}{3}$

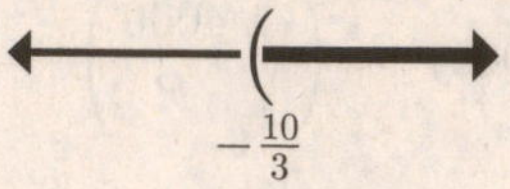

25.
$$\frac{x+3}{4} < \frac{2x-4}{3}$$
$$12 \cdot \frac{x+3}{4} < 12 \cdot \frac{2x-4}{3}$$
$$3(x+3) < 4(2x-4)$$
$$3x+9 < 8x-16$$
$$-5x < -25$$
$$x > 5 \Rightarrow (5, \infty)$$

5

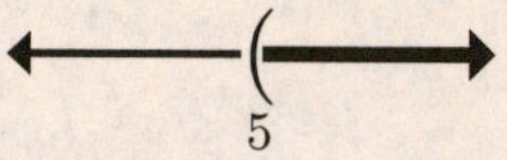

27.
$$\frac{6(x-4)}{5} \ge \frac{3(x+2)}{4}$$
$$20 \cdot \frac{6x-24}{5} \ge 20 \cdot \frac{3x+6}{4}$$
$$4(6x-24) \ge 5(3x+6)$$
$$24x-96 \ge 15x+30$$
$$9x \ge 126$$
$$x \ge 14 \Rightarrow [14, \infty)$$

14

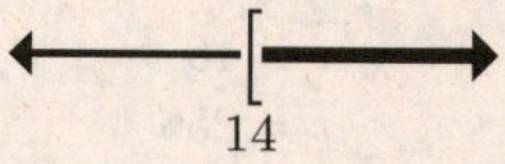

29.
$$\frac{5}{9}(a+3)-a \ge \frac{4}{3}(a-3)-1$$
$$9\left[\frac{5}{9}(a+3)-a\right] \ge 9\left[\frac{4}{3}(a-3)-1\right]$$
$$5(a+3)-9a \ge 12(a-3)-9$$
$$5a+15-9a \ge 12a-36-9$$
$$-16a \ge -60$$
$$a \le \frac{-60}{-16}$$
$$a \le \tfrac{15}{4} \Rightarrow \left(-\infty, \tfrac{15}{4}\right]$$

$\frac{15}{4}$

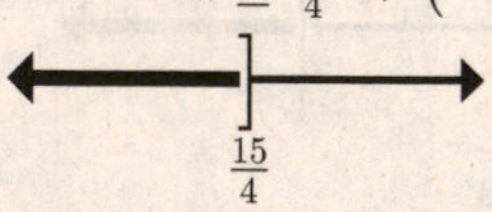

31.
$$\frac{2}{3}a-\frac{3}{4}a < \frac{3}{5}\left(a+\frac{2}{3}\right)+\frac{1}{3}$$
$$60\left(\frac{2}{3}a-\frac{3}{4}a\right) < 60\left[\frac{3}{5}\left(a+\frac{2}{3}\right)+\frac{1}{3}\right]$$
$$40a-45a < 36\left(a+\frac{2}{3}\right)+20$$
$$-5a < 36a+24+20$$
$$-41a < 44$$
$$a > -\tfrac{44}{41} \Rightarrow \left(-\tfrac{44}{41}, \infty\right)$$

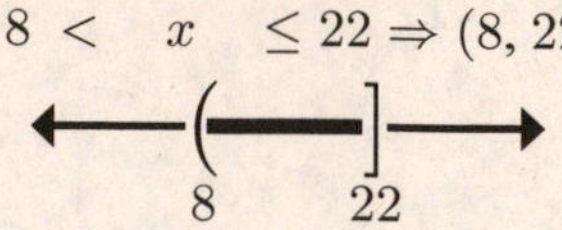

$-\frac{44}{41}$

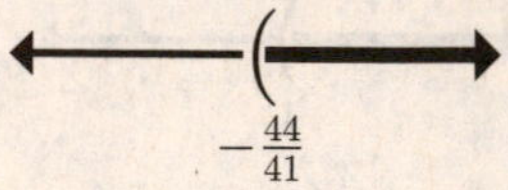

33.
$$4 < 2x-8 \le 10$$
$$12 < 2x \le 18$$
$$6 < x \le 9 \Rightarrow (6, 9]$$

6 9

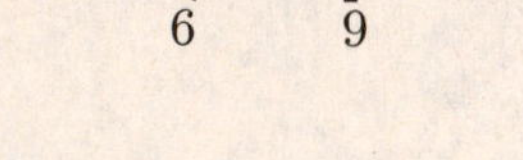

35.
$$9 \ge \frac{x-4}{2} > 2$$
$$18 \ge x-4 > 4$$
$$22 \ge x > 8$$
$$8 < x \le 22 \Rightarrow (8, 22]$$

8 22

EXERCISES 1.7

37.
$$0 \le \frac{4-x}{3} \le 5$$
$$0 \le 4-x \le 15$$
$$-4 \le -x \le 11$$
$$4 \ge x \ge -11$$
$$-11 \le x \le 4 \Rightarrow [-11, 4]$$

39.
$$-2 \ge \frac{1-x}{2} \ge -10$$
$$-4 \ge 1-x \ge -20$$
$$-5 \ge -x \ge -21$$
$$5 \le x \le 21 \Rightarrow [5, 21]$$

41.
$$-3x > -2x > -x$$
$-3x > -2x$ **and** $-2x > -x$

$-x > 0$ $\quad$ $-x > 0$

$x < 0$ $\quad$ $x < 0$

$$x < 0 \Rightarrow (-\infty, 0)$$

43.
$$x < 2x < 3x$$
$x < 2x$ **and** $2x < 3x$

$-x < 0$ $\quad$ $-x < 0$

$x > 0$ $\quad$ $x > 0$

$$x > 0 \Rightarrow (0, \infty)$$

45.
$$2x + 1 < 3x - 2 < 12$$
$2x + 1 < 3x - 2$ **and** $3x - 2 < 12$

$-x < -3$ $\quad$ $3x < 14$

$x > 3$ $\quad$ $x < \frac{14}{3}$

$x > 3$

$x < \frac{14}{3}$

$x > 3$ **and** $x < \frac{14}{3}$

Solution set: $\left(3, \frac{14}{3}\right)$

47.
$$2 + x < 3x - 2 < 5x + 2$$
$2 + x < 3x - 2$ **and** $3x - 2 < 5x + 2$

$-2x < -4$ $\quad$ $-2x < 4$

$x > 2$ $\quad$ $x > -2$

$x > 2$

$x > -2$

$x > 2$ **and** $x > -2$

Solution set: $(2, \infty)$

49. $3 + x > 7x - 2 > 5x - 10$

$3 + x > 7x - 2$ **and** $7x - 2 > 5x - 10$

$-6x > -5$ $\quad$ $2x > -8$

$x < \frac{5}{6}$ $\quad$ $x > -4$

$x < \frac{5}{6}$

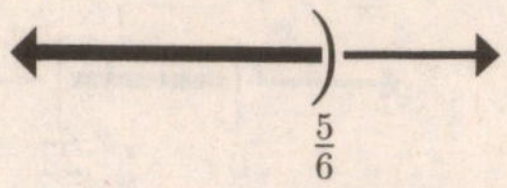

$x > -4$

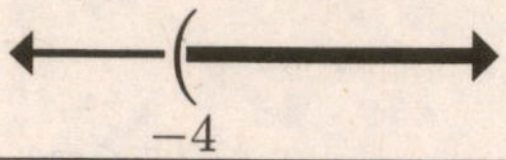

$x < \frac{5}{6}$ **and** $x > -4$

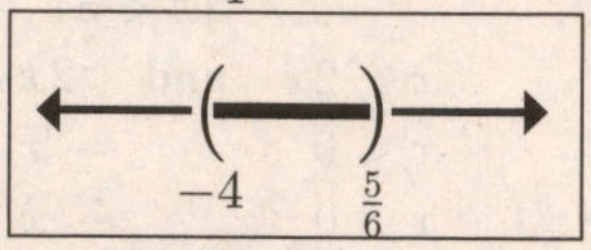

Solution set: $\left(-4, \frac{5}{6}\right)$

51. $x \le x + 1 \le 2x + 3$

$x \le x + 1$ **and** $x + 1 \le 2x + 3$

$0 \le 1$ $\quad$ $-x \le 2$

true for all real numbers x $\quad$ $x \ge -2$

$0 \le 1$

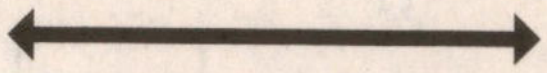

$x \ge -2$

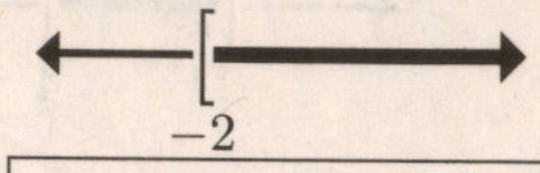

$0 \le 1$ **and** $x \ge -2$

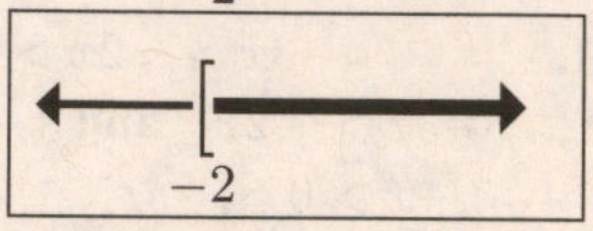

Solution set: $[-2, \infty)$

53. $x^2 + 7x + 12 < 0$

$(x + 3)(x + 4) < 0$

factors = 0: $x = -3, x = -4$

intervals: $(-\infty, -4), (-4, -3), (-3, \infty)$

interval	test number	value of $x^2+7x+12$
$(-\infty, -4)$	-5	$+2$
$(-4, -3)$	-3.5	-0.25
$(-3, \infty)$	0	$+12$

Solution: $(-4, -3)$

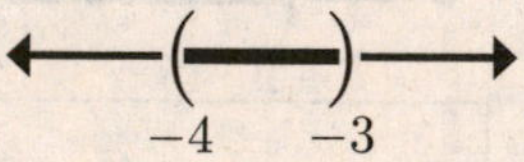

55. $x^2 - 5x + 6 \ge 0$

$(x - 3)(x - 2) \ge 0$

factors = 0: $x = 3, x = 2$

intervals: $(-\infty, 2), (2, 3), (3, \infty)$

interval	test number	value of x^2-5x+6
$(-\infty, 2)$	0	$+6$
$(2, 3)$	2.5	-0.25
$(3, \infty)$	4	$+2$

Solution: $(-\infty, 2] \cup [3, \infty)$

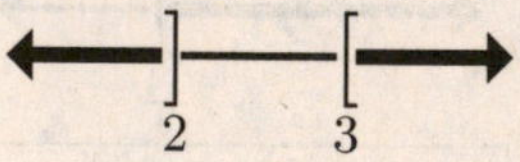

57. $x^2 + 5x + 6 < 0$

$(x + 3)(x + 2) < 0$

factors = 0: $x = -3, x = -2$

intervals: $(-\infty, -3), (-3, -2), (-2, \infty)$

interval	test number	value of x^2+5x+6
$(-\infty, -3)$	-4	$+2$
$(-3, -2)$	-2.5	-0.25
$(-2, \infty)$	0	$+6$

Solution: $(-3, -2)$

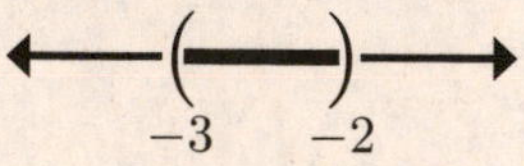

59. $6x^2 + 5x + 1 \ge 0$

$(2x + 1)(3x + 1) \ge 0$

factors = 0: $x = -\frac{1}{2}, x = -\frac{1}{3}$

intervals: $\left(-\infty, -\frac{1}{2}\right), \left(-\frac{1}{2}, -\frac{1}{3}\right), \left(-\frac{1}{3}, \infty\right)$

interval	test number	value of $6x^2+5x+1$
$\left(-\infty, -\frac{1}{2}\right)$	-1	$+2$
$\left(-\frac{1}{2}, -\frac{1}{3}\right)$	-0.4	-0.04
$\left(-\frac{1}{3}, \infty\right)$	0	$+1$

Solution: $\left(-\infty, -\frac{1}{2}\right] \cup \left[-\frac{1}{3}, \infty\right)$

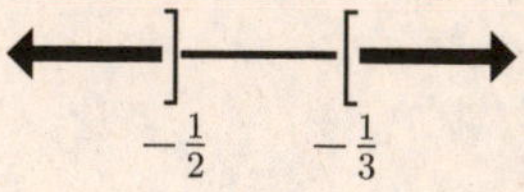

EXERCISES 1.7

61.

$$6x^2 - 5x < -1$$
$$6x^2 - 5x + 1 < 0$$
$$(2x-1)(3x-1) < 0$$

factors $= 0$: $x = \frac{1}{2}, x = \frac{1}{3}$

intervals: $(-\infty, \frac{1}{3}), (\frac{1}{3}, \frac{1}{2}), (\frac{1}{2}, \infty)$

interval	test number	value of $6x^2-5x+1$
$(-\infty, \frac{1}{3})$	0	$+1$
$(\frac{1}{3}, \frac{1}{2})$	0.4	-0.04
$(\frac{1}{2}, \infty)$	1	$+2$

Solution: $(\frac{1}{3}, \frac{1}{2})$

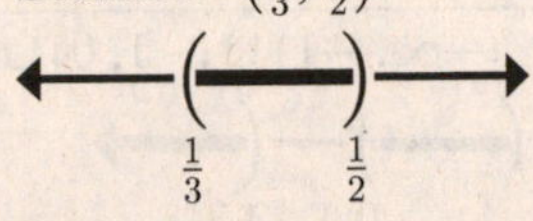

63.

$$2x^2 \geq 3 - x$$
$$2x^2 + x - 3 \geq 0$$
$$(2x+3)(x-1) \geq 0$$

factors $= 0$: $x = -\frac{3}{2}, x = 1$

intervals: $(-\infty, -\frac{3}{2}), (-\frac{3}{2}, 1), (1, \infty)$

interval	test number	value of $2x^2+x-3$
$(-\infty, -\frac{3}{2})$	-2	$+3$
$(-\frac{3}{2}, 1)$	0	-3
$(1, \infty)$	2	$+7$

Solution: $(-\infty, -\frac{3}{2}] \cup [1, \infty)$

$-\frac{3}{2}$ 1

65.

$$x^2 - 3 \geq 0$$
$$x^2 - 3 = 0$$
$$x^2 = 3$$
$$x = \pm\sqrt{3}$$

intervals: $(-\infty, -\sqrt{3}), (-\sqrt{3}, \sqrt{3}), (\sqrt{3}, \infty)$

interval	test number	value of x^2-3
$(-\infty, -\sqrt{3})$	-2	$+1$
$(-\sqrt{3}, \sqrt{3})$	0	-3
$(\sqrt{3}, \infty)$	2	$+1$

Solution: $(-\infty, -\sqrt{3}] \cup [\sqrt{3}, \infty)$

$-\sqrt{3}$ $\sqrt{3}$

67.

$$x^2 - 11 < 0$$
$$x^2 - 11 = 0$$
$$x^2 = 11$$
$$x = \pm\sqrt{11}$$

intervals: $(-\infty, -\sqrt{11}), (-\sqrt{11}, \sqrt{11}), (\sqrt{11}, \infty)$

interval	test number	value of x^2-11
$(-\infty, -\sqrt{11})$	-4	$+5$
$(-\sqrt{11}, \sqrt{11})$	0	-11
$(\sqrt{11}, \infty)$	4	$+5$

Solution: $(-\sqrt{11}, \sqrt{11})$

$-\sqrt{11}$ $\sqrt{11}$

69. $\dfrac{x+3}{x-2} < 0$

factors $= 0$: $x = -3$, $x = 2$

intervals: $(-\infty, -3)$, $(-3, 2)$, $(2, \infty)$

interval	test number	sign of $\frac{x+3}{x-2}$
$(-\infty, -3)$	-4	$+$
$(-3, 2)$	0	$-$
$(2, \infty)$	3	$+$

Solution: $(-3, 2)$

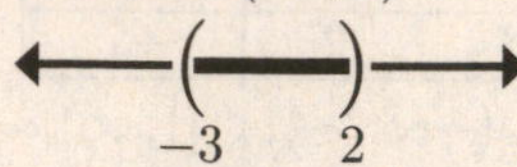

71. $\dfrac{x^2+x}{x^2-1} > 0$

$\dfrac{x(x+1)}{(x+1)(x-1)} > 0$

factors $= 0$: $x = 0$, $x = -1$, $x = 1$

intervals: $(-\infty, -1)$, $(-1, 0)$, $(0, 1)$, $(1, \infty)$

interval	test number	sign of $\frac{x^2+x}{x^2-1}$
$(-\infty, -1)$	-2	$+$
$(-1, 0)$	$-\frac{1}{2}$	$+$
$(0, 1)$	$\frac{1}{2}$	$-$
$(1, \infty)$	2	$+$

Solution: $(-\infty, -1) \cup (-1, 0) \cup (1, \infty)$

73. $\dfrac{x^2+5x+6}{x^2+x-6} \geq 0 \Rightarrow \dfrac{(x+3)(x+2)}{(x+3)(x-2)} \geq 0$

factors $= 0$: $x = -3$, $x = \pm 2$

intervals: $(-\infty, -3)$, $(-3, -2)$, $(-2, 2)$, $(2, \infty)$

interval	test number	sign of $\frac{x^2+5x+6}{x^2+x-6}$
$(-\infty, -3)$	-4	$+$
$(-3, -2)$	-2.5	$+$
$(-2, 2)$	0	$-$
$(2, \infty)$	3	$+$

Include endpoints which make the numerator equal to 0. Do not include endpoints which make the denominator equal to 0.

Solution: $(-\infty, -3) \cup (-3, -2] \cup (2, \infty)$

75. $\dfrac{6x^2-x-1}{x^2+4x+4} > 0 \Rightarrow \dfrac{(2x-1)(3x+1)}{(x+2)(x+2)} > 0$

factors $= 0$: $x = \frac{1}{2}$, $x = -\frac{1}{3}$, $x = -2$

intervals: $(-\infty, -2)$, $\left(-2, -\frac{1}{3}\right)$, $\left(-\frac{1}{3}, \frac{1}{2}\right)$, $\left(\frac{1}{2}, \infty\right)$

interval	test number	sign of $\frac{6x^2-x-1}{x^2+4x+4}$
$(-\infty, -2)$	-3	$+$
$\left(-2, -\frac{1}{3}\right)$	-1	$+$
$\left(-\frac{1}{3}, \frac{1}{2}\right)$	0	$-$
$\left(\frac{1}{2}, \infty\right)$	1	$+$

Solution: $(-\infty, -2) \cup \left(-2, -\frac{1}{3}\right) \cup \left(\frac{1}{2}, \infty\right)$

77. $\dfrac{3}{x} > 2$

$\dfrac{3}{x} - 2 > 0$

$\dfrac{3-2x}{x} > 0$

factors $= 0$: $x = \frac{3}{2}$, $x = 0$

intervals: $(-\infty, 0)$, $\left(0, \frac{3}{2}\right)$, $\left(\frac{3}{2}, \infty\right)$

interval	test number	sign of $\frac{3-2x}{x}$
$(-\infty, 0)$	-1	$-$
$\left(0, \frac{3}{2}\right)$	1	$+$
$\left(\frac{3}{2}, \infty\right)$	2	$-$

Solution: $\left(0, \frac{3}{2}\right)$

EXERCISES 1.7

79. $\frac{6}{x} < 4$

$\frac{6}{x} - 4 < 0$

$\frac{6 - 4x}{x} < 0$

factors = 0: $x = \frac{3}{2}, x = 0$

intervals: $(-\infty, 0)$, $(0, \frac{3}{2})$, $(\frac{3}{2}, \infty)$

interval	test number	sign of $\frac{6-4x}{x}$
$(-\infty, 0)$	-1	$-$
$(0, \frac{3}{2})$	1	$+$
$(\frac{3}{2}, \infty)$	2	$-$

Solution: $(-\infty, 0) \cup (\frac{3}{2}, \infty)$

0 $\quad \frac{3}{2}$

81. $\frac{3}{x-2} \le 5$

$\frac{3}{x-2} - 5 \le 0$

$\frac{3}{x-2} - \frac{5(x-2)}{x-2} \le 0$

$\frac{3 - 5x + 10}{x-2} \le 0$

$\frac{13 - 5x}{x-2} \le 0$

factors = 0: $x = \frac{13}{5}, x = 2$

intervals: $(-\infty, 2)$, $(2, \frac{13}{5})$, $(\frac{13}{5}, \infty)$

interval	test number	sign of $\frac{13-5x}{x-2}$
$(-\infty, 2)$	0	$-$
$(2, \frac{13}{5})$	$\frac{11}{5}$	$+$
$(\frac{13}{5}, \infty)$	3	$-$

Solution: $(-\infty, 2) \cup [\frac{13}{5}, \infty)$

2 $\quad \frac{13}{5}$

Include endpoints which make the numerator equal to 0. Do not include endpoints which make the denominator equal to 0.

83. $\frac{6}{x^2 - 1} < 1$

$\frac{6}{x^2 - 1} - 1 < 0$

$\frac{6}{x^2 - 1} - \frac{x^2 - 1}{x^2 - 1} < 0$

$\frac{7 - x^2}{x^2 - 1} < 0$

$\frac{7 - x^2}{(x+1)(x-1)} < 0$

factors = 0: $x = \pm\sqrt{7}, x = \pm 1$

intervals: $\left(-\infty, -\sqrt{7}\right)$, $\left(-\sqrt{7}, -1\right)$, $(-1, 1)$, $\left(1, \sqrt{7}\right)$, $\left(\sqrt{7}, \infty\right)$

interval	test number	sign of $\frac{7-x^2}{x^2-1}$
$\left(-\infty, -\sqrt{7}\right)$	-3	$-$
$\left(-\sqrt{7}, -1\right)$	-2	$+$
$(-1, 1)$	0	$-$
$\left(1, \sqrt{7}\right)$	2	$+$
$\left(\sqrt{7}, \infty\right)$	3	$-$

Sol'n: $\left(-\infty, -\sqrt{7}\right) \cup (-1, 1) \cup \left(\sqrt{7}, \infty\right)$

$-\sqrt{7} \quad -1 \quad 1 \quad \sqrt{7}$

85. Let x = the number of lessons.

$\boxed{\text{Total cost}} \le 960$

$400 + 80x \le 960$

$80x \le 560$

$x \le 7$

They can take at most 7 lessons.

EXERCISES 1.7

87. Let $x =$ the number of minutes after 3 minutes. The total cost $= 40 + 10x$ cents.

$$\boxed{\text{Total cost}} < 200$$
$$40 + 10x < 200$$
$$10x < 160$$
$$x < 16$$

A person can talk for up to 16 minutes after the initial 3 minutes, for a total of up to 19 minutes for less than \$2.

89. Let $x =$ the number of books. Then the total cost $= 150 + 9.75x$.

$$\boxed{\text{Total cost}} < 275$$
$$150 + 9.75x < 275$$
$$9.75x < 125$$
$$x < 12.8$$

He can buy up to 12 books.

91. Let $p =$ the price of the refrigerator. Then the total cost $= p + 0.065p + 0.0025p$.

$$\boxed{\text{Total cost}} < 1200$$
$$p + 0.065p + 0.0025p < 1200$$
$$1.0675p < 1200$$
$$p < 1124.122$$
$$p \le \$1124.12$$

93.
$$R > C$$
$$26x > 6x + 3660$$
$$20x > 3660$$
$$x > 183$$

95. Let $a =$ the assessed value. Find when Method 1 < Method 2:

$$2200 + 0.04a < 1200 + 0.06a$$
$$1000 < 0.02a$$
$$50000 < a$$

The first method will benefit the taxpayer when $a > \$50,000$.

97. Let $b =$ the hospital bill. Find when Cost of Plan 1 > Cost of Plan 2:

$$200 + 0.30(b - 200) > 400 + 0.20(b - 400)$$
$$200 + 0.30b - 60 > 400 + 0.20b - 80$$
$$0.1b > 180$$
$$b > 1800$$

Plan 2 is better for bills over \$1,800.

99. Let $P =$ the perimeter. Then the length of one side is equal to $\frac{P}{3}$.

$$50 < P < 60$$
$$\tfrac{50}{3} < \tfrac{P}{3} < \tfrac{60}{3}$$
$$16\tfrac{2}{3} < \text{length} < 20$$

The length of a side is between $16\frac{2}{3}$ and 20 cm.

101.
$$-16t^2 + 160t > 144$$
$$-16t^2 + 160t - 144 > 0$$
$$-16(t^2 - 10t + 9) > 0$$
$$t^2 - 10t + 9 < 0$$
$$(t - 1)(t - 9) < 0$$

factors $= 0$: $t = 1, t = 9$

intervals: $(-\infty, 1)$, $(1, 9)$, $(9, \infty)$

interval	test number	value of $t^2-10t+9$
$(-\infty, 1)$	0	+9
$(1, 9)$	2	−7
$(9, \infty)$	10	+9

Solution: $(1, 9)$. It will exceed 144 ft between 1 and 9 seconds.

103. Answers may vary.

105. Answers may vary.

107. False.

$$x^2 - 100 \leq 0$$
$$(x+10)(x-10) \leq 0$$
factors $= 0$: $x = -10, x = 10$
intervals: $(-\infty, -10), (-10, 10), (10, \infty)$

interval	test number	value of x^2-100
$(-\infty, -10)$	-11	$+21$
$(-10, 10)$	0	-100
$(10, \infty)$	11	$+21$

Solution: $[-10, 10]$

109. False. The first step is top subtract 2 from both sides of the equation.

Exercises 1.8 (page 175)

1. x

3. $x = k$ or $x = -k$

5. $-k < x < k$

7. $x \leq -k$ or $x \geq k$

9. $|7| = 7$

11. $|0| = 0$

13. $|5| - |-3| = 5 - 3 = 2$

15. $|\pi - 2| = +(\pi - 2) = \pi - 2$

17. $x \geq 5 \Rightarrow |x - 5| = x - 5$

19. $|x^3| = \begin{cases} x^3 & \text{if } x \geq 0 \\ -x^3 & \text{if } x < 0 \end{cases}$

21.
$$|x+2| = 2$$
$$x + 2 = 2 \quad \textbf{or} \quad x + 2 = -2$$
$$x = 0 \qquad\qquad x = -4$$

23.
$$|3x - 1| - 7 = -2$$
$$|3x - 1| = 5$$
$$3x - 1 = 5 \quad \textbf{or} \quad 3x - 1 = -5$$
$$3x = 6 \qquad\qquad 3x = -4$$
$$x = 2 \qquad\qquad x = -\tfrac{4}{3}$$

25.
$$\left|\frac{3x-4}{2}\right| = 5$$
$$\frac{3x-4}{2} = 5 \quad \textbf{or} \quad \frac{3x-4}{2} = -5$$
$$3x - 4 = 10 \qquad\qquad 3x - 4 = -10$$
$$3x = 14 \qquad\qquad 3x = -6$$
$$x = \tfrac{14}{3} \qquad\qquad x = -2$$

27.
$$\left|\frac{2x-4}{5}\right| + 6 = 8$$
$$\left|\frac{2x-4}{5}\right| = 2$$
$$\frac{2x-4}{5} = 2 \quad \textbf{or} \quad \frac{2x-4}{5} = -2$$
$$2x - 4 = 10 \qquad\qquad 2x - 4 = -10$$
$$2x = 14 \qquad\qquad 2x = -6$$
$$x = 7 \qquad\qquad x = -3$$

29.
$$\left|\frac{x-3}{4}\right| = -2$$
An absolute value can never equal a negative number.
no solution

31.
$$\left|\frac{x-5}{3}\right| = 0$$
$$\frac{x-5}{3} = 0 \quad \textbf{or} \quad \frac{x-5}{3} = -0$$
$$x - 5 = 0 \qquad\qquad x - 5 = 0$$
$$x = 5 \qquad\qquad x = 5$$

EXERCISES 1.8

33. $\left|\dfrac{4x-2}{x}\right| = 3$

$$\begin{aligned} \frac{4x-2}{x} &= 3 \quad \textbf{or} \quad \frac{4x-2}{x} = -3 \\ 4x-2 &= 3x \qquad\quad 4x-2 = -3x \\ x &= 2 \qquad\qquad\quad 7x = 2 \\ x &= 2 \qquad\qquad\quad x = \tfrac{2}{7} \end{aligned}$$

35. $|x| = x$

True for all $x \geq 0$.

37. $|x+3| = |x|$

$$\begin{aligned} x+3 &= x \quad \textbf{or} \quad x+3 = -x \\ 0 &= 3 \qquad\qquad 2x = -3 \\ &\text{not true} \qquad\quad x = -\tfrac{3}{2} \end{aligned}$$

39. $|x-3| = |2x+3|$

$$\begin{aligned} x-3 &= 2x+3 \quad \textbf{or} \quad x-3 = -(2x+3) \\ -x &= 6 \qquad\qquad\qquad x-3 = -2x-3 \\ x &= -6 \qquad\qquad\qquad 3x = 0 \\ & \qquad\qquad\qquad\qquad\quad x = 0 \end{aligned}$$

41. $|x+2| = |x-2|$

$$\begin{aligned} x+2 &= x-2 \quad \textbf{or} \quad x+2 = -(x-2) \\ 0 &= -4 \qquad\qquad\quad x+2 = -x+2 \\ &\text{not true} \qquad\qquad 2x = 0 \\ & \qquad\qquad\qquad\qquad x = 0 \end{aligned}$$

43. $\left|\dfrac{x+3}{2}\right| = |2x-3|$

$$\begin{aligned} \frac{x+3}{2} &= 2x-3 \quad \textbf{or} \quad \frac{x+3}{2} = -(2x-3) \\ x+3 &= 4x-6 \qquad\qquad \frac{x+3}{2} = -2x+3 \\ -3x &= -9 \qquad\qquad\quad x+3 = -4x+6 \\ x &= 3 \qquad\qquad\qquad 5x = 3 \\ & \qquad\qquad\qquad\qquad x = \tfrac{3}{5} \end{aligned}$$

45. $\left|\dfrac{3x-1}{2}\right| = \left|\dfrac{2x+3}{3}\right|$

$$\begin{aligned} \frac{3x-1}{2} &= \frac{2x+3}{3} \quad \textbf{or} \quad \frac{3x-1}{2} = -\frac{2x+3}{3} \\ 6\left(\frac{3x-1}{2} = \frac{2x+3}{3}\right) & \qquad 6\left(\frac{3x-1}{2} = -\frac{2x+3}{3}\right) \\ 3(3x-1) &= 2(2x+3) \qquad 3(3x-1) = -2(2x+3) \\ 9x-3 &= 4x+6 \qquad\quad 9x-3 = -4x-6 \\ 5x &= 9 \qquad\qquad\qquad 13x = -3 \\ x &= \tfrac{9}{5} \qquad\qquad\qquad x = -\tfrac{3}{13} \end{aligned}$$

47. $|x-3| < 6$

$$\begin{aligned} -6 < x-3 &< 6 \\ -3 < \;\; x \;\; &< 9 \end{aligned}$$

$(-3, 9)$

−3 9

49. $|x+3| > 6$

$$\begin{aligned} x+3 > 6 \quad &\textbf{or} \quad x+3 < -6 \\ x > 3 \qquad & \qquad\quad x < -9 \end{aligned}$$

$(-\infty, -9) \cup (3, \infty)$

−9 3

EXERCISES 1.8

51.

$$|2x+4| \geq 10$$
$$2x+4 \geq 10 \quad \textbf{or} \quad 2x+4 \leq -10$$
$$2x \geq 6 \qquad\qquad 2x \leq -14$$
$$x \geq 3 \qquad\qquad x \leq -7$$
$$(-\infty, -7] \cup [3, \infty)$$

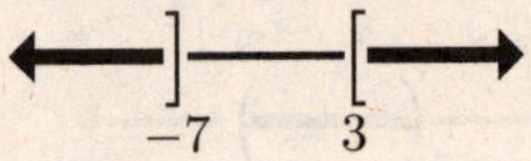

−7 3

53.

$$|3x+5|+1 \leq 9$$
$$|3x+5| \leq 8$$
$$-8 \leq 3x+5 \leq 8$$
$$-13 \leq 3x \leq 3$$
$$-\tfrac{13}{3} \leq x \leq 1$$
$$\left[-\tfrac{13}{3}, 1\right]$$

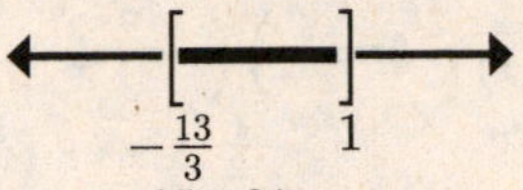

$-\frac{13}{3}$ 1

55.

$$|x+3| > 0$$
$$x+3 > 0 \quad \textbf{or} \quad x+3 < -0$$
$$x > -3 \qquad\qquad x < -3$$
$$(-\infty, -3) \cup (-3, \infty)$$

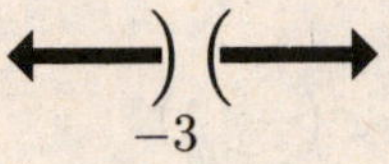

−3

57.

$$\left|\tfrac{5x+2}{3}\right| < 1$$
$$-1 < \tfrac{5x+2}{3} < 1$$
$$-3 < 5x+2 < 3$$
$$-5 < 5x < 1$$
$$-1 < x < \tfrac{1}{5}$$
$$\left(-1, \tfrac{1}{5}\right)$$

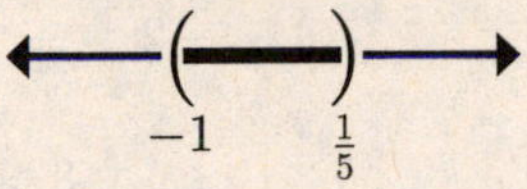

−1 $\frac{1}{5}$

59.

$$3\left|\tfrac{3x-1}{2}\right| > 5$$
$$\left|\tfrac{3x-1}{2}\right| > \tfrac{5}{3}$$
$$\tfrac{3x-1}{2} > \tfrac{5}{3} \quad \textbf{or} \quad \tfrac{3x-1}{2} < -\tfrac{5}{3}$$
$$6 \cdot \tfrac{3x-1}{2} > 6 \cdot \tfrac{5}{3} \qquad 6 \cdot \tfrac{3x-1}{2} < 6\left(-\tfrac{5}{3}\right)$$
$$9x-3 > 10 \qquad 9x-3 < -10$$
$$9x > 13 \qquad 9x < -7$$
$$x > \tfrac{13}{9} \qquad x < -\tfrac{7}{9}$$
$$\left(-\infty, -\tfrac{7}{9}\right) \cup \left(\tfrac{13}{9}, \infty\right)$$

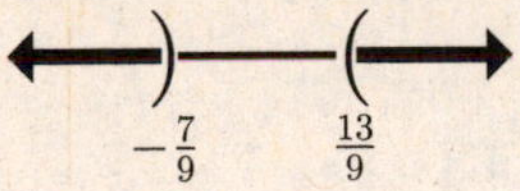

$-\frac{7}{9}$ $\frac{13}{9}$

61.

$$\frac{|x-1|}{-2} > -3$$
$$|x-1| < 6$$
$$-6 < x-1 < 6$$
$$-5 < x < 7$$
$$(-5, 7)$$

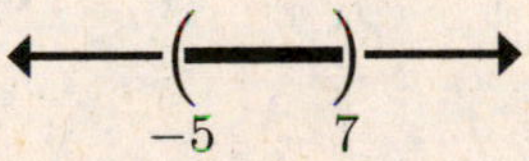

−5 7

63. $0 < |2x+1| < 3$

$0 < |2x+1|$ **and** $|2x+1| < 3$

(1) $|2x+1| > 0$

$2x+1 > 0$ **or** $2x+1 < -0$

$2x > -1$ $\quad$ $2x < -1$

$x > -\frac{1}{2}$ $\quad$ $x < -\frac{1}{2}$

(2) $|2x+1| < 3$

$-3 < 2x+1 < 3$

$-4 < 2x < 2$

$-2 < x < 1$

(1) $-\frac{1}{2}$

(2) -2, 1

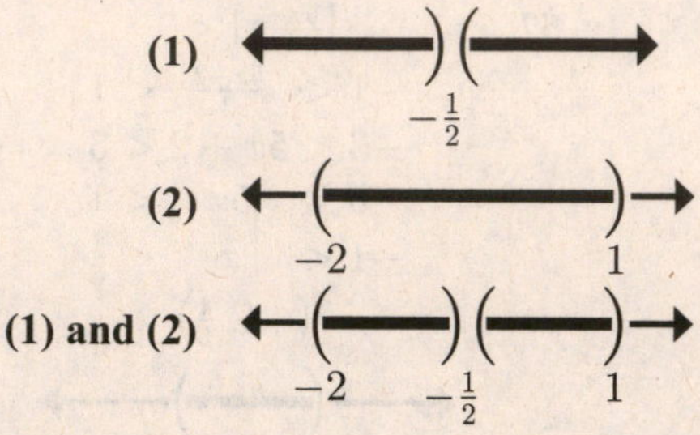

(1) $-\frac{1}{2}$

(2) -2, 1

(1) and (2) -2, $-\frac{1}{2}$, 1 $\Rightarrow \left(-2, -\frac{1}{2}\right) \cup \left(-\frac{1}{2}, 1\right)$

65. $8 > |3x-1| > 3$

$|3x-1| > 3$ **and** $8 > |3x-1|$

(1) $|3x-1| > 3$

$3x-1 > 3$ **or** $3x-1 < -3$

$3x > 4$ $\quad$ $3x < -2$

$x > \frac{4}{3}$ $\quad$ $x < -\frac{2}{3}$

(2) $|3x-1| < 8$

$-8 < 3x-1 < 8$

$-7 < 3x < 9$

$-\frac{7}{3} < x < 3$

(1) $-\frac{2}{3}$, $\frac{4}{3}$

(2) $-\frac{7}{3}$, 3

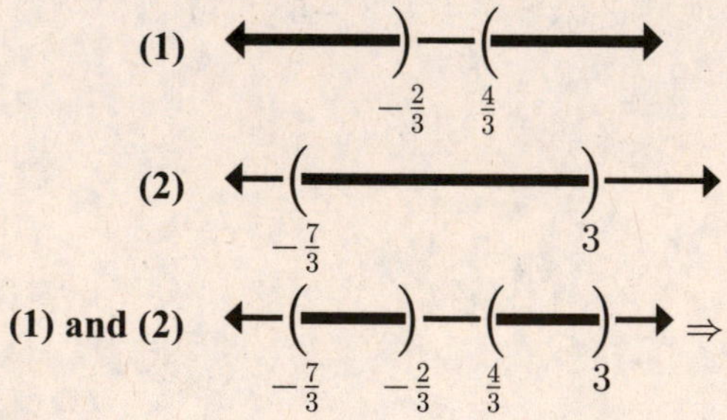

(1) $-\frac{2}{3}$, $\frac{4}{3}$

(2) $-\frac{7}{3}$, 3

(1) and (2) $-\frac{7}{3}$, $-\frac{2}{3}$, $\frac{4}{3}$, 3 $\Rightarrow \left(-\frac{7}{3}, -\frac{2}{3}\right) \cup \left(\frac{4}{3}, 3\right)$

EXERCISES 1.8

67.

$$2 < \left|\frac{x-5}{3}\right| < 4$$

$$2 < \left|\frac{x-5}{3}\right| \quad \textbf{and} \quad \left|\frac{x-5}{3}\right| < 4$$

(1) $\left|\frac{x-5}{3}\right| > 2$ **(2)** $\left|\frac{x-5}{3}\right| < 4$

$\frac{x-5}{3} > 2$ **or** $\frac{x-5}{3} < -2$ $-4 < \frac{x-5}{3} < 4$

$x - 5 > 6$ $x - 5 < -6$ $-12 < x - 5 < 12$

$x > 11$ $x < -1$ $-7 < x < 17$

(1) 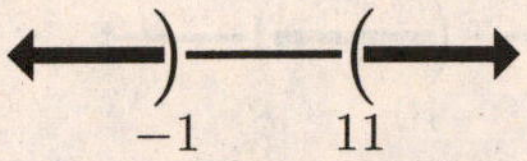 -1 11 **(2)** 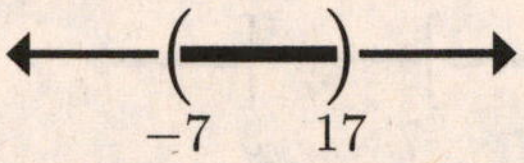 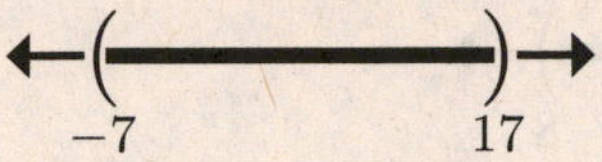 -7 17

(1) 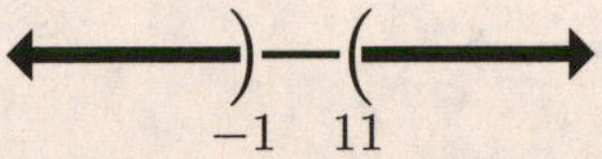 -1 11

(2) -7 17

(1) and (2) 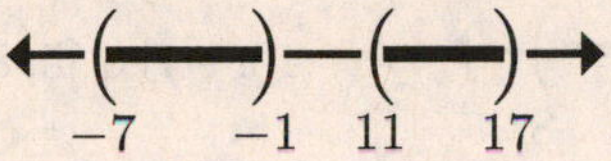 -7 -1 11 17 $\Rightarrow (-7, -1) \cup (11, 17)$

69.

$$10 > \left|\frac{x-2}{2}\right| > 4$$

$$\left|\frac{x-2}{2}\right| > 4 \quad \textbf{and} \quad 10 > \left|\frac{x-2}{2}\right|$$

(1) $\left|\frac{x-2}{2}\right| > 4$ **(2)** $\left|\frac{x-2}{2}\right| < 10$

$\frac{x-2}{2} > 4$ **or** $\frac{x-2}{2} < -4$ $-10 < \frac{x-2}{2} < 10$

$x - 2 > 8$ $x - 2 < -8$ $-20 < x - 2 < 20$

$x > 10$ $x < -6$ $-18 < x < 22$

(1) 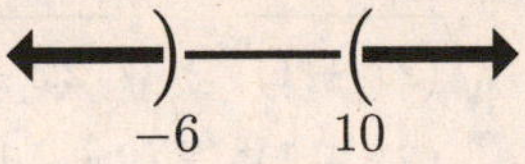 -6 10 **(2)** 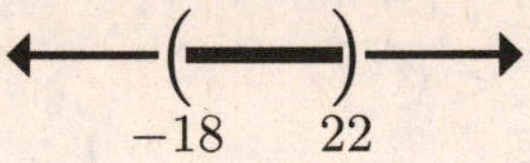 -18 22

(1) 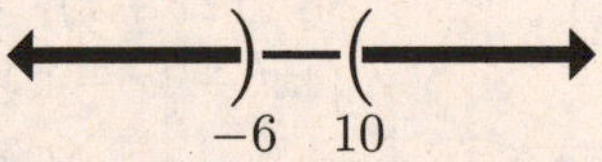 -6 10

(2)  -18 22

(1) and (2) 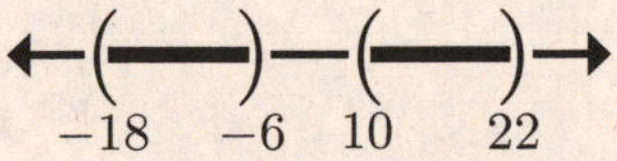 -18 -6 10 22 $\Rightarrow (-18, -6) \cup (10, 22)$

EXERCISES 1.8

71.
$$2 \le \left|\frac{x+1}{3}\right| < 3$$

$$2 \le \left|\frac{x+1}{3}\right| \quad \textbf{and} \quad \left|\frac{x+1}{3}\right| < 3$$

(1) $\left|\frac{x+1}{3}\right| \ge 2$

$$\frac{x+1}{3} \ge 2 \quad \textbf{or} \quad \frac{x+1}{3} \le -2$$
$$x+1 \ge 6 \qquad x+1 \le -6$$
$$x \ge 5 \qquad x \le -7$$

(2) $\left|\frac{x+1}{3}\right| < 3$

$$-3 < \frac{x+1}{3} < 3$$
$$-9 < x+1 < 9$$
$$-10 < x < 8$$

(1) 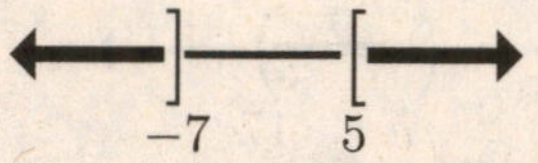

(2) 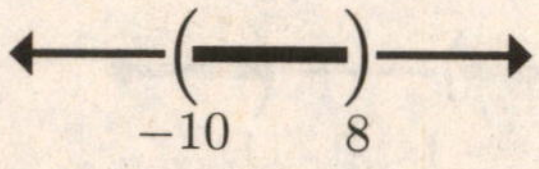

(1) 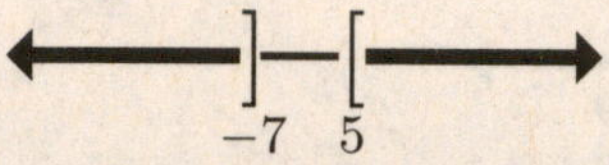

(2) 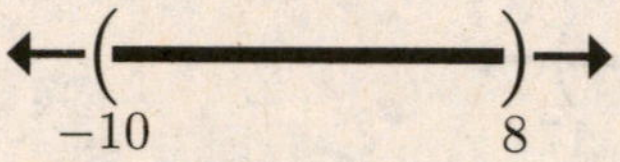

(1) and (2) 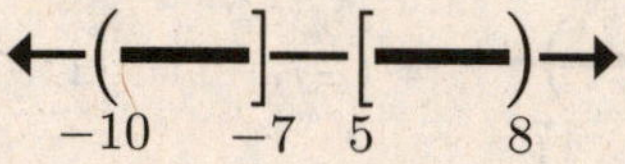

$\Rightarrow (-10, -7] \cup [5, 8)$

73.
$$|x+1| \ge |x|$$
$$\sqrt{(x+1)^2} \ge \sqrt{x^2}$$
$$(x+1)^2 \ge x^2$$
$$x^2 + 2x + 1 \ge x^2$$
$$2x \ge -1$$
$$x \ge -\tfrac{1}{2}$$
Solution: $\left[-\tfrac{1}{2}, \infty\right)$

75.
$$|2x+1| < |2x-1|$$
$$\sqrt{(2x+1)^2} < \sqrt{(2x-1)^2}$$
$$(2x+1)^2 < (2x-1)^2$$
$$4x^2 + 4x + 1 < 4x^2 - 4x + 1$$
$$8x < 0$$
$$x < 0$$
Solution: $(-\infty, 0)$

77.
$$|x+1| < |x|$$
$$\sqrt{(x+1)^2} < \sqrt{x^2}$$
$$(x+1)^2 < x^2$$
$$x^2 + 2x + 1 < x^2$$
$$2x < -1$$
$$x < -\tfrac{1}{2}$$
Solution: $\left(-\infty, -\tfrac{1}{2}\right)$

79.
$$|2x+1| \ge |2x-1|$$
$$\sqrt{(2x+1)^2} \ge \sqrt{(2x-1)^2}$$
$$(2x+1)^2 \ge (2x-1)^2$$
$$4x^2 + 4x + 1 \ge 4x^2 - 4x + 1$$
$$8x \ge 0$$
$$x \ge 0$$
Solution: $[0, \infty)$

81.
$$|t - 78°| \le 8°$$
$$-8° \le t - 78° \le 8°$$
$$70° \le t \le 86°$$

EXERCISES 1.8

83.

$$\begin{aligned} 0.6° + 0.5° &= 1.1° \\ 0.6° - 0.5° &= 0.1° \\ 0.1° \le \quad c \quad &\le 1.1° \\ 0.6° - 0.5° \le \quad c \quad &\le 0.6° + 0.5° \\ -0.5° \le c - 0.6° &\le 0.5° \\ |c - 0.6°| &\le 0.5° \end{aligned}$$

85.

$$\begin{aligned} \frac{38+72}{2} = \frac{110}{2} &= 55 \\ 38 &= 55 - 17 \\ 72 &= 55 + 17 \\ 38 < \quad h \quad &< 72 \\ 55 - 17 < \quad h \quad &< 55 + 17 \\ -17 < h - 55 &< 17 \\ |h - 55| &< 17 \end{aligned}$$

87.

$$\begin{aligned} |p - 25.46| &\le 1.00 \\ -1.00 \le p - 25.46 &\le 1.00 \\ 24.46 \le \quad p \quad &\le 26.46 \end{aligned}$$

a. 24.76% and 26.45% are within the range. **b.** The error is less than 1%.

89-93. Answers may vary.

95. False. Absolute value equations can have zero, one, or two solutions.

97. False.

$$\begin{aligned} &|x| \ge 5 \\ x \ge 5 \quad &\textbf{or} \quad x \le -5 \\ &(-\infty, -5] \cup [5, \infty) \end{aligned}$$

99. True.

$$\begin{aligned} |x| + 555 &< 554 \\ |x| &< -1 \end{aligned}$$

This inequality is never true.

Chapter 1 Review (page 178)

1. $3x + 7 = 4$

no restrictions on x

2. $x + \frac{1}{x} = 2$

restriction: $x \neq 0$

3. $\frac{1}{x-1} = 4$

restriction: $x \neq 1$

4. $\frac{1}{x-2} = \frac{2}{x-3}$

restriction: $x \neq 2, x \neq 3$

5.

$$\begin{aligned} 3(9x+4) &= 28 \\ 27x + 12 &= 28 \\ 27x &= 16 \\ x &= \tfrac{16}{27} \end{aligned}$$

conditional equation

6.

$$\begin{aligned} \tfrac{3}{2}a &= 7(a+11) \\ 2 \cdot \tfrac{3}{2}a &= 2 \cdot 7(a+11) \\ 3a &= 14a + 154 \\ -11a &= 154 \\ a &= -\tfrac{154}{11} = -14 \end{aligned}$$

conditional equation

7.

$$\begin{aligned} 8(3x-5) - 4(x+3) &= 12 \\ 24x - 40 - 4x - 12 &= 12 \\ 20x - 52 &= 12 \\ 20x &= 64 \\ x = \tfrac{64}{20} &= \tfrac{16}{5} \Rightarrow \text{conditional equation} \end{aligned}$$

8.
$$\begin{aligned}
\frac{x+3}{x+4}+\frac{x+3}{x+2}&=2\\
(x+4)(x+2)\left(\frac{x+3}{x+4}+\frac{x+3}{x+2}\right)&=(x+4)(x+2)\cdot 2\\
(x+2)(x+3)+(x+4)(x+3)&=\left(x^2+6x+8\right)\cdot 2\\
x^2+5x+6+x^2+7x+12&=2x^2+12x+16\\
2x^2+12x+18&=2x^2+12x+16\\
18&\neq 16\Rightarrow \text{no solution, contradiction}
\end{aligned}$$

9.
$$\begin{aligned}
\frac{3}{x-1}&=\frac{1}{2}\\
2(x-1)\cdot\frac{3}{x-1}&=2(x-1)\cdot\frac{1}{2}\\
6&=x-1\\
7&=x
\end{aligned}$$
conditional equation

10.
$$\begin{aligned}
\frac{8x^2+72x}{9+x}&=8x\\
(9+x)\cdot\frac{8x^2+72x}{9+x}&=(9+x)\cdot 8x\\
8x^2+72x&=72x+8x^2
\end{aligned}$$
all real numbers except -9, identity

11.
$$\begin{aligned}
\frac{3x}{x-1}-\frac{5}{x+3}&=3\\
(x-1)(x+3)\left(\frac{3x}{x-1}-\frac{5}{x+3}\right)&=(x-1)(x+3)\cdot 3\\
3x(x+3)-5(x-1)&=\left(x^2+2x-3\right)\cdot 3\\
3x^2+9x-5x+5&=3x^2+6x-9\\
4x+5&=6x-9\\
-2x&=-14\\
x&=7\Rightarrow \text{conditional equation}
\end{aligned}$$

12.
$$\begin{aligned}
x+\frac{1}{2x-3}&=\frac{2x^2}{2x-3}\\
(2x-3)\left(x+\frac{1}{2x-3}\right)&=(2x-3)\cdot\frac{2x^2}{2x-3}\\
(2x-3)x+1&=2x^2\\
2x^2-3x+1&=2x^2\\
-3x&=-1\\
x&=\tfrac{1}{3}\Rightarrow \text{conditional equation}
\end{aligned}$$

13.
$$\begin{aligned}
\frac{4}{x^2-13x-48}-\frac{1}{x^2+x-6}&=\frac{2}{x^2-18x+32}\\
\frac{4}{(x-16)(x+3)}-\frac{1}{(x+3)(x-2)}&=\frac{2}{(x-16)(x-2)}\\
4(x-2)-(x-16)&=2(x+3)\qquad\{\text{multiply by common denominator}\}\\
4x-8-x+16&=2x+6\\
3x+8&=2x+6\\
x&=-2\Rightarrow \text{conditional equation}
\end{aligned}$$

14.

$$\frac{a-1}{a+3}+\frac{2a-1}{3-a}=\frac{2-a}{a-3}$$
$$\frac{a-1}{a+3}+\frac{1-2a}{a-3}=\frac{2-a}{a-3}$$
$$(a-1)(a-3)+(1-2a)(a+3)=(2-a)(a+3) \quad \{\text{multiply by common denominator}\}$$
$$a^2-3a-a+3+a+3-2a^2-6a=2a+6-a^2-3a$$
$$-a^2-9a+6=-a^2-a+6$$
$$-9a+6=-a+6$$
$$0=8a$$
$$0=a \Rightarrow \text{conditional equation}$$

15.

$$C=\frac{5}{9}(F-32)$$
$$\frac{9}{5}C=\frac{9}{5}\cdot\frac{5}{9}(F-32)$$
$$\frac{9}{5}C=F-32$$
$$\frac{9}{5}C+32=F$$

16.

$$P_n=l+\frac{si}{f}$$
$$P_n-l=\frac{si}{f}$$
$$f(P_n-l)=f\cdot\frac{si}{f}$$
$$f(P_n-l)=si$$
$$\frac{f(P_n-l)}{P_n-l}=\frac{si}{P_n-l}$$
$$f=\frac{si}{P_n-l}$$

17.

$$\frac{1}{f}=\frac{1}{f_1}+\frac{1}{f_2}$$
$$ff_1f_2\cdot\frac{1}{f}=ff_1f_2\left(\frac{1}{f_1}+\frac{1}{f_2}\right)$$
$$f_1f_2=ff_2+ff_1$$
$$f_1f_2-ff_1=ff_2$$
$$f_1(f_2-f)=ff_2$$
$$\frac{f_1(f_2-f)}{f_2-f}=\frac{ff_2}{f_2-f}$$
$$f_1=\frac{ff_2}{f_2-f}$$

18.

$$S=\frac{a-lr}{1-r}$$
$$S(1-r)=\frac{a-lr}{1-r}(1-r)$$
$$S(1-r)=a-lr$$
$$S-Sr=a-lr$$
$$lr=a-S+Sr$$
$$\frac{lr}{r}=\frac{a-S+Sr}{r}$$
$$l=\frac{a-S+Sr}{r}$$

19. Let $x =$ the score on the first exam. The other scores were $x+4$, $x+8$ and $x+12$.

$$\frac{\boxed{\text{Sum of scores}}}{4} = 66$$
$$\frac{x+x+4+x+8+x+12}{4} = 66$$
$$\frac{4x+24}{4} = 66$$
$$4x+24 = 264$$
$$4x = 240$$
$$x = 60$$

His score on the first test was 60%.

20. Let $w =$ the width. Then $w+5 =$ the length.

$$\boxed{\text{Perimeter}} = 100$$
$$2w + 2(w+5) = 100$$
$$2w + 2w + 10 = 100$$
$$4w + 10 = 100$$
$$4w = 90$$
$$w = 22.5$$

The dimensions are 22.5 ft by 27.5 ft.

21. Let $t =$ the time the cars travel.

$$\boxed{\text{Distance 1st car travels}} + \boxed{\text{Distance 2nd car travels}} = \boxed{\text{Total distance}}$$
$$45t + 50t = 285$$
$$95t = 285$$
$$t = 3 \Rightarrow \text{They will be 285 miles apart after 3 hours.}$$

22. Let $t =$ the time the cars travel.

$$\boxed{\text{Distance 1st car travels}} - \boxed{\text{Distance 2nd car travels}} = \boxed{\text{Distance between them}}$$
$$46t - 40t = 3$$
$$6t = 3$$
$$t = 0.5 \Rightarrow \text{They will be 3 miles apart after 0.5 hours.}$$

23. Let $x =$ the liters of water added.

$$\boxed{\text{Liters of alcohol at start}} + \boxed{\text{Liters of alcohol added}} = \boxed{\text{Liters of alcohol at end}}$$
$$0.50(1) + 0 = 0.20(1+x)$$
$$0.50 = 0.20 + 0.20x$$
$$0.30 = 0.20x$$
$$1.5 = x \Rightarrow \text{1.5 liters of water should be added.}$$

24. Let $x =$ hours for both working together.

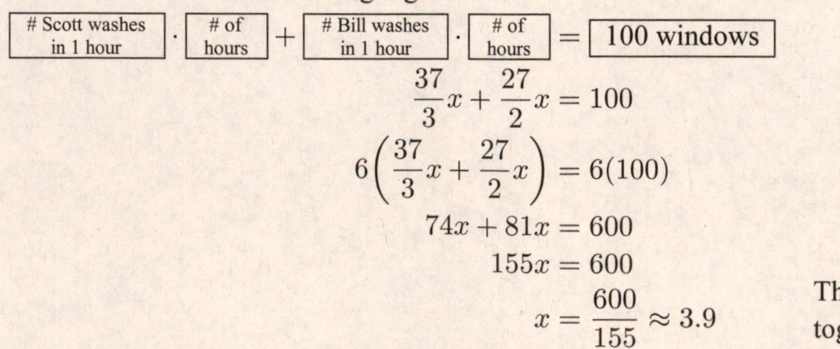

$$\boxed{\text{\# Scott washes in 1 hour}} \cdot \boxed{\text{\# of hours}} + \boxed{\text{\# Bill washes in 1 hour}} \cdot \boxed{\text{\# of hours}} = \boxed{\text{100 windows}}$$
$$\frac{37}{3}x + \frac{27}{2}x = 100$$
$$6\left(\frac{37}{3}x + \frac{27}{2}x\right) = 6(100)$$
$$74x + 81x = 600$$
$$155x = 600$$
$$x = \frac{600}{155} \approx 3.9$$

They can wash 100 windows together in about 3.9 hours.

25. Let $x =$ hours for both pipes to fill the tank.

$\boxed{\text{1st pipe in 1 hour}} + \boxed{\text{2nd pipe in 1 hour}} = \boxed{\text{Total in 1 hour}}$

$$\frac{1}{9} + \frac{1}{12} = \frac{1}{x}$$
$$36x\left(\frac{1}{9} + \frac{1}{12}\right) = 36x\left(\frac{1}{x}\right)$$
$$4x + 3x = 36$$
$$7x = 36$$
$$x = \frac{36}{7} = 5\frac{1}{7}$$

The tank can be filled in $5\frac{1}{7}$ hours.

26. Let $x =$ the ounces of pure zinc added.

$\boxed{\text{Ounces of zinc at start}} + \boxed{\text{Ounces of zinc added}} = \boxed{\text{Ounces of zinc at end}}$

$$0.30(20) + x = 0.40(20 + x)$$
$$6 + x = 8 + 0.40x$$
$$0.60x = 2$$
$$6x = 20$$
$$x = \frac{20}{6} = 3\frac{1}{3}$$

$3\frac{1}{3}$ ounces of zinc should be added.

27. Let $x =$ the amount invested at 11%. Then $10{,}000 - x =$ the amount invested at 14%.

$\boxed{\text{Interest at 11\%}} + \boxed{\text{Interest at 14\%}} = \boxed{\text{Total interest}}$

$$0.11x + 0.14(10{,}000 - x) = 1{,}265$$
$$0.11x + 1{,}400 - 0.14x = 1{,}265$$
$$-0.03x = -135$$
$$x = 4{,}500$$

\$4,500 was invested at 11% and \$5,500 was invested at 14%.

28. Let $x =$ # of rugs for equal costs.

$\boxed{\text{Cost of 1st loom}} = \boxed{\text{Cost of 2nd loom}}$

$$750 + 115x = 950 + 95x$$
$$20x = 200$$
$$x = 10$$

The costs are the same on either loom for 10 rugs.

29. $3\sqrt{-300} = 3\sqrt{-1}\sqrt{100}\sqrt{3} = 30i\sqrt{3}$

30. $-\sqrt{\frac{-45}{4}} = -\sqrt{-1}\cdot\frac{\sqrt{9}\sqrt{5}}{\sqrt{4}} = -\frac{3\sqrt{5}}{2}i$

31. $(2 - 3i) + (-4 + 2i) = 2 - 3i - 4 + 2i = -2 - i$

32. $\left(3 - \sqrt{-36}\right) + \left(\sqrt{-16} + 2\right) = (3 - 6i) + (4i + 2) = 3 - 6i + 4i + 2 = 5 - 2i$

33. $(2 - 3i) - (4 + 2i) = 2 - 3i - 4 - 2i = -2 - 5i$

34. $(5 - 11i)(5 + 11i) = 25 + 55i - 55i - 121i^2 = 25 - 121(-1) = 146 = 146 + 0i$

35. $(8 - 3i)^2 = (8 - 3i)(8 - 3i) = 64 - 24i - 24i + 9i^2 = 64 - 48i + 9(-1) = 55 - 48i$

36. $\left(3 + \sqrt{-9}\right)\left(2 - \sqrt{-25}\right) = (3 + 3i)(2 - 5i) = 6 - 9i - 15i^2 = 6 - 9i + 15 = 21 - 9i$

37. $\frac{3}{i} = \frac{3i}{ii} = \frac{3i}{i^2} = \frac{3i}{-1} = 0 - 3i$

38. $-\frac{5}{6i} = -\frac{5\cdot i}{6i\cdot i} = -\frac{5i}{6i^2} = -\frac{5i}{-6} = 0 + \frac{5}{6}i$

39. $\frac{3}{1+i} = \frac{3(1-i)}{(1+i)(1-i)} = \frac{3(1-i)}{1^2 - i^2} = \frac{3-3i}{2} = \frac{3}{2} - \frac{3}{2}i$

40. $\dfrac{2i}{2-i}=\dfrac{2i(2+i)}{(2-i)(2+i)}=\dfrac{4i+2i^2}{2^2-i^2}=\dfrac{-2+4i}{5}=-\dfrac{2}{5}+\dfrac{4}{5}i$

41. $\dfrac{3+i}{3-i}=\dfrac{(3+i)(3+i)}{(3-i)(3+i)}=\dfrac{9+6i+i^2}{3^2-i^2}=\dfrac{8+6i}{10}=\dfrac{8}{10}+\dfrac{6}{10}i=\dfrac{4}{5}+\dfrac{3}{5}i$

42. $\dfrac{3-2i}{1+i}=\dfrac{(3-2i)(1-i)}{(1+i)(1-i)}=\dfrac{3-5i+2i^2}{1^2-i^2}=\dfrac{1-5i}{2}=\dfrac{1}{2}-\dfrac{5}{2}i$

43. $i^{53}=i^{52}i=\left(i^4\right)^{13}i=1^{13}i=0+i$

44. $i^{103}=i^{100}i^3=\left(i^4\right)^{25}i^3=1^{25}i^3=0-i$

45. $-\dfrac{2}{i^3}=-\dfrac{2\cdot i}{i^3\cdot i}=-\dfrac{2i}{i^4}=-\dfrac{2i}{1}=0-2i$

46. $|3-i|=\sqrt{3^2+(-1)^2}=\sqrt{9+1}$
$=\sqrt{10}+0i$

47. $\left|\dfrac{1+i}{1-i}\right|=\left|\dfrac{(1+i)(1+i)}{(1-i)(1+i)}\right|=\left|\dfrac{1+2i+i^2}{1^2-i^2}\right|=\left|\dfrac{2i}{2}\right|=|0+i|=\sqrt{0^2+1^2}=1$

48. $64r^2+9s^2=64r^2-\left(-9s^2\right)=(8r)^2-(3si)^2=(8r+3si)(8r-3si)$

49.
$$\begin{aligned}2x^2-x-6&=0\\(2x+3)(x-2)&=0\end{aligned}$$
$2x+3=0$ **or** $x-2=0$
$2x=-3$ $\quad x=2$
$x=-\frac{3}{2}$ $\quad x=2$

50.
$$\begin{aligned}12x^2+13x&=4\\12x^2+13x-4&=0\\(4x-1)(3x+4)&=0\end{aligned}$$
$4x-1=0$ **or** $3x+4=0$
$4x=1$ $\quad 3x=-4$
$x=\frac{1}{4}$ $\quad x=-\frac{4}{3}$

51.
$$\begin{aligned}5x^2-8x&=0\\x(5x-8)&=0\end{aligned}$$
$x=0$ **or** $5x-8=0$
$x=0$ $\quad 5x=8$
$x=0$ $\quad x=\frac{8}{5}$

52.
$$\begin{aligned}27x^2&=30x-8\\27x^2-30x+8&=0\\(9x-4)(3x-2)&=0\end{aligned}$$
$9x-4=0$ **or** $3x-2=0$
$9x=4$ $\quad 3x=2$
$x=\frac{4}{9}$ $\quad x=\frac{2}{3}$

53.
$$\begin{aligned}2x^2&=16\\x^2&=8\\\sqrt{x^2}&=\pm\sqrt{8}\\x&=\pm2\sqrt{2}\end{aligned}$$

54.
$$\begin{aligned}12x^2&=-60\\x^2&=-5\\\sqrt{x^2}&=\pm\sqrt{-5}\\x&=\pm i\sqrt{5}\end{aligned}$$

55.

$$\begin{aligned} (4z-5)^2 &= 32 \\ \sqrt{(4z-5)^2} &= \pm\sqrt{32} \\ 4z-5 &= \pm 4\sqrt{2} \\ 4z &= 5 \pm 4\sqrt{2} \\ z &= \frac{5 \pm 4\sqrt{2}}{4} \end{aligned}$$

56.

$$\begin{aligned} (5x-7)^2 &= -45 \\ \sqrt{(5x-7)^2} &= \pm\sqrt{-45} \\ 5x-7 &= \pm 3i\sqrt{5} \\ 5x &= 7 \pm 3i\sqrt{5} \\ x &= \frac{7 \pm 3i\sqrt{5}}{5} = \frac{7}{5} \pm \frac{3\sqrt{5}}{5}i \end{aligned}$$

57.

$$\begin{aligned} x^2 - 8x + 15 &= 0 \\ x^2 - 8x &= -15 \\ x^2 - 8x + 16 &= -15 + 16 \\ (x-4)^2 &= 1 \end{aligned}$$

$$x - 4 = \sqrt{1} \quad \textbf{or} \quad x - 4 = -\sqrt{1}$$

$$x - 4 = 1 \qquad x - 4 = -1$$

$$x = 5 \qquad x = 3$$

58.

$$\begin{aligned} 3x^2 + 18x &= -24 \\ \frac{3x^2 + 18x}{3} &= \frac{-24}{3} \\ x^2 + 6x &= -8 \\ x^2 + 6x + 9 &= -8 + 9 \\ (x+3)^2 &= 1 \end{aligned}$$

$$x + 3 = \sqrt{1} \quad \textbf{or} \quad x + 3 = -\sqrt{1}$$

$$x + 3 = 1 \qquad x + 3 = -1$$

$$x = -2 \qquad x = -4$$

59.

$$\begin{aligned} 5x^2 - x - 1 &= 0 \\ 5x^2 - x &= 1 \\ x^2 - \frac{1}{5}x &= \frac{1}{5} \\ x^2 - \frac{1}{5}x + \frac{1}{100} &= \frac{1}{5} + \frac{1}{100} \\ \left(x - \frac{1}{10}\right)^2 &= \frac{21}{100} \end{aligned}$$

$$x - \frac{1}{10} = \sqrt{\frac{21}{100}} \quad \textbf{or} \quad x - \frac{1}{10} = -\sqrt{\frac{21}{100}}$$

$$x - \frac{1}{10} = \frac{\sqrt{21}}{10} \qquad x - \frac{1}{10} = -\frac{\sqrt{21}}{10}$$

$$x = \frac{1 + \sqrt{21}}{10} \qquad x = \frac{1 - \sqrt{21}}{10}$$

60.
$$5x^2 - x = 0$$
$$x^2 - \frac{1}{5}x = 0$$
$$x^2 - \frac{1}{5}x + \frac{1}{100} = 0 + \frac{1}{100}$$
$$\left(x - \frac{1}{10}\right)^2 = \frac{1}{100}$$
$$x - \frac{1}{10} = \sqrt{\frac{1}{100}} \quad \textbf{or} \quad x - \frac{1}{10} = -\sqrt{\frac{1}{100}}$$
$$x - \frac{1}{10} = \frac{1}{10} \qquad x - \frac{1}{10} = -\frac{1}{10}$$
$$x = \frac{2}{10} = \frac{1}{5} \qquad x = \frac{0}{10} = 0$$

61.
$$3x^2 - 2x + 1 = 0$$
$$3x^2 - 2x = -1$$
$$x^2 - \frac{2}{3}x = -\frac{1}{3}$$
$$x^2 - \frac{2}{3}x + \frac{1}{9} = -\frac{1}{3} + \frac{1}{9}$$
$$\left(x - \frac{1}{3}\right)^2 = -\frac{2}{9}$$
$$x - \frac{1}{3} = \sqrt{-\frac{2}{9}} \quad \textbf{or} \quad x - \frac{1}{3} = -\sqrt{-\frac{2}{9}}$$
$$x - \frac{1}{3} = \frac{\sqrt{2}}{3}i \qquad x - \frac{1}{3} = -\frac{\sqrt{2}}{3}i$$
$$x = \frac{1}{3} + \frac{\sqrt{2}}{3}i \qquad x = \frac{1}{3} - \frac{\sqrt{2}}{3}i$$

62. $x^2 + 5x - 14 = 0 \Rightarrow a = 1, b = 5, c = -14$
$$x = \frac{-b \pm \sqrt{b^2 - 4ac}}{2a} = \frac{-(5) \pm \sqrt{(5)^2 - 4(1)(-14)}}{2(1)} = \frac{-5 \pm \sqrt{25 + 56}}{2} = \frac{-5 \pm \sqrt{81}}{2}$$
$$= \frac{-5 \pm 9}{2}$$
$$x = \frac{-5 + 9}{2} = \frac{4}{2} = 2 \text{ or } x = \frac{-5 - 9}{2} = \frac{-14}{2} = -7$$

63. $3x^2 - 25x = 18 \Rightarrow 3x^2 - 25x - 18 = 0 \Rightarrow a = 3, b = -25, c = -18$

$$x = \frac{-b \pm \sqrt{b^2 - 4ac}}{2a} = \frac{-(-25) \pm \sqrt{(-25)^2 - 4(3)(-18)}}{2(3)} = \frac{25 \pm \sqrt{625 + 216}}{6}$$

$$= \frac{25 \pm \sqrt{841}}{6} = \frac{25 \pm 29}{6}$$

$$x = \frac{25 + 29}{6} = \frac{54}{6} = 9 \text{ or } x = \frac{25 - 29}{6} = \frac{-4}{6} = -\frac{2}{3}$$

64. $5x^2 = 1 - x \Rightarrow 5x^2 + x - 1 = 0 \Rightarrow a = 5, b = 1, c = -1$

$$x = \frac{-b \pm \sqrt{b^2 - 4ac}}{2a} = \frac{-(1) \pm \sqrt{(1)^2 - 4(5)(-1)}}{2(5)} = \frac{-1 \pm \sqrt{1 + 20}}{10} = \frac{-1 \pm \sqrt{21}}{10}$$

65. $5 = a^2 + 2a \Rightarrow a^2 + 2a - 5 = 0 \Rightarrow a = 1, b = 2, c = -5$

$$a = \frac{-b \pm \sqrt{b^2 - 4ac}}{2a} = \frac{-(2) \pm \sqrt{(2)^2 - 4(1)(-5)}}{2(1)} = \frac{-2 \pm \sqrt{4 + 20}}{2} = \frac{-2 \pm \sqrt{24}}{2}$$

$$= \frac{-2 \pm 2\sqrt{6}}{2}$$

$$= -1 \pm \sqrt{6}$$

66. $3x^2 + 4 = 2x \Rightarrow 3x^2 - 2x + 4 = 0 \Rightarrow a = 3, b = -2, c = 4$

$$x = \frac{-b \pm \sqrt{b^2 - 4ac}}{2a} = \frac{-(-2) \pm \sqrt{(-2)^2 - 4(3)(4)}}{2(3)} = \frac{2 \pm \sqrt{4 - 48}}{6} = \frac{2 \pm \sqrt{-44}}{6}$$

$$= \frac{2}{6} \pm \frac{2\sqrt{11}}{6}i$$

$$= \frac{1}{3} \pm \frac{\sqrt{11}}{3}i$$

67. $6x^2 + 5x + 1 = 0$

$a = 6, b = 5, c = 1$

$b^2 - 4ac = (5)^2 - 4(6)(1) = 25 - 24 = 1$

68. two different rational numbers

69. $kx^2 + 4x + 12 = 0$

$a = k, b = 4, c = 12$

Set the discriminant equal to 0:

$$b^2 - 4ac = 0$$
$$4^2 - 4(k)(12) = 0$$
$$16 - 48k = 0$$
$$-48k = -16$$
$$k = \frac{1}{3}$$

70. $4y^2 + (k+2)y = 1 - k$

$4y^2 + (k+2)y - 1 + k = 0$

$a = 4, b = k + 2, c = -1 + k$

Set the discriminant equal to 0:

$$b^2 - 4ac = 0$$
$$(k+2)^2 - 4(4)(-1+k) = 0$$
$$k^2 + 4k + 4 + 16 - 16k = 0$$
$$k^2 - 12k + 20 = 0$$
$$(k-10)(k-2) = 0$$

$k - 10 = 0$ **or** $k - 2 = 0$

$k = 10$ $\quad$ $k = 2$

71. $\dfrac{3x}{2} - \dfrac{2x}{x-1} = x - 3$

$$2(x-1)\left(\frac{3x}{2} - \frac{2x}{x-1}\right) = 2(x-1)(x-3)$$
$$(x-1)\cdot 3x - 2(2x) = 2\left(x^2 - 4x + 3\right)$$
$$3x^2 - 3x - 4x = 2x^2 - 8x + 6$$
$$x^2 + x - 6 = 0$$
$$(x+3)(x-2) = 0$$

$x + 3 = 0$ **or** $x - 2 = 0$

$x = -3$ $\quad$ $x = 2$

72. $\dfrac{4}{a-4} + \dfrac{4}{a-1} = 5$

$$(a-4)(a-1)\left(\tfrac{4}{a-4} + \tfrac{4}{a-1}\right) = (a-4)(a-1)5$$
$$4(a-1) + 4(a-4) = 5\left(a^2 - 5a + 4\right)$$
$$4a - 4 + 4a - 16 = 5a^2 - 25a + 20$$
$$0 = 5a^2 - 33a + 40$$
$$0 = (5a-8)(a-5)$$

$5a - 8 = 0$ **or** $a - 5 = 0$

$5a = 8$ $\quad$ $a = 5$

$a = \frac{8}{5}$ $\quad$ $a = 5$

73. Let $x =$ one side of the garden.

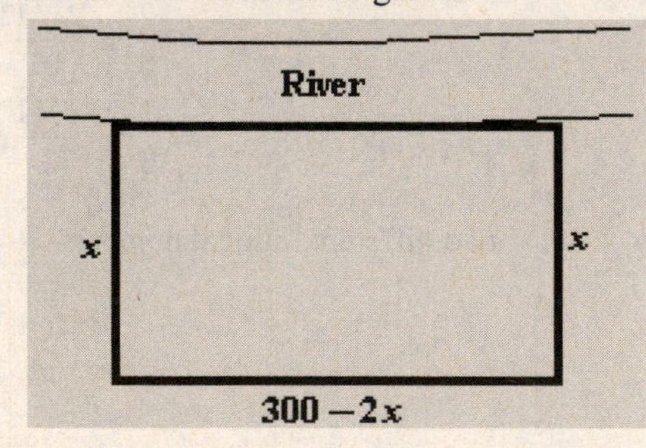

$$\text{Area} = 10450$$
$$x(300 - 2x) = 10450$$
$$-2x^2 + 300x = 10450$$
$$0 = 2x^2 - 300x + 10450$$
$$0 = 2\left(x^2 - 150x + 5225\right)$$
$$0 = 2(x-95)(x-55)$$

$x - 95 = 0$ **or** $x - 55 = 0$

$x = 95$ $\quad$ $x = 55$

The dimensions are 95 yards by 110 yards or 55 yards by 190 yards.

74. Let $r =$ the rate of the propeller-driven plane. Then the rate of the jet plane is $r + 120$.

$\boxed{\text{Jet time}} = \boxed{\text{Propeller time}} - 3$

	Rate	Time	Dist.
Propeller	r	$\frac{3520}{r}$	3520
Jet	$r+120$	$\frac{3520}{r+120}$	3520

$$\frac{3520}{r+120} = \frac{3520}{r} - 3$$
$$r(r+120)\frac{3520}{r+120} = r(r+120)\left(\frac{3520}{r} - 3\right)$$
$$3520r = 3520(r+120) - 3r(r+120)$$
$$3520r = 3520r + 422400 - 3r^2 - 360r$$
$$3r^2 + 360r + 422{,}400 = 0$$
$$3(r-320)(r+440) = 0$$
$$r - 320 = 0 \quad \textbf{or} \quad r + 440 = 0$$
$$r = 320 \qquad\qquad r = -440$$

Since $r = -440$ does not make sense, the solution is $r = 320$. The prop. plane's rate is 320 mph, while the jet plane's rate is 440 mph.

75. Set $h = 48$:

$$h = -16t^2 + 64t$$
$$48 = -16t^2 + 64t$$
$$16t^2 - 64t + 48 = 0$$
$$16(t-1)(t-3) = 0$$
$$t - 1 = 0 \quad \textbf{or} \quad t - 3 = 0$$
$$t = 1 \qquad\qquad t = 3$$

The shortest time required for the ball to reach a height of 48 feet is 1 second.

76. Let $x =$ the width of the walk. Then the total dimensions are $16 + 2x$ by $20 + 2x$.

$\boxed{\text{Total area}} - \boxed{\text{Area of pool}} = \boxed{\text{Area of walk}}$

$$(16+2x)(20+2x) - (16)(20) = 117$$
$$320 + 72x + 4x^2 - 320 = 117$$
$$4x^2 + 72x - 117 = 0$$
$$(2x+39)(2x-3) = 0$$
$$2x + 39 = 0 \quad \textbf{or} \quad 2x - 3 = 0$$
$$x = -\tfrac{39}{2} \qquad\qquad x = \tfrac{3}{2}$$

Since $x = -\frac{39}{2}$ does not make sense, the only solution is $x = \frac{3}{2}$. The walk is $1\frac{1}{2}$ feet wide.

77.

$$x^3 + 4x^2 - 12x = 0$$
$$x(x^2 + 4x - 12) = 0$$
$$x(x+6)(x-2) = 0$$
$$x = 0 \quad \textbf{or} \quad x + 6 = 0 \quad \textbf{or} \quad x - 2 = 0$$
$$x = 0 \qquad x = -6 \qquad x = 2$$

78.

$$3x^3 + 4x^2 - 4x = 0$$
$$x(3x^2 + 4x - 4) = 0$$
$$x(3x-2)(x+2) = 0$$
$$x = 0 \quad \textbf{or} \quad 3x - 2 = 0 \quad \textbf{or} \quad x + 2 = 0$$
$$x = 0 \qquad x = \tfrac{2}{3} \qquad x = -2$$

79.

$$x^4 - 2x^2 + 1 = 0$$
$$(x^2-1)(x^2-1) = 0$$
$$x^2 - 1 = 0 \quad \textbf{or} \quad x^2 - 1 = 0$$
$$x^2 = 1 \qquad\qquad x^2 = 1$$
$$x = \pm 1 \qquad\qquad x = \pm 1$$

80.

$$x^4 - 36 = -35x^2$$
$$x^4 + 35x^2 - 36 = 0$$
$$(x^2+36)(x^2-1) = 0$$
$$x^2 + 36 = 0 \quad \textbf{or} \quad x^2 - 1 = 0$$
$$x^2 = -36 \qquad\qquad x^2 = 1$$
$$x = \pm 6i \qquad\qquad x = \pm 1$$

81.
$$a - a^{1/2} - 6 = 0$$
$$\left(a^{1/2}+2\right)\left(a^{1/2}-3\right) = 0$$
$$a^{1/2}+2=0 \quad \textbf{or} \quad a^{1/2}-3=0$$
$$a^{1/2}=-2 \qquad a^{1/2}=3$$
$$\left(a^{1/2}\right)^2=(-2)^2 \qquad \left(a^{1/2}\right)^2=(3)^2$$
$$a=4 \qquad a=9$$
$a=4$ does not check and is extraneous.

82.
$$x^{2/3}+x^{1/3}-6=0$$
$$\left(x^{1/3}-2\right)\left(x^{1/3}+3\right)=0$$
$$x^{1/3}-2=0 \quad \textbf{or} \quad x^{1/3}+3=0$$
$$x^{1/3}=2 \qquad x^{1/3}=-3$$
$$\left(x^{1/3}\right)^3=(2)^3 \qquad \left(x^{1/3}\right)^3=(-3)^3$$
$$x=8 \qquad x=-27$$
Both answers check.

83.
$$6y^{-2}+13y^{-1}-5=0$$
$$\left(3y^{-1}-1\right)\left(2y^{-1}+5\right)=0$$
$$3y^{-1}-1=0 \quad \textbf{or} \quad 2y^{-1}+5=0$$
$$y^{-1}=\tfrac{1}{3} \qquad y^{-1}=-\tfrac{5}{2}$$
$$\left(y^{-1}\right)^{-1}=\left(\tfrac{1}{3}\right)^{-1} \qquad \left(y^{-1}\right)^{-1}=\left(-\tfrac{5}{2}\right)^{-1}$$
$$y=3 \qquad y=-\frac{2}{5}$$
Both answers check.

84.
$$\sqrt{5x-11}-5=-3$$
$$\sqrt{5x-11}=2$$
$$\left(\sqrt{5x-11}\right)^2=2^2$$
$$5x-11=4$$
$$x=3$$
The solution checks.

85.
$$\sqrt{x-1}+x=7$$
$$\sqrt{x-1}=7-x$$
$$\left(\sqrt{x-1}\right)^2=(7-x)^2$$
$$x-1=49-14x+x^2$$
$$0=x^2-15x+50$$
$$0=(x-5)(x-10)$$
$$x-5=0 \quad \textbf{or} \quad x-10=0$$
$$x=5 \qquad x=10$$
$x=10$ does not check and is extraneous.

86.
$$\sqrt{a+9}-\sqrt{a}=3$$
$$\sqrt{a+9}=3+\sqrt{a}$$
$$\left(\sqrt{a+9}\right)^2=\left(3+\sqrt{a}\right)^2$$
$$a+9=9+6\sqrt{a}+a$$
$$0=6\sqrt{a}$$
$$0^2=\left(6\sqrt{a}\right)^2$$
$$0=36a \Rightarrow a=0$$
The solution checks.

87.
$$\begin{aligned}
\sqrt{5-x}+\sqrt{5+x}&=4\\
\sqrt{5+x}&=4-\sqrt{5-x}\\
\left(\sqrt{5+x}\right)^2&=\left(4-\sqrt{5-x}\right)^2\\
5+x&=16-8\sqrt{5-x}+5-x\\
8\sqrt{5-x}&=16-2x\\
\left(8\sqrt{5-x}\right)^2&=(16-2x)^2\\
64(5-x)&=256-64x+4x^2\\
320-64x&=4x^2-64x+256\\
0&=4x^2-64\\
0&=4(x+4)(x-4)
\end{aligned}$$

$x+4=0$ **or** $x-4=0$

$x=-4$ $\quad x=4$

Both solutions check.

88.
$$\begin{aligned}
\sqrt{y+5}+\sqrt{y}&=1\\
\sqrt{y+5}&=1-\sqrt{y}\\
\left(\sqrt{y+5}\right)^2&=\left(1-\sqrt{y}\right)^2\\
y+5&=1-2\sqrt{y}+y\\
2\sqrt{y}&=-4\\
\left(2\sqrt{y}\right)^2&=(-4)^2\\
4y&=16\\
y&=4
\end{aligned}$$

The solution does not check. $\Rightarrow$ No solution.

89.
$$\begin{aligned}
\sqrt[3]{4x-9}+3&=2\\
\sqrt[3]{4x-9}&=-1\\
\left(\sqrt[3]{4x-9}\right)^3&=(-1)^3\\
4x-9&=-1\\
4x&=8\\
x&=2
\end{aligned}$$

The solution checks.

90.
$$\begin{aligned}
\sqrt[4]{x-2}+3&=5\\
\sqrt[4]{x-2}&=2\\
\left(\sqrt[4]{x-2}\right)^4&=(2)^4\\
x-2&=16\\
x&=18
\end{aligned}$$

The solution checks.

91.
$$\begin{aligned}
2x-9&<5\\
2x&<14\\
x&<7\Rightarrow(-\infty,7)
\end{aligned}$$

7

92.
$$\begin{aligned}
5x+3&\ge 2\\
5x&\ge -1\\
x&\ge -\tfrac{1}{5}\Rightarrow\left[-\tfrac{1}{5},\infty\right)
\end{aligned}$$

$-\frac{1}{5}$

93. $\frac{5(x-1)}{2} < x$

$5(x-1) < 2x$

$5x - 5 < 2x$

$3x < 5$

$x < \frac{5}{3} \Rightarrow \left(-\infty, \frac{5}{3}\right)$

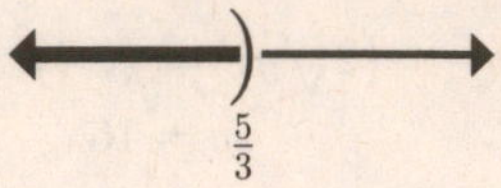

94. $\frac{1}{4}x + \frac{2}{3}x - x > \frac{1}{2} + \frac{1}{2}(x+1)$

$12\left(\frac{1}{4}x + \frac{2}{3}x - x\right) > 12\left(\frac{1}{2} + \frac{1}{2}(x+1)\right)$

$3x + 8x - 12x > 6 + 6(x+1)$

$-x > 6 + 6x + 6$

$-12 > 7x$

$-\frac{12}{7} > x \Rightarrow \left(-\infty, -\frac{12}{7}\right)$

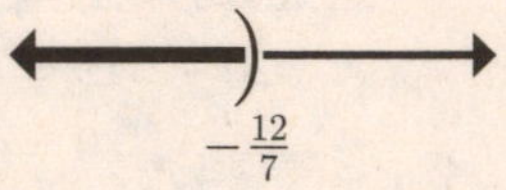

95. $0 \le \frac{3+x}{2} < 4$

$0 \le 3 + x < 8$

$-3 \le x < 5 \Rightarrow [-3, 5)$

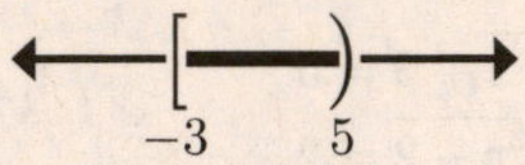

96. $2 + a < 3a - 2 \le 5a + 2$

$2 + a < 3a - 2$ **and** $3a - 2 \le 5a + 2$

$4 < 2a$ $\quad -4 \le 2a$

$a > 2$ $\quad a \ge -2$

$a > 2$

$a > -2$

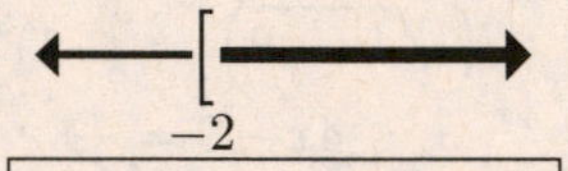

$a > 2$
and
$a \ge -2$

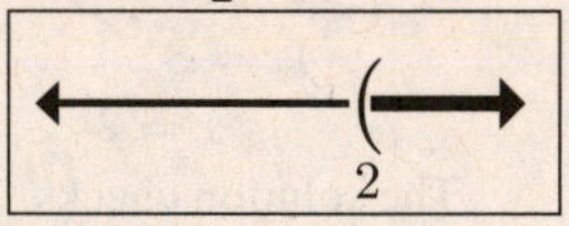

Solution set: $(2, \infty)$

97. $(x+2)(x-4) > 0$

factors = 0: $x = -2, x = 4$

intervals: $(-\infty, -2), (-2, 4), (4, \infty)$

interval	test number	value of $(x+2)(x-4)$
$(-\infty, -2)$	-3	$+7$
$(-2, 4)$	0	-8
$(4, \infty)$	5	$+7$

Solution: $(-\infty, -2) \cup (4, \infty)$

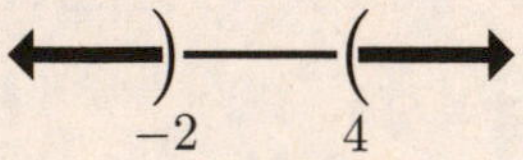

98. $(x-1)(x+4) < 0$

factors = 0: $x = 1, x = -4$

intervals: $(-\infty, -4), (-4, 1), (1, \infty)$

interval	test number	value of $(x-1)(x+4)$
$(-\infty, -4)$	-5	$+6$
$(-4, 1)$	0	-4
$(1, \infty)$	2	$+8$

Solution: $(-4, 1)$

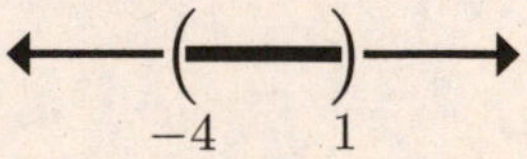

CHAPTER 1 REVIEW

99. $x^2 - 2x - 3 \le 0$

$(x-3)(x+1) \le 0$

factors = 0: $x = 3, x = -1$

intervals: $(-\infty, -1), (-1, 3), (3, \infty)$

interval	test number	value of x^2-2x-3
$(-\infty, -1)$	-2	$+5$
$(-1, 3)$	0	-3
$(3, \infty)$	4	$+5$

Solution: $[-1, 3]$

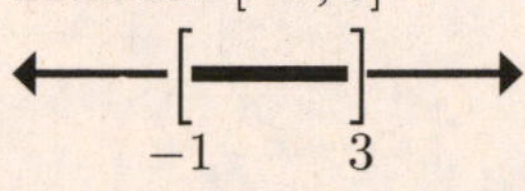

100. $2x^2 + x - 3 > 0$

$(2x+3)(x-1) > 0$

factors = 0: $x = -\frac{3}{2}, x = 1$

intervals: $(-\infty, -\frac{3}{2}), (-\frac{3}{2}, 1), (1, \infty)$

interval	test number	value of $2x^2+x-3$
$(-\infty, -\frac{3}{2})$	-2	$+3$
$(-\frac{3}{2}, 1)$	0	-3
$(1, \infty)$	2	$+7$

Solution: $(-\infty, -\frac{3}{2}) \cup (1, \infty)$

$-\frac{3}{2}$ $\quad$ 1

101. $\dfrac{x+2}{x-3} \ge 0$

factors = 0: $x = -2, x = 3$

intervals: $(-\infty, -2), (-2, 3), (3, \infty)$

interval	test number	sign of $\frac{x+2}{x-3}$
$(-\infty, -2)$	-3	$+$
$(-2, 3)$	0	$-$
$(3, \infty)$	4	$+$

Include endpoints which make the numerator equal to 0. Do not include endpoints which make the denominator equal to 0.

Solution: $(-\infty, -2] \cup (3, \infty)$

-2 $\quad$ 3

102. $\dfrac{x-1}{x+4} \le 0$

factors = 0: $x = 1, x = -4$

intervals: $(-\infty, -4), (-4, 1), (1, \infty)$

interval	test number	sign of $\frac{x-1}{x+4}$
$(-\infty, -4)$	-5	$+$
$(-4, 1)$	0	$-$
$(1, \infty)$	2	$+$

Include endpoints which make the numerator equal to 0. Do not include endpoints which make the denominator equal to 0.

Solution: $(-4, 1]$

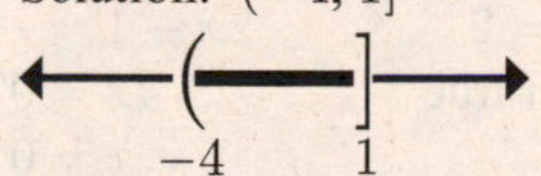

103. $\dfrac{x^2 + x - 2}{x-3} \ge 0$

$\dfrac{(x+2)(x-1)}{x-3} \ge 0$

factors = 0: $x = -2, x = 1, x = 3$

int.: $(-\infty, -2), (-2, 1), (1, 3), (3, \infty)$

interval	test number	sign of $\frac{x^2+x-2}{x-3}$
$(-\infty, -2)$	-3	$-$
$(-2, 1)$	0	$+$
$(1, 3)$	2	$-$
$(3, \infty)$	4	$+$

Include endpoints which make the numerator equal to 0. Do not include endpoints which make the denominator equal to 0.

Solution: $[-2, 1] \cup (3, \infty)$

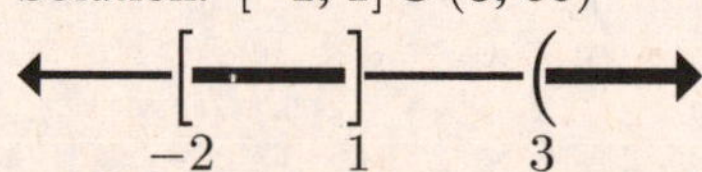

104.

$$\frac{5}{x} < 2$$
$$\frac{5}{x} - 2 < 0$$
$$\frac{5 - 2x}{x} < 0$$

factors $= 0$: $x = \frac{5}{2}, x = 0$

intervals: $(-\infty, 0), \left(0, \frac{5}{2}\right), \left(\frac{5}{2}, \infty\right)$

interval	test number	value of $\frac{5-2x}{x}$
$(-\infty, 0)$	-1	-7
$\left(0, \frac{5}{2}\right)$	1	$+3$
$\left(\frac{5}{2}, \infty\right)$	3	$-\frac{1}{3}$

Solution: $(-\infty, 0) \cup \left(\frac{5}{2}, \infty\right)$

$0 \qquad \frac{5}{2}$

105.

$$|x + 1| = 6$$
$$x + 1 = 6 \quad \textbf{or} \quad x + 1 = -6$$
$$x = 5 \qquad\qquad x = -7$$

106.

$$\left|\frac{3x + 11}{7}\right| - 1 = 0$$
$$\left|\frac{3x + 11}{7}\right| = 1$$
$$\frac{3x + 11}{7} = 1 \quad \textbf{or} \quad \frac{3x + 11}{7} = -1$$
$$3x + 11 = 7 \qquad 3x + 11 = -7$$
$$3x = -4 \qquad 3x = -18$$
$$x = -\frac{4}{3} \qquad x = -6$$

107.

$$\left|\frac{2a - 6}{3a}\right| - 6 = 0$$
$$\left|\frac{2a - 6}{3a}\right| = 6$$
$$\frac{2a - 6}{3a} = 6 \quad \textbf{or} \quad \frac{2a - 6}{3a} = -6$$
$$2a - 6 = 18a \qquad 2a - 6 = -18a$$
$$-16a = 6 \qquad 20a = 6$$
$$a = -\frac{6}{16} \qquad a = \frac{6}{20}$$
$$a = -\frac{3}{8} \qquad a = \frac{3}{10}$$

108.

$$|2x - 1| = |2x + 1|$$
$$2x - 1 = 2x + 1 \quad \textbf{or} \quad 2x - 1 = -(2x + 1)$$
$$0 = 2 \qquad 2x - 1 = -2x - 1$$
never true
$$4x = 0$$
$$x = 0$$

109.

$$|3x - 11| + 16 = 5$$
$$|3x - 11| = -11$$

An absolute value can never equal a negative number.

no solution

110.

$$|x + 3| < 3$$
$$-3 < x + 3 < 3$$
$$-6 < x < 0$$
$$(-6, 0)$$

$-6 \qquad 0$

111.

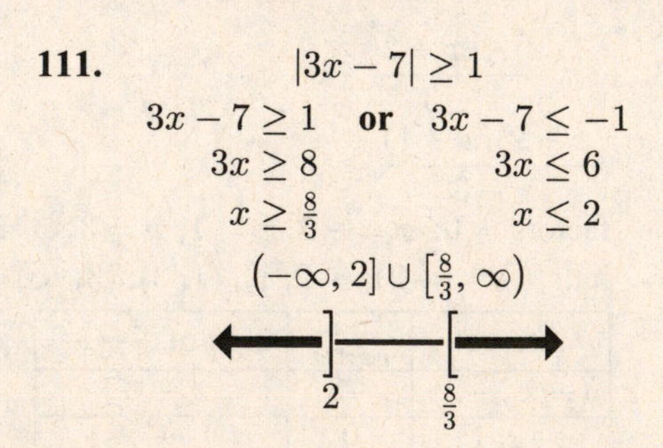

$$|3x - 7| \geq 1$$
$$3x - 7 \geq 1 \quad \textbf{or} \quad 3x - 7 \leq -1$$
$$3x \geq 8 \qquad 3x \leq 6$$
$$x \geq \frac{8}{3} \qquad x \leq 2$$
$$(-\infty, 2] \cup \left[\frac{8}{3}, \infty\right)$$

112.
$$\left|\frac{x+2}{3}\right| + 5 < 6$$
$$\left|\frac{x+2}{3}\right| < 1$$
$$-1 < \frac{x+2}{3} < 1$$
$$-3 < x+2 < 3$$
$$-5 < x < 1$$
$$(-5, 1)$$

−5 1

113.
$$\left|\frac{x-3}{4}\right| > 8$$
$$\frac{x-3}{4} > 8 \quad \textbf{or} \quad \frac{x-3}{4} < -8$$
$$x - 3 > 32 \qquad x - 3 < -32$$
$$x > 35 \qquad x < -29$$
$$(-\infty, -29) \cup (35, \infty)$$

−29 35

114.
$$1 < |2x+3| < 4$$
$1 < |2x+3|$ **and** $|2x+3| < 4$

(1) $|2x+3| > 1$
$$2x + 3 > 1 \quad \textbf{or} \quad 2x + 3 < -1$$
$$2x > -2 \qquad 2x < -4$$
$$x > -1 \qquad x < -2$$

(2) $|2x+3| < 4$
$$-4 < 2x + 3 < 4$$
$$-7 < 2x < 1$$
$$-\frac{7}{2} < x < \frac{1}{2}$$

(1) −2 −1

(2) $-\frac{7}{2}$ $\frac{1}{2}$

(1) −2 −1

(2) $-\frac{7}{2}$ $\frac{1}{2}$

(1) and (2) $-\frac{7}{2}$ −2 −1 $\frac{1}{2}$ $\Rightarrow \left(-\frac{7}{2}, -2\right) \cup \left(-1, \frac{1}{2}\right)$

115.
$$0 < |3x-4| < 7$$
$0 < |3x-4|$ **and** $|3x-4| < 7$

(1) $|3x-4| > 0$
$$3x - 4 > 0 \quad \textbf{or} \quad 3x - 4 < -0$$
$$3x > 4 \qquad 3x < 4$$
$$x > \frac{4}{3} \qquad x < \frac{4}{3}$$

(2) $|3x-4| < 7$
$$-7 < 3x - 4 < 7$$
$$-3 < 3x < 11$$
$$-1 < x < \frac{11}{3}$$

(1) $\frac{4}{3}$

(2) −1 $\frac{11}{3}$

(1) $\frac{4}{3}$

(2) −1 $\frac{11}{3}$

(1) and (2) −1 $\frac{4}{3}$ $\frac{11}{3}$ $\Rightarrow \left(-1, \frac{4}{3}\right) \cup \left(\frac{4}{3}, \frac{11}{3}\right)$

Chapter 1 Test (page 188)

1. $\dfrac{x}{x(x-1)} = 2$

restrictions: $x \neq 0, x \neq 1$

2. $\dfrac{4}{3x-2} + 3 = 7$

restrictions: $x \neq \frac{2}{3}$

3.
$$\begin{aligned} 7(2a+5) - 7 &= 6(a+8) \\ 14a + 35 - 7 &= 6a + 48 \\ 8a &= 20 \\ a &= \frac{20}{8} = \frac{5}{2} \end{aligned}$$

4.
$$\begin{aligned} \frac{1}{a-2} - \frac{1}{5a} &= \frac{3}{2a} \\ 10a(a-2)\left(\frac{1}{a-2} - \frac{1}{5a}\right) &= 10a(a-2) \cdot \frac{3}{2a} \\ 10a(1) - 2(a-2) &= 15(a-2) \\ 10a - 2a + 4 &= 15a - 30 \\ 34 &= 7a \\ \frac{34}{7} &= a \end{aligned}$$

5.
$$\begin{aligned} z &= \frac{x-\mu}{\sigma} \\ \sigma z &= \sigma \cdot \frac{x-\mu}{\sigma} \\ z\sigma &= x - \mu \\ z\sigma + \mu &= x \end{aligned}$$

6.
$$\begin{aligned} \frac{1}{a} &= \frac{1}{b} + \frac{1}{c} \\ abc \cdot \frac{1}{a} &= abc\left(\frac{1}{b} + \frac{1}{c}\right) \\ bc &= ac + ab \\ bc &= a(c+b) \\ \frac{bc}{c+b} &= \frac{a(c+b)}{c+b} \\ \frac{bc}{c+b} &= a \end{aligned}$$

7. Let $x =$ the score on the final.

Note: This score is counted twice.

$$\begin{aligned} \frac{\boxed{\text{Sum of scores}}}{5} &= 80 \\ \frac{75 + 75 + 75 + x + x}{5} &= 80 \\ \frac{2x + 225}{5} &= 80 \\ 2x + 225 &= 400 \\ 2x &= 175 \\ x &= 87.5 \end{aligned}$$

The student needs to score 87.5.

8. Let $x =$ the amount invested at 6%. Then $20{,}000 - x =$ the amount invested at 7%.

$$\boxed{\text{Interest at 6\%}} + \boxed{\text{Interest at 7\%}} = \boxed{\text{Total interest}}$$

$$\begin{aligned} 0.06x + 0.07(20{,}000 - x) &= 1{,}260 \\ 0.06x + 1{,}400 - 0.07x &= 1{,}260 \\ -0.01x &= -140 \\ x &= 14{,}000 \end{aligned}$$

\$14,000 was invested at 6%.

9. $3\sqrt{-96} = 3\sqrt{-1}\sqrt{16}\sqrt{6} = 12i\sqrt{6}$

10. $\sqrt{-\dfrac{18}{5}} = \sqrt{-\dfrac{18 \cdot 5}{5 \cdot 5}} = \sqrt{-\dfrac{90}{25}} = \dfrac{\sqrt{-1}\sqrt{9}\sqrt{10}}{\sqrt{25}} = \dfrac{3\sqrt{10}}{5}i$

11. $(4-5i)-(-3+7i)=4-5i+3-7i$
$=7-12i$

12. $(4-5i)(3-7i)=12-43i+35i^2$
$=12-43i-35$
$=-23-43i$

13. $\dfrac{2}{2-i}=\dfrac{2(2+i)}{(2-i)(2+i)}=\dfrac{4+2i}{2^2-i^2}=\dfrac{4+2i}{5}=\dfrac{4}{5}+\dfrac{2}{5}i$

14. $\dfrac{1+i}{1-i}=\dfrac{(1+i)(1+i)}{(1-i)(1+i)}=\dfrac{1+2i+i^2}{1^2-i^2}=\dfrac{2i}{2}=0+i$

15. $i^{13}=i^{12}i=(i^4)^3i=1^3i=i$

16. $1i^4=1\cdot 1=1$

17.
$$\begin{aligned}4x^2-8x+3&=0\\(2x-3)(2x-1)&=0\end{aligned}$$
$2x-3=0$ **or** $2x-1=0$
$2x=3$ $\quad 2x=1$
$x=\frac{3}{2}$ $\quad x=\frac{1}{2}$

18.
$$\begin{aligned}2b^2-12&=-5b\\2b^2+5b-12&=0\\(2b-3)(b+4)&=0\end{aligned}$$
$2b-3=0$ **or** $b+4=0$
$2b=3$ $\quad b=-4$
$b=\frac{3}{2}$ $\quad b=-4$

19.
$$\begin{aligned}5x^2&=-135\\x^2&=-27\\x&=\pm\sqrt{-27}\\x&=\pm 3i\sqrt{3}\end{aligned}$$

20.
$$\begin{aligned}x^2-14x&=23\\x^2-14x+49&=23+49\\(x-7)^2&=72\end{aligned}$$
$x-7=\sqrt{72}$ **or** $x-7=-\sqrt{72}$
$x-7=6\sqrt{2}$ $\quad x-7=-6\sqrt{2}$
$x=7+6\sqrt{2}$ $\quad x=7-6\sqrt{2}$

21. $3x^2-5x-9=0\Rightarrow a=3,b=-5,c=-9$
$$x=\frac{-b\pm\sqrt{b^2-4ac}}{2a}=\frac{-(-5)\pm\sqrt{(-5)^2-4(3)(-9)}}{2(3)}=\frac{5\pm\sqrt{25+108}}{6}=\frac{5\pm\sqrt{133}}{6}$$

22.
$$\begin{aligned}\frac{3}{x^2-5x-14}&=\frac{4}{x^2+5x+6}\\\frac{3}{(x-7)(x+2)}&=\frac{4}{(x+2)(x+3)}\\(x-7)(x+2)(x+3)\frac{3}{(x-7)(x+2)}&=(x-7)(x+2)(x+3)\frac{4}{(x+2)(x+3)}\\3(x+3)&=4(x-7)\\3x+9&=4x-28\\37&=x\end{aligned}$$

CHAPTER 1 TEST

23. $x^2 + (k+1)x + k + 4 = 0$

$a = 1, b = k+1, c = k+4$

Set the discriminant equal to 0:

$$b^2 - 4ac = 0$$
$$(k+1)^2 - 4(1)(k+4) = 0$$
$$k^2 + 2k + 1 - 4k - 16 = 0$$
$$k^2 - 2k - 15 = 0$$
$$(k-5)(k+3) = 0$$
$$k - 5 = 0 \quad \textbf{or} \quad k + 3 = 0$$
$$k = 5 \qquad\qquad k = -3$$

24. Set $h = 0$:

$$h = -16t^2 + 128t$$
$$0 = -16t^2 + 128t$$
$$0 = -16t(t-8)$$
$$-16t = 0 \quad \textbf{or} \quad t - 8 = 0$$
$$t = 0 \qquad\qquad t = 8$$

The projectile will return after 8 seconds.

25. $|5 - 12i| = \sqrt{5^2 + (-12)^2} = \sqrt{25 + 144} = \sqrt{169} = 13$

26.
$$\left|\frac{1}{3+i}\right| = \left|\frac{1(3-i)}{(3+i)(3-i)}\right| = \left|\frac{3-i}{3^2 - i^2}\right| = \left|\frac{3-i}{10}\right| = \left|\frac{3}{10} - \frac{1}{10}i\right| = \sqrt{\left(\frac{3}{10}\right)^2 + \left(-\frac{1}{10}\right)^2}$$
$$= \sqrt{\frac{9}{100} + \frac{1}{100}}$$
$$= \sqrt{\frac{10}{100}} = \frac{\sqrt{10}}{10}$$

27.
$$z^4 - 13z^2 + 36 = 0 \qquad z^2 - 4 = 0 \quad \textbf{or} \quad z^2 - 9 = 0$$
$$\left(z^2 - 4\right)\left(z^2 - 9\right) = 0 \qquad z^2 = 4 \qquad\qquad z^2 = 9$$
$$z = \pm 2 \qquad\qquad z = \pm 3$$

28.
$$2p^{2/5} - p^{1/5} - 1 = 0 \qquad 2p^{1/5} + 1 = 0 \quad \textbf{or} \quad p^{1/5} - 1 = 0$$
$$\left(2p^{1/5} + 1\right)\left(p^{1/5} - 1\right) = 0 \qquad p^{1/5} = -\tfrac{1}{2} \qquad\qquad p^{1/5} = 1$$
$$\left(p^{1/5}\right)^5 = \left(-\tfrac{1}{2}\right)^5 \qquad\qquad \left(p^{1/5}\right)^5 = (1)^5$$
$$p = -\tfrac{1}{32} \qquad\qquad p = 1$$

Both answers check.

29.

$$\sqrt{x+5} = 12$$
$$\left(\sqrt{x+5}\right)^2 = 12^2$$
$$x+5 = 144$$
$$x = 139$$

The answer checks.

30.

$$\sqrt{2z+3} = 1 - \sqrt{z+1}$$
$$\left(\sqrt{2z+3}\right)^2 = \left(1-\sqrt{z+1}\right)^2$$
$$2z+3 = 1 - 2\sqrt{z+1} + z + 1$$
$$2\sqrt{z+1} = -z-1$$
$$\left(2\sqrt{z+1}\right)^2 = (-z-1)^2$$
$$4(z+1) = z^2+2z+1$$
$$4z+4 = z^2+2z+1$$
$$0 = z^2-2z-3$$
$$0 = (z+1)(z-3)$$

$z+1=0$ **or** $z-3=0$

$z=-1$ $\quad z=3$

The answer $z=3$ is extraneous.

31. $5x-3 \le 7$

$5x \le 10$

$x \le 2 \Rightarrow (-\infty, 2]$

32.

$$\frac{x+3}{4} > \frac{2x-4}{3}$$
$$12 \cdot \tfrac{x+3}{4} > 12 \cdot \tfrac{2x-4}{3}$$
$$3(x+3) > 4(2x-4)$$
$$3x+9 > 8x-16$$
$$-5x > -25$$
$$x < 5 \Rightarrow (-\infty, 5)$$

33. $5 \le 2x-1 < 7$

$6 \le 2x < 8$

$3 \le x < 4 \Rightarrow [3, 4)$

34.

$$1+x < 3x-3 < 4x-2$$

$1+x < 3x-3$ **and** $3x-3 < 4x-2$

$-2x < -4$ $\quad -x < 1$

$x > 2$ $\quad x > -1$

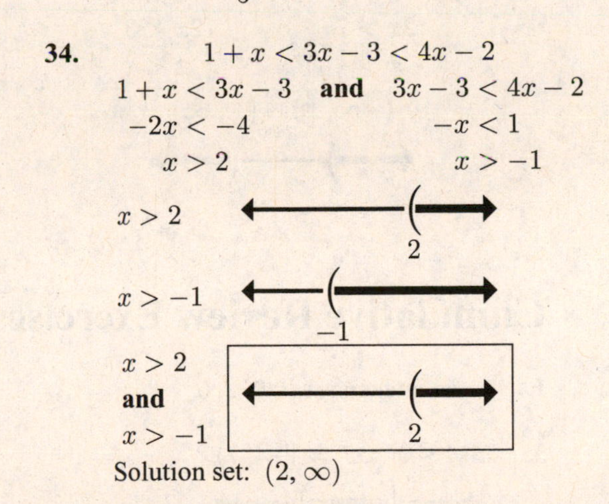

Solution set: $(2, \infty)$

35. $x^2 - 7x - 8 \geq 0$
$(x+1)(x-8) \geq 0$
factors $= 0$: $x = -1, x = 8$
intervals: $(-\infty, -1), (-1, 8), (8, \infty)$

interval	test number	value of x^2-7x-8
$(-\infty, -1)$	-2	$+10$
$(-1, 8)$	0	-8
$(8, \infty)$	9	$+10$

Solution: $(-\infty, -1] \cup [8, \infty)$

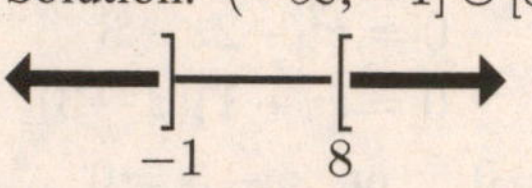

36. $\dfrac{x+2}{x-1} \leq 0$
factors $= 0$: $x = -2, x = 1$
intervals: $(-\infty, -2), (-2, 1), (1, \infty)$

interval	test number	sign of $\frac{x+2}{x-1}$
$(-\infty, -2)$	-3	$+$
$(-2, 1)$	0	$-$
$(1, \infty)$	2	$+$

Include endpoints which make the numerator equal to 0. Do not include endpoints which make the denominator equal to 0.
Solution: $[-2, 1)$

−2 1

37. $\left|\dfrac{3x+2}{2}\right| = 4$

$\frac{3x+2}{2} = 4$ **or** $\frac{3x+2}{2} = -4$
$3x + 2 = 8$ $\quad 3x + 2 = -8$
$3x = 6$ $\quad 3x = -10$
$x = 2$ $\quad x = -\frac{10}{3}$

38. $|x+3| = |x-3|$

$x + 3 = x - 3$ **or** $x + 3 = -(x-3)$
$0 = -6$ $\quad x + 3 = -x + 3$
not true $\quad 2x = 0$
$x = 0$

39. $|2x - 5| > 2$

$2x - 5 > 2$ **or** $2x - 5 < -2$
$2x > 7$ $\quad 2x < 3$
$x > \frac{7}{2}$ $\quad x < \frac{3}{2}$
$(-\infty, \frac{3}{2}) \cup (\frac{7}{2}, \infty)$

$\frac{3}{2}$ $\frac{7}{2}$

40. $\left|\dfrac{2x+3}{3}\right| \leq 5$

$-5 \leq \frac{2x+3}{3} \leq 5$
$-15 \leq 2x + 3 \leq 15$
$-18 \leq 2x \leq 12$
$-9 \leq x \leq 6$
$[-9, 6]$

−9 6

Cumulative Review Exercises (page 189)

1. even integers: $-2, 0, 2, 6$

2. prime numbers: 2, 5, 11

3. $-4 \leq x < 7 \Rightarrow [-4, 7)$

−4 7

4. $x \geq 2$ or $x < 0 \Rightarrow (-\infty, 0) \cup [2, \infty)$

0 2

5. Commutative Property of Addition

6. Transitive Property

CUMULATIVE REVIEW EXERCISES

7. $(81a^4)^{1/2} = \left[(9a^2)^2\right]^{1/2} = 9a^2$

8. $81(a^4)^{1/2} = 81\left[(a^2)^2\right]^{1/2} = 81a^2$

9. $(a^{-3}b^{-2})^{-2} = (a^{-3})^{-2}(b^{-2})^{-2} = a^6b^4$

10. $\left(\dfrac{4x^4}{12x^2y}\right)^{-2} = \left(\dfrac{12x^2y}{4x^4}\right)^2 = \left(\dfrac{3y}{x^2}\right)^2 = \dfrac{9y^2}{x^4}$

11. $\left(\dfrac{4x^0y^2}{x^2y}\right)^{-2} = \left(\dfrac{x^2y}{4x^0y^2}\right)^2 = \left(\dfrac{x^2}{4y}\right)^2 = \dfrac{x^4}{16y^2}$

12. $\left(\dfrac{4x^{-5}y^2}{6x^{-2}y^{-3}}\right)^2 = \left(\dfrac{2y^5}{3x^3}\right)^2 = \dfrac{4y^{10}}{9x^6}$

13. $(a^{1/2}b)^2(ab^{1/2})^2 = (ab^2)(a^2b) = a^3b^3$

14. $(a^{1/2}b^{1/2}c)^2 = abc^2$

15. $\dfrac{3}{\sqrt{3}} = \dfrac{3\sqrt{3}}{\sqrt{3}\sqrt{3}} = \dfrac{3\sqrt{3}}{3} = \sqrt{3}$

16. $\dfrac{2}{\sqrt[3]{4x}} = \dfrac{2\sqrt[3]{2x^2}}{\sqrt[3]{4x}\sqrt[3]{2x^2}} = \dfrac{2\sqrt[3]{2x^2}}{\sqrt[3]{8x^3}} = \dfrac{2\sqrt[3]{2x^2}}{2x} = \dfrac{\sqrt[3]{2x^2}}{x}$

17. $\dfrac{3}{y-\sqrt{3}} = \dfrac{3\left(y+\sqrt{3}\right)}{\left(y-\sqrt{3}\right)\left(y+\sqrt{3}\right)} = \dfrac{3\left(y+\sqrt{3}\right)}{y^2-\left(\sqrt{3}\right)^2} = \dfrac{3\left(y+\sqrt{3}\right)}{y^2-3}$

18. $\dfrac{3x}{\sqrt{x}-1} = \dfrac{3x(\sqrt{x}+1)}{(\sqrt{x}-1)(\sqrt{x}+1)} = \dfrac{3x(\sqrt{x}+1)}{(\sqrt{x})^2-1^2} = \dfrac{3x(\sqrt{x}+1)}{x-1}$

19. $\sqrt{75} - 3\sqrt{5} = \sqrt{25}\sqrt{3} - 3\sqrt{5} = 5\sqrt{3} - 3\sqrt{5}$

20. $\sqrt{18} + \sqrt{8} - 2\sqrt{2} = \sqrt{9}\sqrt{2} + \sqrt{4}\sqrt{2} - 2\sqrt{2} = 3\sqrt{2} + 2\sqrt{2} - 2\sqrt{2} = 3\sqrt{2}$

21. $\left(\sqrt{2}-\sqrt{3}\right)^2 = \left(\sqrt{2}-\sqrt{3}\right)\left(\sqrt{2}-\sqrt{3}\right) = \sqrt{4} - 2\sqrt{6} + \sqrt{9} = 5 - 2\sqrt{6}$

22. $\left(3-\sqrt{5}\right)\left(3+\sqrt{5}\right) = 9 - \sqrt{25} = 9 - 5 = 4$

23. $(3x^2 - 2x + 5) - 3(x^2 + 2x - 1) = 3x^2 - 2x + 5 - 3x^2 - 6x + 3 = -8x + 8$

24. $5x^2(2x^2 - x) + x(x^2 - x^3) = 10x^4 - 5x^3 + x^3 - x^4 = 9x^4 - 4x^3$

25. $(3x - 5)(2x + 7) = 6x^2 + 21x - 10x - 35 = 6x^2 + 11x - 35$

26. $(z + 2)(z^2 - z + 2) = z^3 - z^2 + 2z + 2z^2 - 2z + 4 = z^3 + z^2 + 4$

CUMULATIVE REVIEW EXERCISES

27.
$$\begin{array}{r} 2x^2 - \phantom{x^2+{}} x + 1 \\ 3x+2 \,\overline{)\,6x^3 + x^2 + x + 2} \\ \underline{6x^3 + 4x^2 \phantom{{}+x+2}} \\ -3x^2 + x \phantom{{}+2} \\ \underline{-3x^2 - 2x \phantom{{}+2}} \\ 3x + 2 \\ \underline{3x + 2} \\ 0 \end{array}$$

28.
$$\begin{array}{r} 3x^2 + \phantom{0x^3+{}} 1 + \frac{-x}{x^2+2} \\ x^2+2 \,\overline{)\,3x^4 + 0x^3 + 7x^2 - x + 2} \\ \underline{3x^4 \phantom{{}+0x^3} + 6x^2 \phantom{{}-x+2}} \\ x^2 - x + 2 \\ \underline{x^2 \phantom{{}-x} + 2} \\ -x \phantom{{}+2} \end{array}$$

29. $3t^2 - 6t = 3t(t-2)$

30. $3x^2 - 10x - 8 = (3x+2)(x-4)$

31.
$$\begin{aligned} x^8 - 2x^4 + 1 = \left(x^4-1\right)\left(x^4-1\right) &= \left(x^2+1\right)\left(x^2-1\right)\left(x^2+1\right)\left(x^2-1\right) \\ &= \left(x^2+1\right)^2(x+1)(x-1)(x+1)(x-1) \\ &= \left(x^2+1\right)^2(x+1)^2(x-1)^2 \end{aligned}$$

32. $x^6 - 1 = \left(x^3\right)^2 - 1^2 = (x^3+1)(x^3-1) = (x+1)(x^2-x+1)(x-1)(x^2+x+1)$

33.
$$\frac{x^2-4}{x^2+5x+6} \cdot \frac{x^2-2x-15}{x^2+3x-10} = \frac{(x+2)(x-2)}{(x+2)(x+3)} \cdot \frac{(x-5)(x+3)}{(x+5)(x-2)} = \frac{x-5}{x+5}$$

34.
$$\begin{aligned} \frac{6x^3+x^2-x}{x+2} \div \frac{3x^2-x}{x^2+4x+4} &= \frac{x(6x^2+x-1)}{x+2} \cdot \frac{x^2+4x+4}{3x^2-x} \\ &= \frac{x(2x+1)(3x-1)}{x+2} \cdot \frac{(x+2)(x+2)}{x(3x-1)} = (2x+1)(x+2) \end{aligned}$$

35.
$$\begin{aligned} \frac{2}{x+3} + \frac{5x}{x-3} = \frac{2(x-3)}{(x+3)(x-3)} + \frac{5x(x+3)}{(x-3)(x+3)} &= \frac{2x-6}{(x+3)(x-3)} + \frac{5x^2+15x}{(x+3)(x-3)} \\ &= \frac{5x^2+17x-6}{(x+3)(x-3)} \end{aligned}$$

36.
$$\frac{x-2}{x+3}\left(\frac{x+3}{x^2-4} - 1\right) = \frac{x-2}{x+3}\left(\frac{x+3}{x^2-4} - \frac{x^2-4}{x^2-4}\right) = \frac{x-2}{x+3}\left(\frac{-x^2+x+7}{(x+2)(x-2)}\right) = \frac{-x^2+x+7}{(x+3)(x+2)}$$

37.
$$\frac{\frac{1}{a}+\frac{1}{b}}{\frac{1}{ab}} = \frac{ab\left(\frac{1}{a}+\frac{1}{b}\right)}{ab\left(\frac{1}{ab}\right)} = \frac{b+a}{1} = b+a$$

38.
$$\begin{aligned} \frac{x^{-1}-y^{-1}}{x-y} = \frac{\frac{1}{x}-\frac{1}{y}}{x-y} &= \frac{xy\left(\frac{1}{x}-\frac{1}{y}\right)}{xy(x-y)} \\ &= \frac{y-x}{xy(x-y)} = -\frac{1}{xy} \end{aligned}$$

CUMULATIVE REVIEW EXERCISES

39.
$$\begin{aligned}\frac{3x}{x+5} &= \frac{x}{x-5}\\ 3x(x-5) &= x(x+5)\\ 3x^2-15x &= x^2+5x\\ 2x^2-20x &= 0\\ 2x(x-10) &= 0\end{aligned}$$
$$2x=0 \quad \textbf{or} \quad x-10=0$$
$$x=0 \qquad\qquad x=10$$

40.
$$\begin{aligned}8(2x-3)-3(5x+2) &= 4\\ 16x-24-15x-6 &= 4\\ x &= 34\end{aligned}$$

41.
$$\begin{aligned}\frac{1}{R} &= \frac{1}{R_1}+\frac{1}{R_2}\\ RR_1R_2\cdot\frac{1}{R} &= RR_1R_2\left(\frac{1}{R_1}+\frac{1}{R_2}\right)\\ R_1R_2 &= RR_2+RR_1\\ R_1R_2 &= R(R_2+R_1)\\ \frac{R_1R_2}{R_2+R_1} &= \frac{R(R_2+R_1)}{R_2+R_1}\\ \frac{R_1R_2}{R_2+R_1} &= R\end{aligned}$$

42.
$$\begin{aligned}S &= \frac{a-lr}{1-r}\\ S(1-r) &= \frac{a-lr}{1-r}(1-r)\\ S(1-r) &= a-lr\\ S-Sr &= a-lr\\ S-a &= Sr-lr\\ S-a &= r(S-l)\\ \frac{S-a}{S-l} &= r\end{aligned}$$

43.

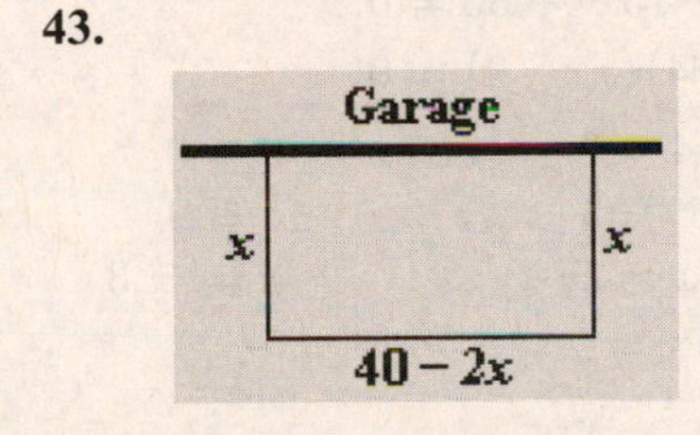

$$\begin{aligned}\text{Area} &= 192\\ x(40-2x) &= 192\\ 40x-2x^2 &= 192\\ 0 &= 2x^2-40x+192\\ 0 &= 2(x-8)(x-12)\end{aligned}$$
$$x-8=0 \quad \textbf{or} \quad x-12=0$$
$$x=8 \qquad\qquad x=12$$

If $x=8$, then the dimensions are 8 feet by 24 feet.
If $x=12$, then the dimensions are 12 feet by 16 feet.

44. Let $x=$ the amount invested at 6%. Then $25{,}000-x=$ the amount invested at 7%.

[Interest at 6%] + [Interest at 7%] = [Total interest]

$$\begin{aligned}0.06x+0.07(25{,}000-x) &= 1{,}670\\ 0.06x+1{,}750-0.07x &= 1{,}670\\ -0.01x &= -80\\ x &= 8{,}000 \Rightarrow \$8{,}000 \text{ was invested at 6\%.}\end{aligned}$$

45.
$$\frac{2+i}{2-i}=\frac{(2+i)(2+i)}{(2-i)(2+i)}=\frac{4+4i+i^2}{4-i^2}=\frac{3+4i}{5}=\frac{3}{5}+\frac{4}{5}i$$

CUMULATIVE REVIEW EXERCISES

46. $$\frac{i(3-i)}{(1+i)(1+i)} = \frac{i(3-i)(1-i)(1-i)}{(1+i)(1-i)(1+i)(1-i)} = \frac{(3i-i^2)(1-2i+i^2)}{(1-i^2)(1-i^2)}$$
$$= \frac{(1+3i)(-2i)}{1-2i^2+i^4} = \frac{-2i-6i^2}{4} = \frac{6-2i}{4} = \frac{3}{2} - \frac{1}{2}i$$

47. $$|3+4i| = \sqrt{3^2+4^2} = \sqrt{9+16} = \sqrt{25} = 5$$

48. $$\frac{5}{i^7} + 5i = \frac{5i}{i^7 i} + 5i = \frac{5i}{i^8} + 5i = \frac{5i}{(i^4)^2} + 5i = \frac{5i}{1^2} + 5i = 5i + 5i = 10i = 0 + 10i$$

49.
$$\begin{aligned} 15x^2 - 16x - 7 &= 0 \\ (5x-7)(3x+1) &= 0 \\ 5x - 7 = 0 \quad &\textbf{or} \quad 3x + 1 = 0 \\ x = \tfrac{7}{5} \qquad & \qquad x = -\tfrac{1}{3} \end{aligned}$$

50.
$$\begin{aligned} (7x-4)^2 &= -8 \\ \sqrt{(7x-4)^2} &= \pm\sqrt{-8} \\ (7x-4) &= \pm 2i\sqrt{2} \\ 7x &= 4 \pm 2i\sqrt{2} \\ x &= \frac{4}{7} \pm \frac{2\sqrt{2}}{7}i \end{aligned}$$

51.
$$\begin{aligned} \frac{x+3}{x-1} - \frac{6}{x} &= 1 \\ x(x-1)\left(\frac{x+3}{x-1} - \frac{6}{x}\right) &= x(x-1)(1) \\ x(x+3) - 6(x-1) &= x^2 - x \\ x^2 + 3x - 6x + 6 &= x^2 - x \\ -2x &= -6 \\ x &= 3 \end{aligned}$$

52.
$$\begin{aligned} x^4 + 36 &= 13x^2 \\ x^4 - 13x^2 + 36 &= 0 \\ (x^2-4)(x^2-9) &= 0 \\ x^2 - 4 = 0 \quad &\textbf{or} \quad x^2 - 9 = 0 \\ x^2 = 4 \qquad & \qquad x^2 = 9 \\ x = \pm 2 \qquad & \qquad x = \pm 3 \end{aligned}$$

53.
$$\begin{aligned} \sqrt{y+2} + \sqrt{11-y} &= 5 \\ \sqrt{y+2} - 5 &= -\sqrt{11-y} \\ \left(\sqrt{y+2} - 5\right)^2 &= \left(-\sqrt{11-y}\right)^2 \\ y + 2 - 10\sqrt{y+2} + 25 &= 11 - y \\ -10\sqrt{y+2} &= -2y - 16 \\ \left(-10\sqrt{y+2}\right)^2 &= (-2y-16)^2 \\ 100(y+2) &= 4y^2 + 64y + 256 \\ 100y + 200 &= 4y^2 + 64y + 256 \\ 0 &= 4y^2 - 36y + 56 \\ 0 &= 4(y-2)(y-7) \\ y - 2 = 0 \quad &\textbf{or} \quad y - 7 = 0 \\ y = 2 \qquad & \qquad y = 7 \end{aligned}$$
Both solutions check.

54.
$$\begin{aligned} z^{2/3} - 13z^{1/3} + 36 &= 0 \\ (z^{1/3}-4)(z^{1/3}-9) &= 0 \\ z^{1/3} - 4 = 0 \quad &\textbf{or} \quad z^{1/3} - 9 = 0 \\ z^{1/3} = 4 \qquad & \qquad z^{1/3} = 9 \\ (z^{1/3})^3 = 4^3 \qquad & \qquad (z^{1/3})^3 = 9^3 \\ z = 64 \qquad & \qquad z = 729 \end{aligned}$$
Both solutions check.

CUMULATIVE REVIEW EXERCISES

55. $5x - 7 \le 4$

$5x \le 11$

$x \le \frac{11}{5} \Rightarrow \left(-\infty, \frac{11}{5}\right]$

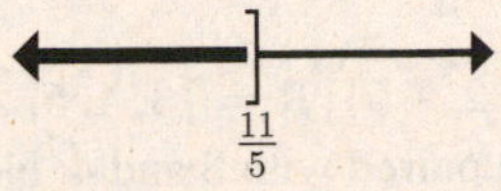

$\frac{11}{5}$

56. $x^2 - 8x + 15 > 0$

$(x-3)(x-5) > 0$

factors $= 0$: $x = 3, x = 5$

intervals: $(-\infty, 3), (3, 5), (5, \infty)$

interval	test number	value of $x^2-8x+15$
$(-\infty, 3)$	0	+15
$(3, 5)$	4	−1
$(5, \infty)$	6	+3

Solution: $(-\infty, 3) \cup (5, \infty)$

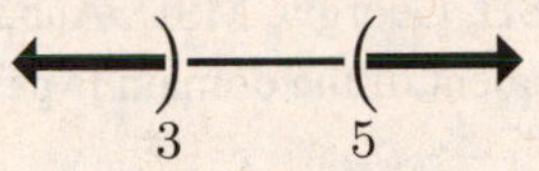

3 5

57. $\dfrac{x^2+4x+3}{x-2} \ge 0$

$\dfrac{(x+3)(x+1)}{x-2} \ge 0$

factors $= 0$: $x = -3, x = -1, x = 2$

int.: $(-\infty, -3), (-3, -1), (-1, 2), (2, \infty)$

interval	test number	sign of $\frac{x^2+4x+3}{x-2}$
$(-\infty, -3)$	−4	−
$(-3, -1)$	−2	+
$(-1, 2)$	0	−
$(2, \infty)$	3	+

Include endpoints which make the numerator equal to 0. Do not include endpoints which make the denominator equal to 0.

Solution: $[-3, -1] \cup (2, \infty)$

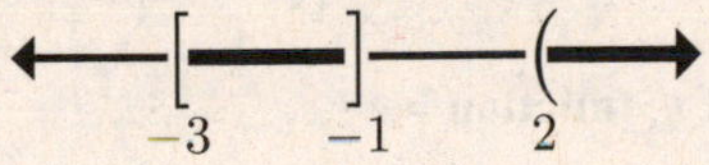

−3 −1 2

58. $\dfrac{9}{x} > x$

$\dfrac{9}{x} - x > 0$

$\dfrac{9-x^2}{x} > 0$

$\dfrac{(3+x)(3-x)}{x} > 0$

factors $= 0$: $x = -3, x = 3, x = 0$

int.: $(-\infty, -3), (-3, 0), (0, 3), (3, \infty)$

interval	test number	sign of $\frac{9-x^2}{x}$
$(-\infty, -3)$	−4	+
$(-3, 0)$	−1	−
$(0, 3)$	1	+
$(3, \infty)$	4	−

Solution: $(-\infty, -3) \cup (0, 3)$

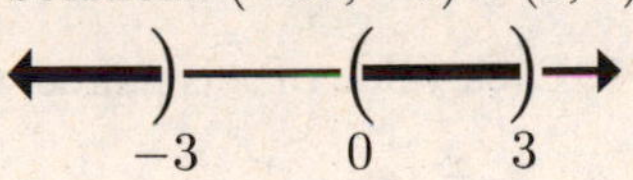

−3 0 3

59. $|2x - 3| \ge 5$

$2x - 3 \ge 5$ **or** $2x - 3 \le -5$

$2x \ge 8$ $\quad$ $2x \le -2$

$x \ge 4$ $\quad$ $x \le -1$

$(-\infty, -1] \cup [4, \infty)$

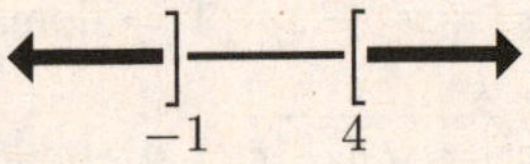

−1 4

60. $\left|\dfrac{3x-5}{2}\right| < 2$

$-2 < \dfrac{3x-5}{2} < 2$

$-4 < 3x - 5 < 4$

$1 < 3x < 9$

$\frac{1}{3} < x < 3$

$\left(\frac{1}{3}, 3\right)$

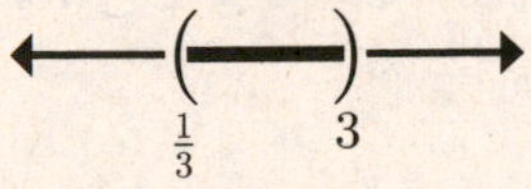

$\frac{1}{3}$ 3

Exercises 2.1 (page 201)

1. function

3. domain

5. $y = f(x)$

7. dependent

9. D = {2, 3, 4, 5}; R = {3, 4, 5, 6}
Each element of the domain is paired with only one element of the range. Function.

11. D = {1, 2, −5}; R = {3, 4, 5, 2}
1 is both paired with 3 and 4. Not a function.

13. {(LSU, Tigers), (Georgia, Bulldogs), (MSU, Bulldogs), (Auburn, Tigers) }
D = {LSU, Georgia, MSU, Auburn}; R = {Tigers, Bulldogs}
Each element of the domain is paired with only one element of the range. Function.

15. {(76, September 9), (76, October 12), (78, May 10), (80, June 1) }
D = {76, 78, 80}; R = {September 9, October 12, May 10, June 1}
76 is paired with September 9 and October 12. Not a function.

17. $y = x$
Each value of x is paired with only one value of y.
function

19. $y^2 = x$
$y = \pm\sqrt{x}$
At least one value of x is paired with more than one value of y. **not a function**

21. $y = x^2$
Each value of x is paired with only one value of y.
function

23. $|y| = x$
$y = \pm x$
At least one value of x is paired with more than one value of y. **not a function**

25. $|x - 2| = y$
$y = |x - 2|$
Each value of x is paired with only one value of y.
function

27. $|x| = |y|$
$|y| = |x|$
$y = \pm|x|$
At least one value of x is paired with more than one value of y. **not a function**

29. $y = 7$; Each value of x is paired with only one value of y. **function**

31. $y - 7 = \sqrt{x}$
$y = \sqrt{x} + 7$
Each value of x is paired with only one value of y. **function**

33. $x^3 + y^2 = 25$
$y^2 = -x^3 + 25$
$y = \pm\sqrt{-x^3 + 25}$
At least one value of x is paired with more than one value of y. **not a function**

35. $f(x) = 3x + 5 \Rightarrow \text{domain} = (-\infty, \infty)$

37. $f(x) = x^2 - x + 1 \Rightarrow \text{domain} = (-\infty, \infty)$

39. $f(x) = \sqrt{x - 2} \Rightarrow x - 2 \geq 0$
domain $= [2, \infty)$

41. $f(x) = \sqrt{4 - x} \Rightarrow 4 - x \geq 0$
domain $= (-\infty, 4]$

EXERCISES 2.1

43. $f(x)=\sqrt{x^2-1}\Rightarrow x^2-1\geq 0$
domain $=(-\infty,-1]\cup[1,\infty)$

45. $f(x)=\sqrt[3]{x+1}\Rightarrow$ domain $=(-\infty,\infty)$

47. $f(x)=\dfrac{3}{x+1}\Rightarrow x\neq -1$
domain $=(-\infty,-1)\cup(-1,\infty)$

49. $f(x)=\dfrac{x}{x-3}\Rightarrow x\neq 3$
domain $=(-\infty,3)\cup(3,\infty)$

51. $f(x)=\dfrac{x}{x^2-4}=\dfrac{x}{(x+2)(x-2)}$
$x\neq -2, x\neq 2$
domain $=(-\infty,-2)\cup(-2,2)\cup(2,\infty)$

53. $f(x)=\dfrac{1}{x^2-4x-5}=\dfrac{1}{(x+1)(x-5)}$
$x\neq -1, x\neq 5$
domain $=(-\infty,-1)\cup(-1,5)\cup(5,\infty)$

55. $f(x)=|x|+3\Rightarrow$ domain $=(-\infty,\infty)$

57. $f(x)=3x-2$

$f(2)=3(2)-2$ $=6-2$ $=4$	$f(-3)=3(-3)-2$ $=-9-2$ $=-11$	$f(k)=3k-2$	$f(k^2-1)=3(k^2-1)-2$ $=3k^2-3-2$ $=3k^2-5$

59. $f(x)=\frac{1}{2}x+3$

$f(2)=\frac{1}{2}(2)+3$ $=1+3$ $=4$	$f(-3)=\frac{1}{2}(-3)+3$ $=-\frac{3}{2}+3$ $=\frac{3}{2}$	$f(k)=\frac{1}{2}k+3$	$f(k^2-1)=\frac{1}{2}(k^2-1)+3$ $=\frac{1}{2}k^2-\frac{1}{2}+3$ $=\frac{1}{2}k^2+\frac{5}{2}$

61. $f(x)=x^2$

$f(2)=2^2$ $=4$	$f(-3)=(-3)^2$ $=9$	$f(k)=k^2$	$f(k^2-1)=(k^2-1)^2$ $=(k^2-1)(k^2-1)$ $=k^4-2k^2+1$

63. $f(x)=x^2+3x-1$

$f(2)=2^2+3(2)-1$ $=4+6-1$ $=9$	$f(-3)=(-3)^2+3(-3)-1$ $=9-9-1$ $=-1$	$f(k)=k^2+3k-1$

$f(k^2-1)=(k^2-1)^2+3(k^2-1)-1$ $=k^4-2k^2+1+3k^2-3-1$ $=k^4+k^2-3$

65. $$f(x) = x^3 - 2$$

$f(2) = 2^3 - 2$	$f(-3) = (-3)^3 - 2$	$f(k) = k^3 - 2$	$f(k^2-1) = (k^2-1)^3 - 2$
$= 8 - 2$	$= -27 - 2$		$= (k^2-1)(k^2-1)(k^2-1) - 2$
$= 6$	$= -29$		$= k^6 - 3k^4 + 3k^2 - 1 - 2$
			$= k^6 - 3k^4 + 3k^2 - 3$

67. $$f(x) = \left|x^2 + 1\right|$$

$f(2) = \left|2^2 + 1\right|$	$f(-3) = \left|(-3)^2 + 1\right|$	$f(k) = \left|k^2 + 1\right|$	$f(k^2-1) = \left|(k^2-1)^2 + 1\right|$
$= \lvert 5 \rvert$	$= \lvert 10 \rvert$	$= k^2 + 1$	$= (k^2-1)^2 + 1$
$= 5$	$= 10$	$[k^2 + 1 \geq 0]$	$= k^4 - 2k^2 + 1 + 1$
			$= k^4 - 2k^2 + 2$
			$\left[(k^2-1)^2 + 1 \geq 0\right]$

69. $$f(x) = \frac{2}{x+4}$$

$f(2) = \dfrac{2}{2+4}$	$f(-3) = \dfrac{2}{-3+4}$	$f(k) = \dfrac{2}{k+4}$	$f(k^2-1) = \dfrac{2}{k^2-1+4}$
$= \dfrac{2}{6} = \dfrac{1}{3}$	$= \dfrac{2}{1} = 2$		$= \dfrac{2}{k^2+3}$

71. $$f(x) = \frac{1}{x^2-1}$$

$f(2) = \dfrac{1}{2^2-1}$	$f(-3) = \dfrac{1}{(-3)^2-1}$	$f(k) = \dfrac{1}{k^2-1}$	$f(k^2-1) = \dfrac{1}{(k^2-1)^2-1}$
$= \dfrac{1}{4-1}$	$= \dfrac{1}{9-1}$		$= \dfrac{1}{k^4-2k^2+1-1}$
$= \dfrac{1}{3}$	$= \dfrac{1}{8}$		$= \dfrac{1}{k^4-2k^2}$

73. $$f(x) = \sqrt{x^2+1}$$

$f(2) = \sqrt{2^2+1}$	$f(-3) = \sqrt{(-3)^2+1}$	$f(k) = \sqrt{k^2+1}$	$f(k^2-1) = \sqrt{(k^2-1)^2+1}$
$= \sqrt{4+1}$	$= \sqrt{9+1}$		$= \sqrt{k^4-2k^2+1+1}$
$= \sqrt{5}$	$= \sqrt{10}$		$= \sqrt{k^4-2k^2+2}$

75. $$f(x) = \sqrt[3]{x} - 1$$

$f(2) = \sqrt[3]{2} - 1$	$f(-3) = \sqrt[3]{-3} - 1$	$f(k) = \sqrt[3]{k} - 1$	$f(k^2-1) = \sqrt[3]{k^2-1} - 1$
	$= -\sqrt[3]{3} - 1$		

EXERCISES 2.1

77. $$\frac{f(x+h)-f(x)}{h}=\frac{[3(x+h)+1]-[3x+1]}{h}=\frac{[3x+3h+1]-[3x+1]}{h}$$
$$=\frac{3x+3h+1-3x-1}{h}=\frac{3h}{h}=3$$

79. $$\frac{f(x+h)-f(x)}{h}=\frac{[-7(x+h)+8]-[-7x+8]}{h}=\frac{[-7x-7h+8]-[-7x+8]}{h}$$
$$=\frac{-7x-7h+8+7x-8}{h}=\frac{-7h}{h}=-7$$

81. $$\frac{f(x+h)-f(x)}{h}=\frac{\left[(x+h)^2+1\right]-[x^2+1]}{h}=\frac{[x^2+2xh+h^2+1]-[x^2+1]}{h}$$
$$=\frac{x^2+2xh+h^2+1-x^2-1}{h}$$
$$=\frac{2xh+h^2}{h}=\frac{h(2x+h)}{h}=2x+h$$

83. $$\frac{f(x+h)-f(x)}{h}=\frac{\left[4(x+h)^2-6\right]-[4x^2-6]}{h}=\frac{[4x^2+8xh+4h^2-6]-[4x^2-6]}{h}$$
$$=\frac{4x^2+8xh+4h^2-6-4x^2+6}{h}$$
$$=\frac{8xh+4h^2}{h}=\frac{h(8x+4h)}{h}=8x+4h$$

85. $$\frac{f(x+h)-f(x)}{h}=\frac{\left[(x+h)^2+3(x+h)-7\right]-[x^2+3x-7]}{h}$$
$$=\frac{[x^2+2xh+h^2+3x+3h-7]-[x^2+3x-7]}{h}$$
$$=\frac{x^2+2xh+h^2+3x+3h-7-x^2-3x+7}{h}$$
$$=\frac{2xh+h^2+3h}{h}=\frac{h(2x+h+3)}{h}=2x+h+3$$

87. $$\frac{f(x+h)-f(x)}{h}=\frac{\left[2(x+h)^2-4(x+h)+2\right]-[2x^2-4x+2]}{h}$$
$$=\frac{[2x^2+4xh+2h^2-4x-4h+2]-[2x^2-4x+2]}{h}$$
$$=\frac{2x^2+4xh+2h^2-4x-4h+2-2x^2+4x-2}{h}$$
$$=\frac{4xh+2h^2-4h}{h}=\frac{h(4x+2h-4)}{h}=4x+2h-4$$

89. $$\frac{f(x+h)-f(x)}{h} = \frac{\left[-(x+h)^2+(x+h)-3\right]-\left[-x^2+x-3\right]}{h}$$
$$= \frac{\left[-x^2-2xh-h^2+x+h-3\right]-\left[-x^2+x-3\right]}{h}$$
$$= \frac{-x^2-2xh-h^2+x+h-3+x^2-x+3}{h}$$
$$= \frac{-2xh-h^2+h}{h} = \frac{h(-2x-h+1)}{h} = -2x-h+1$$

91. $$\frac{f(x+h)-f(x)}{h} = \frac{(x+h)^3-x^3}{h} = \frac{\left[x^3+3x^2h+3xh^2+h^3\right]-\left[x^3\right]}{h}$$
$$= \frac{3x^2h+3xh^2+h^3}{h}$$
$$= \frac{h(3x^2+3xh+h^2)}{h} = 3x^2+3xh+h^2$$

93. $$\frac{f(x+h)-f(x)}{h} = \frac{\frac{1}{x+h}-\frac{1}{x}}{h} = \frac{\left(\frac{1}{x+h}-\frac{1}{x}\right)\cdot x(x+h)}{h\cdot x(x+h)}$$
$$= \frac{x-(x+h)}{xh(x+h)} = \frac{-h}{xh(x+h)} = -\frac{1}{x(x+h)}$$

95. $f(x) = -0.6x+132$
$f(25) = -0.6(25)+132 = 117$

97. $s(t) = -16t^2+10t+300$
$s(3) = -16(3)^2+10(3)+300 = 186$ ft

99. $g(d) = 300d$
$g(365) = 300(365) = 109{,}500$ gallons

101. Let x = the length. Then $x+5$ = the width.
$A(x) = x(x+5) = x^2+5x$

103. **a.** $C(x) = 8x+75$
b. $C(85) = 8(85)+75 = \$755$

105. **a.** $C(x) = 0.07x+9.99$
b. $C(20) = 0.07(20)+9.99 = \$11.39$

107-109. Answers may vary.

111. They are different. 10 is in the domain of $f(x)$, but not in the domain of $g(x)$.

113. Answers may vary.

115. $\frac{f(x+h)-f(x)}{h} = \frac{5-5}{h} = \frac{0}{h} = 0.$

117. d

119. e

121. c

123. a

Exercises 2.2 (page 216)

1. quadrants

3. to the right

5. first

7. linear

9. x-intercept

11. horizontal

13. midpoint

15. $A(2,3)$

17. $C(-2,-3)$

19. $E(0,0)$

21. $G(-5,-5)$

EXERCISES 2.2

23, 25, 27, 29.

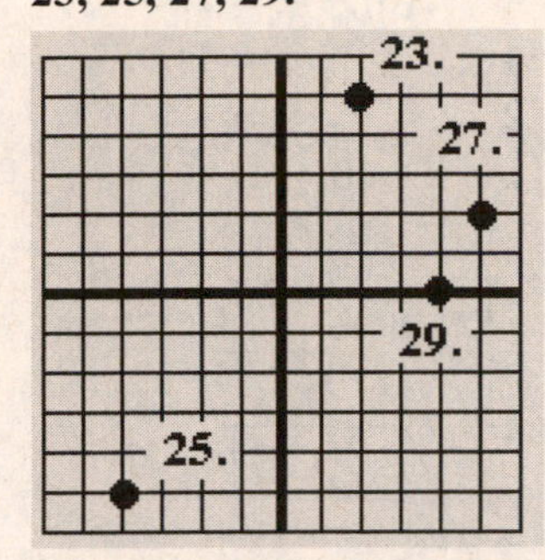

23. QI

25. QIII

27. QI

29. $+\,x$-axis

31. $y - 2x = 7$

$y = 2x + 7$

x	y
0	7
-2	3

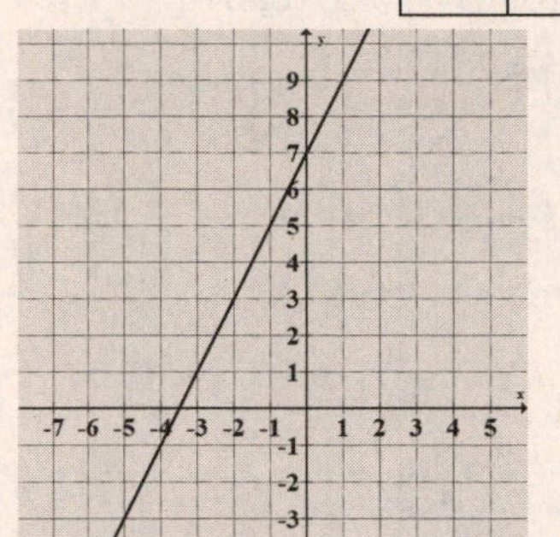

33. $y + 5x = 5$

$y = -5x + 5$

x	y
0	5
1	0

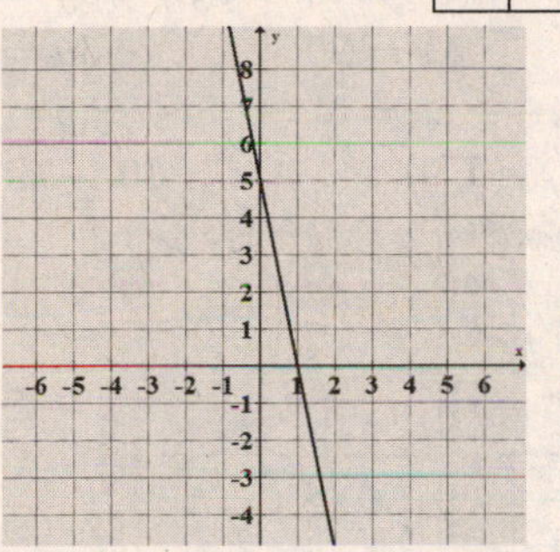

35. $6x - 3y = 10$

$-3y = -6x + 10$

$y = 2x - \frac{10}{3}$

x	y
0	$-\frac{10}{3}$
2	$\frac{2}{3}$

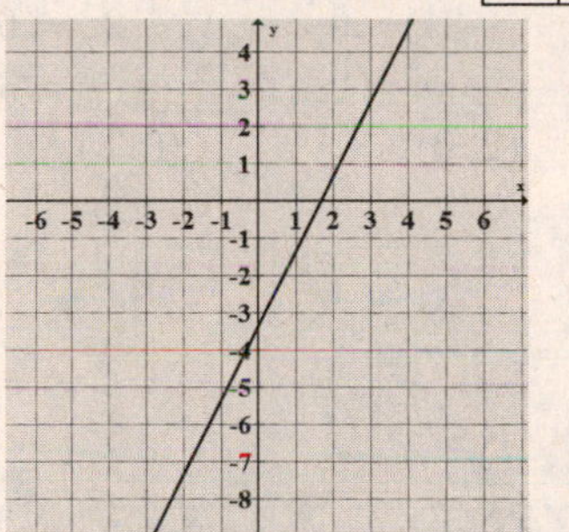

37. $3x = 6y - 1$

$-6y = -3x - 1$

$y = \frac{1}{2}x + \frac{1}{6}$

x	y
0	$\frac{1}{6}$
-2	$-\frac{5}{6}$

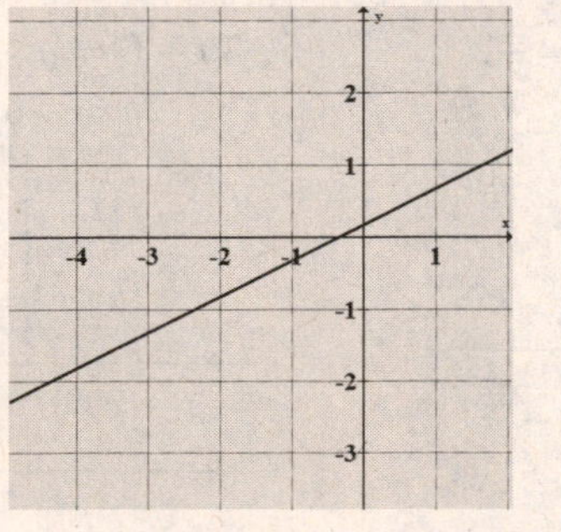

39. $2(x + y + 1) = x + 2$

$2x + 2y + 2 = x + 2$

$2y = -x$

$y = -\frac{1}{2}x$

x	y
0	0
-2	1

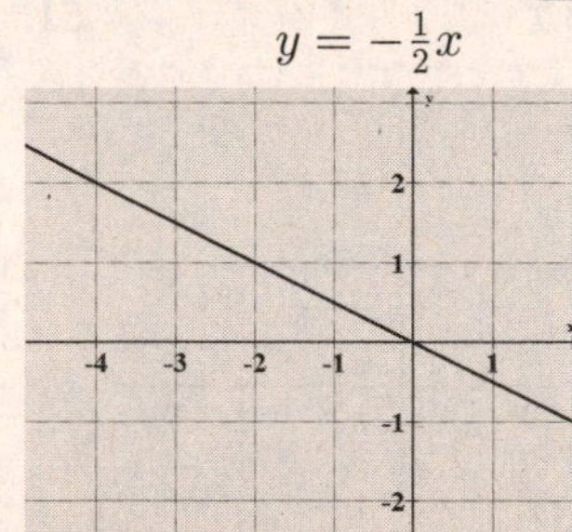

EXERCISES 2.2

41.
$$x + y = 5 \qquad x + y = 5$$
$$x + 0 = 5 \qquad 0 + y = 5$$
$$x = 5 \qquad y = 5$$
$$(5, 0) \qquad (0, 5)$$

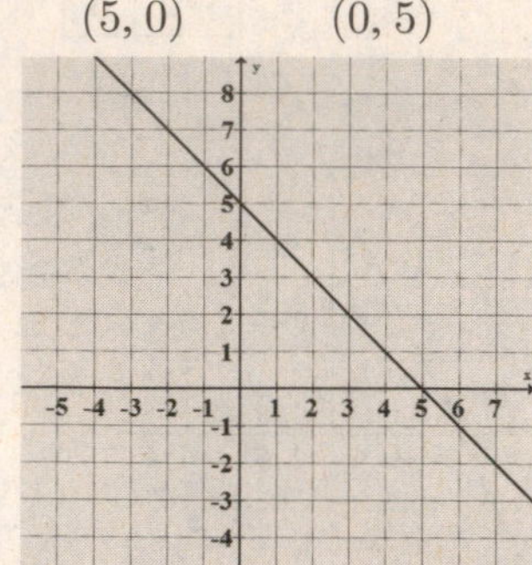

43.
$$2x - y = 4 \qquad 2x - y = 4$$
$$2x - 0 = 4 \qquad 2(0) - y = 4$$
$$2x = 4 \qquad -y = 4$$
$$x = 2 \qquad y = -4$$
$$(2, 0) \qquad (0, -4)$$

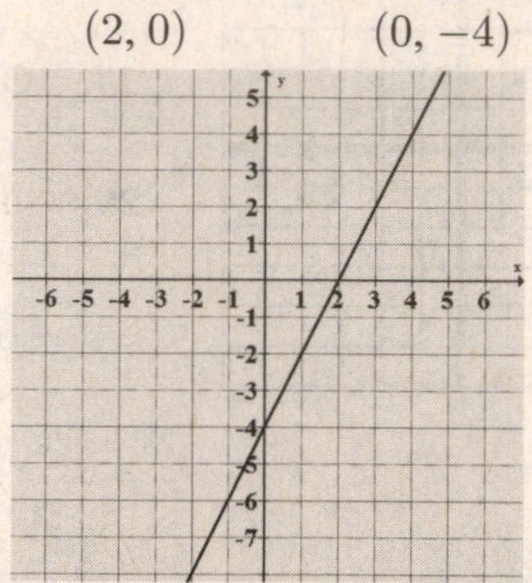

45.
$$3x + 2y = 6 \qquad 3x + 2y = 6$$
$$3x + 2(0) = 6 \qquad 3(0) + 2y = 6$$
$$3x = 6 \qquad 2y = 6$$
$$x = 2 \qquad y = 3$$
$$(2, 0) \qquad (0, 3)$$

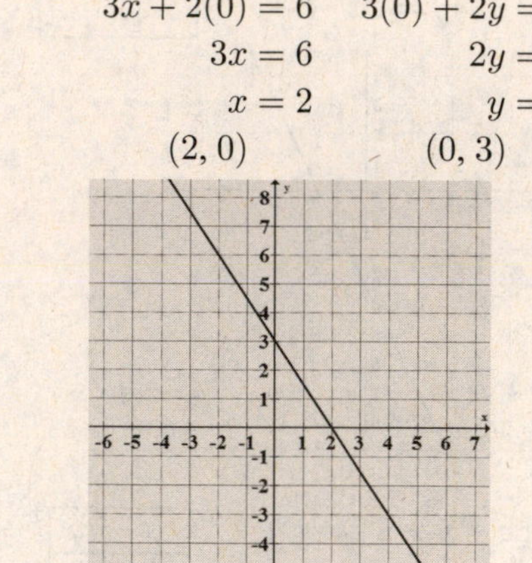

47.
$$4x - 5y = 20 \qquad 4x - 5y = 20$$
$$4x - 5(0) = 20 \qquad 4(0) - 5y = 20$$
$$4x = 20 \qquad -5y = 20$$
$$x = 5 \qquad y = -4$$
$$(5, 0) \qquad (0, -4)$$

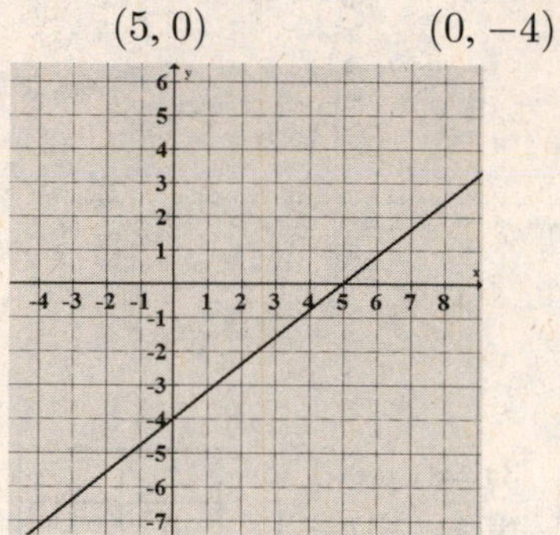

49. $y = 3$

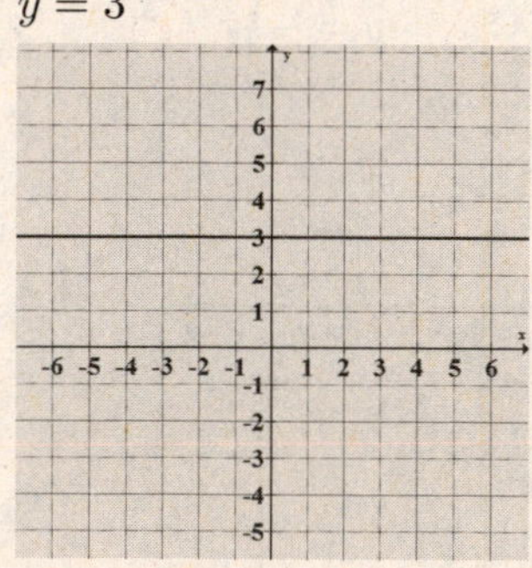

51. $3x + 5 = -1$

$3x = -6 \Rightarrow x = -2$

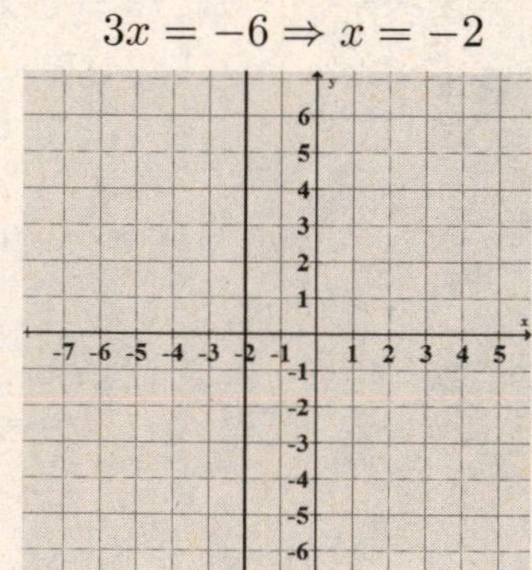

53. $3(y + 2) = y$

$3y + 6 = y$

$2y = -6 \Rightarrow y = -3$

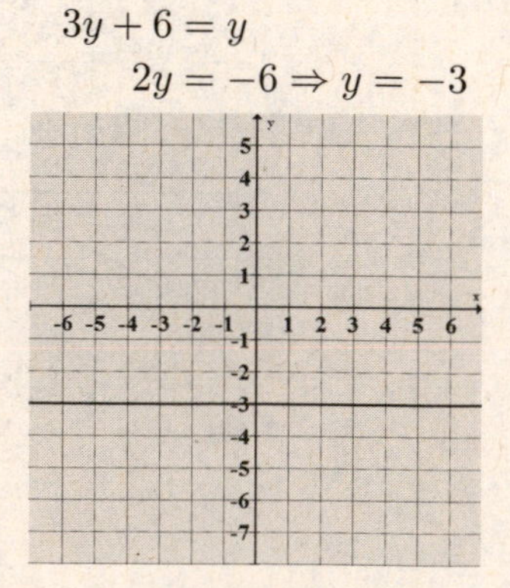

EXERCISES 2.2

55. $3(y + 2x) = 6x + y$
$3y + 6x = 6x + y$
$2y = 0 \Rightarrow y = 0$

57. $y = 3.7x - 4.5$

x-int: $x = 1.22$

59. $1.5x - 3y = 7$
$-3y = -1.5x + 7$
$y = 0.5x - \frac{7}{3}$

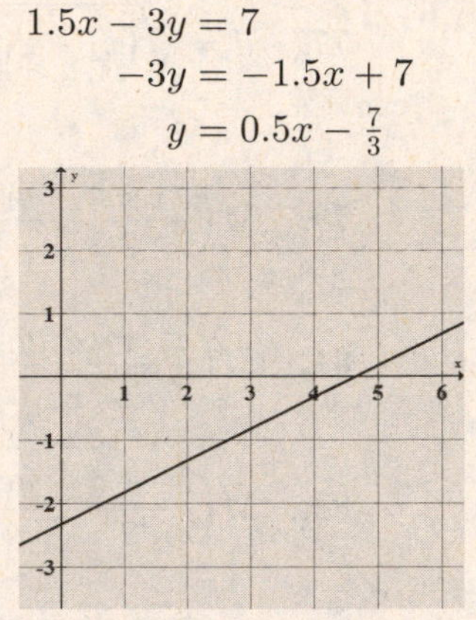

x-int: $x = 4.67$

61.
$$\begin{aligned} d &= \sqrt{(x_2 - x_1)^2 + (y_2 - y_1)^2} \\ &= \sqrt{(4 - 0)^2 + (-3 - 0)^2} \\ &= \sqrt{4^2 + (-3)^2} \\ &= \sqrt{16 + 9} = \sqrt{25} = 5 \end{aligned}$$

63.
$$\begin{aligned} d &= \sqrt{(x_2 - x_1)^2 + (y_2 - y_1)^2} \\ &= \sqrt{(-3 - 0)^2 + (2 - 0)^2} \\ &= \sqrt{(-3)^2 + (2)^2} \\ &= \sqrt{9 + 4} = \sqrt{13} \end{aligned}$$

65.
$$\begin{aligned} d &= \sqrt{(x_2 - x_1)^2 + (y_2 - y_1)^2} \\ &= \sqrt{(1 - 0)^2 + (1 - 0)^2} \\ &= \sqrt{(1)^2 + (1)^2} \\ &= \sqrt{1 + 1} = \sqrt{2} \end{aligned}$$

67.
$$\begin{aligned} d &= \sqrt{(x_2 - x_1)^2 + (y_2 - y_1)^2} \\ &= \sqrt{\left(\sqrt{3} - 0\right)^2 + (1 - 0)^2} \\ &= \sqrt{\left(\sqrt{3}\right)^2 + (1)^2} \\ &= \sqrt{3 + 1} = \sqrt{4} = 2 \end{aligned}$$

69.
$$\begin{aligned} d &= \sqrt{(x_2 - x_1)^2 + (y_2 - y_1)^2} \\ &= \sqrt{(3 - 6)^2 + (7 - 3)^2} \\ &= \sqrt{(-3)^2 + (4)^2} \\ &= \sqrt{9 + 16} = \sqrt{25} = 5 \end{aligned}$$

71.
$$\begin{aligned} d &= \sqrt{(x_2 - x_1)^2 + (y_2 - y_1)^2} \\ &= \sqrt{[4 - (-1)]^2 + [-6 - 6]^2} \\ &= \sqrt{(5)^2 + (-12)^2} \\ &= \sqrt{25 + 144} = \sqrt{169} = 13 \end{aligned}$$

73.
$$\begin{aligned} d &= \sqrt{(x_2 - x_1)^2 + (y_2 - y_1)^2} \\ &= \sqrt{[-2 - (-6)]^2 + [-15 - (-21)]^2} \\ &= \sqrt{(4)^2 + (6)^2} \\ &= \sqrt{16 + 36} = \sqrt{52} = 2\sqrt{13} \end{aligned}$$

75.
$$\begin{aligned} d &= \sqrt{(x_2 - x_1)^2 + (y_2 - y_1)^2} \\ &= \sqrt{[3 - (-5)]^2 + [-3 - 5]^2} \\ &= \sqrt{(8)^2 + (-8)^2} \\ &= \sqrt{64 + 64} = \sqrt{128} = 8\sqrt{2} \end{aligned}$$

EXERCISES 2.2

77. $d = \sqrt{(x_2 - x_1)^2 + (y_2 - y_1)^2} = \sqrt{[\pi - \pi]^2 + [-2 - 5]^2} = \sqrt{(0)^2 + (-7)^2} = \sqrt{0 + 49} = \sqrt{49}$
$= 7$

79. $M\left(\frac{x_1 + x_2}{2}, \frac{y_1 + y_2}{2}\right) = M\left(\frac{2+6}{2}, \frac{4+8}{2}\right) = M\left(\frac{8}{2}, \frac{12}{2}\right) = M(4, 6)$

81. $M\left(\frac{x_1 + x_2}{2}, \frac{y_1 + y_2}{2}\right) = M\left(\frac{2+(-2)}{2}, \frac{-5+7}{2}\right) = M\left(\frac{0}{2}, \frac{2}{2}\right) = M(0, 1)$

83. $M\left(\frac{x_1 + x_2}{2}, \frac{y_1 + y_2}{2}\right) = M\left(\frac{-8+6}{2}, \frac{5+(-4)}{2}\right) = M\left(\frac{-2}{2}, \frac{1}{2}\right) = M\left(-1, \frac{1}{2}\right)$

85. $M\left(\frac{x_1 + x_2}{2}, \frac{y_1 + y_2}{2}\right) = M\left(\frac{0+\sqrt{5}}{2}, \frac{0+\sqrt{5}}{2}\right) = M\left(\frac{\sqrt{5}}{2}, \frac{\sqrt{5}}{2}\right)$

87. Let Q have coordinates (x, y):
$M\left(\frac{x_1+x_2}{2}, \frac{y_1+y_2}{2}\right) = (3, 5)$
$\frac{x_1 + x_2}{2} = 3 \qquad \frac{y_1 + y_2}{2} = 5$
$\frac{1 + x}{2} = 3 \qquad \frac{4 + y}{2} = 5$
$1 + x = 6 \qquad 4 + y = 10$
$x = 5 \qquad y = 6$
$Q(5, 6)$

89. Let Q have coordinates (x, y):
$M\left(\frac{x_1+x_2}{2}, \frac{y_1+y_2}{2}\right) = (5, 5)$
$\frac{x_1 + x_2}{2} = 5 \qquad \frac{y_1 + y_2}{2} = 5$
$\frac{5 + x}{2} = 5 \qquad \frac{-5 + y}{2} = 5$
$5 + x = 10 \qquad -5 + y = 10$
$x = 5 \qquad y = 15$
$Q(5, 15)$

91. Let the points be identified as $A(13, -2)$, $B(9, -8)$ and $C(5, -2)$.
$AB = \sqrt{(x_2 - x_1)^2 + (y_2 - y_1)^2} = \sqrt{(13 - 9)^2 + (-2 - (-8))^2} = \sqrt{16 + 36} = \sqrt{52} = 2\sqrt{13}$
$BC = \sqrt{(x_2 - x_1)^2 + (y_2 - y_1)^2} = \sqrt{(9 - 5)^2 + (-8 - (-2))^2} = \sqrt{16 + 36} = \sqrt{52} = 2\sqrt{13}$
Since AB and BC have the same length, the triangle is isosceles.

93. $M = \left(\frac{2+6}{2}, \frac{4+10}{2}\right) = \left(\frac{8}{2}, \frac{14}{2}\right) = (4, 7)$; $N = \left(\frac{4+6}{2}, \frac{6+10}{2}\right) = \left(\frac{10}{2}, \frac{16}{2}\right) = (5, 8)$
$MN = \sqrt{(x_2 - x_1)^2 + (y_2 - y_1)^2} = \sqrt{(4 - 5)^2 + (7 - 8)^2} = \sqrt{1 + 1} = \sqrt{2}$

95. $M = \left(\frac{0+a}{2}, \frac{b+0}{2}\right) = \left(\frac{a}{2}, \frac{b}{2}\right)$; $L = \left(\frac{a}{2}, 0\right)$; $N = \left(0, \frac{b}{2}\right)$
Area of $AOB = \frac{1}{2} \cdot \text{base} \cdot \text{height} = \frac{1}{2}(OA)(OB) = \frac{1}{2}(a)(b) = \frac{1}{2}ab$
Area of $OLMN = \text{length} \cdot \text{width} = (OL)(ON) = \frac{a}{2} \cdot \frac{b}{2} = \frac{1}{4}ab = \frac{1}{2}(\text{Area of } AOB)$

97. $y = 17500x + 325000$
$y = 17500(5) + 325000$
$y = 87500 + 325000$
$y = 412500$
The value will be \$412,500.

99. $p = -\frac{1}{10}q + 170$
$150 = -\frac{1}{10}q + 170$
$\frac{1}{10}q = 20$
$q = 200$
200 scanners will be sold.

101. $V = \frac{nv}{N}$
$60 = \frac{12v}{20}$
$1200 = 12v$
$100 = v$
The smaller gear is spinning at 100 rpm.

103. $d = \sqrt{(x_2 - x_1)^2 + (y_2 - y_1)^2}$
$= \sqrt{[0 - 30]^2 + [10 - 25]^2}$
$= \sqrt{(-30)^2 + (-15)^2}$
$= \sqrt{900 + 225} = \sqrt{1125} = 15\sqrt{5}$ yd

105. $M\left(\frac{x_1 + x_2}{2}, \frac{y_1 + y_2}{2}\right) = M\left(\frac{30 + 0}{2}, \frac{25 + 10}{2}\right) = M\left(\frac{30}{2}, \frac{35}{2}\right) = M(15, 17.5)$

107.

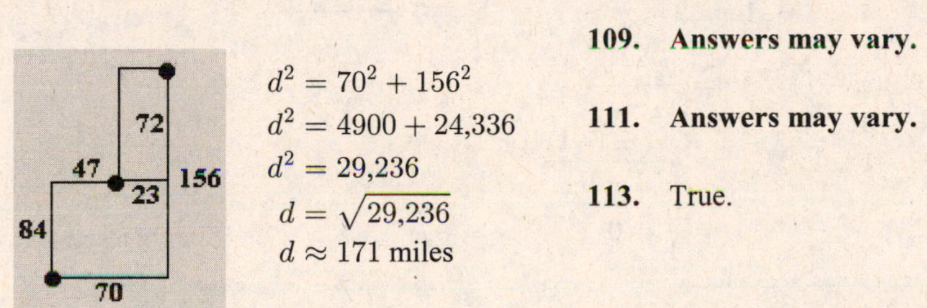

$d^2 = 70^2 + 156^2$
$d^2 = 4900 + 24{,}336$
$d^2 = 29{,}236$
$d = \sqrt{29{,}236}$
$d \approx 171$ miles

109. **Answers may vary.**

111. **Answers may vary.**

113. True.

115. False. Vertical lines have equations that are not functions.

117. False. The vertical line $x = 0$ has infinitely many y-intercepts.

119. True.

Exercises 2.3 (page 229)

1. divided

3. run

5. the change in

7. vertical

9. perpendicular

11. $m = \frac{y_2 - y_1}{x_2 - x_1} = \frac{2 - (-1)}{2 - (-1)} = \frac{3}{3} = 1$

13. $m = \frac{y_2 - y_1}{x_2 - x_1} = \frac{-2 - 3}{6 - (-6)} = \frac{-5}{12} = -\frac{5}{12}$

15. $m = \frac{y_2 - y_1}{x_2 - x_1} = \frac{5 - (-2)}{-1 - 3} = \frac{7}{-4} = -\frac{7}{4}$

17. $m = \frac{y_2 - y_1}{x_2 - x_1} = \frac{1 - (-7)}{4 - 8} = \frac{8}{-4} = -2$

EXERCISES 2.3

19. $m = \dfrac{y_2 - y_1}{x_2 - x_1} = \dfrac{-14 - (-14)}{2 - (-7)} = \dfrac{0}{9} = 0$

21. $m = \dfrac{y_2 - y_1}{x_2 - x_1} = \dfrac{-2 - 3}{-5 - (-5)} = \dfrac{-5}{0} \Rightarrow$ und.

23. $m = \dfrac{y_2 - y_1}{x_2 - x_1} = \dfrac{\frac{7}{3} - \frac{2}{3}}{\frac{5}{2} - \frac{3}{2}} = \dfrac{\frac{5}{3}}{\frac{2}{2}} = \dfrac{\frac{5}{3}}{1} = \dfrac{5}{3}$

25. $m = \dfrac{y_2 - y_1}{x_2 - x_1} = \dfrac{a - c}{(b + c) - (a + b)}$
$= \dfrac{a - c}{c - a} = -1$

27. $y = 3x + 2$

x	y
0	2
1	5

$m = \dfrac{y_2 - y_1}{x_2 - x_1} = \dfrac{5 - 2}{1 - 0}$
$= \dfrac{3}{1} = 3$

29. $y = 4x - 6$

x	y
0	-6
1	-2

$m = \dfrac{y_2 - y_1}{x_2 - x_1} = \dfrac{-2 - (-6)}{1 - 0}$
$= \dfrac{4}{1} = 4$

31. $5x - 10y = 3$

x	y
0	$-\frac{3}{10}$
1	$\frac{1}{5}$

$m = \dfrac{y_2 - y_1}{x_2 - x_1}$
$= \dfrac{\frac{1}{5} - \left(-\frac{3}{10}\right)}{1 - 0}$
$= \dfrac{\frac{5}{10}}{1} = \dfrac{1}{2}$

33. $3(y + 2) = 2x - 3$
$3y - 2x = -9$

x	y
0	-3
3	-1

$m = \dfrac{y_2 - y_1}{x_2 - x_1}$
$= \dfrac{-1 - (-3)}{3 - 0}$
$= \dfrac{2}{3}$

35. $3(y + x) = 3(x - 1)$
$3y = -3$
$y = -1$

x	y
0	-1
1	-1

$m = \dfrac{y_2 - y_1}{x_2 - x_1}$
$= \dfrac{-1 - (-1)}{1 - 0}$
$= \dfrac{0}{1} = 0$

37. horizontal $\Rightarrow m = 0$

39. $f(x) = \dfrac{1}{4} \Rightarrow y = \dfrac{1}{4}$
horizontal $\Rightarrow m = 0$

41. $x = -\dfrac{1}{2}$
vertical $\Rightarrow m$ is undefined.

43. The slope is negative.

45. The slope is positive.

47. The slope is undefined.

49. $m_1 m_2 = 3\left(-\frac{1}{3}\right) = -1$
perpendicular

51. $m_1 = \sqrt{8} = 2\sqrt{2} = m_2$
parallel

53. $m_1 m_2 = -\sqrt{2}\left(\frac{\sqrt{2}}{2}\right) = -1$
perpendicular

55. $m_1 m_2 = -0.125(8) = -1$
perpendicular

57. $m_1 m_2 = ab^{-1}\left(-a^{-1}b\right) = -a^0 b^0 = -1$
perpendicular

EXERCISES 2.3

For Exercises 59-63 use the slope of line through R and S calculated below:

$$m_{RS} = \frac{y_2 - y_1}{x_2 - x_1} = \frac{7-5}{2-(-3)} = \frac{2}{5}$$

59. $m_{PQ} = \frac{y_2 - y_1}{x_2 - x_1} = \frac{6-4}{7-2} = \frac{2}{5} = m_{RS} \Rightarrow$ parallel

61. $m_{PQ} = \frac{y_2 - y_1}{x_2 - x_1} = \frac{1-6}{-2-(-4)} = \frac{-5}{2} = -\frac{5}{2} \Rightarrow$ perpendicular

63. $m_{PQ} = \frac{y_2 - y_1}{x_2 - x_1} = \frac{6a-a}{3a-a} = \frac{5a}{2a} = \frac{5}{2} \Rightarrow$ neither

65. $m_{PQ} = \frac{y_2 - y_1}{x_2 - x_1} = \frac{9-7}{2-(-3)} = \frac{2}{5}$; $m_{RS} = \frac{y_2 - y_1}{x_2 - x_1} = \frac{-6-(-4)}{x-10} = \frac{-2}{x-10}$

$\frac{2}{5} \cdot \frac{-1}{-1} = \frac{-2}{-5}$; $x - 10 = -5 \Rightarrow \boxed{x = 5}$

67. $m_{PQ} = \frac{y_2 - y_1}{x_2 - x_1} = \frac{0-(-7)}{1-2} = \frac{7}{-1} = -7$; $m_{RS} = \frac{y_2 - y_1}{x_2 - x_1} = \frac{y-5}{-2-(-9)} = \frac{y-5}{7}$

$-7 = \frac{-7}{1}$; Perp. slope $= \frac{1}{7}$; $y - 5 = 1 \Rightarrow \boxed{y = 6}$

69. $m_{PQ} = \frac{y_2 - y_1}{x_2 - x_1} = \frac{9-8}{-6-(-2)} = \frac{1}{-4} = -\frac{1}{4}$

$m_{PR} = \frac{y_2 - y_1}{x_2 - x_1} = \frac{5-8}{2-(-2)} = \frac{-3}{4} = -\frac{3}{4} \Rightarrow$ not on same line

71. $m_{PQ} = \frac{y_2 - y_1}{x_2 - x_1} = \frac{0-a}{0-(-a)} = \frac{-a}{a} = -1$

$m_{PR} = \frac{y_2 - y_1}{x_2 - x_1} = \frac{-a-a}{a-(-a)} = \frac{-2a}{2a} = -1 \Rightarrow$ on same line

73. $m_{PQ} = \frac{y_2 - y_1}{x_2 - x_1} = \frac{-5-4}{2-5} = \frac{-9}{-3} = 3$

$m_{PR} = \frac{y_2 - y_1}{x_2 - x_1} = \frac{-3-4}{8-5} = \frac{-7}{3} = -\frac{7}{3}$

$m_{QR} = \frac{y_2 - y_1}{x_2 - x_1} = \frac{-3-(-5)}{8-2} = \frac{2}{6} = \frac{1}{3} \Rightarrow$ None are perpendicular.

75. $m_{PQ} = \frac{y_2 - y_1}{x_2 - x_1} = \frac{9-3}{1-1} = \frac{6}{0} \Rightarrow$ undefined $\Rightarrow$ vertical

$m_{PR} = \frac{y_2 - y_1}{x_2 - x_1} = \frac{3-3}{7-1} = \frac{0}{6} = 0 \Rightarrow$ horizontal

$m_{QR} = \frac{y_2 - y_1}{x_2 - x_1} = \frac{3-9}{7-1} = \frac{-6}{6} = -1 \Rightarrow PQ$ and PR are perpendicular.

77. $m_{PQ} = \dfrac{y_2 - y_1}{x_2 - x_1} = \dfrac{b-0}{a-0} = \dfrac{b}{a}$

$m_{PR} = \dfrac{y_2 - y_1}{x_2 - x_1} = \dfrac{a-0}{-b-0} = \dfrac{a}{-b} = -\dfrac{a}{b}$

$m_{QR} = \dfrac{y_2 - y_1}{x_2 - x_1} = \dfrac{a-b}{-b-a} = \dfrac{a-b}{-b-a} \Rightarrow PQ$ and PR are perpendicular.

79. $m_{AB} = \dfrac{y_2 - y_1}{x_2 - x_1} = \dfrac{4-(-1)}{-3-(-1)} = \dfrac{5}{-2} = -\dfrac{5}{2}$

$m_{AC} = \dfrac{y_2 - y_1}{x_2 - x_1} = \dfrac{1-(-1)}{4-(-1)} = \dfrac{2}{5} \Rightarrow AB$ and AC are perpendicular. $\Rightarrow$ right triangle

81. $m_{AB} = \dfrac{y_2 - y_1}{x_2 - x_1} = \dfrac{0-(-1)}{3-1} = \dfrac{1}{2}$; $d(A, B) = \sqrt{(1-3)^2 + (-1-0)^2} = \sqrt{5}$

$m_{BC} = \dfrac{y_2 - y_1}{x_2 - x_1} = \dfrac{2-0}{2-3} = \dfrac{2}{-1} = -2$; $d(B, C) = \sqrt{(3-2)^2 + (0-2)^2} = \sqrt{5}$

$m_{CD} = \dfrac{y_2 - y_1}{x_2 - x_1} = \dfrac{1-2}{0-2} = \dfrac{-1}{-2} = \dfrac{1}{2}$; $d(C, D) = \sqrt{(2-0)^2 + (2-1)^2} = \sqrt{5}$

$m_{DA} = \dfrac{y_2 - y_1}{x_2 - x_1} = \dfrac{1-(-1)}{0-1} = \dfrac{2}{-1} = -2$; $d(D, A) = \sqrt{(1-0)^2 + (-1-1)^2} = \sqrt{5}$

Adjacent sides are perpendicular and congruent, so the figure is a square.

83. $m_{AB} = \dfrac{y_2 - y_1}{x_2 - x_1} = \dfrac{3-(-2)}{3-(-2)} = \dfrac{5}{5} = 1$; $m_{BC} = \dfrac{y_2 - y_1}{x_2 - x_1} = \dfrac{6-3}{2-3} = \dfrac{3}{-1} = -3$

$m_{CD} = \dfrac{y_2 - y_1}{x_2 - x_1} = \dfrac{1-6}{-3-2} = \dfrac{-5}{-5} = 1$; $m_{DA} = \dfrac{y_2 - y_1}{x_2 - x_1} = \dfrac{1-(-2)}{-3-(-2)} = \dfrac{3}{-1} = -3$

Opposite sides are parallel, so the figure is a parallelogram.

85. $M\left(\dfrac{5+7}{2}, \dfrac{9+5}{2}\right) = M\left(\dfrac{12}{2}, \dfrac{14}{2}\right) = M(6, 7)$; $N\left(\dfrac{1+7}{2}, \dfrac{3+5}{2}\right) = N\left(\dfrac{8}{2}, \dfrac{8}{2}\right) = N(4, 4)$

$m_{MN} = \dfrac{y_2 - y_1}{x_2 - x_1} = \dfrac{4-7}{4-6} = \dfrac{-3}{-2} = \dfrac{3}{2}$; $m_{AC} = \dfrac{y_2 - y_1}{x_2 - x_1} = \dfrac{9-3}{5-1} = \dfrac{6}{4} = \dfrac{3}{2} \Rightarrow MN \parallel AC$

87. $m = \dfrac{y_2 - y_1}{x_2 - x_1} = \dfrac{42-14}{5-1} = \dfrac{28}{4} = 7$

The rate of growth was 7 students per year.

89. $m = \dfrac{y_2 - y_1}{x_2 - x_1} = \dfrac{6700-2200}{10-3}$

$= \dfrac{4500}{7} \approx 642.86$

The cost decreased about \$642.86 per year.

91. $\dfrac{\Delta T}{\Delta t}$ = the hourly rate of change of temperature.
(Let $t = x$ and $T = y$.)

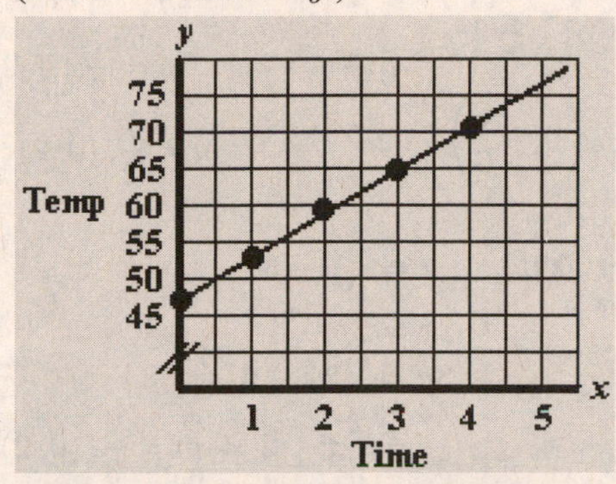

93. $D = 590t$; The slope is the speed of the plane.

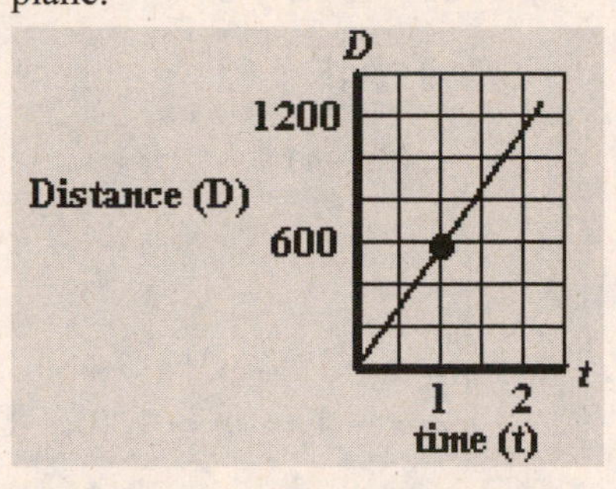

95. Answers may vary.

97. Answers may vary.

99. False. $m = \dfrac{y_2 - y_1}{x_2 - x_1}$.

101. True. $(\Delta y = 0.)$

103. False. The line will be horizontal, so the slope is 0.

105. $m = \dfrac{y_2 - y_1}{x_2 - x_1} = \dfrac{10.25 - 6.95}{2014 - 2008} = \dfrac{3.30}{6} = 0.55$. True.

Exercises 2.4 (page 243)

1. slope-intercept

3. y-intercept

5. $Ax + By = C$

7. $y = mx + b$
$y = 3x - 2$

9. $y = mx + b$
$y = 5x - \dfrac{1}{5}$

11. $y = mx + b$
$y = ax + \dfrac{1}{a}$

13. $y = mx + b$
$y = ax + a$

15.
$$\begin{aligned} y &= mx + b \\ 0 &= \tfrac{3}{2}(0) + b \\ 0 &= b \end{aligned} \qquad \begin{aligned} y &= mx + b \\ y &= \tfrac{3}{2}x + 0 \\ 2y &= 3x \\ -3x + 2y &= 0 \\ 3x - 2y &= 0 \end{aligned}$$

17.
$$\begin{aligned} y &= mx + b \\ 5 &= -3(-3) + b \\ 5 &= 9 + b \\ -4 &= b \end{aligned} \qquad \begin{aligned} y &= mx + b \\ y &= -3x - 4 \\ 3x + y &= -4 \end{aligned}$$

19.
$$\begin{aligned} y &= mx + b \\ \sqrt{2} &= \sqrt{2}(0) + b \\ \sqrt{2} &= b \end{aligned} \qquad \begin{aligned} y &= mx + b \\ y &= \sqrt{2}x + \sqrt{2} \\ -\sqrt{2}x + y &= \sqrt{2} \\ \sqrt{2}x - y &= -\sqrt{2} \end{aligned}$$

EXERCISES 2.4

21.
$$3x - 2y = 8$$
$$-2y = -3x + 8$$
$$y = \frac{3}{2}x - 4$$
$$m = \frac{3}{2}, (0, -4)$$

23.
$$-2(x + 3y) = 5$$
$$-2x - 6y = 5$$
$$-6y = 2x + 5$$
$$y = -\frac{1}{3}x - \frac{5}{6}$$
$$m = -\frac{1}{3}, \left(0, -\frac{5}{6}\right)$$

25.
$$x = \frac{2y - 4}{7}$$
$$7x = 2y - 4$$
$$-2y = -7x - 4$$
$$y = \frac{7}{2}x + 2$$
$$m = \frac{7}{2}, (0, 2)$$

27.
$$x - y = 1$$
$$y = x - 1 \Rightarrow m = 1, (0, -1)$$

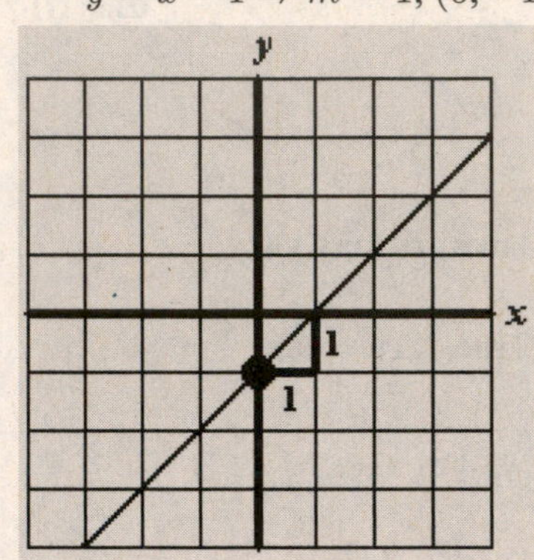

29.
$$x = \frac{3}{2}y - 3$$
$$2x = 3y - 6$$
$$-3y = -2x - 6$$
$$y = \frac{2}{3}x + 2 \Rightarrow m = \frac{2}{3}, (0, 2)$$

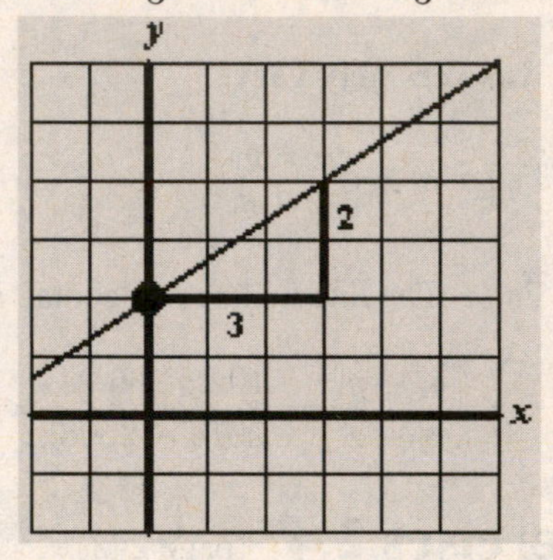

31.
$$3(y - 4) = -2(x - 3)$$
$$3y - 12 = -2x + 6$$
$$3y = -2x + 18$$
$$y = -\frac{2}{3}x + 6 \Rightarrow m = -\frac{2}{3}, (0, 6)$$

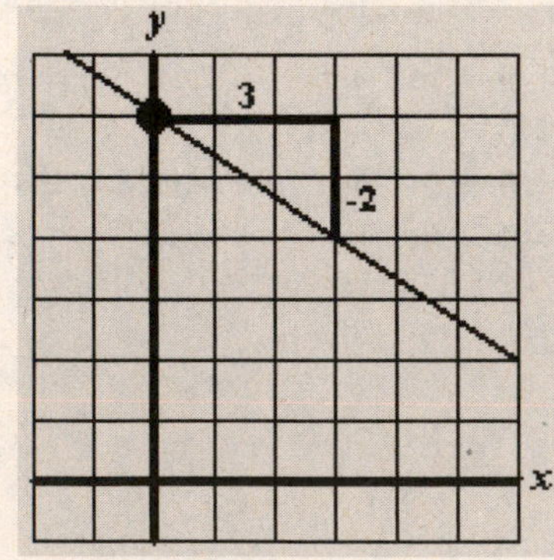

EXERCISES 2.4

33. $y = 3x + 4 \quad y = 3x - 7$
$m = 3 \quad m = 3$
The lines are parallel.

35. $x + y = 2 \quad y = x + 5$
$y = -x + 2 \quad m = 1$
$m = -1$
The lines are perpendicular.

37. $y = 3x + 7 \quad 2y = 6x - 9$
$m = 3 \quad y = 3x - \frac{9}{2}$
$m = 3$
The lines are parallel.

39. $3x + 6y = 1 \quad y = \frac{1}{2}x$
$6y = -3x + 1 \quad m = \frac{1}{2}$
$y = -\frac{1}{2}x + \frac{1}{6}$
$m = -\frac{1}{2}$
The lines are neither.

41. $y = 3 \quad x = 4$
horizontal vertical
The lines are perpendicular.

43. $x = \frac{y-2}{3} \quad 3(y-3) + x = 0$
$3x = y - 2 \quad 3y - 9 + x = 0$
$-y = -3x - 2 \quad 3y = -x + 9$
$y = 3x + 2 \quad y = -\frac{1}{3}x + 3$
$m = 3 \quad m = -\frac{1}{3}$
The lines are perpendicular.

45.
$$\begin{aligned} y - y_1 &= m(x - x_1) \\ y - 4 &= 2(x - 2) \\ y - 4 &= 2x - 4 \\ -2x + y &= 0 \\ 2x - y &= 0 \end{aligned}$$

47.
$$\begin{aligned} y - y_1 &= m(x - x_1) \\ y - \frac{1}{2} &= 2\left(x + \frac{3}{2}\right) \\ y - \frac{1}{2} &= 2x + 3 \\ 2y - 1 &= 4x + 6 \\ -4x + 2y &= 7 \\ 4x - 2y &= -7 \end{aligned}$$

49.
$$\begin{aligned} y - y_1 &= m(x - x_1) \\ y - 1 &= \frac{2}{5}(x + 1) \\ 5(y - 1) &= 2(x + 1) \\ 5y - 5 &= 2x + 2 \\ -2x + 5y &= 7 \\ 2x - 5y &= -7 \end{aligned}$$

51.
$$\begin{aligned} y - y_1 &= m(x - x_1) \\ y + 3 &= 0(x + 6) \\ y + 3 &= 0 \\ y &= -3 \end{aligned}$$

53. m is und $\Rightarrow$ vertical
$x =$ constant
$x = -6$

55.
$$\begin{aligned} y - y_1 &= m(x - x_1) \\ y - 0 &= \pi(x - \pi) \\ y &= \pi x - \pi^2 \\ -\pi x + y &= -\pi^2 \\ \pi x - y &= \pi^2 \end{aligned}$$

EXERCISES 2.4

57. From the graph, $m = \frac{2}{3}$ and the line passes through $(2, 5)$.

$$\begin{aligned} y - y_1 &= m(x - x_1) \\ y - 5 &= \frac{2}{3}(x - 2) \\ 3(y - 5) &= 3 \cdot \frac{2}{3}(x - 2) \\ 3y - 15 &= 2(x - 2) \\ 3y - 15 &= 2x - 4 \\ -2x + 3y &= 11 \\ 2x - 3y &= -11 \end{aligned}$$

59. $m = \frac{y_2 - y_1}{x_2 - x_1} = \frac{4 - 0}{4 - 0} = \frac{4}{4} = 1$

$$\begin{aligned} y - y_1 &= m(x - x_1) \\ y - 0 &= 1(x - 0) \\ y &= x \end{aligned}$$

61. $m = \frac{y_2 - y_1}{x_2 - x_1} = \frac{-3 - 4}{0 - 3} = \frac{-7}{-3} = \frac{7}{3}$

$$\begin{aligned} y - y_1 &= m(x - x_1) \\ y + 3 &= \frac{7}{3}(x - 0) \\ y &= \frac{7}{3}x - 3 \end{aligned}$$

63. From the graph, $m = -\frac{9}{5}$ and the line passes through $(-2, 4)$

$$\begin{aligned} y - y_1 &= m(x - x_1) \\ y - 4 &= -\frac{9}{5}(x + 2) \\ y - 4 &= -\frac{9}{5}x - \frac{18}{5} \\ y &= -\frac{9}{5}x - \frac{18}{5} + 4 \\ y &= -\frac{9}{5}x + \frac{2}{5} \end{aligned}$$

65. $y = 4x - 7$ $y - y_1 = m(x - x_1)$

$m = 4$ $y - 0 = 4(x - 0)$

Use $m = 4$. $\boxed{y = 4x}$

67. $4x - y = 7$ $y - y_1 = m(x - x_1)$

$-y = -4x + 7$ $y - 5 = 4(x - 2)$

$y = 4x - 7$ $y - 5 = 4x - 8$

$m = 4$ $\boxed{y = 4x - 3}$

Use $m = 4$.

69.

$$\begin{aligned} x &= \frac{5}{4}y - 2 \\ 4x &= 5y - 8 \\ -5y &= -4x - 8 \\ y &= \frac{4}{5}x + \frac{8}{5} \\ m &= \frac{4}{5} \end{aligned}$$

Use $m = \frac{4}{5}$.

$$\begin{aligned} y - y_1 &= m(x - x_1) \\ y + 2 &= \frac{4}{5}(x - 4) \\ y + 2 &= \frac{4}{5}x - \frac{16}{5} \\ &\boxed{y = \frac{4}{5}x - \frac{26}{5}} \end{aligned}$$

71. $y = 4x - 7$ $y - y_1 = m(x - x_1)$

$m = 4$ $y - 0 = -\frac{1}{4}(x - 0)$

Use $m = -\frac{1}{4}$. $\boxed{y = -\frac{1}{4}x}$

73.

$$4x - y = 7$$
$$-y = -4x + 7$$
$$y = 4x - 7$$
$$m = 4$$

Use $m = -\frac{1}{4}$.

$$y - y_1 = m(x - x_1)$$
$$y - 5 = -\frac{1}{4}(x - 2)$$
$$y - 5 = -\frac{1}{4}x + \frac{1}{2}$$
$$\boxed{y = -\frac{1}{4}x + \frac{11}{2}}$$

75.

$$x = \frac{5}{4}y - 2$$
$$4x = 5y - 8$$
$$-5y = -4x - 8$$
$$y = \frac{4}{5}x + \frac{8}{5}$$
$$m = \frac{4}{5}$$

Use $m = -\frac{5}{4}$.

$$y - y_1 = m(x - x_1)$$
$$y + 2 = -\frac{5}{4}(x - 4)$$
$$y + 2 = -\frac{5}{4}x + 5$$
$$\boxed{y = -\frac{5}{4}x + 3}$$

77. Since $y = 3$ is the equation of a horizontal line, any perpendicular line will be vertical. Find the midpoint:

$$x = \frac{2 + (-6)}{2} = -2; y = \frac{4 + 10}{2} = 7$$

The vertical line through $(-2, 7)$ is $x = -2$.

79. Since $x = 3$ is the equation of a vertical line, any parallel line will be vertical. Find the midpoint:

$$x = \frac{2 + 8}{2} = 5; y = \frac{-4 + 12}{2} = 4$$

The vertical line through $(5, 4)$ is $x = 5$.

81. Let $x =$ the number of years the truck has been owned and let $y =$ the value of the truck. Then two points on the line are given: $(0, 24300)$ and $(7, 1900)$.

$$m = \frac{24300 - 1900}{0 - 7} = \frac{22400}{-7} = -3200$$
$$y - y_1 = m(x - x_1)$$
$$y - 24300 = -3200(x - 0)$$
$$y - 24300 = -3200x$$
$$y = -3200x + 24300$$

83. Let $x =$ the number of years the building has been owned and let $y =$ the value of the building. Then two points on the line are given: $(0, 475000)$ and $(10, 950000)$.

$$m = \frac{950000 - 475000}{10 - 0} = \frac{475000}{10} = 47500$$
$$y - y_1 = m(x - x_1)$$
$$y - 475000 = 47500(x - 0)$$
$$y - 475000 = 47500x$$
$$y = 47500x + 475000$$

85. Let $x =$ the number of years the TV has been owned and let $y =$ the value of the TV. Then two points on the line are given: $(0, 1900)$ and $(3, 1190)$.

$$m = \frac{1900 - 1190}{0 - 3} = \frac{710}{-3} = -\frac{710}{3}$$
$$y - y_1 = m(x - x_1)$$
$$y - 1900 = -\frac{710}{3}(x - 0)$$
$$y - 1900 = -\frac{710}{3}x$$
$$y = -\frac{710}{3}x + 1900$$

87. Let $x =$ the number of years the copier has been owned and let $y =$ the value of the copier. Then one point on the line is given: $(0, 1050)$. Since the copier depreciates by \$120 per year, $m = -120$.

$$y - y_1 = m(x - x_1)$$
$$y - 1050 = -120(x - 0)$$
$$y - 1050 = -120x$$
$$y = -120x + 1050$$

Let $x = 8$ and find the value of y:

$$y = -120x + 1050$$
$$= -120(8) + 1050 = 90$$

The salvage value will be \$90.

89. Let $x =$ the number of years the table has been owned and let $y =$ the value of the table. Then one point on the line is given: $(2, 450)$. Since the table appreciates by \$40 per year, $m = 40$.

$$y - y_1 = m(x - x_1)$$
$$y - 450 = 40(x - 2)$$
$$y - 450 = 40x - 80$$
$$y = 40x + 370$$

Let $x = 13$ and find the value of y:

$$y = 40x + 370$$
$$= 40(13) + 370 = 890$$

The value will be \$890.

91. Let $x =$ the number of years the cottage has been owned and let $y =$ the value of the cottage. Then one point on the line is given: $(3, 47700)$. Since the cottage appreciates by \$3500 per year, $m = 3500$.

$$y - y_1 = m(x - x_1)$$
$$y - 47700 = 3500(x - 3)$$
$$y - 47700 = 3500x - 10500$$
$$y = 3500x + 37200$$

Let $x = 0$ and find the value of y:

$$y = 3500x + 37200$$
$$= 3500(0) + 37200 = 37200$$

The purchase price was \$37,200.

93. Let $x =$ the hours of labor and let $y =$ the labor charge. Then $m =$ the hourly charge.

$$y = mx \qquad y = 46x$$
$$69 = m(1.5) \qquad y = 46(5) = 230$$
$$46 = m$$

The charge will be \$230.

95. Let $x =$ the number of fires and let $y =$ the population. Then two points on the line are given: $(300, 57000)$ and $(325, 59000)$.

$$m = \frac{59000 - 57000}{325 - 300} = \frac{2000}{25} = 80$$
$$y - y_1 = m(x - x_1)$$
$$y - 57000 = 80(x - 300)$$
$$y - 57000 = 80x - 24000$$
$$y = 80x + 33000$$

Let $y = 100000$ and find the value of x:

$$y = 80x + 33000$$
$$100000 = 80x + 33000$$
$$67000 = 80x$$
$$837.5 = x \Rightarrow$$ There will be about 838 fires when the population is 100,000.

97. Let F replace x and C replace y. Then two points on the line are given: $(32, 0)$ and $(212, 100)$.

$$m = \frac{100 - 0}{212 - 32} = \frac{100}{180} = \frac{5}{9}$$
$$C - C_1 = m(F - F_1)$$
$$C - 0 = \frac{5}{9}(F - 32)$$
$$C = \frac{5}{9}(F - 32)$$

99. Let $y =$ the percent who smoke and let $x =$ the # of years since 1974. Two points are given: $(0, 47)$ and $(20, 29)$.

$$m = \frac{29 - 47}{20 - 0} = \frac{-18}{20} = -\frac{9}{10}$$
$$y - y_1 = m(x - x_1)$$
$$y - 47 = -\frac{9}{10}(x - 0)$$
$$\boxed{y = -\frac{9}{10}x + 47}$$

Let $x = 50$:

$$y = -\frac{9}{10}(50) + 47 = -45 + 47 = 2$$

2% will smoke in 2024.

101. Two points on the line are given: $(0, 37.5)$ and $(2, 45)$.

$$m = \frac{45 - 37.5}{2 - 0} = \frac{7.5}{2} = 3.75$$

$$y - y_1 = m(x - x_1)$$
$$y - 37.5 = 3.75(x - 0)$$
$$y = 3.75x + 37.5$$

Let $x = 10$ and find the value of y:

$$y = 3.75x + 37.5$$
$$= 3.75(10) + 37.5$$
$$= 37.5 + 37.5 = 75$$

The price will be $75 in the year 2020.

103. The equation describing the production is $y = -70x + 1900$, where x represents the number of years and y is the level of production. Let $x = 3\frac{1}{2} = \frac{7}{2}$.

$$y = -70x + 1900$$
$$= -70\left(\frac{7}{2}\right) + 1900 = 1655$$

The production will be 1655 barrels per day.

105. a.

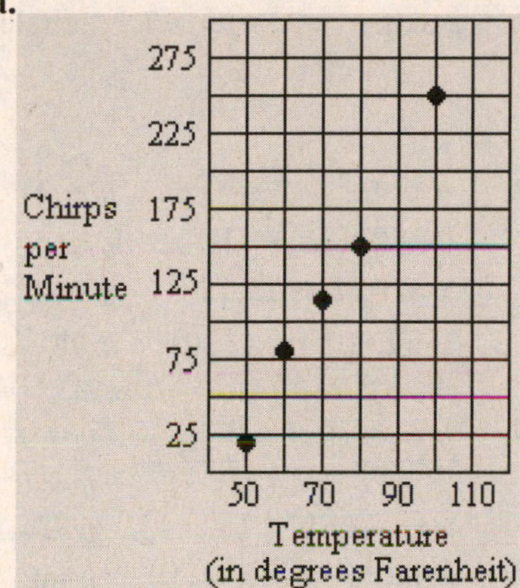

b. Use (50, 20) and (100, 250) for the regression line.

$$m = \frac{250 - 20}{100 - 50} = \frac{230}{50} = \frac{23}{5}$$

$$y - y_1 = m(x - x_1)$$
$$y - 20 = \frac{23}{5}(x - 50)$$
$$y - 20 = \frac{23}{5}x - 230$$
$$y = \frac{23}{5}x - 210$$

c. $y = \frac{23}{5}(90) - 210 = 204$

The rate will be about 204 chirps per minute.

107. $y = 4.44x - 196.62$

109. Answers may vary.

111. Answers may vary.

113. $m = \frac{b - 0}{0 - a} = -\frac{b}{a}$

$$y - y_1 = m(x - x_1)$$
$$y - b = -\frac{b}{a}(x - 0)$$
$$y - b = -\frac{b}{a}x$$
$$ay - ab = -bx$$
$$bx + ay = ab$$
$$\frac{bx + ay}{ab} = \frac{ab}{ab}$$
$$\frac{x}{a} + \frac{y}{b} = 1$$

115. Answers may vary.

117. Answers may vary.

119. $Ax + By = C$
$By = -Ax + C$
$y = -\frac{A}{B}x + \frac{C}{B}$
False. $m = -\frac{A}{B}$

121. Both are horizontal. True.

123. $x = 99$ is vertical, so the parallel line must be vertical too ($x = -99$). False.

125. $\sqrt{5}x + \sqrt{10}y = \sqrt{15}$
$\frac{\sqrt{5}x + \sqrt{10}y}{\sqrt{5}} = \frac{\sqrt{15}}{\sqrt{5}}$
$x + \sqrt{2}y = \sqrt{3}$; True.

Exercises 2.5 (page 265)

1. x-intercept

3. axis of symmetry

5. x-axis

7. circle, center

9. $x^2 + y^2 = r^2$

11. $y = x^2 - 4$
$0 = (x+2)(x-2)$
$x = -2, x = 2$
x-int: $(-2, 0), (2, 0)$

$y = x^2 - 4$
$y = 0^2 - 4$
$y = -4$
y-int: $(0, -4)$

13. $y = 4x^2 - 2x$
$0 = 2x(2x - 1)$
$x = 0, x = \frac{1}{2}$
x-int: $(0, 0), \left(\frac{1}{2}, 0\right)$

$y = 4x^2 - 2x$
$y = 4(0)^2 - 2(0)$
$y = 0$
y-int: $(0, 0)$

15. $y = x^2 - 4x - 5$
$0 = (x+1)(x-5)$
$x = -1, x = 5$
x-int: $(-1, 0), (5, 0)$

$y = x^2 - 4x - 5$
$y = 0^2 - 4(0) - 5$
$y = -5$
y-int: $(0, -5)$

17. $y = x^2 + x - 2$
$0 = (x+2)(x-1)$
$x = -2, x = 1$
x-int: $(-2, 0), (1, 0)$

$y = x^2 + x - 2$
$y = 0^2 + 0 - 2$
$y = -2$
y-int: $(0, -2)$

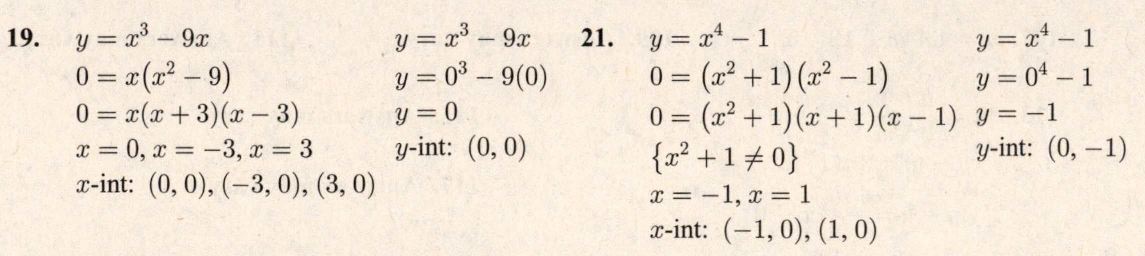

19. $y = x^3 - 9x$
$0 = x(x^2 - 9)$
$0 = x(x+3)(x-3)$
$x = 0, x = -3, x = 3$
x-int: $(0, 0), (-3, 0), (3, 0)$

$y = x^3 - 9x$
$y = 0^3 - 9(0)$
$y = 0$
y-int: $(0, 0)$

21. $y = x^4 - 1$
$0 = \left(x^2 + 1\right)\left(x^2 - 1\right)$
$0 = \left(x^2 + 1\right)(x+1)(x-1)$
$\left\{x^2 + 1 \neq 0\right\}$
$x = -1, x = 1$
x-int: $(-1, 0), (1, 0)$

$y = x^4 - 1$
$y = 0^4 - 1$
$y = -1$
y-int: $(0, -1)$

23. 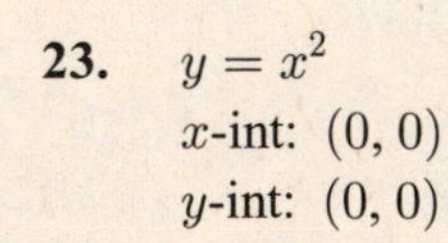
$y = x^2$
x-int: $(0, 0)$
y-int: $(0, 0)$

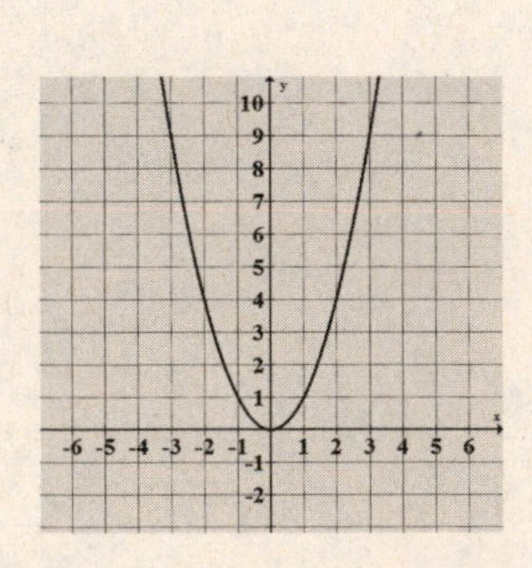

EXERCISES 2.5

25. $y = -x^2 + 2$

x-int: $\left(\sqrt{2}, 0\right), \left(-\sqrt{2}, 0\right)$

y-int: $(0, 2)$

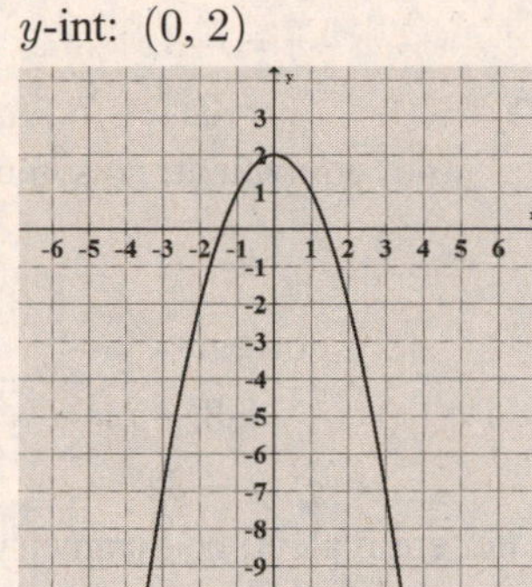

27. $y = x^2 - 4x$

x-int: $(0, 0), (4, 0)$

y-int: $(0, 0)$

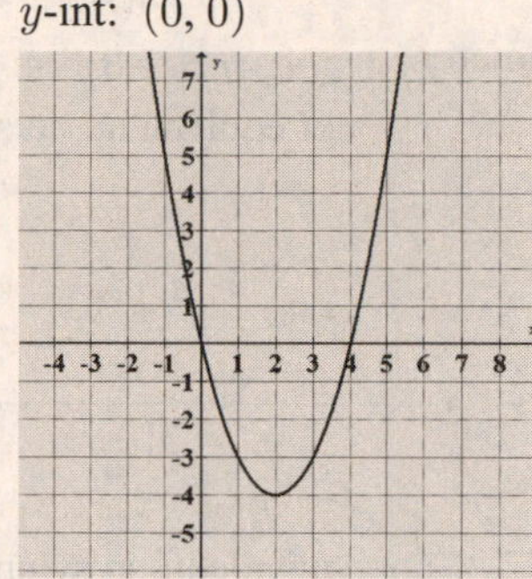

29. $y = \frac{1}{2}x^2 - 2x$

x-int: $(0, 0), (4, 0)$

y-int: $(0, 0)$

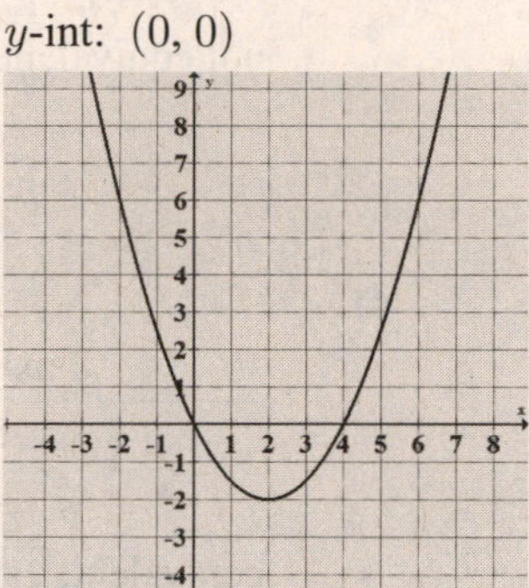

31. $y = x^2 + 2$

x-axis	y-axis	origin
$-y = x^2 + 2$	$y = (-x)^2 + 2$	$-y = (-x)^2 + 2$
not equivalent: no symmetry	$y = x^2 + 2$	$-y = x^2 + 2$
	equivalent: symmetry	not equivalent: no symmetry

33. $y^2 + 1 = x$

x-axis	y-axis	origin
$(-y)^2 + 1 = x$	$y^2 + 1 = -x$	$(-y)^2 + 1 = -x$
$y^2 + 1 = x$	not equivalent: no symmetry	$y^2 + 1 = -x$
equivalent: symmetry		not equivalent: no symmetry

35. $y^2 = x^2$

x-axis	y-axis	origin
$(-y)^2 = x^2$	$y^2 = (-x)^2$	$(-y)^2 = (-x)^2$
$y^2 = x^2$	$y^2 = x^2$	$y^2 = x^2$
equivalent: symmetry	equivalent: symmetry	equivalent: symmetry

37. $y = 3x^2 + 7$

x-axis	y-axis	origin
$-y = 3x^2 + 7$	$y = 3(-x)^2 + 7$	$-y = 3(-x)^2 + 7$
not equivalent: no symmetry	$y = 3x^2 + 7$	$-y = 3x^2 + 7$
	equivalent: symmetry	not equivalent: no symmetry

EXERCISES 2.5

39. $y = 3x^3 + 7$

x-axis	y-axis	origin
$-y = 3x^3 + 7$	$y = 3(-x)^3 + 7$	$-y = 3(-x)^3 + 7$
not equivalent: no symmetry	$y = -3x^3 + 7$	$-y = -3x^3 + 7$
	not equivalent: no symmetry	$y = 3x^3 - 7$
		not equivalent: no symmetry

41. $y^2 = 3x$

x-axis	y-axis	origin
$(-y)^2 = 3x$	$y^2 = 3(-x)$	$(-y)^2 = 3(-x)$
$y^2 = 3x$	$y^2 = -3x$	$y^2 = -3x$
equivalent: symmetry	not equivalent: no symmetry	not equivalent: no symmetry

43. $y = \lvert x \rvert$

x-axis	y-axis	origin
$-y = \lvert x \rvert$	$y = \lvert -x \rvert$	$-y = \lvert -x \rvert$
not equivalent: no symmetry	$y = \lvert -1 \rvert \lvert x \rvert$	$-y = \lvert -1 \rvert \lvert x \rvert$
	$y = \lvert x \rvert$	$-y = \lvert x \rvert$
	equivalent: symmetry	not equivalent: no symmetry

45. $\lvert y \rvert = x$

x-axis	y-axis	origin
$\lvert -y \rvert = x$	$\lvert y \rvert = -x$	$\lvert -y \rvert = -x$
$\lvert -1 \rvert \lvert y \rvert = x$	not equivalent: no symmetry	$\lvert -1 \rvert \lvert y \rvert = -x$
$\lvert y \rvert = x$		$\lvert y \rvert = -x$
equivalent: symmetry		not equivalent: no symmetry

47. $y = x^2 + 4x$
x-int: $(0, 0)$, $(-4, 0)$
y-int: $(0, 0)$
symmetry: none

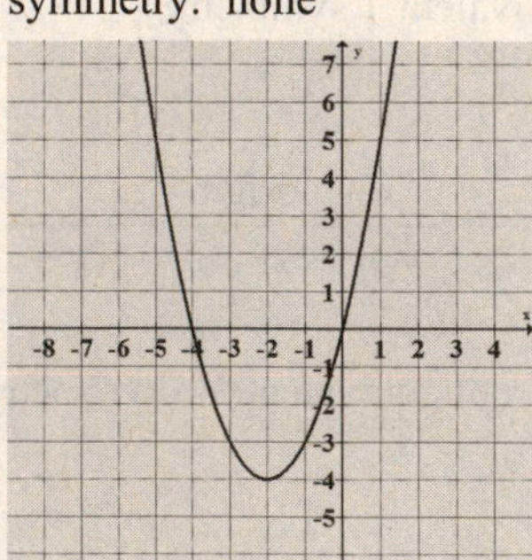

49. $y = x^3$
x-int: $(0, 0)$
y-int: $(0, 0)$
symmetry: origin

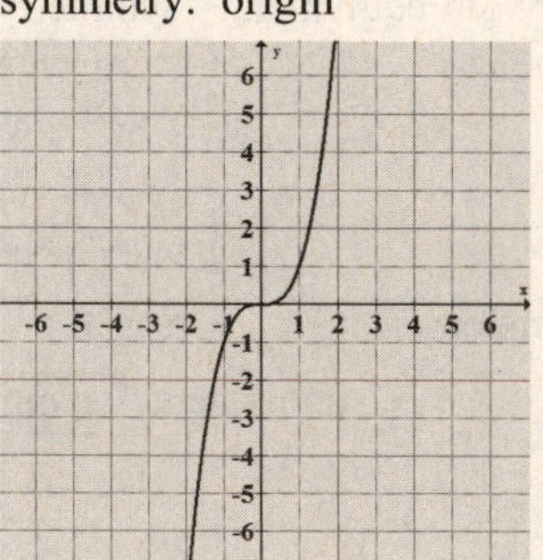

51. $y = \lvert x - 2 \rvert$
x-int: $(2, 0)$
y-int: $(0, 2)$
symmetry: none

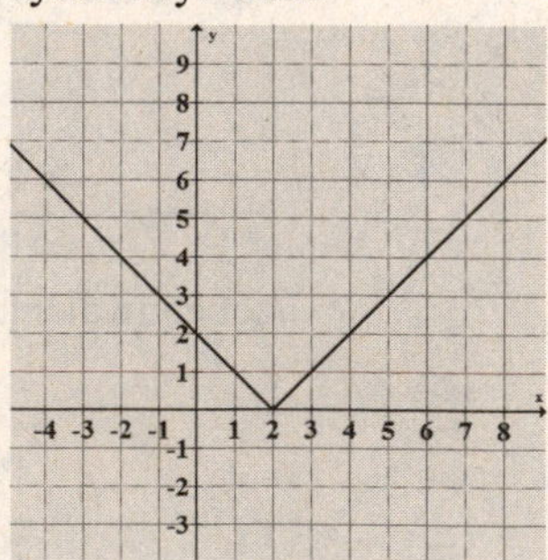

EXERCISES 2.5

53. $y = -|x| + 3$
x-int: $(-3, 0), (3, 0)$
y-int: $(0, 3)$
symmetry: y-axis

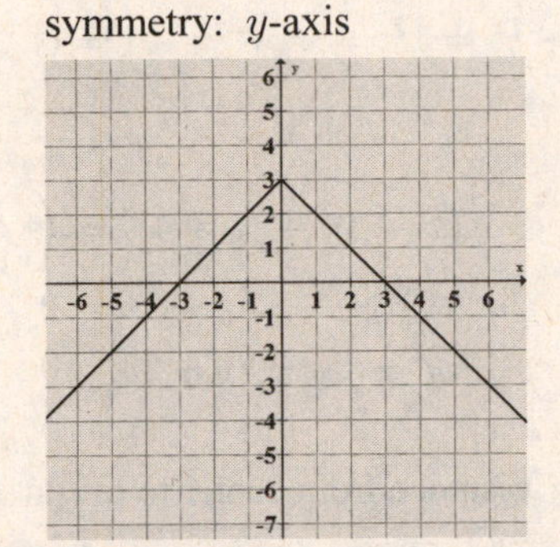

55. $y^2 = -x$
x-int: $(0, 0)$
y-int: $(0, 0)$
symmetry: x-axis

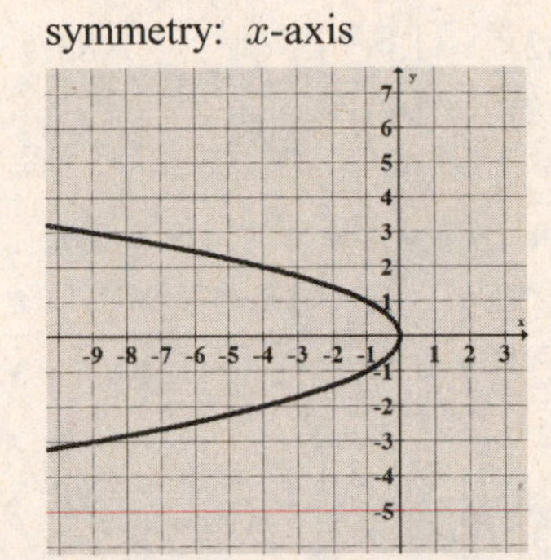

57. $y^2 = 9x$
x-int: $(0, 0)$
y-int: $(0, 0)$
symmetry: x-axis

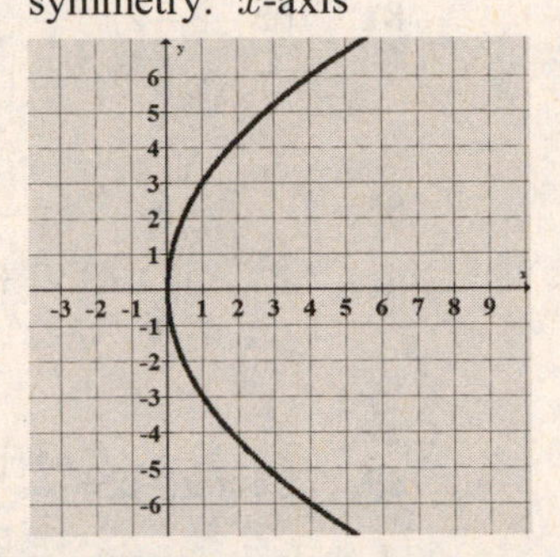

59. $y = \sqrt{x} - 1$
x-int: $(1, 0)$
y-int: $(0, -1)$
symmetry: none

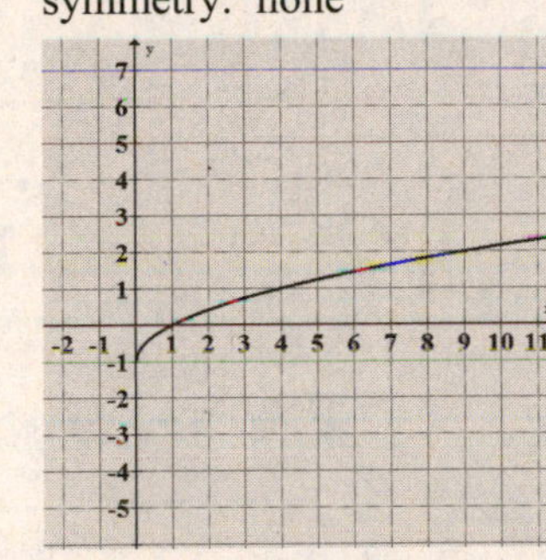

61. $xy = 4$
x-int: none
y-int: none
symmetry: origin

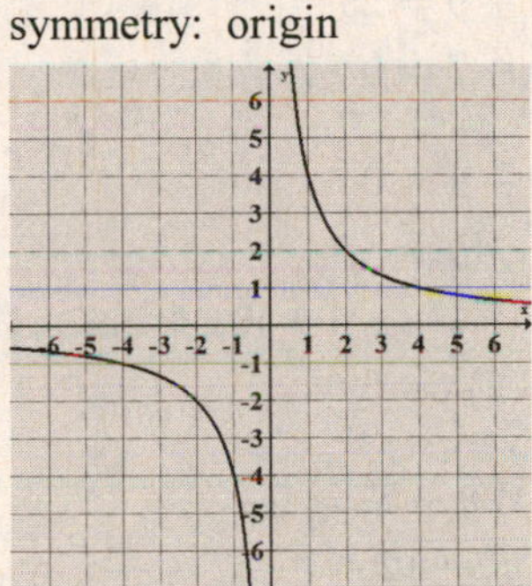

63. $y = \sqrt[3]{x}$
x-int: $(0, 0)$
y-int: $(0, 0)$
symmetry: origin

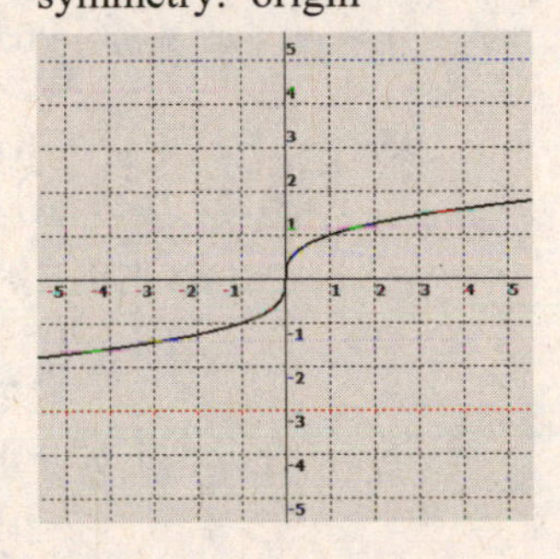

65.
$$x^2 + y^2 = 100$$
$$(x-0)^2 + (y-0)^2 = 10^2$$
$$\text{C: } (0, 0);\ r = 10$$

67.
$$x^2 + (y-5)^2 = 49$$
$$(x-0)^2 + (y-5)^2 = 7^2$$
$$\text{C: } (0, 5);\ r = 7$$

69.
$$(x+6)^2 + y^2 = \tfrac{1}{4}$$
$$(x-(-6))^2 + (y-0)^2 = \left(\tfrac{1}{2}\right)^2$$
$$\text{C: } (-6, 0);\ r = \tfrac{1}{2}$$

71.
$$(x-4)^2 + (y-1)^2 = 9$$
$$(x-4)^2 + (y-1)^2 = 3^2$$
$$\text{C: } (4, 1);\ r = 3$$

73.
$$\left(x - \tfrac{1}{4}\right)^2 + (y+2)^2 = 45$$
$$\left(x - \tfrac{1}{4}\right)^2 + (y-(-2))^2 = \left(\sqrt{45}\right)^2$$
$$\left(x - \tfrac{1}{4}\right)^2 + (y-(-2))^2 = \left(3\sqrt{5}\right)^2$$
$$\text{C: } \left(\tfrac{1}{4}, -2\right);\ r = 3\sqrt{5}$$

75.
$$(x-0)^2 + (y-0)^2 = 5^2$$
$$x^2 + y^2 = 25$$

EXERCISES 2.5

77. $(x-0)^2+(y-(-6))^2=6^2$
$$x^2+(y+6)^2=36$$

79. $(x-8)^2+(y-0)^2=\left(\frac{1}{5}\right)^2$
$$(x-8)^2+y^2=\frac{1}{25}$$

81. $(x-(-2))^2+(y-12)^2=13^2$
$$(x+2)^2+(y-12)^2=169$$

83. $x^2+y^2=1^2 \Rightarrow x^2+y^2-1=0$

85.
$$\begin{aligned}(x-6)^2+(y-8)^2&=4^2\\ x^2-12x+36+y^2-16y+64&=16\\ x^2+y^2-12x-16y+84&=0\end{aligned}$$

87.
$$\begin{aligned}(x-3)^2+(y+4)^2&=\left(\sqrt{2}\right)^2\\ x^2-6x+9+y^2+8y+16&=2\\ x^2+y^2-6x+8y+23&=0\end{aligned}$$

89. Center: $x=\dfrac{3+3}{2}=3,\ y=\dfrac{-2+8}{2}=3$

$r=$ distance from center to endpoint
$$\begin{aligned}&=\sqrt{(3-3)^2+(3-8)^2}=5\\ (x-3)^2+(y-3)^2&=5^2\\ x^2-6x+9+y^2-6y+9&=25\\ x^2+y^2-6x-6y-7&=0\end{aligned}$$

91. $r=$ distance from center to origin
$$\begin{aligned}&=\sqrt{(0-(-3))^2+(0-4)^2}=5\\ (x+3)^2+(y-4)^2&=5^2\\ x^2+6x+9+y^2-8y+16&=25\\ x^2+y^2+6x-8y&=0\end{aligned}$$

93.
$$\begin{aligned}x^2+y^2-6x+4y+4&=0\\ x^2-6x+y^2+4y&=-4\\ x^2-6x+9+y^2+4y+4&=-4+9+4\\ (x-3)^2+(y+2)^2&=9\end{aligned}$$

95.
$$\begin{aligned}x^2+y^2-10x-12y+57&=0\\ x^2-10x+y^2-12y&=-57\\ x^2-10x+25+y^2-12y+36&=-57+25+36\\ (x-5)^2+(y-6)^2&=4\end{aligned}$$

97.
$$\begin{aligned}2x^2+2y^2-8x-16y+22&=0\\ x^2+y^2-4x-8y+11&=0\\ x^2-4x+y^2-8y&=-11\\ x^2-4x+4+y^2-8y+16&=-11+4+16\\ (x-2)^2+(y-4)^2&=9\end{aligned}$$

EXERCISES 2.5

99. $x^2 + y^2 - 25 = 0$

$$x^2 + y^2 = 25$$

$C(0, 0), r = 5$

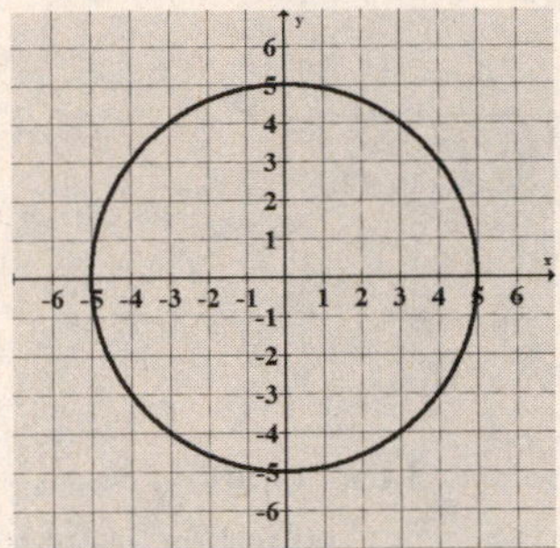

101. $(x - 1)^2 + (y + 2)^2 = 4$

$C(1, -2), r = 2$

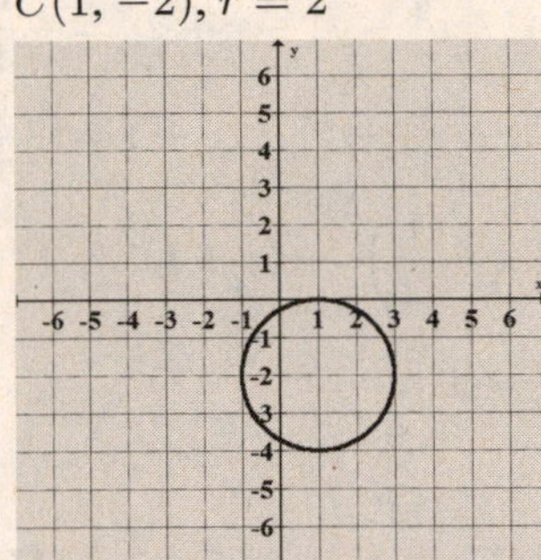

103. $x^2 + y^2 + 2x - 24 = 0$

$$x^2 + 2x + y^2 = 24$$

$$x^2 + 2x + 1 + y^2 = 24 + 1$$

$$(x + 1)^2 + y^2 = 25$$

$C(-1, 0), r = 5$

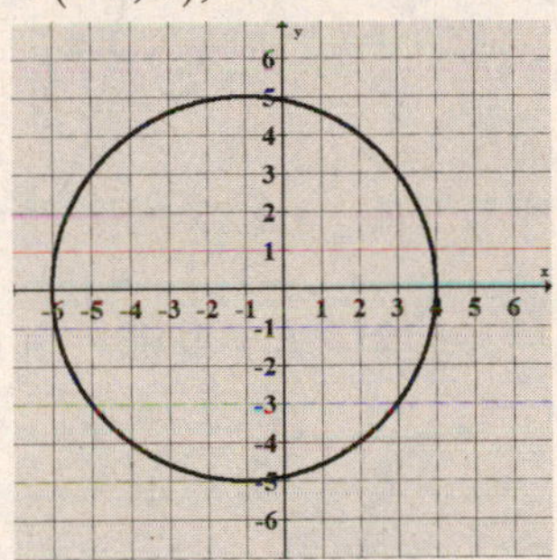

105. $x^2 + y^2 + 4x + 2y - 11 = 0$

$$x^2 + 4x + y^2 + 2y = 11$$

$$x^2 + 4x + 4 + y^2 + 2y + 1 = 11 + 4 + 1$$

$$(x + 2)^2 + (y + 1)^2 = 16$$

$C(-2, -1), r = 4$

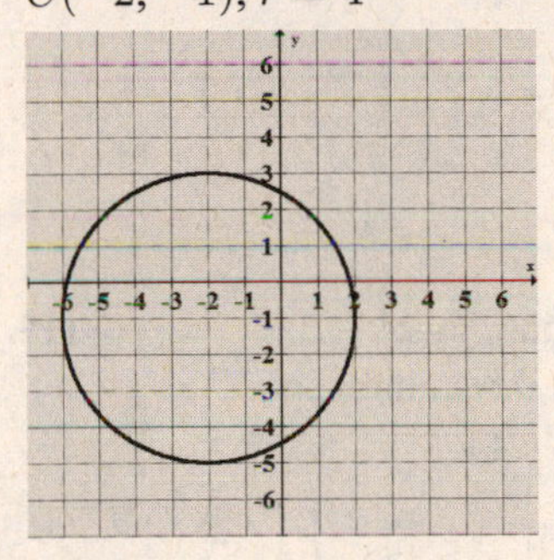

107. $9x^2 + 9y^2 - 12y = 5$

$$x^2 + y^2 - \frac{4}{3}y = \frac{5}{9}$$

$$x^2 + y^2 - \frac{4}{3}y + \frac{4}{9} = \frac{5}{9} + \frac{4}{9}$$

$$x^2 + \left(y - \frac{2}{3}\right)^2 = 1$$

$C\left(0, \frac{2}{3}\right), r = 1$

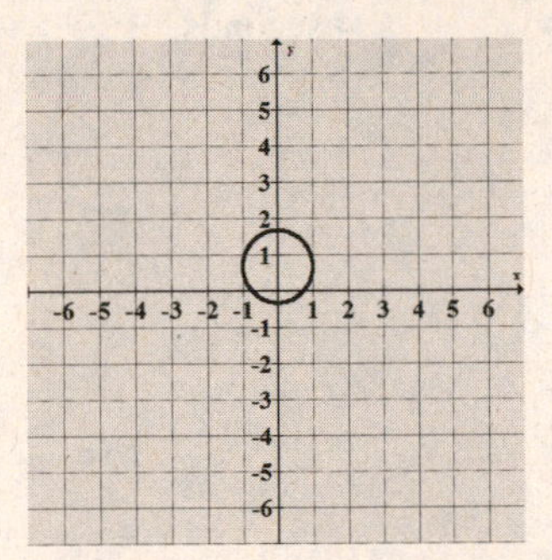

109. $4x^2 + 4y^2 - 4x + 8y + 1 = 0$

$$x^2 + y^2 - x + 2y = -\frac{1}{4}$$

$$x^2 - x + \frac{1}{4} + y^2 + 2y + 1 = -\frac{1}{4} + \frac{1}{4} + 1$$

$$\left(x - \frac{1}{2}\right)^2 + (y+1)^2 = 1$$

$$C\left(\frac{1}{2}, -1\right), r = 1$$

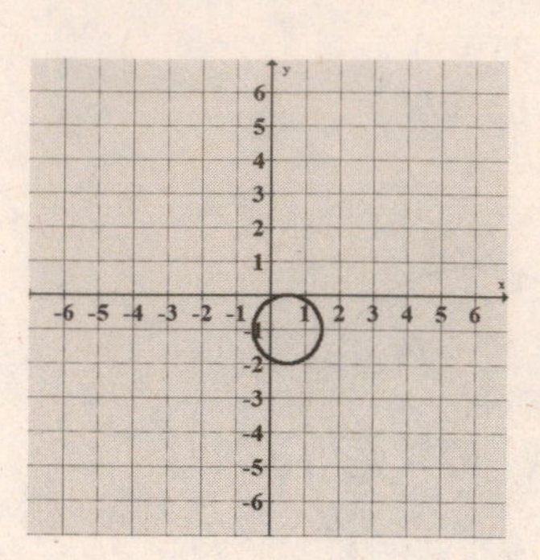

111. $y = 2x^2 - x + 1$
Vertex: (0.25, 0.88)

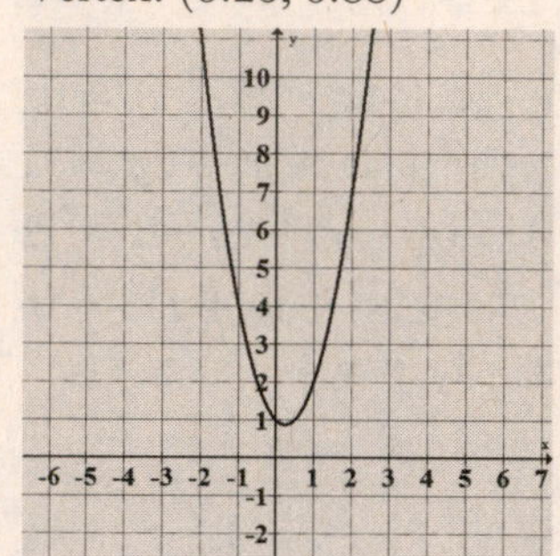

113. $y = 7 + x - x^2$
Vertex: (0.50, 7.25)

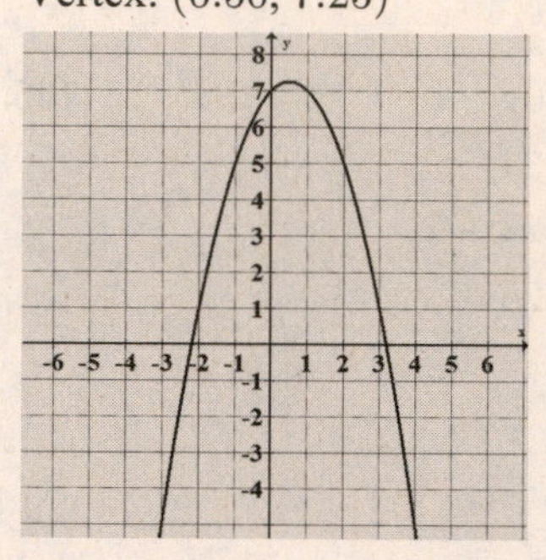

115. Graph $y = x^2 - 7$.
Find the x-intercepts.
$x = -2.65, \; x = 2.65$

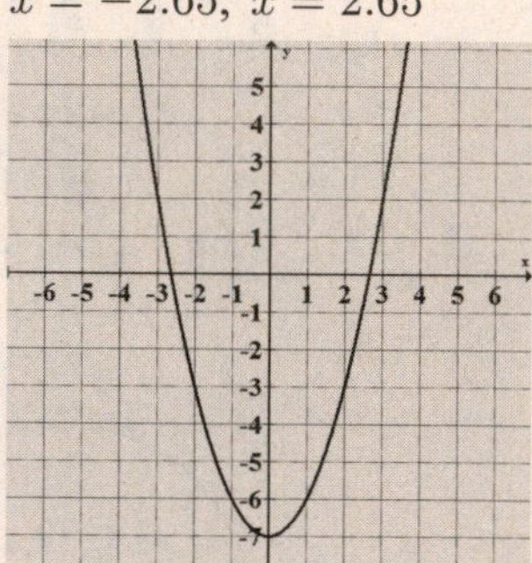

117. Graph $y = x^3 - 3$.
Find the x-intercepts.
$x = 1.44$

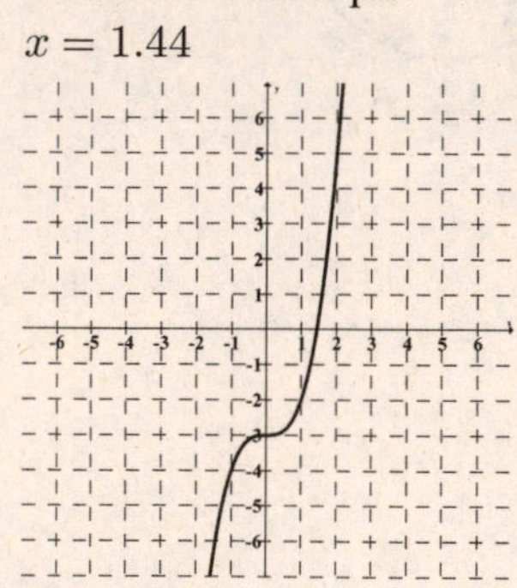

119. Let $y = 0$:
$y = 64t - 16t^2$
$0 = 16t(4 - t)$
$t = 0$ or $t = 4$
It strikes the ground after 4 seconds.

121. $D = 0.08V^2 + 0.9V$;

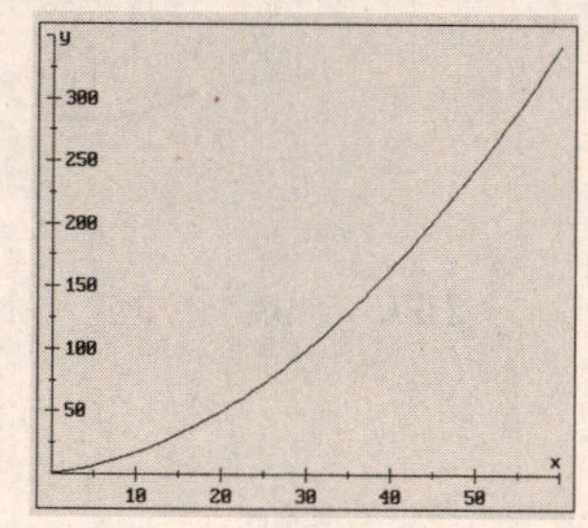

123. $r = \frac{12}{2} = 6$

$$(x-0)^2 + (y-0)^2 = 6^2$$
$$x^2 + y^2 = 36$$

125. $r = \frac{60}{2} = 30$

$$(x-0)^2 + (y-35)^2 = 30^2$$
$$x^2 + (y-35)^2 = 900$$

127. $r = \sqrt{(10-7)^2 + (0-4)^2} = 5$

$$(x-7)^2 + (y-4)^2 = 5^2$$
$$x^2 - 14x + 49 + y^2 - 8y + 16 = 25$$
$$x^2 + y^2 - 14x - 8y + 40 = 0$$

129. **Answers may vary.**

131. **Answers may vary.**

133.
$$x^2 - 4x + y^2 - 6y + 13 = 0$$
$$x^2 - 4x + 4 + y^2 - 6y + 9 = -13 + 4 + 9$$
$$(x+2)^2 + (y-3)^2 = 0 \Rightarrow \text{a single point}$$

135. False. The graphs are symmetric with respect to the y-axis.

137. True.

139. True.

141. False. The graph is the single point $\left(4, -\frac{1}{7}\right)$.

Exercises 2.6 (page 276)

1. quotient

3. means

5. extremes, means

7. inverse

9. joint

11.
$$\frac{4}{x} = \frac{2}{7}$$
$$4 \cdot 7 = 2 \cdot x$$
$$28 = 2x$$
$$14 = x$$

13.
$$\frac{x}{2} = \frac{3}{x+1}$$
$$x(x+1) = 3 \cdot 2$$
$$x^2 + x = 6$$
$$x^2 + x - 6 = 0$$
$$(x+3)(x-2) = 0 \Rightarrow x = -3 \text{ or } x = 2$$

15. Let $x =$ the number of women.

$$\frac{3}{5} = \frac{x}{30}$$
$$3 \cdot 30 = 5 \cdot x$$
$$90 = 5x$$
$$18 = x \Rightarrow \text{There are 18 women.}$$

17.
$$y = kx$$
$$15 = k(30)$$
$$\frac{1}{2} = k$$

19.
$$I = \frac{k}{R}$$
$$50 = \frac{k}{20}$$
$$1000 = k$$

21.
$$E = kIR$$
$$125 = k(5)(25)$$
$$125 = 125k$$
$$1 = k$$

23.
$$y = kx \qquad y = \frac{15}{4}x$$
$$15 = k(4) \qquad y = \frac{15}{4} \cdot \frac{7}{5}$$
$$\frac{15}{4} = k \qquad y = \frac{21}{4}$$

25.
$$w = \frac{k}{z} \qquad w = \frac{30}{z}$$
$$10 = \frac{k}{3} \qquad w = \frac{30}{5}$$
$$30 = k \qquad w = 6$$

EXERCISES 2.6

27. $$\begin{aligned} P &= krs \\ 16 &= k(5)(-8) \\ 16 &= -40k \\ -\tfrac{16}{40} &= k \\ -\tfrac{2}{5} &= k \end{aligned} \qquad \begin{aligned} P &= -\tfrac{2}{5}rs \\ P &= -\tfrac{2}{5}(2)(10) \\ P &= -8 \end{aligned}$$

29. direct

31. neither

33. Let $x =$ the amount of caffeine.

$$\begin{aligned} \tfrac{55}{12} &= \tfrac{x}{44} \\ 55 \cdot 44 &= 12 \cdot x \\ 2420 &= 12x \\ 202 \text{ mg} &\approx x \end{aligned} \qquad \begin{aligned} \tfrac{47}{12} &= \tfrac{x}{44} \\ 47 \cdot 44 &= 12 \cdot x \\ 2068 &= 12x \\ 172 \text{ mg} &\approx x \end{aligned} \qquad \begin{aligned} \tfrac{37}{12} &= \tfrac{x}{44} \\ 37 \cdot 44 &= 12 \cdot x \\ 1628 &= 12x \\ 136 \text{ mg} &\approx x \end{aligned}$$

35. Let $x =$ the amount of adhesive needed.

$$\begin{aligned} \tfrac{\frac{1}{2}}{140} &= \tfrac{x}{500} \\ \frac{1}{2} \cdot 500 &= 140 \cdot x \\ 250 &= 140x \\ 1.79 &\approx x \end{aligned}$$

About 2 gallons of adhesive will be needed.

37. $$\begin{aligned} V &= \frac{kT}{P} \\ 20 &= \frac{k(330)}{40} \\ 800 &= 330k \\ \frac{800}{330} &= k \\ \frac{80}{33} &= k \end{aligned} \qquad \begin{aligned} V &= \frac{\frac{80}{33}T}{P} \\ V &= \frac{\frac{80}{33}(300)}{50} \\ V &= \frac{\frac{8000}{11}}{50} \\ V &= \frac{160}{11} = 14\frac{6}{11} \text{ ft}^3 \end{aligned}$$

39. $$\begin{aligned} d &= kt^2 \\ 16 &= k(1)^2 \\ 16 &= k \end{aligned} \qquad \begin{aligned} d &= 16t^2 \\ 144 &= 16t^2 \\ 9 &= t^2 \\ 3 &= t \Rightarrow 3 \text{ seconds} \end{aligned}$$

41. $$\begin{aligned} t &= kl^2 \\ 1 &= k(1)^2 \\ 1 &= k \end{aligned} \qquad \begin{aligned} t &= l^2 \\ 2 &= l^2 \\ \sqrt{2} &= l \Rightarrow \sqrt{2} \text{ meters} \end{aligned}$$

43. $$\begin{aligned} I &= \frac{k}{d^2} \\ 60 &= \frac{k}{10^2} \\ 60 &= \frac{k}{100} \\ 6000 &= k \end{aligned} \qquad \begin{aligned} I &= \frac{6000}{d^2} \\ I &= \frac{6000}{20^2} \\ I &= \frac{6000}{400} \\ I &= 15 \Rightarrow 15 \text{ lumens} \end{aligned}$$

45. $$\begin{aligned} E = kmv^2 &= k(2m)(3v)^2 \\ &= k(2m)\left(9v^2\right) \\ &= 18 \cdot kmv^2 \end{aligned}$$

The energy is multiplied by 18.

47. $$\begin{aligned} G = \frac{km_1m_2}{d^2} &= \frac{k(3m_1)(3m_2)}{(2d)^2} \\ &= \frac{k \cdot 9m_1m_2}{4d^2} \\ &= \frac{9}{4} \cdot \frac{km_1m_2}{d^2} \end{aligned}$$

The force is multiplied by $\frac{9}{4}$.

49. Consider this figure:

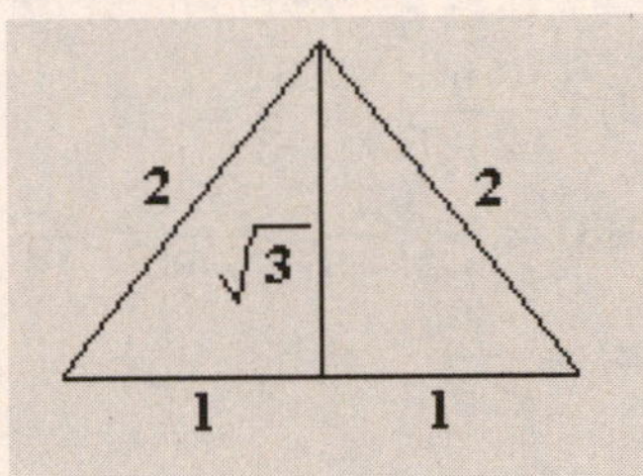

$h = \sqrt{3}$ can be computed using the Pythagorean Theorem.

$$A = \frac{1}{2}bh = \frac{1}{2}(2)\sqrt{3} = \sqrt{3}$$
$$A = ks^2$$
$$\sqrt{3} = k(2)^2$$
$$\sqrt{3} = 4k$$
$$\frac{\sqrt{3}}{4} = k$$

51-57. Answers may vary.

59. d

61. b

63. c

65. a

Chapter 2 Review (page 279)

1. D = {3, 4, 5, 6}; R = {4, 5, 6, 7}
Each element of the domain is paired with only one element of the range. Function.

2. D = {2, 3, −4}; R = {4, 5, 6, 3}
2 is both paired with 4 and 5. Not a function.

3. $y = 3$
Each value of x is paired with only one value of y.
function

4. $y + 5x^2 = 2$
$y = -5x^2 + 2$
Each value of x is paired with only one value of y.
function

5. $y^2 - x = 5$
$y^2 = x + 5$
$y = \pm\sqrt{x+5}$
Each value of x is paired with more than one value of y. **not a function**

6. $y = |x| + x$
Each value of x is paired with only one value of y.
function

7. $f(x) = y = 3x^2 - 5$
domain = $(-\infty, \infty)$

8. $f(x) = y = \dfrac{3x}{x-5}$
domain = $(-\infty, 5) \cup (5, \infty)$

9. $f(x) = \dfrac{3x}{4x^2 - 16} = \dfrac{3x}{4(x+2)(x-2)}$
$x \neq -2, x \neq 2$
domain = $(-\infty, -2) \cup (-2, 2) \cup (2, \infty)$

10. $f(x) = y = \sqrt{x-1}$
domain = $[1, \infty)$

11. $f(x) = y = \sqrt{5-x}$
domain = $(-\infty, 5]$

12. $f(x) = y = \sqrt{x^2+1}$
$x^2 + 1 \geq 0 \Rightarrow$ domain = $(-\infty, \infty)$

13. $f(x) = 5x - 2$
$f(2) = 5(2) - 2 = 8$
$f(-3) = 5(-3) - 2 = -17$
$f(k) = 5k - 2$

14. $f(x) = \dfrac{6}{x-5}$
$f(2) = \dfrac{6}{2-5} = \dfrac{6}{-3} = -2$
$f(-3) = \dfrac{6}{-3-5} = \dfrac{6}{-8} = -\dfrac{3}{4}$
$f(k) = \dfrac{6}{k-5}$

15. $f(x) = |x-2|$
$f(2) = |2-2| = |0| = 0$
$f(-3) = |-3-2| = |-5| = 5$
$f(k) = |k-2|$

16. $f(x) = \dfrac{x^2-3}{x^2+3}$
$f(2) = \dfrac{2^2-3}{2^2+3} = \dfrac{1}{7}$
$f(-3) = \dfrac{(-3)^2-3}{(-3)^2+3} = \dfrac{6}{12} = \dfrac{1}{2}$
$f(k) = \dfrac{k^2-3}{k^2+3}$

17.
$$\frac{f(x+h)-f(x)}{h} = \frac{[5(x+h)-6]-[5x-6]}{h} = \frac{[5x+5h-6]-[5x-6]}{h} = \frac{5x+5h-6-5x+6}{h} = \frac{5h}{h} = 5$$

18.
$$\begin{aligned}\frac{f(x+h)-f(x)}{h} &= \frac{\left[2(x+h)^2-7(x+h)+3\right]-[2x^2-7x+3]}{h} \\ &= \frac{[2x^2+4xh+2h^2-7x-7h+3]-[2x^2-7x+3]}{h} \\ &= \frac{2x^2+4xh+2h^2-7x-7h+3-2x^2+7x-3}{h} \\ &= \frac{4xh+2h^2-7h}{h} = \frac{h(4x+2h-7)}{h} = 4x+2h-7\end{aligned}$$

19. $f(x) = -0.6x + 132$
$f(45) = -0.6(45) + 132 = 105$

20. **a.** $I(h) = 3.5h - 50$
b. $I(200) = 3.5(200) - 50 = \$650$

21. $A(2, 0)$ **22.** $B(-2, 1)$ **23.** $C(0, -1)$ **24.** $D(3, -1)$

25. $A(-3, 5)$: QII **26.** $B(5, -3)$: QIV

27. $C(0, -7)$: negative y-axis **28.** $D\left(-\frac{1}{2}, 0\right)$: negative x-axis

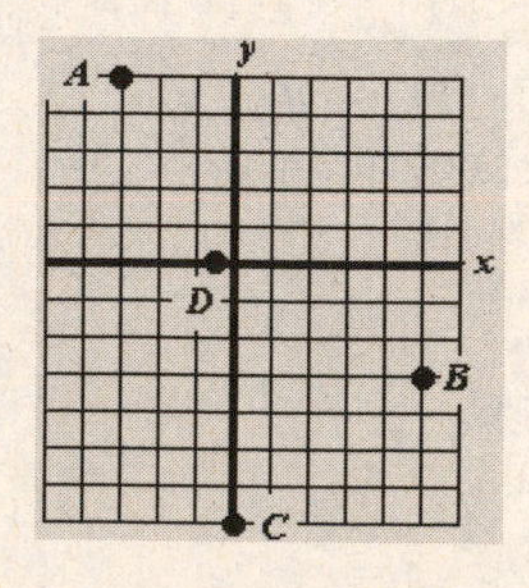

29. $2x - y = 6$

$-y = -2x + 6$

$y = 2x - 6$

x	y
0	−6
2	−2

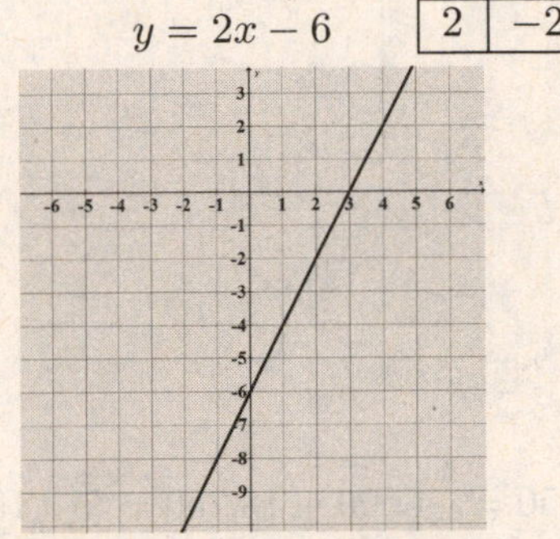

30. $2x + 5y = -10$

$5y = -2x - 10$

$y = -\frac{2}{5}x - 2$

x	y
0	−2
−5	0

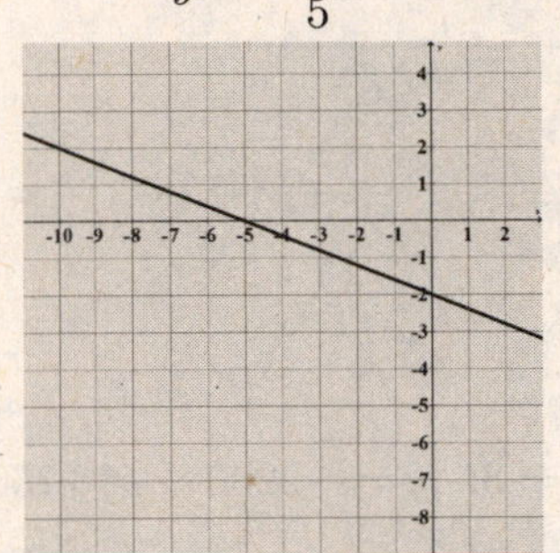

31.

$$\begin{aligned} 3x - 5y &= 15 \\ 3x - 5(0) &= 15 \\ 3x &= 15 \\ x &= 5 \end{aligned} \qquad \begin{aligned} 3x - 5y &= 15 \\ 3(0) - 5y &= 15 \\ -5y &= 15 \\ y &= -3 \end{aligned}$$

(5, 0) (0, −3)

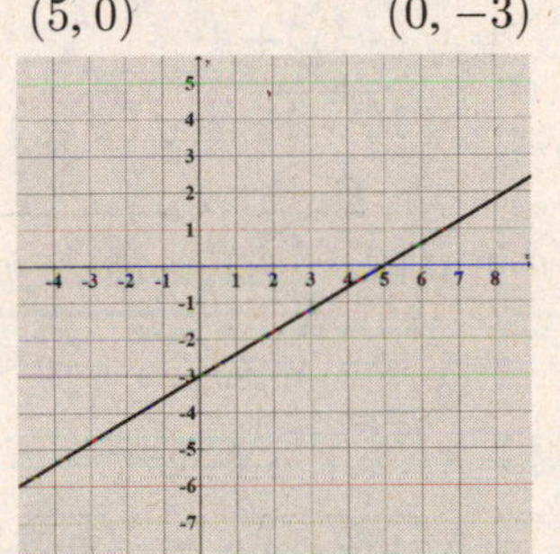

32.

$$\begin{aligned} x + y &= 7 \\ x + 0 &= 7 \\ x &= 7 \end{aligned} \qquad \begin{aligned} x + y &= 7 \\ 0 + y &= 7 \\ y &= 7 \end{aligned}$$

(7, 0) (0, 7)

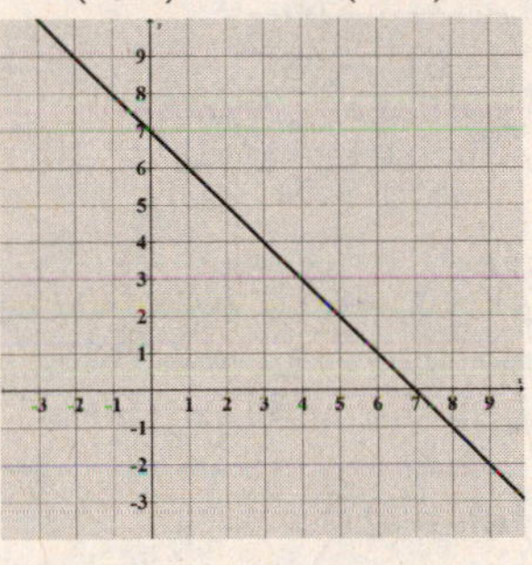

33.

$$\begin{aligned} x + y &= -7 \\ x + 0 &= -7 \\ x &= -7 \end{aligned} \qquad \begin{aligned} x + y &= -7 \\ 0 + y &= -7 \\ y &= -7 \end{aligned}$$

(−7, 0) (0, −7)

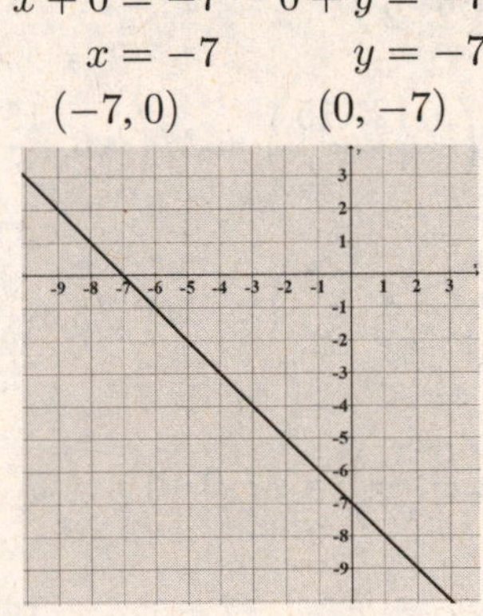

34.

$$\begin{aligned} x - 5y &= 5 \\ x - 5(0) &= 5 \\ x &= 5 \end{aligned} \qquad \begin{aligned} x - 5y &= 5 \\ 0 - 5y &= 5 \\ -5y &= 5 \\ y &= -1 \end{aligned}$$

(5, 0) (0, −1)

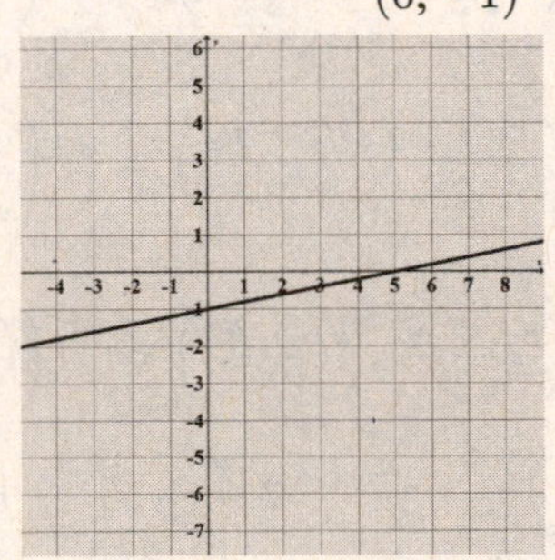

35. $y = 4 \Rightarrow$ horizontal

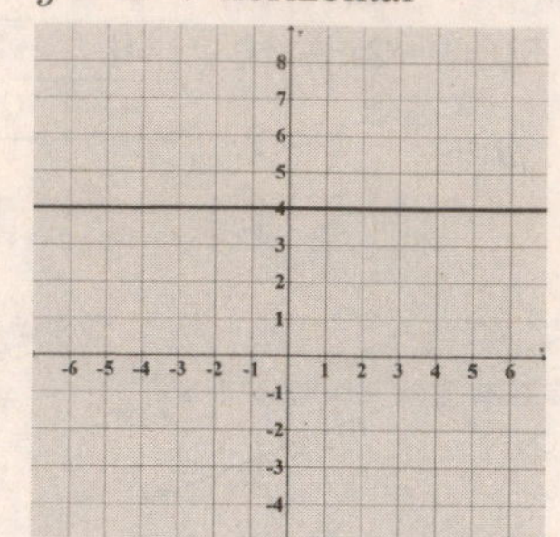

36. $x = -2 \Rightarrow$ vertical

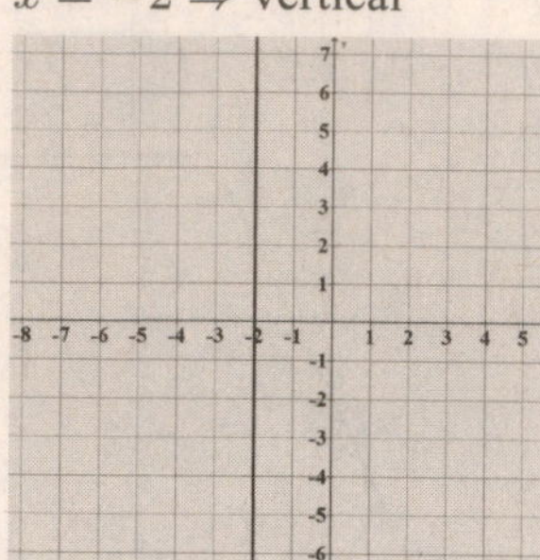

37. Let $x = 3$: $y = -2200x + 18{,}750 = -2200(3) + 18{,}750 = -6600 + 18{,}750 = \$12{,}150$

38. Let $x = 5$: $y = 16{,}500x + 250{,}000 = 16{,}500(5) + 250{,}000 = 82{,}500 + 250{,}000 = \$332{,}500$

39.
$$\begin{aligned} d &= \sqrt{(x_2 - x_1)^2 + (y_2 - y_1)^2} \\ &= \sqrt{(-3-3)^2 + (7-(-1))^2} \\ &= \sqrt{(-6)^2 + (8)^2} \\ &= \sqrt{36+64} = \sqrt{100} = 10 \end{aligned}$$

40.
$$\begin{aligned} d &= \sqrt{(x_2 - x_1)^2 + (y_2 - y_1)^2} \\ &= \sqrt{(-12-(-8))^2 + (10-6)^2} \\ &= \sqrt{(-4)^2 + 4^2} \\ &= \sqrt{16+16} = \sqrt{32} = 4\sqrt{2} \end{aligned}$$

41.
$$\begin{aligned} d &= \sqrt{(x_2 - x_1)^2 + (y_2 - y_1)^2} \\ &= \sqrt{\left(\sqrt{3}-\sqrt{3}\right)^2 + (9-7)^2} \\ &= \sqrt{0^2 + (2)^2} \\ &= \sqrt{0+4} = \sqrt{4} = 2 \end{aligned}$$

42.
$$\begin{aligned} d &= \sqrt{(x_2 - x_1)^2 + (y_2 - y_1)^2} \\ &= \sqrt{(a-(-a))^2 + (-a-a)^2} \\ &= \sqrt{(2a)^2 + (-2a)^2} \\ &= \sqrt{4a^2+4a^2} = \sqrt{8a^2} = 2\sqrt{2}|a| \end{aligned}$$

43. $M\left(\frac{x_1+x_2}{2}, \frac{y_1+y_2}{2}\right) = M\left(\frac{-3+3}{2}, \frac{7+(-1)}{2}\right) = M\left(\frac{0}{2}, \frac{6}{2}\right) = M(0,3)$

44. $M\left(\frac{x_1+x_2}{2}, \frac{y_1+y_2}{2}\right) = M\left(\frac{0+(-12)}{2}, \frac{5+10}{2}\right) = M\left(\frac{-12}{2}, \frac{15}{2}\right) = M\left(-6, \frac{15}{2}\right)$

45. $M\left(\frac{x_1+x_2}{2}, \frac{y_1+y_2}{2}\right) = M\left(\frac{\sqrt{3}+\sqrt{3}}{2}, \frac{9+7}{2}\right) = M\left(\frac{2\sqrt{3}}{2}, \frac{16}{2}\right) = M\left(\sqrt{3}, 8\right)$

46. $M\left(\frac{x_1+x_2}{2}, \frac{y_1+y_2}{2}\right) = M\left(\frac{a+(-a)}{2}, \frac{-a+a}{2}\right) = M\left(\frac{0}{2}, \frac{0}{2}\right) = M(0,0)$

47. $m = \frac{y_2-y_1}{x_2-x_1} = \frac{7-(-5)}{1-3} = \frac{12}{-2} = -6$

48. $m = \frac{y_2-y_1}{x_2-x_1} = \frac{-7-7}{-5-2} = \frac{-14}{-7} = 2$

CHAPTER 2 REVIEW

49. $m = \frac{y_2 - y_1}{x_2 - x_1} = \frac{\frac{1}{2} - (-8)}{5 - 5} = \frac{8\frac{1}{2}}{0}$: und.

50. $m = \frac{y_2 - y_1}{x_2 - x_1} = \frac{-8 - (-8)}{-1 - \frac{2}{3}} = \frac{0}{-1\frac{2}{3}} = 0$

51. $m = \frac{y_2 - y_1}{x_2 - x_1} = \frac{b - a}{a - b} = -1$

52. $m = \frac{y_2 - y_1}{x_2 - x_1} = \frac{(b - a) - b}{b - (a + b)} = \frac{-a}{-a} = 1$

53. $y = 3x + 6$ $\quad m = \frac{y_2 - y_1}{x_2 - x_1} = \frac{9 - 6}{1 - 0} = \frac{3}{1} = 3$

x	y
0	6
1	9

54. $y = -\frac{1}{5}x - 6$ $\quad m = \frac{y_2 - y_1}{x_2 - x_1} = \frac{-7 - (-6)}{5 - 0} = \frac{-1}{5} = -\frac{1}{5}$

x	y
0	-6
5	-7

55. The slope is zero.

56. The slope is undefined.

57. The slope is negative.

58. The slope is positive.

59. $m_1 m_2 = 5\left(-\frac{1}{5}\right) = -1$
perpendicular

60. $m_1 \neq m_2$; $m_1 m_2 = \frac{2}{7} \cdot \frac{7}{2} = 1 \neq -1$
neither

61.
$$m = \frac{y_2 - y_1}{x_2 - x_1} = \frac{10 - 5}{6 - (-2)} = \frac{5}{8}$$
$$m = \frac{y_2 - y_1}{x_2 - x_1} = \frac{y - 2}{10 - 2} = \frac{5}{8}$$
$$8(y - 2) = 5(8)$$
$$8y - 16 = 40$$
$$8y = 56$$
$$y = 7$$

62.
$$m = \frac{y_2 - y_1}{x_2 - x_1} = \frac{10 - 5}{6 - (-2)} = \frac{5}{8}$$
$$m = \frac{y_2 - y_1}{x_2 - x_1} = \frac{-3 - 5}{x - (-2)} = \frac{-8}{5}$$
$$5(-8) = -8(x + 2)$$
$$-40 = -8x - 16$$
$$8x = 24$$
$$x = 3$$

63. $m = \frac{\Delta y}{\Delta x} = \frac{3000}{15} = 200$ ft per minute

64. $m = \frac{\Delta y}{\Delta x} = \frac{147{,}500 - 50{,}000}{3 - 1} = \frac{97{,}500}{2} = \$48{,}750$ per year

65. $y = mx + b$
$y = \frac{2}{3}x + 3$

66. $y = mx + b$
$y = -\frac{3}{2}x - 5$

67.
$$3x - 2y = 10$$
$$-2y = -3x + 10$$
$$y = \tfrac{3}{2}x - 5$$
$m = \frac{3}{2}$, $(0, -5)$

68.
$$2x + 4y = -8$$
$$4y = -2x - 8$$
$$y = -\tfrac{1}{2}x - 2$$
$m = -\frac{1}{2}$, $(0, -2)$

69.
$$-2y = -3x + 10$$
$$y = \tfrac{3}{2}x - 5$$
$m = \frac{3}{2}$, $(0, -5)$

70. $2x = -4y - 8$
$4y = -2x - 8$
$y = -\frac{1}{2}x - 2$
$m = -\frac{1}{2}, (0, -2)$

71. $5x + 2y = 7$
$2y = -5x + 7$
$y = -\frac{5}{2}x + \frac{7}{2}$
$m = -\frac{5}{2}, (0, \frac{7}{2})$

72. $3x - 4y = 14$
$-4y = -3x + 14$
$y = \frac{3}{4}x - \frac{7}{2}$
$m = \frac{3}{4}, (0, -\frac{7}{2})$

73. $y = \dfrac{3}{5}x - 2$
$m = \dfrac{3}{5}, b = -2$

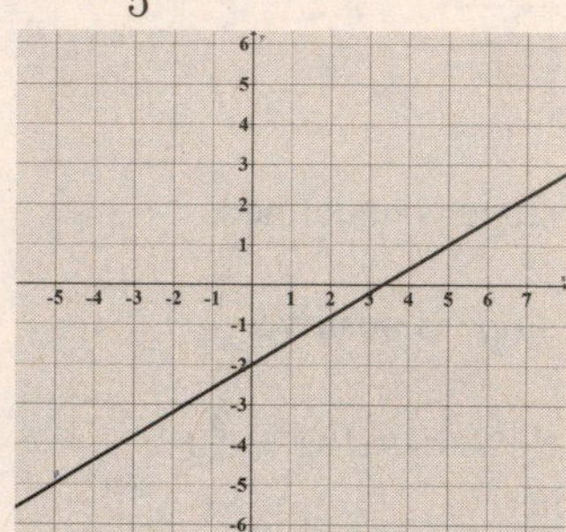

74. $y = -\dfrac{4}{3}x + 3$
$m = -\dfrac{4}{3}, b = 3$

75. $y = 3x + 8$
$m = 3$

$2y = 6x - 19$
$y = 3x - \frac{19}{2}$
$m = 3$

The lines are parallel.

76. $2x + 3y = 6$
$3y = -2x + 6$
$y = -\frac{2}{3}x + 2$
$m = -\frac{2}{3}$

$3x - 2y = 15$
$-2y = -3x + 15$
$y = \frac{3}{2}x - \frac{15}{2}$
$m = \frac{3}{2}$

The lines are perpendicular.

77. $m = \dfrac{y_2 - y_1}{x_2 - x_1} = \dfrac{7 - 0}{-5 - 0} = -\dfrac{7}{5}$
$y - y_1 = m(x - x_1)$
$y - 0 = -\dfrac{7}{5}(x - 0)$
$y = -\dfrac{7}{5}x$
$5y = 5\left(-\dfrac{7}{5}x\right)$
$5y = -7x$
$7x + 5y = 0$

78. $y - y_1 = m(x - x_1)$
$y - 1 = -4(x + 2)$
$y - 1 = -4x - 8$
$4x + y = -7$

79.
$$\begin{aligned} y - y_1 &= m(x - x_1) \\ y + 1 &= -\frac{1}{5}(x - 2) \\ 5(y + 1) &= 5 \cdot \left[-\frac{1}{5}(x - 2)\right] \\ 5y + 5 &= -(x - 2) \\ 5y + 5 &= -x + 2 \\ x + 5y &= -3 \end{aligned}$$

80. $m = \dfrac{y_2 - y_1}{x_2 - x_1} = \dfrac{1 - (-5)}{4 - 7} = \dfrac{6}{-3} = -2$
$$\begin{aligned} y - y_1 &= m(x - x_1) \\ y + 5 &= -2(x - 7) \\ y + 5 &= -2x + 14 \\ 2x + y &= 9 \end{aligned}$$

81. $m = 0 \Rightarrow$ horizontal
$y = 17$

82. m is undefined $\Rightarrow$ vertical
$x = -5$

83.
$$\begin{aligned} 3x - 4y &= 7 \\ -4y &= -3x + 7 \\ y &= \frac{3}{4}x - \frac{7}{4} \end{aligned}$$
$m = \dfrac{3}{4}$

Use $m = \dfrac{3}{4}$.
$$\begin{aligned} y - y_1 &= m(x - x_1) \\ y - 0 &= \frac{3}{4}(x - 2) \\ y &= \frac{3}{4}x - \frac{3}{2} \end{aligned}$$

84. $m = \dfrac{y_2 - y_1}{x_2 - x_1} = \dfrac{-10 - 4}{4 - 2} = -7$
$$\begin{aligned} y - y_1 &= m(x - x_1) \\ y + 2 &= -7(x - 7) \\ y + 2 &= -7x + 49 \\ y &= -7x + 47 \end{aligned}$$

85.
$$\begin{aligned} x + 3y &= 4 \\ 3y &= -x + 4 \\ y &= -\frac{1}{3}x + \frac{4}{3} \end{aligned}$$
$m = -\dfrac{1}{3}$

Use $m = 3$.
$$\begin{aligned} y - y_1 &= m(x - x_1) \\ y - 5 &= 3(x - 0) \\ y - 5 &= 3x \\ y &= 3x + 5 \end{aligned}$$

86. $m = \dfrac{y_2 - y_1}{x_2 - x_1} = \dfrac{-10 - 4}{4 - 2} = -7$

Use $m = \dfrac{1}{7}$.
$$\begin{aligned} y - y_1 &= m(x - x_1) \\ y + 2 &= \frac{1}{7}(x - 7) \\ y + 2 &= \frac{1}{7}x - 1 \\ y &= \frac{1}{7}x - 3 \end{aligned}$$

87. Let $x =$ the number of rolls hung and let $y =$ the total charge. Then two points on the line are given: $(11, 177)$ and $(20, 294)$

$$m = \frac{294 - 177}{20 - 11} = \frac{117}{9} = 13$$

$$y - y_1 = m(x - x_1)$$
$$y - 177 = 13(x - 11)$$
$$y - 177 = 13x - 143$$
$$y = 13x + 34$$

Let $x = 27$:

$y = 13(27) + 34 = 385$. The charge is \$385.

88. $14x + 18y = 5040$

Let $x = 180$:

$$14(180) + 18y = 5040$$
$$2520 + 18y = 5040$$
$$18y = 2520$$
$$y = 140$$

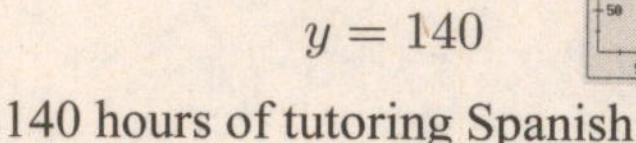

140 hours of tutoring Spanish

89.

$y = 4x - 8x^2$
$0 = 4x(1 - 2x)$
$x = 0, x = \frac{1}{2}$
x-int: $(0, 0), \left(\frac{1}{2}, 0\right)$

$y = 4x - 8x^2$
$y = 4(0) - 8(0)^2$
$y = 0$
y-int: $(0, 0)$

90.

$y = x^2 - 10x - 24$
$0 = (x - 12)(x + 2)$
$x = 12, x = -2$
x-int: $(12, 0), (-2, 0)$

$y = x^2 - 10x - 24$
$y = 0^2 - 10(0) - 24$
$y = -24$
y-int: $(0, -24)$

91. $y^2 = 8x$

x-axis	y-axis	origin
$(-y)^2 = 8x$	$y^2 = 8(-x)$	$(-y)^2 = 8(-x)$
$y^2 = 8x$	$y^2 = -8x$	$y^2 = -8x$
equivalent: **symmetry**	not equivalent: no symmetry	not equivalent: no symmetry

92. $y = 3x^4 + 6$

x-axis	y-axis	origin
$-y = 3x^4 + 6$	$y = 3(-x)^4 + 6$	$-y = 3(-x)^4 + 6$
not equivalent: no symmetry	$y = 3x^4 + 6$	$-y = 3x^4 + 6$
	equivalent: **symmetry**	not equivalent: no symmetry

93. $y = -2|x|$

x-axis	y-axis	origin
$-y = -2\|x\|$	$y = -2\|-x\|$	$-y = -2\|-x\|$
$y = 2\|x\|$	$y = -2\|-1\|\|x\|$	$y = 2\|-x\|$
not equivalent: no symmetry	$y = -2\|x\|$	$y = 2\|-1\|\|x\|$
	equivalent: **symmetry**	$y = 2\|x\|$
		not equivalent: no symmetry

94. $y = |x + 2|$

x-axis	y-axis	origin
$-y = \|x + 2\|$	$y = \|-x + 2\|$	$-y = \|-x + 2\|$
not equivalent: no symmetry	not equivalent: no symmetry	not equivalent: no symmetry

CHAPTER 2 REVIEW

95. $y = x^2 + 2$
x-int: none, y-int: $(0, 2)$
symmetry: y-axis

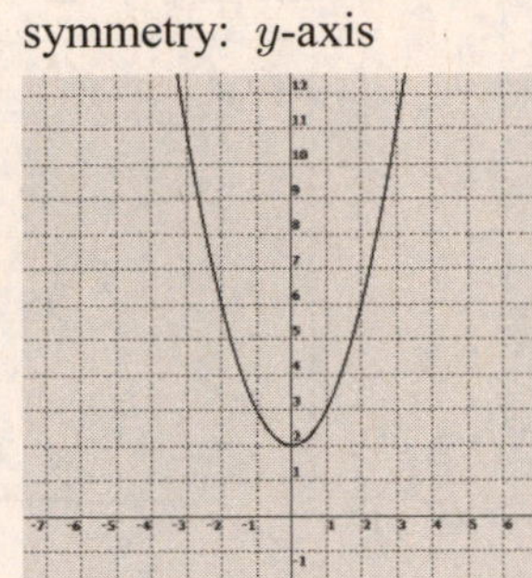

96. $y = -x^2 + 9$
x-int: $(\pm 3, 0)$, y-int: $(0, 9)$
symmetry: y-axis

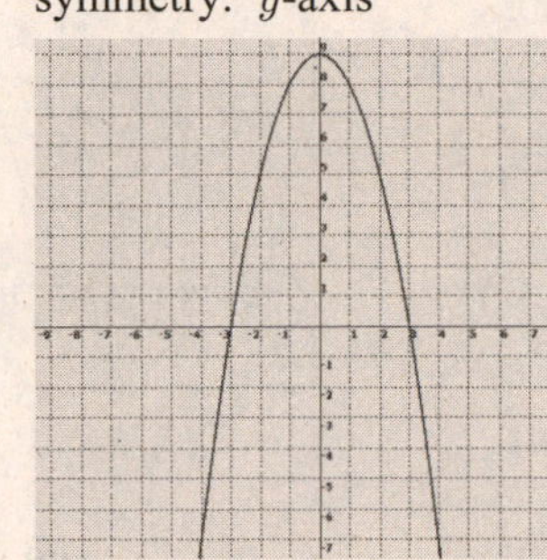

97. $y = x^3 - 2$
x-int: $\left(\sqrt[3]{2}, 0\right)$,
y-int: $(0, -2)$
symmetry: none

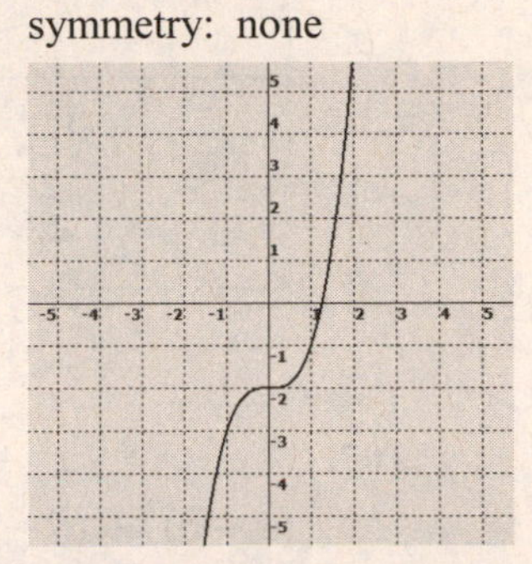

98. $y = \sqrt{x} + 2$
x-int: none, y-int: $(0, 2)$
symmetry: none

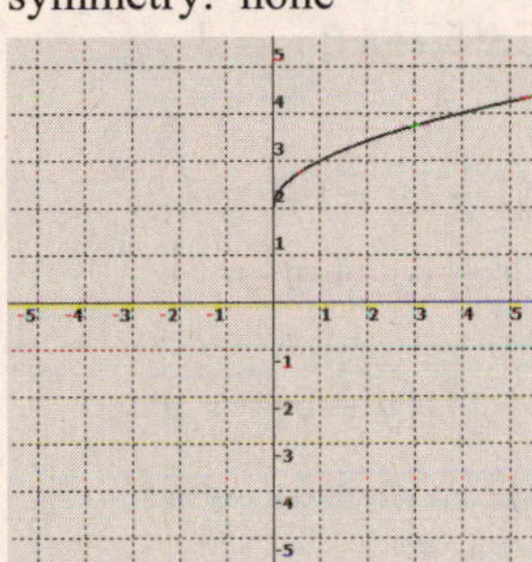

99. $y = -\sqrt{x - 4}$
x-int: $(4, 0)$, y-int: none
symmetry: none

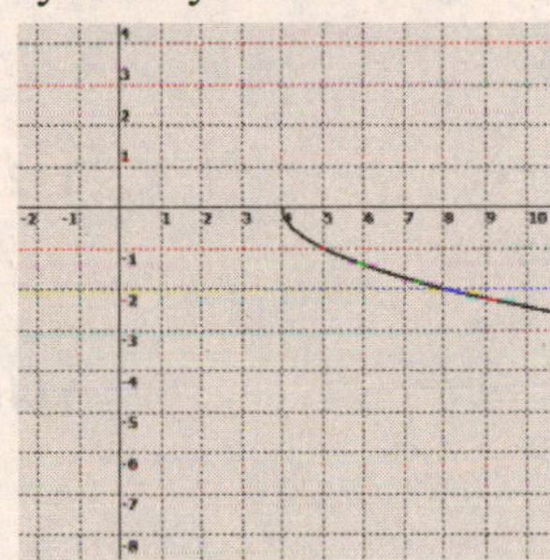

100. $y = \frac{1}{2}|x|$
x-int: $(0, 0)$, y-int: $(0, 0)$
symmetry: y-axis

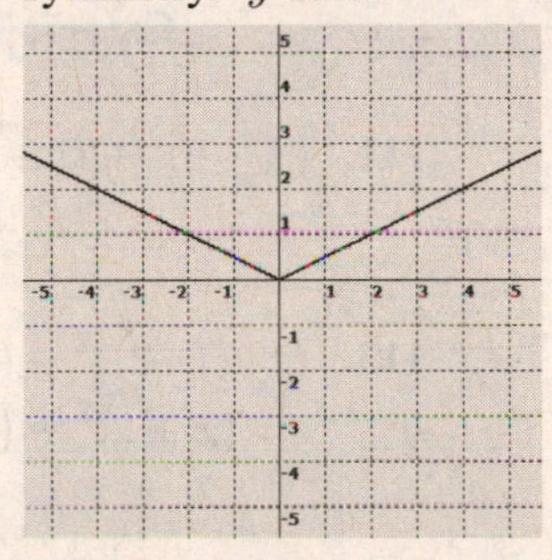

101. $y = |x + 1| + 2$
x-int: none, y-int: $(0, 3)$
symmetry: none

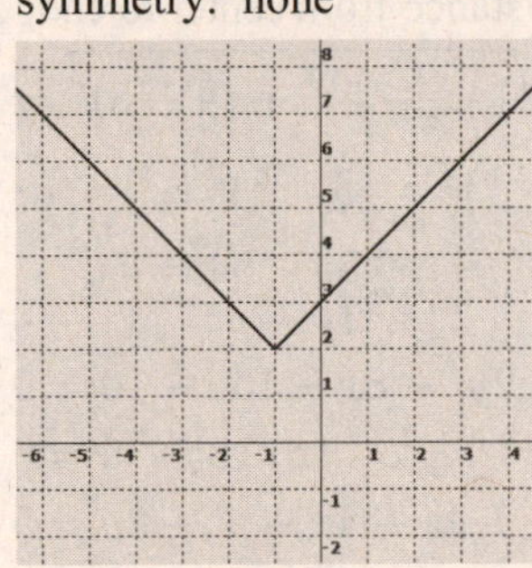

102. $y = \sqrt[3]{x} - 1$
x-int: $(1, 0)$, y-int: $(0, -1)$
symmetry: none

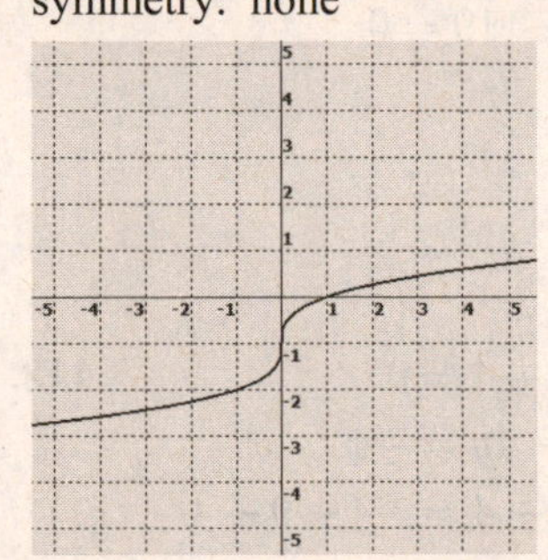

103. $y = |x - 4| + 2$

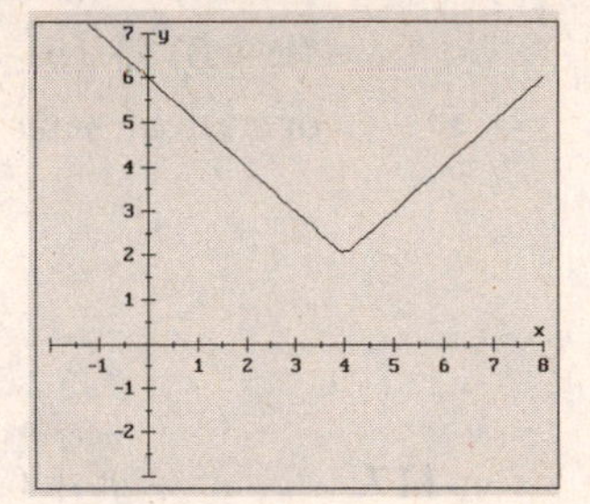

CHAPTER 2 REVIEW

104. $y = -\sqrt{x+2} + 3$

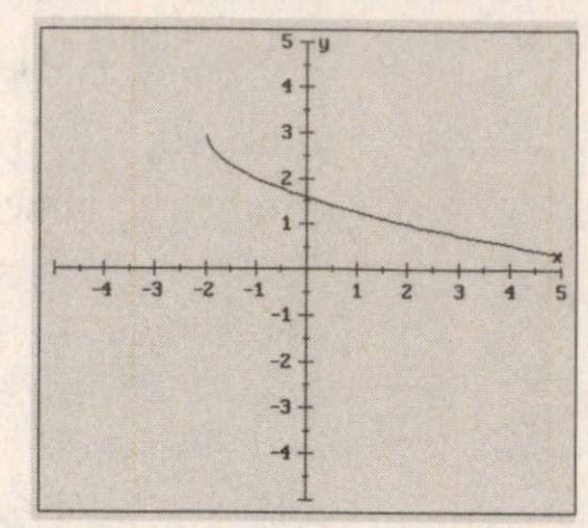

105. $y = x + 2|x|$

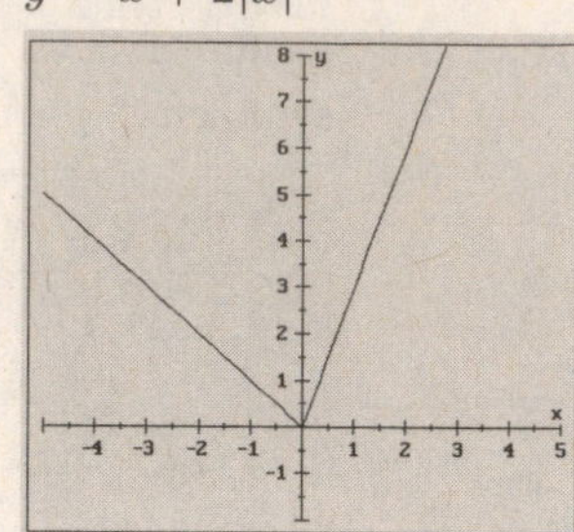

106. $y^2 = x - 3$

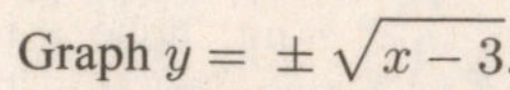

Graph $y = \pm\sqrt{x-3}$.

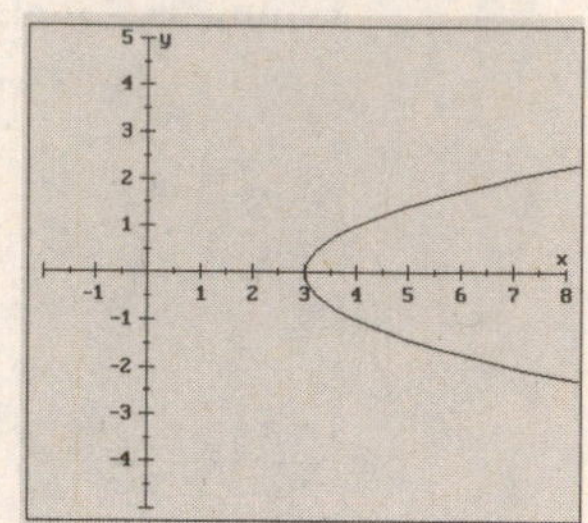

107.
$$x^2 + y^2 = 64$$
$$(x-0)^2 + (y-0)^2 = 8^2$$
C: $(0, 0)$; $r = 8$

108.
$$x^2 + (y-6)^2 = 100$$
$$(x-0)^2 + (y-6)^2 = 10^2$$
C: $(0, 6)$; $r = 10$

109.
$$(x+7)^2 + y^2 = \tfrac{1}{4}$$
$$(x-(-7))^2 + (y-0)^2 = \left(\tfrac{1}{2}\right)^2$$
C: $(-7, 0)$; $r = \frac{1}{2}$

110.
$$(x-5)^2 + (y+1)^2 = 9$$
$$(x-5)^2 + (y-(-1))^2 = 3^2$$
C: $(5, -1)$; $r = 3$

111.
$$(x-0)^2 + (y-0)^2 = 7^2$$
$$x^2 + y^2 = 49$$

112.
$$(x-3)^2 + (y-0)^2 = \left(\tfrac{1}{5}\right)^2$$
$$(x-3)^2 + y^2 = \tfrac{1}{25}$$

113.
$$(x-(-2))^2 + (y-12)^2 = 5^2$$
$$(x+2)^2 + (y-12)^2 = 25$$

114.
$$\left(x - \tfrac{2}{7}\right)^2 + (y-5)^2 = 9^2$$
$$\left(x - \tfrac{2}{7}\right)^2 + (y-5)^2 = 81$$

115. $C(-3, 4)$; $r = 12$
$$(x-h)^2 + (y-k)^2 = r^2$$
$$(x+3)^2 + (y-4)^2 = 144$$
or $x^2 + y^2 + 6x - 8y - 119 = 0$

116. Center: $x = \dfrac{-6+5}{2} = -\dfrac{1}{2}$
$$y = \frac{-3+8}{2} = \frac{5}{2}$$
$r =$ distance from center to endpoint
$$= \sqrt{\left(-\tfrac{1}{2} - 5\right)^2 + \left(\tfrac{5}{2} - 8\right)^2} = \sqrt{\tfrac{121}{2}}$$
$$\left(x + \tfrac{1}{2}\right)^2 + \left(y - \tfrac{5}{2}\right)^2 = \tfrac{121}{2}, \text{ or}$$
$$x^2 + y^2 + x - 5y - 54 = 0$$

117.
$$x^2 + y^2 + 6x - 4y + 4 = 0$$
$$x^2 + 6x + y^2 - 4y = -4$$
$$x^2 + 6x + 9 + y^2 - 4y + 4 = -4 + 9 + 4$$
$$(x+3)^2 + (y-2)^2 = 9$$

118.
$$2x^2 + 2y^2 - 8x - 16y - 10 = 0$$
$$x^2 + y^2 - 4x - 8y - 5 = 0$$
$$x^2 - 4x + y^2 - 8y = 5$$
$$x^2 - 4x + 4 + y^2 - 8y + 16 = 5 + 4 + 16$$
$$(x-2)^2 + (y-4)^2 = 25$$

CHAPTER 2 REVIEW

119.
$$x^2 + y^2 - 16 = 0$$
$$(x-0)^2 + (y-0)^2 = 16$$
$C(0, 0), r = 4$

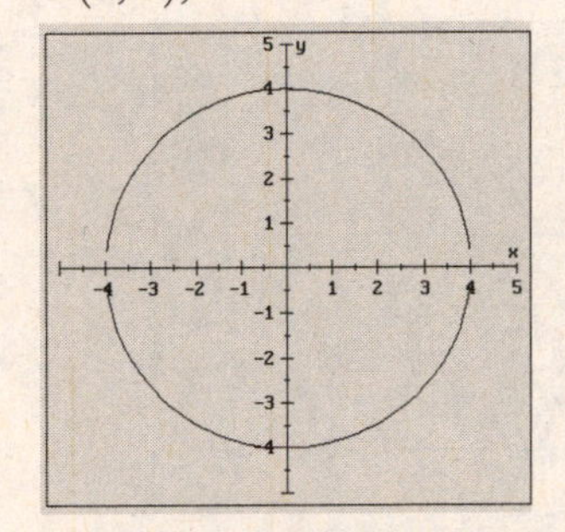

120.
$$x^2 + y^2 - 4x = 5$$
$$x^2 - 4x + y^2 = 5$$
$$x^2 - 4x + 4 + y^2 = 5 + 4$$
$$(x-2)^2 + y^2 = 9$$
$C(2, 0), r = 3$

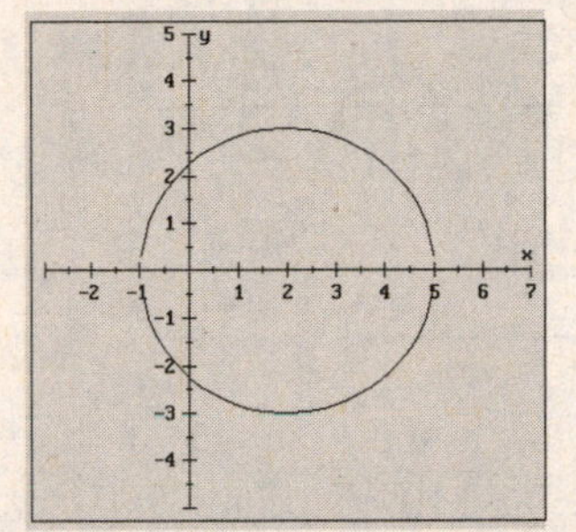

121.
$$x^2 + y^2 - 2y = 15$$
$$x^2 + y^2 - 2y + 1 = 15 + 1$$
$$x^2 + (y-1)^2 = 16$$
$C(0, 1), r = 4$

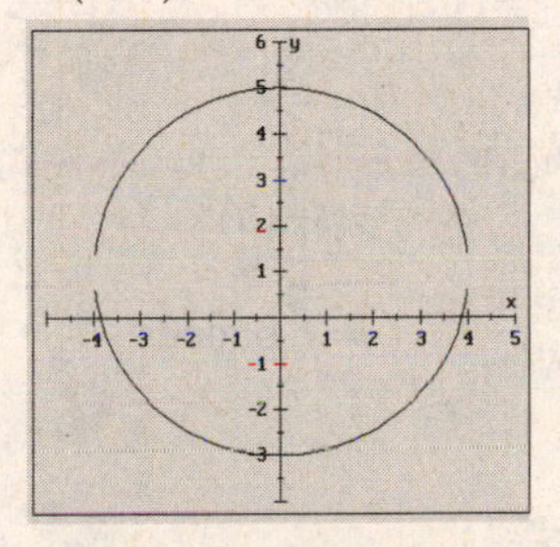

122.
$$x^2 + y^2 - 4x + 2y = 4$$
$$x^2 - 4x + 4 + y^2 + 2y + 1 = 4 + 4 + 1$$
$$(x-2)^2 + (y+1)^2 = 9$$
$C(2, -1), r = 3$

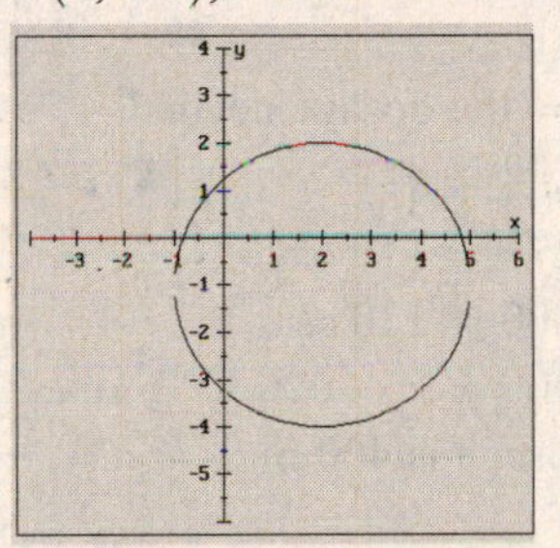

123. Graph $y = x^2 - 11$.
Find the x-intercepts.
$x = -3.32,\ x = 3.32$

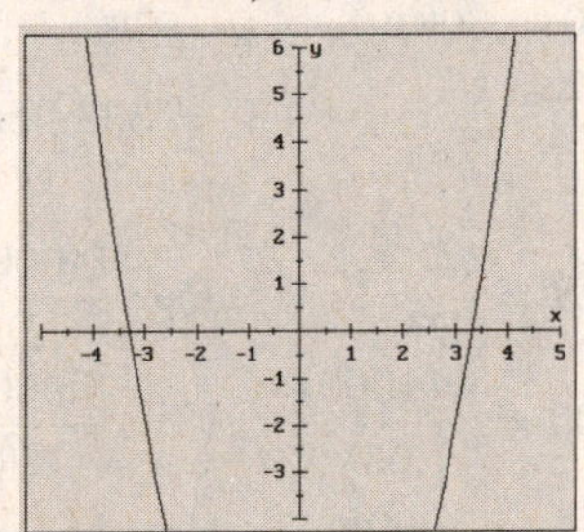

124. Graph $y = x^3 - x$.
Find the x-intercepts.
$x = -1,\ x = 0, x = 1$

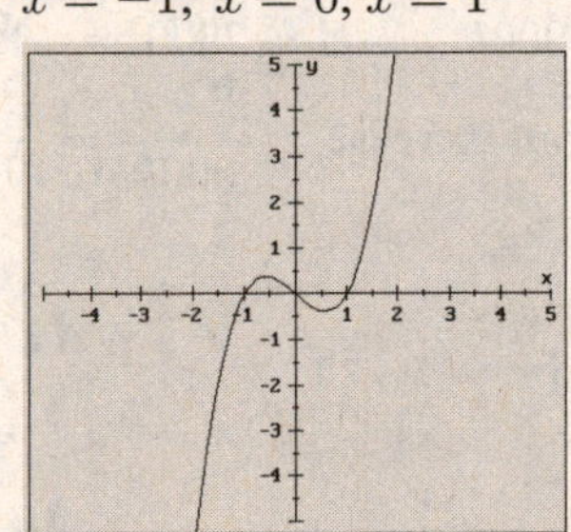

CHAPTER 2 REVIEW

125. Graph $y = |x^2 - 2| - 1$.
Find the x-intercepts.
$x = -1.73,\ x = -1,\ x = 1,\ x = 1.73$

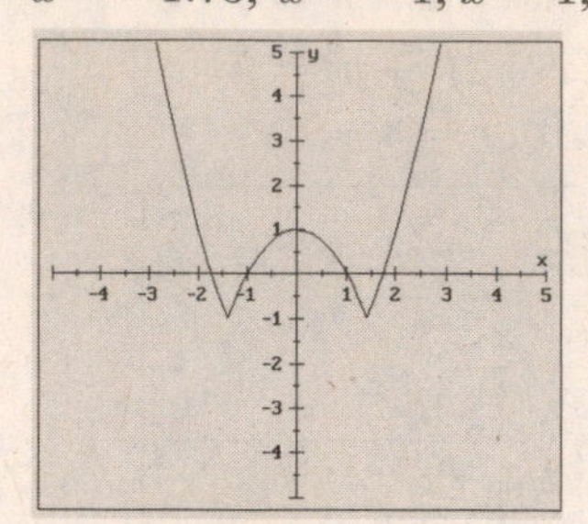

126. Graph $y = x^2 - 3x - 5$.
Find the x-intercepts.
$x = -1.19,\ x = 4.19$

127.

$$\begin{aligned} \frac{x+3}{10} &= \frac{x-1}{x} \\ x(x+3) &= 10(x-1) \\ x^2 + 3x &= 10x - 10 \\ x^2 - 7x + 10 &= 0 \\ (x-5)(x-2) &= 0 \\ x = 5 &\text{ or } x = 2 \end{aligned}$$

128.

$$\begin{aligned} \frac{x-1}{2} &= \frac{12}{x+1} \\ (x+1)(x-1) &= 2(12) \\ x^2 - 1 &= 24 \\ x^2 &= 25 \\ x &= \pm 5 \end{aligned}$$

129. Let $x =$ the dosage needed.

$$\begin{aligned} \tfrac{250}{110} &= \tfrac{x}{176} \\ 250 \cdot 176 &= 110 \cdot x \\ 44000 &= 110x \\ 400 &= x \end{aligned}$$

The dosage is 400 mg.

130.

$$\begin{aligned} f &= ks & f &= \frac{3}{5}s \\ 3 &= k(5) & f &= \frac{3}{5}(3) \\ \frac{3}{5} &= k & f &= \frac{9}{5}\text{pounds} \end{aligned}$$

131.

$$E = kv^2$$

30 mph	50 mph
$E = k(30)^2$	$E = k(50)^2$
$E = 900k$	$E = 2500k$

$$\text{Factor of increase} = \frac{2500k}{900k} = \frac{25}{9}$$

132.

$$\begin{aligned} V &= \frac{kT}{P} & V &= \frac{\frac{100}{3}T}{P} \\ 400 &= \frac{k(300)}{25} & V &= \frac{\frac{100}{3}(200)}{20} \\ 10000 &= 300k & V &= \tfrac{1000}{3} \\ \tfrac{100}{3} &= k & V &= 333\tfrac{1}{3}\text{ cm}^3 \end{aligned}$$

133. $A = klw$
$A = 1lw \Rightarrow k = 1$

134.

$$\begin{aligned} R &= \frac{kL}{D^2} & R &= \frac{0.0005L}{D^2} \\ 200 &= \frac{k(1000)}{(.05)^2} & V &= \frac{0.0005(1500)}{(0.08)^2} \\ 200 &= \frac{1000k}{.0025} & V &\approx 117\text{ ohms} \\ 0.0005 &= k \end{aligned}$$

Chapter 2 Test (page 291)

1. $f(x) = \dfrac{3}{2x-5}$

domain $= \left(-\infty, \dfrac{5}{2}\right) \cup \left(\dfrac{5}{2}, \infty\right)$

2. $f(x) = \sqrt{x+3}$: domain $= [-3, \infty)$

3. $f(-1) = \dfrac{-1}{-1-1} = \dfrac{-1}{-2} = \dfrac{1}{2}$

$f(2) = \dfrac{2}{2-1} = \dfrac{2}{1} = 2$

4. $f(-1) = \sqrt{-1+7} = \sqrt{6}$

$f(2) = \sqrt{2+7} = \sqrt{9} = 3$

5.
$$\begin{aligned}\frac{f(x+h)-f(x)}{h} &= \frac{\left[(x+h)^2-(x+h)+5\right]-[x^2-x+5]}{h}\\ &= \frac{[x^2+2xh+h^2-x-h+5]-[x^2-x+5]}{h}\\ &= \frac{x^2+2xh+h^2-x-h+5-x^2+x-5}{h}\\ &= \frac{2xh+h^2-h}{h} = \frac{h(2x+h-1)}{h} = 2x+h-1\end{aligned}$$

6. $(-3, \pi) \Rightarrow$ QII

7. $(0, -8) \Rightarrow$ negative y-axis

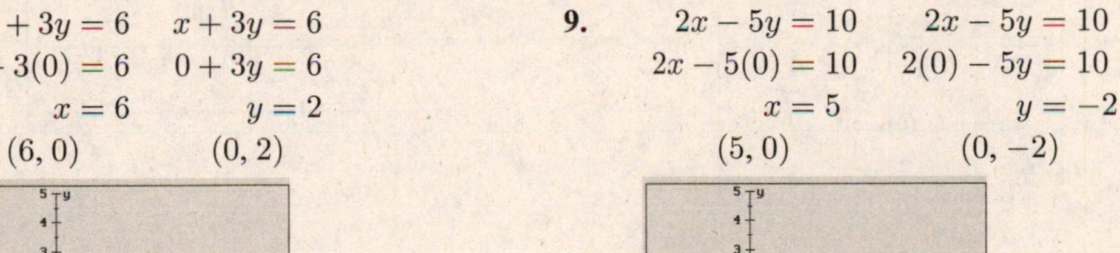

8.
$$\begin{array}{rr} x+3y=6 & x+3y=6\\ x+3(0)=6 & 0+3y=6\\ x=6 & y=2\\ (6,0) & (0,2)\end{array}$$

9.
$$\begin{array}{rr} 2x-5y=10 & 2x-5y=10\\ 2x-5(0)=10 & 2(0)-5y=10\\ x=5 & y=-2\\ (5,0) & (0,-2)\end{array}$$

CHAPTER 2 TEST

10. $2(x+y)=3x+5$
$2x+2y=3x+5$
$2y=x+5$
$y=\frac{1}{2}x+\frac{5}{2}$

x	y
0	$\frac{5}{2}$
1	3

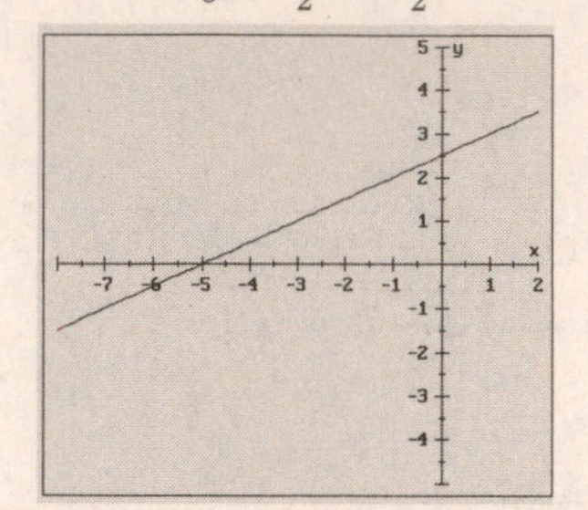

11. $3x-5y=3(x-5)$
$3x-5y=3x-15$
$-5y=-15$
$y=3$

x	y
0	3
-2	3

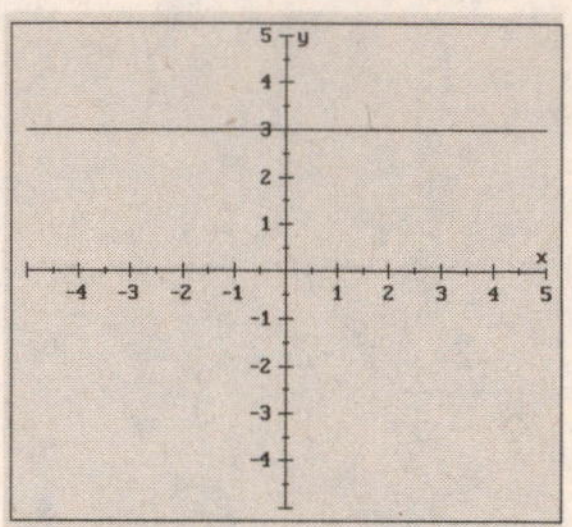

12. $\frac{1}{2}(x-2y)=y-1$
$\frac{1}{2}x-y=y-1$
$x-2y=2y-2$
$-4y=-x-2$
$y=\frac{1}{4}x+\frac{1}{2}$

x	y
0	$\frac{1}{2}$
2	1

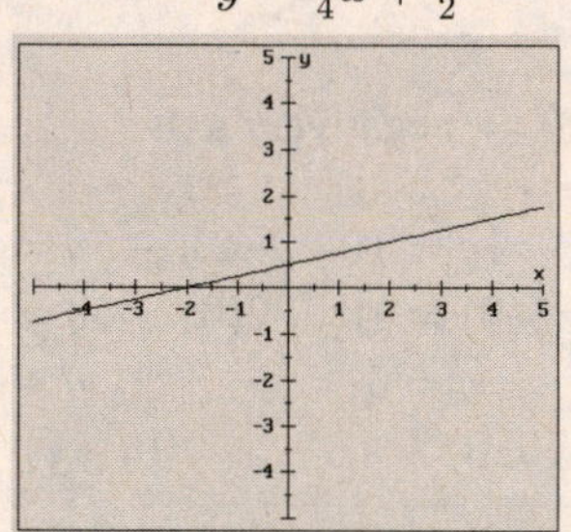

13. $\dfrac{x+y-5}{7}=3x$
$x+y-5=21x$
$y=20x+5$

x	y
0	5
$-\frac{1}{4}$	0

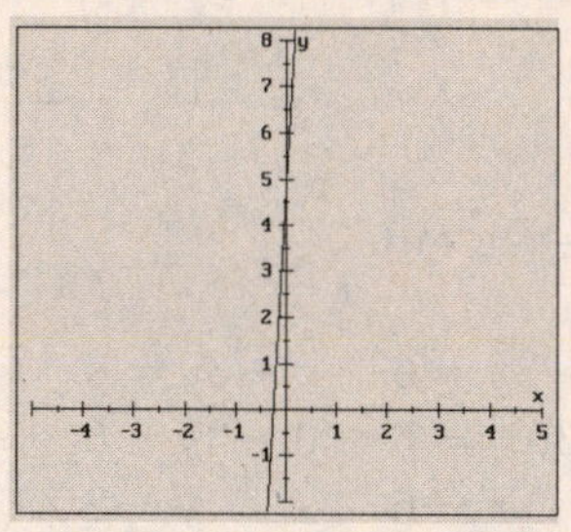

14. $$d=\sqrt{(x_2-x_1)^2+(y_2-y_1)^2}=\sqrt{(1-(-3))^2+(-1-4)^2}=\sqrt{(4)^2+(-5)^2}=\sqrt{16+25}=\sqrt{41}$$

15. $$d=\sqrt{(x_2-x_1)^2+(y_2-y_1)^2}=\sqrt{(0-(-\pi))^2+(\pi-0)^2}=\sqrt{\pi^2+\pi^2}=\sqrt{2\pi^2}=\pi\sqrt{2}\approx 4.44$$

16. $$M\left(\frac{x_1+x_2}{2},\frac{y_1+y_2}{2}\right)=M\left(\frac{3+(-3)}{2},\frac{-7+7}{2}\right)=M\left(\frac{0}{2},\frac{0}{2}\right)=M(0,0)$$

17. $$M\left(\frac{x_1+x_2}{2},\frac{y_1+y_2}{2}\right)=M\left(\frac{0+\sqrt{8}}{2},\frac{\sqrt{2}+\sqrt{18}}{2}\right)=M\left(\frac{2\sqrt{2}}{2},\frac{4\sqrt{2}}{2}\right)=M\left(\sqrt{2},2\sqrt{2}\right)$$

18. $$m=\frac{y_2-y_1}{x_2-x_1}=\frac{1-(-9)}{-5-3}=\frac{10}{-8}=-\frac{5}{4}$$

CHAPTER 2 TEST

19. $m = \frac{y_2 - y_1}{x_2 - x_1} = \frac{0 - 3}{-\sqrt{12} - \sqrt{3}} = \frac{-3}{-3\sqrt{3}} = \frac{1}{\sqrt{3}} = \frac{\sqrt{3}}{3}$

20. $y = 3x - 2$ $\quad y = 2x - 3$

$m = 3$ $\quad m = 2$

neither

21. $2x - 3y = 5$ $\quad 3x + 2y = 7$

$-3y = -2x + 5$ $\quad 2y = -3x + 7$

$y = \frac{2}{3}x - \frac{5}{3}$ $\quad y = -\frac{3}{2}x + \frac{7}{2}$

$m = \frac{2}{3}$ $\quad m = -\frac{3}{2}$

perpendicular

22. $y - y_1 = m(x - x_1)$

$y + 5 = 2(x - 3)$

$y + 5 = 2x - 6$

$y = 2x - 11$

23. $y = mx + b$

$y = 3x + \frac{1}{2}$

24. $2x - y = 3$

$-y = -2x + 3$

$y = 2x - 3$

$m = 2$ $\quad y = 2x + 5$

25. $2x - y = 3$

$-y = -2x + 3$

$y = 2x - 3$

$m = 2$ $\quad y = -\frac{1}{2}x + 5$

26. $m = \frac{y_2 - y_1}{x_2 - x_1} = \frac{\frac{1}{2} - \left(-\frac{3}{2}\right)}{3 - 2} = \frac{\frac{4}{2}}{1} = 2$

$y - y_1 = m(x - x_1)$

$y - \frac{1}{2} = 2(x - 3)$

$y - \frac{1}{2} = 2x - 6$

$y = 2x - \frac{11}{2}$

27. If the line is parallel to the y-axis, then it is a vertical line: $x = 3$

28. $y = x^3 - 16x$

$0 = x(x^2 - 16)$

$0 = x(x + 4)(x - 4)$

$x = 0, x = -4, x = 4$

x-int: $(0, 0), (-4, 0), (4, 0)$

$y = x^3 - 16x$

$y = 0^3 - 16(0)$

$y = 0$

y-int: $(0, 0)$

29. $y = |x - 4|$

$0 = |x - 4|$

$0 = x - 4$

$4 = x$

x-int: $(4, 0)$

$y = |x - 4|$

$y = |0 - 4|$

$y = |-4|$

$y = 4$

y-int: $(0, 4)$

30. $y^2 = x - 1$

x-axis	y-axis	origin
$(-y)^2 = x - 1$	$y^2 = -x - 1$	$(-y)^2 = -x - 1$
$y^2 = x - 1$	not equivalent: no symmetry	$y^2 = -x - 1$
equivalent: **symmetry**		not equivalent: no symmetry

CHAPTER 2 TEST

31. $y = x^4 + 1$

x-axis	y-axis	origin
$-y = x^4 + 1$	$y = (-x)^4 + 1$	$-y = (-x)^4 + 1$
not equivalent: no symmetry	$y = x^4 + 1$	$-y = x^4 + 1$
	equivalent: **symmetry**	not equivalent: no symmetry

32. $y = x^2 - 9$
x-int: $(3, 0), (-3, 0)$
y-int: $(0, -9)$
symmetry: y-axis

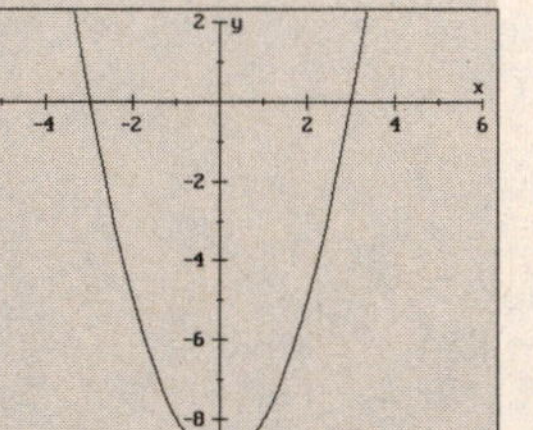

33. $x = |y|$
x-int: $(0, 0)$
y-int: $(0, 0)$
symmetry: x-axis

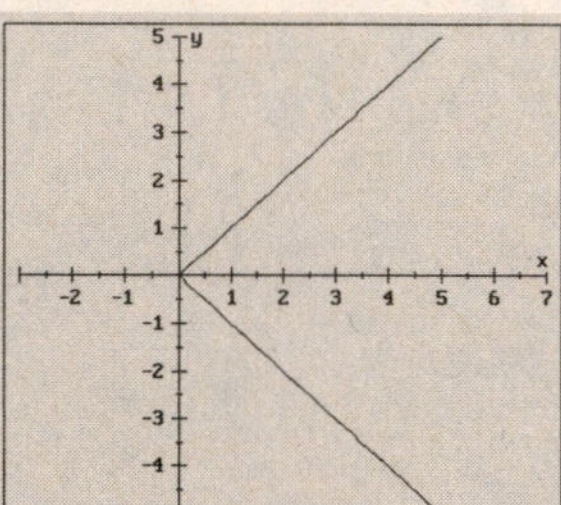

34. 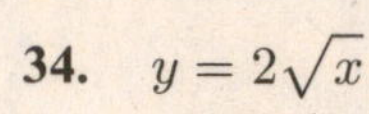 $y = 2\sqrt{x}$
x-int: $(0, 0)$
y-int: $(0, 0)$
symmetry: none

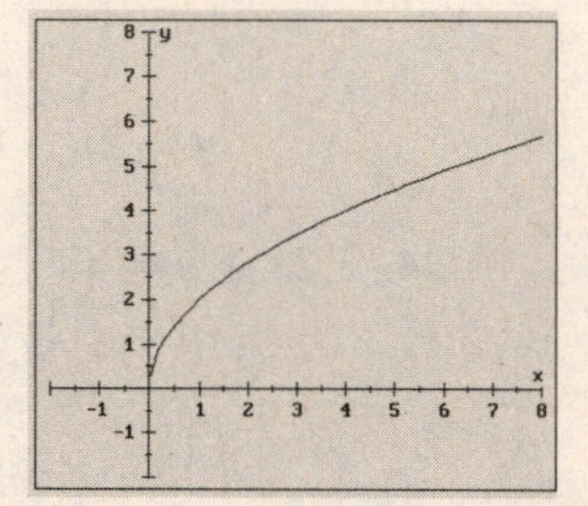

35. $x = y^3$
x-int: $(0, 0)$
y-int: $(0, 0)$
symmetry: origin

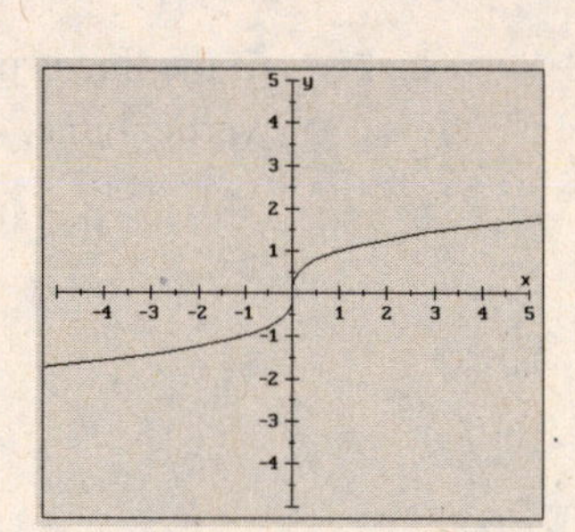

36. $C(5, 7); r = 8$
$(x - h)^2 + (y - k)^2 = r^2$
$(x - 5)^2 + (y - 7)^2 = 64$

37. $r = \sqrt{(2-6)^2 + (4-8)^2}$
$= \sqrt{32}$
$(x - h)^2 + (y - k)^2 = r^2$
$(x - 2)^2 + (y - 4)^2 = 32$

38. $x^2 + y^2 = 9$
$C(0, 0), r = 3$

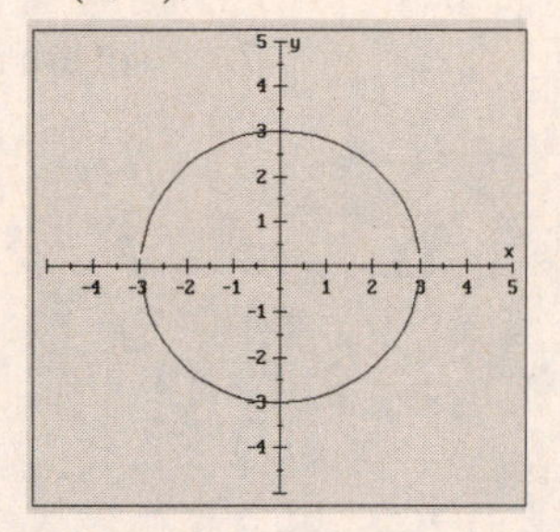

39. $x^2 - 4x + y^2 + 3 = 0$
$x^2 - 4x + y^2 = -3$
$x^2 - 4x + 4 + y^2 = -3 + 4$
$(x - 2)^2 + y^2 = 1$
$C(2, 0), r = 1$

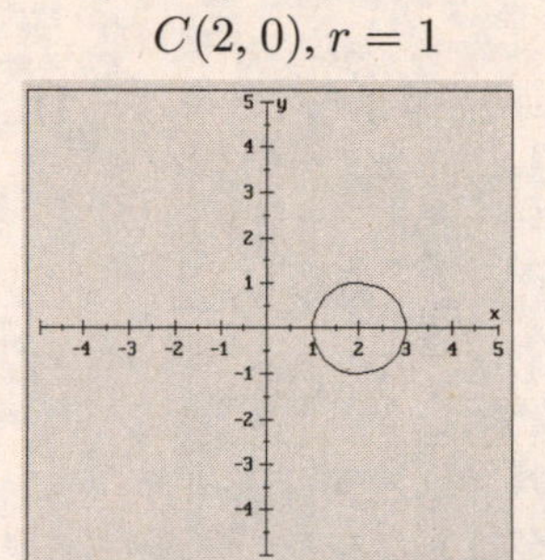

40. $y = kz^2$

41. $w = krs^2$

42. $P = kQ$
$7 = k(2)$
$\frac{7}{2} = k$
$P = \frac{7}{2}Q$
$P = \frac{7}{2}(5)$
$P = \frac{35}{2}$

43. $y = \frac{kx}{z^2}$
$16 = \frac{k(3)}{2^2}$
$16 = \frac{3k}{4}$
$\frac{64}{3} = k$
$y = \frac{\frac{64}{3}x}{z^2}$
$2 = \frac{\frac{64}{3}x}{3^2}$
$18 = \frac{64}{3}x$
$\frac{3}{64} \cdot 18 = \frac{3}{64} \cdot \frac{64}{3}x$
$\frac{27}{32} = x$

44. Graph $y = x^2 - 7$.
Find any positive x-intercept.
$x = 2.65$

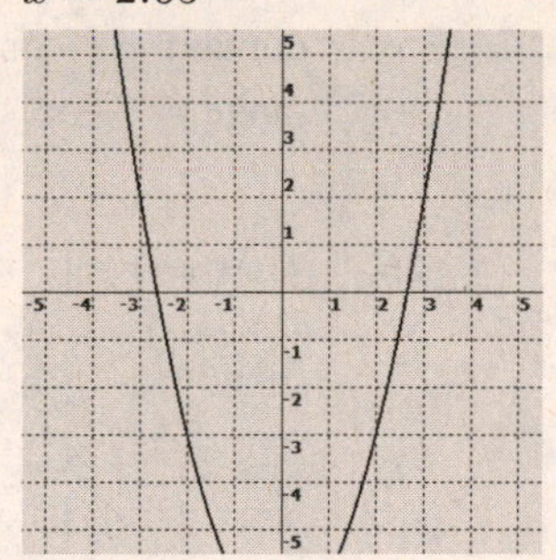

45. Graph $y = x^2 - 5x - 5$.
Find any plosive x-intercept.
$x = 5.85$

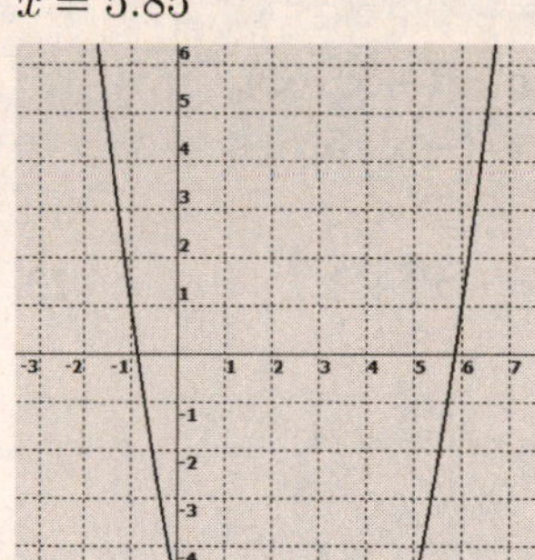

Exercises 3.1 (page 302)

1. (x, y), domain range

3. identity

5. cubing

7. square root

9. $f(x) = 2x + 3$

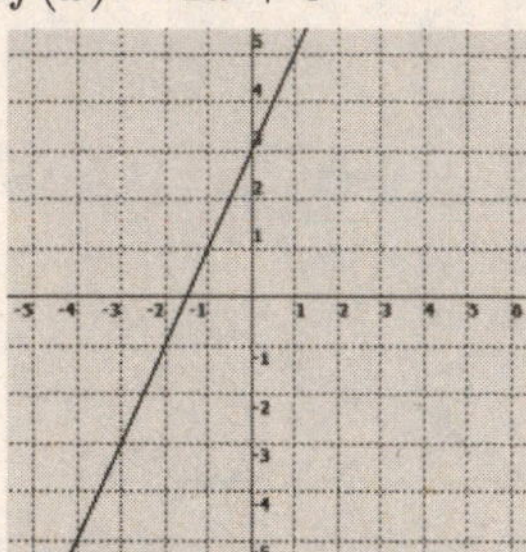

domain $= (-\infty, \infty)$
range $= (-\infty, \infty)$

11. $f(x) = -\dfrac{3}{4}x + 4$

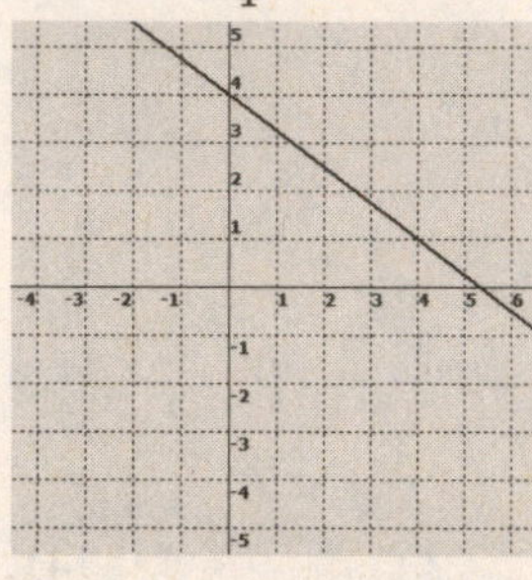

domain $= (-\infty, \infty)$
range $= (-\infty, \infty)$

13. $f(x) = x^2 - 4$

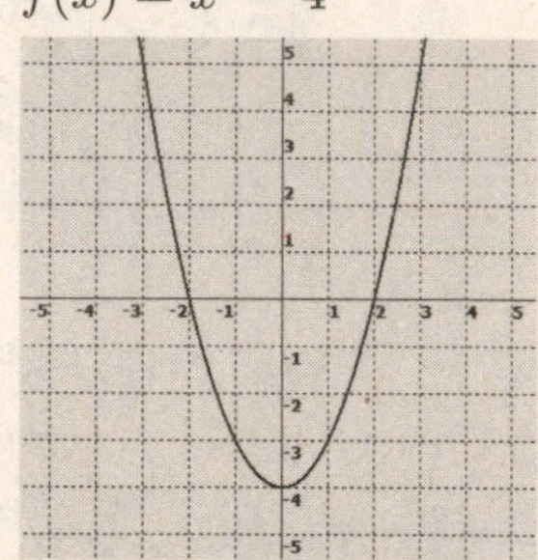

domain $= (-\infty, \infty)$
range $= [-4, \infty)$

15. $f(x) = -\dfrac{1}{2}x^2 + 5$

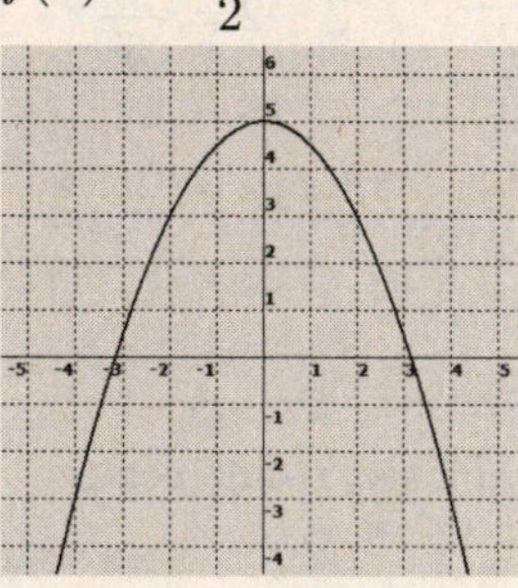

domain $= (-\infty, \infty)$
range $= (-\infty, 5]$

17. $f(x) = 3(x + 2)^2$

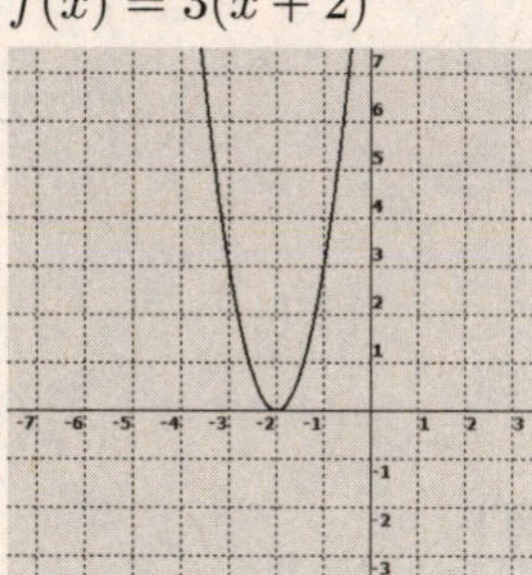

domain $= (-\infty, \infty)$
range $= [0, \infty)$

19. $f(x) = x^3 - 2$

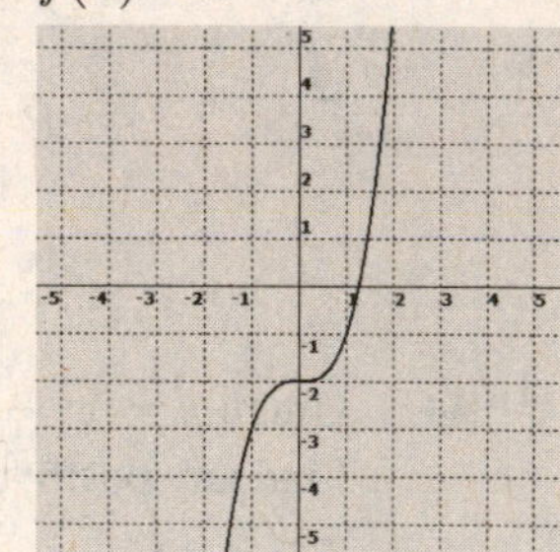

domain $= (-\infty, \infty)$
range $= (-\infty, \infty)$

21. $f(x) = -x^3 + 1$

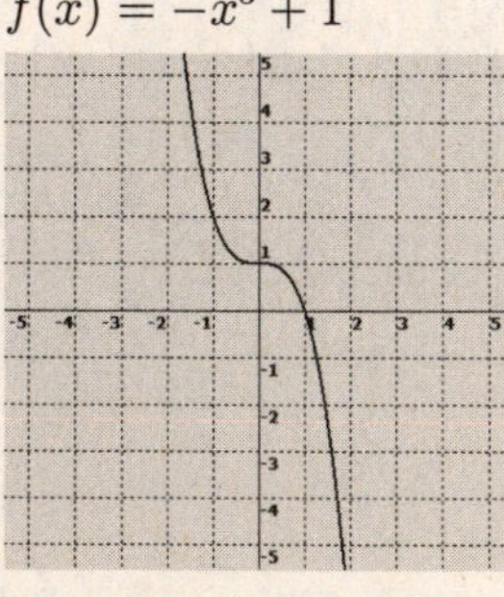

domain $= (-\infty, \infty)$
range $= (-\infty, \infty)$

23. $f(x) = -\dfrac{1}{2}x^3 - 4$

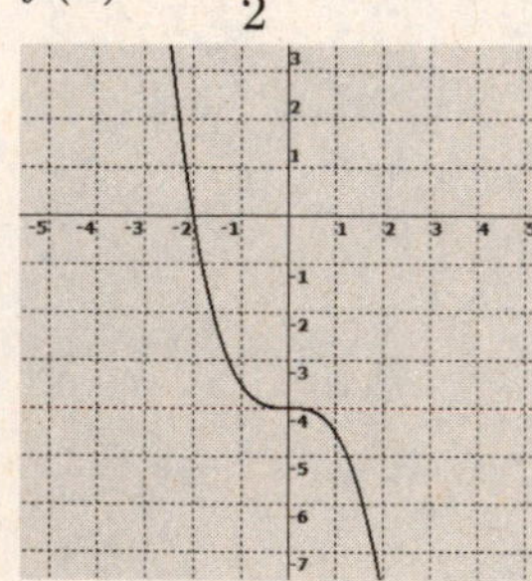

domain $= (-\infty, \infty)$
range $= (-\infty, \infty)$

25. $f(x) = -|x|$

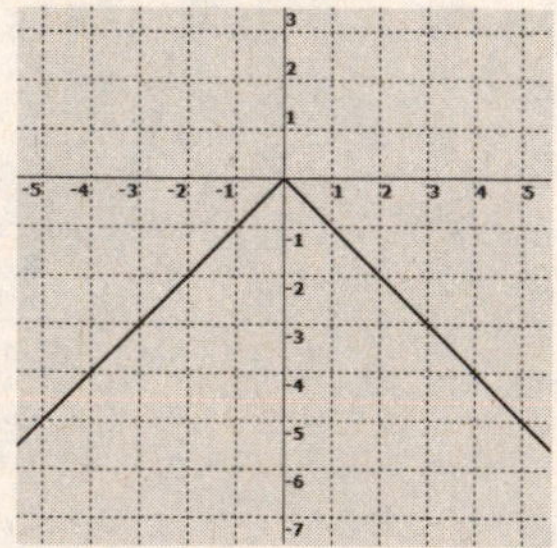

domain $= (-\infty, \infty)$
range $= (-\infty, 0]$

EXERCISES 3.1

27. $f(x) = |x - 2|$

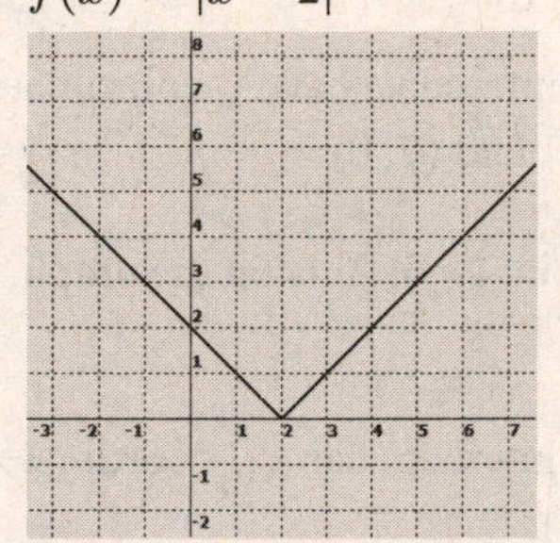

domain $= (-\infty, \infty)$
range $= [0, \infty)$

29. $f(x) = \left|\frac{1}{2}x + 3\right|$

domain $= (-\infty, \infty)$
range $= [0, \infty)$

31. $f(x) = 4|x| + 1$

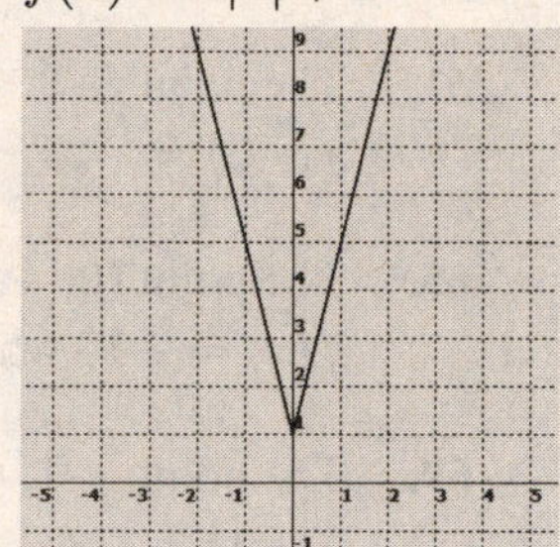

domain $= (-\infty, \infty)$
range $= [1, \infty)$

33. $f(x) = \sqrt{x} + 2$

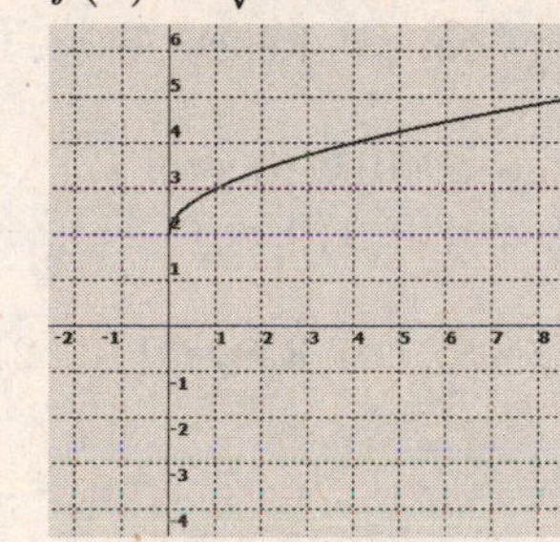

domain $= [0, \infty)$
range $= [2, \infty)$

35. $f(x) = 2\sqrt{x} - 3$

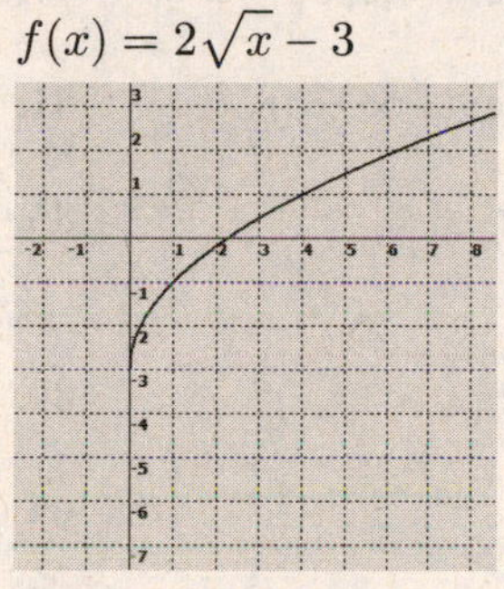

domain $= [0, \infty)$
range $= [-3, \infty)$

37. $f(x) = \sqrt{2x - 4}$

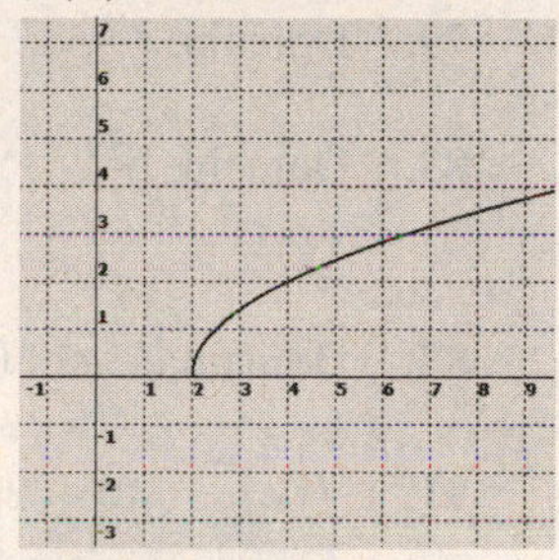

domain $= [2, \infty)$
range $= [0, \infty)$

39. $f(x) = \sqrt[3]{x} + 2$

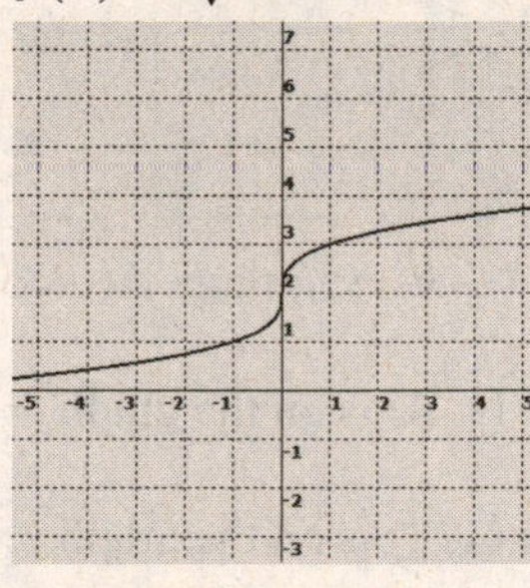

domain $= (-\infty, \infty)$
range $= (-\infty, \infty)$

41. $f(x) = 3\sqrt[3]{x}$

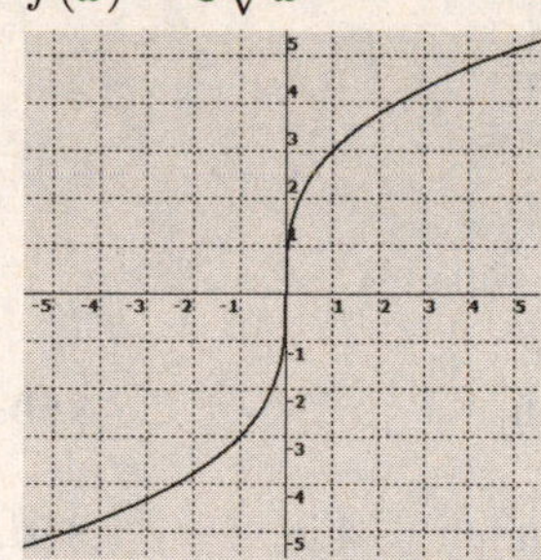

domain $= (-\infty, \infty)$
range $= (-\infty, \infty)$

43. $f(x) = -2\sqrt[3]{x + 1}$

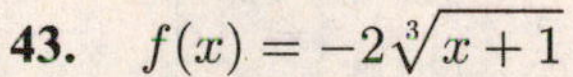

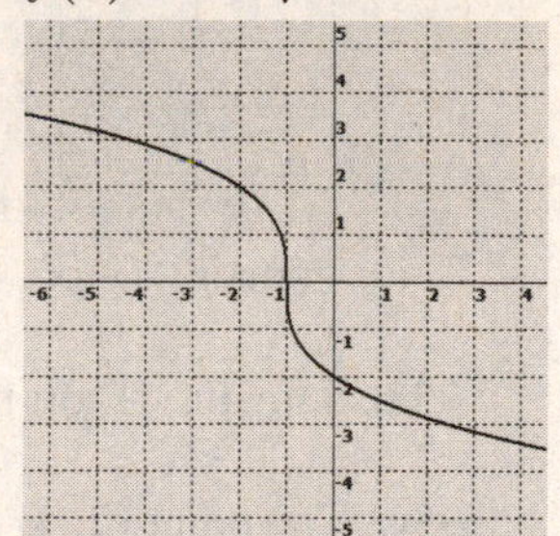

domain $= (-\infty, \infty)$
range $= (-\infty, \infty)$

45. function

47. not a function

49. function

51. function

53. The point $(-11, 2)$ is on the graph, so $f(-11) = 2$.

55. The point $(2, 0)$ is on the graph, so $f(2) = 0$.

57. The x-intercepts have y-coordinates of 0: $(2, 0)$ and $(8, 0)$

59. The point $(-6, 6)$ is on the graph, so $f(x) = 6$ when $x = -6$.

61. The point $(-3, -25)$ is on the graph, so $f(-3) = -25$.

63. The x-intercepts have y-coordinates of 0: $(-2, 0)$ and $(2, 0)$

65. The point $(0, -16)$ is on the graph, so $f(x) = -16$ when $x = 0$.

67. The point $(-6, 2)$ is on the graph, so $f(-6) = 2$.

69. The point $(-1, -3)$ is on the graph, so $f(-1) = -3$.

71. The x-intercept has a y-coordinate of 0: $(-4, 0)$

73. The point $(-5, 1)$ is on the graph, so $f(x) = 1$ when $x = -5$.

75. Domain: $(-\infty, \infty)$, Range: $(-\infty, \infty)$

77. Domain: $(-\infty, \infty)$, Range: $[-4, \infty)$

79. Domain: $(-\infty, \infty)$, Range: $(-\infty, \infty)$

81. Domain: $(-\infty, \infty)$, Range: $(-\infty, 1]$

83. Domain: $[-2, 1)$, Range: $(-3, 3]$

85. Domain: $(-\infty, 0) \cup (0, \infty)$
Range: $(-\infty, 0) \cup (0, \infty)$

87. Domain: $(-\infty, 0]$, Range: $[2, \infty)$

89. Domain: $(-\infty, \infty)$, Range: $\{-2, 2\}$

91. $f(x) = |3x + 2|$

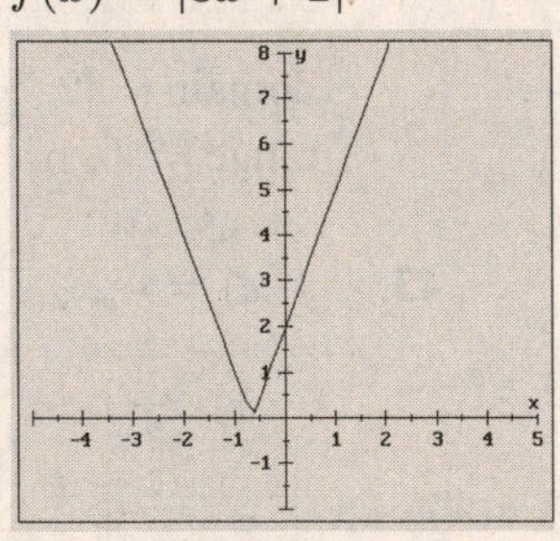

domain: $(-\infty, \infty)$; range: $[0, \infty)$

93. $f(x) = \sqrt[3]{5x - 1}$

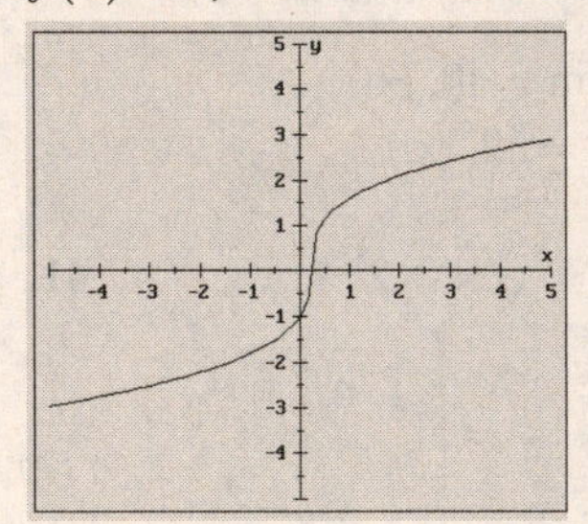

domain: $(-\infty, \infty)$; range: $(-\infty, \infty)$

95a. domain: $[1, 6]$; range: $[2, 5]$

95b. The point $(1, 5)$ is on the graph $\Rightarrow$ 5 in.

95c. The point $(2, 4)$ is on the graph $\Rightarrow$ 4 in.

95d. The point $(4, 2)$ is on the graph $\Rightarrow$ August

97-99. Answers may vary.

101. $\sqrt{5} \approx 2.236$

103. False. Functions always pass the Vertical Line Test.

105. False. The range does not have to be the same as the domain.

107. b

109. a

Exercises 3.2 (page 322)

1. upward **3.** to the right **5.** 2, downward **7.** y-axis

9. horizontally

11. $g(x) = x^2 - 2$
Shift $f(x) = x^2$ D 2

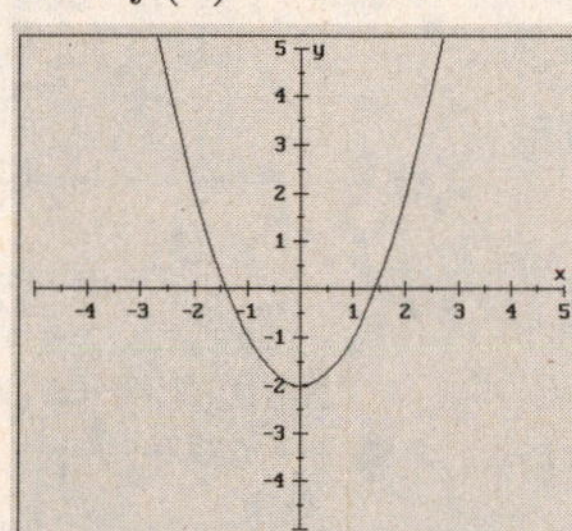

13. $g(x) = (x + 3)^2$
Shift $f(x) = x^2$ L 3

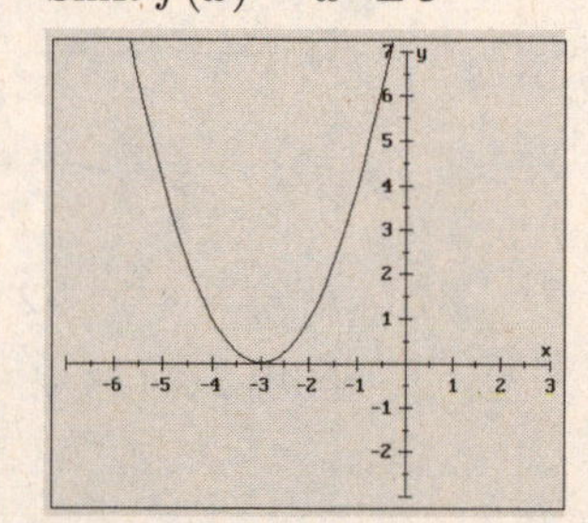

15. $h(x) = (x + 1)^2 + 2$
Shift $f(x) = x^2$ U 2, L 1

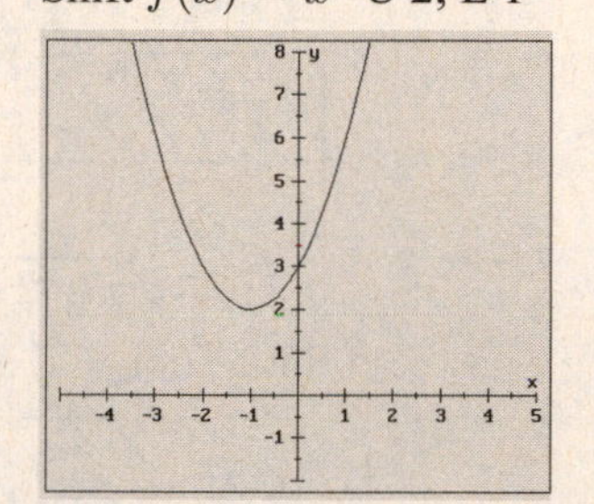

17. $h(x) = \left(x + \frac{1}{2}\right)^2 - \frac{1}{2}$
Shift $f(x) = x^2$ D $\frac{1}{2}$, L $\frac{1}{2}$

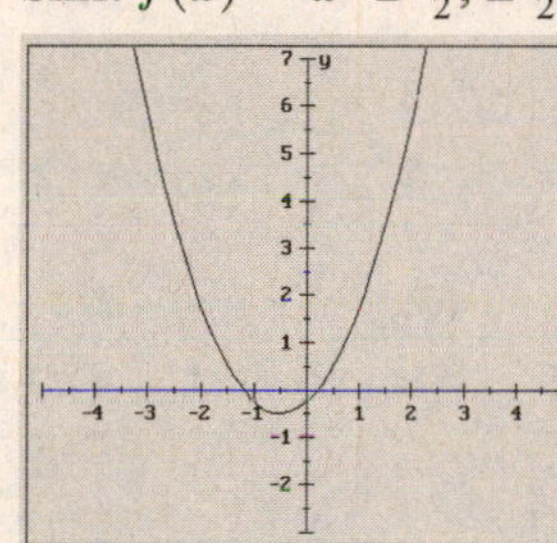

19. $g(x) = x^3 + 1$
Shift $f(x) = x^3$ U 1

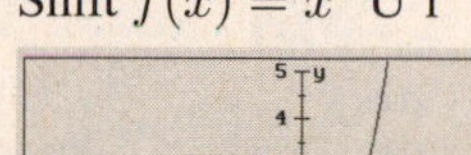

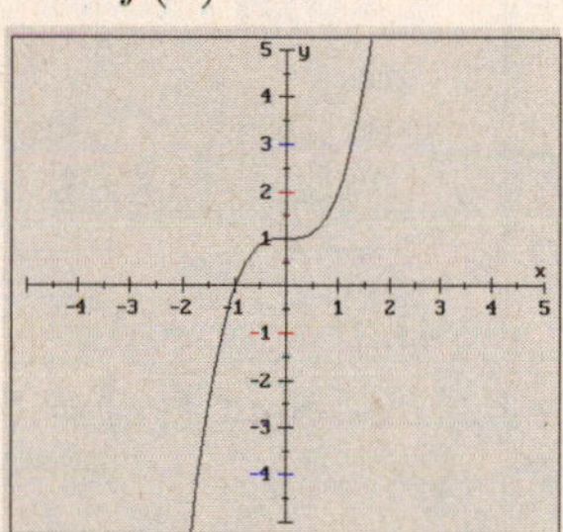

21. $g(x) = (x - 2)^3$
Shift $f(x) = x^3$ R 2

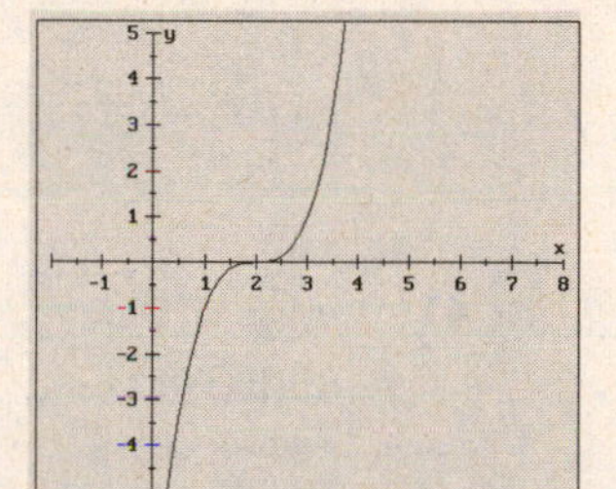

23. $h(x) = (x - 2)^3 - 3$
Shift $f(x) = x^3$ D 3, R 2

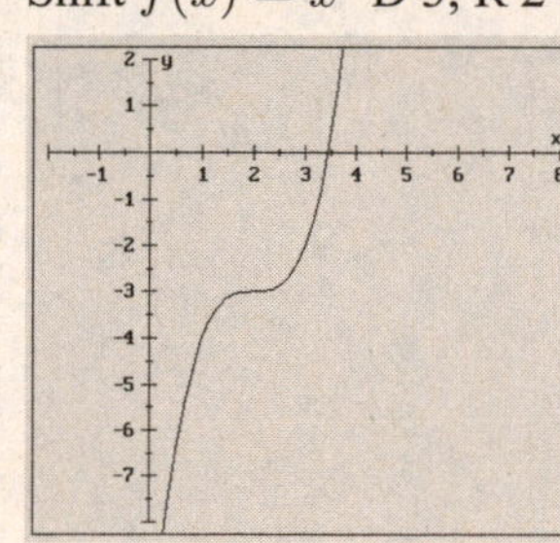

25. $y + 2 = x^3$
$y = x^3 - 2$
Shift $f(x) = x^3$ D 2

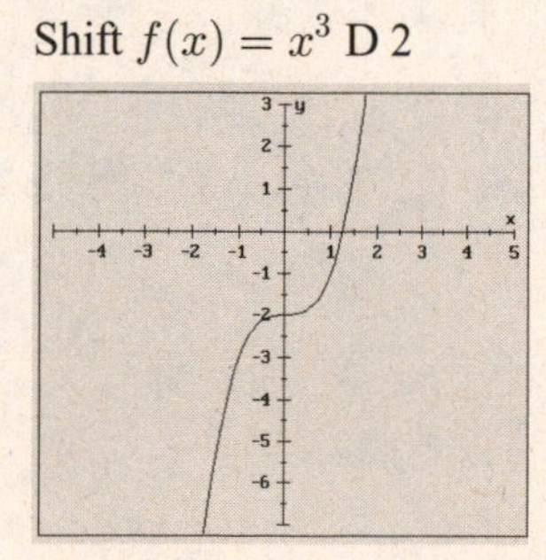

27. $g(x) = |x| + 2$
Shift $f(x) = |x|$ U 2

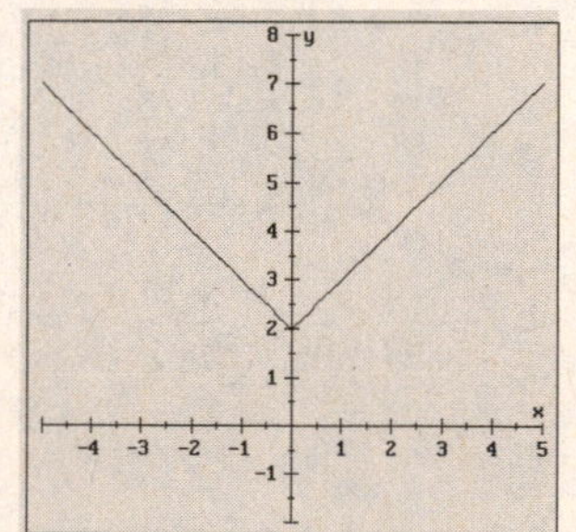

EXERCISES 3.2

29. $g(x) = |x - 5|$

Shift $f(x) = |x|$ R 5

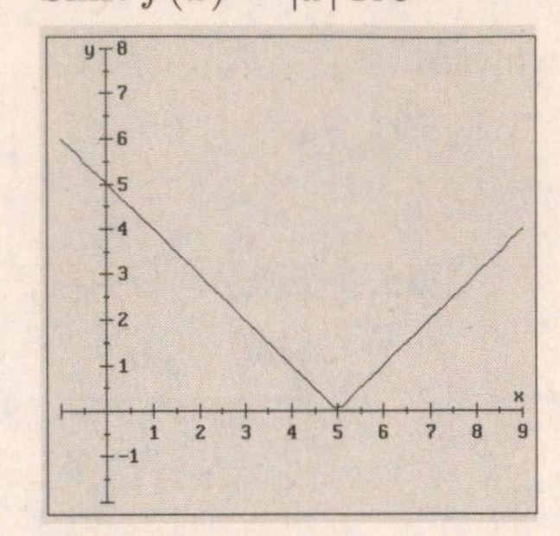

31. $f(x) = |x + 2| - 1$

Shift $f(x) = |x|$ D 1, L 2

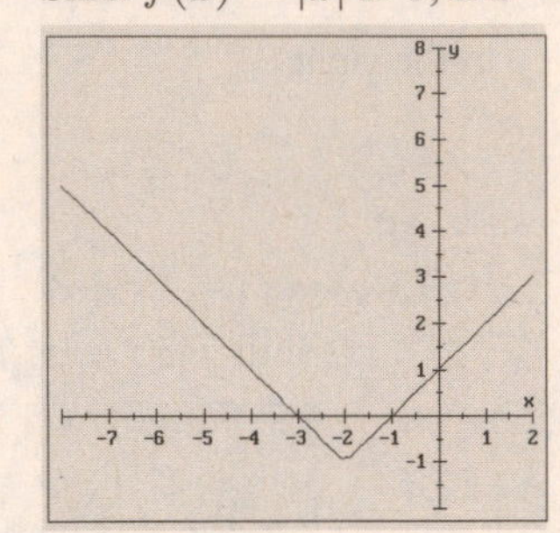

33. $g(x) = \sqrt{x} + 1$

Shift $f(x) = \sqrt{x}$ U 1

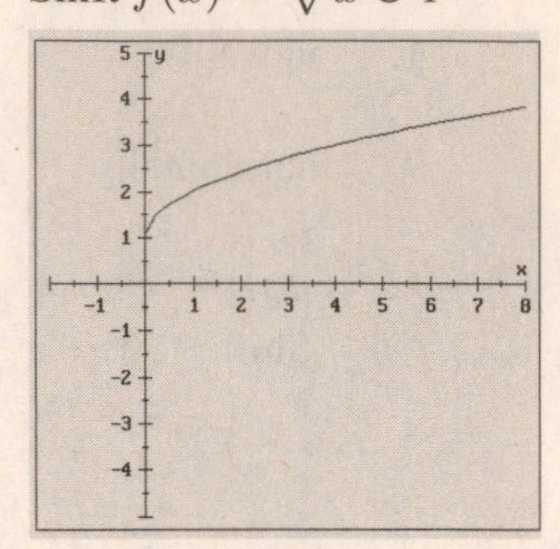

35. $g(x) = \sqrt{x + 2}$

Shift $f(x) = \sqrt{x}$ L 2

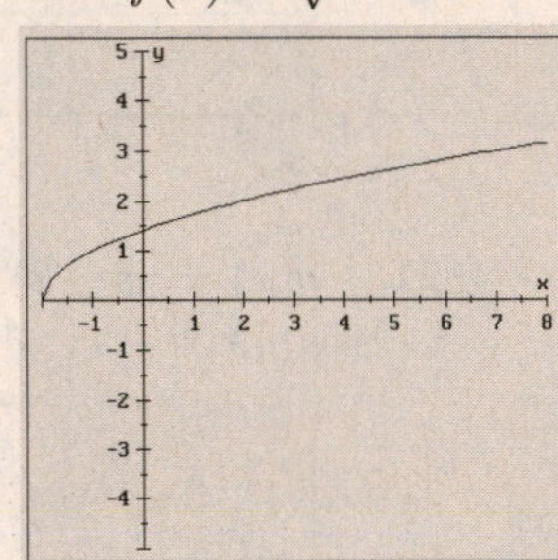

37. $h(x) = \sqrt{x - 2} - 1$

Shift $f(x) = \sqrt{x}$ D 1, R 2

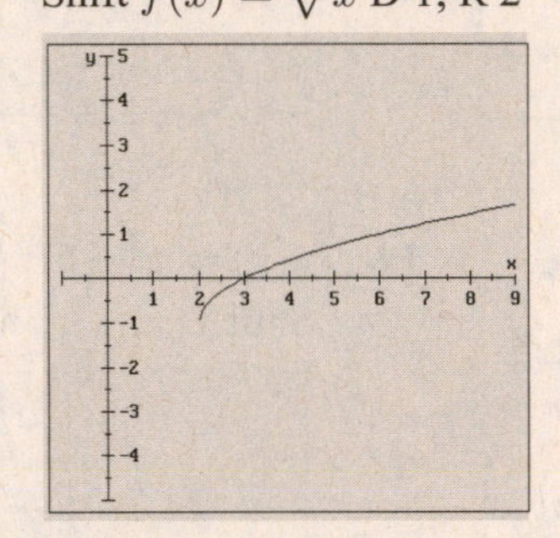

39. $g(x) = \sqrt[3]{x} - 4$

Shift $f(x) = \sqrt[3]{x}$ D 4

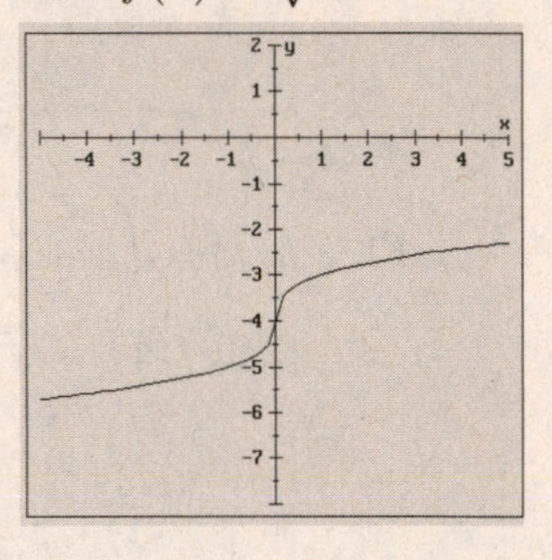

41. $g(x) = \sqrt[3]{x - 2}$

Shift $f(x) = \sqrt[3]{x}$ R 2

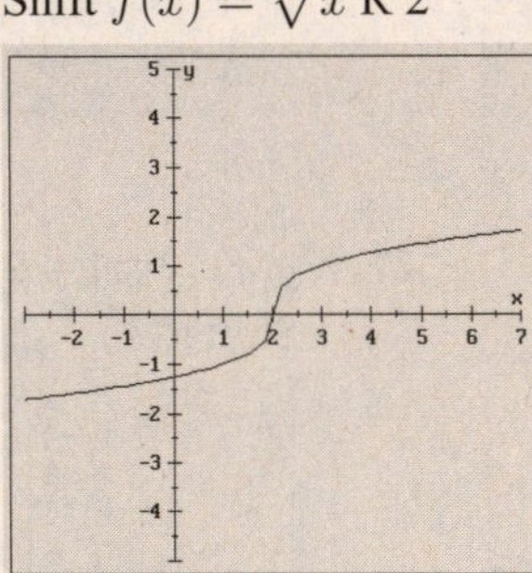

43. $h(x) = \sqrt[3]{x + 1} - 1$

Shift $f(x) = \sqrt[3]{x}$ D 1, L 1

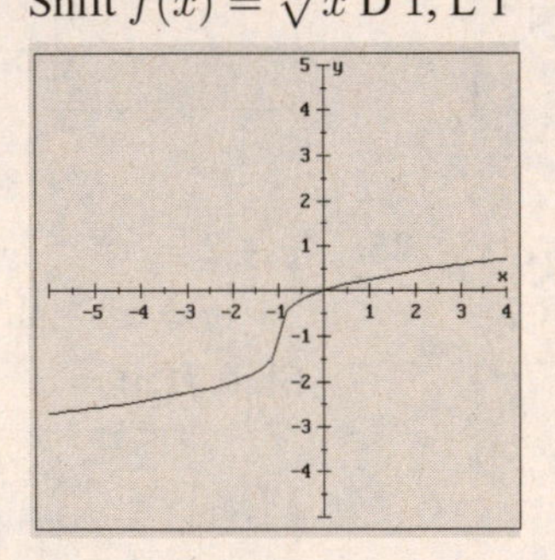

45. $f(x) = -x^2$

Reflect $y = x^2$ about x

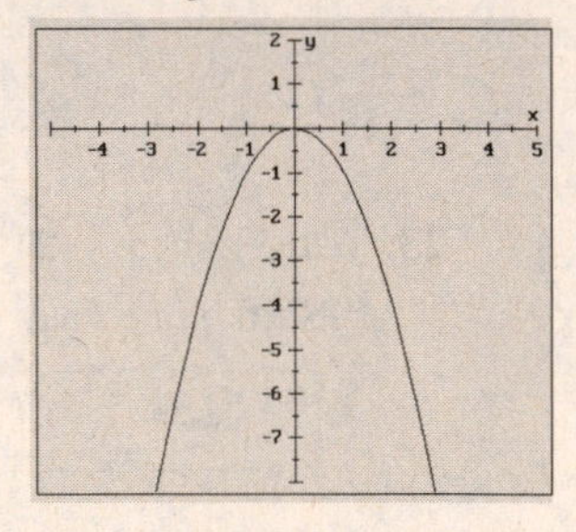

EXERCISES 3.2

47. $h(x) = -x^3$
Reflect $y = x^3$ about x

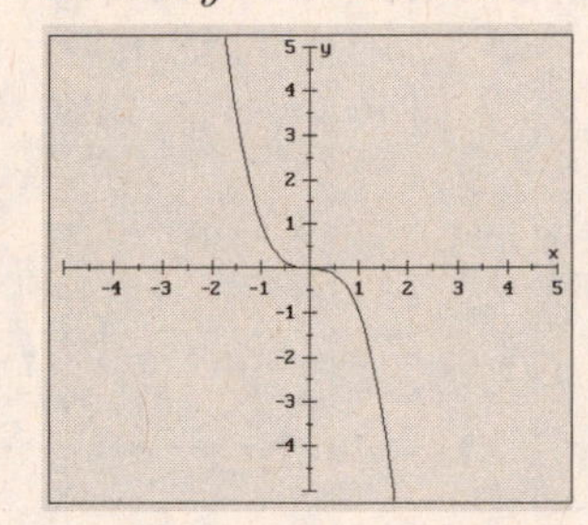

49. $f(x) = -|x|$
Reflect $y = |x|$ about x

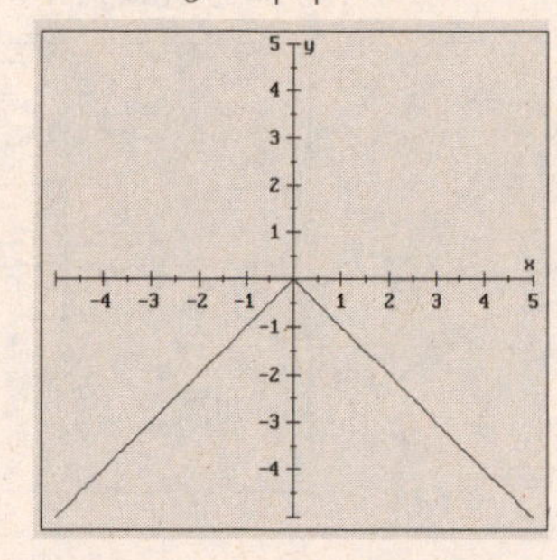

51. $f(x) = -\sqrt{x}$
Reflect $y = \sqrt{x}$ about x

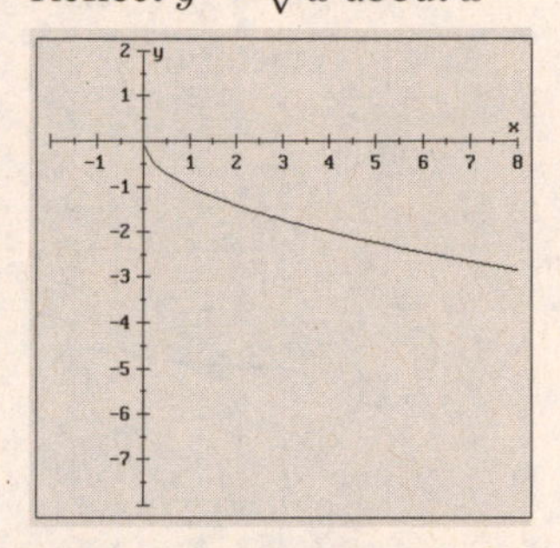

53. $f(x) = -\sqrt[3]{x}$
Reflect $y = \sqrt[3]{x}$ about x

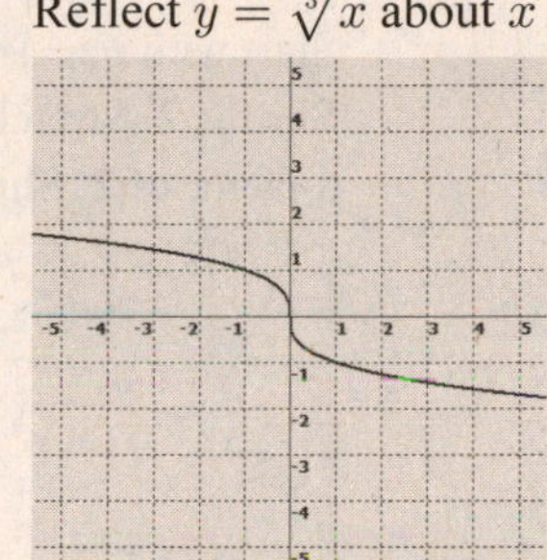

55. $f(x) = 2x^2$: Stretch
$y = x^2$ vert. by a factor of 2

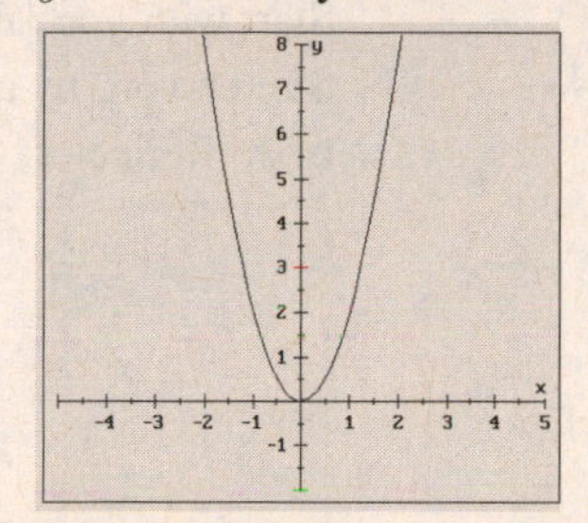

57. $h(x) = -3x^2$: Stretch
$y = x^2$ vert. by a factor of 3
Reflect about x

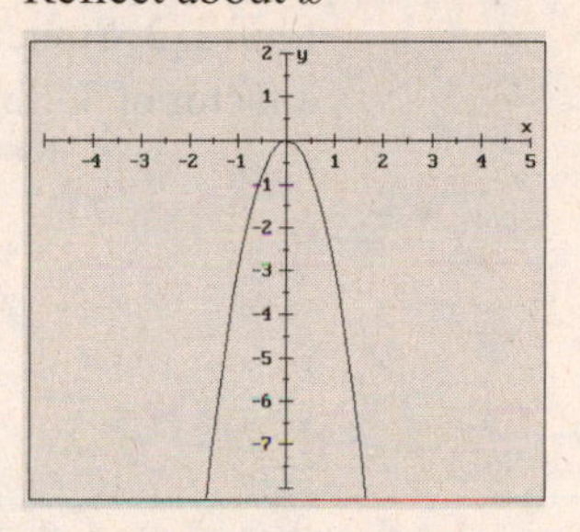

59. $f(x) = \frac{1}{2}x^3$: Shrink
$y = x^3$ vert. by a factor of $\frac{1}{2}$

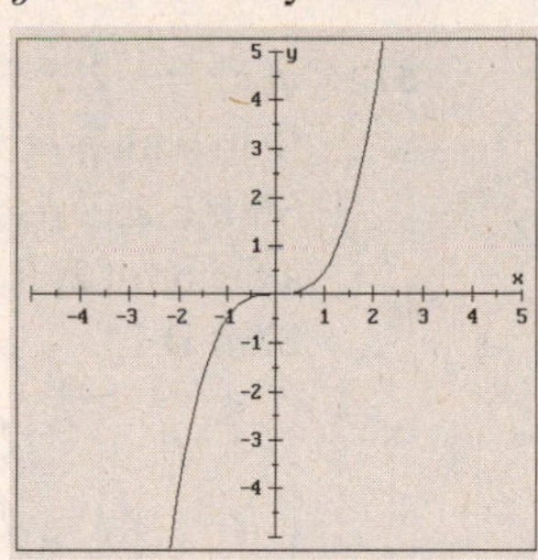

61. $h(x) = -3|x|$: Stretch
$y = |x|$ vert. by a factor of 3
Reflect about x

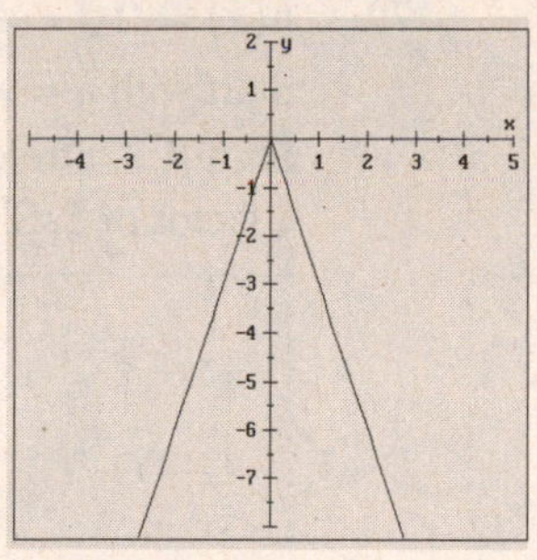

63. $f(x) = 3\sqrt{x}$: Stretch
$y = \sqrt{x}$ vert. by a
factor of 3

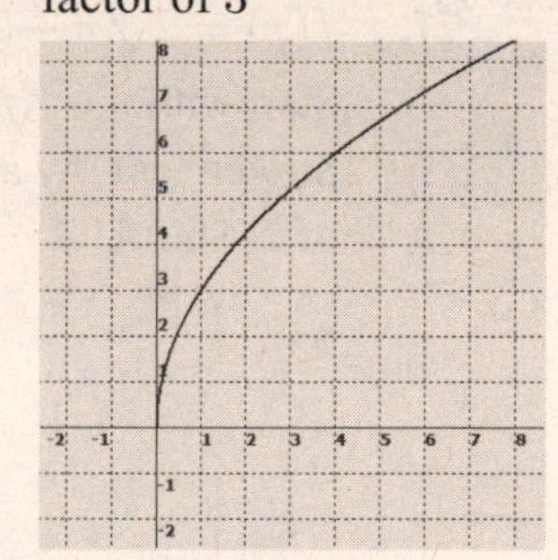

65. $f(x) = \frac{1}{2}\sqrt[3]{x}$: Shrink $y = \sqrt[3]{x}$ vert. by a factor of $\frac{1}{2}$

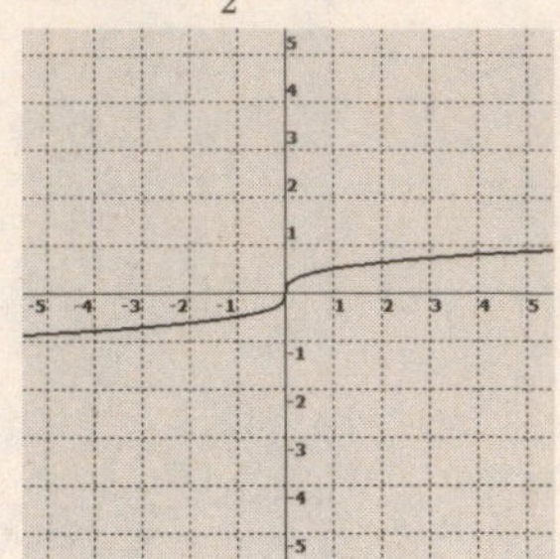

67. $f(x) = \left(\frac{1}{2}x\right)^3$: Stretch $y = x^3$ hor. by a factor of 2

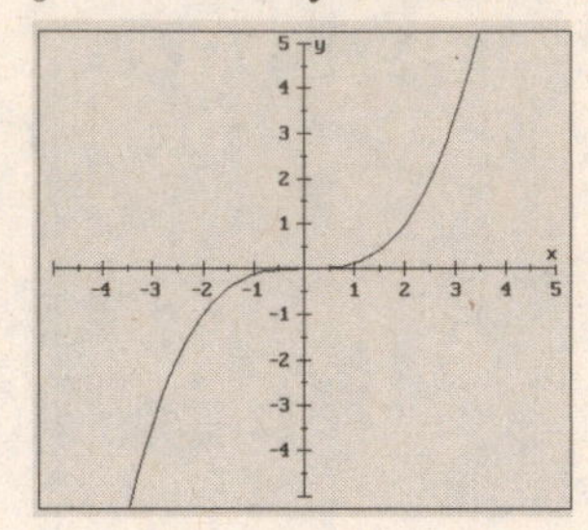

69. $f(x) = (2x)^2$: Shrink $y = x^2$ hor. by a factor of 2

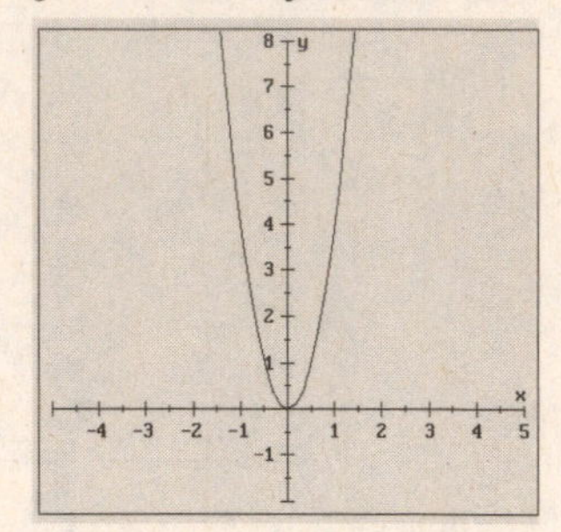

71. $g(x) = 3(x + 2)^2 - 1$
Start with $y = x^2$
Shift L 2, Stretch vert. by a factor of 3, Shift D 1

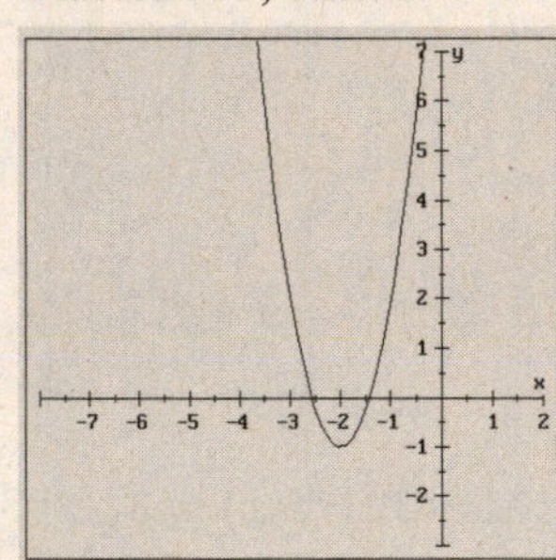

73. $h(x) = -2|x| + 3$
Start with $y = |x|$
Stretch vert. by a factor of 2; Reflect x; Shift U 3

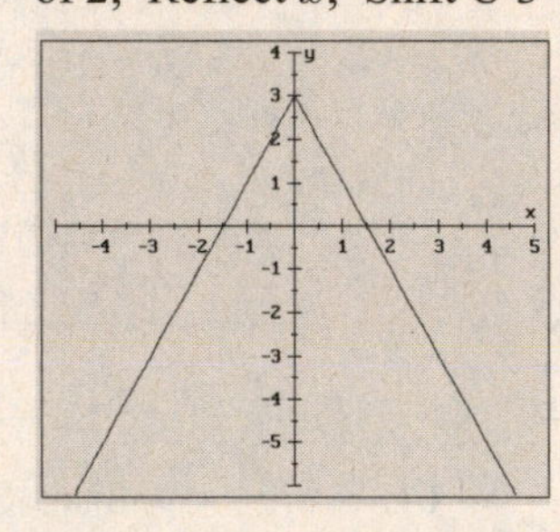

75. $f(x) = 2|x - 2| + 1$
Start with $y = |x|$
Shift R 2, Stretch vert. by a factor of 2, Shift U 1

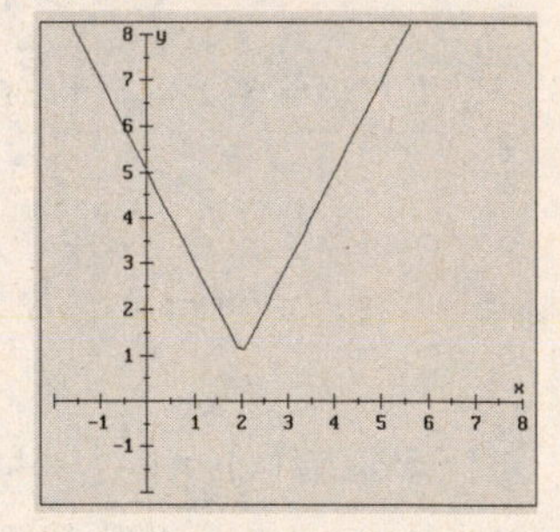

77. $f(x) = 2\sqrt{x} + 3$
Start with $y = \sqrt{x}$
Stretch vert. by a factor of 2, Shift U 3

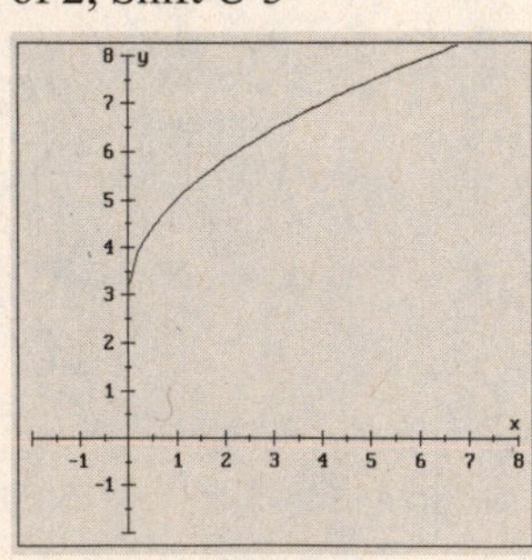

79. $h(x) = 2\sqrt{x - 2} + 1$
Start with $y = \sqrt{x}$
Shift R 2, Stretch vert. by a factor of 2, Shift U 1

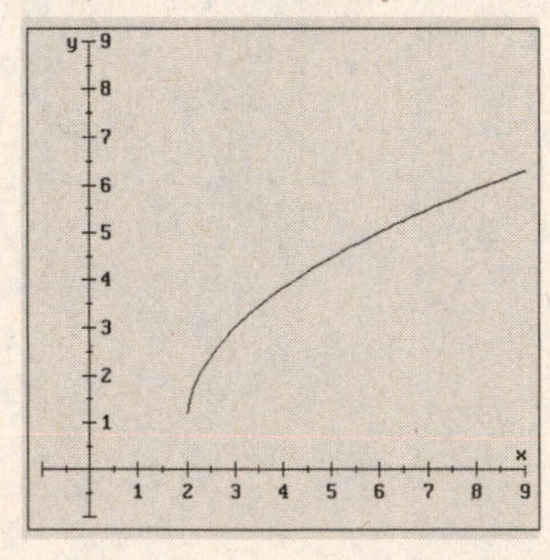

81. $g(x) = -2(x + 2)^3 - 1$
Start with $y = x^3$
Shift L 2, Stretch vert. by a factor of 2, Reflect x, Shift D 1

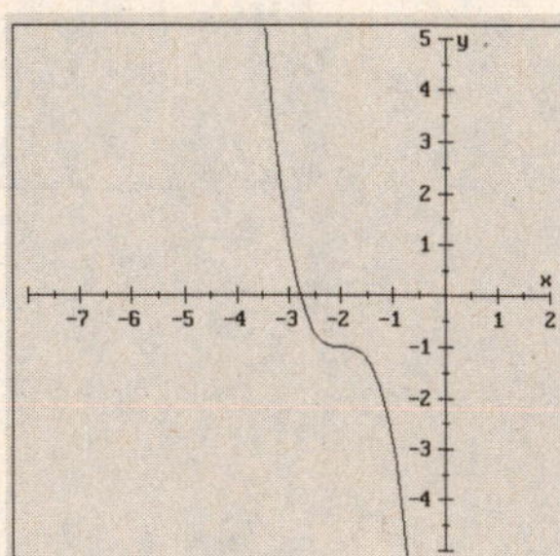

EXERCISES 3.2

83. $f(x) = 2\sqrt[3]{x} + 4$
Start with $y = \sqrt[3]{x}$
Stretch vert. by a factor of 2, Shift U 4

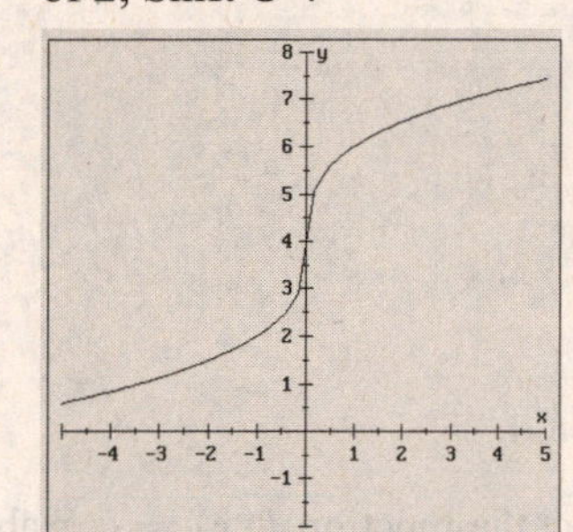

85. Shift $y = f(x)$ U 1

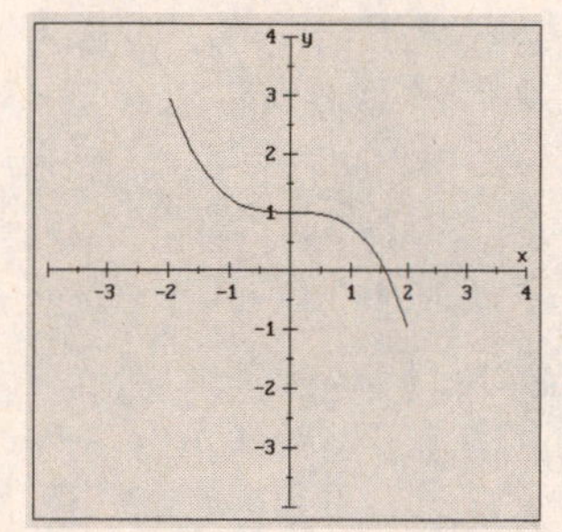

87. Stretch $y = f(x)$ vert. by a factor of 2

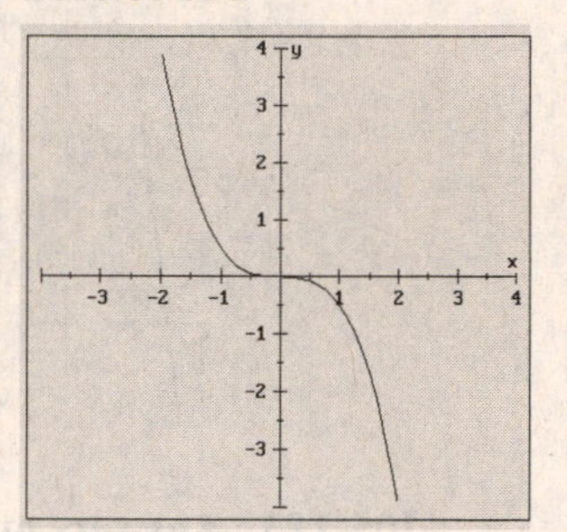

89. Shift $y = f(x)$ U 1, R 2

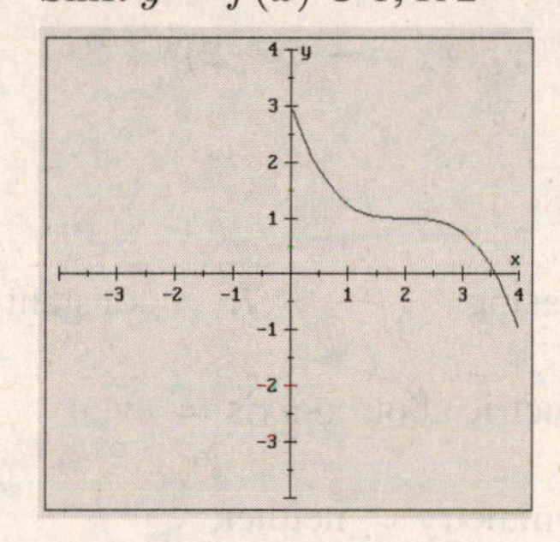

91. Stretch $y = f(x)$ vert. by a factor of 2, reflect about y

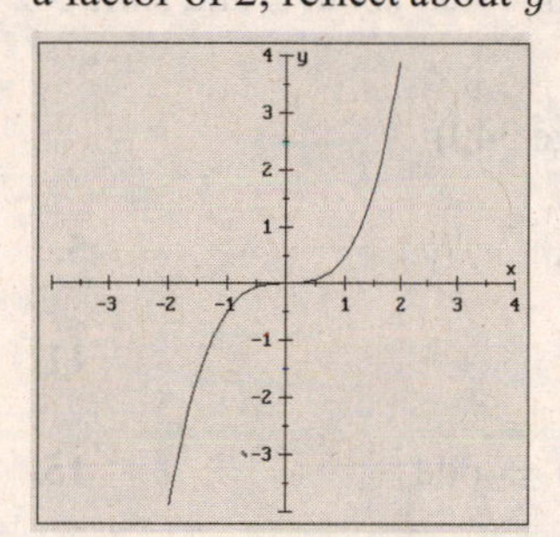

93. Shift $y = f(x)$ U 2

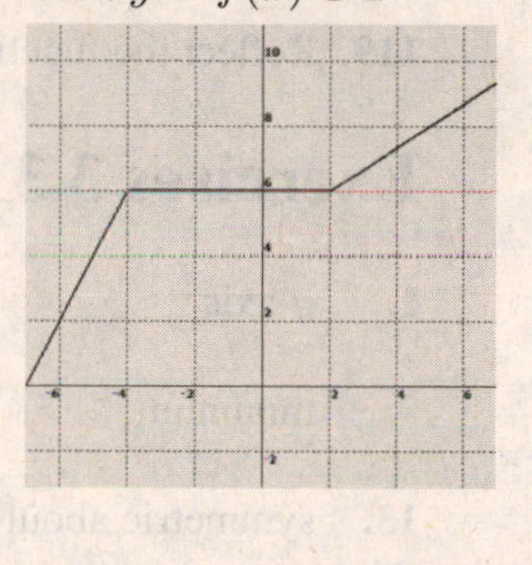

95. Shift $y = f(x)$ L 2

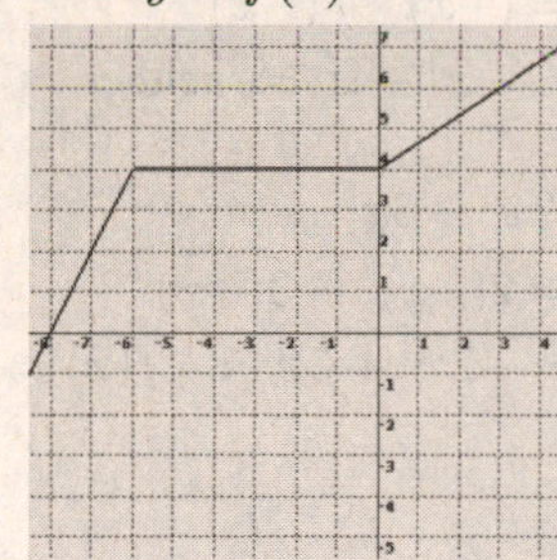

97. Shift $y = f(x)$ D 2, R 4

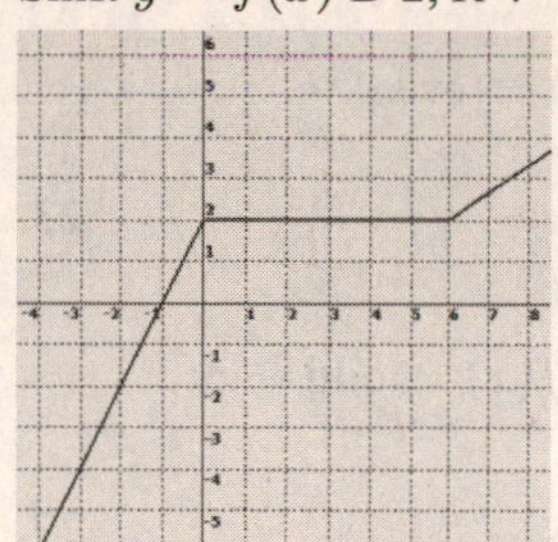

99. Reflect $y = f(x)$ about y

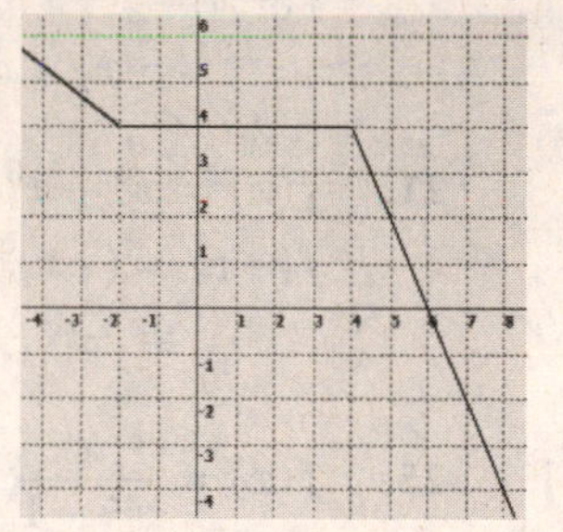

101. Stretch $y = f(x)$ vert. by a factor of 4

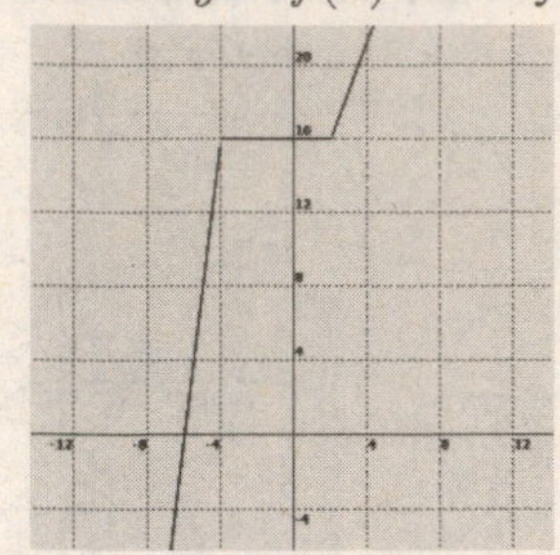

103. Shrink $y = f(x)$ horiz. by a factor of $\frac{1}{4}$

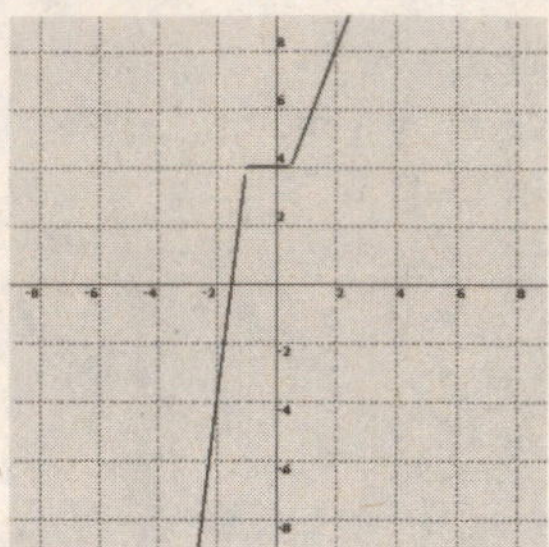

105-113. Answers may vary.

115. Reflect the function $f(x) = \sqrt{x}$ about x and shift U 1. $g(x) = -\sqrt{x} + 1$

117. Reflect the function $f(x) = \sqrt[3]{x}$ about x and stretch vert. by a factor of 2. $g(x) = -2\sqrt[3]{x}$

119. Reflect the function $f(x) = x^3$ about x and shift U 2 and R 2. $g(x) = -(x-2)^3 + 2$

Exercises 3.3 (page 340)

1. y-axis

3. $f(x)$

5. increasing

7. constant

9. minimum

11. symmetric about y-axis $\Rightarrow$ even

13. symmetric about origin $\Rightarrow$ odd

15. no symmetry $\Rightarrow$ neither

17. $f(x) = x^4 + x^2$
$$f(-x) = (-x)^4 + (-x)^2 = x^4 + x^2 = f(x) \Rightarrow \text{even}$$

19. $f(x) = x^3 + x^2$
$$f(-x) = (-x)^3 + (-x)^2 = -x^3 + x^2 \Rightarrow \text{neither}$$

21. $f(x) = x^5 + x^3$
$$f(-x) = (-x)^5 + (-x)^3 = -x^5 - x^3 = -f(x) \Rightarrow \text{odd}$$

23. $f(x) = 2x^3 - 3x$
$$f(-x) = 2(-x)^3 - 3(-x) = -2x^3 + 3x = -f(x) \Rightarrow \text{odd}$$

25. $f(x) = \dfrac{x}{x^2 - 1}$
$$f(-x) = \frac{-x}{(-x)^2 - 1} = \frac{-x}{x^2 - 1} = -f(x) \Rightarrow \text{odd}$$

27. $f(x) = \dfrac{1}{x^4}$
$$f(-x) = \frac{1}{(-x)^4} = \frac{1}{x^4} = f(x) \Rightarrow \text{even}$$

29. $f(x) = \sqrt{x+1}$
$$f(-x) = \sqrt{-x+1} \Rightarrow \text{neither}$$

31. $f(x) = \dfrac{|x|}{x}$
$$f(-x) = \frac{|-x|}{-x} = \frac{|x|}{-x} = -f(x) \Rightarrow \text{odd}$$

EXERCISES 3.3

33. decreasing: $(-\infty, 0)$; increasing: $(0, \infty)$

35. increasing: $(-\infty, 0)$; decreasing: $(0, \infty)$

37. decreasing: $(-\infty, -2)$; constant: $(-2, 2)$; increasing: $(2, \infty)$

39. local min. is 2

41. local max. is 5

43. local max. is 2, local min. is 1

45. local max. is 1, local min. is 0

47. local max. is 3, local min. is -3

49. **a.** $f(-2) = 2(-2) + 2 = -2$
b. $f(0) = 3$

51. **a.** $f(-1) = 2$
b. $f(1) = 2 - 1 = 1$
c. $f(2) = 2 + 1 = 3$

53. $f(x) = \begin{cases} x + 2 & \text{if } x < 0 \\ 2 & \text{if } x \geq 0 \end{cases}$

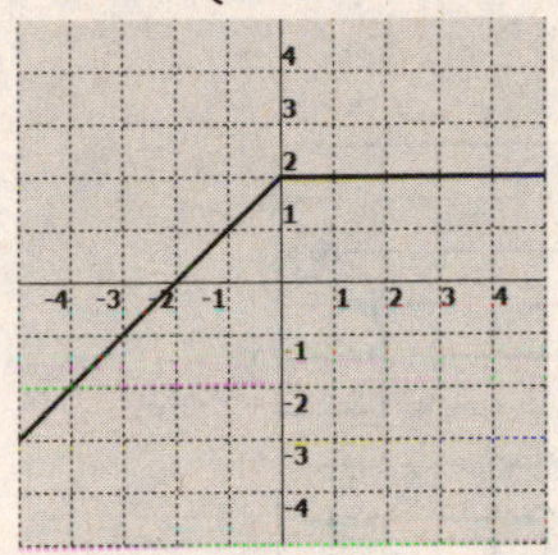

55. $f(x) = \begin{cases} x & \text{if } x \leq 0 \\ 2 & \text{if } x > 0 \end{cases}$

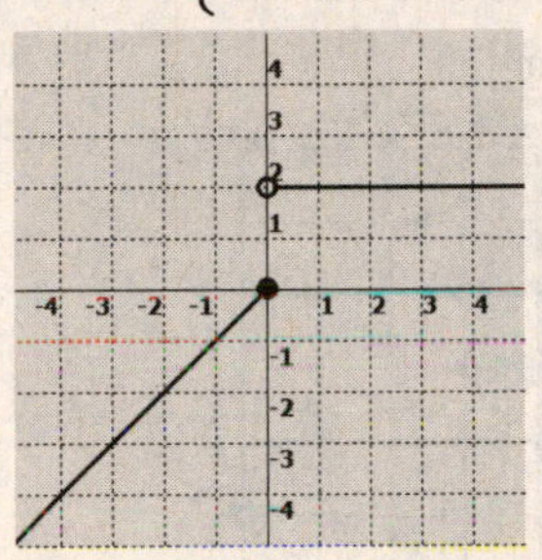

57. $f(x) = \begin{cases} -4 - x & \text{if } x < 1 \\ 3 & \text{if } x \geq 1 \end{cases}$

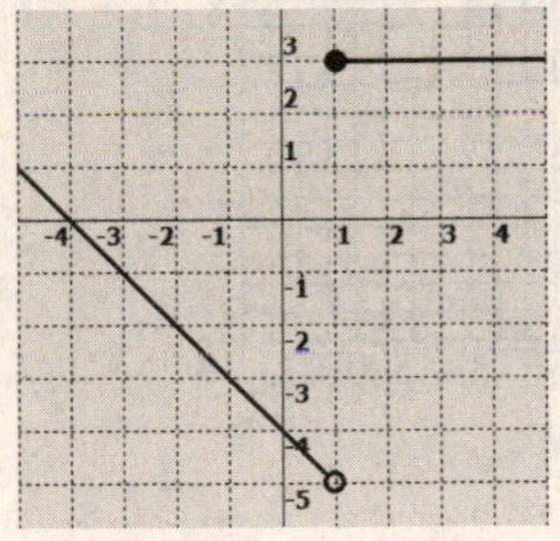

59. $f(x) = \begin{cases} -x & \text{if } x < 0 \\ x^2 & \text{if } x \geq 0 \end{cases}$

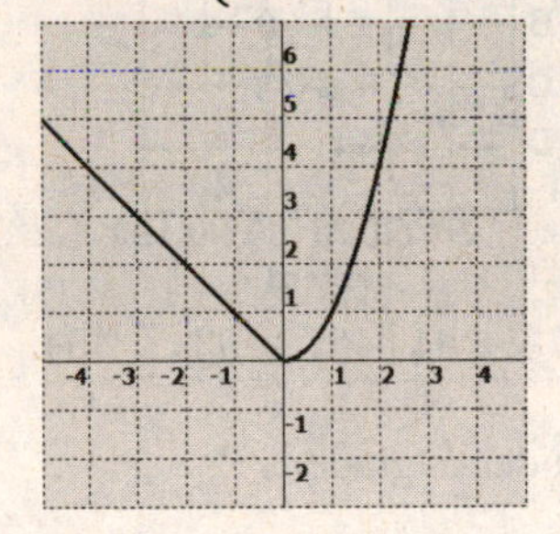

EXERCISES 3.3

61. $f(x) = \begin{cases} 0 & \text{if } x < 0 \\ x^2 & \text{if } 0 \le x \le 2 \\ 4 - 2x & \text{if } x > 2 \end{cases}$

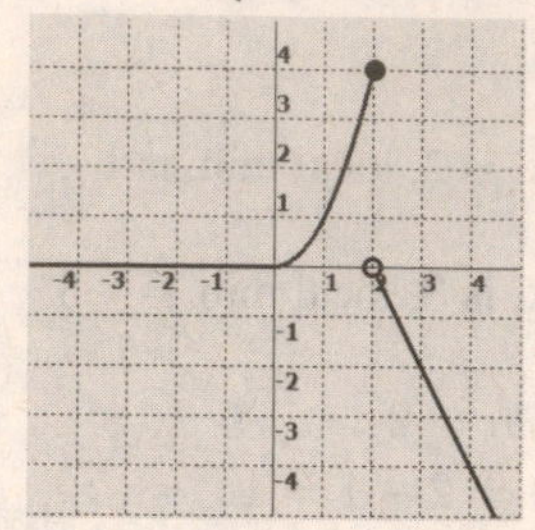

63. **a.** $f(3) = [[3]] = 3$
b. $f(-4) = [[-4]] = -4$
c. $f(-2.3) = [[-2.3]] = -3$

65. **a.** $f(-1) = [[-1 + 3]] = [[2]] = 2$ **b.** $f(\frac{2}{3}) = [[\frac{2}{3} + 3]] = [[3\frac{2}{3}]] = 3$
c. $f(1.3) = [[1.3 + 3]] = [[4.3]] = 4$

67. $y = [[2x]]$

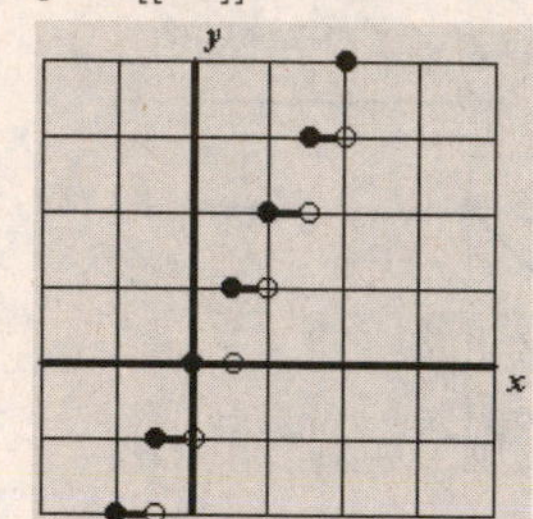

69. $y = [[x]] - 1$

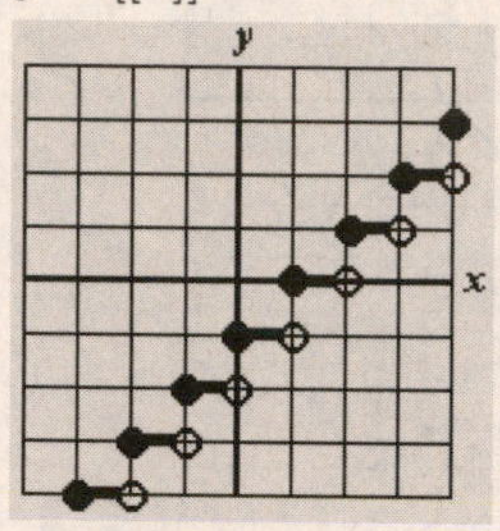

71.

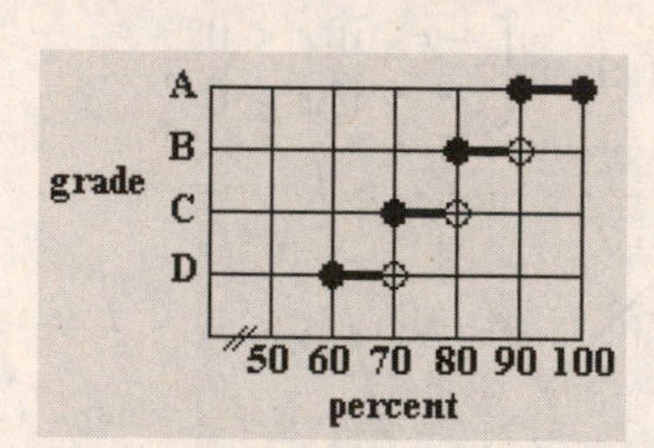

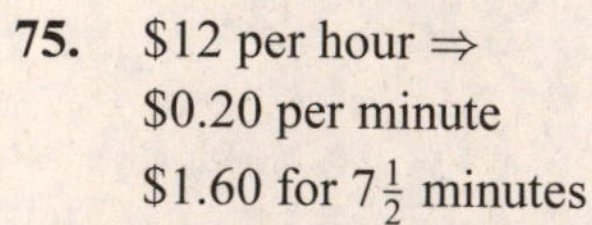

$$\frac{67 + 73 + 84 + 87 + 93}{5} = \frac{404}{5} = 80.8$$

The student's grade is B.

73. \$32 for 275 miles

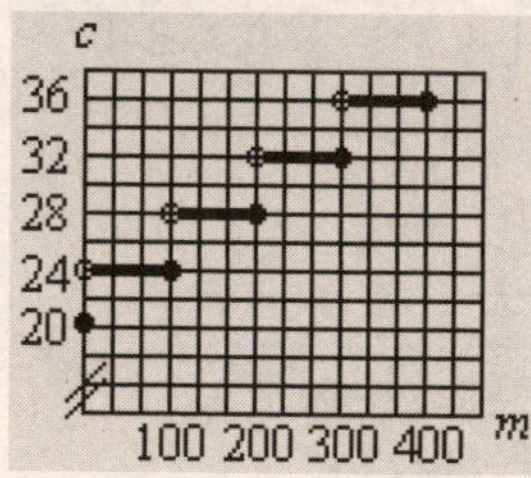

75. \$12 per hour ⇒
\$0.20 per minute
\$1.60 for $7\frac{1}{2}$ minutes

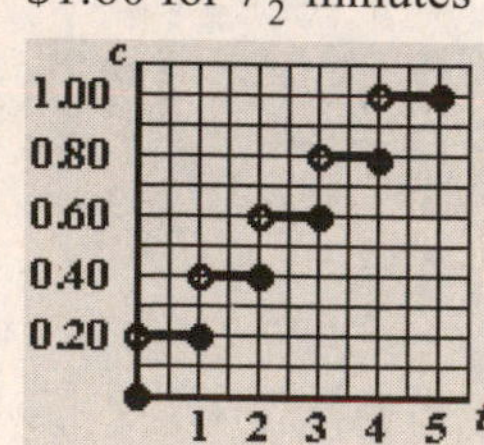

77.

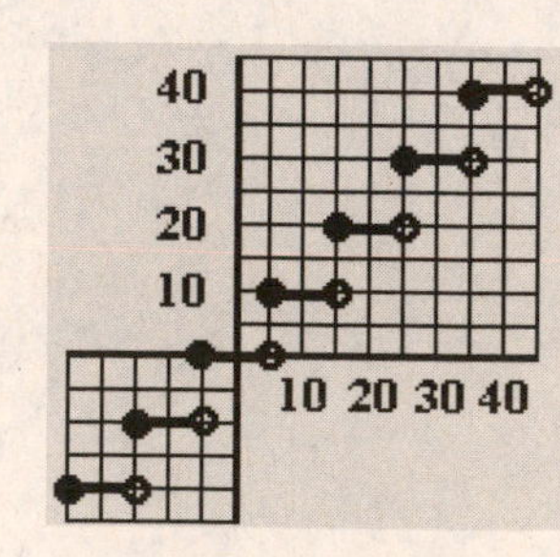

79. $y = \frac{|x|}{x}$, Not defined at $x = 0$, so not the same

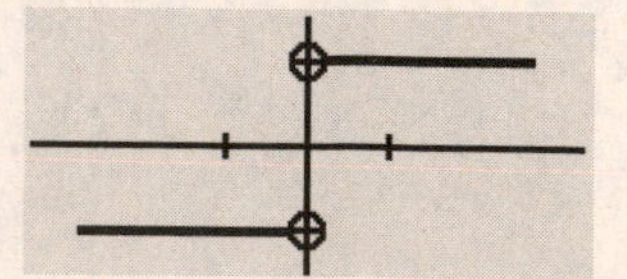

81-89. Answers may vary.

91. $f(-x) = (-x)^{100} + (-x)^{50}$
$= x^{100} + x^{50} = f(x)$: True.

93. $f(-x) = \sqrt[7]{-x} = -\sqrt[7]{x} = -f(x)$: True.

95. True.

97. True.

99. False. $f(x_1)$ is a local maximum value.

Exercises 3.4 (page 359)

1. $f(x) + g(x)$

3. $f(x)g(x)$

5. intersection

7. $g(f(x))$

9. commutative

11. $(f+g)(x) = f(x) + g(x) = (2x+1) + (3x-2) = 5x - 1$; domain $= (-\infty, \infty)$

13. $(f \cdot g)(x) = f(x)g(x) = (2x+1)(3x-2) = 6x^2 - x - 2$; domain $= (-\infty, \infty)$

15. $(f-g)(x) = f(x) - g(x) = (x^2 + x) - (x^2 - 1) = x + 1$; domain $= (-\infty, \infty)$

17. $(f/g)(x) = \dfrac{f(x)}{g(x)} = \dfrac{x^2+x}{x^2-1} = \dfrac{x(x+1)}{(x+1)(x-1)} = \dfrac{x}{x-1}$; domain $= (-\infty, -1) \cup (-1, 1) \cup (1, \infty)$

19. $(f+g)(x) = f(x) + g(x) = (x^2 - 7x + 3) + (x^2 - 5x + 6) = 2x^2 - 12x + 9$; domain $= (-\infty, \infty)$

21. $(f \cdot g)(x) = f(x)g(x) = (x^2 - 7x + 3)(x^2 - 5x + 6) = x^4 - 12x^3 + 44x^2 - 57x + 18$;
domain $= (-\infty, \infty)$

23. $(f+g)(x) = f(x) + g(x) = (x^2 - 7) + \left(\sqrt{x}\right) = x^2 + \sqrt{x} - 7$; domain $= [0, \infty)$

25. $(f/g)(x) = \dfrac{f(x)}{g(x)} = \dfrac{x^2 - 7}{\sqrt{x}}$; domain $= (0, \infty)$

27. $(f+g)(2) = f(2) + g(2) = \left[(2)^2 - 1\right] + [3(2) - 2] = 3 + 4 = 7$

29. $(f-g)(0) = f(0) - g(0) = \left[(0)^2 - 1\right] - [3(0) - 2] = -1 - (-2) = 1$

31. $(f \cdot g)(2) = f(2) \cdot g(2) = \left[(2)^2 - 1\right] \cdot [3(2) - 2] = (3)(4) = 12$

33. $(f/g)\left(\frac{2}{3}\right) = \dfrac{f\left(\frac{2}{3}\right)}{g\left(\frac{2}{3}\right)} = \dfrac{\left[\left(\frac{2}{3}\right)^2 - 1\right]}{\left[3\left(\frac{2}{3}\right) - 2\right]} = \dfrac{-\frac{5}{9}}{0} \Rightarrow$ undefined

35. $(f+g)(8) = f(8) + g(8) = [2(8) - 5] + \sqrt[3]{8} = 11 + 2 = 13$

37. $(f-g)(-27)=f(-27)-g(-27)=[2(-27)-5]-\sqrt[3]{-27}=-59-(-3)=-56$

39. $(f\cdot g)(-1)=f(-1)\cdot g(-1)=[2(-1)-5]\cdot\sqrt[3]{-1}=(-7)(-1)=7$

41. $(f/g)\left(\frac{1}{8}\right)=\dfrac{f\left(\frac{1}{8}\right)}{g\left(\frac{1}{8}\right)}=\dfrac{\left[2\left(\frac{1}{8}\right)-5\right]}{\sqrt[3]{\frac{1}{8}}}=\dfrac{-\frac{19}{4}}{\frac{1}{2}}=-\dfrac{19}{2}$

43. Let $f(x)=3x^2$ and $g(x)=2x$.
Then $(f+g)(x)=3x^2+2x=h(x)$.

45. Let $f(x)=3x^2$ and $g(x)=x^2-1$.
Then $(f/g)(x)=\dfrac{3x^2}{x^2-1}=h(x)$.

47. Let $f(x)=3x^3$ and $g(x)=-x$.
Then $(f-g)(x)=3x^3+x$
$=x\left(3x^2+1\right)=h(x)$.

49. Let $f(x)=x+9$ and $g(x)=x-2$.
Then $(f\cdot g)(x)=(x+9)(x-2)$
$=x^2+7x-18=h(x)$.

51. The domain of $f\circ g$ is the set of all real numbers in the domain of $g(x)$ such that $g(x)$ is in the domain of $f(x)$. Domain of $g(x)$: $(-\infty,\infty)$. Domain of $f(x)=(-\infty,\infty)$. Thus, all values of $g(x)$ are in the domain of $f(x)$. Domain of $f\circ g$: $(-\infty,\infty)$
$(f\circ g)(x)=f(g(x))=f(x+1)=3(x+1)=3x+3$

53. The domain of $f\circ f$ is the set of all real numbers in the domain of $f(x)$ such that $f(x)$ is in the domain of $f(x)$. Domain of $f(x)$: $(-\infty,\infty)$. Thus, all values of $f(x)$ are in the domain of $f(x)$.
Domain of $f\circ f$: $(-\infty,\infty)$ $(f\circ f)(x)=f(f(x))=f(3x)=3(3x)=9x$

55. The domain of $g\circ f$ is the set of all real numbers in the domain of $f(x)$ such that $f(x)$ is in the domain of $g(x)$. Domain of $f(x)$: $(-\infty,\infty)$. Domain of $g(x)=(-\infty,\infty)$. Thus, all values of $f(x)$ are in the domain of $g(x)$. Domain of $g\circ f$: $(-\infty,\infty)$
$(g\circ f)(x)=g(f(x))=g\left(x^2\right)=2x^2$

57. The domain of $g\circ g$ is the set of all real numbers in the domain of $g(x)$ such that $g(x)$ is in the domain of $g(x)$. Domain of $g(x)$: $(-\infty,\infty)$. Thus, all values of $g(x)$ are in the domain of $g(x)$.
Domain of $g\circ g$: $(-\infty,\infty)$ $(g\circ g)(x)=g(g(x))=g(2x)=2(2x)=4x$

59. The domain of $f\circ g$ is the set of all real numbers in the domain of $g(x)$ such that $g(x)$ is in the domain of $f(x)$. Domain of $g(x)$: $(-\infty,\infty)$. Domain of $f(x)=(-\infty,\infty)$. Thus, all values of $g(x)$ are in the domain of $f(x)$. Domain of $f\circ g$: $(-\infty,\infty)$
$(f\circ g)(x)=f(g(x))=f(4x-1)=2(4x-1)^2-3(4x-1)+7$
$=2\left(16x^2-8x+1\right)-12x+3+7=32x^2-16x+2-12x+10=32x^2-28x+12$

EXERCISES 3.4

61. The domain of $f \circ f$ is the set of all real numbers in the domain of $f(x)$ such that $f(x)$ is in the domain of $f(x)$. Domain of $f(x)$: $(-\infty, \infty)$. Thus, all values of $f(x)$ are in the domain of $f(x)$.

$\boxed{\text{Domain of } f \circ f\text{: } (-\infty, \infty)}$

$$(f \circ f)(x) = f(f(x)) = f\left(2x^2 - 3x + 7\right) = 2\left(2x^2 - 3x + 7\right)^2 - 3\left(2x^2 - 3x + 7\right) + 7$$
$$= 2\left(4x^4 - 12x^3 + 37x^2 - 42x + 49\right) - 6x^2 + 9x - 21 + 7$$
$$= 8x^4 - 24x^3 + 74x^2 - 84x + 98 - 6x^2 + 9x - 14 = 8x^4 - 24x^3 + 68x^2 - 75x + 84$$

63. The domain of $f \circ g$ is the set of all real numbers in the domain of $g(x)$ such that $g(x)$ is in the domain of $f(x)$. Domain of $g(x)$: $(-\infty, \infty)$. Domain of $f(x) = [0, \infty)$. Thus, we must have $g(x) \geq 0 \Rightarrow x + 1 \geq 0 \Rightarrow x \geq -1$. $\boxed{\text{Domain of } f \circ g\text{: } [-1, \infty)}$

$$(f \circ g)(x) = f(g(x)) = f(x + 1) = \sqrt{x + 1}$$

65. The domain of $f \circ f$ is the set of all real numbers in the domain of $f(x)$ such that $f(x)$ is in the domain of $f(x)$. Domain of $f(x)$: $[0, \infty)$. Thus, we must have $f(x) \geq 0 \Rightarrow \sqrt{x} \geq 0$. This is true for all real values of x. $\boxed{\text{Domain of } f \circ f\text{: } [0, \infty)}$

$$(f \circ f)(x) = f(f(x)) = f\left(\sqrt{x}\right) = \sqrt{\sqrt{x}} = \left((x)^{1/2}\right)^{1/2} = x^{1/4} = \sqrt[4]{x}$$

67. The domain of $g \circ f$ is the set of all real numbers in the domain of $f(x)$ such that $f(x)$ is in the domain of $g(x)$. Domain of $f(x)$: $[-1, \infty)$. Domain of $g(x) = (-\infty, \infty)$. Thus, all values of $f(x)$ are in the domain of $g(x)$. $\boxed{\text{Domain of } g \circ f\text{: } [-1, \infty)}$

$$(g \circ f)(x) = g(f(x)) = g\left(\sqrt{x + 1}\right) = \left(\sqrt{x + 1}\right)^2 - 1 = x$$

69. The domain of $g \circ g$ is the set of all real numbers in the domain of $g(x)$ such that $g(x)$ is in the domain of $g(x)$. Domain of $g(x)$: $(-\infty, \infty)$. Thus, all values of $g(x)$ are in the domain of $g(x)$.

$\boxed{\text{Domain of } g \circ g\text{: } (-\infty, \infty)}$ $(g \circ g)(x) = g(g(x)) = g\left(x^2 - 1\right) = \left(x^2 - 1\right)^2 - 1 = x^4 - 2x^2$

71. The domain of $f \circ g$ is the set of all real numbers in the domain of $g(x)$ such that $g(x)$ is in the domain of $f(x)$. Domain of $g(x)$: $(-\infty, 2) \cup (2, \infty)$. Domain of $f(x) = (-\infty, 1) \cup (1, \infty)$. Thus, we must have $g(x) \neq 1 \Rightarrow \dfrac{1}{x - 2} \neq 1 \Rightarrow 1 \neq x - 2 \Rightarrow x \neq 3$

$\boxed{\text{Domain of } f \circ g\text{: } (-\infty, 2) \cup (2, 3) \cup (3, \infty)}$

$$(f \circ g)(x) = f(g(x)) = f\left(\frac{1}{x - 2}\right) = \frac{1}{\frac{1}{x-2} - 1} = \frac{1}{\frac{1}{x-2} - 1} \cdot \frac{x - 2}{x - 2} = \frac{x - 2}{1 - (x - 2)} = \frac{x - 2}{3 - x}$$

73. The domain of $f \circ f$ is the set of all real numbers in the domain of $f(x)$ such that $f(x)$ is in the domain of $f(x)$. Domain of $f(x)$: $(-\infty, 1) \cup (1, \infty)$. Thus, we must have $f(x) \neq 1 \Rightarrow$ $\dfrac{1}{x - 1} \neq 1 \Rightarrow 1 \neq x - 1 \Rightarrow x \neq 2$ $\boxed{\text{Domain of } f \circ f\text{: } (-\infty, 1) \cup (1, 2) \cup (2, \infty)}$

$$(f \circ f)(x) = f(f(x)) = f\left(\frac{1}{x - 1}\right) = \frac{1}{\frac{1}{x-1} - 1} = \frac{1}{\frac{1}{x-1} - 1} \cdot \frac{x - 1}{x - 1} = \frac{x - 1}{1 - (x - 1)} = \frac{x - 1}{2 - x}$$

EXERCISES 3.4

75. $(f \circ g)(2) = f(g(2)) = f(5(2) - 2) = f(8) = 2(8) - 5 = 11$

77. $(g \circ f)(-3) = g(f(-3)) = g(2(-3) - 5) = g(-11) = 5(-11) - 2 = -57$

79. $(f \circ f)\left(-\frac{1}{2}\right) = f\left(f\left(-\frac{1}{2}\right)\right) = f\left(2\left(-\frac{1}{2}\right) - 5\right) = f(-6) = 2(-6) - 5 = -17$

81. $(f \circ g)(-3) = f(g(-3)) = f(4(-3) + 4) = f(-8) = 3(-8)^2 - 2 = 190$

83. $(g \circ f)(3) = g(f(3)) = g\left(3(3)^2 - 2\right) = g(25) = 4(25) + 4 = 104$

85. $(f \circ f)\left(\sqrt{3}\right) = f\left(f\left(\sqrt{3}\right)\right) = f\left(3\left(\sqrt{3}\right)^2 - 2\right) = f(7) = 3(7)^2 - 2 = 145$

87. $(f \circ g)(100) = f(g(100)) = f\left(\sqrt{100}\right) = f(10) = \dfrac{2}{10} = \dfrac{1}{5}$

89. $(g \circ f)\left(\dfrac{1}{32}\right) = g\left(f\left(\dfrac{1}{32}\right)\right) = g\left(\dfrac{2}{\frac{1}{32}}\right) = g(64) = \sqrt{64} = 8$

91. $(g \circ g)\left(\dfrac{81}{256}\right) = g\left(g\left(\dfrac{81}{256}\right)\right) = g\left(\sqrt{\dfrac{81}{256}}\right) = g\left(\dfrac{9}{16}\right) = \sqrt{\dfrac{9}{16}} = \dfrac{3}{4}$

93. Let $f(x) = x - 2$ and $g(x) = 3x$.
Then $(f \circ g)(x) = f(g(x))$
$= f(3x) = 3x - 2.$

95. Let $f(x) = x - 2$ and $g(x) = x^2$.
Then $(f \circ g)(x) = f(g(x))$
$= f\left(x^2\right) = x^2 - 2.$

97. Let $f(x) = x^2$ and $g(x) = x - 2$.
Then $(f \circ g)(x) = f(g(x))$
$= f(x - 2) = (x - 2)^2.$

99. Let $f(x) = \sqrt{x}$ and $g(x) = x + 2$.
Then $(f \circ g)(x) = f(g(x))$
$= f(x + 2) = \sqrt{x + 2}.$

101. Let $f(x) = x + 2$ and $g(x) = \sqrt{x}$.
Then $(f \circ g)(x) = f(g(x))$
$= f\left(\sqrt{x}\right) = \sqrt{x} + 2.$

103. Let $f(x) = x$ and $g(x) = x$.
Then $(f \circ g)(x) = f(g(x))$
$= f(x) = x.$

105. $(f + g)(-4) = f(-4) + g(-4)$
$= -2 + 2 = 0$

107. $(f \cdot g)(5) = f(5) \cdot g(5) = -2(0) = 0$

109. $(f \circ g)(3) = f(g(3)) = f(2) = 1$

111. $(f \circ f)(-2) = f(f(-2)) = f(0) = 1$

113. $(f + g)(2) = f(2) + g(2) = 4 + 4 = 8$

115. $(f \circ g)(2) = f(g(2)) = f(4) = 9$

117. a. $(R - C)(x) = 300x - (60{,}000 + 40x)$
$= 260x - 60{,}000$

b. $(R - C)(500) = 260(500) - 60{,}000$
$= 70{,}000$

EXERCISES 3.4

119. $r(t) = \dfrac{d(t)}{2} = \dfrac{3t}{2}$; $A(t) = \pi(r(t))^2 = \pi\left(\dfrac{3t}{2}\right)^2 = \dfrac{9}{4}\pi t^2$

$A(120) = \dfrac{9}{4}\pi(120)^2 \approx 101{,}787.6$ square inches

121. If the area is A and the length of a side is s, then $s^2 = A \Rightarrow s = \sqrt{A}$. Then $P = 4s = 4\sqrt{A}$.

123. Answers may vary.

125. $(f+f)(x) = f(x) + f(x) = 3x + 3x = 6x$
$f(x+x) = f(2x) = 3(2x) = 6x$

127. $(f \circ f)(x) = f(f(x)) = f\left(\dfrac{x-1}{x+1}\right) = \dfrac{\frac{x-1}{x+1} - 1}{\frac{x-1}{x+1} + 1} = \dfrac{(x+1)\left(\frac{x-1}{x+1} - 1\right)}{(x+1)\left(\frac{x-1}{x+1} + 1\right)} = \dfrac{x - 1 - (x+1)}{x - 1 + x + 1}$
$= \dfrac{-2}{2x} = -\dfrac{1}{x}$

129. Answers may vary.

131. Answers may vary.

133. False. $(f-g)(x) = -(-f+g)(x)$
$= -(g-f)(x)$

135. True.

137. False. $(g \circ g \circ g)(x) = g(g(g(x))) = g(g(-x^3)) = g\left(-(-x^3)^3\right) = g(x^9) = -(x^9)^3 = -x^{27}$

139. True. $(f \circ g)(-1) = f(g(-1)) = f\left((-1)^{864}\right) = f(1) = 1^{975} = 1.$

Exercises 3.5 (page 372)

1. one-to-one

3. identity

5. $f(x) = 5$
$f(1) = f(-1) = 5$
not one-to-one

7. $f(x) = 3x$
one-to-one

9. $f(x) = x^2 + 3$
$f(1) = f(-1) = 4$
not one-to-one

11. $f(x) = x^3 + 5$
one-to-one

13. $f(x) = x^3 - x$
$f(1) = f(-1) = 0$
not one-to-one

15. $f(x) = |x|$
$f(1) = f(-1) = 1$
not one-to-one

17. $f(x) = \sqrt{x}$
one-to-one

19. $f(x) = \sqrt[3]{x}$
one-to-one

21. $f(x) = (x-2)^2, x \geq 2$
one-to-one

23. one-to-one

25. not one-to-one
(not a function)

27. one-to-one

EXERCISES 3.5

29. $(f \circ g)(x) = f(g(x)) = f\left(\frac{1}{5}x\right) = 5\left(\frac{1}{5}x\right) = x$

$(g \circ f)(x) = g(f(x)) = g(5x) = \frac{1}{5}(5x) = x$

31. $(f \circ g)(x) = f(g(x)) = f\left(\sqrt[3]{x-8}\right) = \left(\sqrt[3]{x-8}\right)^3 + 8 = x - 8 + 8 = x$

$(g \circ f)(x) = g(f(x)) = g(x^3 + 8) = \sqrt[3]{x^3 + 8 - 8} = \sqrt[3]{x^3} = x$

33. $(f \circ g)(x) = f(g(x)) = f((x+1)^5) = \left(\sqrt[5]{(x+1)^5}\right) - 1 = x + 1 - 1 = x$

$(g \circ f)(x) = g(f(x)) = g(\sqrt[5]{x} - 1) = (\sqrt[5]{x} - 1 + 1)^5 = (\sqrt[5]{x})^5 = x$

35. $(f \circ g)(x) = f(g(x)) = f\left(\frac{1}{x-1}\right) = \frac{\frac{1}{x-1} + 1}{\frac{1}{x-1}} = \frac{(x-1)\left(\frac{1}{x-1} + 1\right)}{(x-1)\frac{1}{x-1}} = \frac{1 + x - 1}{1} = x$

$(g \circ f)(x) = g(f(x)) = g\left(\frac{x+1}{x}\right) = \frac{1}{\frac{x+1}{x} - 1} = \frac{x(1)}{x\left(\frac{x+1}{x} - 1\right)} = \frac{x}{x + 1 - x} = \frac{x}{1} = x$

37.

$$\begin{aligned} y = f(x) &= 3x \\ x &= 3y \\ \frac{x}{3} &= y \\ f^{-1}(x) &= \frac{x}{3} = \frac{1}{3}x \end{aligned}$$

$$\begin{aligned} (f \circ f^{-1})(x) &= f(f^{-1}(x)) \\ &= f\left(\frac{x}{3}\right) \\ &= 3\left(\frac{x}{3}\right) \\ &= x \end{aligned}$$

$$\begin{aligned} (f^{-1} \circ f)(x) &= f^{-1}(f(x)) \\ &= f^{-1}(3x) \\ &= \frac{3x}{3} \\ &= x \end{aligned}$$

39.

$$\begin{aligned} y = f(x) &= 3x + 2 \\ x &= 3y + 2 \\ x - 2 &= 3y \\ \frac{x-2}{3} &= y \\ f^{-1}(x) &= \frac{x-2}{3} \end{aligned}$$

$$\begin{aligned} (f \circ f^{-1})(x) &= f(f^{-1}(x)) \\ &= f\left(\frac{x-2}{3}\right) \\ &= 3\left(\frac{x-2}{3}\right) + 2 \\ &= x - 2 + 2 = x \end{aligned}$$

$$\begin{aligned} (f^{-1} \circ f)(x) &= f^{-1}(f(x)) \\ &= f^{-1}(3x + 2) \\ &= \frac{(3x+2) - 2}{3} \\ &= \frac{3x}{3} = x \end{aligned}$$

41.

$$\begin{aligned} y = f(x) &= x^3 + 2 \\ x &= y^3 + 2 \\ x - 2 &= y^3 \\ \sqrt[3]{x-2} &= y \\ f^{-1}(x) &= \sqrt[3]{x-2} \end{aligned}$$

$$\begin{aligned} (f \circ f^{-1})(x) &= f(f^{-1}(x)) \\ &= f\left(\sqrt[3]{x-2}\right) \\ &= \left(\sqrt[3]{x-2}\right)^3 + 2 \\ &= x - 2 + 2 = x \end{aligned}$$

$$\begin{aligned} (f^{-1} \circ f)(x) &= f^{-1}(f(x)) \\ &= f^{-1}(x^3 + 2) \\ &= \sqrt[3]{(x^3 + 2) - 2} \\ &= \sqrt[3]{x^3} = x \end{aligned}$$

43.

$$\begin{aligned} y = f(x) &= \sqrt[5]{x} \\ x &= \sqrt[5]{y} \\ x^5 &= y \\ f^{-1}(x) &= x^5 \end{aligned}$$

$$\begin{aligned} (f \circ f^{-1})(x) &= f(f^{-1}(x)) \\ &= f(x^5) \\ &= \sqrt[5]{x^5} = x \end{aligned}$$

$$\begin{aligned} (f^{-1} \circ f)(x) &= f^{-1}(f(x)) \\ &= f^{-1}(\sqrt[5]{x}) \\ &= (\sqrt[5]{x})^5 = x \end{aligned}$$

EXERCISES 3.5

45.
$$\begin{aligned} y = f(x) &= \frac{1}{x+3} \\ x &= \frac{1}{y+3} \\ x(y+3) &= 1 \\ y+3 &= \frac{1}{x} \\ f^{-1}(x) &= \frac{1}{x} - 3 \end{aligned}$$
$$\begin{aligned} (f \circ f^{-1})(x) &= f\big(f^{-1}(x)\big) \\ &= f\left(\frac{1}{x} - 3\right) \\ &= \frac{1}{\frac{1}{x} - 3 + 3} \\ &= \frac{1}{\frac{1}{x}} \\ &= x \end{aligned}$$
$$\begin{aligned} (f^{-1} \circ f)(x) &= f^{-1}(f(x)) \\ &= f^{-1}\left(\frac{1}{x+3}\right) \\ &= \frac{1}{\frac{1}{x+3}} - 3 \\ &= x + 3 - 3 \\ &= x \end{aligned}$$

47.
$$\begin{aligned} y = f(x) &= \frac{1}{2x} \\ x &= \frac{1}{2y} \\ x(2y) &= 1 \\ 2xy &= 1 \\ y &= \frac{1}{2x} \\ f^{-1}(x) &= \frac{1}{2x} \end{aligned}$$
$$\begin{aligned} (f \circ f^{-1})(x) &= f\big(f^{-1}(x)\big) \\ &= f\left(\frac{1}{2x}\right) \\ &= \frac{1}{2\left(\frac{1}{2x}\right)} \\ &= \frac{1}{\frac{1}{x}} \\ &= x \end{aligned}$$
$$\begin{aligned} (f^{-1} \circ f)(x) &= f^{-1}(f(x)) \\ &= f^{-1}\left(\frac{1}{2x}\right) \\ &= \frac{1}{2\left(\frac{1}{2x}\right)} \\ &= \frac{1}{\frac{1}{x}} \\ &= x \end{aligned}$$

49.
$$\begin{aligned} y = f(x) &= 5x \\ x &= 5y \\ \frac{x}{5} &= y \\ f^{-1}(x) &= \frac{1}{5}x \end{aligned}$$

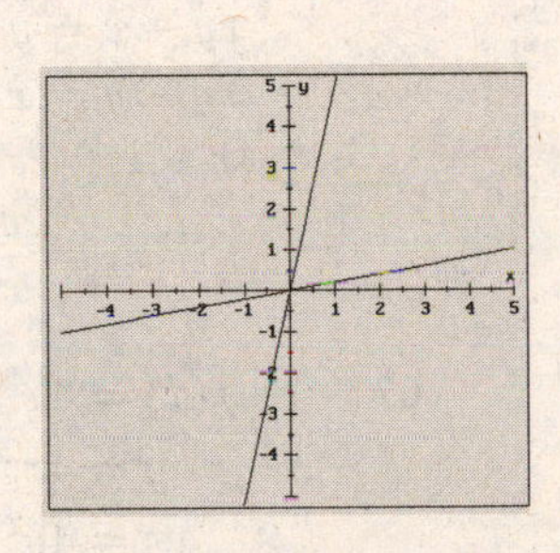

51.
$$\begin{aligned} y = f(x) &= 2x - 4 \\ x &= 2y - 4 \\ x + 4 &= 2y \\ \frac{x+4}{2} &= y \\ f^{-1}(x) &= \frac{x+4}{2} \end{aligned}$$

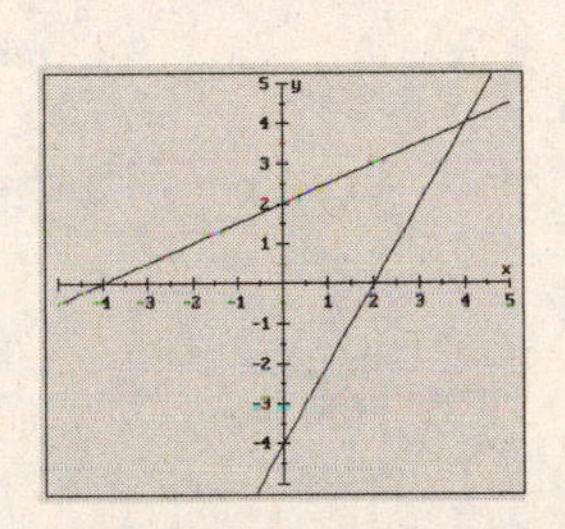

53.
$$\begin{aligned} x - y &= 2 \\ y - x &= 2 \\ y &= x + 2 \\ f^{-1}(x) &= x + 2 \end{aligned}$$

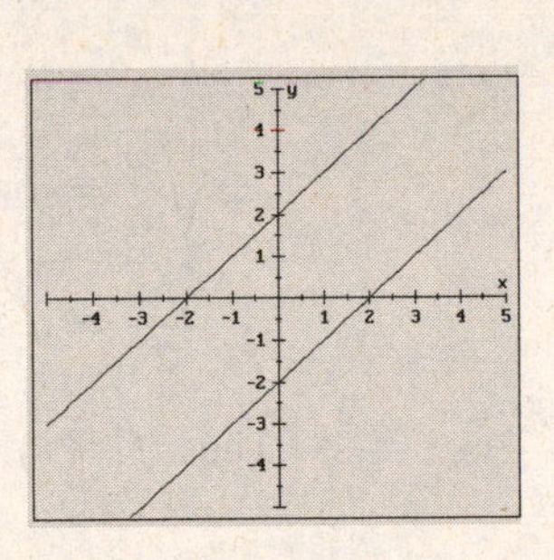

55.
$$\begin{aligned} 2x + y &= 4 \\ 2y + x &= 4 \\ 2y &= 4 - x \\ y &= \frac{4-x}{2} \\ f^{-1}(x) &= \frac{4-x}{2} \end{aligned}$$

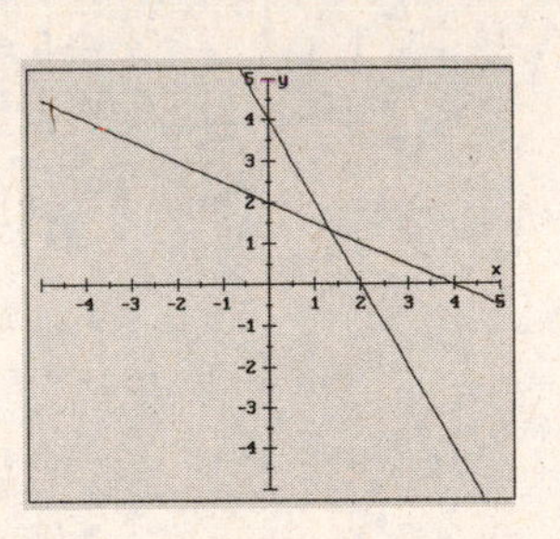

EXERCISES 3.5

57.
$$y = f(x) = \sqrt[3]{x-4}$$
$$x = \sqrt[3]{y-4}$$
$$x^3 = y - 4$$
$$x^3 + 4 = y$$
$$f^{-1}(x) = x^3 + 4$$

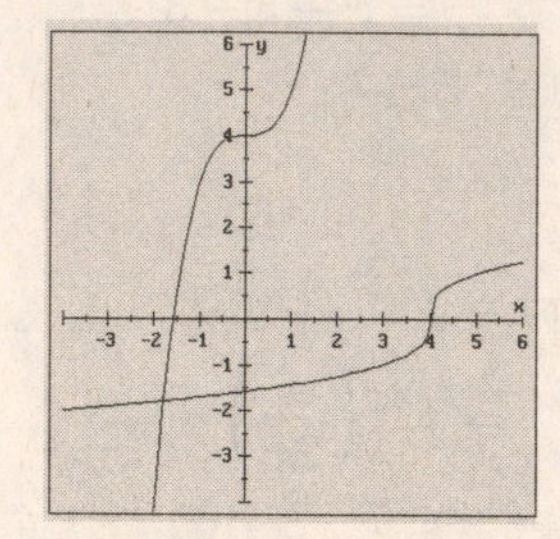

59.
$$y = f(x) = (x-6)^3$$
$$x = (y-6)^3$$
$$\sqrt[3]{x} = y - 6$$
$$\sqrt[3]{x} + 6 = y$$
$$f^{-1}(x) = \sqrt[3]{x} + 6$$

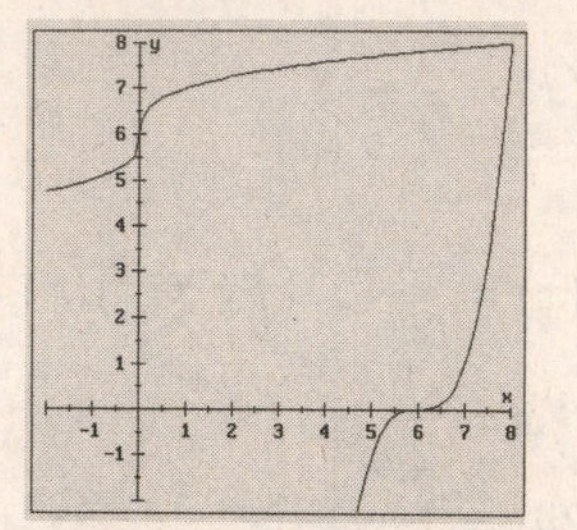

61.
$$y = f(x) = \frac{1}{2x}$$
$$x = \frac{1}{2y}$$
$$2xy = 1$$
$$y = \frac{1}{2x}$$
$$f^{-1}(x) = \frac{1}{2x}$$

63.
$$y = f(x) = \frac{x+1}{x-1}$$
$$x = \frac{y+1}{y-1}$$
$$x(y-1) = y+1$$
$$xy - x = y + 1$$
$$xy - y = x + 1$$
$$y(x-1) = x+1$$
$$f^{-1}(x) = \frac{x+1}{x-1}$$

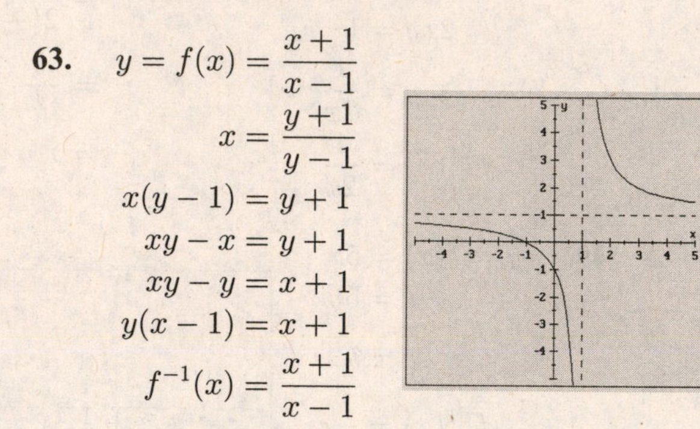

65.
$$f(x) = x^2 + 5 \quad x \ge 0$$
$$y = x^2 + 5 \quad x \ge 0$$
$$x = y^2 + 5 \quad y \ge 0$$
$$x - 5 = y^2 \quad y \ge 0$$
$$\pm\sqrt{x-5} = y \quad y \ge 0$$
Thus, $f^{-1}(x) = \sqrt{x-5}$ $(x \ge 5)$.

67.
$$f(x) = 4x^2 \quad x \ge 0$$
$$y = 4x^2 \quad x \ge 0$$
$$x = 4y^2 \quad y \ge 0$$
$$\frac{x}{4} = y^2 \quad y \ge 0$$
$$\pm\sqrt{\frac{x}{4}} = y \quad y \ge 0$$
$$\frac{\sqrt{x}}{2} = y \quad y \ge 0$$
Thus, $f^{-1}(x) = \dfrac{\sqrt{x}}{2}$ $(x \ge 0)$.

69.
$$f(x) = x^2 - 3 \quad x \le 0$$
$$y = x^2 - 3 \quad x \le 0$$
$$x = y^2 - 3 \quad y \le 0$$
$$x + 3 = y^2 \quad y \le 0$$
$$\pm\sqrt{x+3} = y \quad y \le 0$$
Thus, $f^{-1}(x) = -\sqrt{x+3}$ $(x \ge -3)$.

71.
$$f(x) = x^4 - 8 \quad x \ge 0$$
$$y = x^4 - 8 \quad x \ge 0$$
$$x = y^4 - 8 \quad y \ge 0$$
$$x + 8 = y^4 \quad y \ge 0$$
$$\pm\sqrt[4]{x+8} = y \quad y \ge 0$$
Thus, $f^{-1}(x) = \sqrt[4]{x+8}$ $(x \ge -8)$.

EXERCISES 3.5

73.
$$\begin{aligned} f(x) &= \sqrt{4-x^2} && 0 \le x \le 2 \\ y &= \sqrt{4-x^2} && 0 \le x \le 2 \\ x &= \sqrt{4-y^2} && 0 \le y \le 2 \\ x^2 &= 4-y^2 && 0 \le y \le 2 \\ y^2 &= 4-x^2 && 0 \le y \le 2 \\ y &= \pm\sqrt{4-x^2} && 0 \le y \le 2 \end{aligned}$$
Thus, $f^{-1}(x) = \sqrt{4-x^2}$ $(0 \le x \le 2)$.

75. $f(x) = \dfrac{x}{x-2}$

Domain of $f = \boxed{(-\infty, 2) \cup (2, \infty)}$

$f^{-1}(x) = \dfrac{2x}{x-1}$

Range of f = Domain of f^{-1}

$= \boxed{(-\infty, 1) \cup (1, \infty)}$

77. $f(x) = \dfrac{1}{x} - 2$

Domain of $f = \boxed{(-\infty, 0) \cup (0, \infty)}$

$f^{-1}(x) = \dfrac{1}{x+2}$

Range of f = Domain of f^{-1}

$= \boxed{(-\infty, -2) \cup (-2, \infty)}$

79. **a.** $y = 0.75x + 8.50$

b. $y = 0.75(4) + 8.50$
$= \$11.50$

c.
$$\begin{aligned} y &= 0.75x + 8.50 \\ x &= 0.75y + 8.50 \\ x - 8.50 &= 0.75y \\ \frac{x-8.50}{0.75} &= y \end{aligned}$$

d.
$$\begin{aligned} y &= \frac{x-8.50}{0.75} \\ &= \frac{10-8.50}{0.75} \\ &= \frac{1.50}{0.75} = 2 \end{aligned}$$

81-83. Answers may vary.

85. $f(0) = 3$, so $f^{-1}(3) = 0$.

87. $a \ge 0$

89. False. Only one-to-one functions have inverses.

91. False. The squaring function is not one-to-one and does not have an inverse.

93. True.

95. False. The graph of a function and its inverse are symmetric about the line $y = x$.

Chapter 3 Review (page 375)

1. $f(x) = \dfrac{1}{2}(x-2)^2$

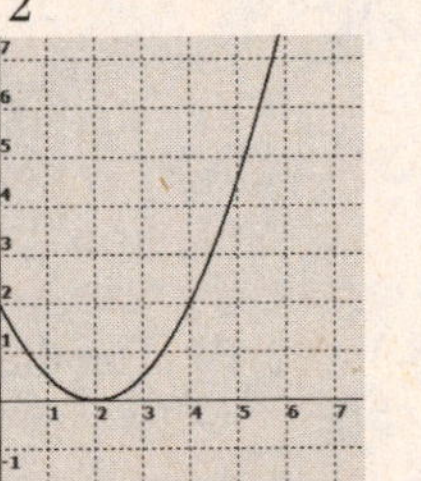

domain $= (-\infty, \infty)$
range $= [0, \infty)$

2.

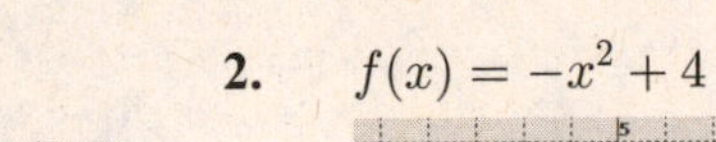

$f(x) = -x^2 + 4$

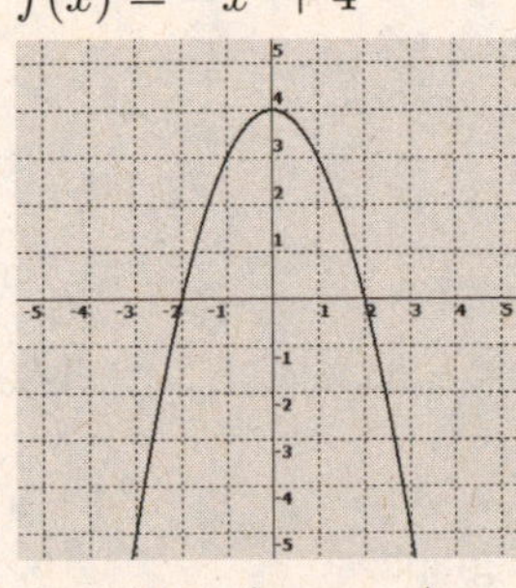

domain $= (-\infty, \infty)$
range $= (-\infty, 4]$

3.

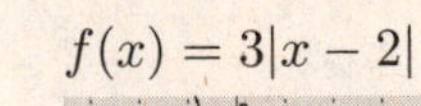

$f(x) = 3|x-2|$

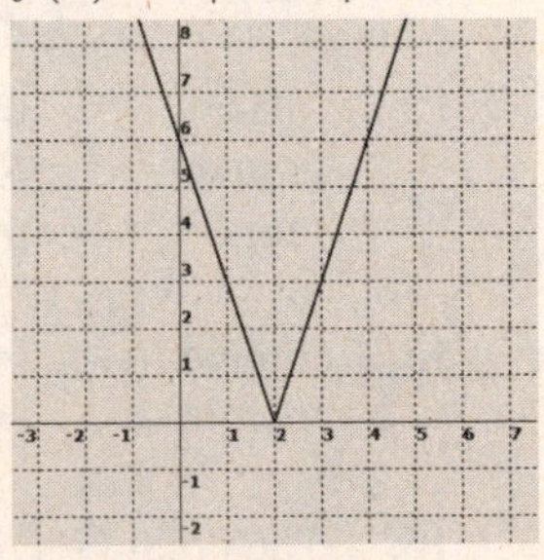

domain $= (-\infty, \infty)$
range $= [0, \infty)$

CHAPTER 3 REVIEW

4. $f(x) = -\frac{1}{2}|x| + 3$

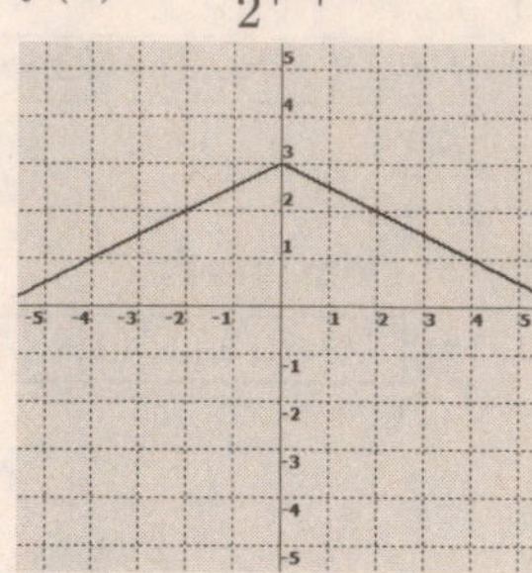

domain $= (-\infty, \infty)$
range $= (-\infty, 3]$

5. $f(x) = 2x^3 + 2$

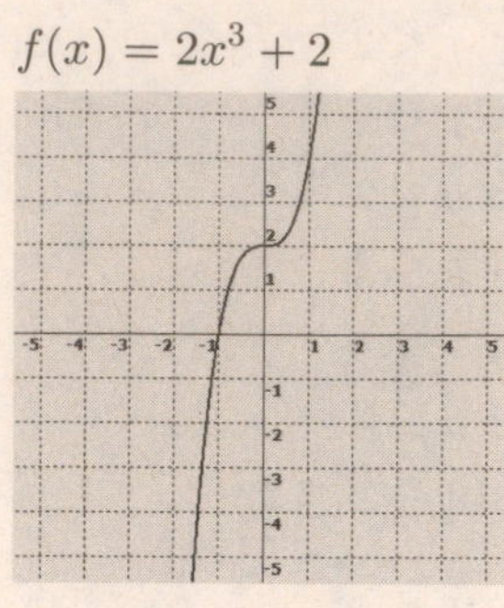

domain $= (-\infty, \infty)$
range $= (-\infty, \infty)$

6. $f(x) = -(x-4)^3$

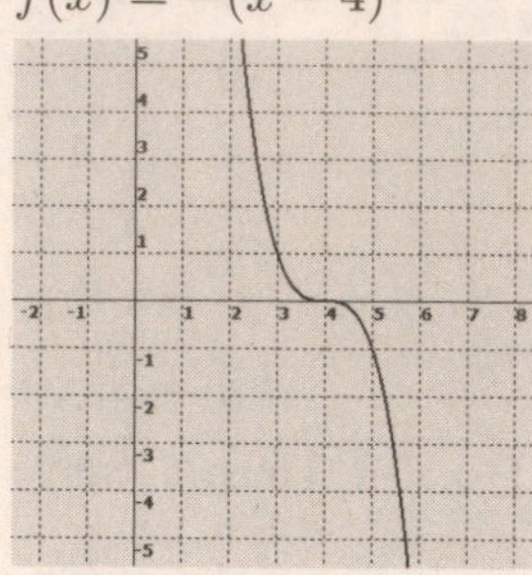

domain $= (-\infty, \infty)$
range $= (-\infty, \infty)$

7. $f(x) = \sqrt{x+5} + 1$

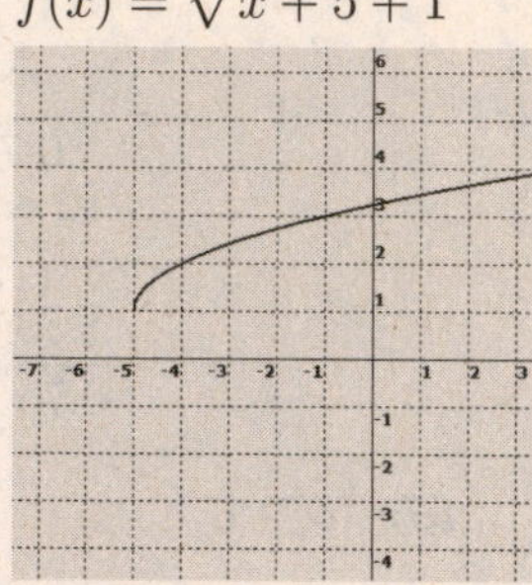

domain $= [-5, \infty)$
range $= [1, \infty)$

8. $f(x) = -\sqrt{x} - 4$

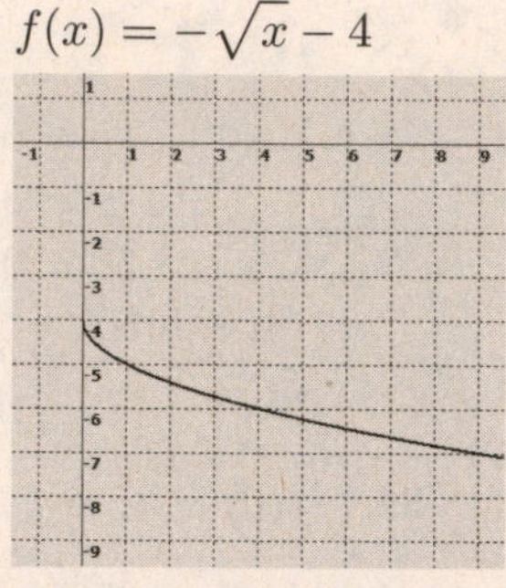

domain $= [0, \infty)$
range $= (-\infty, -4]$

9. $f(x) = 2\sqrt[3]{x}$

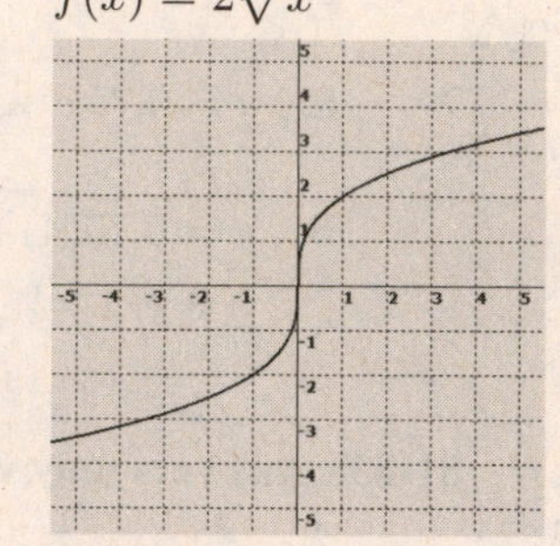

domain $= (-\infty, \infty)$
range $= (-\infty, \infty)$

10. $f(x) = -\sqrt[3]{x-1}$

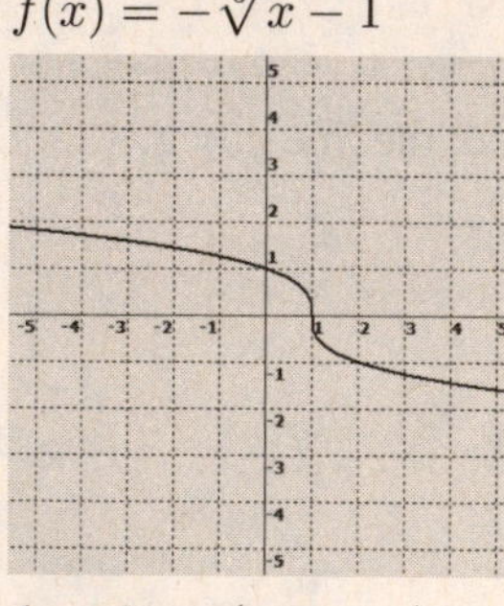

domain $= (-\infty, \infty)$
range $= (-\infty, \infty)$

11. function

12. not a function

13. domain: $(-\infty, \infty)$; range $= (-\infty, 4]$

14. The point $(2, 0)$ is on the graph, so $f(2) = 0$.

15. The point $(-1, 3)$ is on the graph, so $f(-1) = 3$.

16. The points $(-2, 0)$, $(0, 0)$, and $(2, 0)$ are on the graph, so $f(x) = 0$ when $x = -2, 0,$ and 2.

CHAPTER 3 REVIEW

17. domain: $(-\infty, \infty)$; range $= (-\infty, -2]$

18. The point $(-1, -6)$ is on the graph, so $f(-1) = -6$.

19. The point $(-4, -4)$ is on the graph, so $f(-4) = -4$.

20. The points $(-6, -8)$ and $(0, -8)$ are on the graph, so $f(x) = -8$ when $x = -6$ and 0.

21. $f(x) = -0.6x + 132$
$f(19) = -0.6(19) + 132 \approx 121$ bpm

22. $s(t) = -16t^2 + 10t + 300$
$s(2.5) = -16(2.5)^2 + 10(2.5) + 300 = 225$ ft

23. $g(x) = x^2 + 5$
Shift $y = x^2$ U 5

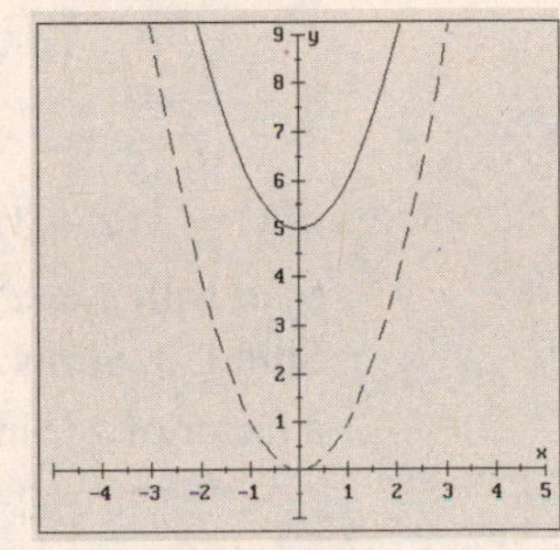

24. $g(x) = (x - 7)^3$
Shift $y = x^3$ R 7

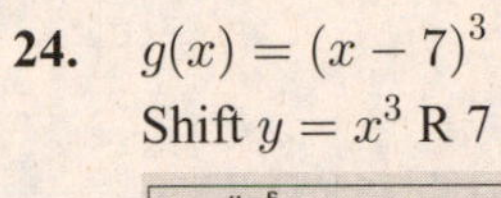

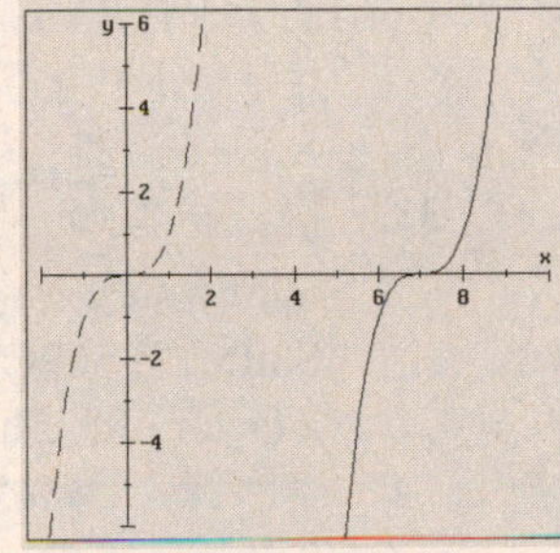

25. $g(x) = \sqrt{x + 2} + 3$
Shift $y = \sqrt{x}$ U 3, L 2

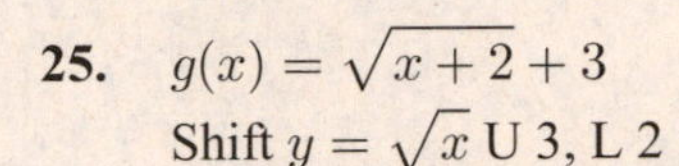

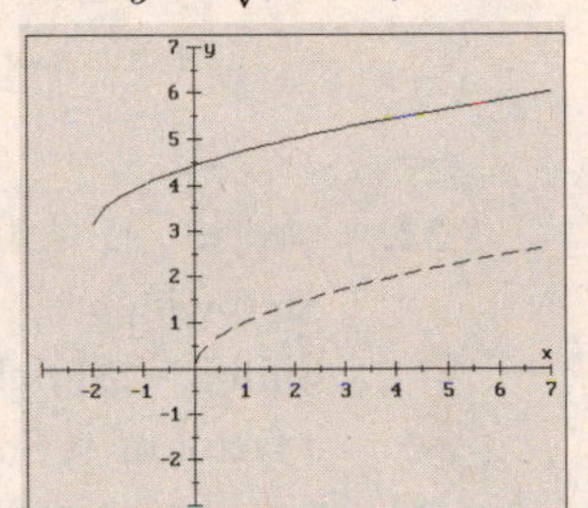

26. $g(x) = |x - 4| + 2$
Shift $y = |x|$ U 2, R 4

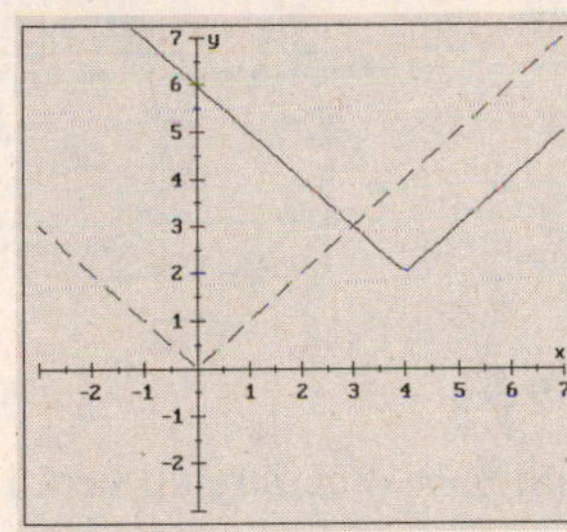

27. $g(x) = \frac{1}{3}x^3$: Shrink $y = x^3$ vert. by a factor of $\frac{1}{3}$

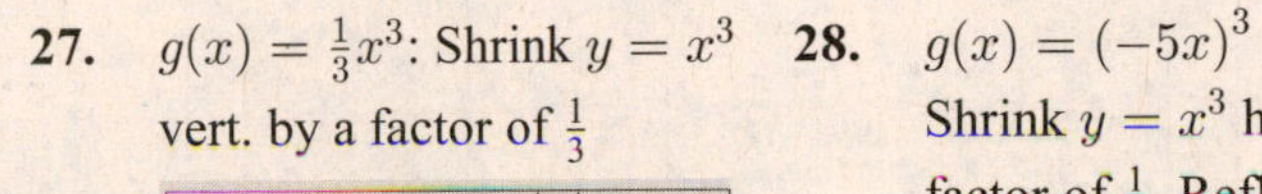

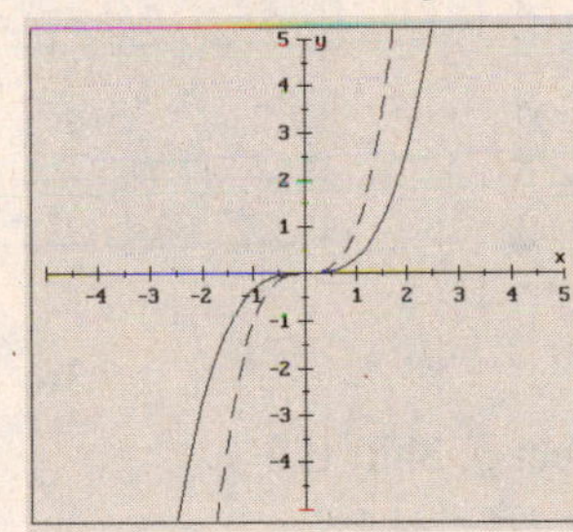

28. $g(x) = (-5x)^3$
Shrink $y = x^3$ horiz. by a factor of $\frac{1}{5}$. Reflect about y.

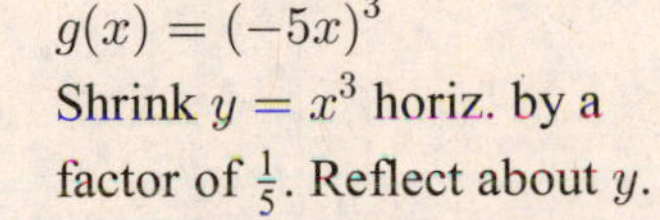

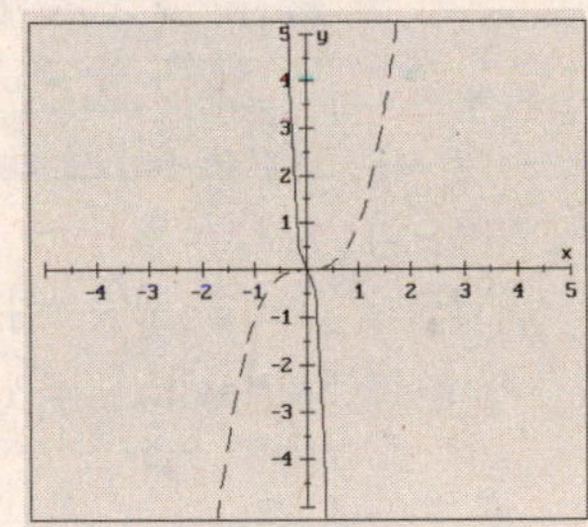

29. $f(x) = 2(x - 6)^2 - 8$
Start with $y = x^2$.
Shift R 6, Stretch vert. by a factor of 2, Shift D 8

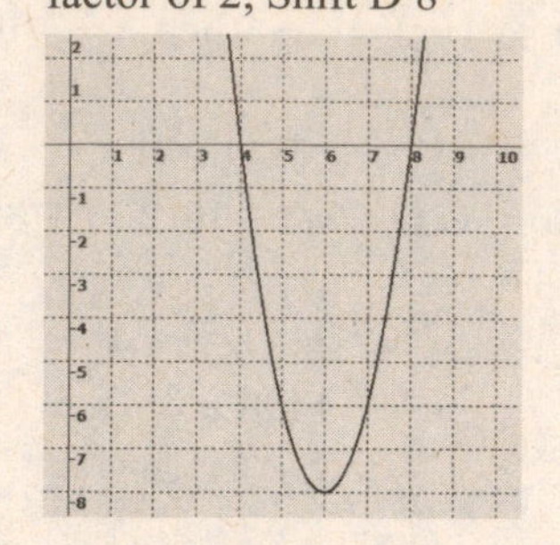

30. $f(x) = \frac{1}{2}(x + 2)^2 + 6$
Start with $y = x^2$.
Shift L 2, Shrink vert. by a factor of $\frac{1}{2}$, Shift U 6

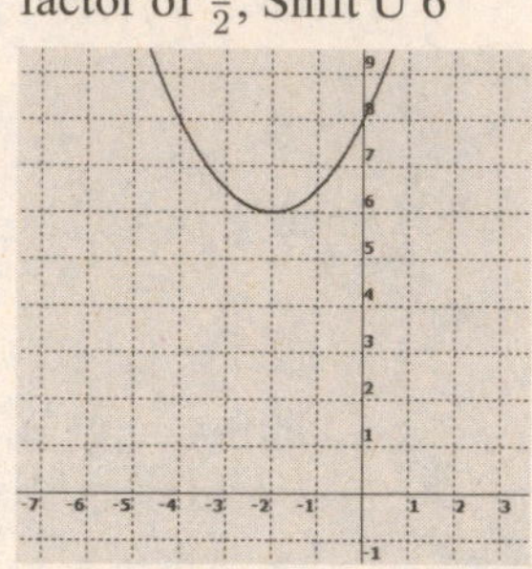

31. $g(x) = -|x - 4| + 3$
Start with $y = |x|$
Shift R 4, Reflect x, Shift U 3

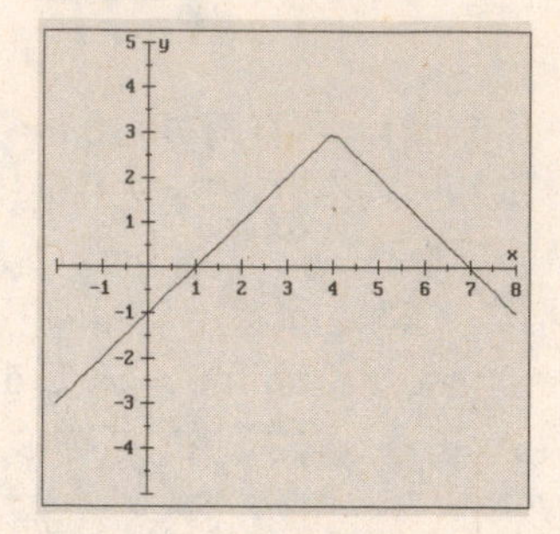

32. $g(x) = \frac{1}{4}|x - 4| + 1$
Start with $y = |x|$
Shift R 4, Shrink vert. by a factor of $\frac{1}{4}$, Shift U 1

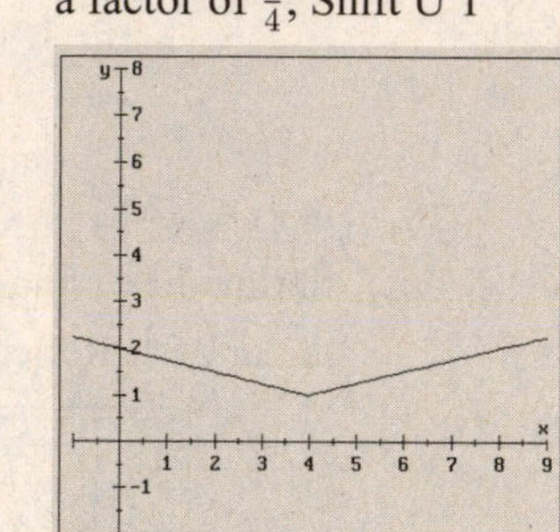

33. $g(x) = 3\sqrt{x + 3} + 2$
Start with $y = \sqrt{x}$
Shift L 3, Stretch vert. by a factor of 3, Shift U 2

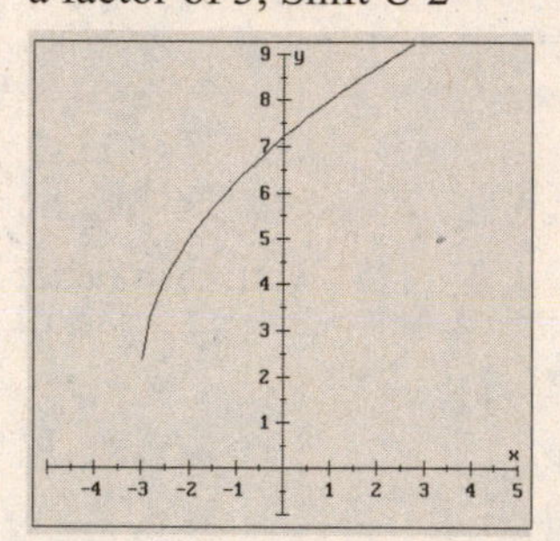

34. $g(x) = \frac{1}{3}(x + 3)^3 + 2$
Start with $y = x^3$
Shift L 3, Shrink vert. by a factor of $\frac{1}{3}$, Shift U 2

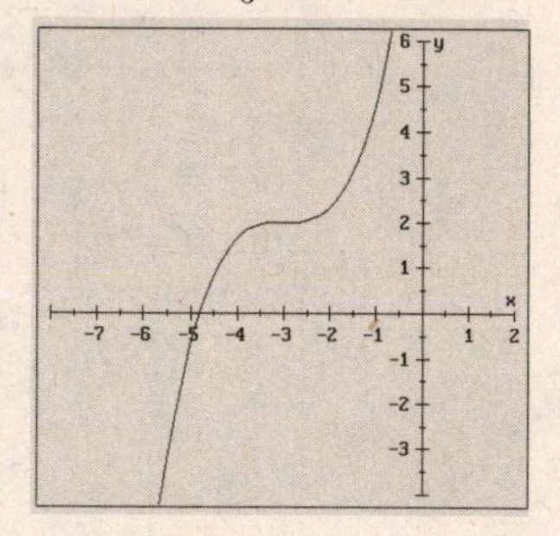

35. $f(x) = \sqrt{-x} + 3$
Start with $y = \sqrt{x}$. Reflect y, Shift U 3

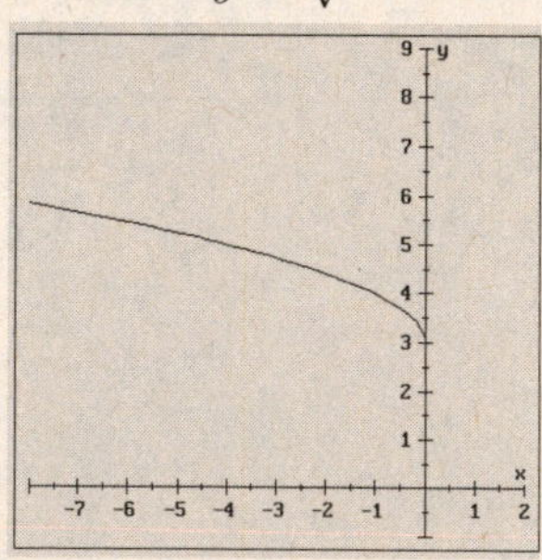

36. $g(x) = 2\sqrt[3]{x} - 5$
Start with $y = \sqrt[3]{x}$. Stretch vert. by a factor of 2, Shift D 5

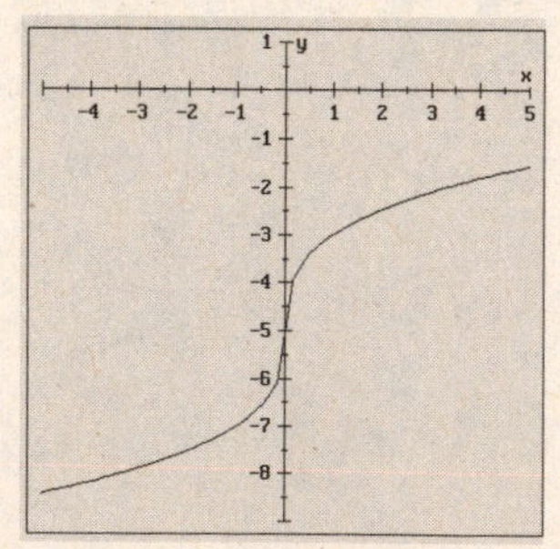

37. symmetric about y-axis $\Rightarrow$ even

38. symmetric about origin $\Rightarrow$ odd

39. no symmetry $\Rightarrow$ neither

40. symmetric about origin $\Rightarrow$ odd

41. $y = f(x) = x^3 - x$
$f(-x) = (-x)^3 - (-x)$
$= -x^3 + x$
$= -f(x) \Rightarrow$ odd

42. $f(x) = x^2 - 4x$
$f(-x) = (-x)^2 - 4(-x)$
$= x^2 + 4x$
$\Rightarrow$ neither even nor odd

43. $f(x) = x^3 - x^2$
$f(-x) = (-x)^3 - (-x)^2$
$= -x^3 - x^2$
$\Rightarrow$ neither even nor odd

44. $y = f(x) = 1 - x^4$
$f(-x) = 1 - (-x)^4$
$= 1 - x^4 = f(x) \Rightarrow$ even

45. inc: $(-\infty, 4)$; dec: $(4, \infty)$

46. inc: $(-\infty, -2) \cup (2, \infty)$; dec: $(-2, 2)$

47. local max. is 2, local min. is 0

48. local max. is 2, local min. is -2

49. **a.** $f(-2) = -2 - 2 = -4$
b. $f(3) = 3^2 = 9$

50. **a.** $f\left(\frac{3}{2}\right) = 2 - \frac{3}{2} = \frac{1}{2}$
b. $f(2) = 2 + 1 = 3$

51. $f(x) = \begin{cases} x + 3 & \text{if } x \leq 0 \\ 3 & \text{if } x > 0 \end{cases}$

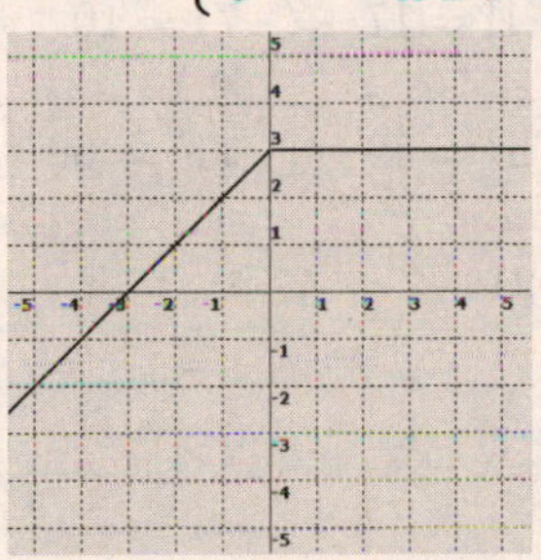

inc: $(-\infty, 0)$; const: $(0, \infty)$

52. $f(x) = \begin{cases} x + 5 & \text{if } x \leq 0 \\ 5 - x & \text{if } x > 0 \end{cases}$

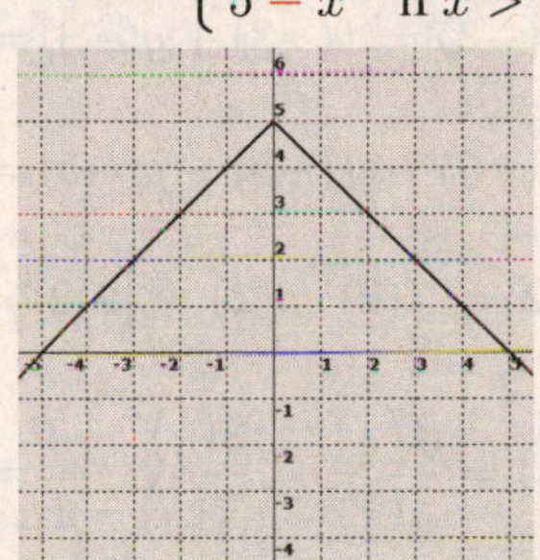

inc: $(-\infty, 0)$; dec: $(0, \infty)$

53. $f(x) = \begin{cases} -3x + 1 & \text{if } x < 0 \\ \frac{1}{3}x^2 - 4 & \text{if } x \geq 0 \end{cases}$

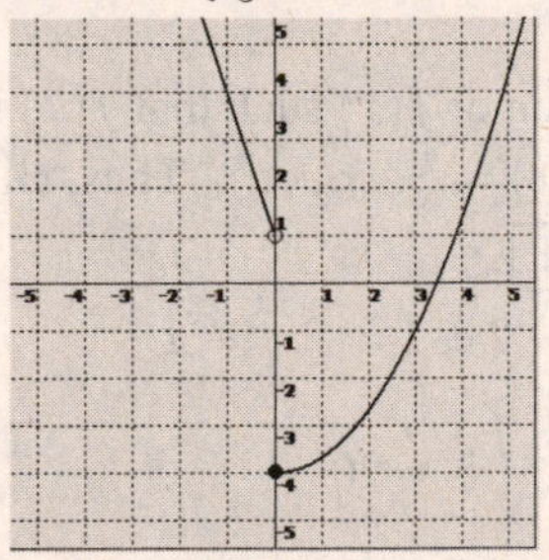

dec: $(-\infty, 0)$; inc: $(0, \infty)$

54. $f(x) = \begin{cases} \frac{1}{2}(x + 1)^2 + 2 & \text{if } x \leq -1 \\ -2 & \text{if } -1 < x < 1 \\ 2x + \frac{1}{2} & \text{if } x \geq 1 \end{cases}$

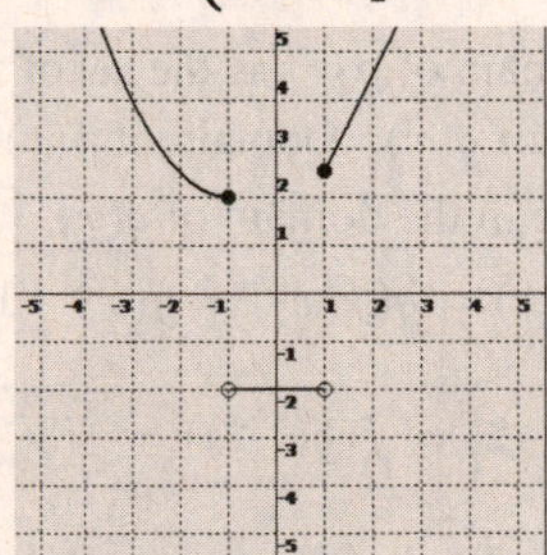

dec: $(-\infty, -1)$; const: $(-1, 1)$ inc: $(1, \infty)$

55. $f(1.7) = [[2(1.7)]] = [[3.4]] = 3$

56. $f(4.99) = [[4.99 - 5]] = [[-0.01]] = -1$

57. $f(x) = [[x]] + 2$

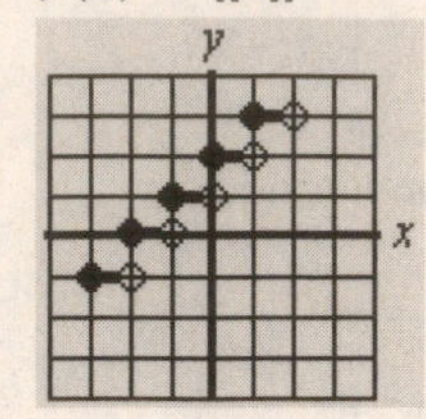

58. $f(x) = [[x - 1]]$

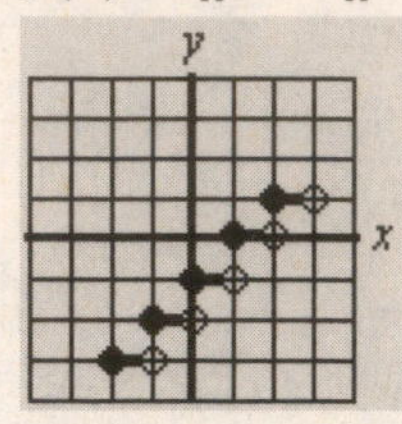

59. $20 + 3(8) = \$44$

60. $4 + 11(2) = \$26$

61. $(f + g)(x) = f(x) + g(x) = (x^2 - 1) + (2x + 1) = x^2 + 2x$; domain $= (-\infty, \infty)$

62. $(f \cdot g)(x) = f(x)g(x) = (x^2 - 1)(2x + 1) = 2x^3 + x^2 - 2x - 1$; domain $= (-\infty, \infty)$

63. $(f - g)(x) = f(x) - g(x) = (x^2 - 1) - (2x + 1) = x^2 - 2x - 2$; domain $= (-\infty, \infty)$

64. $(f/g)(x) = \dfrac{f(x)}{g(x)} = \dfrac{x^2 - 1}{2x + 1}$; domain $= \left(-\infty, -\frac{1}{2}\right) \cup \left(-\frac{1}{2}, \infty\right)$

65. $(f + g)(-3) = f(-3) + g(-3) = \left[2(-3)^2 - 1\right] + \left[2(-3) - 1\right] = 17 + (-7) = 10$

66. $(f - g)(-5) = f(-5) - g(-5) = \left[2(-5)^2 - 1\right] - \left[2(-5) - 1\right] = 49 - (-11) = 60$

67. $(f \cdot g)(2) = f(2) \cdot g(2) = \left[2(2)^2 - 1\right] \cdot \left[2(2) - 1\right] = 7 \cdot 3 = 21$

68. $(f/g)\left(\frac{1}{2}\right) = \dfrac{f\left(\frac{1}{2}\right)}{g\left(\frac{1}{2}\right)} = \dfrac{2\left(\frac{1}{2}\right)^2 - 1}{2\left(\frac{1}{2}\right) - 1} = \dfrac{-\frac{1}{2}}{0} \Rightarrow$ undefined

69. The domain of $f \circ g$ is the set of all real numbers in the domain of $g(x)$ such that $g(x)$ is in the domain of $f(x)$. Domain of $g(x)$: $(-\infty, \infty)$. Domain of $f(x) = (-\infty, \infty)$. Thus, all values of $g(x)$ are in the domain of $f(x)$. Domain of $f \circ g$: $(-\infty, \infty)$
$(f \circ g)(x) = f(g(x)) = f(2x + 1) = (2x + 1)^2 - 1 = 4x^2 + 4x + 1 - 1 = 4x^2 + 4x$

70. The domain of $g \circ f$ is the set of all real numbers in the domain of $f(x)$ such that $f(x)$ is in the domain of $g(x)$. Domain of $f(x)$: $(-\infty, \infty)$. Domain of $g(x) = (-\infty, \infty)$. Thus, all values of $f(x)$ are in the domain of $g(x)$. Domain of $g \circ f$: $(-\infty, \infty)$
$(g \circ f)(x) = g(f(x)) = g(x^2 - 1) = 2(x^2 - 1) + 1 = 2x^2 - 2 + 1 = 2x^2 - 1$

71. $(f \circ g)(-2) = f(g(-2)) = f(3(-2) + 1) = f(-5) = (-5)^2 - 5 = 20$

72. $(g \circ f)(-2) = g(f(-2)) = g\left((-2)^2 - 5\right) = g(-1) = 3(-1) + 1 = -2$

73. Let $f(x) = \sqrt{x}$ and $g(x) = x - 5$.
Then $(f \circ g)(x) = f(g(x))$
$= f(x - 5) = \sqrt{x - 5}$.

74. Let $f(x) = x^3$ and $g(x) = x + 6$.
Then $(f \circ g)(x) = f(g(x))$
$= f(x + 6) = (x + 6)^3$.

75. $f(x) = x^2 + 7$
$f(1) = f(-1) = 8$
not one-to-one

76. $f(x) = x^3$
one-to-one

77. one-to-one

78. not one-to-one

79. $(f \circ g)(x) = f(g(x)) = f\left(\frac{x+3}{8}\right) = 8\left(\frac{x+3}{8}\right) - 3 = x + 3 - 3 = x$

$(g \circ f)(x) = g(f(x)) = g(8x - 3) = \frac{(8x-3)+3}{8} = \frac{8x}{8} = x$

80. $(f \circ g)(x) = f(g(x)) = f\left(2 - \frac{1}{x}\right) = \frac{1}{2 - \left(2 - \frac{1}{x}\right)} = \frac{1}{\frac{1}{x}} = x$

$(g \circ f)(x) = g(f(x)) = g\left(\frac{1}{2 - x}\right) = 2 - \frac{1}{\frac{1}{2-x}} = 2 - (2 - x) = x$

81.
$$\begin{aligned} y = f(x) &= 7x - 1 \\ x &= 7y - 1 \\ x + 1 &= 7y \\ \frac{x+1}{7} &= y \\ f^{-1}(x) &= \frac{x+1}{7} \end{aligned}$$

82.
$$\begin{aligned} y = f(x) &= 5x - 8 \\ x &= 5y - 8 \\ x + 8 &= 5y \\ \frac{x+8}{5} &= y \\ f^{-1}(x) &= \frac{x+8}{5} \end{aligned}$$

83.
$$\begin{aligned} y = f(x) &= x^3 - 10 \\ x &= y^3 - 10 \\ x - 10 &= y^3 \\ \sqrt[3]{x - 10} &= y \\ f^{-1}(x) &= \sqrt[3]{x - 10} \end{aligned}$$

84.
$$\begin{aligned} y = f(x) &= \sqrt[3]{x + 5} \\ x &= \sqrt[3]{y + 5} \\ x^3 &= y + 5 \\ x^3 - 5 &= y \\ f^{-1}(x) &= x^3 - 5 \end{aligned}$$

85.
$$\begin{aligned} y = f(x) &= \frac{5}{x} \\ x &= \frac{5}{y} \\ xy &= 5 \\ y &= \frac{5}{x} \\ f^{-1}(x) &= \frac{5}{x} \end{aligned}$$

86.
$$\begin{aligned} y = f(x) &= \frac{1}{2 - x} \\ x &= \frac{1}{2 - y} \\ x(2 - y) &= 1 \\ 2 - y &= \frac{1}{x} \\ 2 - \frac{1}{x} &= y \\ f^{-1}(x) &= 2 - \frac{1}{x} \end{aligned}$$

87. $y = f(x) = \dfrac{x}{1-x}$

$x = \dfrac{y}{1-y}$

$x(1-y) = y$

$x - xy = y$

$x = xy + y$

$x = y(x+1)$

$\dfrac{x}{x+1} = y$

$f^{-1}(x) = \dfrac{x}{x+1}$

88. $y = f(x) = \dfrac{3}{x^3}$

$x = \dfrac{3}{y^3}$

$xy^3 = 3$

$y^3 = \dfrac{3}{x}$

$y = \sqrt[3]{\dfrac{3}{x}}$

$f^{-1}(x) = \sqrt[3]{\dfrac{3}{x}} = \dfrac{\sqrt[3]{3x^2}}{x}$

89. $y = f(x) = 2x - 5$

$x = 2y - 5$

$x + 5 = 2y$

$\dfrac{x+5}{2} = y$

$f^{-1}(x) = \dfrac{x+5}{2}$

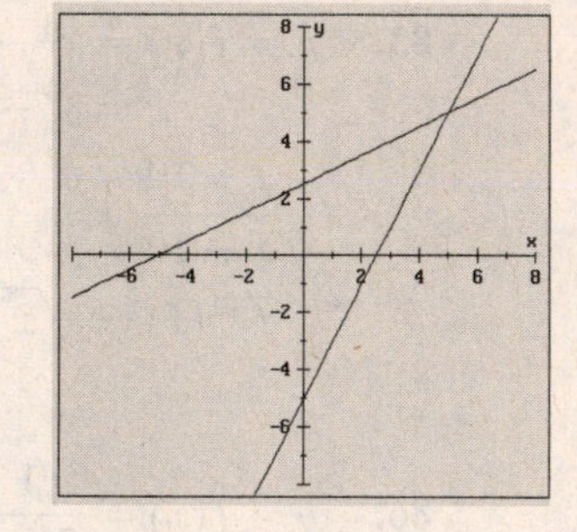

90. $y = \dfrac{2x+3}{5x-10}$

$x = \dfrac{2y+3}{5y-10}$

$x(5y-10) = 2y+3$

$5xy - 10x = 2y + 3$

$5xy - 2y = 10x + 3$

$y(5x-2) = 10x+3$

$y = \dfrac{10x+3}{5x-2}$

Range of f = Domain of f^{-1}

$= \left(-\infty, \dfrac{2}{5}\right) \cup \left(\dfrac{2}{5}, \infty\right)$

Chapter 3 Test (page 385)

1. $f(x) = 2|x+1| + 2$

2. $f(x) = -2x^3 - 4$

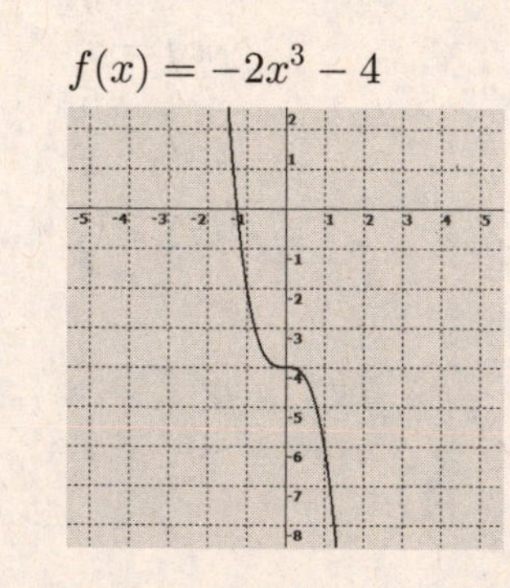

3. domain: $(-\infty, \infty)$
range: $(-\infty, 5]$

4. The point $(1, 4)$ is on the graph, so $f(1) = 4$.

5. $f(x) = (x-3)^2 + 1$
Shift $y = x^2$ U 1, R 3

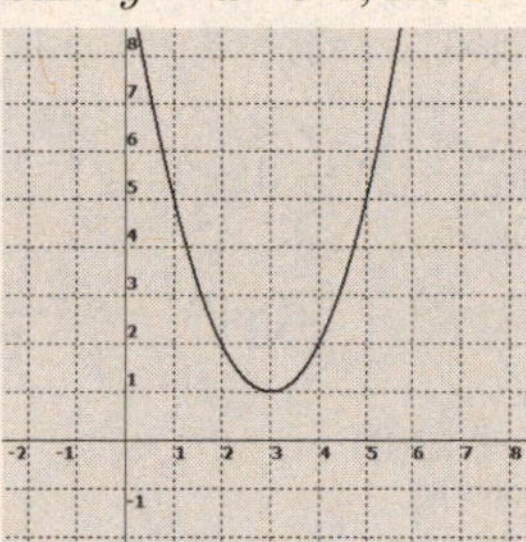

6. $f(x) = \sqrt{x-1} + 5$
Shift $y = \sqrt{x}$ U 5, R 1

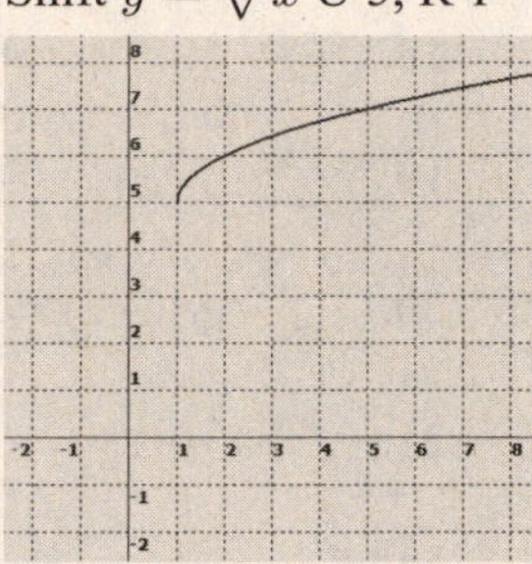

7. $f(x) = -(x-1)^3 + 3$
Shift $y = x^3$ R 1, reflect about x, shift U 3.

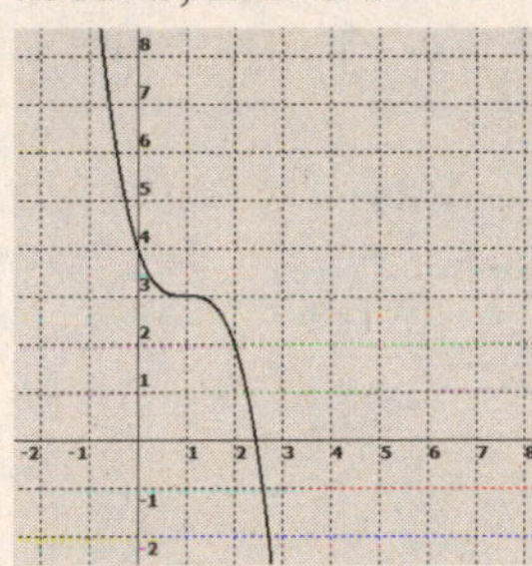

8. $f(x) = -\frac{1}{2}|x+5| - 2$
Shift $y = |x|$ L 5, reflect about x, shrink vert. by a factor of $\frac{1}{2}$, shift D 2

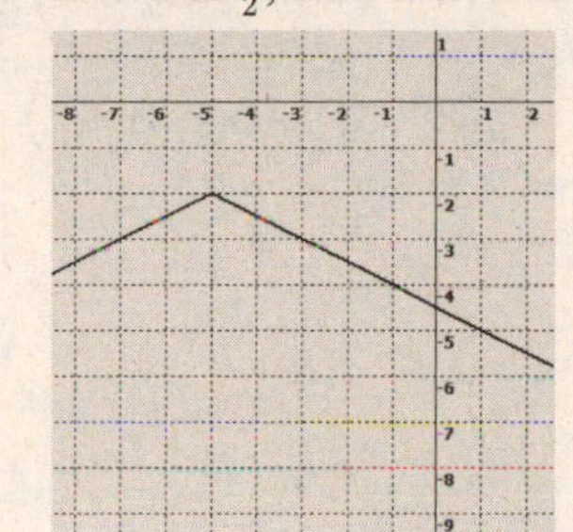

9. $f(x) = 2\sqrt[3]{x-6} - 1$
Shift $y = \sqrt[3]{x}$ R 6, stretch vert. by a factor of 2, shift D 1.

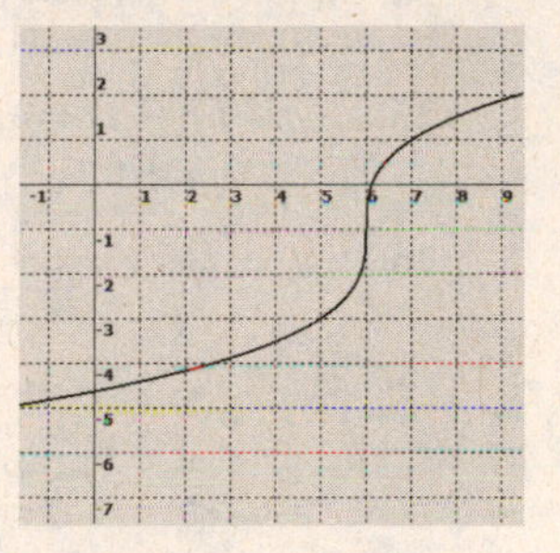

10. symmetric about origin $\Rightarrow$ odd

11. $y = f(x) = 2x^4 - 3x^2 - 7$

$$f(-x) = 2(-x)^4 - 3(-x)^2 - 7 = 2x^4 - 3x^2 - 7 = f(x) \Rightarrow \text{even}$$

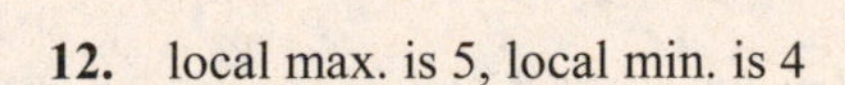

12. local max. is 5, local min. is 4

13. $f\left(\frac{3}{2}\right) = 3 - \frac{3}{2} = \frac{3}{2}$

14. $f(5) = |5| = 5$

15. $f(x) = \begin{cases} -x-1 & \text{if } x < 1 \\ 4 & \text{if } x \geq 1 \end{cases}$

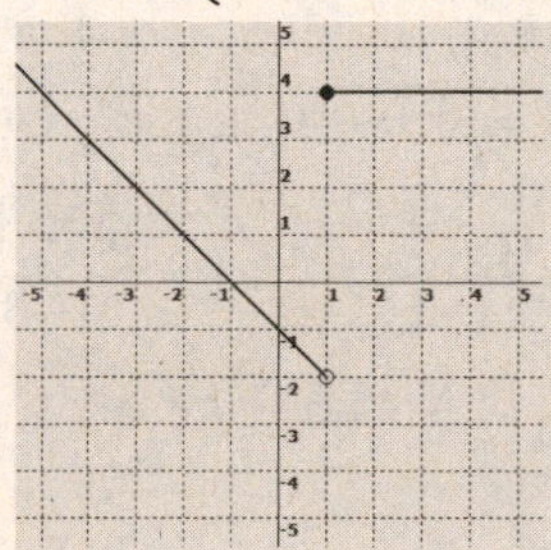

16. $(f+g)(x) = f(x) + g(x) = (3x) + (x^2 + 2) = x^2 + 3x + 2$

CHAPTER 3 TEST

17. $(f/g)(x) = \frac{f(x)}{g(x)} = \frac{3x}{x^2+2}$

18. $(g \circ f)(x) = g(f(x)) = g(3x) = (3x)^2 + 2 = 9x^2 + 2$

19. $(f \circ g)(x) = f(g(x)) = f(x^2 + 2) = 3(x^2 + 2) = 3x^2 + 6$

20. $(f + g)(-2) = f(-2) + g(-2) = \left[2(-2)^2 - 5(-2) + 1\right] + [5(-2) + 1] = 19 + (-9) = 10$

21. $(f - g)(2) = f(2) - g(2) = \left[2(2)^2 - 5(2) + 1\right] - [5(2) + 1] = -1 - 11 = -12$

22. $(f \cdot g)(-1) = f(-1) \cdot g(-1) = \left[2(-1)^2 - 5(-1) + 1\right] \cdot [5(-1) + 1] = 8 \cdot (-4) = -32$

23. $(f/g)(0) = \dfrac{f(0)}{g(0)} = \dfrac{2(0)^2 - 5(0) + 1}{5(0) + 1} = \dfrac{1}{1} = 1$

24. $(f \circ g)(-1) = f(g(-1)) = f(5(-1) + 1) = f(-4) = 2(-4)^2 - 5(-4) + 1 = 53$

25. $(g \circ f)(-3) = g(f(-3)) = g\left(2(-3)^2 - 5(-3) + 1\right) = g(34) = 5(34) + 1 = 171$

26.
$$\begin{aligned} y = f(x) &= 5x - 2 \\ x &= 5y - 2 \\ x + 2 &= 5y \\ \frac{x+2}{5} &= y \\ f^{-1}(x) &= \frac{x+2}{5} \end{aligned}$$

27.
$$\begin{aligned} y = f(x) &= \frac{x+1}{x-1} \\ x &= \frac{y+1}{y-1} \\ x(y-1) &= y+1 \\ xy - x &= y+1 \\ xy - y &= x+1 \\ y(x-1) &= x+1 \\ f^{-1}(x) &= \frac{x+1}{x-1} \end{aligned}$$

28.
$$\begin{aligned} y &= x^3 - 3 \\ x &= y^3 - 3 \\ x + 3 &= y^3 \\ \sqrt[3]{x+3} &= y \\ f^{-1}(x) &= \sqrt[3]{x+3} \end{aligned}$$

29. $f(x) = \dfrac{3}{x} - 2$; $f^{-1}(x) = \dfrac{3}{x+2}$
Range of f = Domain of f^{-1}
$= \boxed{(-\infty, -2) \cup (-2, \infty)}$

30. $f(x) = \dfrac{3x-1}{x-3}$; $f^{-1}(x) = \dfrac{3x-1}{x-3}$; Range of f = Domain of f^{-1} = $\boxed{(-\infty, 3) \cup (3, \infty)}$

Cumulative Review Exercises (page 386)

1.
$$\begin{aligned} 5x - 3y &= 15 \\ 5x - 3(0) &= 15 \\ 5x &= 15 \\ x &= 3 \\ &(3, 0) \end{aligned} \qquad \begin{aligned} 5x - 3y &= 15 \\ 5(0) - 3y &= 15 \\ -3y &= 15 \\ y &= -5 \\ &(0, -5) \end{aligned}$$

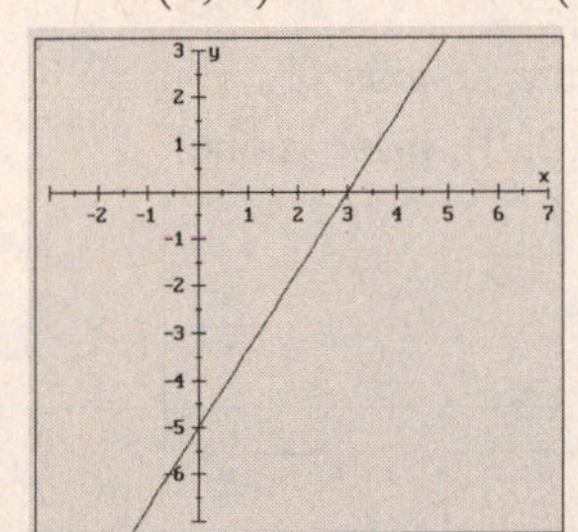

2.
$$\begin{aligned} 3x + 2y &= 12 \\ 3x + 2(0) &= 12 \\ 3x &= 12 \\ x &= 4 \\ &(4, 0) \end{aligned} \qquad \begin{aligned} 3x + 2y &= 12 \\ 3(0) + 2y &= 12 \\ 2y &= 12 \\ y &= 6 \\ &(0, 6) \end{aligned}$$

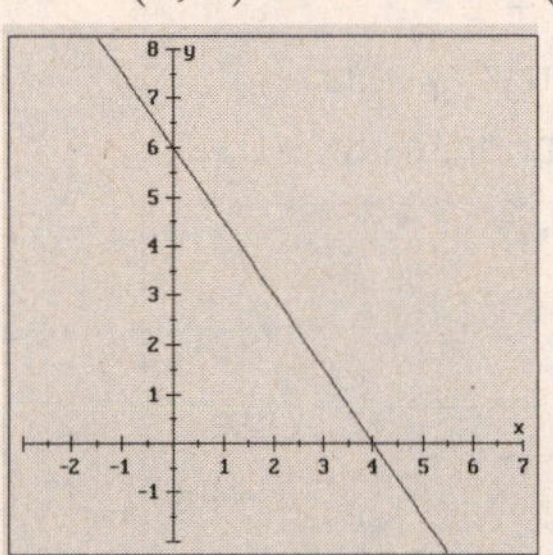

3. **a.** $d = \sqrt{(x_2 - x_1)^2 + (y_2 - y_1)^2} = \sqrt{(-2 - 3)^2 + \left(\frac{7}{2} - \left(-\frac{1}{2}\right)\right)^2} = \sqrt{25 + 16} = \sqrt{41}$

b. $x = \dfrac{x_1 + x_2}{2} = \dfrac{-2 + 3}{2} = \dfrac{1}{2}; y = \dfrac{\frac{7}{2} + \left(-\frac{1}{2}\right)}{2} = \dfrac{\frac{6}{2}}{2} = \dfrac{3}{2}$ $\left(\dfrac{1}{2}, \dfrac{3}{2}\right)$

c. $m = \dfrac{-\frac{1}{2} - \frac{7}{2}}{3 - (-2)} = \dfrac{-\frac{8}{2}}{5} = -\dfrac{4}{5}$

4. **a.** $d = \sqrt{(x_2 - x_1)^2 + (y_2 - y_1)^2} = \sqrt{(3 - (-7))^2 + (7 - 3)^2} = \sqrt{100 + 16} = \sqrt{116} = 2\sqrt{29}$

b. $x = \dfrac{x_1 + x_2}{2} = \dfrac{3 + (-7)}{2} = \dfrac{-4}{2} = -2; y = \dfrac{7 + 3}{2} = \dfrac{10}{2} = 5$ $(-2, 5)$

c. $m = \dfrac{3 - 7}{-7 - 3} = \dfrac{-4}{-10} = \dfrac{2}{5}$

5. $m = \dfrac{y_2 - y_1}{x_2 - x_1} = \dfrac{-6 - 9}{-4 - (-1)} = \dfrac{-15}{-3} = 5$

6. $m = \dfrac{y_2 - y_1}{x_2 - x_1} = \dfrac{-\frac{1}{3} - \left(-\frac{1}{3}\right)}{5 - 2} = \dfrac{0}{3} = 0$

7. $m = \dfrac{y_2 - y_1}{x_2 - x_1} = \dfrac{-7 - 5}{3 - (-3)} = \dfrac{-12}{6} = -2$
$$\begin{aligned} y - y_1 &= m(x - x_1) \\ y - 5 &= -2(x + 3) \\ y &= -2x - 1 \end{aligned}$$

8.
$$\begin{aligned} y - y_1 &= m(x - x_1) \\ y - \tfrac{5}{2} &= \tfrac{7}{2}\left(x - \tfrac{3}{2}\right) \\ y &= \tfrac{7}{2}x - \tfrac{21}{4} + \tfrac{5}{2} \\ y &= \tfrac{7}{2}x - \tfrac{11}{4} \end{aligned}$$

9. $3x - 5y = 7$
$-5y = -3x + 7$
$y = \frac{3}{5}x - \frac{7}{5}$
$m = \frac{3}{5}$
Use $m = \frac{3}{5}$.

$y - y_1 = m(x - x_1)$
$y - 3 = \frac{3}{5}(x + 5)$
$y - 3 = \frac{3}{5}x + 3$
$\boxed{y = \frac{3}{5}x + 6}$

10. $x - 4y = 12$
$-4y = -x + 12$
$y = \dfrac{1}{4}x - 3$
$m = \frac{1}{4}$
Use $m = -4$.

$y - y_1 = m(x - x_1)$
$y - 0 = -4(x - 0)$
$\boxed{y = -4x}$

11. $x^2 = y - 2$
symmetry: y-axis
x-int: none, y-int: $(0, 2)$

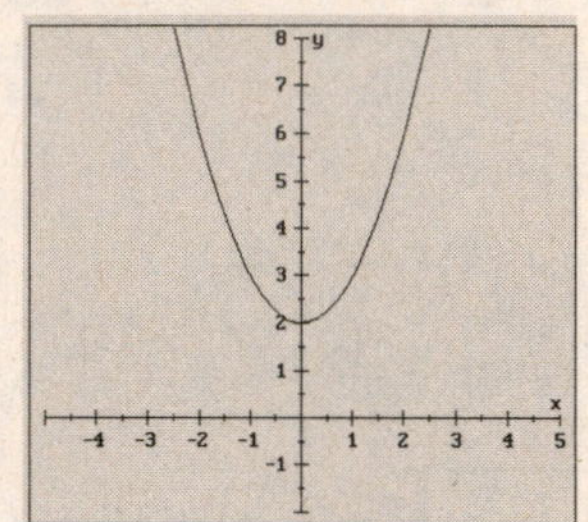

12. $y^2 = x - 2$
symmetry: x-axis
x-int: $(2, 0)$, y-int: none

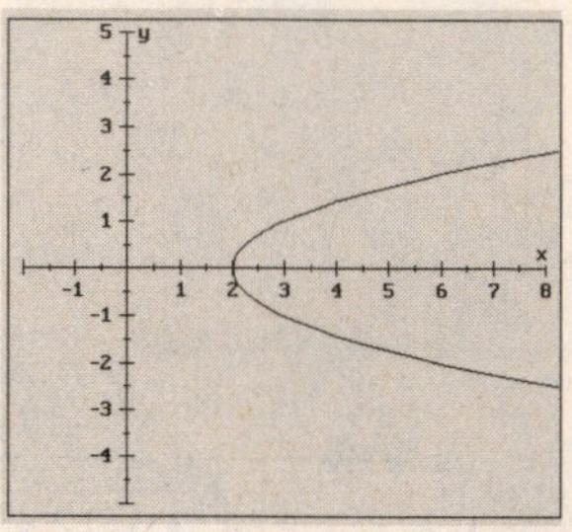

13. $x^2 + (y - 7)^2 = \dfrac{1}{4}$
$C(0, 7);\ r = \sqrt{\dfrac{1}{4}} = \dfrac{1}{2}$

14.

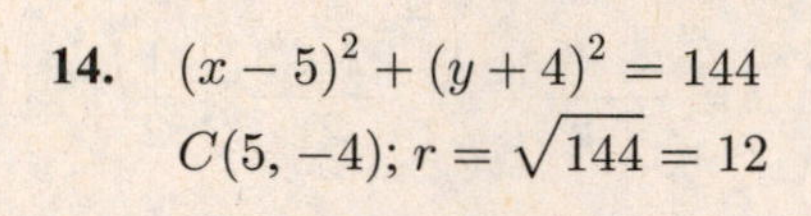

$(x - 5)^2 + (y + 4)^2 = 144$
$C(5, -4);\ r = \sqrt{144} = 12$

15. $x^2 + y^2 = 100$
Circle: $C(0, 0);\ r = 10$

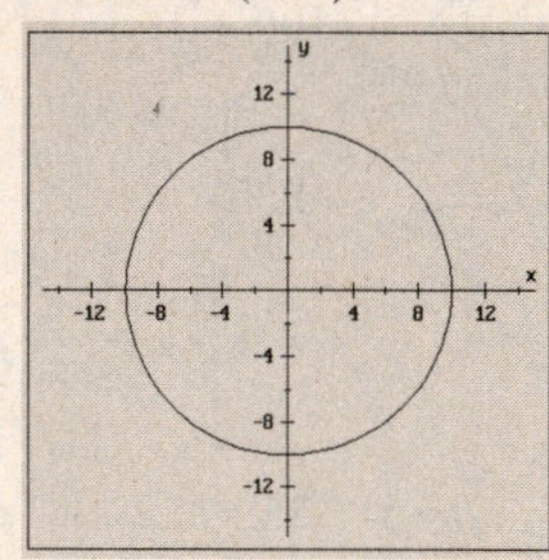

16. $x^2 - 2x + y^2 = 8$
$(x - 1)^2 + y^2 = 9$
Circle: $C(1, 0);\ r = 3$

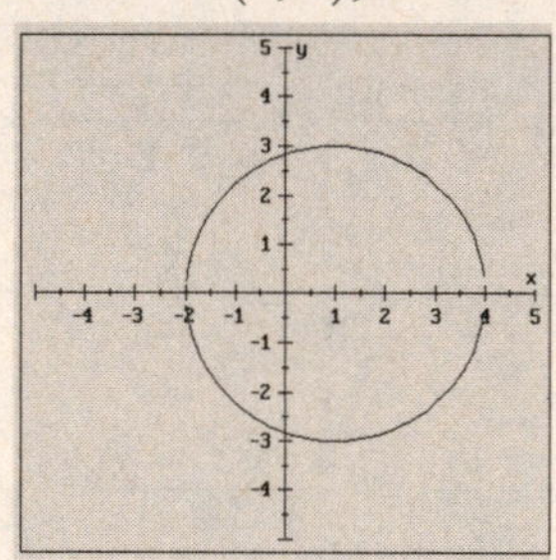

17. $\dfrac{x - 2}{x} = \dfrac{x - 6}{5}$
$5(x - 2) = x(x - 6)$
$5x - 10 = x^2 - 6x$
$0 = x^2 - 11x + 10$
$0 = (x - 10)(x - 1)$
$x = 10$ or $x = 1$

18.

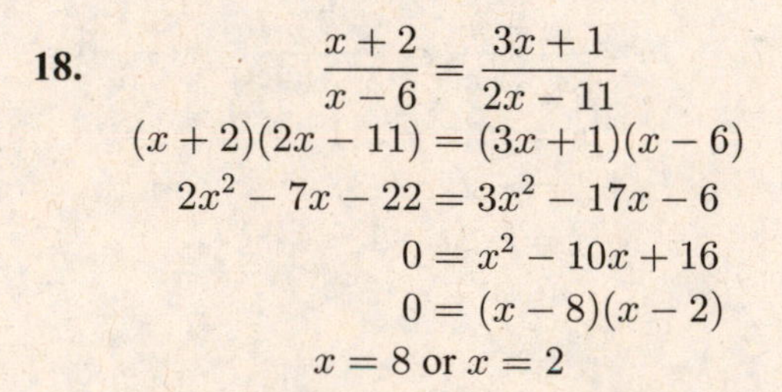

$\dfrac{x + 2}{x - 6} = \dfrac{3x + 1}{2x - 11}$
$(x + 2)(2x - 11) = (3x + 1)(x - 6)$
$2x^2 - 7x - 22 = 3x^2 - 17x - 6$
$0 = x^2 - 10x + 16$
$0 = (x - 8)(x - 2)$
$x = 8$ or $x = 2$

CUMULATIVE REVIEW EXERCISES

19. $m = \dfrac{54-37}{4-2} = \dfrac{17}{2} = 8.5$

$y = mx + b$
$y = 8.5x + b$
$37 = 8.5(2) + b$
$37 = 17 + b$
$20 = b$

$y = 8.5x + 20$
$y = 8.5(5) + 20$
$y = 62.50$
It will cost \$62.50.

20. $E = ks^2$

$E = k(50)^2 \quad E = k(20)^2$
$E = 2500k \quad E = 400k$
$\dfrac{50 \text{ mph } E}{20 \text{ mph } E} = \dfrac{2500k}{400k} = \dfrac{25}{4}$

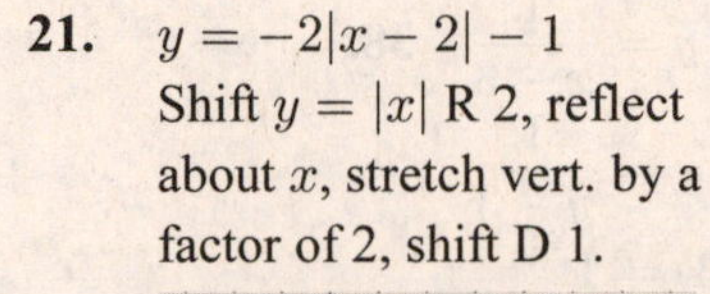

21. $y = -2|x-2| - 1$
Shift $y = |x|$ R 2, reflect about x, stretch vert. by a factor of 2, shift D 1.

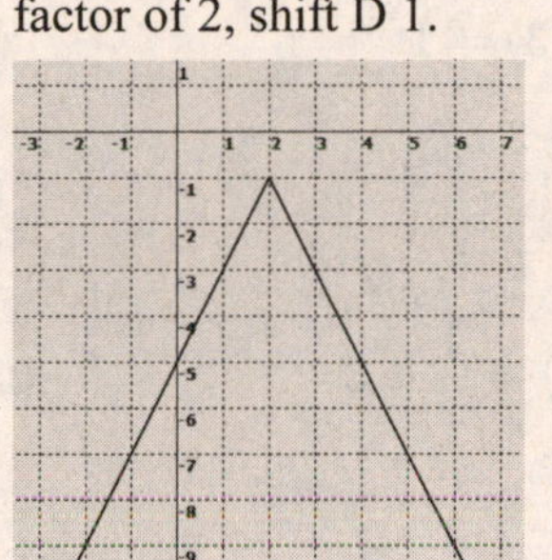

22. $y = x^2 - 4$
Shift $y = x^2$ D 4.

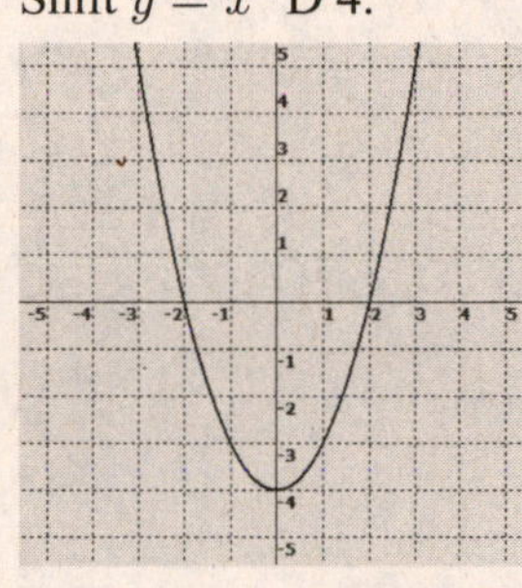

23. $y = -x^2 + 4$
Reflect $y = x^2$ about x, shift U 4.

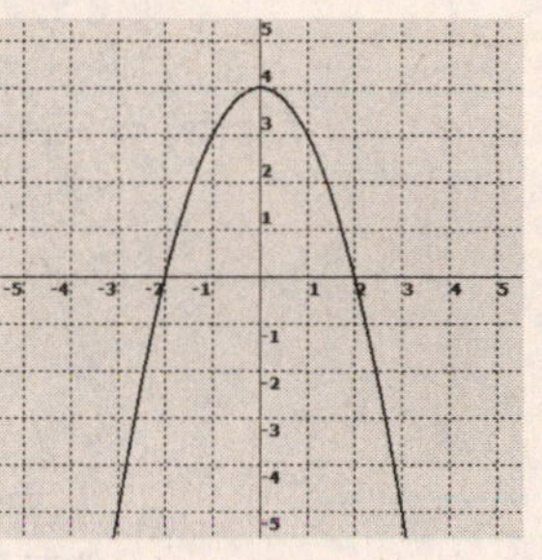

24. $y = -x^3 - 5$
Reflect $y = x^3$ about x, shift D 5.

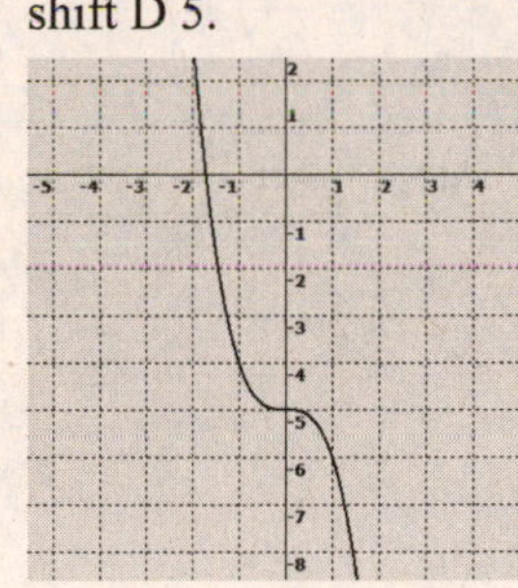

25. $y = 2\sqrt{x+4} - 1$
Shift $y = \sqrt{x}$ L 4, stretch vert. by a factor of 2, shift D 1.

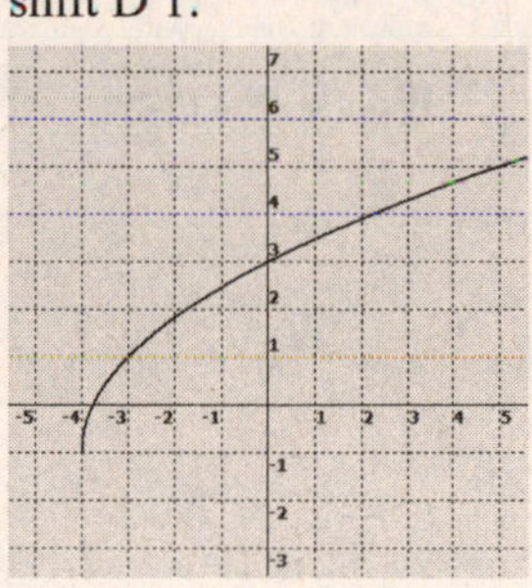

26. $y = \sqrt[3]{x-1} - 3$
Shift $y = \sqrt[3]{x}$ R 1 and D 3.

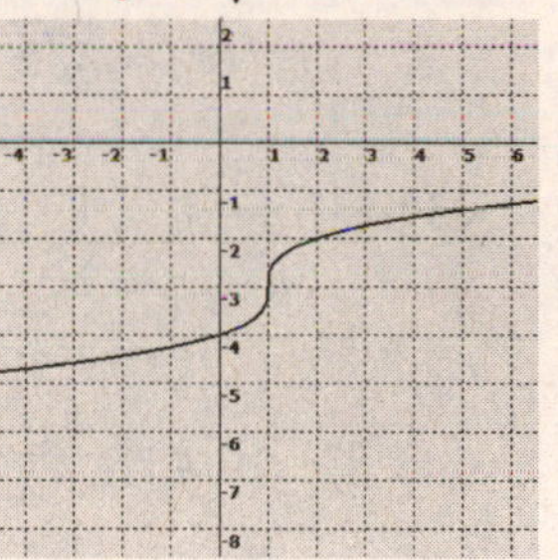

27. domain: $(-\infty, \infty)$; range: $[0, \infty)$

28. $(f+g)(x) = f(x) + g(x) = (3x-4) + (x^2+1) = x^2 + 3x - 3$; domain $= (-\infty, \infty)$

29. $(f-g)(x) = f(x) - g(x) = (3x-4) - (x^2+1) = -x^2 + 3x - 5$; domain $= (-\infty, \infty)$

30. $(f \cdot g)(x) = f(x)g(x) = (3x-4)(x^2+1) = 3x^3 - 4x^2 + 3x - 4$; domain $= (-\infty, \infty)$

31. $(f/g)(x) = \dfrac{f(x)}{g(x)} = \dfrac{3x-4}{x^2+1}$; domain $= (-\infty, \infty)$

CUMULATIVE REVIEW EXERCISES

32. $(f \circ g)(2) = f(g(2)) = f((2^2 + 1)) = f(5) = 3(5) - 4 = 11$

33. $(g \circ f)(2) = g(f(2)) = g(3(2) - 4) = g(2) = 2^2 + 1 = 5$

34. $(f \circ g)(x) = f(g(x)) = f(x^2 + 1) = \ = 3(x^2 + 1) - 4 = 3x^2 - 1$

35. $(g \circ f)(x) = g(f(x)) = g(3x - 4) = (3x - 4)^2 + 1 = 9x^2 - 24x + 17$

36.
$$\begin{aligned} y &= 3x + 2 \\ x &= 3y + 2 \\ x - 2 &= 3y \\ \tfrac{x-2}{3} &= y \\ f^{-1}(x) &= \tfrac{x-2}{3} \end{aligned}$$

37.
$$\begin{aligned} y &= \frac{1}{x-3} \\ x &= \frac{1}{y-3} \\ x(y-3) &= 1 \\ y - 3 &= \frac{1}{x} \\ y &= \frac{1}{x} + 3 \\ f^{-1}(x) &= \frac{1}{x} + 3 \end{aligned}$$

38.
$$\begin{aligned} y &= x^2 + 5 \\ x &= y^2 + 5 \\ x - 5 &= y^2 \\ \pm\sqrt{x-5} &= y \\ \sqrt{x-5} &= y \ \ (y \geq 0) \\ f^{-1}(x) &= \sqrt{x-5} \end{aligned}$$

39.
$$\begin{aligned} 3x - y &= 1 \\ 3y - x &= 1 \\ 3y &= x + 1 \\ y &= \frac{x+1}{3} \\ f^{-1}(x) &= \frac{x+1}{3} \end{aligned}$$

40. $y = kwz$

41. $y = \dfrac{kx}{t^2}$

Exercises 4.1 (page 402)

1. $f(x) = ax^2 + bx + c$

3. $(3, 5)$

5. upward

7. $-\dfrac{b}{2a}$

9. $f(x) = \frac{1}{2}x^2 + 3$
$a = \frac{1}{2} \Rightarrow a > 0$
up, minimum

11. $f(x) = -3(x+1)^2 + 2$
$a = -3 \Rightarrow a < 0$
down, maximum

13. $f(x) = -2x^2 + 5x - 1$
$a = -2 \Rightarrow a < 0$
down, maximum

15. $y = x^2 - 1 = (x-0)^2 - 1$
Vertex: $(0, -1)$

17. $f(x) = (x-3)^2 + 5$
Vertex: $(3, 5)$

19. $f(x) = -2(x+6)^2 - 4$; Vertex: $(-6, -4)$

21. $f(x) = \frac{2}{3}(x-3)^2$; Vertex: $(3, 0)$

23. $f(x) = x^2 - 4x + 4$; $a = 1, b = -4, c = 4$
$$x = -\frac{b}{2a} = -\frac{-4}{2(1)} = 2$$
$$y = x^2 - 4x + 4 = 2^2 - 4(2) + 4 = 0$$
Vertex: $(2, 0)$

25. $y = x^2 + 6x - 3$; $a = 1, b = 6, c = -3$
$$x = -\frac{b}{2a} = -\frac{6}{2(1)} = -3$$
$$y = x^2 + 6x - 3 = (-3)^2 + 6(-3) - 3 = -12$$
Vertex: $(-3, -12)$

27. $y = -2x^2 + 12x - 17$;
$a = -2, b = 12, c = -17$
$$x = -\frac{b}{2a} = -\frac{12}{2(-2)} = 3$$
$$y = -2x^2 + 12x - 17 = -2(3)^2 + 12(3) - 17 = 1$$
Vertex: $(3, 1)$

29. $y = 3x^2 - 4x + 5$;
$a = 3, b = -4, c = 5$
$$x = -\frac{b}{2a} = -\frac{-4}{2(3)} = \frac{4}{6} = \frac{2}{3}$$
$$y = 3x^2 - 4x + 5 = 3\left(\frac{2}{3}\right)^2 - 4\left(\frac{2}{3}\right) + 5 = \frac{11}{3}$$
Vertex: $\left(\frac{2}{3}, \frac{11}{3}\right)$

31. $y = \dfrac{1}{2}x^2 + 4x - 3$;
$a = \dfrac{1}{2}, b = 4, c = -3$
$$x = -\frac{b}{2a} = -\frac{4}{2\left(\frac{1}{2}\right)} = -4$$
$$y = \frac{1}{2}x^2 + 4x - 3 = \frac{1}{2}(-4)^2 + 4(-4) - 3 = -11$$
Vertex: $(-4, -11)$

33. $f(x) = x^2 - 4 = (x - 0)^2 - 4$

$a = 1 \Rightarrow$ up, vertex: $(0, -4)$

$0 = x^2 - 4$
$0 = (x + 2)(x - 2)$
$x = -2, x = 2 \Rightarrow (-2, 0), (2, 0)$

$f(0) = -4 \Rightarrow (0, -4)$

axis of symmetry: $x = 0$

$f(1) = -3 \Rightarrow (1, -3)$
$(-1, -3)$ on graph by symmetry

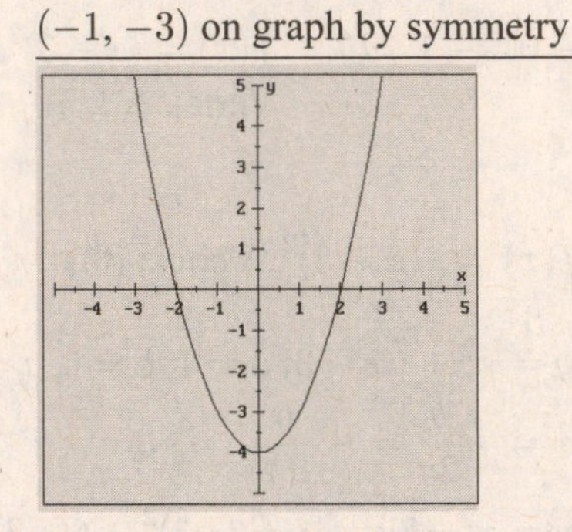

35. $f(x) = -3x^2 + 6 = -3(x - 0)^2 + 6$

$a = -3 \Rightarrow$ down, vertex: $(0, 6)$

$0 = -3x^2 + 6$
$x^2 = 2$
$x = \pm\sqrt{2} \Rightarrow \left(-\sqrt{2}, 0\right), \left(\sqrt{2}, 0\right)$

$f(0) = 6 \Rightarrow (0, 6)$

axis of symmetry: $x = 0$

$f(1) = 3 \Rightarrow (1, 3)$
$(-1, 3)$ on graph by symmetry

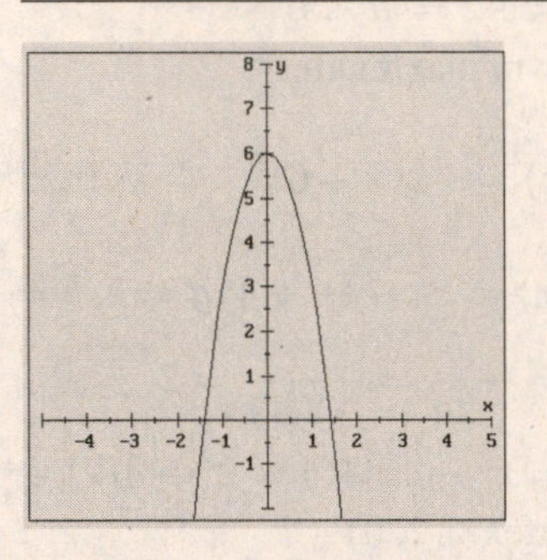

37. $f(x) = -\frac{1}{2}x^2 + 8 = -\frac{1}{2}(x - 0)^2 + 8$

$a = -\frac{1}{2} \Rightarrow$ down, vertex: $(0, 8)$

$0 = -\frac{1}{2}x^2 + 8$
$x^2 = 16$
$x = \pm 4 \Rightarrow (-4, 0), (4, 0)$

$f(0) = 8 \Rightarrow (0, 8)$

axis of symmetry: $x = 0$

$f(2) = 6 \Rightarrow (2, 6)$
$(-2, 6)$ on graph by symmetry

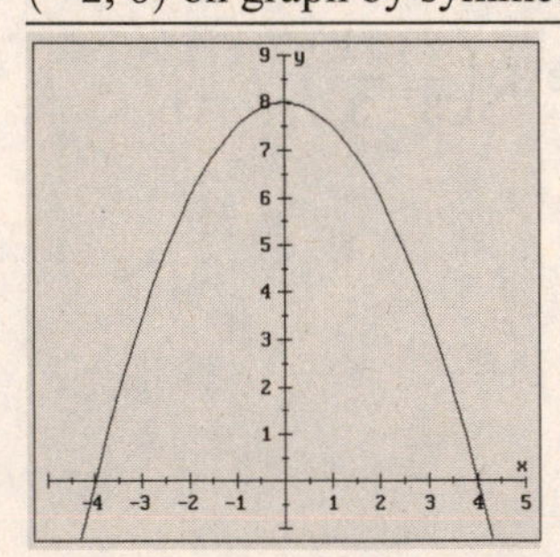

39. $f(x) = (x - 3)^2 - 1$

$a = 1 \Rightarrow$ up, vertex: $(3, -1)$

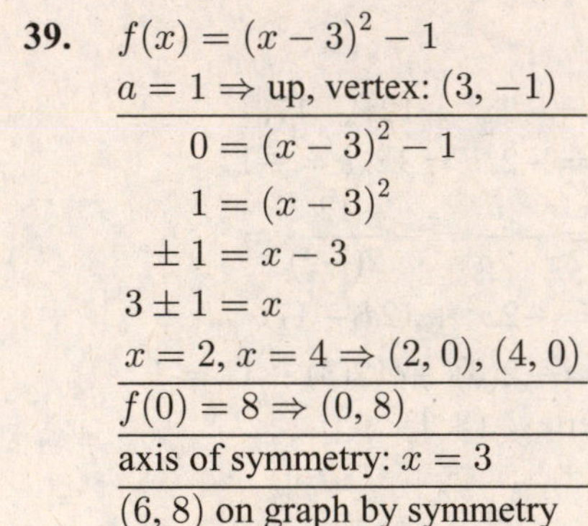

$0 = (x - 3)^2 - 1$
$1 = (x - 3)^2$
$\pm 1 = x - 3$
$3 \pm 1 = x$
$x = 2, x = 4 \Rightarrow (2, 0), (4, 0)$

$f(0) = 8 \Rightarrow (0, 8)$

axis of symmetry: $x = 3$

$(6, 8)$ on graph by symmetry

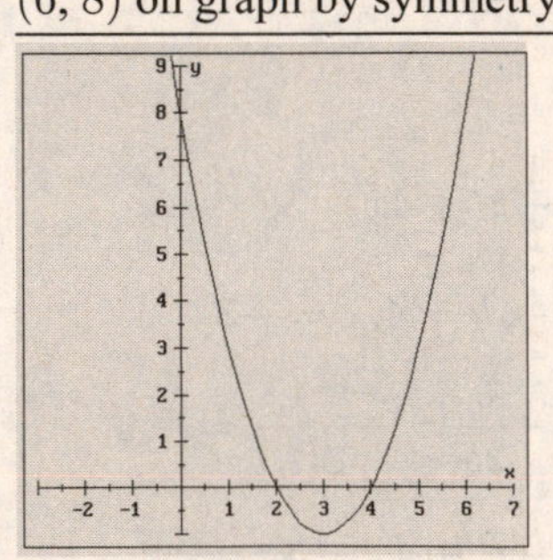

41. $f(x) = 2(x+1)^2 - 2$

$a = 2 \Rightarrow$ up, vertex: $(-1, -2)$

$$0 = 2(x+1)^2 - 2$$
$$2 = 2(x+1)^2$$
$$1 = (x+1)^2$$
$$\pm 1 = x+1$$
$$-1 \pm 1 = x$$

$x = -2, x = 0 \Rightarrow (-2, 0), (0, 0)$

$f(0) = 0 \Rightarrow (0, 0)$

axis of symmetry: $x = -1$

$f(1) = 6 \Rightarrow (1, 6)$

$(-3, 6)$ on graph by symmetry

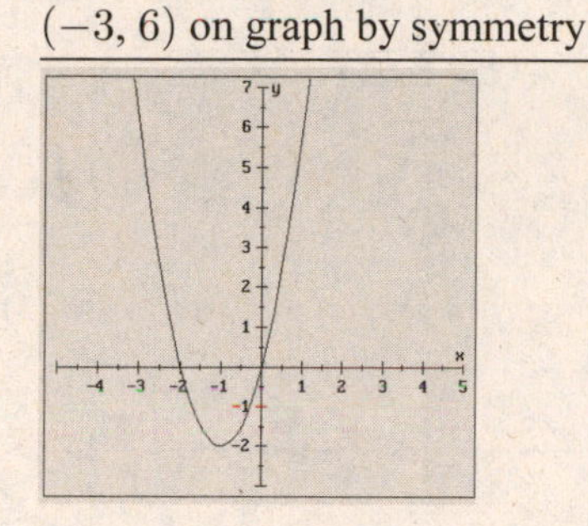

43. $f(x) = -(x+4)^2 + 1$

$a = -1 \Rightarrow$ down, vertex: $(-4, 1)$

$$0 = -(x+4)^2 + 1$$
$$(x+4)^2 = 1$$
$$x + 4 = \pm 1$$
$$x = -4 \pm 1$$

$x = -5, x = -3 \Rightarrow (-5, 0), (-3, 0)$

$f(0) = -15 \Rightarrow (0, -15)$

axis of symmetry: $x = -4$

$(-8, -15)$ on graph by symmetry

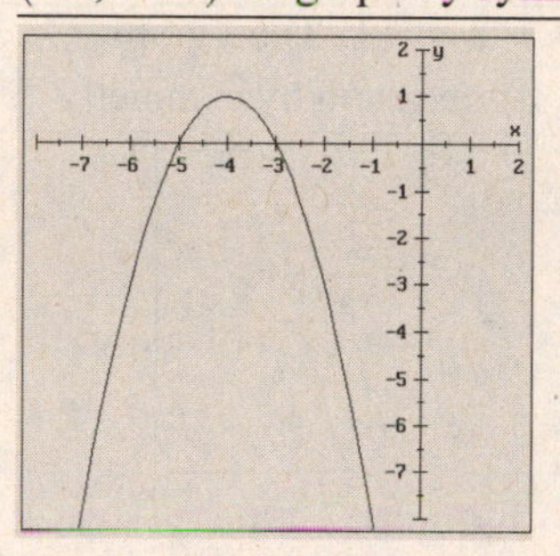

45. $f(x) = -3(x-2)^2 + 6$

$a = -3 \Rightarrow$ down, vertex: $(2, 6)$

$$0 = -3(x-2)^2 + 6$$
$$3(x-2)^2 = 6$$
$$(x-2)^2 = 2$$
$$x - 2 = \pm\sqrt{2}$$
$$x = 2 \pm \sqrt{2}$$

$\left(2 - \sqrt{2}, 0\right), \left(2 + \sqrt{2}, 0\right)$

$f(0) = -6 \Rightarrow (0, -6)$

axis of symmetry: $x = 2$

$(4, -6)$ on graph by symmetry

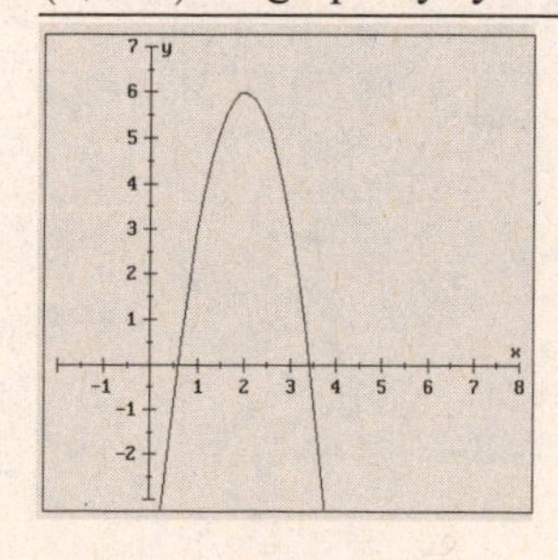

47. $f(x) = \frac{1}{3}(x-1)^2 - 3$

$a = \frac{1}{3} \Rightarrow$ up, vertex: $(1, -3)$

$$0 = \tfrac{1}{3}(x-1)^2 - 3$$
$$0 = (x-1)^2 - 9$$
$$9 = (x-1)^2$$
$$\pm 3 = x - 1$$
$$x = 1 \pm 3$$

$x = 4, x = -2 \Rightarrow (4, 0), (-2, 0)$

$f(0) = -\frac{8}{3} \Rightarrow \left(0, -\frac{8}{3}\right)$

axis of symmetry: $x = 1$

$\left(2, -\frac{8}{3}\right)$ on graph by symmetry

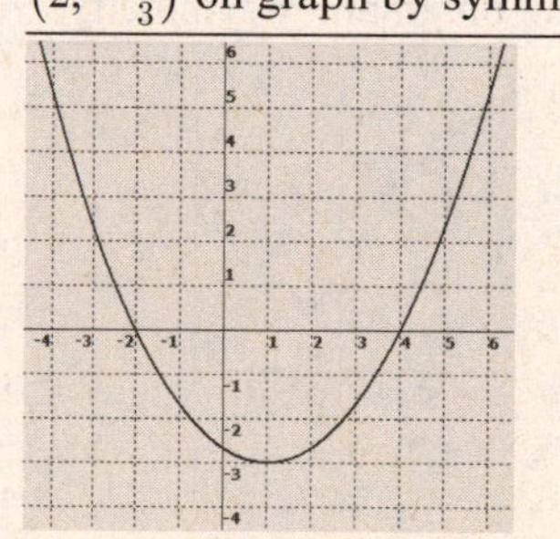

EXERCISES 4.1

49. $f(x) = x^2 + 2x;\ a = 1, b = 2, c = 0$

$x = -\dfrac{b}{2a} = -\dfrac{2}{2(1)} = -1$

$y = x^2 + 2x = (-1)^2 + 2(-1) = -1$

vertex: $(-1, -1)$, $a = 1 \Rightarrow$ up

$0 = x^2 + 2x$

$0 = x(x + 2)$

$x = 0$ or $x = -2 \Rightarrow (0, 0), (-2, 0)$

$f(0) = 0 \Rightarrow (0, 0)$

axis of symmetry: $x = -1$

$f(1) = 3 \Rightarrow (1, 3)$ on graph

$(-3, 3)$ on graph by symmetry

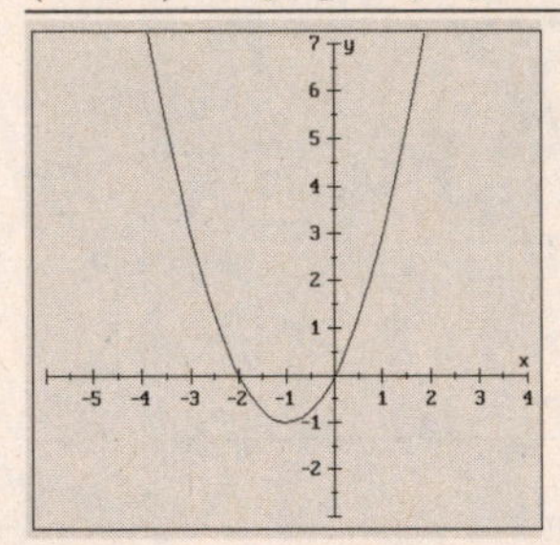

51. $f(x) = x^2 - 6x - 7;\ a = 1, b = -6, c = -7$

$x = -\dfrac{b}{2a} = -\dfrac{-6}{2(1)} = 3$

$y = x^2 - 6x - 7 = 3^2 - 6(3) - 7 = -16$

vertex: $(3, -16)$, $a = 1 \Rightarrow$ up

$0 = x^2 - 6x - 7$

$0 = (x + 1)(x - 7)$

$x = -1$ or $x = 7 \Rightarrow (-1, 0), (7, 0)$

$f(0) = -7 \Rightarrow (0, -7)$

axis of symmetry: $x = 3$

$(6, -7)$ on graph by symmetry

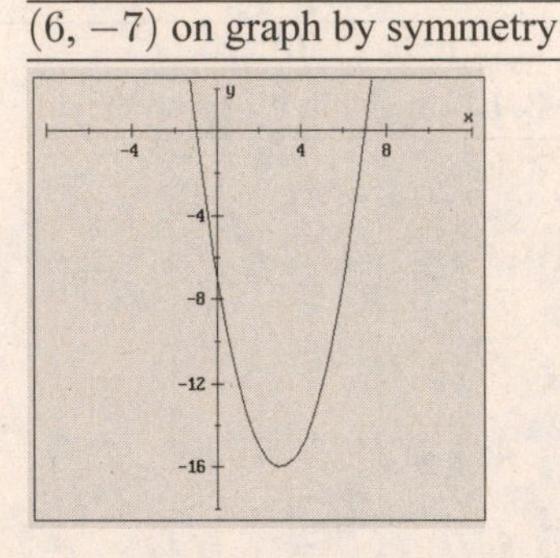

53. $f(x) = -x^2 - 4x + 1$

$a = -1, b = -4, c = 1$

$x = -\frac{b}{2a} = -\frac{-4}{2(-1)} = -2$

$y = -(-2)^2 - 4(-2) + 1 = 5$

vertex: $(-2, 5)$, $a = -1 \Rightarrow$ down

$0 = -x^2 - 4x + 1$

$x = -2 \pm \sqrt{5}$ by quadratic formula

$\left(-2 - \sqrt{5}, 0\right), \left(-2 + \sqrt{5}, 0\right)$

$f(0) = 0 \Rightarrow (0, 1)$

axis of symmetry: $x = -2$

$(-4, 1)$ on graph by symmetry

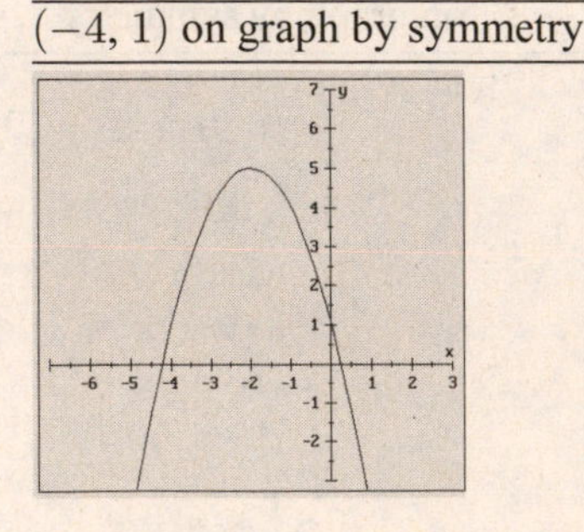

55. $f(x) = 2x^2 - 12x + 10$

$a = 2, b = -12, c = 10$

$x = -\frac{b}{2a} = -\frac{-12}{2(2)} = 3$

$y = 2(3)^2 - 12(3) + 10 = -8$

vertex: $(3, -8)$, $a = 2 \Rightarrow$ up

$0 = 2x^2 - 12x + 10$

$0 = 2(x - 1)(x - 5)$

$x = 1$ or $x = 5 \Rightarrow (1, 0), (5, 0)$

$f(0) = 10 \Rightarrow (0, 10)$

axis of symmetry: $x = 3$

$(6, 10)$ on graph by symmetry

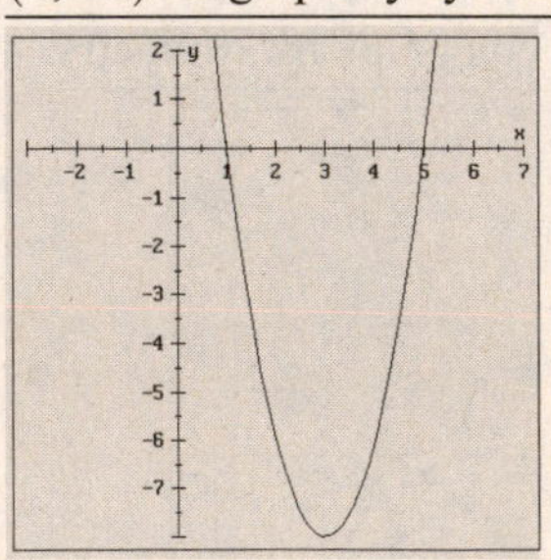

57. $f(x) = -3x^2 - 6x - 9$
$a = -3, b = -6, c = -9$
$x = -\frac{b}{2a} = -\frac{-6}{2(-3)} = -1$
$y = -3(-1)^2 - 6(-1) - 9 = -6$
vertex: $(-1, -6)$, $a = -3 \Rightarrow$ down

$0 = -3x^2 - 6x - 9$
$0 = -3(x^2 + 2x + 3)$
impossible $\Rightarrow$ no x-intercepts

$f(0) = -9 \Rightarrow (0, -9)$

axis of symmetry: $x = -1$

$(-2, -9)$ on graph by symmetry

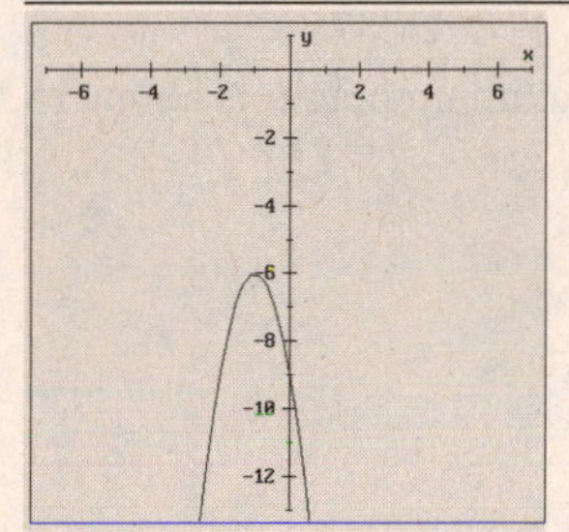

59. $f(x) = \frac{1}{2}x^2 - 2x - \frac{5}{2}$
$a = \frac{1}{2}, b = -2, c = -\frac{5}{2}$
$x = -\frac{b}{2a} = -\frac{-2}{2(\frac{1}{2})} = 2$
$y = \frac{1}{2}(2)^2 - 2(2) - \frac{5}{2} = -\frac{9}{2}$
vertex: $(2, -\frac{9}{2})$, $a = \frac{1}{2} \Rightarrow$ up

$0 = \frac{1}{2}x^2 - 2x - \frac{5}{2}$
$0 = x^2 - 4x - 5$
$0 = (x + 1)(x - 5)$
$x = -1, x = 5 \Rightarrow (-1, 0), (5, 0)$

$f(0) = -\frac{5}{2} \Rightarrow (0, -\frac{5}{2})$

axis of symmetry: $x = 2$

$(4, -\frac{5}{2})$ on graph by symmetry

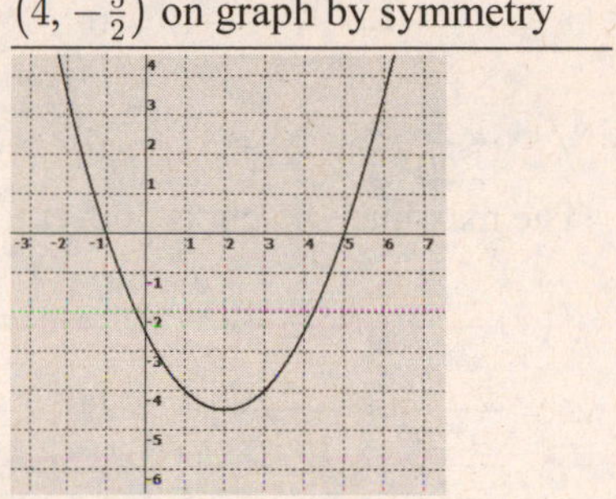

61. Let $x =$ the width of the region.
Then $\dfrac{300 - 2x}{2} = 150 - x =$ the length.
Area = width · length
$y = x(150 - x)$
$y = -x^2 + 150x$
$a = -1, b = 150, c = 0$
$x = -\dfrac{b}{2a} = -\dfrac{150}{2(-1)} = 75$
$150 - x = 150 - 75 = 75$
$y = 75(150 - 75) = 5625$
The dimensions are 75 ft by 75 ft, with an area of 5625 ft^2.

63. Let $x =$ the width.
Then $800 - 2x =$ the length.
Area $= lw$
$y = (800 - 2x)x$
$y = -2x^2 + 800x$
$a = -2, b = 800, c = 0$
$x = -\dfrac{b}{2a} = -\dfrac{800}{2(-2)} = 200$
$y = (800 - 400)200 = 80{,}000$
The maximum area of 80,000 ft^2 has dimensions 200 ft by 400 ft.

65. Let $x =$ the width.
Then $50 - x =$ the length.
Area $= lw$
$y = (50 - x)x$
$y = -x^2 + 50x$
$a = -1, b = 50, c = 0$
$x = -\dfrac{b}{2a} = -\dfrac{50}{2(-1)} = 25$
$y = (50 - 25)25 = 625$
The maximum area occurs when the dimensions are 25 ft by 25 ft.

67. Set up the variables :

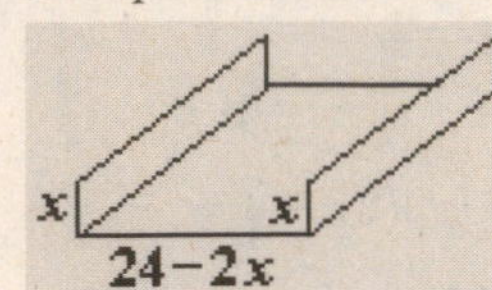

Area $= lw$

$y = x(24 - 2x)$

$y = 24x - 2x^2$

$y = -2x^2 + 24x$

$a = -2, b = 24, c = 0$

$x = -\frac{b}{2a} = -\frac{24}{2(-2)} = 6$

$y = -2x^2 + 24x = -2(6)^2 + 24(6) = 72$

The maximum area occurs when the depth is 6 inches and the width is 12 inches.

69. $x^2 + 20y - 400 = 0$

$20y = -x^2 + 400$

$y = -\frac{1}{20}x^2 + 20$

$a = -\frac{1}{20}, b = 0, c = 20$

$x = -\frac{b}{2a} = -\frac{0}{2(-\frac{1}{20})} = 0$

$y = -\frac{1}{20}x^2 + 20 = -\frac{1}{20}(0)^2 + 20 = 20$

The maximum height is 20 feet.

71. $f(x) = -0.06x^2 + 1.5x + 6$

$a = -0.06, b = 1.5, c = 6$

$x = -\frac{b}{2a} = -\frac{1.5}{2(-0.06)} = 12.5$

$f(12.5) = -0.06(12.5)^2 + 1.5(12.5) + 6$

$= 15.375$

The maximum height is about 15.4 ft.

73. $f(x) = -\frac{1}{20}x^2 + 3x - 19 \Rightarrow a = -\frac{1}{20}, b = 3, c = -19$

$x = -\frac{b}{2a} = -\frac{3}{2(-\frac{1}{20})} = 30$

$y = -\frac{1}{20}x^2 + 3x - 19 = -\frac{1}{20}(30)^2 + 3(30) - 19 = 26$

The maximum height is 26 ft.

75. Revenue = Price · # Sold

$y = x(1200 - x)$

$y = 1200x - x^2$

$y = -x^2 + 1200x$

$a = -1, b = 1200, c = 0$

Find the vertex:

$x = -\frac{b}{2a} = -\frac{1200}{2(-1)} = 600$

The maximum revenue occurs when the price is $600.

77. $C(x) = 1.5x^2 - 144x + 5856 \Rightarrow a = 1.5, b = -144, c = 5856$

$x = -\frac{b}{2a} = -\frac{-144}{2(1.5)} = 48$

$C(48) = 1.5(48)^2 - 144(48) + 5856 = 2400$

48 cameras should be made, for a minimum cost of $2400.

EXERCISES 4.1

79. Let $x = \#$ of penny decreases.
Then Fare $= 180 - x$ (cents)
Riders $= 150{,}000 + 1000x$
Revenue $=$ Fare $\cdot$ # Riders

$$y = (180 - x)(150{,}000 + 1000x)$$
$$y = 27{,}000{,}000 + 30{,}000x - 1000x^2$$
$$a = -1000,\ b = 30{,}000,\ c = 27{,}000{,}000$$

Find the vertex:

$$x = -\frac{b}{2a} = -\frac{30{,}000}{2(-1000)} = 15$$

The maximum revenue occurs when the fare is decreased by 15 pennies, or when the fare is decreased to \$1.65.

81. Let $x = \#$ of \$5 increases.
Then Rate $= 90 + 5x$
Rooms $= 200 - 10x$
Revenue $=$ Rate $\cdot$ # Rooms

$$y = (90 + 5x)(200 - 10x)$$
$$y = 18{,}000 + 100x - 50x^2$$
$$a = -50,\ b = 100,\ c = 18{,}000$$

Find the vertex:

$$x = -\frac{b}{2a} = -\frac{100}{2(-50)} = 1$$

The maximum revenue occurs when the room rate increases by 1 five-dollar increment, or when the rate is \$95.

83. $s = -16t^2 + 80t$
$a = -16,\ b = 80,\ c = 0$
Find the x-coord. of the vertex:

$$x = t = -\frac{b}{2a} = -\frac{80}{2(-16)} = \frac{5}{2} = 2.5$$

The max. height occurs after 2.5 seconds.

85. $s = -16t^2 + 80t$
$a = -16,\ b = 80,\ c = 0$
Find the y-coord. of the vertex.
Note: The x-coord. was found in **#83**.

$$y = s = -16t^2 + 80t$$
$$= -16(2.5)^2 + 80(2.5) = 100$$

The max. height is 100 ft.

87. $y = 2x^2 + 9x - 56$
Vertex: $(-2.25, -66.13)$

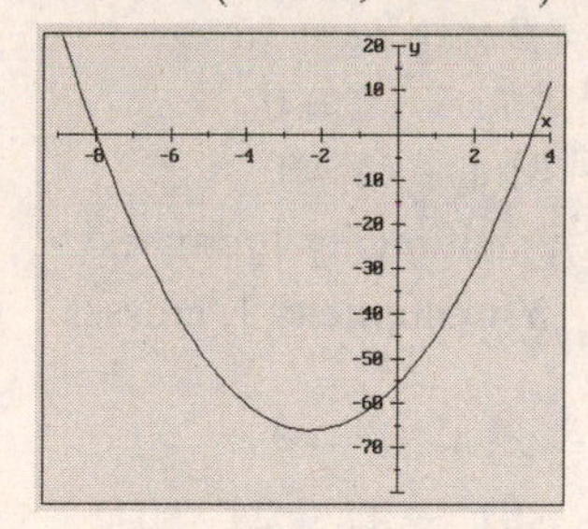

89. $y = (x - 7)(5x + 2)$
Vertex: $(3.3, -68.5)$

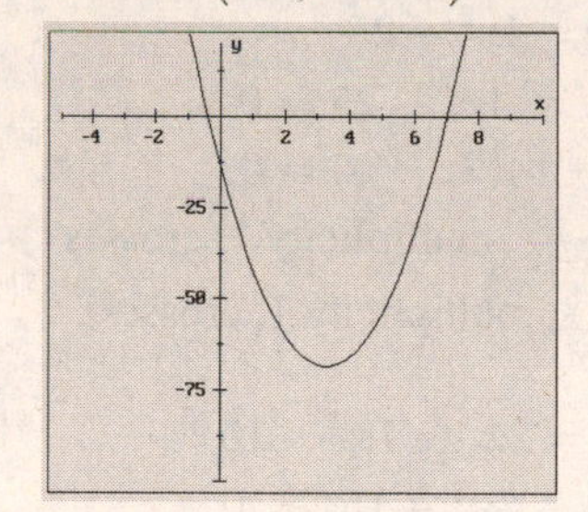

91. $f(x) = 1.679x^2 - 3.907x - 0.229$

93. $f(x) = 0.086616x^2 - 11.317553x + 410.484123$
$f(130) = 0.086616(130)^2 - 11.317553(130) + 410.484123 \approx 403$ lb

95. **Answers may vary.**

97. **Answers may vary.**

99. The equation of the line is $y = -\frac{3}{4}x + 9$.
Thus the point $(x, y) = \left(x, -\frac{3}{4}x + 9\right)$.
Area $= x\left(-\frac{3}{4}x + 9\right)$
$y = -\frac{3}{4}x^2 + 9x$

$a = -\frac{3}{4}, b = 9, c = 0$
Find the x-coord. of the vertex:
$$x = -\frac{b}{2a} = -\frac{9}{2\left(-\frac{3}{4}\right)} = 6$$
Thus, the dimensions are 6 by $4\frac{1}{2}$ units.

101. Let $x =$ one number.
Then $6 - x =$ the other number.
Sum of squares $= x^2 + (6 - x)^2$
$y = x^2 + 36 - 12x + x^2$
$y = 2x^2 - 12x + 36$

$a = 2, b = -12, c = 36$
Find the x-coord. of the vertex:
$$x = -\frac{b}{2a} = -\frac{-12}{2(2)} = 3$$
Thus, the numbers are both 3.

103. True.

105. False. The graph of a quadratic function always has exactly one y-intercept.

107. False. The range is $[k, \infty)$.

109. True.

Exercises 4.2 (page 421)

1. 4

3. zeros

5. falls, rises

7. rises, rises

9. multiplicity

11. $P(a)$ and $P(b)$

13. polynomial
degree = 5

15. polynomial
degree = 7

17. not a polynomial

19. not a polynomial

21. polynomial

23. not a polynomial

25. $f(x) = 4x^2 - 25$
$4x^2 - 25 = 0$
$(2x + 5)(2x - 5) = 0$
$x = -\frac{5}{2}$, multiplicity 1, crosses
$x = \frac{5}{2}$, multiplicity 1, crosses

27. $f(x) = 2x^2 + 7x - 15$
$2x^2 + 7x - 15 = 0$
$(2x - 3)(x + 5) = 0$
$x = \frac{3}{2}$, multiplicity 1, crosses
$x = -5$, multiplicity 1, crosses

29. $g(x) = 2x^3 - 7x^2 - 15x$
$2x^3 - 7x^2 - 15x = 0$
$x\left(2x^2 - 7x - 15\right) = 0$
$x(2x + 3)(x - 5) = 0$
$x = 0$, multiplicity 1, crosses
$x = -\frac{3}{2}$, multiplicity 1, crosses
$x = 5$, multiplicity 1, crosses

31. $g(x) = x^3 + 6x^2 - 4x - 24$
$x^3 + 6x^2 - 4x - 24 = 0$
$x^2(x + 6) - 4(x + 6) = 0$
$(x + 6)\left(x^2 - 4\right) = 0$
$(x + 6)(x + 2)(x - 2) = 0$
$x = -6$, multiplicity 1, crosses
$x = -2$, multiplicity 1, crosses
$x = 2$, multiplicity 1, crosses

EXERCISES 4.2

33. $f(x) = x^4 + 2x^3 - 3x^2$

$$x^4 + 2x^3 - 3x^2 = 0$$
$$x^2(x^2 + 2x - 3) = 0$$
$$x^2(x+3)(x-1) = 0$$

$x = 0$, multiplicity 2, touches
$x = -3$, multiplicity 1, crosses
$x = 1$, multiplicity 1, crosses

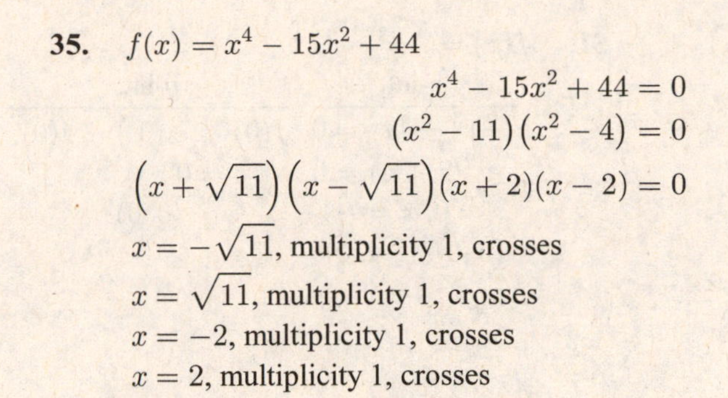

35. $f(x) = x^4 - 15x^2 + 44$

$$x^4 - 15x^2 + 44 = 0$$
$$(x^2 - 11)(x^2 - 4) = 0$$
$$\left(x + \sqrt{11}\right)\left(x - \sqrt{11}\right)(x+2)(x-2) = 0$$

$x = -\sqrt{11}$, multiplicity 1, crosses
$x = \sqrt{11}$, multiplicity 1, crosses
$x = -2$, multiplicity 1, crosses
$x = 2$, multiplicity 1, crosses

37. $h(x) = 3x^2(x+4)^2(x-5)$

$x = 0$, multiplicity 2, touches
$x = -4$, multiplicity 2, touches
$x = 5$, multiplicity 1, crosses

39. $h(x) = (2x-5)(x+3)(x-1)^2$

$x = \frac{5}{2}$, multiplicity 1, crosses
$x = -3$, multiplicity 1, crosses
$x = 1$, multiplicity 2, touches

41. $f(x) = \boldsymbol{\sqrt{5}x^7} + 10x^3 - 2x$

Degree = 7 (odd); Lead Coef: pos.
falls left, rises right

43. $g(x) = \boldsymbol{-\frac{1}{2}x^5} + 3x^4 + 2x^2 - 4$

Degree = 5 (odd); Lead Coef: neg.
rises left, falls right

45. $f(x) = \boldsymbol{7x^4} - 2x^2 + 1$

Degree = 4 (even); Lead Coef: pos.
rises left, rises right

47. $h(x) = \boldsymbol{-3x^4} - 5x - 1$

Degree = 4 (even); Lead Coef: neg.
falls left, falls right

49. $f(x) = x^3 - 9x$

x-int.	y-int.
$x^3 - 9x = 0$	$f(0) = 0^3 - 9(0)$
$x(x^2 - 9) = 0$	$y = 0$
$x(x+3)(x-3) = 0$	$(0, 0)$
$x = 0, x = -3, x = 3$	

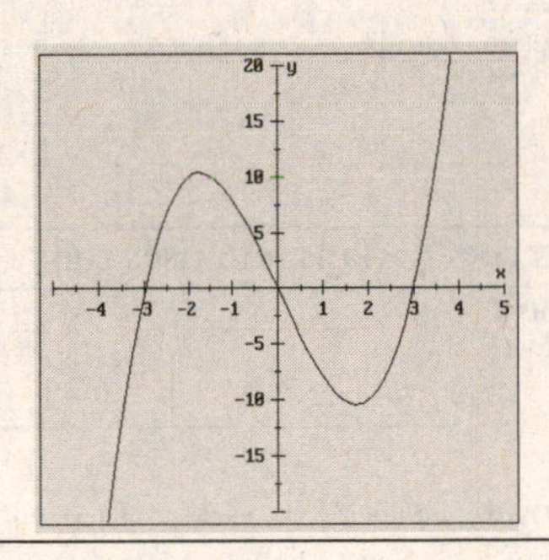

odd deg, pos coef $\Rightarrow$ falls left, rises right

	$(-\infty, -3)$	$(-3, 0)$	$(0, 3)$	$(3, \infty)$
Sign of $f(x)=x^3-9x$	$-$	$+$	$-$	$+$
Test point	$f(-4) = -28$	$f(-1) = 8$	$f(1) = -8$	$f(4) = 28$
Graph of $f(x)$	below axis	above axis	below axis	above axis

(Interval endpoints: -3, 0, 3)

$f(-x) = (-x)^3 - 9(-x) = -x^3 + 9x = -f(x) \Rightarrow$ odd, symmetric about origin

51. $f(x) = -x^3 - 4x^2$

x-int.	y-int.
$-x^3 - 4x^2 = 0$	$f(0) = -(0^3) - 4(0)^2$
$-x^2(x+4) = 0$	$y = 0$
$x = 0, x = -4$	$(0, 0)$
$(0, 0), (-4, 0)$	

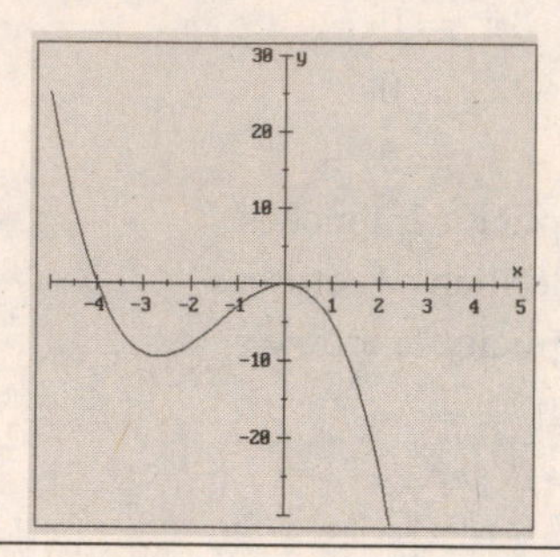

odd deg, neg coef $\Rightarrow$ rises left, falls right

Sign of $f(x) = -x^3 - 4x^2$	$+$	$-$	$-$
	$(-\infty, -4)$	$(-4, 0)$	$(0, \infty)$
	-4	0	
Test point	$f(-5) = 25$	$f(-1) = -3$	$f(1) = -5$
Graph of $f(x)$	above axis	below axis	below axis

$f(-x) = -(-x)^3 - 4(-x)^2 = x^3 - 4x^2 \Rightarrow$ neither even nor odd, no symmetry

53. $f(x) = x^3 + x^2$

x-int.	y-int.
$x^3 + x^2 = 0$	$f(0) = 0^3 + 0^2$
$x^2(x+1) = 0$	$y = 0$
$x = 0, x = -1$	$(0, 0)$
$(0, 0), (-1, 0)$	

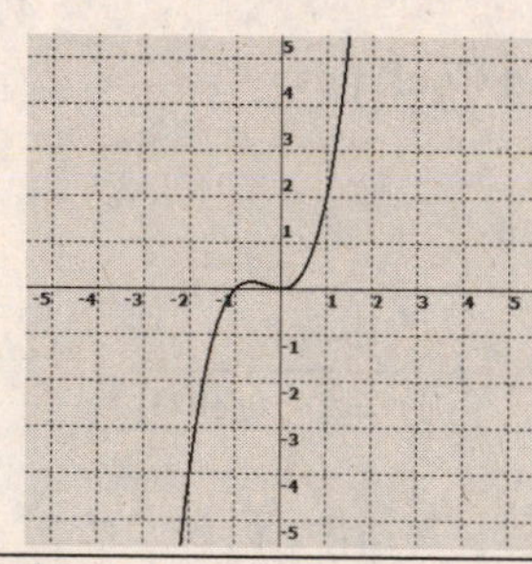

odd deg, pos coef $\Rightarrow$ falls left, rises right

Sign of $f(x) = x^3 + x^2$	$-$	$+$	$+$
	$(-\infty, -1)$	$(-1, 0)$	$(0, \infty)$
	-1	0	
Test point	$f(-2) = -4$	$f\left(-\frac{1}{2}\right) = \frac{1}{8}$	$f(1) = 2$
Graph of $f(x)$	below axis	above axis	above axis

$f(-x) = (-x)^3 + (-x)^2 = -x^3 + x^2 \Rightarrow$ neither even nor odd, no symmetry

EXERCISES 4.2

55. $f(x) = x^3 - 9x^2 + 18x$

x-int.	y-int.
$x^3 - 9x^2 + 18x = 0$	$f(0) = 0^3 - 9(0)^2 + 18(0)$
$x(x^2 - 9x + 18) = 0$	$y = 0$
$x(x-3)(x-6) = 0$	$(0, 0)$
$x = 0, x = 3, x = 6$	
$(0, 0), (3, 0), (6, 0)$	

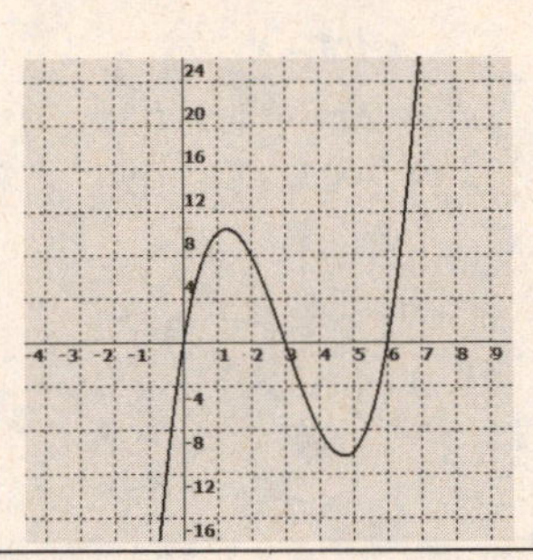

odd deg, pos coef $\Rightarrow$ falls left, rises right

Sign of $f(x)=x^3-9x^2+18x$	$-$	$+$	$-$	$+$
	$(-\infty, 0)$	$(0, 3)$	$(3, 6)$	$(6, \infty)$
	0	3	6	
Test point	$f(-1) = -28$	$f(1) = 10$	$f(4) = -8$	$f(7) = 28$
Graph of $f(x)$	below axis	above axis	below axis	above axis

$f(-x) = (-x)^3 - 9(-x)^2 + 18(-x) = -x^3 - 9x^2 - 18x \Rightarrow$ neither even nor odd, no symmetry

57. $f(x) = x^3 - x^2 - 4x + 4$

x-int.	y-int.
$x^3 - x^2 - 4x + 4 = 0$	$f(0) = (0)^3 - (0)^2 - 4(0) + 4$
$x^2(x-1) - 4(x-1) = 0$	$y = 4$
$(x-1)(x^2-4) = 0$	$(0, 4)$
$(x-1)(x+2)(x-2) = 0$	
$x = 1, x = -2, x = 2$	
$(1, 0), (-2, 0), (2, 0)$	

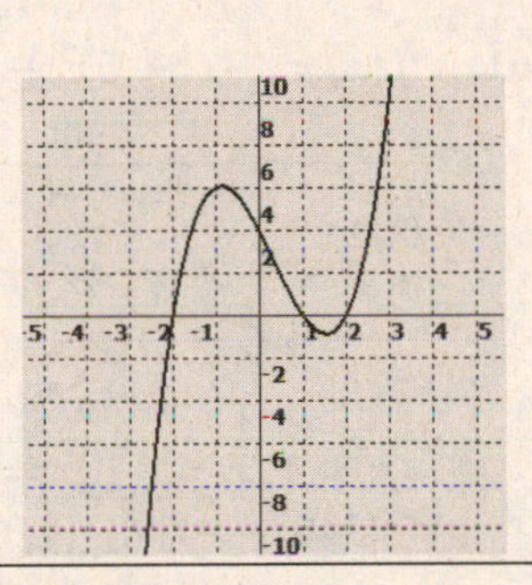

odd deg, pos coef $\Rightarrow$ falls left, rises right

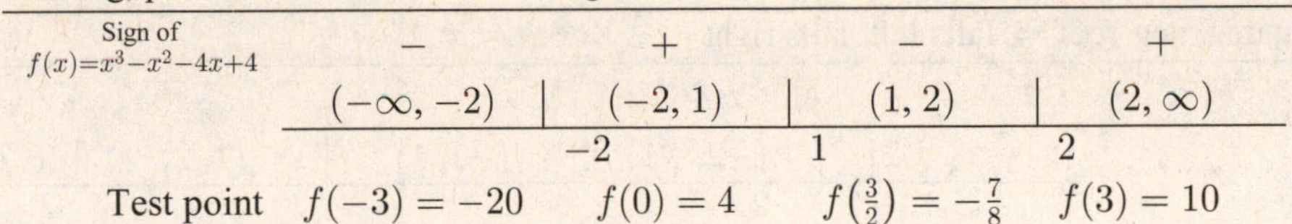

Sign of $f(x)=x^3-x^2-4x+4$	$-$	$+$	$-$	$+$
	$(-\infty, -2)$	$(-2, 1)$	$(1, 2)$	$(2, \infty)$
	-2	1	2	
Test point	$f(-3) = -20$	$f(0) = 4$	$f\left(\frac{3}{2}\right) = -\frac{7}{8}$	$f(3) = 10$
Graph of $f(x)$	below axis	above axis	below axis	above axis

$f(-x) = (-x)^3 - (-x)^2 - 4(-x) + 4 = -x^3 - x^2 + 4x + 4 \Rightarrow$ neither even nor odd, no symmetry

59. $f(x) = x^4 - 2x^2 + 1$

x-int.	y-int.
$x^4 - 2x^2 + 1 = 0$	$f(0) = (0)^4 - 2(0)^2 + 1$
$(x^2 - 1)(x^2 - 1) = 0$	$y = 1$
$x^2 = 1$	$(0, 1)$
$x = 1, x = -1$	
$(1, 0), (-1, 0)$	

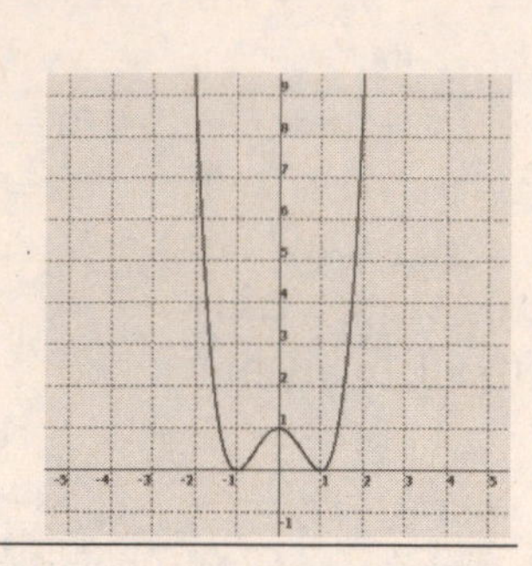

even deg, pos coef $\Rightarrow$ rises left, rises right

Sign of $f(x)=x^4-2x^2+1$	$+$	$+$	$+$
	$(-\infty, -1)$	$(-1, 1)$	$(1, \infty)$
	-1	1	
Test point	$f(-2) = 9$	$f(0) = 1$	$f(2) = 9$
Graph of $f(x)$	above axis	above axis	above axis

$f(-x) = (-x)^4 - 2(-x)^2 + 1 = x^4 - 2x^2 + 1 = f(x)$

$\Rightarrow$ even, symmetric about y-axis

61. $f(x) = -x^4 + 5x^2 - 4$

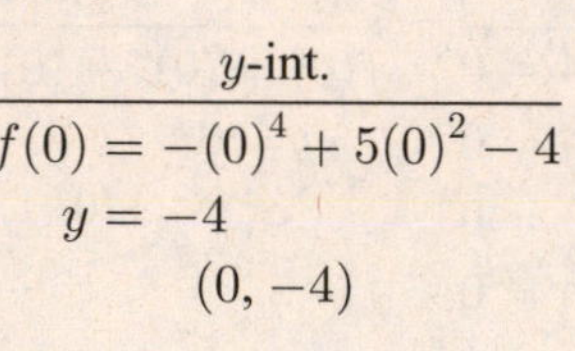

x-int.	y-int.
$-x^4 + 5x^2 - 4 = 0$	$f(0) = -(0)^4 + 5(0)^2 - 4$
$x^4 - 5x^2 + 4 = 0$	$y = -4$
$(x^2 - 1)(x^2 - 4) = 0$	$(0, -4)$
$x^2 = 1$ or $x^2 = 4$	
$x = 1, x = -1, x = 2, x = -2$	
$(1, 0), (-1, 0), (2, 0), (-2, 0)$	

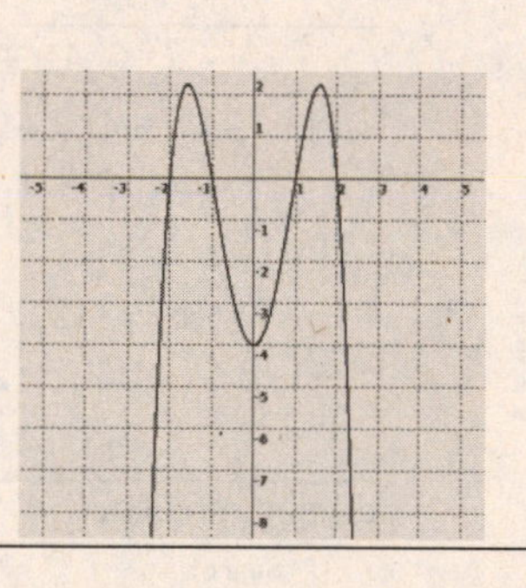

even deg, neg coef $\Rightarrow$ falls left, falls right

Sign of $f(x)=-x^4+5x^2-4$	$-$	$+$	$-$	$+$	$-$
	$(-\infty, -2)$	$(-2, -1)$	$(-1, 1)$	$(1, 2)$	$(2, \infty)$
	-2	-1	1	2	
Test point	$f(-3) = -40$	$f\left(-\frac{3}{2}\right) = \frac{35}{16}$	$f(0) = -4$	$f\left(\frac{3}{2}\right) = \frac{35}{16}$	$f(3) = -40$
Graph of $f(x)$	below axis	above axis	below axis	above axis	below axis

$f(-x) = -(-x)^4 + 5(-x)^2 - 4 = -x^4 + 5x^2 - 4 = f(x)$

$\Rightarrow$ even, symmetric about y-axis

63. $f(x) = -x^4 + 6x^3 - 8x^2$

x-int.	y-int.
$-x^4 + 6x^3 - 8x^2 = 0$	$f(0) = -(0)^4 + 6(0)^3 - 8(0)^2$
$-x^2(x^2 - 6x + 8) = 0$	$y = 0$
$-x^2(x-2)(x-4) = 0$	$(0, 0)$
$x = 0, x = 2, x = 4$	
$(0,0), (2,0), (4,0)$	

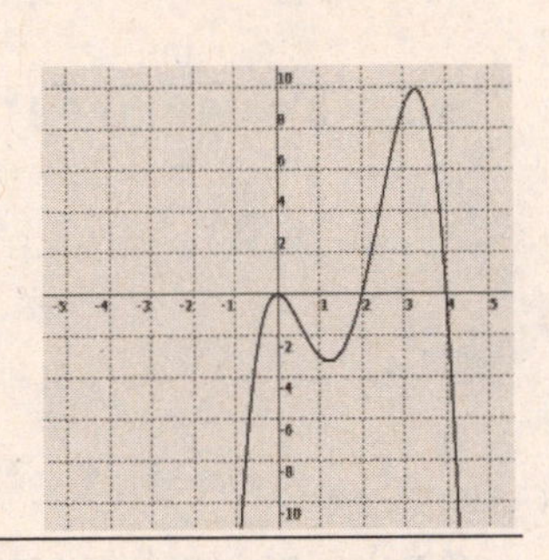

even deg, neg coef $\Rightarrow$ falls left, falls right

Sign of $f(x)=-x^4+6x^3-8x^2$	$-$	$-$	$+$	$-$
	$(-\infty, 0)$	$(0, 2)$	$(2, 4)$	$(4, \infty)$
	0	2	4	
Test point	$f(-1) = -15$	$f(1) = -3$	$f(3) = 9$	$f(5) = -75$
Graph of $f(x)$	below axis	below axis	above axis	below axis

$f(-x) = -(-x)^4 + 6(-x)^3 - 8(-x)^2 = -x^4 - 6x^3 - 8x^2$

$\Rightarrow$ neither even nor odd, no symmetry

65. $f(x) = \frac{1}{2}x^4 - \frac{9}{2}x^2$

x-int.	y-int.
$\frac{1}{2}x^4 - \frac{9}{2}x^2 = 0$	$f(0) = \frac{1}{2}(0)^4 - \frac{9}{2}(0)^2$
$x^4 - 9x^2 = 0$	$y = 0$
$x^2(x^2 - 9) = 0$	$(0, 0)$
$x^2(x+3)(x-3) = 0$	
$x = 0, x = -3, x = 3$	
$(0,0), (-3,0), (3,0)$	

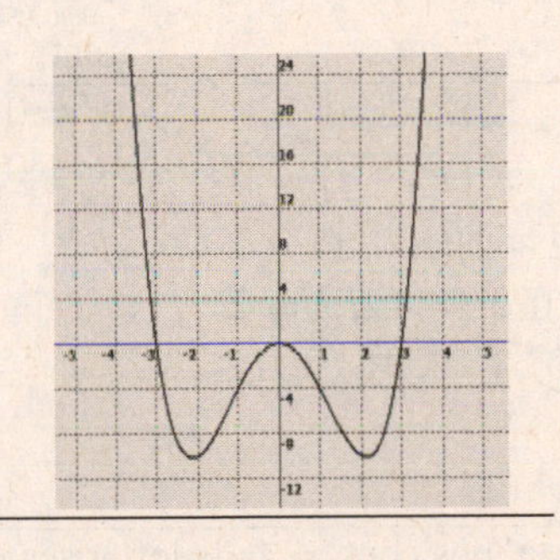

even deg, pos coef $\Rightarrow$ rises left, rises right

Sign of $f(x)=\frac{1}{2}x^4-\frac{9}{2}x^2$	$+$	$-$	$-$	$+$
	$(-\infty, -3)$	$(-3, 0)$	$(0, 3)$	$(3, \infty)$
	-3	0	3	
Test point	$f(-4) = 56$	$f(-1) = -4$	$f(1) = -4$	$f(4) = 56$
Graph of $f(x)$	above axis	below axis	below axis	above axis

$f(-x) = \frac{1}{2}(-x)^4 - \frac{9}{2}(-x)^2 = \frac{1}{2}x^4 - \frac{9}{2}x^2 = f(x)$

$\Rightarrow$ even, symmetric about y-axis

67. $f(x) = x(x-3)(x-2)(x+1) = x^4 - 4x^3 + x^2 + 6x$

x-int.	y-int.
$x(x-3)(x-2)(x+1) = 0$	$f(0) = 0$
$x = 0, x = 3, x = 2, x = -1$	$y = 0$
$(0, 0), (3, 0), (2, 0), (-1, 0)$	$(0, 0)$

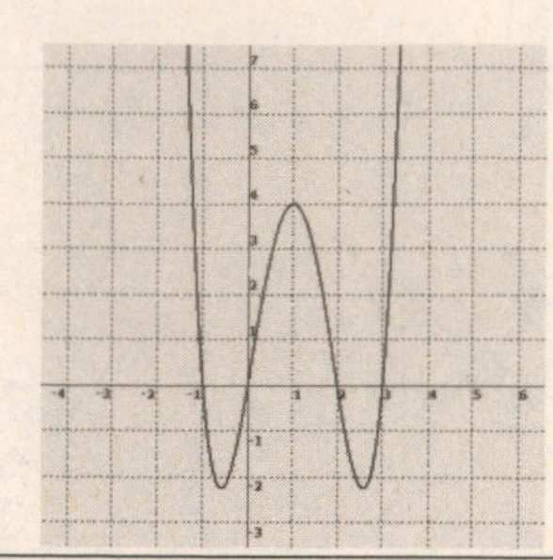

even deg, pos coef $\Rightarrow$ rises left, rises right

Sign of $f(x)= x(x-3)(x-2)(x+1)$	$+$	$-$	$+$	$-$	$+$
	$(-\infty, -1)$	$(-1, 0)$	$(0, 2)$	$(2, 3)$	$(3, \infty)$
	-1	0	2	3	
Test point	$f(-2) = 40$	$f\left(-\frac{1}{2}\right) = -\frac{35}{16}$	$f(1) = 4$	$f\left(\frac{5}{2}\right) = -\frac{35}{16}$	$f(4) = 40$
Graph of $f(x)$	above axis	below axis	above axis	below axis	above axis

$f(-x) = (-x)^4 - 4(-x)^3 + (-x)^2 + 6(-x) = x^4 + 4x^3 + x^2 - 6x$

$\Rightarrow$ neither even nor odd, no symmetry

69. $f(x) = x^5 - 4x^3$

x-int.	y-int.
$x^5 - 4x^3 = 0$	$f(0) = 0$
$x^3(x^2 - 4) = 0$	$y = 0$
$x^3(x+2)(x-2) = 0$	$(0, 0)$
$x = 0, x = -2, x = 2$	
$(0, 0), (-2, 0), (2, 0)$	

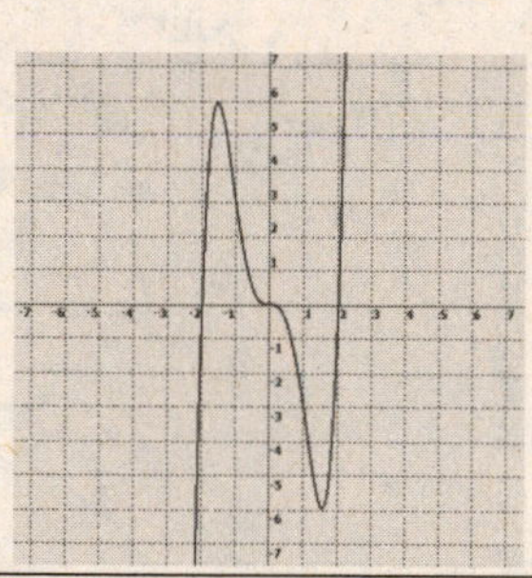

odd deg, pos coef $\Rightarrow$ falls left, rises right

Sign of $f(x)=x^5-4x^3$	$-$	$+$	$-$	$+$
	$(-\infty, -2)$	$(-2, 0)$	$(0, 2)$	$(2, \infty)$
	-2	0	2	
Test point	$f(-3) = -135$	$f(-1) = 3$	$f(1) = -3$	$f(3) = 135$
Graph of $f(x)$	below axis	above axis	below axis	above axis

$f(-x) = (-x)^5 - 4(-x)^3 = -x^5 + 4x^3 = -f(x)$

$\Rightarrow$ odd, symmetric about origin

71. $P(x) = 2x^2 + x - 3$
$P(-2) = 3$; $P(-1) = -2$
Thus, there is a zero between -2 and -1.

73. $P(x) = 3x^3 - 11x^2 - 14x$
$P(4) = -40$; $P(5) = 30$
Thus, there is a zero between 4 and 5.

EXERCISES 4.2

75. $P(x) = x^4 - 8x^2 + 15$
$P(1) = 8$; $P(2) = -1$
Thus, there is a zero between 1 and 2.

77. $P(x) = 30x^3 - 61x^2 - 39x + 10$
$P(2) = -72$; $P(3) = 154$
Thus, there is a zero between 2 and 3.

79. $P(x) = 30x^3 - 61x^2 - 39x + 10$; $P(0) = 10$; $P(1) = -60$; Thus, there is a zero between 0 and 1.

81. **a.** $V(x) = x(20 - 2x)(24 - 2x) = 4x^3 - 88x^2 + 480x$

b. $V(x) = 4x^3 - 88x^2 + 480x = 4x\left(x^2 - 22x + 120\right) = 4x(x - 10)(x - 12)$

x-int.	y-int.
$4x(x-10)(x-12)$	$f(0) = 0$
$x = 0, x = 10, x = 12$	$y = 0$
$(0, 0), (10, 0), (12, 0)$	$(0, 0)$

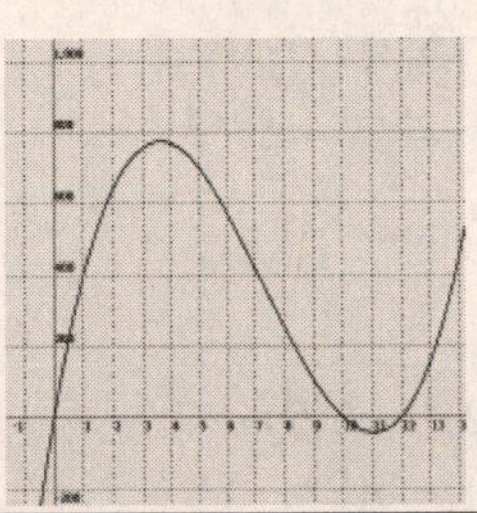

odd deg, pos coef $\Rightarrow$ falls left, rises right

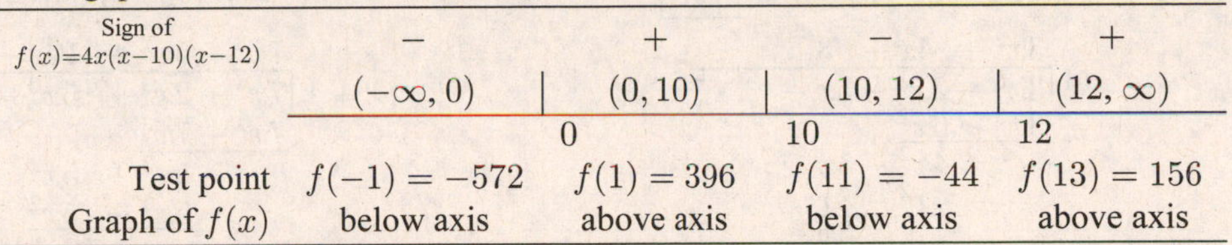

Sign of $f(x)=4x(x-10)(x-12)$	$-$	$+$	$-$	$+$
	$(-\infty, 0)$	$(0, 10)$	$(10, 12)$	$(12, \infty)$
	0	10	12	
Test point	$f(-1) = -572$	$f(1) = 396$	$f(11) = -44$	$f(13) = 156$
Graph of $f(x)$	below axis	above axis	below axis	above axis

c. $x \geq 0$, $20 - 2x \geq 0 \Rightarrow x \leq 10$, $24 - 2x \geq 0 \Rightarrow x \leq 12$; Domain $= [0, 10]$

d. $x \approx 3.6$ in, $V(x) \approx 774.2$ in^3

83. **a.** # Trees $= 270 + x$; Prod/tree $= 840 - 0.1x^2$
$P(x) = $ (# Trees)(Prod/tree) $= (270 + x)\left(840 - 0.1x^2\right) = -0.1x^3 - 27x^2 + 840x + 226{,}800$

b. $x \approx 14$ trees

85. **a.** $P(12) = 10(12)^3 - 100(12)^2 + 210(12) = \$5400$

b.
$$P(x) = 0$$
$$10x^3 - 100x^2 + 210x = 0$$
$$10x\left(x^2 - 10x + 21\right) = 0$$
$10x(x - 3)(x - 7) = 0$; $x = 0$ (not in domain); $x = 3$ (March); $x = 7$ (July)

87. Let $x =$ the height. Then the volume $= V(x) = lwh = 16(10)x$
$V(4) = 16(10)(4) = 640$ in^3; $V(8) = 16(10)(8) = 1280$ in^3
By the Intermediate Value Theorem, there is at least one value of x between 4 and 8 such that $V(x)$ is between $V(4) = 640$ and $V(8) = 1280$, so there is a value of x with $V(x) = 1000$ in^3.

89-99. Answers may vary.

101. b

103. d

105. True.

Exercises 4.3 (page 435)

1. whole **3.** any **5.** factor

7.
$$\begin{array}{r l}
 & 4x^2 + 2x + 1 + \frac{2}{x-1} \\
x-1 & \overline{)\,4x^3 - 2x^2 - x + 1} \\
 & \underline{4x^3 - 4x^2} \\
 & \quad 2x^2 - x \\
 & \quad \underline{2x^2 - 2x} \\
 & \qquad x + 1 \\
 & \qquad \underline{x - 1} \\
 & \qquad\quad 2
\end{array}$$

9.
$$\begin{array}{r l}
 & 2x^3 - 3x^2 + 8x - 1 + \frac{-3}{x+2} \\
x+2 & \overline{)\,2x^4 + x^3 + 2x^2 + 15x - 5} \\
 & \underline{2x^4 + 4x^3} \\
 & \; -3x^3 + 2x^2 \\
 & \; \underline{-3x^3 - 6x^2} \\
 & \qquad 8x^2 + 15x \\
 & \qquad \underline{8x^2 + 16x} \\
 & \qquad\quad -x - 5 \\
 & \qquad\quad \underline{-x - 2} \\
 & \qquad\qquad -3
\end{array}$$

11.
$$\begin{aligned}
P(2) &= 3(2)^3 - 2(2)^2 - 5(2) - 7 \\
&= 3(8) - 2(4) - 5(2) - 7 \\
&= 24 - 8 - 10 - 7 = \boxed{-1}
\end{aligned}$$

$$\begin{array}{r l}
 & 3x^2 + 4x + 3 \\
x-2 & \overline{)\,3x^3 - 2x^2 - 5x - 7} \\
 & \underline{3x^3 - 6x^2} \\
 & \quad 4x^2 - 5x \\
 & \quad \underline{4x^2 - 8x} \\
 & \qquad 3x - 7 \\
 & \qquad \underline{3x - 6} \\
 & \qquad\quad \boxed{-1}
\end{array}$$

13.
$$\begin{aligned}
P(-1) &= 7(-1)^4 + 2(-1)^3 + 5(-1)^2 - 1 \\
&= 7(1) + 2(-1) + 5(1) - 1 \\
&= 7 - 2 + 5 - 1 = \boxed{9}
\end{aligned}$$

$$\begin{array}{r l}
 & 7x^3 - 5x^2 + 10x - 10 \\
x+1 & \overline{)\,7x^4 + 2x^3 + 5x^2 + 0x - 1} \\
 & \underline{7x^4 + 7x^3} \\
 & \; -5x^3 + 5x^2 \\
 & \; \underline{-5x^3 - 5x^2} \\
 & \qquad 10x^2 + 0x \\
 & \qquad \underline{10x^2 + 10x} \\
 & \qquad\quad -10x - 1 \\
 & \qquad\quad \underline{-10x - 10} \\
 & \qquad\qquad \boxed{9}
\end{array}$$

15.
$$\begin{aligned}
P(1) &= 2(1)^5 + (1)^4 - (1)^3 - 2(1) + 3 \\
&= 2(1) + (1) - (1) - 2 + 3 \\
&= 2 + 1 - 1 - 2 + 3 = \boxed{3}
\end{aligned}$$

$$\begin{array}{r l}
 & 2x^4 + 3x^3 + 2x^2 + 2x \\
x-1 & \overline{)\,2x^5 + x^4 - x^3 + 0x^2 - 2x + 3} \\
 & \underline{2x^5 - 2x^4} \\
 & \quad 3x^4 - x^3 \\
 & \quad \underline{3x^4 - 3x^3} \\
 & \qquad 2x^3 + 0x^2 \\
 & \qquad \underline{2x^3 - 2x^2} \\
 & \qquad\quad 2x^2 - 2x \\
 & \qquad\quad \underline{2x^2 - 2x} \\
 & \qquad\qquad 0x + \boxed{3}
\end{array}$$

17. remainder $= P(-2) = 3(-2)^4 + 5(-2)^3 - 4(-2)^2 - 2(-2) + 1$
$= 3(16) + 5(-8) - 4(4) - 2(-2) + 1 = 48 - 40 - 16 + 4 + 1 = -3$

19. remainder $= P(2) = 3(2)^4 + 5(2)^3 - 4(2)^2 - 2(2) + 1$
$= 3(16) + 5(8) - 4(4) - 2(2) + 1 = 48 + 40 - 16 - 4 + 1 = 69$

21. remainder $= P(-3) = 3(-3)^4 + 5(-3)^3 - 4(-3)^2 - 2(-3) + 1$
$= 3(81) + 5(-27) - 4(9) - 2(-3) + 1 = 243 - 135 - 36 + 6 + 1 = 79$

EXERCISES 4.3

23. remainder $= P(4) = 3(4)^4 + 5(4)^3 - 4(4)^2 - 2(4) + 1$
$= 3(256) + 5(64) - 4(16) - 2(4) + 1 = 768 + 320 - 64 - 8 + 1 = 1017$

25. $(x-1)$ is a factor if $P(1) = 0$.
$P(1) = 1^7 - 1 = 0$: true

27. $(x-1)$ is a factor if $P(1) = 0$.
$P(1) = 3(1)^5 + 4(1)^2 - 7 = 0$: true

29. $(x+3)$ is a factor if $P(-3) = 0$. $P(-3) = 2(-3)^3 - 2(-3)^2 + 1 = -71$: false

31. $(x-1)$ is a factor if $P(1) = 0$. $P(1) = (1)^{1984} - (1)^{1776} + (1)^{1492} - (1)^{1066} = 0$: true

33.
$$\begin{array}{r|rrrr} 1 & 3 & -2 & -6 & -4 \\ & & 3 & 1 & -5 \\ \hline & 3 & 1 & -5 & -9 \end{array}$$
$(x-1)(3x^2 + x - 5) - 9$

35.
$$\begin{array}{r|rrrr} 3 & 3 & -2 & -6 & -4 \\ & & 9 & 21 & 45 \\ \hline & 3 & 7 & 15 & 41 \end{array}$$
$(x-3)(3x^2 + 7x + 15) + 41$

37.
$$\begin{array}{r|rrrr} -1 & 3 & -2 & -6 & -4 \\ & & -3 & 5 & 1 \\ \hline & 3 & -5 & -1 & -3 \end{array}$$
$(x+1)(3x^2 - 5x - 1) - 3$

39.
$$\begin{array}{r|rrrr} -3 & 3 & -2 & -6 & -4 \\ & & -9 & 33 & -81 \\ \hline & 3 & -11 & 27 & -85 \end{array}$$
$(x+3)(3x^2 - 11x + 27) - 85$

41.
$$\begin{array}{r|rrrr} 1 & 1 & 1 & 1 & -3 \\ & & 1 & 2 & 3 \\ \hline & 1 & 2 & 3 & 0 \end{array}$$
$x^2 + 2x + 3$

43.
$$\begin{array}{r|rrrr} -1 & 7 & -3 & -5 & 1 \\ & & -7 & 10 & -5 \\ \hline & 7 & -10 & 5 & -4 \end{array}$$
$7x^2 - 10x + 5 + \dfrac{-4}{x+1}$

45.
$$\begin{array}{r|rrrrr} 3 & 4 & -3 & 0 & -1 & 5 \\ & & 12 & 27 & 81 & 240 \\ \hline & 4 & 9 & 27 & 80 & 245 \end{array}$$
$4x^3 + 9x^2 + 27x + 80 + \dfrac{245}{x-3}$

47.
$$\begin{array}{r|rrrrrr} 4 & 3 & 0 & 0 & 0 & -768 & 0 \\ & & 12 & 48 & 192 & 768 & 0 \\ \hline & 3 & 12 & 48 & 192 & 0 & 0 \end{array}$$
$3x^4 + 12x^3 + 48x^2 + 192x$

49.
$$\begin{array}{r|rrrr} 2 & 5 & 2 & -1 & 1 \\ & & 10 & 24 & 46 \\ \hline & 5 & 12 & 23 & 47 \end{array} \quad P(2) = 47$$

51.
$$\begin{array}{r|rrrr} -5 & 5 & 2 & -1 & 1 \\ & & -25 & 115 & -570 \\ \hline & 5 & -23 & 114 & -569 \end{array}$$
$P(-5) = -569$

53.
$$\begin{array}{r|rrrr} i & 5 & 2 & -1 & 1 \\ & & 5i & -5+2i & -2-6i \\ \hline & 5 & 2+5i & -6+2i & -1-6i \end{array}$$
$P(i) = -1 - 6i$

55.
$$\begin{array}{r|rrrrr} \frac{1}{2} & 2 & 0 & -1 & 0 & 2 \\ & & 1 & \frac{1}{2} & -\frac{1}{4} & -\frac{1}{8} \\ \hline & 2 & 1 & -\frac{1}{2} & -\frac{1}{4} & \frac{15}{8} \end{array} \quad P\left(\tfrac{1}{2}\right) = \tfrac{15}{8}$$

EXERCISES 4.3

57. $$\begin{array}{r|rrrrr} i & 2 & 0 & -1 & 0 & 2 \\ & & 2i & -2 & -3i & 3 \\ \hline & 2 & 2i & -3 & -3i & 5 \end{array} \quad P(i) = 5$$

59. $$\begin{array}{r|rrrrr} 1 & 1 & -8 & 14 & 8 & -15 \\ & & 1 & -7 & 7 & 15 \\ \hline & 1 & -7 & 7 & 15 & 0 \end{array} \quad P(1) = 0$$

61. $$\begin{array}{r|rrrrr} -3 & 1 & -8 & 14 & 8 & -15 \\ & & -3 & 33 & -141 & 399 \\ \hline & 1 & -11 & 47 & -133 & 384 \end{array} \quad P(-3) = 384$$

63. $$\begin{array}{r|rrrrr} -i & 1 & -8 & 14 & 8 & -15 \\ & & -i & -1+8i & 8-13i & -13-16i \\ \hline & 1 & -8-i & 13+8i & 16-13i & -28-16i \end{array} \quad P(-i) = -28-16i$$

65. $$\begin{array}{r|rrrrrr} i & 1 & 0 & -1 & -8 & 0 & 8 \\ & & i & -1 & -2i & 2-8i & 8+2i \\ \hline & 1 & i & -2 & -8-2i & 2-8i & 16+2i \end{array} \quad P(i) = 16+2i$$

67. $$\begin{array}{r|rrrrrr} -2i & 1 & 0 & -1 & -8 & 0 & 8 \\ & & -2i & -4 & 10i & 20+16i & 32-40i \\ \hline & 1 & -2i & -5 & -8+10i & 20+16i & 40-40i \end{array} \quad P(-2i) = 40-40i$$

69. $x+2$ is a factor if $P(-2) = 0$.

$$\begin{array}{r|rrrr} -2 & 3 & -13 & -10 & 56 \\ & & -6 & 38 & -56 \\ \hline & 3 & -19 & -28 & 0 \end{array} \quad \text{yes}$$

71. $x-1$ is a factor if $P(1) = 0$.

$$\begin{array}{r|rrrrr} 1 & 1 & -3 & 4 & -2 & 4 \\ & & 1 & -2 & 2 & 0 \\ \hline & 1 & -2 & 2 & 0 & 4 \end{array} \quad \text{no}$$

73. $x+3$ is a factor if $P(-3) = 0$.

$$\begin{array}{r|rrrrrr} -3 & 3 & 0 & -22 & 15 & 3 & 9 \\ & & -9 & 27 & -15 & 0 & -9 \\ \hline & 3 & -9 & 5 & 0 & 3 & 0 \end{array} \quad \text{yes}$$

75. 4 is a zero if $x-4$ is a factor.

$$\begin{array}{r|rrrr} 4 & -3 & 13 & 10 & -56 \\ & & -12 & 4 & 56 \\ \hline & -3 & 1 & 14 & 0 \end{array} \quad \text{yes}$$

77. -2 is a zero if $x+2$ is a factor.

$$\begin{array}{r|rrrrr} -2 & 4 & 1 & 20 & 0 & -4 \\ & & -8 & 14 & -68 & 136 \\ \hline & 4 & -7 & 34 & -68 & 132 \end{array} \quad \text{no}$$

79. $\frac{1}{2}$ is a zero if $x - \frac{1}{2}$ is a factor.

$$\begin{array}{r|rrrrrr} \frac{1}{2} & 4 & -2 & 6 & 5 & -6 & 1 \\ & & 2 & 0 & 3 & 4 & -1 \\ \hline & 4 & 0 & 6 & 8 & -2 & 0 \end{array} \quad \text{yes}$$

EXERCISES 4.3

81. $x = -1$ is a solution, so $(x+1)$ is a factor. Use synthetic division to divide by $(x+1)$.

$$\begin{array}{r|rrrr} -1 & 1 & 3 & -13 & -15 \\ & & -1 & -2 & 15 \\ \hline & 1 & 2 & -15 & 0 \end{array}$$

$$x^3 + 3x^2 - 13x - 15 = 0$$
$$(x+1)(x^2 + 2x - 15) = 0$$
$$(x+1)(x+5)(x-3) = 0$$

Solution set: $\{-1, -5, 3\}$

83. $x = -\frac{1}{2}$ is a solution, so $\left(x + \frac{1}{2}\right)$ is a factor. Use synthetic division to divide by $\left(x + \frac{1}{2}\right)$.

$$\begin{array}{r|rrrr} -\frac{1}{2} & 2 & 1 & -18 & -9 \\ & & -1 & 0 & 9 \\ \hline & 2 & 0 & -18 & 0 \end{array}$$

$$2x^3 + x^2 - 18x - 9 = 0$$
$$\left(x + \tfrac{1}{2}\right)(2x^2 - 18) = 0$$
$$2\left(x + \tfrac{1}{2}\right)(x^2 - 9) = 0$$
$$2\left(x + \tfrac{1}{2}\right)(x+3)(x-3) = 0$$

Solution set: $\left\{-\frac{1}{2}, -3, 3\right\}$

85. $x = 2$ is a solution, so $(x-2)$ is a factor. Use synthetic division to divide by $(x-2)$.

$$\begin{array}{r|rrrr} 2 & 1 & -6 & 7 & 2 \\ & & 2 & -8 & -2 \\ \hline & 1 & -4 & -1 & 0 \end{array}$$

$$x^3 - 6x^2 + 7x + 2 = 0$$
$$(x-2)(x^2 - 4x - 1) = 0$$

Use the quadratic formula to finish.

Solution set: $\left\{2, 2+\sqrt{5}, 2-\sqrt{5}\right\}$

87. $x = -3$ is a solution, so $(x+3)$ is a factor. Use synthetic division to divide by $(x+3)$.

$$\begin{array}{r|rrrr} -3 & 1 & -3 & 1 & 57 \\ & & -3 & 18 & -57 \\ \hline & 1 & -6 & 19 & 0 \end{array}$$

$$x^3 - 3x^2 + x + 57 = 0$$
$$(x+3)(x^2 - 6x + 19) = 0$$

Use the quadratic formula to finish.

Solution set: $\left\{-3, 3 \pm \sqrt{10}\,i\right\}$

89. $x = 1$ is a solution, so $(x-1)$ is a factor. Use synthetic division to divide by $(x-1)$.

$$\begin{array}{r|rrrrr} 1 & 1 & -2 & -2 & 6 & -3 \\ & & 1 & -1 & -3 & 3 \\ \hline & 1 & -1 & -3 & 3 & 0 \end{array}$$

$$x^4 - 2x^3 - 2x^2 + 6x - 3 = 0$$
$$(x-1)(x^3 - x^2 - 3x + 3) = 0$$

Use the fact that $x = 1$ is a double root and divide the depressed polynomial by $(x-1)$:

$$\begin{array}{r|rrrr} 1 & 1 & -1 & -3 & 3 \\ & & 1 & 0 & -3 \\ \hline & 1 & 0 & -3 & 0 \end{array}$$

$$(x-1)(x^3 - x^2 - 3x + 3) = 0$$
$$(x-1)(x-1)(x^2 - 3) = 0$$

Solution set: $\left\{1, 1, \sqrt{3}, -\sqrt{3}\right\}$

91. $x = 2$ is a solution, so $(x-2)$ is a factor. Use synthetic division to divide by $(x-2)$.

$$\begin{array}{r|rrrrr} 2 & 1 & -5 & 7 & -5 & 6 \\ & & 2 & -6 & 2 & -6 \\ \hline & 1 & -3 & 1 & -3 & 0 \end{array}$$

$$x^4 - 5x^3 + 7x^2 - 5x + 6 = 0$$
$$(x-2)(x^3 - 3x^2 + x - 3) = 0$$

$x = 3$ is a root, so $(x-3)$ is a factor. Use synthetic division to divide by $(x-3)$.

$$\begin{array}{r|rrrr} 3 & 1 & -3 & 1 & -3 \\ & & 3 & 0 & 3 \\ \hline & 1 & 0 & 1 & 0 \end{array}$$

$$(x-2)(x^3 - 3x^2 + x - 3) = 0$$
$$(x-2)(x-3)(x^2 + 1) = 0$$

$x^2 + 1 = 0 \Rightarrow x^2 = -1 \Rightarrow x = \pm i$

Solution set: $\{2, 3, \pm i\}$

93. $(x-4)(x-5) = x^2 - 9x + 20$

95. $(x-1)(x-1)(x-1) = (x^2 - 2x + 1)(x-1) = x^3 - 3x^2 + 3x - 1$

97. $(x-2)(x-4)(x-5) = (x^2 - 6x + 8)(x-5) = x^3 - 11x^2 + 38x - 40$

99. $(x-1)(x+1)\left(x - \sqrt{2}\right)\left(x + \sqrt{2}\right) = (x^2-1)(x^2-2) = x^4 - 3x^2 + 2$

101. $\left(x - \sqrt{2}\right)(x-i)(x+i) = \left(x - \sqrt{2}\right)(x^2 - i^2) = \left(x - \sqrt{2}\right)(x^2+1) = x^3 - \sqrt{2}x^2 + x - \sqrt{2}$

103.
$$\begin{aligned}(x-0)[x-(1+i)][x-(1-i)] &= x\left[x^2 - (1-i)x - (1+i)x + (1+i)(1-i)\right] \\ &= x\left[x^2 - x + ix - x - ix + 1 - i^2\right] \\ &= x\left[x^2 - 2x + 2\right] = x^3 - 2x^2 + 2x\end{aligned}$$

105. Answers may vary.

107. Answers may vary.

109.
$$\begin{aligned} P(0) &= 0 \\ a_n(0)^n + a_{n-1}(0)^{n-1} + \cdots + a_1(0) + a_0 &= 0 \\ 0 + 0 + \cdots + 0 + a_0 &= 0 \\ a_0 &= 0 \end{aligned}$$

111. $P(2) = 0 \Rightarrow (x-2)$ is a factor. $P(-2) = 0 \Rightarrow (x+2)$ is a factor. The product of two factors will also be a factor, so $(x-2)(x+2) = x^2 - 4$ is a factor of the polynomial $P(x)$.

113. True.

115. $P(-1) = 0$. True.

117. False. $P(i) = 3$.

119. False. $\frac{246}{135}$ is a zero.

Exercises 4.4 (page 445)

1. zero

3. conjugate

5.
$$\begin{aligned} &(-x)^3 - (-x)^2 - 4 \\ &= -x^3 - x^2 - 4 \end{aligned}$$
0 variations

7. $7x^4 + 5x^3 - 2x + 1 \Rightarrow 7(-x)^4 + 5(-x)^3 - 2(-x) + 1$
$\Rightarrow 7x^4 - 5x^3 + 2x + 1 \Rightarrow$ 2 variations $\Rightarrow$ at most 2 negative roots

9. lower bound

11. $P(x) = x^{10} - 1 \Rightarrow$ 10 zeros

13. $P(x) = 3x^4 - 4x^2 - 2x + 7 \Rightarrow$ 4 zeros

15.
$$\begin{aligned} P(x) &= x\left(3x^4 - 2\right) - 12x \\ &= 3x^5 - 14x \end{aligned}$$
5 total zeros $\Rightarrow$ $\boxed{\text{4 other zeros}}$

EXERCISES 4.4

17. $P(x) = x^4 - 81$
4 linear factors, 4 zeros

19. $P(x) = 4x^5 + 8x^3$
5 linear factors, 5 zeros

21. If $2i$ is a zero, then $-2i$ is a zero also:
$P(x) = (x - 2i)(x + 2i) = x^2 - 4i^2 = x^2 + 4$

23. If $3 - i$ is a zero, then $3 + i$ is a zero also:
$$\begin{aligned} P(x) = [x - (3 - i)][x - (3 + i)] &= x^2 - (3 + i)x - (3 - i)x + (3 - i)(3 + i) \\ &= x^2 - 3x - ix - 3x + ix + 9 - i^2 \\ &= x^2 - 6x + 10 \end{aligned}$$

25. If $-i$ is a zero, then i is a zero also:
$$\begin{aligned} P(x) &= (x - 3)(x + i)(x - i) \\ &= (x - 3)\left(x^2 - i^2\right) \\ &= (x - 3)\left(x^2 + 1\right) \\ &= x^3 - 3x^2 + x - 3 \end{aligned}$$

27. If $2 + i$ is a zero, then $2 - i$ is a zero also:
$$\begin{aligned} P(x) &= (x - 2)[x - (2 + i)][x - (2 - i)] \\ &= (x - 2)\left[x^2 - (2 - i)x - (2 + i)x + (2 + i)(2 - i)\right] \\ &= (x - 2)\left[x^2 - 2x + ix - 2x - ix + 4 - i^2\right] \\ &= (x - 2)\left[x^2 - 4x + 5\right] = x^3 - 6x^2 + 13x - 10 \end{aligned}$$

29. If i is a zero, then $-i$ is a zero also:
$$\begin{aligned} P(x) &= (x - 3)(x - 2)(x - i)(x + i) \\ &= \left(x^2 - 5x + 6\right)\left(x^2 - i^2\right) \\ &= \left(x^2 - 5x + 6\right)\left(x^2 + 1\right) = x^4 - 5x^3 + 7x^2 - 5x + 6 \end{aligned}$$

31. If i and $1 - i$ are zeros, then $-i$ and $1 + i$ are zeros also:
$$\begin{aligned} P(x) &= (x - i)(x + i)[x - (1 - i)][x - (1 + i)] \\ &= \left(x^2 - i^2\right)\left[x^2 - (1 + i)x - (1 - i)x + (1 - i)(1 + i)\right] \\ &= \left(x^2 + 1\right)\left[x^2 - x - ix - x + ix + 1 - i^2\right] \\ &= \left(x^2 + 1\right)\left[x^2 - 2x + 2\right] = x^4 - 2x^3 + 3x^2 - 2x + 2 \end{aligned}$$

33. If $2i$ is a double zero, then there are two factors of $(x - 2i)$. [The problem does not specify real coefficients, so we do not include $-2i$ as a zero.]

$$P(x) = (x - 2i)(x - 2i)$$
$$= x^2 - 4ix + 4i^2 = x^2 - 4ix - 4$$

35. $P(x) = 3x^3 + 5x^2 - 4x + 3$

2 sign variations $\Rightarrow$ 2 or 0 positive zeros

$$P(-x) = 3(-x)^3 + 5(-x)^2 - 4(-x) + 3$$
$$= -3x^3 + 5x^2 + 4x + 3$$

1 sign variation $\Rightarrow$ 1 negative zero

# pos	# neg	# nonreal
2	1	0
0	1	2

37. $P(x) = 2x^3 + 7x^2 + 5x + 5$

0 sign variations $\Rightarrow$ 0 positive zeros

$$P(-x) = 2(-x)^3 + 7(-x)^2 + 5(-x) + 3$$
$$= -2x^3 + 7x^2 - 5x + 3$$

3 sign variations $\Rightarrow$ 3 or 1 negative zeros

# pos	# neg	# nonreal
0	3	0
0	1	2

39. $P(x) = 8x^4 + 5$

0 sign variations $\Rightarrow$ 0 positive zeros

$$P(-x) = 8(-x)^4 + 5$$
$$= 8x^4 + 5$$

0 sign variations $\Rightarrow$ 0 negative zeros

# pos	# neg	# nonreal
0	0	4

41. $P(x) = x^4 + 8x^2 - 5x - 10$: 1 sign variation $\Rightarrow$ 1 positive zero

$$P(-x) = (-x)^4 + 8(-x)^2 - 5(-x) - 10$$
$$= x^4 + 8x^2 + 5x - 10\text{: 1 sign variation} \Rightarrow \text{1 negative zero}$$

# pos	# neg	# nonreal
1	1	2

43. $P(x) = -x^{10} - x^8 - x^6 - x^4 - x^2 - 1$: 0 sign variations $\Rightarrow$ 0 positive zeros

$$P(-x) = -(-x)^{10} - (-x)^8 - (-x)^6 - (-x)^4 - (-x)^2 - 1$$
$$= -x^{10} - x^8 - x^6 - x^4 - x^2 - 1\text{: 0 sign variations} \Rightarrow \text{0 negative zeros}$$

# pos	# neg	# nonreal
0	0	10

45. $P(x) = x^9 + x^7 + x^5 + x^3 + x = x\left(x^8 + x^6 + x^4 + x^2 + 1\right)$: 0 sign variations $\Rightarrow$ 0 positive zeros

$$P(-x) = (-x)\left[(-x)^8 + (-x)^6 + (-x)^4 + (-x)^2 + 1\right]$$
$$= -x\left[x^8 + x^6 + x^4 + x^2 + 1\right]\text{: 0 sign variations} \Rightarrow \text{0 negative zeros}$$

# pos	# neg	# zero	# nonreal
0	0	1	8

EXERCISES 4.4

47. $P(x) = -2x^4 - 3x^2 + 2x + 3$: 1 sign variation $\Rightarrow$ 1 positive zero

$P(-x) = -2(-x)^4 - 3(-x)^2 + 2(-x) + 3$

$= -2x^4 - 3x^2 - 2x + 3$: 1 sign variation $\Rightarrow$ 1 negative zero

# pos	# neg	# nonreal
1	1	2

49. $P(x) = x^2 - 2x - 4$

$$\begin{array}{r|rrr} 4 & 1 & -2 & -4 \\ & & 4 & 8 \\ \hline & 1 & 2 & 4 \end{array} \qquad \begin{array}{r|rrr} -2 & 1 & -2 & -4 \\ & & -2 & 8 \\ \hline & 1 & -4 & 4 \end{array}$$

Upper bound: 4 Lower bound: -2

51. $P(x) = 18x^2 - 6x - 1$

$$\begin{array}{r|rrr} 1 & 18 & -6 & -1 \\ & & 18 & 12 \\ \hline & 18 & 12 & 11 \end{array} \qquad \begin{array}{r|rrr} -1 & 18 & -6 & -1 \\ & & -18 & 24 \\ \hline & 18 & -24 & 23 \end{array}$$

Upper bound: 1 Lower bound: -1

53. $P(x) = 6x^3 - 13x^2 - 110x$

$$\begin{array}{r|rrrr} 6 & 6 & -13 & -110 & 0 \\ & & 36 & 138 & 168 \\ \hline & 6 & 23 & 28 & 168 \end{array}$$

Upper bound: 6

$$\begin{array}{r|rrrr} -4 & 6 & -13 & -110 & 0 \\ & & -24 & 148 & -152 \\ \hline & 6 & -37 & 38 & -152 \end{array}$$

Lower bound: -4

55. $P(x) = x^5 + x^4 - 8x^3 - 8x^2 + 15x + 15$

$$\begin{array}{r|rrrrrr} 3 & 1 & 1 & -8 & -8 & 15 & 15 \\ & & 3 & 12 & 12 & 12 & 81 \\ \hline & 1 & 4 & 4 & 4 & 27 & 96 \end{array}$$

Upper bound: 3

$$\begin{array}{r|rrrrrr} -4 & 1 & 1 & -8 & -8 & 15 & 15 \\ & & -4 & 12 & -16 & 96 & -444 \\ \hline & 1 & -3 & 4 & -24 & 111 & -429 \end{array}$$

Lower bound: -4

57. $P(x) = 3x^5 - 11x^4 - 2x^3 + 38x^2 - 21x - 15$

$$\begin{array}{r|rrrrrr} 4 & 3 & -11 & -2 & 38 & -21 & -15 \\ & & 12 & 4 & 8 & 184 & 652 \\ \hline & 3 & 1 & 2 & 46 & 163 & 637 \end{array} \qquad \begin{array}{r|rrrrrr} -2 & 3 & -11 & -2 & 38 & -21 & -15 \\ & & -6 & 34 & -64 & 52 & -62 \\ \hline & 3 & -17 & 32 & -26 & 31 & -77 \end{array}$$

Upper bound: 4 Lower bound: -2

59. **Answers may vary.**

61. **Answers may vary.**

63. The number of nonreal zeros must occur in conjugate pairs, so the number of nonreal zeros will always be even. Since a polynomial of odd degree has an odd number of zeros, at least one zero must not be nonreal. Thus, at least one zero of such a polynomial will be real.

65. False. The theorem states that every polynomial function has at least one complex zero.

67. False. You need to know that the coefficients are real numbers to use the Conjugate Pairs Theorem.

69. False. It will have 789 zeros.

Exercises 4.5 (page 456)

1. -7

3. zero

5. num: $\pm 1, \pm 2, \pm 3, \pm 4, \pm 6, \pm 12$; den: ± 1
possible zeros: $\pm 1, \pm 2, \pm 3, \pm 4, \pm 6, \pm 12$

7. num: $\pm 1, \pm 2, \pm 3, \pm 6$; den: $\pm 1, \pm 2$
possible zeros: $\pm 1, \pm 2, \pm 3, \pm 6, \pm \frac{1}{2}, \pm \frac{3}{2}$

9. num: $\pm 1, \pm 2, \pm 5, \pm 10$; den: $\pm 1, \pm 2, \pm 4$
possible zeros: $\pm 1, \pm 2, \pm 5, \pm 10, \pm \frac{1}{2}, \pm \frac{5}{2}, \pm \frac{1}{4}, \pm \frac{5}{4}$

11. $P(x) = x^3 - 5x^2 - x + 5$

Possible rational zeros: $\pm 1, \pm 5$

Descartes' Rule of Signs:

# pos	# neg	# nonreal
2	1	0
0	1	2

Test $x = -1$:

$$\begin{array}{r|rrrr} -1 & 1 & -5 & -1 & 5 \\ & & -1 & 6 & -5 \\ \hline & 1 & -6 & 5 & 0 \end{array}$$

$$\begin{aligned} P(x) &= x^3 - 5x^2 - x + 5 \\ &= (x+1)(x^2 - 6x + 5) \\ &= (x+1)(x-5)(x-1) \end{aligned}$$

Zeros: $\{-1, 5, 1\}$

13. $P(x) = x^3 - 2x^2 - x + 2$

Possible rational zeros

$\pm 1, \pm 2$

Descartes' Rule of Signs

# pos	# neg	# nonreal
2	1	0
0	1	2

Test $x = -1$:

$$\begin{array}{r|rrrr} -1 & 1 & -2 & -1 & 2 \\ & & -1 & 3 & -2 \\ \hline & 1 & -3 & 2 & 0 \end{array}$$

$$\begin{aligned} P(x) &= x^3 - 2x^2 - x + 2 \\ &= (x+1)(x^2 - 3x + 2) \\ &= (x+1)(x-1)(x-2) \end{aligned}$$

Zeros: $\{-1, 1, 2\}$

15. $P(x) = x^3 - x^2 - 4x + 4$

Possible rational zeros

$\pm 1, \pm 2, \pm 4$

Descartes' Rule of Signs

# pos	# neg	# nonreal
2	1	0
0	1	2

Test $x = 1$:

$$\begin{array}{r|rrrr} 1 & 1 & -1 & -4 & 4 \\ & & 1 & 0 & -4 \\ \hline & 1 & 0 & -4 & 0 \end{array}$$

$$\begin{aligned} P(x) &= x^3 - x^2 - 4x + 4 \\ &= (x-1)(x^2 - 4) \\ &= (x-1)(x+2)(x-2) \end{aligned}$$

Zeros: $\{1, -2, 2\}$

EXERCISES 4.5

17. $P(x)=x^3-2x^2-9x+18$

Possible rational zeros:

$\pm 1, \pm 2, \pm 3, \pm 6,$
$\pm 9, \pm 18$

Descartes' Rule of Signs:

# pos	# neg	# nonreal
2	1	0
0	1	2

Test $x=2$:

$$\begin{array}{r|rrrr} 2 & 1 & -2 & -9 & 18 \\ & & 2 & 0 & -18 \\ \hline & 1 & 0 & -9 & 0 \end{array}$$

$$\begin{aligned} P(x) &= x^3-2x^2-9x+18 \\ &= (x-2)\left(x^2-9\right) \\ &= (x-2)(x+3)(x-3) \end{aligned}$$

Zeros: $\{2, -3, 3\}$

19. $P(x)=2x^3-x^2-2x+1$

Possible rational zeros:

$\pm 1, \pm \frac{1}{2}$

Descartes' Rule of Signs

# pos	# neg	# nonreal
2	1	0
0	1	2

Test $x=1$:

$$\begin{array}{r|rrrr} 1 & 2 & -1 & -2 & 1 \\ & & 2 & 1 & -1 \\ \hline & 2 & 1 & -1 & 0 \end{array}$$

$$\begin{aligned} P(x) &= 2x^3-x^2-2x+1 \\ &= (x-1)\left(2x^2+x-1\right) \\ &= (x-1)(2x-1)(x+1) \end{aligned}$$

Zeros: $\left\{1, \frac{1}{2}, -1\right\}$

21. $3x^3+5x^2+x-1=0$

Possible rational zeros:

$\pm 1, \pm \frac{1}{3}$

Descartes' Rule of Signs:

# pos	# neg	# nonreal
1	2	0
1	0	2

Test $x=-1$:

$$\begin{array}{r|rrrr} -1 & 3 & 5 & 1 & -1 \\ & & -3 & -2 & 1 \\ \hline & 3 & 2 & -1 & 0 \end{array}$$

$$\begin{aligned} P(x) &= 3x^3+5x^2+x-1 \\ &= (x+1)\left(3x^2+2x-1\right) \\ &= (x+1)(3x-1)(x+1) \end{aligned}$$

Zeros: $\left\{-1, \frac{1}{3}, -1\right\}$

23. Possible rational zeros

$\pm 1, \pm 2, \pm 3, \pm 6,$
$\pm 9, \pm 18, \pm \frac{1}{2}, \pm \frac{3}{2},$
$\pm \frac{9}{2}, \pm \frac{1}{3}, \pm \frac{2}{3}, \pm \frac{1}{5},$
$\pm \frac{2}{5}, \pm \frac{3}{5}, \pm \frac{6}{5}, \pm \frac{9}{5},$
$\pm \frac{18}{5}, \pm \frac{1}{6}, \pm \frac{1}{10}, \pm \frac{3}{10},$
$\pm \frac{9}{10}, \pm \frac{1}{15}, \pm \frac{2}{15}, \pm \frac{1}{30}$

Descartes' Rule of Signs

# pos	# neg	# nonreal
2	1	0
0	1	2

Test $x=\frac{2}{3}$:

$$\begin{array}{r|rrrr} \frac{2}{3} & 30 & -47 & -9 & 18 \\ & & 20 & -18 & -18 \\ \hline & 30 & -27 & -27 & 0 \end{array}$$

$$\begin{aligned} P(x) &= 30x^3-47x^2-9x+18 \\ &= \left(x-\tfrac{2}{3}\right)\left(30x^2-27x-27\right) \\ &= 3\left(x-\tfrac{2}{3}\right)\left(10x^2-9x-9\right) \\ &= (3x-2)(2x-3)(5x+3) \end{aligned}$$

Zeros: $\left\{\frac{2}{3}, \frac{3}{2}, -\frac{3}{5}\right\}$

25. Possible rational zeros

$\pm 1, \pm 2, \pm 3, \pm 4, \pm 6, \pm 8, \pm 12, \pm 24, \pm \frac{1}{3}, \pm \frac{2}{3}, \pm \frac{4}{3}, \pm \frac{8}{3}, \pm \frac{1}{5}, \pm \frac{2}{5}, \pm \frac{3}{5}, \pm \frac{4}{5}, \pm \frac{6}{5}, \pm \frac{8}{5}, \pm \frac{12}{5}, \pm \frac{24}{5}, \pm \frac{1}{15}, \pm \frac{2}{15}, \pm \frac{4}{15}, \pm \frac{8}{15}$

Descartes' Rule of Signs

# pos	# neg	# nonreal
2	1	0
0	1	2

Test $x = 4$:

$$\begin{array}{r|rrrr} 4 & 15 & -61 & -2 & 24 \\ & & 60 & -4 & -24 \\ \hline & 15 & -1 & -6 & 0 \end{array}$$

$P(x) = 15x^3 - 61x^2 - 2x + 24$
$= (x-4)(15x^2 - x - 6)$
$= (x-4)(3x-2)(5x+3)$

Zeros: $\{4, \frac{2}{3}, -\frac{3}{5}\}$

27. Possible rational zeros

$\pm 1, \pm 2, \pm 3, \pm 5, \pm 6, \pm 10, \pm 15, \pm 30, \pm \frac{1}{2}, \pm \frac{3}{2}, \pm \frac{5}{2}, \pm \frac{15}{2}, \pm \frac{1}{3}, \pm \frac{2}{3}, \pm \frac{5}{3}, \pm \frac{10}{3}, \pm \frac{1}{4}, \pm \frac{3}{4}, \pm \frac{5}{4}, \pm \frac{15}{4}, \pm \frac{1}{6}, \pm \frac{5}{6}, \pm \frac{1}{8}, \pm \frac{3}{8}, \pm \frac{5}{8}, \pm \frac{15}{8}, \pm \frac{1}{12}, \pm \frac{5}{12}, \pm \frac{1}{24}, \pm \frac{5}{24}$

Descartes' Rule of Signs

# pos	# neg	# nonreal
3	0	0
1	0	2

Test $x = \frac{3}{2}$:

$$\begin{array}{r|rrrr} \frac{3}{2} & 24 & -82 & 89 & -30 \\ & & 36 & -69 & 30 \\ \hline & 24 & -46 & 20 & 0 \end{array}$$

$P(x) = 24x^3 - 82x^2 + 89x - 30$
$= (x - \frac{3}{2})(24x^2 - 46x + 20)$
$= 2(x - \frac{3}{2})(12x^2 - 23x + 10)$
$= (2x-3)(4x-5)(3x-2)$

Zeros: $\{\frac{3}{2}, \frac{5}{4}, \frac{2}{3}\}$

29. Possible rational zeros

$\pm 1, \pm 2, \pm 3, \pm 4, \pm 6, \pm 8, \pm 12, \pm 24$

Descartes' Rule of Signs

# pos	# neg	# nonreal
4	0	0
2	0	2
0	0	4

$P(x) = x^4 - 10x^3 + 35x^2 - 50x + 24$
$= (x-1)(x^3 - 9x^2 + 26x - 24)$
$= (x-1)(x-2)(x^2 - 7x + 12)$
$= (x-1)(x-2)(x-3)(x-4)$

Zeros: $\{1, 2, 3, 4\}$

Test $x = 1$:

$$\begin{array}{r|rrrrr} 1 & 1 & -10 & 35 & -50 & 24 \\ & & 1 & -9 & 26 & -24 \\ \hline & 1 & -9 & 26 & -24 & 0 \end{array}$$

Test $x = 2$:

$$\begin{array}{r|rrrr} 2 & 1 & -9 & 26 & -24 \\ & & 2 & -14 & 24 \\ \hline & 1 & -7 & 12 & 0 \end{array}$$

31. Possible rational zeros

$\pm 1, \pm 2, \pm 3, \pm 5, \pm 6, \pm 10, \pm 15, \pm 30$

Descartes' Rule of Signs

# pos	# neg	# nonreal
2	2	0
2	0	2
0	2	2
0	0	4

$P(x) = x^4 + 3x^3 - 13x^2 - 9x + 30$
$= (x-2)(x^3 + 5x^2 - 3x - 15)$
$= (x-2)(x+5)(x^2 - 3)$

$x^2 - 3$ does not factor rationally.

Rational zeros: $\{2, -5\}$

Test $x = 2$:

$$\begin{array}{r|rrrrr} 2 & 1 & 3 & -13 & -9 & 30 \\ & & 2 & 10 & -6 & -30 \\ \hline & 1 & 5 & -3 & -15 & 0 \end{array}$$

Test $x = -5$:

$$\begin{array}{r|rrrr} -5 & 1 & 5 & -3 & -15 \\ & & -5 & 0 & 15 \\ \hline & 1 & 0 & -3 & 0 \end{array}$$

EXERCISES 4.5

33. Possible rational zeros

$\pm 1, \pm 3, \pm \frac{1}{2}, \pm \frac{3}{2}, \pm \frac{1}{4}, \pm \frac{3}{4}$

Descartes' Rule of Signs

# pos	# neg	# nonreal
3	1	0
1	1	2

$$\begin{aligned} P(x) &= 4x^4 - 8x^3 - x^2 + 8x - 3 \\ &= (x+1)\left(4x^3 - 12x^2 + 11x + 3\right) \\ &= (x+1)(x-1)\left(4x^2 - 8x + 3\right) \\ &= (x+1)(x-1)(2x-3)(2x-1) \end{aligned}$$

Zeros: $\left\{-1, 1, \frac{3}{2}, \frac{1}{2}\right\}$

Test $x = -1$:

$$\begin{array}{r|rrrrr} -1 & 4 & -8 & -1 & 8 & -3 \\ & & -4 & 12 & -11 & 3 \\ \hline & 4 & -12 & 11 & -3 & 0 \end{array}$$

Test $x = 1$:

$$\begin{array}{r|rrrr} 1 & 4 & -12 & 11 & -3 \\ & & 4 & -8 & 3 \\ \hline & 4 & -8 & 3 & 0 \end{array}$$

35. Possible rational zeros

$\pm 1, \pm 2, \pm 4, \pm 5, \pm 8, \pm 10, \pm 20, \pm 40, \pm \frac{1}{2}, \pm \frac{5}{2}$

Descartes' Rule of Signs

# pos	# neg	# nonreal
1	3	0
1	1	2

$$\begin{aligned} P(x) &= 2x^4 - x^3 - 2x^2 - 4x - 40 \\ &= (x+2)\left(2x^3 - 5x^2 + 8x - 20\right) \\ &= (x+2)\left(x - \tfrac{5}{2}\right)\left(2x^2 + 8\right) \end{aligned}$$

$2x^2 + 8$ does not factor rationally.

Rational zeros: $\left\{-2, \frac{5}{2}\right\}$

Test $x = -2$:

$$\begin{array}{r|rrrrr} -2 & 2 & -1 & -2 & -4 & -40 \\ & & -4 & 10 & -16 & 40 \\ \hline & 2 & -5 & 8 & -20 & 0 \end{array}$$

Test $x = \frac{5}{2}$:

$$\begin{array}{r|rrrr} \frac{5}{2} & 2 & -5 & 8 & -20 \\ & & 5 & 0 & 20 \\ \hline & 2 & 0 & 8 & 0 \end{array}$$

37. Possible rational zeros

$\pm 1, \pm \frac{1}{2}, \pm \frac{1}{3}, \pm \frac{1}{4}, \pm \frac{1}{6}, \pm \frac{1}{9}, \pm \frac{1}{12}, \pm \frac{1}{18}, \pm \frac{1}{36}$

Descartes' Rule of Signs

# pos	# neg	# nonreal
3	1	0
1	1	2

$$\begin{aligned} P(x) &= 36x^4 - x^2 + 2x - 1 \\ &= \left(x + \tfrac{1}{2}\right)\left(36x^3 - 18x^2 + 8x - 2\right) \\ &= \left(x + \tfrac{1}{2}\right)\left(x - \tfrac{1}{3}\right)\left(36x^2 - 6x + 6\right) \end{aligned}$$

$36x^2 - 6x + 6$ does not factor rationally. Rational zeros: $\left\{-\frac{1}{2}, \frac{1}{3}\right\}$

Test $x = -\frac{1}{2}$:

$$\begin{array}{r|rrrrr} -\frac{1}{2} & 36 & 0 & -1 & 2 & -1 \\ & & -18 & 9 & -4 & 1 \\ \hline & 36 & -18 & 8 & -2 & 0 \end{array}$$

Test $x = \frac{1}{3}$:

$$\begin{array}{r|rrrr} \frac{1}{3} & 36 & -18 & 8 & -2 \\ & & 12 & -2 & 2 \\ \hline & 36 & -6 & 6 & 0 \end{array}$$

EXERCISES 4.5

39. Possible rational zeros: $\pm 1,\ \pm 2,\ \pm 3,\ \pm 4,\ \pm 6,\ \pm 12$

Descartes' Rule of Signs

# pos	# neg	# nonreal
2	3	0
2	1	2
0	3	2
0	1	4

$$\begin{aligned} P(x) &= x^5+3x^4-5x^3-15x^2+4x+12 \\ &= (x+1)\left(x^4+2x^3-7x^2-8x+12\right) \\ &= (x+1)(x-1)\left(x^3+3x^2-4x-12\right) \\ &= (x+1)(x-1)(x-2)\left(x^2+5x+6\right) \\ &= (x+1)(x-1)(x-2)(x+2)(x+3) \end{aligned}$$

Zeros: $\{-1, 1, 2, -2, -3\}$

Test $x=-1$:

$$\begin{array}{r|rrrrrr} -1 & 1 & 3 & -5 & -15 & 4 & 12 \\ & & -1 & -2 & 7 & 8 & -12 \\ \hline & 1 & 2 & -7 & -8 & 12 & 0 \end{array}$$

Test $x=1$:

$$\begin{array}{r|rrrrr} 1 & 1 & 2 & -7 & -8 & 12 \\ & & 1 & 3 & -4 & -12 \\ \hline & 1 & 3 & -4 & -12 & 0 \end{array}$$

Test $x=2$:

$$\begin{array}{r|rrrr} 2 & 1 & 3 & -4 & -12 \\ & & 2 & 10 & 12 \\ \hline & 1 & 5 & 6 & 0 \end{array}$$

41. Possible rational zeros: $\pm 1,\ \pm 2,\ \pm 3,\ \pm 4,\ \pm 6,\ \pm 12,\ \pm \frac{1}{2},\ \pm \frac{3}{2},\ \pm \frac{1}{4},\ \pm \frac{3}{4}$

Descartes' Rule of Signs

# pos	# neg	# nonreal
4	1	0
2	1	2
0	1	4

Test $x=3$:

$$\begin{array}{r|rrrrrr} 3 & 4 & -12 & 15 & -45 & -4 & 12 \\ & & 12 & 0 & 45 & 0 & -12 \\ \hline & 4 & 0 & 15 & 0 & -4 & 0 \end{array}$$

Test $x=-\frac{1}{2}$:

$$\begin{array}{r|rrrrr} -\frac{1}{2} & 4 & 0 & 15 & 0 & -4 \\ & & -2 & 1 & -8 & 4 \\ \hline & 4 & -2 & 16 & -8 & 0 \end{array}$$

Test $x=\frac{1}{2}$:

$$\begin{array}{r|rrrr} \frac{1}{2} & 4 & -2 & 16 & -8 \\ & & 2 & 0 & 8 \\ \hline & 4 & 0 & 16 & 0 \end{array}$$

$$\begin{aligned} P(x) &= 4x^5-12x^4+15x^3-45x^2-4x+12 \\ &= (x-3)\left(4x^4+15x^2-4\right) \\ &= (x-3)\left(x+\tfrac{1}{2}\right)\left(4x^3-2x^2+16x-8\right) \\ &= (x-3)\left(x+\tfrac{1}{2}\right)\left(x-\tfrac{1}{2}\right)\left(4x^2+16\right) \end{aligned}$$

$4x^2+16$ does not factor rationally. Rational zeros: $\left\{3, -\frac{1}{2}, \frac{1}{2}\right\}$

43. First, factor out the common factor of x: $x^7-12x^5+48x^3-64x = x\left(x^6-12x^4+48x^2-64\right)$

Possible rational zeros: $\pm 1,\ \pm 2,\ \pm 4,\ \pm 8,\ \pm 16,\ \pm 32,\ \pm 64$

Descartes' Rule of Signs

# pos	# neg	# zero	# nonreal
3	3	1	0
3	1	1	2
1	3	1	2
1	1	1	4

Test $x=2$:

$$\begin{array}{r|rrrrrrr} 2 & 1 & 0 & -12 & 0 & 48 & 0 & -64 \\ & & 2 & 4 & -16 & -32 & 32 & 64 \\ \hline & 1 & 2 & -8 & -16 & 16 & 32 & 0 \end{array}$$

Test $x=2$:

$$\begin{array}{r|rrrrrr} 2 & 1 & 2 & -8 & -16 & 16 & 32 \\ & & 2 & 8 & 0 & -32 & -32 \\ \hline & 1 & 4 & 0 & -16 & -16 & 0 \end{array}$$

Test $x=2$:

$$\begin{array}{r|rrrrr} 2 & 1 & 4 & 0 & -16 & -16 \\ & & 2 & 12 & 24 & 16 \\ \hline & 1 & 6 & 12 & 8 & 0 \end{array}$$

Test $x=-2$:

$$\begin{array}{r|rrrr} -2 & 1 & 6 & 12 & 8 \\ & & -2 & -8 & -8 \\ \hline & 1 & 4 & 4 & 0 \end{array}$$

continued on next page...

43. continued

$$\begin{aligned}P(x) &= x^7+7x^6+21x^5+35x^4+35x^3+21x^2+7x+1\\&=(x+1)(x^6+6x^5+15x^4+20x^3+15x^2+6x+1)\\&=(x+1)(x+1)(x^5+5x^4+10x^3+10x^2+5x+1)\\&=(x+1)(x+1)(x+1)(x^4+4x^3+6x^2+4x+1)\\&=(x+1)(x+1)(x+1)(x+1)(x^3+3x^2+3x+1)\\&=(x+1)(x+1)(x+1)(x+1)(x+1)(x^2+2x+1)\\&=(x+1)(x+1)(x+1)(x+1)(x+1)(x+1)(x+1)\end{aligned}$$

Solution set $= \{-1,-1,-1,-1,-1,-1,-1\}$

45. Possible rational zeros

$\pm 1,\ \pm 2,\ \pm 3,\ \pm 6$

Test $x=3$:

$$\begin{array}{r|rrrr} 3 & 1 & -3 & -2 & 6 \\ & & 3 & 0 & -6 \\ \hline & 1 & 0 & -2 & 0 \end{array}$$

Descartes' Rule of Signs

# pos	# neg	# nonreal
2	1	0
0	1	2

$$\begin{aligned}P(x)&=x^3-3x^2-2x+6\\&=(x-3)(x^2-2)\end{aligned}$$

$x-3=0$ **or** $x^2-2=0$

$x=3 \qquad x=\pm\sqrt{2}$

Zeros: $\left\{3,-\sqrt{2},\sqrt{2}\right\}$

47. Possible rational zeros

$\pm 1,\ \pm \frac{1}{2}$

Test $x=\frac{1}{2}$:

$$\begin{array}{r|rrrr} \frac{1}{2} & 2 & -1 & 2 & -1 \\ & & 1 & 0 & 1 \\ \hline & 2 & 0 & 2 & 0 \end{array}$$

Descartes' Rule of Signs

# pos	# neg	# nonreal
3	0	0
1	0	2

$$\begin{aligned}P(x)&=2x^3-x^2+2x-1\\&=\left(x-\tfrac{1}{2}\right)(2x^2+2)\end{aligned}$$

$x-\frac{1}{2}=0$ **or** $2x^2+2=0$

$x=\frac{1}{2} \qquad x=\pm\sqrt{-1}$

$x=\pm i$

Zeros: $\left\{\frac{1}{2},-i,i\right\}$

49. Possible rational zeros

$\pm 1,\ \pm 2,\ \pm 4,\ \pm 8$
± 16

Descartes' Rule of Signs

# pos	# neg	# nonreal
2	2	0
2	0	2
0	2	2
0	0	4

$$\begin{aligned}P(x)&=x^4-2x^3-8x^2+8x+16\\&=(x-2)(x^3-8x-8)\\&=(x-2)(x+2)(x^2-2x-4)\end{aligned}$$

Use the quadratic formula.

Zeros: $\left\{2,-2,1\pm\sqrt{5}\right\}$

Test $x=2$:

$$\begin{array}{r|rrrrr} 2 & 1 & -2 & -8 & 8 & 16 \\ & & 2 & 0 & -16 & -16 \\ \hline & 1 & 0 & -8 & -8 & 0 \end{array}$$

Test $x=-2$:

$$\begin{array}{r|rrrr} -2 & 1 & 0 & -8 & -8 \\ & & -2 & 4 & 8 \\ \hline & 1 & -2 & -4 & 0 \end{array}$$

EXERCISES 4.5

51. Possible rational zeros

$\pm 1, \pm 3, \pm 9, \pm \frac{1}{2}$
$\pm \frac{3}{2}, \pm \frac{9}{2}$

Descartes' Rule of Signs

# pos	# neg	# nonreal
1	3	0
1	1	2

$$\begin{aligned} P(x) &= 2x^4 + x^3 + 17x^2 + 9x - 9 \\ &= (x+1)(2x^3 - x^2 + 18x - 9) \\ &= (x+1)\left(x - \tfrac{1}{2}\right)(2x^2 + 18) \end{aligned}$$

$2x^2 + 18 = 0 \Rightarrow x = \pm\sqrt{-9}$

Zeros: $\left\{-1, \frac{1}{2}, \pm 3i\right\}$

Test $x = -1$:

$$\begin{array}{r|rrrrr} -1 & 2 & 1 & 17 & 9 & -9 \\ & & -2 & 1 & -18 & 9 \\ \hline & 2 & -1 & 18 & -9 & 0 \end{array}$$

Test $x = \frac{1}{2}$:

$$\begin{array}{r|rrrr} \frac{1}{2} & 2 & -1 & 18 & -9 \\ & & 1 & 0 & 9 \\ \hline & 2 & 0 & 18 & 0 \end{array}$$

53. Possible rational zeros

$\pm 1, \pm 5,$
± 25

Descartes' Rule of Signs

# pos	# neg	# nonreal
5	0	0
3	0	2
1	0	4

$$\begin{aligned} P(x) &= x^5 - 3x^4 + 28x^3 - 76x^2 + 75x - 25 \\ &= (x-1)(x^4 - 2x^3 + 26x^2 - 50x + 25) \\ &= (x-1)(x-1)(x^3 - x^2 + 25x - 25) \\ &= (x-1)(x-1)(x-1)(x^2 + 25) \end{aligned}$$

$x^2 + 25 = 0 \Rightarrow x = \pm\sqrt{-25}$

Zeros: $\{1, 1, 1, \pm 5i\}$

Test $x = 1$:

$$\begin{array}{r|rrrrrr} 1 & 1 & -3 & 28 & -76 & 75 & -25 \\ & & 1 & -2 & 26 & -50 & 25 \\ \hline & 1 & -2 & 26 & -50 & 25 & 0 \end{array}$$

Test $x = 1$:

$$\begin{array}{r|rrrrr} 1 & 1 & -2 & 26 & -50 & 25 \\ & & 1 & -1 & 25 & -25 \\ \hline & 1 & -1 & 25 & -25 & 0 \end{array}$$

Test $x = 1$:

$$\begin{array}{r|rrrr} 1 & 1 & -1 & 25 & -25 \\ & & 1 & 0 & 25 \\ \hline & 1 & 0 & 25 & 0 \end{array}$$

55. Possible rational zeros

$\pm 1, \pm 2, \pm 3,$
$\pm 4, \pm 6,$
$\pm 12, \pm \frac{1}{2}, \pm \frac{3}{2}$

Descartes' Rule of Signs

# pos	# neg	# nonreal
4	1	0
2	1	2
0	1	4

$$\begin{aligned} P(x) &= 2x^5 - 3x^4 + 6x^3 - 9x^2 - 8x + 12 \\ &= (x-1)(2x^4 - x^3 + 5x^2 - 4x - 12) \\ &= (x-1)(x+1)(2x^3 - 3x^2 + 8x - 12) \\ &= (x-1)(x+1)\left(x - \tfrac{3}{2}\right)(2x^2 + 8) \end{aligned}$$

$2x^2 + 8 = 0 \Rightarrow x = \pm\sqrt{-4}$

Zeros: $\left\{1, -1, \frac{3}{2}, \pm 2i\right\}$

Test $x = 1$:

$$\begin{array}{r|rrrrrr} 1 & 2 & -3 & 6 & -9 & -8 & 12 \\ & & 2 & -1 & 5 & -4 & -12 \\ \hline & 2 & -1 & 5 & -4 & -12 & 0 \end{array}$$

Test $x = -1$:

$$\begin{array}{r|rrrrr} -1 & 2 & -1 & 5 & -4 & -12 \\ & & -2 & 3 & -8 & 12 \\ \hline & 2 & -3 & 8 & -12 & 0 \end{array}$$

Test $x = \frac{3}{2}$:

$$\begin{array}{r|rrrr} \frac{3}{2} & 2 & -3 & 8 & -12 \\ & & 3 & 0 & 12 \\ \hline & 2 & 0 & 8 & 0 \end{array}$$

57. If $(1+i)$ is a zero, then so is $(1-i)$, and $x-(1+i)$ and $x-(1-i)$ are factors.
Then $[x-(1+i)][x-(1-i)] = x^2-2x+2$ is a factor. Divide it out:

$$\begin{array}{r|l} & x - \quad 3 \\ \hline x^2-2x+2 & x^3-5x^2+8x-6 \\ & \underline{x^3-2x^2+2x} \\ & -3x^2+6x-6 \\ & \underline{-3x^2+6x-6} \\ & 0 \end{array}$$

$$P(x) = x^3-5x^2+8x-6 = (x^2-2x+2)(x-3)$$

Zeros: $\{1+i, 1-i, 3\}$

59. If $(1+i)$ is a zero, then so is $(1-i)$, and $x-(1+i)$ and $x-(1-i)$ are factors.
Then $[x-(1+i)][x-(1-i)] = x^2-2x+2$ is a factor. Divide it out:

$$\begin{array}{r|l} & x^2 - \quad 9 \\ \hline x^2-2x+2 & x^4-2x^3-7x^2+18x-18 \\ & \underline{x^4-2x^3+2x^2} \\ & -9x^2+18x-18 \\ & \underline{-9x^2+18x-18} \\ & 0 \end{array}$$

$$P(x) = x^4-2x^3-7x^2+18x-18 = (x^2-2x+2)(x^2-9) = (x^2-2x+2)(x+3)(x-3)$$

Zeros: $\{1+i, 1-i, -3, 3\}$

61. Possible rational zeros: $\pm 1, \pm 2, \pm 3, \pm 6, \pm\frac{1}{3}, \pm\frac{2}{3}$

Descartes' Rule of Signs

# pos	# neg	# nonreal
1	2	0
1	0	2

Test $x=-1$:

$$\begin{array}{r|rrrr} -1 & 3 & -4 & -13 & -6 \\ & & -3 & 7 & 6 \\ \hline & 3 & -7 & -6 & 0 \end{array}$$

$$x^3-\frac{4}{3}x^2-\frac{13}{3}x-2=0$$
$$3x^3-4x^2-13x-6=0$$
$$(x+1)(3x^2-7x-6)=0$$
$$(x+1)(3x+2)(x-3)=0$$

Solutions: $\{-1, -\frac{2}{3}, 3\}$

63.
$$x^{-5}-8x^{-4}+25x^{-3}-38x^{-2}+28x^{-1}-8=0$$
$$x^5(x^{-5}-8x^{-4}+25x^{-3}-38x^{-2}+28x^{-1}-8)=x^5(0)$$
$$1-8x+25x^2-38x^3+28x^4-8x^5=0$$

Possible rat. zeros: $\pm 1, \pm\frac{1}{2}, \pm\frac{1}{4}, \pm\frac{1}{8}$

Descartes' Rule of Signs

# pos	# neg	# nonreal
5	0	0
3	0	2
1	0	4

Test $x=1$:

$$\begin{array}{r|rrrrrr} 1 & 8 & -28 & 38 & -25 & 8 & -1 \\ & & 8 & -20 & 18 & -7 & 1 \\ \hline & 8 & -20 & 18 & -7 & 1 & 0 \end{array}$$

Test $x=1$:

$$\begin{array}{r|rrrrr} 1 & 8 & -20 & 18 & -7 & 1 \\ & & 8 & -12 & 6 & -1 \\ \hline & 8 & -12 & 6 & -1 & 0 \end{array}$$

Test $x=\frac{1}{2}$:

$$\begin{array}{r|rrrr} \frac{1}{2} & 8 & -12 & 6 & -1 \\ & & 4 & -4 & 1 \\ \hline & 8 & -8 & 2 & 0 \end{array}$$

continued on next page...

63. continued

$$1 - 8x + 25x^2 - 38x^3 + 28x^4 - 8x^5 = 0$$
$$8x^5 - 28x^4 + 38x^3 - 25x^2 + 8x - 1 = 0$$
$$(x-1)(8x^4 - 20x^3 + 18x^2 - 7x + 1) = 0$$
$$(x-1)(x-1)(8x^3 - 12x^2 + 6x - 1) = 0$$
$$(x-1)(x-1)\left(x-\tfrac{1}{2}\right)(8x^2 - 8x + 2) = 0$$
$$(x-1)(x-1)\left(x-\tfrac{1}{2}\right)(4x-2)(2x-1) = 0$$

Solutions: $\left\{1, 1, \frac{1}{2}, \frac{1}{2}, \frac{1}{2}\right\}$

65. Let $x = R_1$. Then $x + 10 = R_2$ and $x + 50 = R_3$.

$$\tfrac{1}{R} = \tfrac{1}{R_1} + \tfrac{1}{R_2} + \tfrac{1}{R_3}$$
$$\tfrac{1}{6} = \tfrac{1}{x} + \tfrac{1}{x+10} + \tfrac{1}{x+50}$$
$$6x(x+10)(x+50) \cdot \tfrac{1}{6} = 6x(x+10)(x+50)\left(\tfrac{1}{x} + \tfrac{1}{x+10} + \tfrac{1}{x+50}\right)$$
$$x(x+10)(x+50) = 6(x+10)(x+50) + 6x(x+50) + 6x(x+10)$$
$$x^3 + 60x^2 + 500x = 6x^2 + 360x + 3000 + 6x^2 + 300x + 6x^2 + 60x$$
$$x^3 + 42x^2 - 220x - 3000 = 0$$
$$(x-10)(x^2 + 52x + 300) = 0$$

10	1	42	−220	−3000
		10	520	3000
	1	52	300	0

Use the quadratic formula on the second factor. The two solutions from that factor are negative. The only solution that makes sense is $x = 10$. The resistances are 10, 20 and 60 ohms.

67. Let $x =$ the height. Then $x + 7 =$ the length, and $x + 4 =$ the width.
The volume is $x(x+7)(x+4) = x^3 + 11x^2 + 28x$.

$$x^3 + 11x^2 + 28x = 4420$$
$$x^3 + 11x^2 + 28x - 4420 = 0$$
$$(x-13)(x^2 + 24x + 340) = 0$$

Test $x = 13$:

13	1	11	28	−4420
		13	312	4420
	1	24	340	0

The only real solution is $x = 13$.
The height is 13 inches.

69. Possible rat'l zeros

$\pm 1, \pm 2, \pm 3,$
$\pm 5, \pm 6, \pm 9,$
$\pm 10, \pm 15, \pm 18,$
$\pm 25, \pm 30, \pm 45,$
$\pm 50, \pm 75, \pm 90,$
$\pm 150, \pm 225,$
± 450

Descartes' Rule of Signs

# pos	# neg	# nonreal
3	1	0
1	1	2

$$-x^4 + 5x^3 + 91x^2 - 545x + 550 = 100$$
$$x^4 - 5x^3 - 91x^2 + 545x - 450 = 0$$
$$(x-1)(x^3 - 4x^2 - 95x + 450) = 0$$
$$(x-1)(x-5)(x^2 + x - 90) = 0$$
$$(x-1)(x-5)(x-9)(x+10) = 0$$

Solutions: $\{1, 5, 9, -10\}$

Test $x = 1$:

$$\begin{array}{r|rrrrr} 1 & 1 & -5 & -91 & 545 & -450 \\ & & 1 & -4 & -95 & 450 \\ \hline & 1 & -4 & -95 & 450 & 0 \end{array}$$

Test $x = 5$:

$$\begin{array}{r|rrrr} 5 & 1 & -4 & -95 & 450 \\ & & 5 & 5 & -450 \\ \hline & 1 & 1 & -90 & 0 \end{array}$$

The only solutions between 0 and 9 are 1, 5, and 9 miles.

71. **Answers may vary.**

73. **Answers may vary.**

75. A coordinate of a point on the parabola has coordinates $(x, 16 - x^2)$.
Thus, the area of the rectangle is $A = 2x(16 - x^2) = 32x - 2x^3$.

$$32x - 2x^3 = 42$$
$$-2x^3 + 32x - 42 = 0$$
$$x^3 - 16x + 21 = 0$$
$$(x-3)(x^2 + 3x - 7) = 0$$

Using the quadratic formula on the second factor yields the solutions $x = \dfrac{-3 \pm \sqrt{37}}{2} \approx 1.54$ or -4.54.

The only solutions that make sense are $x = 3$ or $x = 1.54$.
Points: $(3, 7)$ or $(1.54, 13.63)$

Test $x = 3$:

$$\begin{array}{r|rrrr} 3 & 1 & 0 & -16 & 21 \\ & & 3 & 9 & -21 \\ \hline & 1 & 3 & -7 & 0 \end{array}$$

77. False. The possible rational zeros are ± 1 and $\pm \frac{1}{5}$.

79. True.

81. False. It must have at least one real zero, but that zero need not be rational.

Exercises 4.6 (page 476)

1. asymptote

3. vertical

5. x-intercept

7. same

9. horizontal or slant; vertical

11. vertical: $x = 2$, horizontal: $y = 1$
domain: $(-\infty, 2) \cup (2, \infty)$
range: $(-\infty, 1) \cup (1, \infty)$

13. $t = f(30) = \frac{600}{30} = 20$ hr

15. $t = f(50) = \frac{600}{50} = 12$ hr

EXERCISES 4.6

17. $c = f(10) = \dfrac{50{,}000(10)}{100-10} \approx \$5555.56$

19. $c = f(50) = \dfrac{50{,}000(50)}{100-50} = \$50{,}000.00$

21. $f(x) = \dfrac{x^2}{x-2}$; den $= 0 \Rightarrow x = 2$
domain $= (-\infty, 2) \cup (2, \infty)$

23. $f(x) = \dfrac{2x^2+7x-2}{x^2-25} = \dfrac{2x^2+7x-2}{(x+5)(x-5)}$
den $= 0 \Rightarrow x = -5, x = 5$
domain $= (-\infty, -5) \cup (-5, 5) \cup (5, \infty)$

25. $f(x) = \dfrac{x-1}{x^3-x} = \dfrac{x-1}{x(x+1)(x-1)}$; den $= 0 \Rightarrow x = 0, x = -1, x = 1$
domain $= (-\infty, -1) \cup (-1, 0) \cup (0, 1) \cup (1, \infty)$

27. $f(x) = \dfrac{3x^2+5}{x^2+1}$; den $= 0 \Rightarrow$ never true
domain $= (-\infty, \infty)$

29. $f(x) = \dfrac{x}{x-3}$; den $= 0 \Rightarrow x = 3$
vertical: $x = 3$

31. $f(x) = \dfrac{x+2}{x^2-1} = \dfrac{x+2}{(x+1)(x-1)}$
den $= 0 \Rightarrow x = -1, x = 1$
vertical: $x = -1, x = 1$

33. $f(x) = \dfrac{1}{x^2-x-6} = \dfrac{1}{(x+2)(x-3)}$
den $= 0 \Rightarrow x = -2, x = 3$
vertical: $x = -2, x = 3$

35. $f(x) = \dfrac{x^2}{x^2+5}$; den $= 0 \Rightarrow$ never true
vertical: none

37. $f(x) = \dfrac{2x-1}{x}$; deg(num) $=$ deg(den)
horizontal: $y = \dfrac{2}{1}$, or $y = 2$

39. $f(x) = \dfrac{x^2+x-2}{2x^2-4}$; deg(num) $=$ deg(den)
horizontal: $y = \dfrac{1}{2}$

41. $f(x) = \dfrac{x+1}{x^3-4x}$; deg(num) $<$ deg(den)
horizontal: $y = 0$

43. $f(x) = \dfrac{x^2}{x-2}$; deg(num) $>$ deg(den)
horizontal: none

45. $f(x) = \dfrac{x^2-5x-6}{x-2} = x - 3 + \dfrac{-12}{x-2}$
slant: $y = x - 3$

47. $f(x) = \dfrac{2x^2-5x+1}{x-4} = 2x + 3 + \dfrac{13}{x-4}$
slant: $y = 2x + 3$

49. $f(x) = \dfrac{x^3+2x^2-x-1}{x^2-1}$
$= x + 2 + \dfrac{1}{x^2-1}$
slant: $y = x + 2$

EXERCISES 4.6

51. $y = \dfrac{1}{x-2}$

Vert: $x = 2$; Horiz: $y = 0$

Slant: none; x-intercepts: none

y-intercepts: $\left(0, -\frac{1}{2}\right)$; Symmetry: none

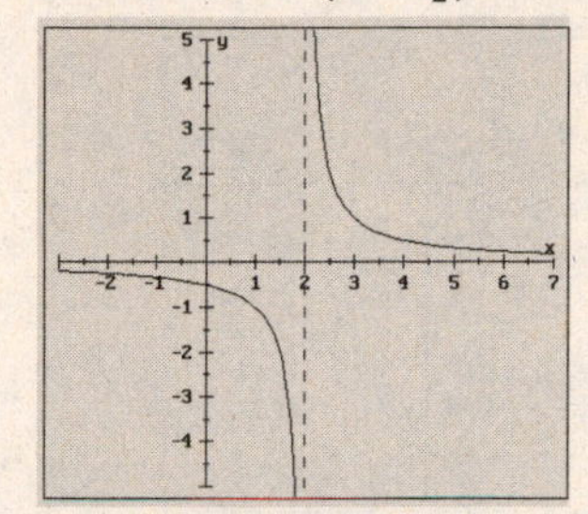

53. $y = \dfrac{x}{x-1}$

Vert: $x = 1$; Horiz: $y = \frac{1}{1} = 1$

Slant: none; x-intercepts: $(0, 0)$

y-intercepts: $(0, 0)$; Symmetry: none

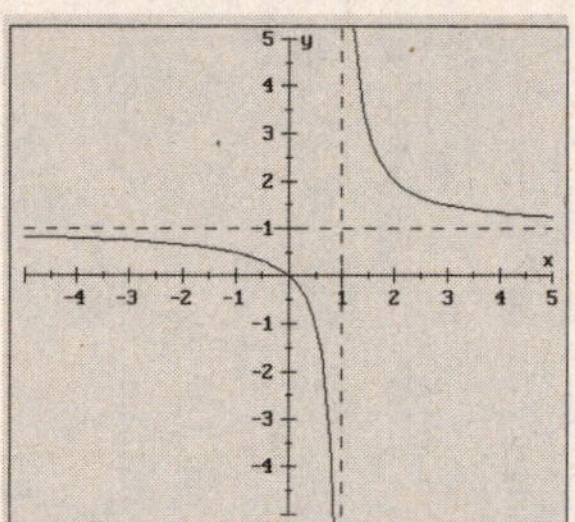

55. $f(x) = \dfrac{x+1}{x+2}$

Vert: $x = -2$; Horiz: $y = \frac{1}{1} = 1$

Slant: none; x-intercepts: $(-1, 0)$

y-intercepts: $\left(0, \frac{1}{2}\right)$; Symmetry: none

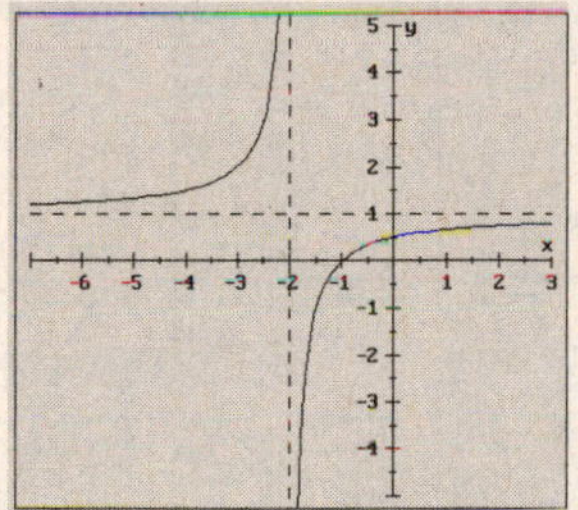

57. $f(x) = \dfrac{2x-1}{x-1}$

Vert: $x = 1$; Horiz: $y = \frac{2}{1} = 2$

Slant: none; x-intercepts: $\left(\frac{1}{2}, 0\right)$

y-intercepts: $(0, 1)$; Symmetry: none

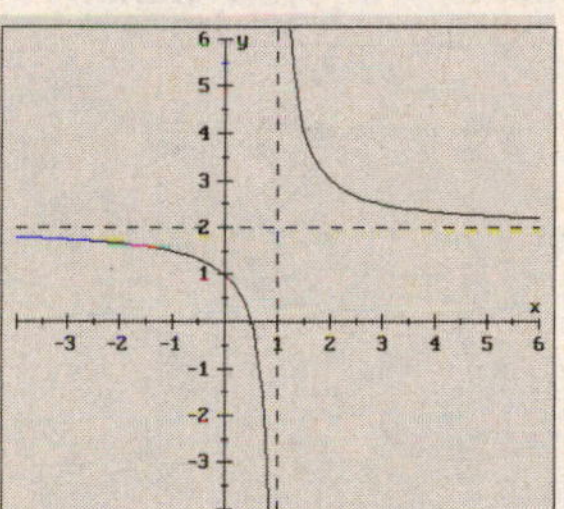

59. $g(x) = \dfrac{x^2-9}{x^2-4} = \dfrac{(x+3)(x-3)}{(x+2)(x-2)}$

Vert: $x = -2, x = 2$; Horiz: $y = \frac{1}{1} = 1$

Slant: none; x-intercepts: $(-3, 0)$, $(3, 0)$

y-intercepts: $\left(0, \frac{9}{4}\right)$; Symmetry: y-axis

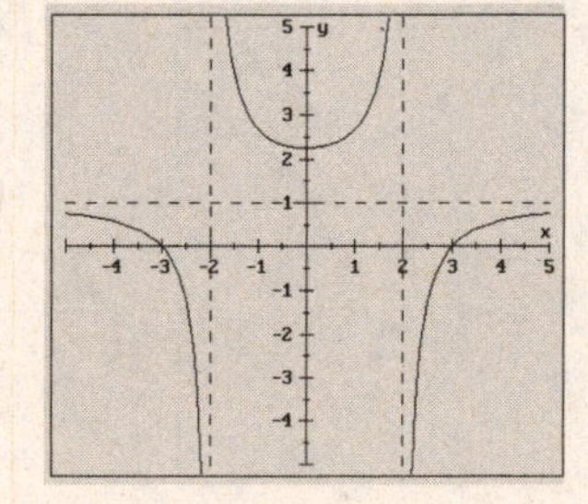

61. $g(x) = \dfrac{x^2-x-2}{x^2-4x+3} = \dfrac{(x+1)(x-2)}{(x-3)(x-1)}$

Vert: $x = 3, x = 1$; Horiz: $y = \frac{1}{1} = 1$

Slant: none; x-intercepts: $(-1, 0)$, $(2, 0)$

y-intercepts: $\left(0, -\frac{2}{3}\right)$; Symmetry: none

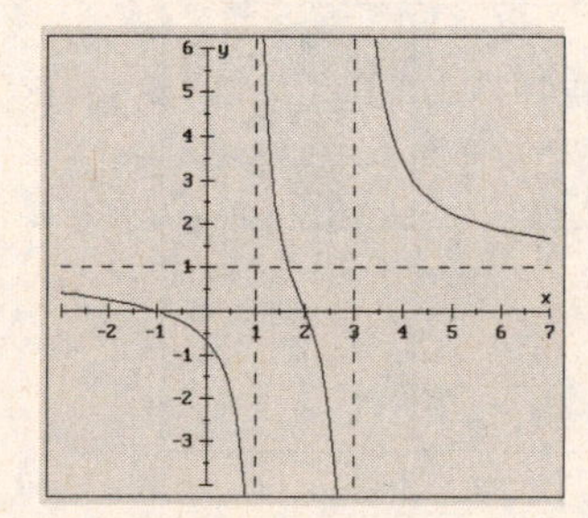

63. $y = \dfrac{x^2 + 2x - 3}{x^3 - 4x} = \dfrac{(x-1)(x+3)}{x(x+2)(x-2)}$

Vert: $x = 0, x = -2, x = 2$; Horiz: $y = 0$
Slant: none; x-intercepts: $(1, 0), (-3, 0)$
y-intercepts: none; Symmetry: none

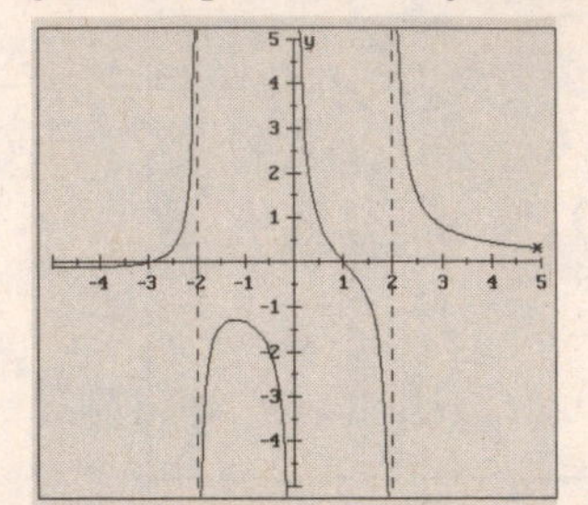

65. $y = \dfrac{x^2 - 9}{x^2} = \dfrac{(x+3)(x-3)}{x^2}$

Vert: $x = 0$; Horiz: $y = \frac{1}{1} = 1$
Slant: none; x-intercepts: $(3, 0), (-3, 0)$
y-intercepts: none; Symmetry: y-axis

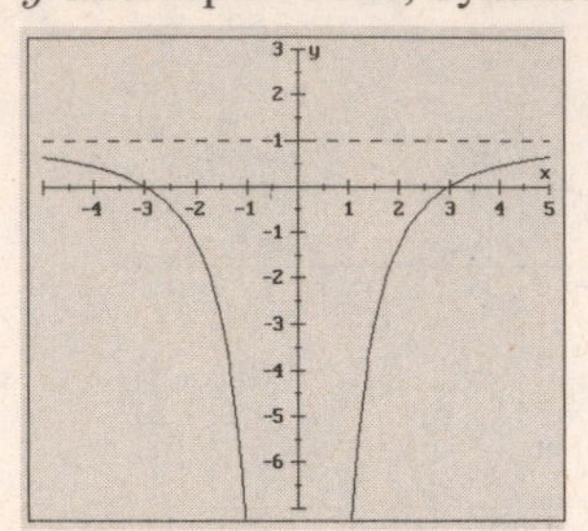

67. $f(x) = \dfrac{x}{(x+3)^2}$

Vert: $x = -3$; Horiz: $y = 0$
Slant: none; x-intercepts: $(0, 0)$
y-intercepts: $(0, 0)$; Symmetry: none

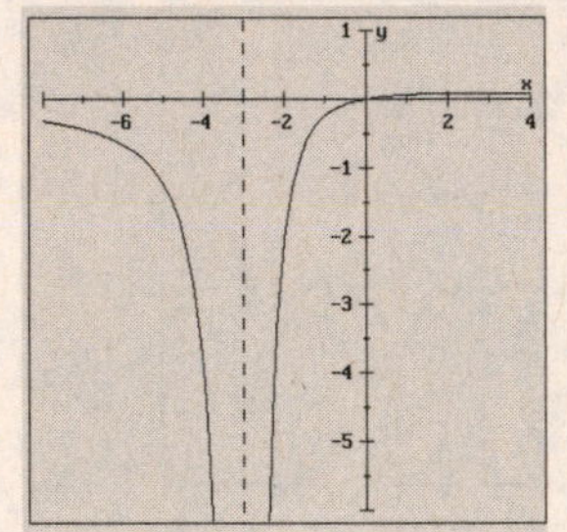

69. $f(x) = \dfrac{x+1}{x^2(x-2)}$

Vert: $x = 0, x = 2$; Horiz: $y = 0$
Slant: none; x-intercepts: $(-1, 0)$
y-intercepts: none; Symmetry: none

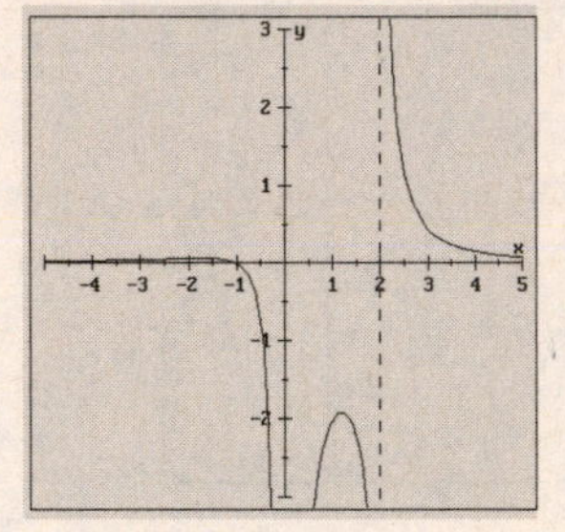

71. $y = \dfrac{x}{x^2 + 1}$

Vert: none; Horiz: $y = 0$
Slant: none; x-intercepts: $(0, 0)$
y-intercepts: $(0, 0)$; Symmetry: origin

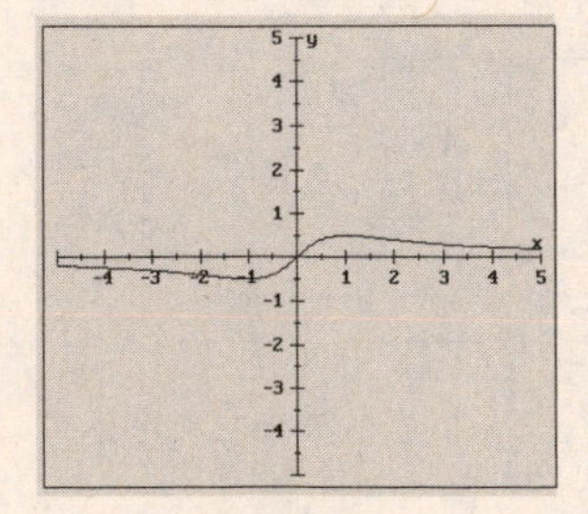

73. $y = \dfrac{3x^2}{x^2 + 1}$

Vert: none; Horiz: $y = \frac{3}{1} = 3$
Slant: none; x-intercepts: $(0, 0)$
y-intercepts: $(0, 0)$; Symmetry: y-axis

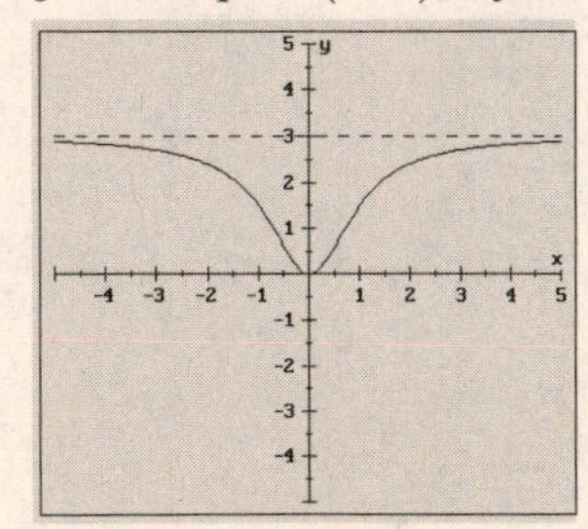

EXERCISES 4.6

75. $h(x) = \dfrac{x^2 - 2x - 8}{x - 1} = \dfrac{(x+2)(x-4)}{x-1}$
$= x - 1 + \dfrac{-9}{x-1}$

Vert: $x = 1$; Horiz: none; Slant: $y = x - 1$
x-intercepts: $(4, 0)$, $(-2, 0)$
y-intercepts: $(0, 8)$; Symmetry: none

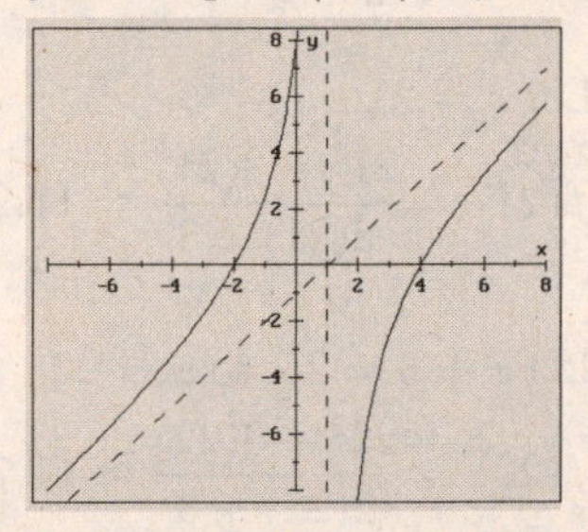

77. $f(x) = \dfrac{x^3 + x^2 + 6x}{x^2 - 1} = \dfrac{x(x^2 + x + 6)}{(x+1)(x-1)}$
$= x + 1 + \dfrac{7x + 1}{x^2 - 1}$

Vert: $x = -1$, $x = 1$; Horiz: none
Slant: $y = x + 1$; x-intercepts: $(0, 0)$
y-intercepts: $(0, 0)$; Symmetry: none

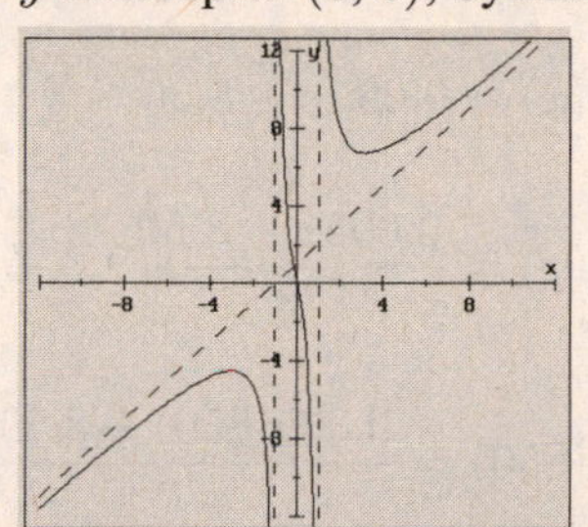

79. $f(x) = \dfrac{x^2}{x} = x$ (if $x \neq 0$)

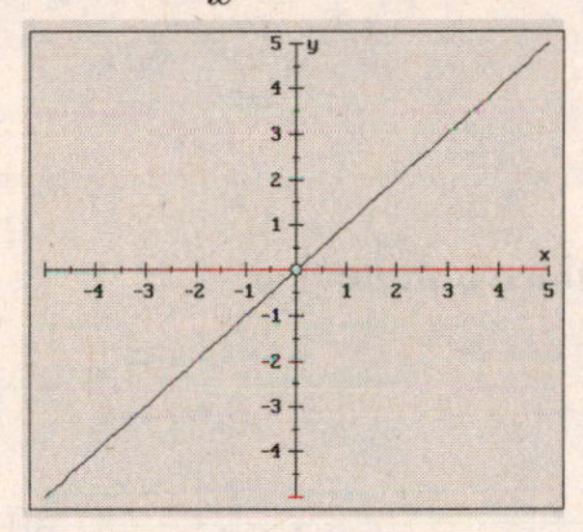

81. $f(x) = \dfrac{x^3 + x}{x} = \dfrac{x(x^2 + 1)}{x} = x^2 + 1$
(if $x \neq 0$)

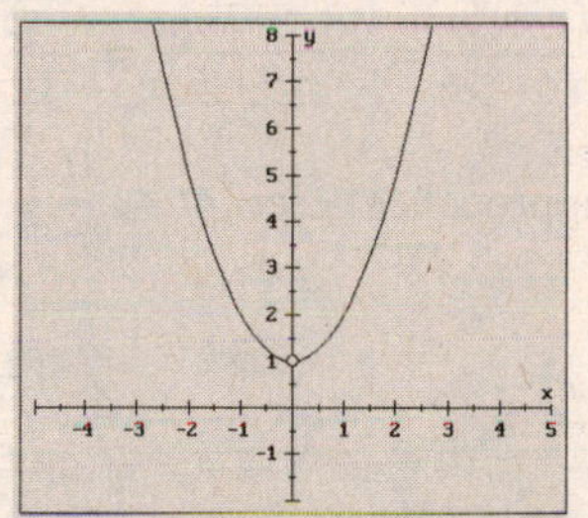

83. $f(x) = \dfrac{x^2 - 2x + 1}{x - 1} = \dfrac{(x-1)(x-1)}{x-1}$
$= x - 1$ (if $x \neq 1$)

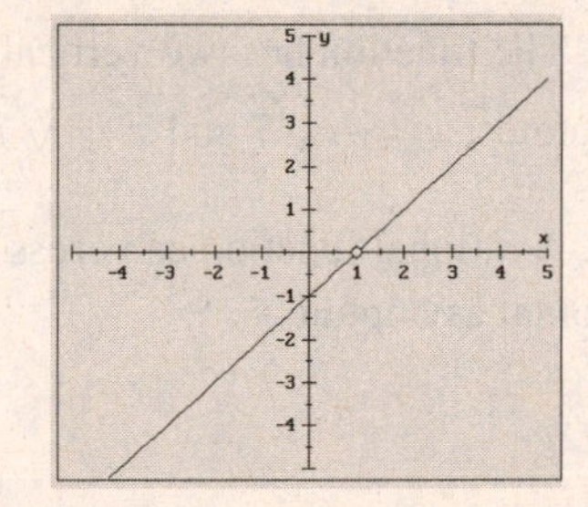

85. $f(x) = \dfrac{x^3 - 1}{x - 1} = \dfrac{(x-1)(x^2 + x + 1)}{x - 1}$
$= x^2 + x + 1$ (if $x \neq 1$)

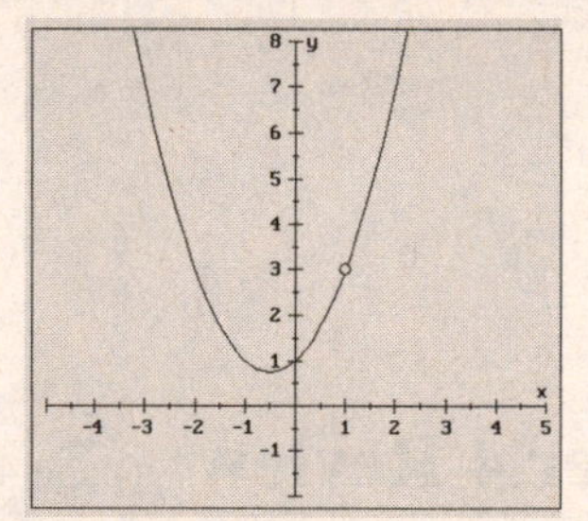

87. **a.** $c(x) = 3.25x + 700$

b. $c(500) = 3.25(500) + 700 = \$2325$

c. $\overline{c}(x) = \dfrac{3.25x + 700}{x}$

d. $\overline{c}(500) = \dfrac{3.25(500) + 700}{500} = \$4.65$

e. $\overline{c}(1000) = \dfrac{3.25(1000) + 700}{1000} = \$3.95$

f. $\overline{c}(2000) = \dfrac{3.25(2000) + 700}{2000} = \$3.60$

89. **a.** $c(x) = 0.095x + 8.50$

b. $\overline{c}(x) = \dfrac{0.095x + 8.50}{x}$

c. $\overline{c}(850) = \dfrac{0.095(850) + 8.50}{850}$
$= \$0.105 = 10.5¢$

90. **a.** Let $x = 21$:
$f(21) = \dfrac{21^2 + 3(21)}{2(21) + 3} = 11.2$ days

b. Let $x + 3 = 25$, so $x = 22$:
$f(22) = \dfrac{22^2 + 3(22)}{2(22) + 3} \approx 11.7$ days

91. **Answers may vary.**

93. **Answers may vary.**

95. $$y = \frac{ax + b}{cx^2 + d} = \frac{\frac{ax+b}{x^2}}{\frac{cx^2+d}{x^2}} = \frac{\frac{ax}{x^2} + \frac{b}{x^2}}{\frac{cx^2}{x^2} + \frac{d}{x^2}} = \frac{\frac{a}{x} + \frac{b}{x^2}}{c + \frac{d}{x^2}}$$

As x approaches $\pm\infty$, $y \approx \dfrac{0 + 0}{c + 0} = 0$. Thus the horizontal asymptote is $y = 0$.

97. $$y = \frac{ax^2 + b}{cx^2 + d} = \frac{\frac{ax^2+b}{x^2}}{\frac{cx^2+d}{x^2}} = \frac{\frac{ax^2}{x^2} + \frac{b}{x^2}}{\frac{cx^2}{x^2} + \frac{d}{x^2}} = \frac{a + \frac{b}{x^2}}{c + \frac{d}{x^2}}$$

As x approaches $\pm\infty$, $y \approx \dfrac{a + 0}{c + 0} = \dfrac{a}{c}$. Thus the horizontal asymptote is $y = \dfrac{a}{c}$.

99. **Answers may vary.**

101. **Answers may vary.**

103. True.

105. False. The function has two vertical asymptotes, $x = -\sqrt{7}$ and $x = \sqrt{7}$.

107. True.

109. False. A rational function can cross a horizontal asymptote.

Chapter 4 Review (page 481)

1. $f(x) = \frac{1}{2}x^2 + 4$
$a = \frac{1}{2} \Rightarrow a > 0$
upward, minimum

2. $f(x) = -4(x + 1)^2 + 5$
$a = -4 \Rightarrow a < 0$
downward, maximum

3. $f(x) = 2(x-1)^2 + 6$
Vertex: $(1, 6)$

4. $f(x) = -2(x+4)^2 - 5$
Vertex: $(-4, -5)$

5. $y = x^2 + 6x - 4;\ a = 1, b = 6, c = -4$

$$x = -\frac{b}{2a} = -\frac{6}{2(1)} = -3$$

$$y = x^2 + 6x - 4 = (-3)^2 + 6(-3) - 4 = -13$$

Vertex: $(-3, -13)$

6. $y = -4x^2 + 4x - 9;\ a = -4, b = 4, c = -9$

$$x = -\frac{b}{2a} = -\frac{4}{2(-4)} = \frac{1}{2}$$

$$y = -4x^2 + 4x - 9 = -4\left(\tfrac{1}{2}\right)^2 + 4\left(\tfrac{1}{2}\right) - 9 = -8$$

Vertex: $\left(\tfrac{1}{2}, -8\right)$

7. $f(x) = (x-2)^2 - 3$
$a = 1 \Rightarrow$ up, vertex: $(2, -3)$

$$0 = (x-2)^2 - 3$$
$$3 = (x-2)^2$$
$$\pm\sqrt{3} = x - 2$$
$$2 \pm \sqrt{3} = x$$
$$\left(2-\sqrt{3}, 0\right), \left(2+\sqrt{3}, 0\right)$$

$f(0) = 1 \Rightarrow (0, 1)$

$f(1) = -2 \Rightarrow (1, -2)$

$(3, -2)$ on graph by symmetry

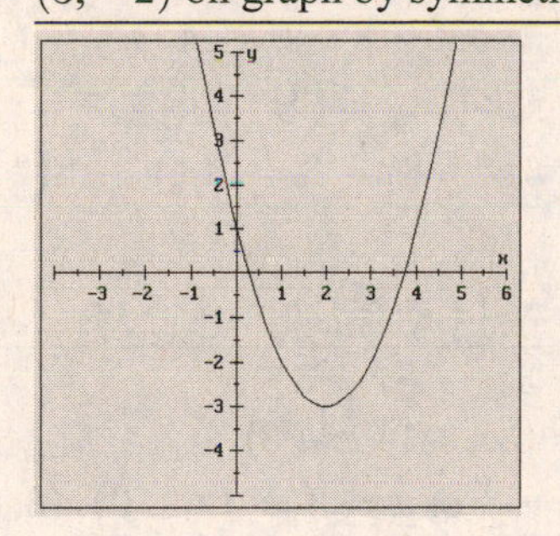

8. $f(x) = -(x-4)^2 + 4$
$a = -1 \Rightarrow$ down, vertex: $(4, 4)$

$$0 = -(x-4)^2 + 4$$
$$(x-4)^2 = 4$$
$$x - 4 = \pm 2$$
$$x = 4 \pm 2$$

$x = 2$ or $x = 6 \Rightarrow (2, 0), (6, 0)$

$f(0) = -12 \Rightarrow (0, -12)$

$f(3) = 3 \Rightarrow (3, 3)$

$(5, 3)$ on graph by symmetry

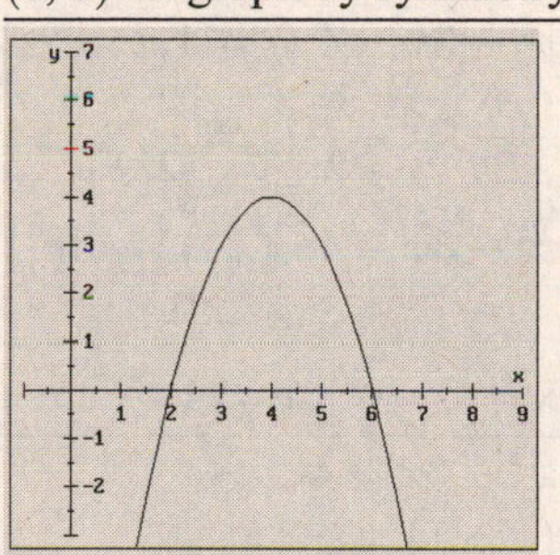

9. $f(x) = x^2 - x;\ a = 1, b = -1, c = 0$

$x = -\frac{b}{2a} = -\frac{-1}{2(1)} = \frac{1}{2}$

$y = x^2 - x = \left(\frac{1}{2}\right)^2 - \frac{1}{2} = -\frac{1}{4}$

vertex: $\left(\frac{1}{2}, -\frac{1}{4}\right)$, $a = 1 \Rightarrow$ up

$0 = x^2 - x$

$0 = x(x-1)$

$x = 0$ or $x = 1 \Rightarrow (0, 0), (1, 0)$

$f(0) = 0 \Rightarrow (0, 0)$

$f(2) = 2 \Rightarrow (2, 2)$ on graph

$(-1, 2)$ on graph by symmetry

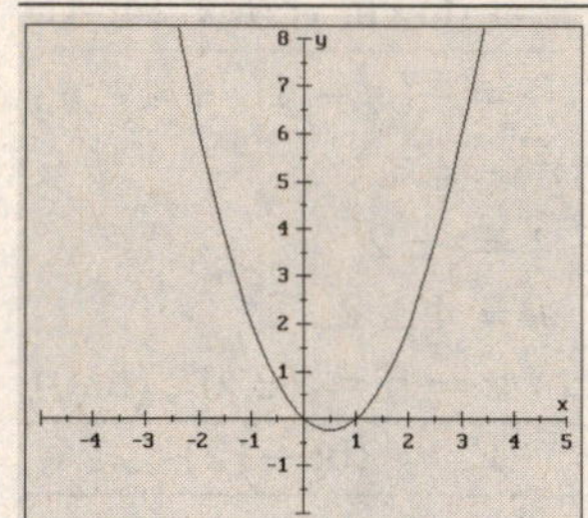

10. $f(x) = x - x^2;\ a = -1, b = 1, c = 0$

$x = -\frac{b}{2a} = -\frac{1}{2(-1)} = \frac{1}{2}$

$y = x - x^2 = \frac{1}{2} - \left(\frac{1}{2}\right)^2 = \frac{1}{4}$

vertex: $\left(\frac{1}{2}, \frac{1}{4}\right)$, $a = -1 \Rightarrow$ down

$0 = x - x^2$

$0 = x(1-x)$

$x = 0$ or $x = 1 \Rightarrow (0, 0), (1, 0)$

$f(0) = 0 \Rightarrow (0, 0)$

$f(2) = -2 \Rightarrow (2, -2)$ on graph

$(-1, -2)$ on graph by symmetry

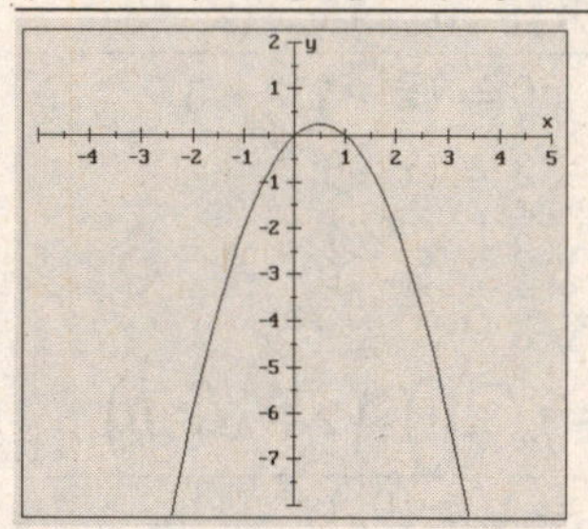

11. $y = x^2 - 3x - 4;\ a = 1, b = -3, c = -4$

$x = -\frac{b}{2a} = -\frac{-3}{2(1)} = \frac{3}{2}$

$y = x^2 - 3x - 4 = \left(\frac{3}{2}\right)^2 - 3\left(\frac{3}{2}\right) - 4$
$= -\frac{25}{4}$

vertex: $\left(\frac{3}{2}, -\frac{25}{4}\right)$, $a = 1 \Rightarrow$ up

$0 = x^2 - 3x - 4$

$0 = (x+1)(x-4)$

$x = -1$ or $x = 4 \Rightarrow (-1, 0), (4, 0)$

$f(0) = -4 \Rightarrow (0, -4)$

$f(2) = -6 \Rightarrow (2, -6)$ on graph

$(1, -6)$ on graph by symmetry

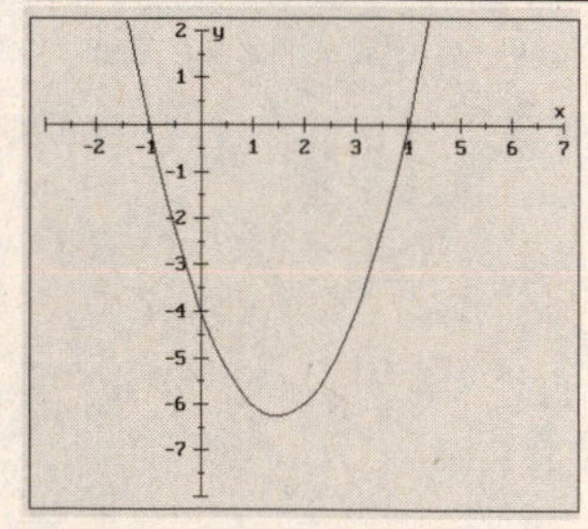

12. $y = 3x^2 - 8x - 3;\ a = 3, b = -8, c = -3$

$x = -\frac{b}{2a} = -\frac{-8}{2(3)} = \frac{4}{3}$

$y = 3x^2 - 8x - 3 = 3\left(\frac{4}{3}\right)^2 - 8\left(\frac{4}{3}\right) - 3$
$= -\frac{25}{3}$

vertex: $\left(\frac{4}{3}, -\frac{25}{3}\right)$, $a = 3 \Rightarrow$ up

$0 = 3x^2 - 8x - 3$

$0 = (3x+1)(x-3)$

$x = -\frac{1}{3}$ or $x = 3 \Rightarrow \left(-\frac{1}{3}, 0\right), (3, 0)$

$f(0) = -3 \Rightarrow (0, -3)$

$f(1) = -8 \Rightarrow (1, -8)$ on graph

$\left(\frac{5}{3}, -8\right)$ on graph by symmetry

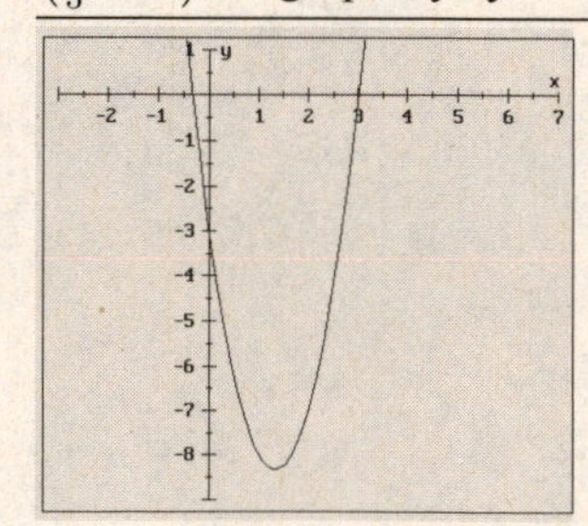

13. $3x^2 + y - 300 = 0$

$y = -3x^2 + 300$

$a = -3, b = 0, c = 300$

$x = -\frac{b}{2a} = -\frac{0}{2(-3)} = 0$

$y = -3(0)^2 + 300 = 300$

The maximum height is 300 units.

14. Let the numbers be x and $1 - x$.

Product $= x(1 - x)$

$y = x - x^2$: $a = -1, b = 1, c = 0$

vertex: $x = -\frac{b}{2a} = -\frac{1}{2(-1)} = \frac{1}{2}$

Both numbers are $\frac{1}{2}$.

15. Let $x =$ the width of the region.

Then $\dfrac{1400 - 2x}{2} = 700 - x =$ the length.

Area = width · length

$y = x(700 - x)$

$y = -x^2 + 700x$

$a = -1, b = 700, c = 0$

$x = -\dfrac{b}{2a} = -\dfrac{700}{2(-1)} = 350$

$700 - x = 700 - 350 = 350$

$y = 350(700 - 350) = 122{,}500$

The dimensions are 350 ft by 350 ft, with an area of 122,500 ft^2.

16. $C(x) = 1.5x^2 - 150x + 4850$

$a = 1.5, b = -150, c = 4850$

$x = -\dfrac{b}{2a} = -\dfrac{-150}{2(1.5)} = 50$

$C(50) = 1.5(50)^2 - 150(50) + 4850 = 1100$

50 cameras should be made, for a minimum cost of $1100.

17. $g(x) = x^3 - 6x^2 + 9x$

$x^3 - 6x^2 + 9x = 0$

$x(x^2 - 6x + 9) = 0$

$x(x - 3)^2 = 0$

$x = 0$, multiplicity 1, crosses

$x = 3$, multiplicity 2, touches

18. $g(x) = x^3 + 7x^2 - 4x - 28$

$x^3 + 7x^2 - 4x - 28 = 0$

$x^2(x + 7) - 4(x + 7)$

$(x + 7)(x^2 - 4) = 0$

$(x + 7)(x + 2)(x - 2) = 0$

$x = -7$, multiplicity 1, crosses

$x = -2$, multiplicity 1, crosses

$x = 2$, multiplicity 1, crosses

19. $f(x) = x^4 - 4x^3 + 3x^2$

$x^4 - 4x^3 + 3x^2 = 0$

$x^2(x^2 - 4x + 3) = 0$

$x^2(x - 1)(x - 3) = 0$

$x = 0$, multiplicity 2, touches

$x = 1$, multiplicity 1, crosses

$x = 3$, multiplicity 1, crosses

20. $f(x) = x^4 - 10x^2 + 24$

$x^4 - 10x^2 + 24 = 0$

$(x^2 - 6)(x^2 - 4) = 0$

$\left(x + \sqrt{6}\right)\left(x - \sqrt{6}\right)(x + 2)(x - 2) = 0$

$x = -\sqrt{6}$, multiplicity 1, crosses

$x = \sqrt{6}$, multiplicity 1, crosses

$x = -2$, multiplicity 1, crosses

$x = 2$, multiplicity 1, crosses

21. $f(x) = \sqrt{2}x^5 + 9x^3 - 7x$
Degree = 5 (odd); Lead Coef: pos.
falls left, rises right

22. $f(x) = -\frac{1}{2}x^7 + 5x^4 + 6x^2 - 7$
Degree = 7 (odd); Lead Coef: neg.
rises left, falls right

23. $f(x) = 7x^6 - 5x^2 + 4$
Degree = 4 (even); Lead Coef: pos.
rises left, rises right

24. $f(x) = -2x^4 - 3x - 8$
Degree = 4 (even); Lead Coef: neg.
falls left, falls right

25. $f(x) = x^3 - x$

x-int.	y-int.
$x^3 - x = 0$	$f(0) = 0^3 - 0$
$x(x^2 - 1) = 0$	$y = 0$
$x(x+1)(x-1) = 0$	$(0, 0)$
$x = 0, x = -1, x = 1$	

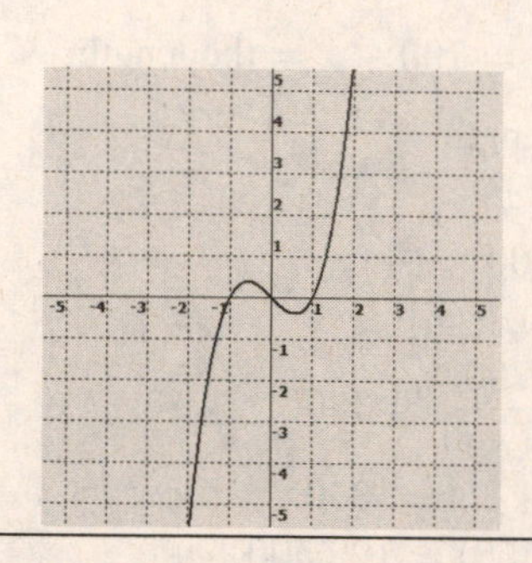

odd deg, pos coef ⇒ falls left, rises right

Sign of $f(x)=x^3-x$	$-$	$+$	$-$	$+$
	$(-\infty, -1)$	$(-1, 0)$	$(0, 1)$	$(1, \infty)$
	-1	0	1	
Test point	$f(-2) = -6$	$f\left(-\frac{1}{2}\right) = \frac{3}{8}$	$f\left(\frac{1}{2}\right) = -\frac{3}{8}$	$f(2) = 6$
Graph of $f(x)$	below axis	above axis	below axis	above axis

$f(-x) = (-x)^3 - (-x) = -x^3 + x = -f(x)$ ⇒ odd, symmetric about origin

26. $f(x) = x^3 - x^2$

x-int.	y-int.
$x^3 - x^2 = 0$	$f(0) = 0^3 - 0^2$
$x^2(x - 1) = 0$	$y = 0$
$x = 0, x = 1$	$(0, 0)$

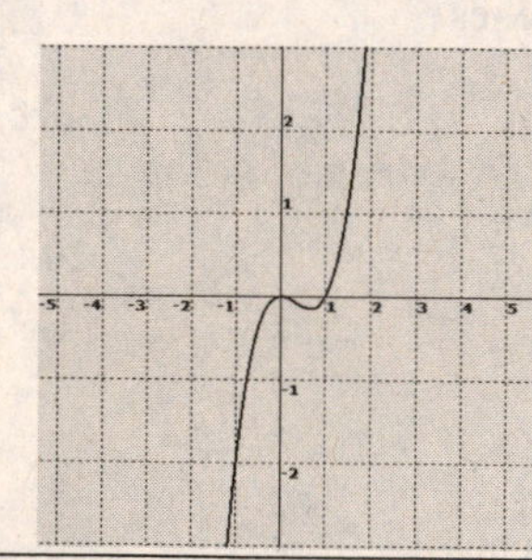

odd deg, pos coef ⇒ falls left, rises right

Sign of $f(x)=x^3-x^2$	$-$	$-$	$+$
	$(-\infty, 0)$	$(0, 1)$	$(1, \infty)$
	0	1	
Test point	$f(-1) = -2$	$f\left(\frac{1}{2}\right) = -\frac{1}{8}$	$f(2) = 4$
Graph of $f(x)$	below axis	below axis	above axis

$f(-x) = (-x)^3 - (-x)^2 = -x^3 - x^2$ ⇒ neither even nor odd, no symmetry

27. $f(x) = -x^3 - 7x^2 - 10x$

x-int.	y-int.
$-x^3 - 7x^2 - 10x = 0$	$f(0) = -(0)^3 - 7(0)^2 - 10(0)$
$-x(x^2 + 7x + 10) = 0$	$y = 0$
$-x(x+2)(x+5) = 0$	$(0, 0)$
$x = 0, x = -2, x = -5$	

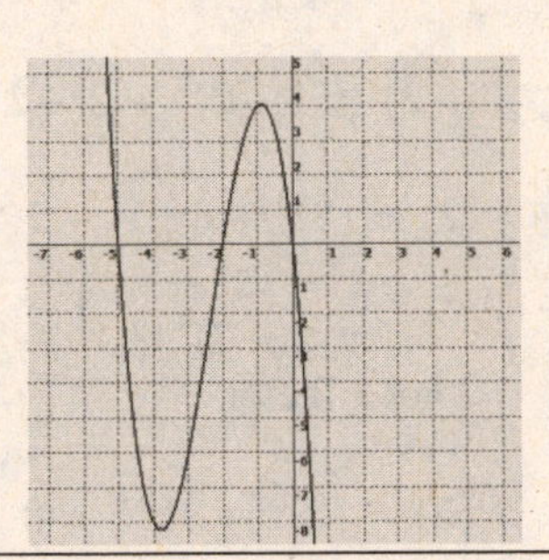

odd deg, neg coef $\Rightarrow$ rises left, falls right

Sign of $f(x)=-x^3-7x^2-10x$	$+$	$-$	$+$	$-$
	$(-\infty, -5)$	$(-5, -2)$	$(-2, 0)$	$(0, \infty)$
		-5	-2	0
Test point	$f(-6) = 24$	$f(-3) = -6$	$f(-1) = 4$	$f(1) = -18$
Graph of $f(x)$	above axis	below axis	above axis	below axis

$f(-x) = -(-x)^3 - 7(-x)^2 - 10(-x) = x^3 - 7x^2 + 10x \Rightarrow$ neither even nor odd, no symmetry

28. $f(x) = -x^4 + 18x^2 - 32$

x-int.	y-int.
$-x^4 + 18x^2 - 32 = 0$	$f(0) = -32$
$-(x^4 - 18x^2 + 32) = 0$	$y = -32$
$-(x^2 - 2)(x^2 - 16) = 0$	$(0, -32)$
$-\left(x + \sqrt{2}\right)\left(x - \sqrt{2}\right)(x+4)(x-4) = 0$	
$x = -\sqrt{2}, x = \sqrt{2}, x = -4, x = 4$	

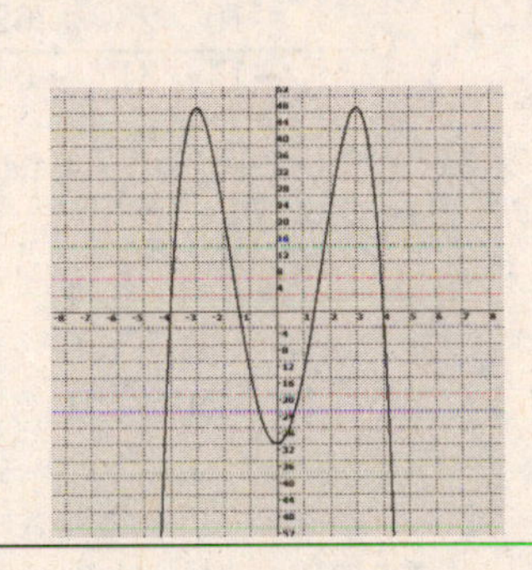

even deg, neg coef $\Rightarrow$ falls left, falls right

Sign of $f(x)=-x^4+18x^2-32$	$-$	$+$	$-$	$+$	$-$
	$(-\infty, -4)$	$\left(-4, -\sqrt{2}\right)$	$\left(-\sqrt{2}, \sqrt{2}\right)$	$\left(\sqrt{2}, 4\right)$	$(4, \infty)$
		-4	$-\sqrt{2}$	$\sqrt{2}$	4
Test point	$f(-5) = -207$	$f(-2) = 24$	$f(0) = -32$	$f(2) = 24$	$f(5) = -207$
Graph of $f(x)$	below axis	above axis	below axis	above axis	below axis

$f(-x) = -(-x)^4 + 18(-x)^2 - 32 = -x^4 + 18x^2 - 32 = f(x) \Rightarrow$ even, symmetric about y-axis

29. $f(x) = 5x^3 + 37x^2 + 59x + 18$
$f(-1) = -9$; $f(0) = 18$
Thus, there is a zero between -1 and 0.

30. $f(x) = 6x^3 - x^2 - 10x - 3$
$f(1) = -8$; $f(2) = 21$
Thus, there is a zero between 1 and 2.

31. $P(1) = 4(1)^4 + 2(1)^3 - 3(1)^2 - 2 = 1$
The remainder is 1.

32. $P(2) = 4(2)^4 + 2(2)^3 - 3(2)^2 - 2 = 66$
The remainder is 66.

33. $P(-3) = 4(-3)^4 + 2(-3)^3 - 3(-3)^2 - 2 = 241 \Rightarrow$ The remainder is 241.

34. $P(-2) = 4(-2)^4 + 2(-2)^3 - 3(-2)^2 - 2 = 34 \Rightarrow$ The remainder is 34.

35. $P(2) = (2)^3 + 4(2)^2 - 2(2) + 4 = 24 \Rightarrow$ The remainder is 24. $\Rightarrow$ not a factor

36. $P(-3) = 2(-3)^4 + 10(-3)^3 + 4(-3)^2 + 7(-3) + 21 = -72 \Rightarrow$ The remainder is -72. $\Rightarrow$ not a factor

37. $P(5) = (5)^5 - 3125 = 0$
The remainder is 0. $\Rightarrow$ factor

38. $P(6) = (6)^5 - 6(6)^4 - 4(6) + 24 = 0$
The remainder is 0. $\Rightarrow$ factor

39.
$$\begin{array}{r|rrrrr} 3 & 3 & 0 & 2 & 3 & 7 \\ & & 9 & 27 & 87 & 270 \\ \hline & 3 & 9 & 29 & 90 & 277 \end{array}$$
$$3x^3 + 9x^2 + 29x + 90 + \frac{277}{x-3}$$

40.
$$\begin{array}{r|rrrrr} 2 & 2 & 0 & -3 & 3 & -1 \\ & & 4 & 8 & 10 & 26 \\ \hline & 2 & 4 & 5 & 13 & 25 \end{array}$$
$$2x^3 + 4x^2 + 5x + 13 + \frac{25}{x-2}$$

41.
$$\begin{array}{r|rrrrrr} -2 & 5 & -4 & 3 & -2 & 1 & -1 \\ & & -10 & 28 & -62 & 128 & -258 \\ \hline & 5 & -14 & 31 & -64 & 129 & -259 \end{array}$$
$$5x^4 - 14x^3 + 31x^2 - 64x + 129 + \frac{-259}{x+2}$$

42.
$$\begin{array}{r|rrrrrr} -1 & 4 & 2 & -1 & 3 & 2 & 1 \\ & & -4 & 2 & -1 & -2 & 0 \\ \hline & 4 & -2 & 1 & 2 & 0 & 1 \end{array}$$
$$4x^4 - 2x^3 + x^2 + 2x + \frac{1}{x+1}$$

43.
$$\begin{array}{r|rrrr} 3 & 5 & 2 & -1 & 1 \\ & & 15 & 51 & 150 \\ \hline & 5 & 17 & 50 & 151 \end{array} \quad P(3) = 151$$

44.
$$\begin{array}{r|rrrr} -3 & 5 & 2 & -1 & 1 \\ & & -15 & 39 & -114 \\ \hline & 5 & -13 & 38 & -113 \end{array}$$
$$P(-3) = -113$$

45.
$$\begin{array}{r|rrrr} \frac{1}{2} & 5 & 2 & -1 & 1 \\ & & \frac{5}{2} & \frac{9}{4} & \frac{5}{8} \\ \hline & 5 & \frac{9}{2} & \frac{5}{4} & \frac{13}{8} \end{array} \quad P\left(\tfrac{1}{2}\right) = \tfrac{13}{8}$$

46.
$$\begin{array}{r|rrrr} i & 5 & 2 & -1 & 1 \\ & & 5i & -5+2i & -2-6i \\ \hline & 5 & 2+5i & -6+2i & -1-6i \end{array}$$
$$P(i) = -1 - 6i$$

47. $x = 3$ is a solution, so $(x-3)$ is a factor.
Use synthetic division to divide by $(x-3)$.
$$\begin{array}{r|rrrr} 3 & 2 & -3 & -11 & 6 \\ & & 6 & 9 & -6 \\ \hline & 2 & 3 & -2 & 0 \end{array}$$
$$2x^3 - 3x^2 - 11x + 6 = 0$$
$$(x-3)(2x^2 + 3x - 2) = 0$$
$$(x-3)(2x-1)(x+2) = 0$$
Solution set: $\left\{3, \frac{1}{2}, -2\right\}$

48. $x = -2$ is a solution, so $(x+2)$ is a factor. Use synthetic division to divide by $(x+2)$.

$$\begin{array}{r|rrrrr} -2 & 1 & 4 & -1 & -20 & -20 \\ & & -2 & -4 & 10 & 20 \\ \hline & 1 & 2 & -5 & -10 & 0 \end{array}$$

$$x^4 + 4x^3 - x^2 - 20x - 20 = 0$$
$$(x+2)\left(x^3 + 2x^2 - 5x - 10\right) = 0$$

Use the fact that $x = -2$ is a double root and divide the depressed polynomial by $(x+2)$:

$$\begin{array}{r|rrrr} -2 & 1 & 2 & -5 & -10 \\ & & -2 & 0 & 10 \\ \hline & 1 & 0 & -5 & 0 \end{array}$$

$$(x+2)\left(x^3 + 2x^2 - 5x - 10\right) = 0$$
$$(x+2)(x+2)\left(x^2 - 5\right) = 0$$

Solution set: $\left\{-2, -2, \sqrt{5}, -\sqrt{5}\right\}$

49. $2(x+1)(x-2)\left(x - \frac{3}{2}\right) = 2(x^2 - x - 2)\left(x - \frac{3}{2}\right) = 2\left(x^3 - \frac{5}{2}x^2 - \frac{1}{2}x + 3\right)$
$= 2x^3 - 5x^2 - x + 6$

50. $2(x-1)(x+3)\left(x - \frac{1}{2}\right) = 2(x^2 + 2x - 3)\left(x - \frac{1}{2}\right) = 2\left(x^3 + \frac{3}{2}x^2 - 4x + \frac{3}{2}\right)$
$= 2x^3 + 3x^2 - 8x + 3$

51. $(x-2)(x+5)(x-i)(x+i) = \left(x^2 + 3x - 10\right)\left(x^2 - i^2\right) = \left(x^2 + 3x - 10\right)\left(x^2 + 1\right)$
$= x^4 + 3x^3 - 9x^2 + 3x - 10$

52. $(x+3)(x-2)(x-i)(x+i) = \left(x^2 + x - 6\right)\left(x^2 - i^2\right) = \left(x^2 + x - 6\right)\left(x^2 + 1\right)$
$= x^4 + x^3 - 5x^2 + x - 6$

53. $3x^6 - 4x^5 + 3x + 2 = 0$
6 zeros

54. $2x^6 - 5x^4 + 5x^3 - 4x^2 + x - 12 = 0$
6 zeros

55. $3x^{65} - 4x^{50} + 3x^{17} + 2x = 0$
65 zeros

56. $x^{1984} - 12 = 0$
1984 zeros

57. $P(x) = x^4 - 16$
4 linear factors, 4 zeros

58. $P(x) = x^{40} + x^{30}$
40 linear factors, 40 zeros

59. $P(x) = 4x^5 + 2x^3$
5 linear factors, 5 zeros

60. $P(x) = x^3 - 64x$
3 linear factors, 3 zeros

61. $2 - i$ is also a zero.

62. $-i = 0 - i$, so $0 + i = i$ is also a zero.

63. If $-i$ is a zero, then i is a zero also:
$$(x-4)(x+i)(x-i) = 0$$
$$(x-4)\left(x^2 - i^2\right) = 0$$
$$(x-4)\left(x^2 + 1\right) = 0$$
$$x^3 - 4x^2 + x - 4 = 0$$

64. If i is a zero, then $-i$ is a zero also:
$$(x+5)(x-i)(x+i) = 0$$
$$(x+5)\left(x^2 - i^2\right) = 0$$
$$(x+5)\left(x^2 + 1\right) = 0$$
$$x^3 + 5x^2 + x + 5 = 0$$

65. $P(x) = 3x^4 + 2x^3 - 4x + 2$: 2 sign variations $\Rightarrow$ 2 or 0 positive zeros
$P(-x) = 3(-x)^4 + 2(-x)^3 - 4(-x) + 2$
$= 3x^4 - 2x^3 + 4x + 2$: 2 sign variations $\Rightarrow$ 2 or 0 negative zeros

# pos	# neg	# nonreal
2	2	0
2	0	2
0	2	2
0	0	4

66. $P(x) = 2x^4 - 3x^3 + 5x^2 + x - 5$: 3 sign variations $\Rightarrow$ 3 or 1 positive zeros
$P(-x) = 2(-x)^4 - 3(-x)^3 + 5(-x)^2 + (-x) - 5$
$= 2x^4 + 3x^3 + 5x^2 - x - 5$: 1 sign variation $\Rightarrow$ 1 negative zero

# pos	# neg	# nonreal
3	1	0
1	1	2

67. $P(x) = 4x^5 + 3x^4 + 2x^3 + x^2 + x - 7$: 1 sign variation $\Rightarrow$ 1 positive zero
$P(-x) = 4(-x)^5 + 3(-x)^4 + 2(-x)^3 + (-x)^2 + (-x) - 7$
$= -4x^5 + 3x^4 - 2x^3 + x^2 - x - 7$: 4 sign variations $\Rightarrow$ 4 or 2 or 0 negative zeros

# pos	# neg	# nonreal
1	4	0
1	2	2
1	0	4

68. $P(x) = 3x^7 - 4x^5 + 3x^3 + x - 4$: 3 sign variations $\Rightarrow$ 3 or 1 positive zeros
$P(-x) = 3(-x)^7 - 4(-x)^5 + 3(-x)^3 + (-x) - 4$
$= -3x^7 + 4x^5 - 3x^3 - x - 4$: 2 sign variations $\Rightarrow$ 2 or 0 negative zeros

# pos	# neg	# nonreal
3	2	2
3	0	4
1	2	4
1	0	6

69. $P(x) = x^4 + x^2 + 24{,}567$: 0 sign variations $\Rightarrow$ 0 positive zeros
$P(-x) = (-x)^4 + (-x)^2 + 24{,}567$
$= x^4 + x^2 + 24{,}567$: 0 sign variations $\Rightarrow$ 0 negative zeros

# pos	# neg	# nonreal
0	0	4

CHAPTER 4 REVIEW

70. $P(x) = -x^7 - 5$: 0 sign variations $\Rightarrow$ 0 positive zeros

$P(-x) = -(-x)^7 - 5$

$= x^7 - 5$: 1 sign variation $\Rightarrow$ 1 negative zero

# pos	# neg	# nonreal
0	1	6

71. $P(x) = 5x^3 - 4x^2 - 2x + 4$

$$\begin{array}{r|rrrr} 2 & 5 & -4 & -2 & 4 \\ & & 10 & 12 & 20 \\ \hline & 5 & 6 & 10 & 24 \end{array}$$

Upper bound: 2

$$\begin{array}{r|rrrr} -1 & 5 & -4 & -2 & 4 \\ & & -5 & 9 & -7 \\ \hline & 5 & -9 & 7 & -3 \end{array}$$

Lower bound: -1

72. $P(x) = x^4 + 3x^3 - 5x^2 - 9x + 1$

$$\begin{array}{r|rrrrr} 2 & 1 & 3 & -5 & -9 & 1 \\ & & 2 & 10 & 10 & 2 \\ \hline & 1 & 5 & 5 & 1 & 3 \end{array}$$

Upper bound: 2

$$\begin{array}{r|rrrrr} -5 & 1 & 3 & -5 & -9 & 1 \\ & & -5 & 10 & -25 & 170 \\ \hline & 1 & -2 & 5 & -34 & 171 \end{array}$$

Lower bound: -5

73. num: $\pm 1, \pm 2, \pm 3, \pm 6$; den: $\pm 1, \pm 2$

possible zeros: $\pm 1, \pm 2, \pm 3, \pm 6, \pm \frac{1}{2}, \pm \frac{3}{2}$

74. num: $\pm 1, \pm 2, \pm 5, \pm 10$; den: $\pm 1, \pm 2, \pm 4$

possible zeros: $\pm 1, \pm 2, \pm 5, \pm 10, \pm \frac{1}{2}, \pm \frac{5}{2}, \pm \frac{1}{4}, \pm \frac{5}{4}$

75. $P(x) = x^3 - 10x^2 + 29x - 20$

Possible rational zeros

$\pm 1, \pm 2, \pm 4, \pm 5, \pm 10, \pm 20$

Descartes' Rule of Signs

# pos	# neg	# nonreal
3	0	0
1	0	2

Test $x = 1$:

$$\begin{array}{r|rrrr} 1 & 1 & -10 & 29 & -20 \\ & & 1 & -9 & 20 \\ \hline & 1 & -9 & 20 & 0 \end{array}$$

$$\begin{aligned} P(x) &= x^3 - 10x^2 + 29x - 20 \\ &= (x-1)(x^2 - 9x + 20) \\ &= (x-1)(x-5)(x-4) \end{aligned}$$

Solution set: $\{1, 5, 4\}$

76. $P(x) = x^3 - 8x^2 - x + 8$

Possible rational zeros

$\pm 1, \pm 2, \pm 4, \pm 8$

Descartes' Rule of Signs

# pos	# neg	# nonreal
2	1	0
0	1	2

Test $x = 1$:

$$\begin{array}{r|rrrr} 1 & 1 & -8 & -1 & 8 \\ & & 1 & -7 & -8 \\ \hline & 1 & -7 & -8 & 0 \end{array}$$

$$\begin{aligned} P(x) &= x^3 - 8x^2 - x + 8 \\ &= (x-1)(x^2 - 7x - 8) \\ &= (x-1)(x-8)(x+1) \end{aligned}$$

Solution set: $\{1, 8, -1\}$

77. Possible rational zeros

$\pm 1, \pm 2, \pm 3, \pm 5,$
$\pm 6, \pm 10, \pm 15, \pm 30,$
$\pm \frac{1}{2}, \pm \frac{3}{2}, \pm \frac{5}{2}, \pm \frac{15}{2}$

Test $x = -2$:

$$\begin{array}{r|rrrr} -2 & 2 & 17 & 41 & 30 \\ & & -4 & -26 & -30 \\ \hline & 2 & 13 & 15 & 0 \end{array}$$

Descartes' Rule of Signs

# pos	# neg	# nonreal
0	3	0
0	1	2

$P(x) = 2x^3 + 17x^2 + 41x + 30$
$= (x+2)(2x^2 + 13x + 15)$
$= (x+2)(2x+3)(x+5)$

Solution set: $\{-2, -\frac{3}{2}, -5\}$

78. Possible rational zeros

$\pm 1, \pm \frac{1}{3}$

Test $x = \frac{1}{3}$:

$$\begin{array}{r|rrrr} \frac{1}{3} & 3 & 2 & 2 & -1 \\ & & 1 & 1 & 1 \\ \hline & 3 & 3 & 3 & 0 \end{array}$$

Descartes' Rule of Signs

# pos	# neg	# nonreal
1	2	0
1	0	2

$P(x) = 3x^3 + 2x^2 + 2x - 10$
$= (x - \frac{1}{3})(3x^2 + 3x + 3)$

$3x^2 + 3x + 3$ does not factor rationally.

Rational solutions: $\{\frac{1}{3}\}$

79. Possible rat. zeros

$\pm 1, \pm 2, \pm 3, \pm 4,$
$\pm 6, \pm 9, \pm 12,$
$\pm 18, \pm 36, \pm \frac{1}{2},$
$\pm \frac{3}{2}, \pm \frac{9}{2}, \pm \frac{1}{4},$
$\pm \frac{3}{4}, \pm \frac{9}{4}$

Descartes' Rule of Signs

# pos	# neg	# nonreal
2	2	0
2	0	2
0	2	2
0	0	4

$P(x) = 4x^4 - 25x^2 + 36$
$= (x-2)(4x^3 + 8x^2 - 9x - 18)$
$= (x-2)(x+2)(4x^2 - 9)$
$= (x-2)(x+2)(2x+3)(2x-3)$

Solution set: $\{2, -2, -\frac{3}{2}, \frac{3}{2}\}$

Test $x = 2$:

$$\begin{array}{r|rrrrr} 2 & 4 & 0 & -25 & 0 & 36 \\ & & 8 & 16 & -18 & -36 \\ \hline & 4 & 8 & -9 & -18 & 0 \end{array}$$

Test $x = -2$:

$$\begin{array}{r|rrrr} -2 & 4 & 8 & -9 & -18 \\ & & -8 & 0 & 18 \\ \hline & 4 & 0 & -9 & 0 \end{array}$$

80. Possible rat. zeros

$\pm 1, \pm 2, \pm 4,$
$\pm 8, \pm 16, \pm 32,$
$\pm \frac{1}{2}$

Descartes' Rule of Signs

# pos	# neg	# nonreal
2	2	0
2	0	2
0	2	2
0	0	4

$P(x) = 2x^4 - 11x^3 - 6x^2 + 64x + 32$
$= (x-4)(2x^3 - 3x^2 - 18x - 8)$
$= (x-4)(x-4)(2x^2 + 5x + 2)$
$= (x-4)(x-4)(2x+1)(x+2)$

Solution set: $\{4, 4, -\frac{1}{2}, -2\}$

Test $x = 4$:

$$\begin{array}{r|rrrrr} 4 & 2 & -11 & -6 & 64 & 32 \\ & & 8 & -12 & -72 & -32 \\ \hline & 2 & -3 & -18 & -8 & 0 \end{array}$$

Test $x = 4$:

$$\begin{array}{r|rrrr} 4 & 2 & -3 & -18 & -8 \\ & & 8 & 20 & 8 \\ \hline & 2 & 5 & 2 & 0 \end{array}$$

81. Possible rational zeros

$\pm 1, \pm 2, \pm 4, \pm 8$
$\pm 16, \pm \frac{1}{3}, \pm \frac{2}{3}, \pm \frac{4}{3}$
$\pm \frac{8}{3}, \pm \frac{16}{3}$

Descartes' Rule of Signs

# pos	# neg	# nonreal
3	0	0
1	0	2

Test $x = \frac{1}{3}$:

$$\begin{array}{r|rrrr} \frac{1}{3} & 3 & -1 & 48 & -16 \\ & & 1 & 0 & 16 \\ \hline & 3 & 0 & 48 & 0 \end{array}$$

$P(x) = 3x^3 - x^2 + 48x - 16$
$= \left(x - \frac{1}{3}\right)\left(3x^2 + 48\right)$
$x = \frac{1}{3}$ **or** $x = \pm\sqrt{-16}$
$x = \pm 4i$

Solution set: $\left\{\frac{1}{3}, -4i, 4i\right\}$

82. Possible rational zeros

$\pm 1, \pm 2, \pm 4, \pm 5$
$\pm 10, \pm 20$

Descartes' Rule of Signs

# pos	# neg	# nonreal
2	2	0
2	0	2
0	2	2
0	0	4

$P(x) = x^4 - 2x^3 - 9x^2 + 8x + 20$
$= (x-2)\left(x^3 - 9x - 10\right)$
$= (x-2)(x-2)\left(x^2 - 2x - 5\right)$

Use the quadratic formula.

Solution set: $\left\{2, -2, 1 \pm \sqrt{6}\right\}$

Test $x = 2$:

$$\begin{array}{r|rrrrr} 2 & 1 & -2 & -9 & 8 & 20 \\ & & 2 & 0 & -18 & -20 \\ \hline & 1 & 0 & -9 & -10 & 0 \end{array}$$

Test $x = -2$:

$$\begin{array}{r|rrrr} -2 & 1 & 0 & -9 & -10 \\ & & -2 & 4 & 10 \\ \hline & 1 & -2 & -5 & 0 \end{array}$$

83. $f(x) = \dfrac{3x^2 + x - 2}{x^2 - 25} = \dfrac{(3x-2)(x+1)}{(x+5)(x-5)}$

den $= 0 \Rightarrow x = -5$ or $x = 5$

domain $= (-\infty, -5) \cup (-5, 5) \cup (5, \infty)$

84. $f(x) = \dfrac{2x^2 + 1}{x^2 + 7}$

den $= 0 \Rightarrow$ never true

domain $= (-\infty, \infty)$

85. $f(x) = \dfrac{x+5}{x^2 - 1} = \dfrac{x+5}{(x+1)(x-1)}$

den $= 0 \Rightarrow x = -1$ or $x = 1$

vertical: $x = -1$ or $x = 1$

86. $f(x) = \dfrac{x-7}{x^2 - 49} = \dfrac{x-7}{(x+7)(x-7)}$
$= \dfrac{1}{x+7}$

den $= 0 \Rightarrow x = -7$

vertical: $x = -7$

87. $f(x) = \dfrac{x}{x^2 + x - 6} = \dfrac{x}{(x+3)(x-2)}$

den $= 0 \Rightarrow x = -3$ or $x = 2$

vertical: $x = -3$ or $x = 2$

88. $f(x) = \dfrac{5x+2}{2x^2 - 6x - 8} = \dfrac{5x+2}{(2x+2)(x-4)}$

den $= 0 \Rightarrow x = -1$ or $x = 4$

vertical: $x = -1$ or $x = 4$

89. $f(x) = \dfrac{2x^2 + x - 2}{4x^2 - 4}$; deg(num) = deg(den)

horizontal: $y = \dfrac{2}{4}$, or $y = \dfrac{1}{2}$

90. $f(x) = \dfrac{5x^2 + 4}{4 - x^2}$; deg(num) = deg(den)

horizontal: $y = \dfrac{5}{-1}$, or $y = -5$

91. $f(x) = \dfrac{x+1}{x^3 - 4x}$; deg(num) < deg(den)
horizontal: $y = 0$

92. $f(x) = \dfrac{x^3}{2x^2 - 6x - 8}$; deg(num) > deg(den)
horizontal: none

93. $f(x) = \dfrac{2x^2 - 5x + 1}{x - 4} = 2x + 3 + \dfrac{13}{x - 4}$
slant: $y = 2x + 3$

94. $f(x) = \dfrac{5x^3 + 1}{x + 5}$; deg(num) $= 3$,
deg(den) $= 1 \Rightarrow$ slant: none

95. $f(x) = \dfrac{2x}{x - 4}$
Vert: $x = 4$; Horiz: $y = \frac{2}{1} = 2$
Slant: none; x-intercepts: $(0, 0)$
y-intercepts: $(0, 0)$; Symmetry: none

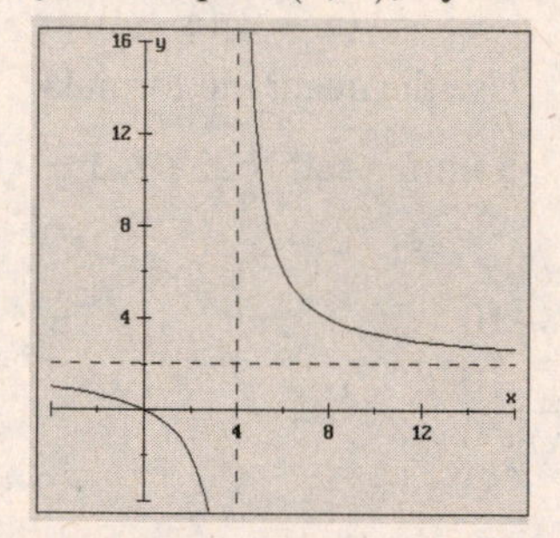

96. $f(x) = \dfrac{-4x}{x + 4}$
Vert: $x = -4$; Horiz: $y = \frac{-4}{1} = -4$
Slant: none; x-intercepts: $(0, 0)$
y-intercepts: $(0, 0)$; Symmetry: none

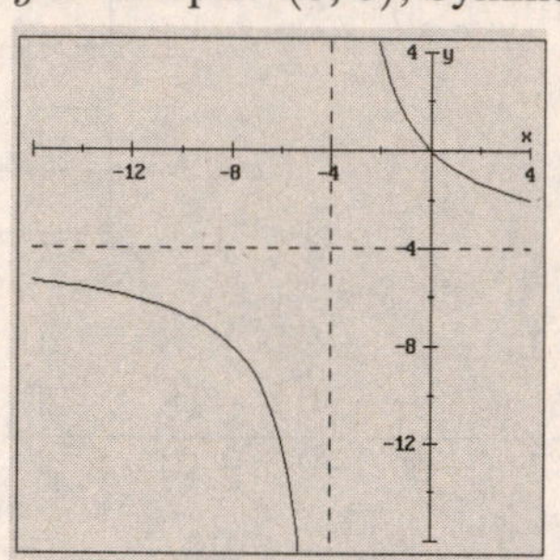

97. $f(x) = \dfrac{x}{(x-1)^2}$
Vert: $x = 1$; Horiz: $y = 0$
Slant: none; x-intercepts: $(0, 0)$
y-intercepts: $(0, 0)$; Symmetry: none

98. $f(x) = \dfrac{(x-1)^2}{x}$: Vert: $x = 0$
Horiz: none; Slant: $y = x - 2$
x-intercepts: $(1, 0)$
y-intercepts: none; Symmetry: none

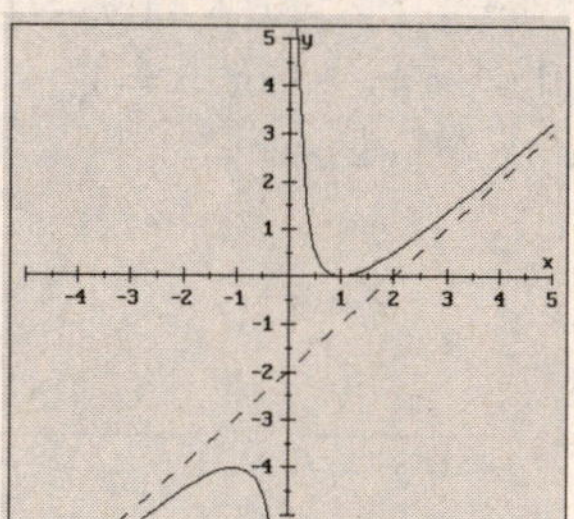

99. $y = \frac{x^2 - x - 2}{x^2 + x - 2} = \frac{(x+1)(x-2)}{(x+2)(x-1)}$

Vert: $x = -2, x = 1$; Horiz: $y = 1$

Slant: none; x-intercepts: $(-1, 0), (2, 0)$

y-intercepts: $(0, 1)$; Symmetry: none

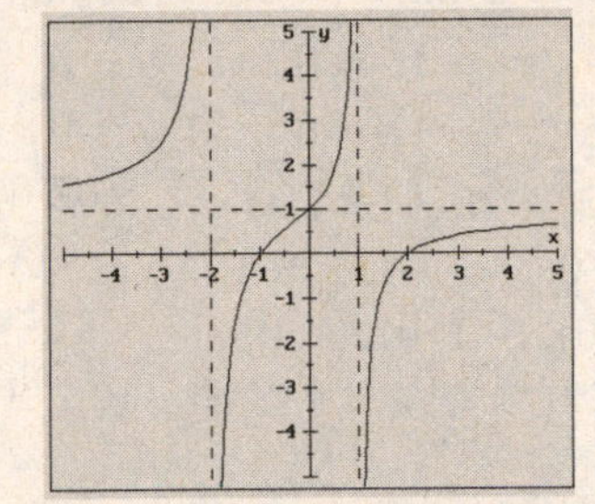

100. $y = \frac{x^3 + x}{x^2 - 4} = \frac{x(x^2+1)}{(x+2)(x-2)}$

Vert: $x = -2, x = 2$;

Slant: $y = x$;

x-int: $(0, 0)$$y$-int: $(0, 0)$;

Symmetry: none

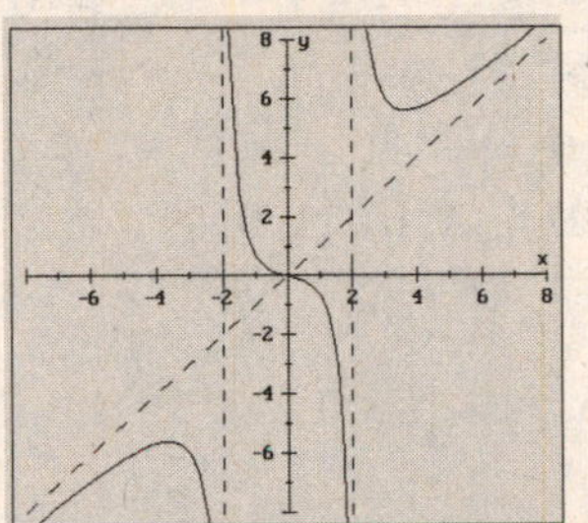

Chapter 4 Test (page 497)

1. $y = 3(x-7)^2 - 3$; Vertex: $(7, -3)$

2. $y = 3x^2 - 24x + 38$

$a = 3, b = -24, c = 38$

vertex: $x = -\frac{b}{2a} = -\frac{-24}{2(3)} = 4$

$$y = 3x^2 - 24x + 38 = 3(4)^2 - 24(4) + 38 = -10$$

3. $y = (x-3)^2 + 1$; Shift $y = x^2$ U 1, R 3.

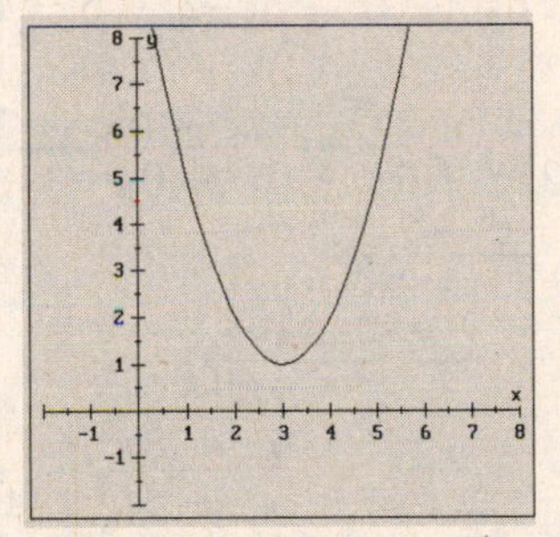

4. $h = 100t - 16t^2$: $a = -16, b = 100, c = 0$

$$x = -\frac{b}{2a} = -\frac{100}{2(-16)} = \frac{25}{8} \text{ seconds}$$

5. $h = 100t - 16t^2$: $a = -16, b = 100, c = 0$

From **#4**, $x = \frac{25}{8}$.

$y = 100\left(\frac{25}{8}\right) - 16\left(\frac{25}{8}\right)^2 = \frac{625}{4}$ feet

6. The roadway is at $y = 0$, so the distance to the lowest point will be the y-coord. of the vertex.

$$x^2 - 2500y + 25000 = 0$$
$$x^2 + 25000 = 2500y$$
$$\frac{1}{2500}x^2 + 10 = y$$
$$y = c - \frac{b^2}{4a} = 10 - \frac{0^2}{4\left(\frac{1}{2500}\right)} = 10$$

The lowest point is 10 ft above.

7. $y = f(x) = x^4 - x^2$

$f(-x) = (-x)^4 - (-x)^2$

$= x^4 - x^2 = f(x) \Rightarrow$ even

x-int.	y-int.
$x^4 - x^2 = 0$	$y = 0^4 - 0^2$
$x^2(x^2 - 1) = 0$	$y = 0$
$x^2 = 0, x^2 = 1$	$(0, 0)$
$x = 0, x = -1, x = 1$	
$(0, 0), (-1, 0), (1, 0)$	

8. $y = f(x) = x^5 - x^3$

$f(-x) = (-x)^5 - (-x)^3$

$= -x^5 - (-x)^3$

$= -x^5 + x^3 = -f(x) \Rightarrow$ odd

x-int.	y-int.
$x^5 - x^3 = 0$	$y = 0^5 - 0^3$
$x^3(x^2 - 1) = 0$	$y = 0$
$x^3 = 0, x^2 = 1$	$(0, 0)$
$x = 0, x = -1, x = 1$	
$(0, 0), (-1, 0), (1, 0)$	

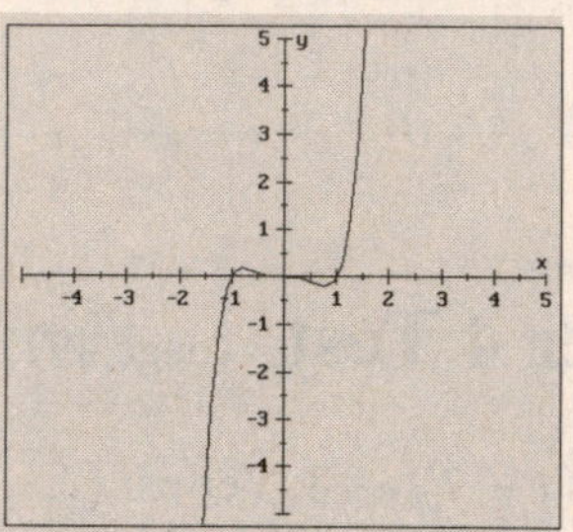

9. $P(-2) = (-2)^2 + 5(-2) + 6 = 4 - 10 + 6 = 0$; -2 is a zero of $P(x)$.

10.

$$\begin{array}{r l}
 & x^4 - 2x^3 + 4x^2 - 8x + 16 \\
x + 2 & \overline{)\,x^5 + 0x^4 + 0x^3 + 0x^2 + 0x + 2} \\
 & \underline{x^5 + 2x^4} \\
 & -2x^4 + 0x^3 \\
 & \underline{-2x^4 - 4x^3} \\
 & 4x^3 + 0x^2 \\
 & \underline{4x^3 + 8x^2} \\
 & -8x^2 + 0x \\
 & \underline{-8x^2 - 16x} \\
 & 16x + 2 \\
 & \underline{16x + 32} \\
 & \boxed{-30}
\end{array}$$

11. $P(3) = 2(3)^4 - 10(3)^3 + 4(3)^2 + 7(3) + 21$

$= -30 \Rightarrow$ remainder $= -30$

$x - 3$ is not a factor of the polynomial.

12.

$$\begin{array}{r|rrrr}
2 & 2 & -3 & -4 & -1 \\
 & & 4 & 2 & -4 \\ \hline
 & 2 & 1 & -2 & -5
\end{array}$$

$(x - 2)(2x^2 + x - 2) - 5$

13.

$$\begin{array}{r|rrr}
5 & 2 & -7 & -15 \\
 & & 10 & 15 \\ \hline
 & 2 & 3 & 0
\end{array} \quad \boxed{2x + 3}$$

14.

$$\begin{array}{r|rrrr}
-2 & 3 & 7 & 2 & 0 \\
 & & -6 & -2 & 0 \\ \hline
 & 3 & 1 & 0 & 0
\end{array} \quad \boxed{3x^2 + x}$$

15.

$$\begin{array}{r|rrrr}
-\frac{1}{3} & 3 & -2 & 0 & 4 \\
 & & -1 & 1 & -\frac{1}{3} \\ \hline
 & 3 & -3 & 1 & \frac{11}{3}
\end{array} \quad P\left(-\tfrac{1}{3}\right) = \tfrac{11}{3}$$

CHAPTER 4 TEST

16.

$$\begin{array}{r|rrrr} i & 3 & -2 & 0 & 4 \\ & & 3i & -3-2i & 2-3i \\ \hline & 3 & -2+3i & -3-2i & 6-3i \end{array}$$

$P(i) = 6 - 3i$

17. $(x-5)(x+1)(x-0) = (x^2 - 4x - 5)x$
$= x^3 - 4x^2 - 5x$

18. $(x-i)(x+i)\left(x-\sqrt{3}\right)\left(x+\sqrt{3}\right) = (x^2 - i^2)(x^2-3) = (x^2+1)(x^2-3) = x^4 - 2x^2 - 3$

19. 3 linear factors, 3 zeros

20. $3+2i$ must also be a zero.

21. $P(x) = 3x^5 - 2x^4 + 2x^2 - x - 3$: 3 sign variations $\Rightarrow$ 3 or 1 positive zeros
$P(-x) = 3(-x)^5 - 2(-x)^4 + 2(-x)^2 - (-x) - 3$
$= -3x^5 - 2x^4 + 2x^2 + x - 4$: 2 sign variations $\Rightarrow$ 2 or 0 negative zeros

# pos	# neg	# nonreal
3	2	0
3	0	2
1	2	2
1	0	4

22. $P(x) = x^5 - x^4 - 5x^3 + 5x^2 + 4x - 5$

$$\begin{array}{r|rrrrrr} 3 & 1 & -1 & -5 & 5 & 4 & -5 \\ & & 3 & 6 & 3 & 24 & 84 \\ \hline & 1 & 2 & 1 & 8 & 28 & 79 \end{array}$$

Upper bound: 3

$$\begin{array}{r|rrrrrr} -3 & 1 & -1 & -5 & 5 & 4 & -5 \\ & & -3 & 12 & -21 & 48 & -156 \\ \hline & 1 & -4 & 7 & -16 & 52 & -161 \end{array}$$

Lower bound: -3

23. num: $\pm 1, \pm 2$; den: $\pm 1, \pm 5$; possible zeros: $\pm 1, \pm 2, \pm \frac{1}{5}, \pm \frac{2}{5}$

24. Possible rational zeros

$\pm 1, \pm 2, \pm 3, \pm 6,$
$\pm \frac{1}{2}, \pm \frac{3}{2}$

Test $x = 2$:

$$\begin{array}{r|rrrr} 2 & 2 & 3 & -11 & -6 \\ & & 4 & 14 & 6 \\ \hline & 2 & 7 & 3 & 0 \end{array}$$

Descartes' Rule of Signs

# pos	# neg	# nonreal
1	2	0
1	0	2

$2x^3 + 3x^2 - 11x - 6 = 0$
$(x-2)(2x^2 + 7x + 3) = 0$
$(x-2)(2x+1)(x+3) = 0$
Solution set: $\{2, -\frac{1}{2}, -3\}$

25.

Possible rational zeros
$\pm 1, \pm 2, \pm 3, \pm 6,$
$\pm 9, \pm 18, \pm 27, \pm 54$

Descartes' Rule of Signs

# pos	# neg	# nonreal
1	3	0
1	1	2

$$x^4 + x^3 + 3x^2 + 9x - 54 = 0$$
$$(x-2)\left(x^3 + 3x^2 + 9x + 27\right) = 0$$
$$(x-2)(x+3)\left(x^2+9\right) = 0$$
$$x^2 + 9 = 0 \Rightarrow x^2 = -9 \Rightarrow x = \pm 3i$$

Solution set: $\{2, -3, \pm 3i\}$

Test $x = 2$:

$$\begin{array}{r|rrrrr} 2 & 1 & 1 & 3 & 9 & -54 \\ & & 2 & 6 & 18 & 54 \\ \hline & 1 & 3 & 9 & 27 & 0 \end{array}$$

Test $x = -3$:

$$\begin{array}{r|rrrr} -3 & 1 & 3 & 9 & 27 \\ & & -3 & 0 & -27 \\ \hline & 1 & 0 & 9 & 0 \end{array}$$

26. $P(1) = 5$, $P(2) = 28$; The Intermediate Value Theorem does not guarantee a zero between 1 and 2.

27. $y = \dfrac{x-1}{x^2-9} = \dfrac{x-1}{(x+3)(x-3)}$

Vert: $x = -3, x = 3$; Horiz: $y = 0$

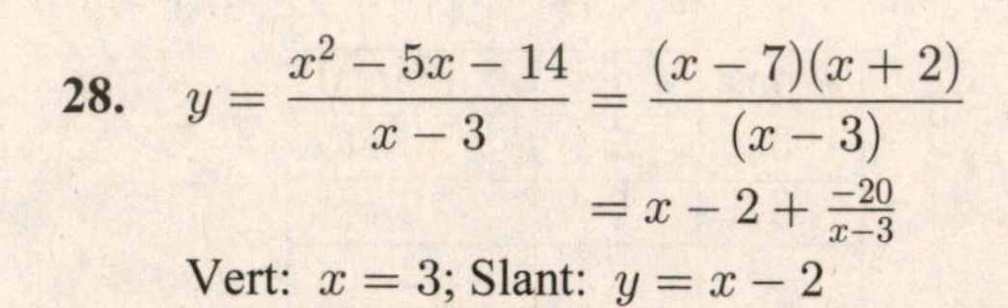

28. $y = \dfrac{x^2 - 5x - 14}{x-3} = \dfrac{(x-7)(x+2)}{(x-3)}$

$= x - 2 + \frac{-20}{x-3}$

Vert: $x = 3$; Slant: $y = x - 2$

29. $y = \dfrac{x^2}{x^2-9} = \dfrac{x^2}{(x+3)(x-3)}$

Vert: $x = -3, x = 3$; Horiz: $y = 1$

Slant: none; x-intercepts: $(0, 0)$

y-intercepts: $(0, 0)$; Symmetry: y-axis

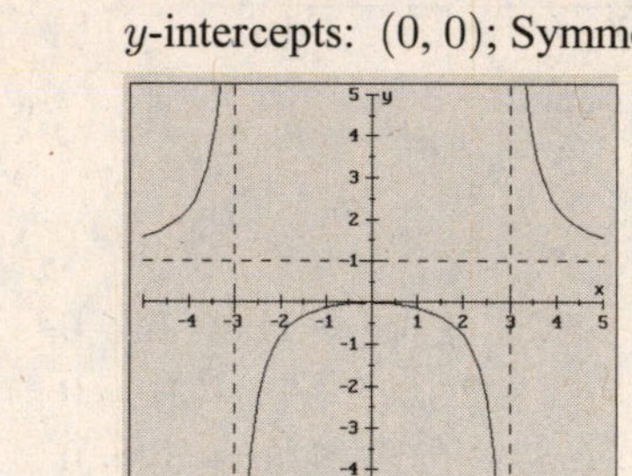

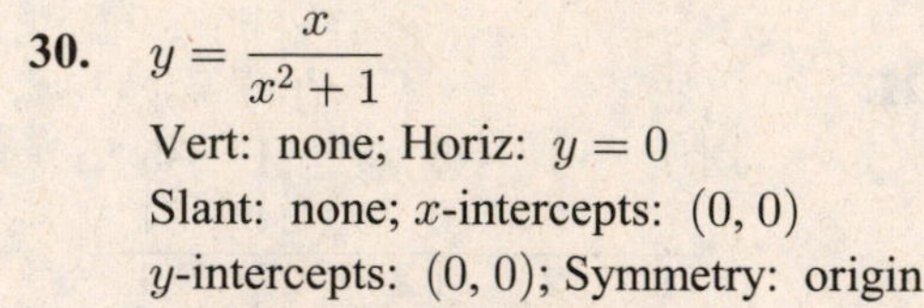

30. $y = \dfrac{x}{x^2+1}$

Vert: none; Horiz: $y = 0$

Slant: none; x-intercepts: $(0, 0)$

y-intercepts: $(0, 0)$; Symmetry: origin

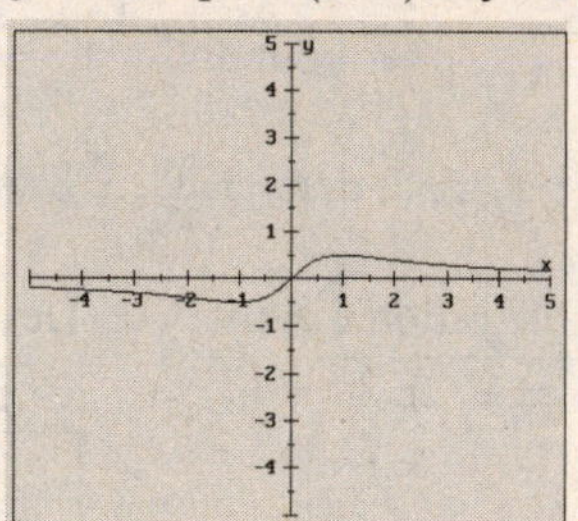

Exercises 5.1 (page 511)

1. exponential

3. $(-\infty, \infty)$

5. $(0, \infty)$

7. asymptote

9. increasing

11. 2.72

13. increasing

15. $4^{\sqrt{3}} \approx 11.0357$

17. $7^{\pi} \approx 451.8079$

19. $5^{\sqrt{2}}5^{\sqrt{2}} = 5^{\sqrt{2}+\sqrt{2}} = 5^{2\sqrt{2}} = \left(5^2\right)^{\sqrt{2}} = 25^{\sqrt{2}}$

21. $\left(a^{\sqrt{8}}\right)^{\sqrt{2}} = a^{\sqrt{8}\cdot\sqrt{2}} = a^{\sqrt{16}} = a^4$

23. $f(0) = 5^0 = 1, f(2) = 5^2 = 25$

25. $f(0) = \left(\frac{1}{3}\right)^{-0} = 1, f(2) = \left(\frac{1}{3}\right)^{-2} = 9$

EXERCISES 5.1

27. $f(x) = 3^x$

points: $(0, 1)$, $(1, 3)$

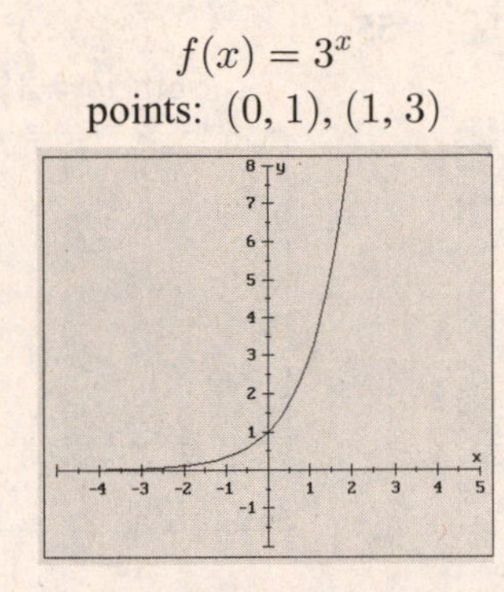

29. $f(x) = \left(\frac{1}{5}\right)^x$

points: $(0, 1)$, $\left(1, \frac{1}{5}\right)$

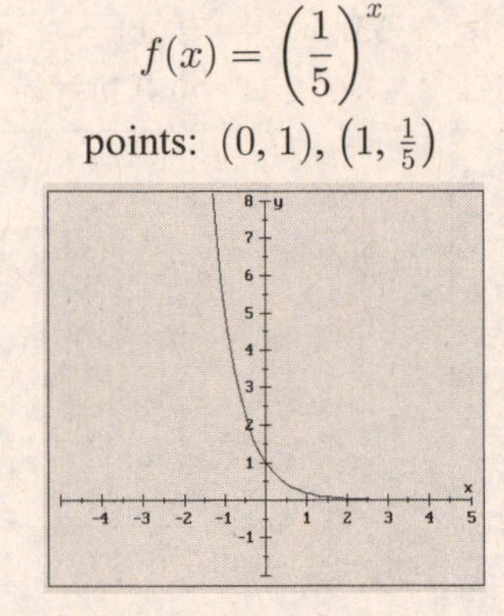

31. $f(x) = \left(\frac{3}{4}\right)^x$

points: $(0, 1)$, $\left(1, \frac{3}{4}\right)$

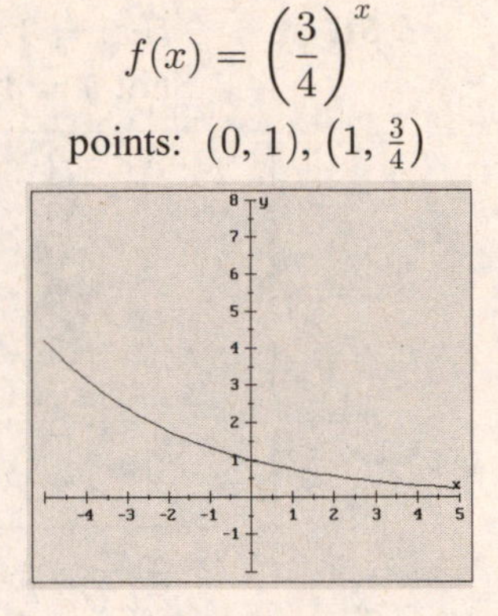

33. $f(x) = (1.5)^x$

points: $(0, 1)$, $(1, 1.5)$

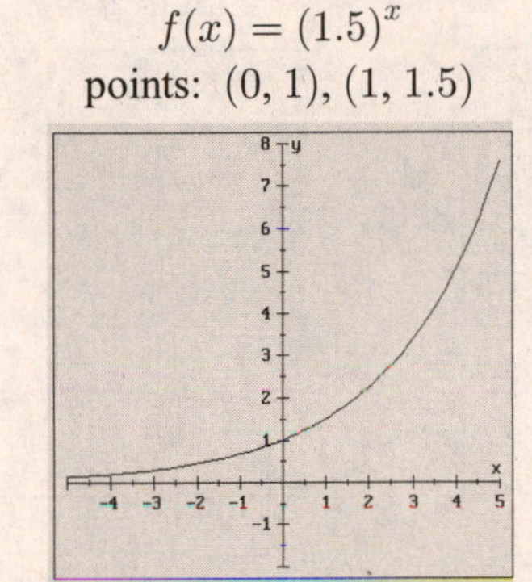

35. $f(x) = 3^{-x}$

points: $(0, 1)$, $\left(1, \frac{1}{3}\right)$

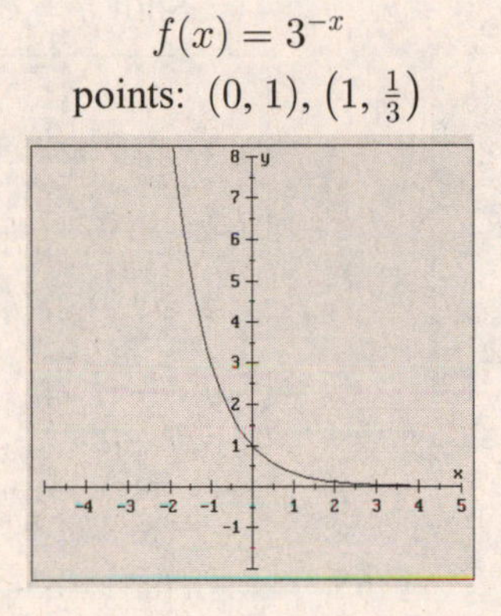

37. $f(x) = -\left(\frac{1}{5}\right)^x$

points: $(0, -1)$, $\left(1, -\frac{1}{5}\right)$

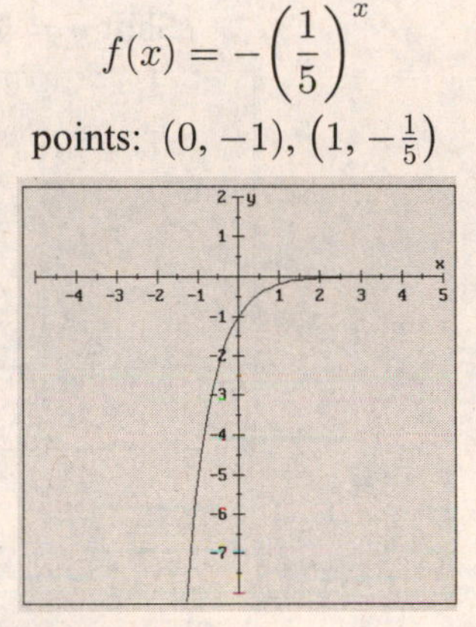

39. The graph passes through $(0, 1)$ and has the x-axis as an asymptote. YES

41. The graph does not pass through $(0, 1)$. NO

43. The graph passes through $(0, 1)$ and has the x-axis as an asymptote, so it could be an exponential function. It passes through the point $\left(1, \frac{1}{2}\right) = (1, b)$. $b = \frac{1}{2}$

45. The graph does not pass through $(0, 1)$. It is not an exponential function.

47. The graph passes through $(0, 1)$ and has the x-axis as an asymptote, so it could be an exponential function. It passes through the point $(1, 2) = (1, b)$. $b = 2$

49. The graph passes through $(0, 1)$ and has the x-axis as an asymptote, so it could be an exponential function.

$y = b^x$

$e^2 = b^2$

$e = b$

EXERCISES 5.1

51. $f(x) = 3^x - 1$
Shift $y = 3^x$ D1.

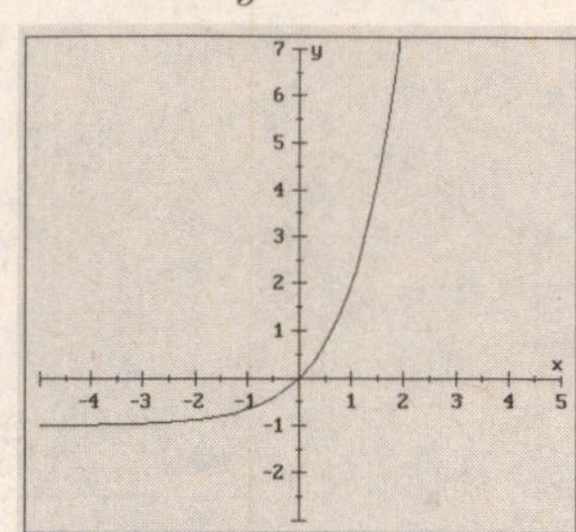

53. $f(x) = 2^x + 1$
Shift $y = 2^x$ U1.

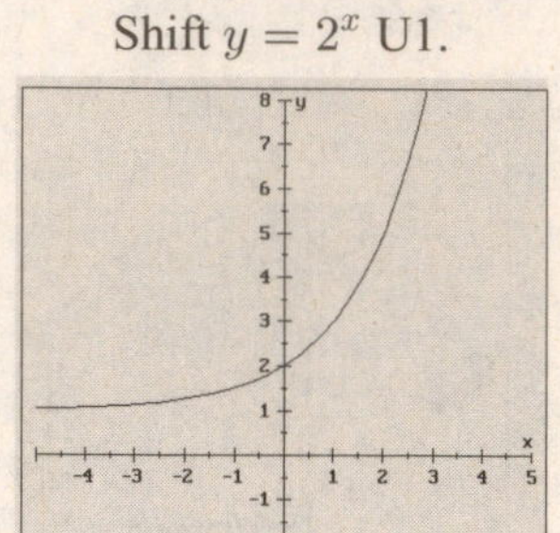

55. $f(x) = 3^{x-1}$
Shift $y = 3^x$ R1.

57. $f(x) = 3^{x+1}$
Shift $y = 3^x$ L1.

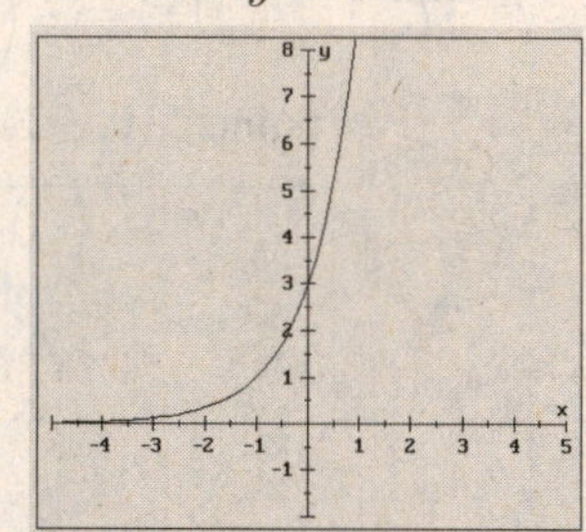

59. $f(x) = e^x - 4$
Shift $y = e^x$ D4.

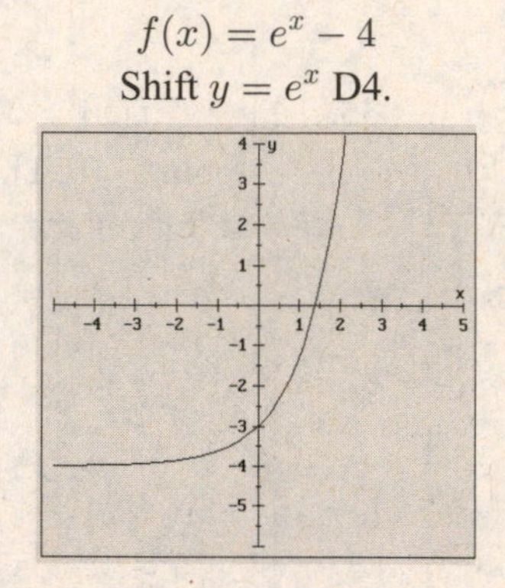

61. $f(x) = e^{x-2}$
Shift $y = e^x$ R2.

63. $f(x) = 2^{x+1} - 2$
Shift $y = 2^x$ L1, D2.

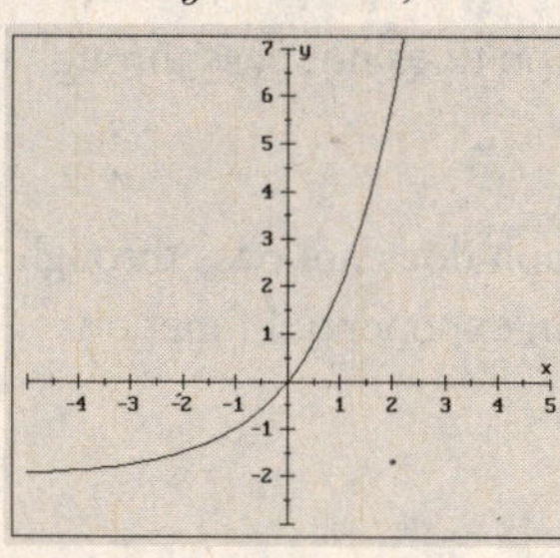

65. $f(x) = 3^{x-2} + 1$
Shift $y = 3^x$ R2, U1.

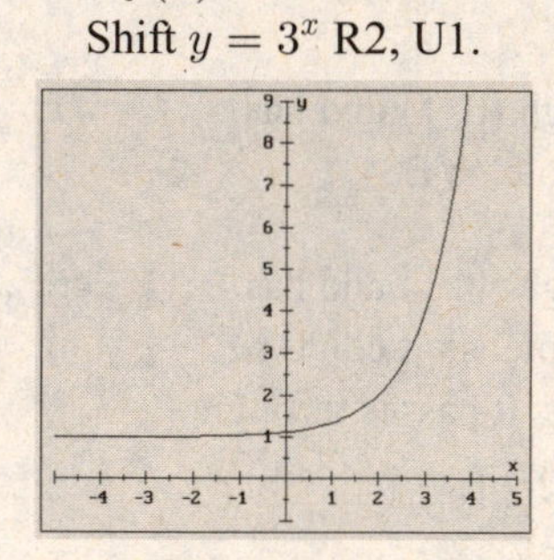

67. $f(x) = -3^x + 1$
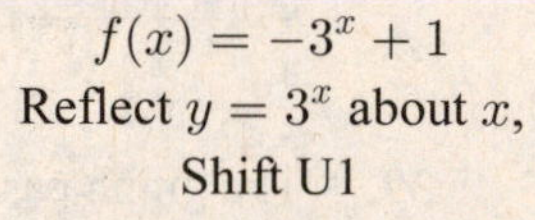
Reflect $y = 3^x$ about x,
Shift U1

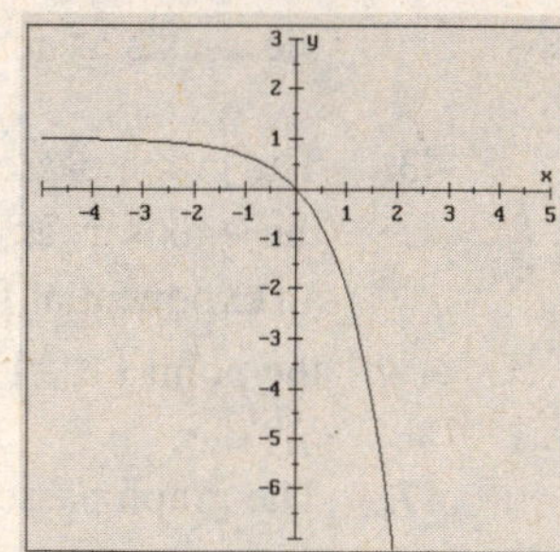

EXERCISES 5.1

69. $f(x) = 2^{-x} - 3$
Reflect $y = 2^x$ about y,
Shift D3

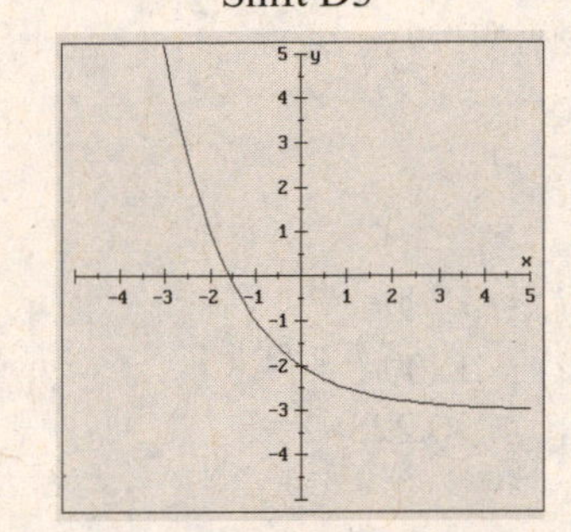

71. $f(x) = -e^x + 2$
Reflect $y = e^x$ about x,
Shift U2

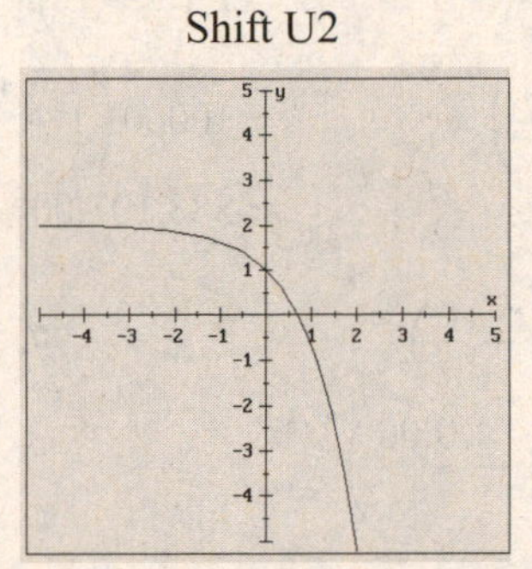

73. $y = 5(2^x)$

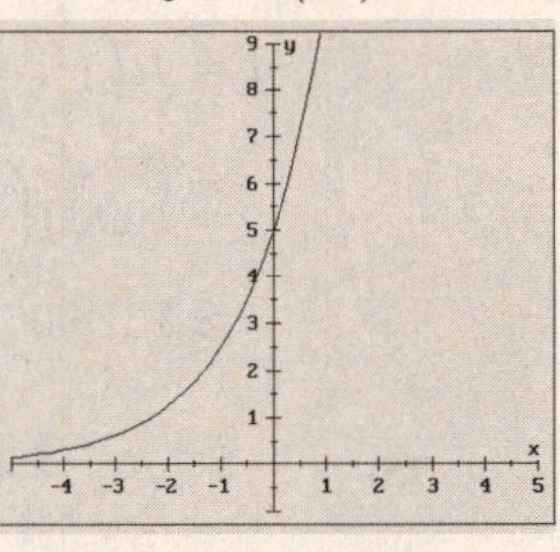

75. $y = 3^{-x}$

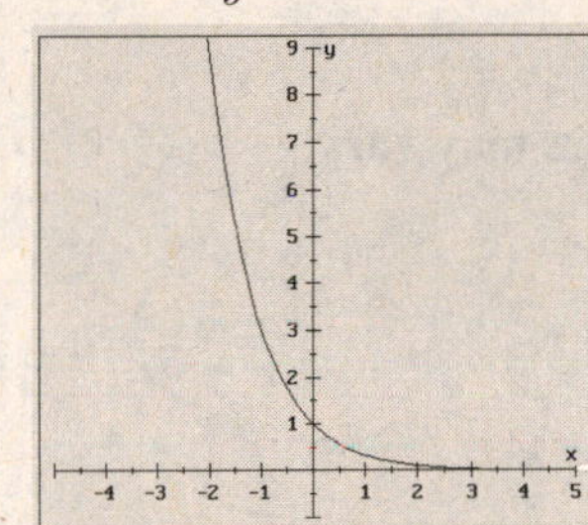

77. $y = 2e^x$

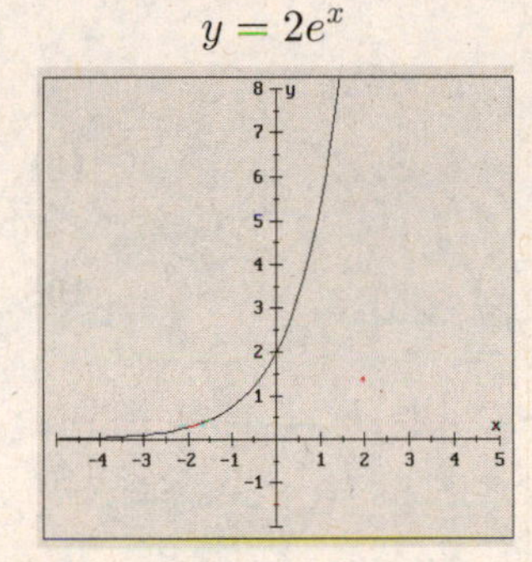

79. $y = 5e^{-0.5x}$

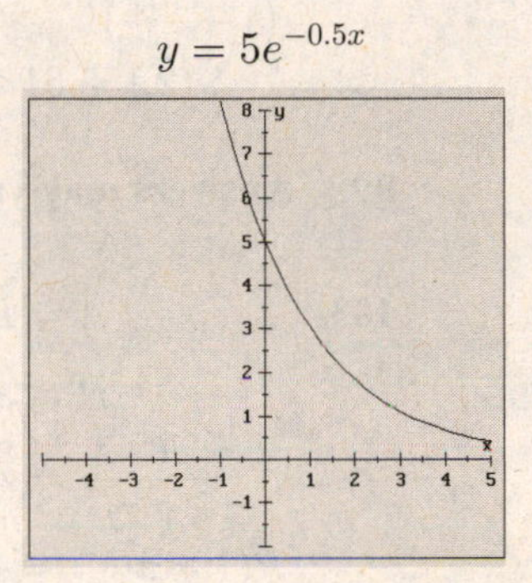

81. $A = P\left(1 + \dfrac{r}{n}\right)^{nt} = 10000\left(1 + \dfrac{0.08}{4}\right)^{4(10)} \approx \$22{,}080.40$

83. 5% interest:

$$A = P\left(1 + \frac{r}{n}\right)^{nt} = 500\left(1 + \frac{0.05}{2}\right)^{2(5)} \approx \$640.04$$

$5\frac{1}{2}$% interest:

$$A = P\left(1 + \frac{r}{n}\right)^{nt} = 500\left(1 + \frac{0.055}{2}\right)^{2(5)} \approx \$655.83$$

Difference $= 655.83 - 640.04$
$= \$15.79$ more

85. $$A = P\left(1 + \frac{r}{n}\right)^{nt} = 1\left(1 + \frac{0.05}{1}\right)^{1(300)} \approx \$2{,}273{,}996.13$$

87. $$A = P\left(1 + \frac{r}{n}\right)^{nt} = 1100\left(1 + \frac{0.0175}{1}\right)^{1(8)} \approx \$1263.77$$

89. $$A = Pe^{rt} = 5000e^{0.082(12)} \approx \$13{,}375.68$$

91. Continuous:

$$A = Pe^{rt} = 5000e^{0.085(5)} \approx \$7647.95$$

Annually:

$$A = P\left(1 + \frac{r}{n}\right)^{nt} = 5000\left(1 + \frac{0.085}{1}\right)^{1(5)} \approx \$7518.28$$

93. Quarterly:

$$\begin{aligned} A &= P\left(1+\frac{r}{n}\right)^{nt} \\ &= 10000\left(1+\frac{0.06}{4}\right)^{4(20)} \\ &\approx \$32{,}906.63 \end{aligned}$$

Daily:

$$\begin{aligned} A &= P\left(1+\frac{r}{n}\right)^{nt} \\ &= 10000\left(1+\frac{0.06}{365}\right)^{365(20)} \\ &\approx \$33{,}197.90 \end{aligned}$$

Difference

$$33{,}197.90 - 32{,}906.63 = \boxed{\$291.27}$$

95.

$$\begin{aligned} A &= P\left(1+\frac{r}{n}\right)^{nt} \\ 40{,}000 &= P\left(1+\frac{0.06}{4}\right)^{4(20)} \\ \frac{40{,}000}{\left(1+\frac{0.06}{4}\right)^{4(20)}} &= P \\ \$12{,}155.61 &\approx P \end{aligned}$$

97.

$$\begin{aligned} P(t) &= 1200e^{0.2t} \\ P(12) &= 1200e^{0.2(12)} \\ &\approx 13{,}228 \end{aligned}$$

99. Answers may vary.

101. Answers may vary.

103.

$$\begin{aligned} A &= P\left(1+\frac{r}{n}\right)^{nt} \\ P &= A(1+r)^n \\ \frac{P}{(1+r)^n} &= A \\ P(1+r)^{-n} &= A \end{aligned}$$

105.

$$\begin{aligned} 5^{3t} &= k^t \\ \left(5^3\right)^t &= k^t \\ 125^t &= k^t \\ 125 &= k \end{aligned}$$

107. True.

109. False. The y-intercept is $(0,-1)$.

111. True.

113. False. They intersect at $(0,1)$.

Exercises 5.2 (page 521)

1. birth, death

3.

$$\begin{aligned} A &= A_0 2^{-t/h} \\ &= 50 \cdot 2^{-100/(12.4)} \\ &\approx 0.1868 \text{ grams} \end{aligned}$$

5.

$$\begin{aligned} A &= A_0 2^{-t/h} \\ &= 1000 \cdot 2^{-200/(30.17)} \\ &\approx 10 \text{ kg} \end{aligned}$$

7. $A = A_0 2^{-t/h} = A_0 \cdot 2^{-60/40} \approx A_0(0.354) \Rightarrow$ About 35.4% will remain.

9.

$$\begin{aligned} A &= A_0 2^{-t/h} \\ &= 1 \cdot 2^{-12/(4.5)} \\ &\approx 0.1575 \text{ unit} \end{aligned}$$

11.

$$\begin{aligned} I &= I_0 k^x \\ &= 8(0.5)^2 \\ &= 2 \text{ lumens} \end{aligned}$$

13.

$$\begin{aligned} I &= I_0 k^x \\ 1 &= I_0 (0.5)^3 \\ \frac{1}{(0.5)^3} &= I_0 \\ 8 \text{ lumens} &= I_0 \end{aligned}$$

EXERCISES 5.2

15. $P = P_0 2^{t/2}$
$= 10{,}000 \cdot 2^{5/2}$
$\approx 56{,}570$ fish

17. $T = 40 + 60(0.75)^t$
$= 40 + 60(0.75)^{3.5}$
$\approx 61.9°$ C

19. $P = 173e^{0.03t}$
$= 173e^{0.03(20)}$
≈ 315

21. $P = P_0 e^{0.27t}$
$= 2e^{0.27(7)}$
≈ 13 cases

23. $P = P_0 e^{kt}$
$= 6e^{0.019(30)}$
≈ 10.6 billion

25. $P = P_0 e^{kt}$
$= 6e^{0.019(50)}$
≈ 15.5 billion
$\frac{15.5}{6} \approx$ a factor of 2.6

27. $P = e^{-0.3t}$
$= e^{-0.3(24)}$
≈ 0.0007
$= 0.07\%$

29. Let $t = 0$:
$x = 0.08\left(1 - e^{-0.1t}\right)$
$= 0.08\left(1 - e^{-0.1(0)}\right)$
$= 0.08\left(1 - e^{0}\right)$
$= 0.08(1 - 1) = 0$

31. $N = P\left(1 - e^{-0.1t}\right)$
$= 50{,}000\left(1 - e^{-0.1(10)}\right)$
$\approx 31{,}606$
$50{,}000 - 31{,}606 = 18{,}394$

33. $P = \dfrac{1{,}200{,}000}{1 + (1200 - 1)e^{-0.4t}}$
$= \dfrac{1{,}200{,}000}{1 + (1200 - 1)e^{-0.4(8)}}$
$\approx 24{,}060$ people

35. $P = \dfrac{55{,}000}{1 + (550 - 1)e^{-0.8t}}$
$= \dfrac{55{,}000}{1 + (550 - 1)e^{-0.8(2)}}$
≈ 492 people

37. $w = 1.54e^{0.503n}$
$= 1.54e^{0.503(5)}$
≈ 19.0 mm

39. $v = 50\left(1 - e^{-0.2t}\right)$
$= 50\left(1 - e^{-0.2(20)}\right)$
≈ 49 meters/second

41.

Males	Females
$P = P_0 e^{kt}$	$P = P_0 e^{kt}$
$= 133e^{0.01t}$	$= 139e^{0.01t}$
$= 133e^{0.01(20)}$	$= 139e^{0.01(20)}$
≈ 162.4 million	≈ 169.8 million

There will be about 7 million more females.

43. Find where these graphs meet:
$y = 1000e^{0.02t}$, $y = 31x + 2000$

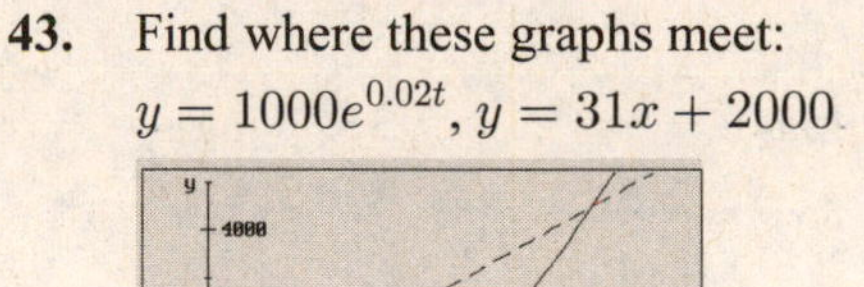

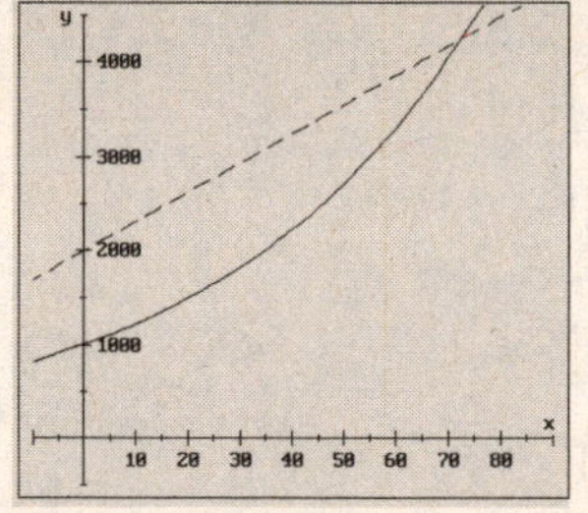

It will take about 72.2 years.

45. $y = 1.035264924^x$

47. $y = \dfrac{e^x + e^{-x}}{2}$

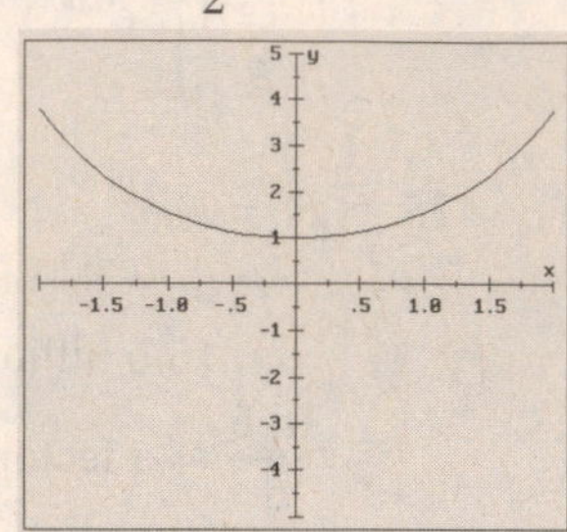

49. $P = \dfrac{1{,}200{,}000}{1 + 1199e^{-0.4t}}$

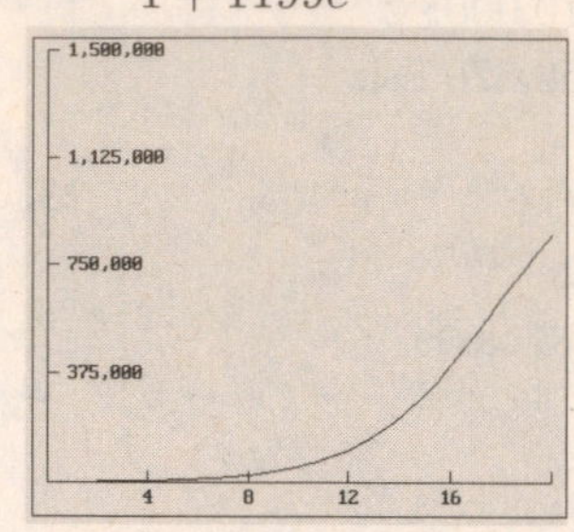

51. $1 + 1 + \frac{1}{2} + \frac{1}{2\cdot 3} + \frac{1}{2\cdot 3\cdot 4} + \frac{1}{2\cdot 3\cdot 4\cdot 5} \approx 2.71\overline{6}$; $e \approx 2.718$: accurate to 2 places

Exercises 5.3 (page 534)

1. $x = b^y$

3. range

5. inverse

7. exponent

9. $(b, 1)$, $(1, 0)$

11. $\log_e x$

13. $(-\infty, \infty)$

15. 10

17. $8^2 = 64$
$\log_8 64 = 2$

19. $4^{-2} = \dfrac{1}{16}$
$\log_4 \dfrac{1}{16} = -2$

21. $\left(\dfrac{1}{2}\right)^{-5} = 32$
$\log_{1/2} 32 = -5$

23. $x^y = z$
$\log_x z = y$

25. $\log_3 81 = 4$
$3^4 = 81$

27. $\log_{1/2} \dfrac{1}{8} = 3$
$\left(\dfrac{1}{2}\right)^3 = \dfrac{1}{8}$

29. $\log_4 \dfrac{1}{64} = -3$
$4^{-3} = \dfrac{1}{64}$

31. $\log_\pi \pi = 1$
$\pi^1 = \pi$

33. $\log_2 8 = x$
$2^x = 8$
$x = 3$

35. $\log_4 \dfrac{1}{64} = x$
$4^x = \dfrac{1}{64}$
$x = -3$

37. $\log_{1/2} \dfrac{1}{8} = x$
$\left(\dfrac{1}{2}\right)^x = \dfrac{1}{8}$
$x = 3$

39. $\log_9 3 = x$
$9^x = 3$
$x = \dfrac{1}{2}$

41. $\log_{1/2} 8 = x$
$\left(\dfrac{1}{2}\right)^x = 8$
$x = -3$

43. $\log_8 x = 2$
$8^2 = x$
$64 = x$

45. $\log_7 x = 1$
$7^1 = x$
$7 = x$

47. $\log_{25} x = \dfrac{1}{2}$
$25^{1/2} = x$
$5 = x$

EXERCISES 5.3

49. $\log_5 x = -2$
$$5^{-2} = x$$
$$\frac{1}{5^2} = x$$
$$\frac{1}{25} = x$$

51. $\log_{36} x = -\frac{1}{2}$
$$36^{-1/2} = x$$
$$\frac{1}{36^{1/2}} = x$$
$$\frac{1}{6} = x$$

53. $\log_x 5^3 = 3$
$$x^3 = 5^3$$
$$x = 5$$

55. $\log_x \frac{9}{4} = 2$
$$x^2 = \frac{9}{4}$$
$$x = \frac{3}{2}$$

57. $\log_x \frac{1}{64} = -3$
$$x^{-3} = \frac{1}{64}$$
$$\frac{1}{x^3} = \frac{1}{4^3}$$
$$x = 4$$

59. $\log_x \frac{9}{4} = -2$
$$x^{-2} = \frac{9}{4}$$
$$(x^{-2})^{-1} = \left(\frac{9}{4}\right)^{-1}$$
$$x^2 = \frac{4}{9}$$
$$x = \frac{2}{3}$$

61. From the definition: $2^{\log_2 5} = 5$

63. From the definition: $x^{\log_4 6} = 6$
$\Rightarrow x = 4$

65. $\log 3.25 \approx 0.5119$

67. $\log 0.00467 \approx -2.3307$

69. $\ln 45.7 \approx 3.8221$

71. $\ln \frac{2}{3} \approx -0.4055$

73. $\ln 35.15 \approx 3.5596$

75. $\ln 7.896 \approx 2.0664$

77. $\log(\ln 1.7) \approx -0.2752$

79. $\ln(\log 0.1)$: undefined

81. $\log y = 1.4023$
$$y \approx 25.2522$$

83. $\log y = -3.71$
$$y \approx 1.9498 \times 10^{-4}$$

85. $\ln y = 1.4023$
$$y \approx 4.0645$$

87. $\ln y = 4.24$
$$y \approx 69.4079$$

89. $\ln y = -3.71$
$$y \approx 0.0245$$

91. $\log y = \ln 8$
$$y \approx 120.0719$$

93. $\log 10{,}000 = x$
$$10^x = 10{,}000$$
$$x = 4$$
$$\log 10{,}000 = 4$$

95. $\log 0.001 = x$
$$10^x = 0.001$$
$$x = -3$$
$$\log 0.001 = -3$$

97. From the definition: $e^{\ln 7} = 7$

99. From the definition: $\ln\left(e^4\right) = 4$

101. The graph passes through the point $(b, 1) = (2, 1)$.
$\Rightarrow b = 2$

103.
$$y = \log_b x$$
$$b^y = x$$
$$b^{-1} = \tfrac{1}{2}$$
$$\left(b^{-1}\right)^{-1} = \left(\tfrac{1}{2}\right)^{-1}$$
$$b = 2$$

EXERCISES 5.3

105. $f(x) = \log_3 x$
points: $(1, 0)$, $(3, 1)$

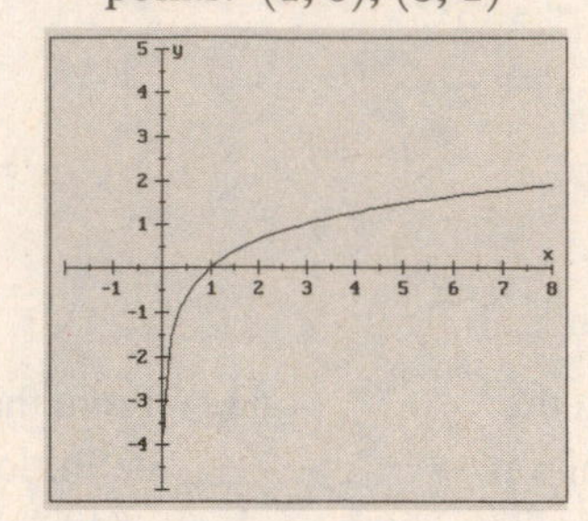

107. $f(x) = \log_{1/3} x$
points: $(1, 0)$, $\left(\frac{1}{3}, 1\right)$

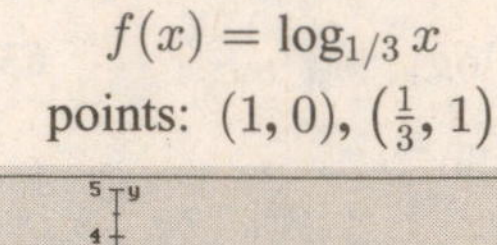

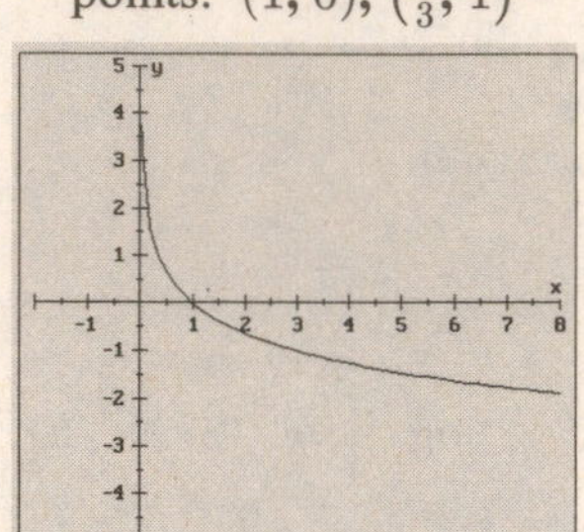

109. $f(x) = -\log_5 x$
Reflect $y = \log_5 x$ about x

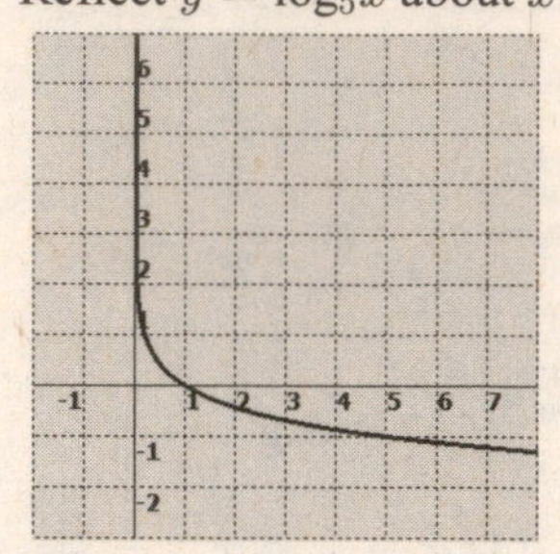

111. $f(x) = 2 + \log_2 x$
Shift $y = \log_2 x$ U2.

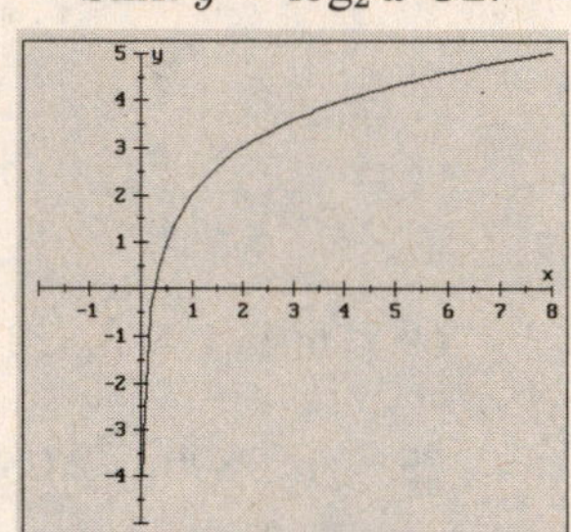

113. $f(x) = \log_3 (x + 2)$
Shift $y = \log_3 x$ L2.

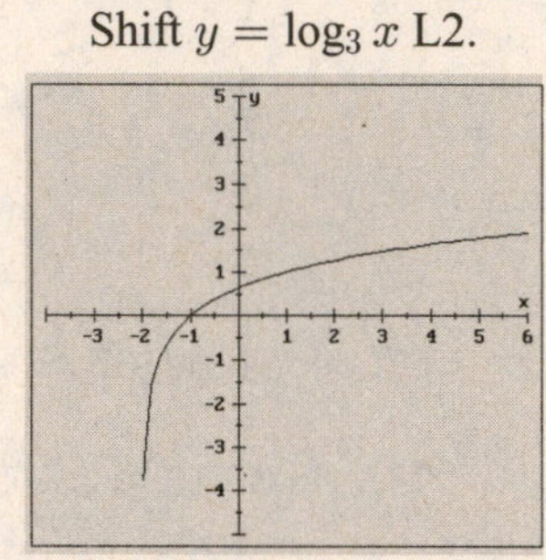

115. $f(x) = 3 + \log_3(x + 1)$
Shift $y = \log_3 x$ U3, L1.

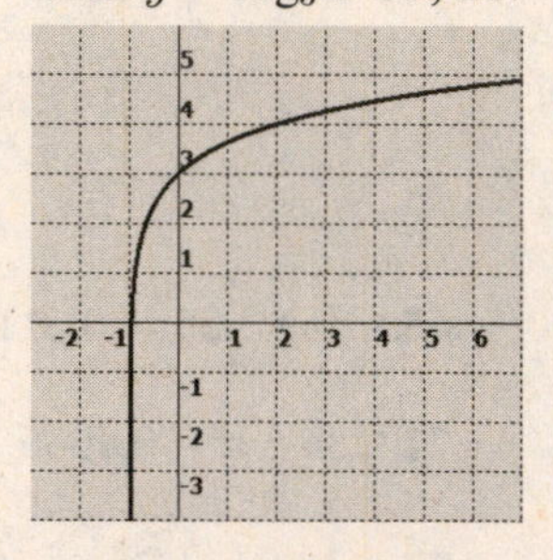

117. $f(x) = -3 + \ln x$
Shift $y = \ln x$ D3.

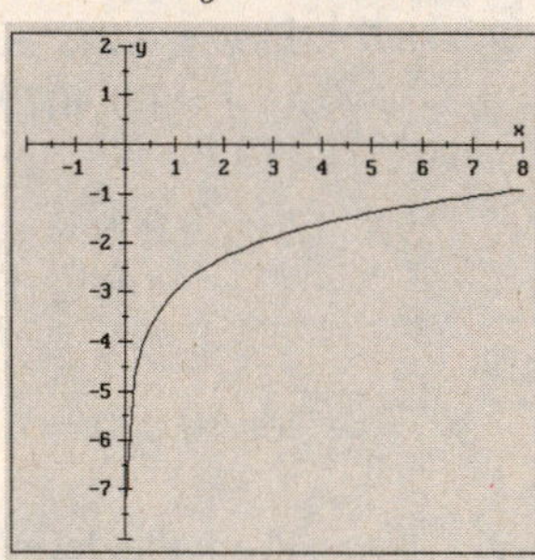

119. $f(x) = \ln (x - 4)$
Shift $y = \ln x$ R4.

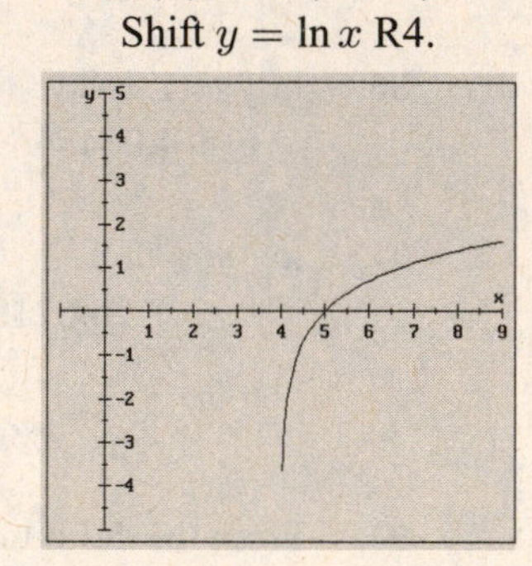

121. $f(x) = 1 - \ln x$
Reflect $y = \ln x$
about x, shift U1.

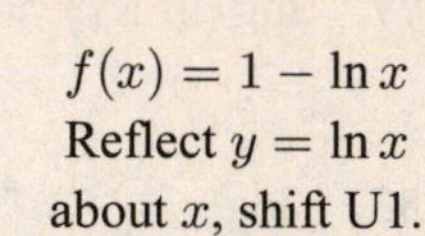

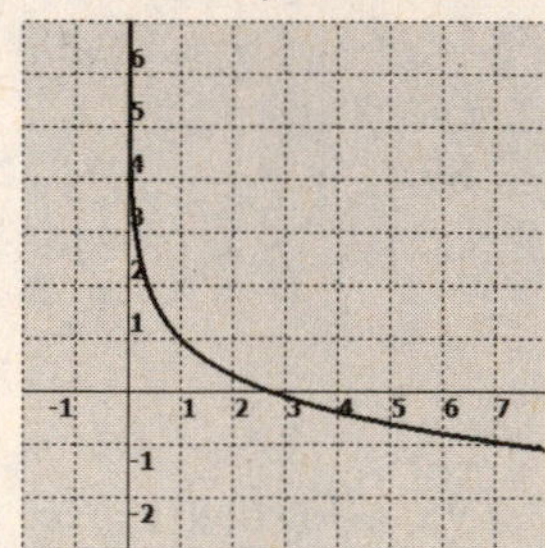

EXERCISES 5.3

123. $f(x) = \log(3x)$

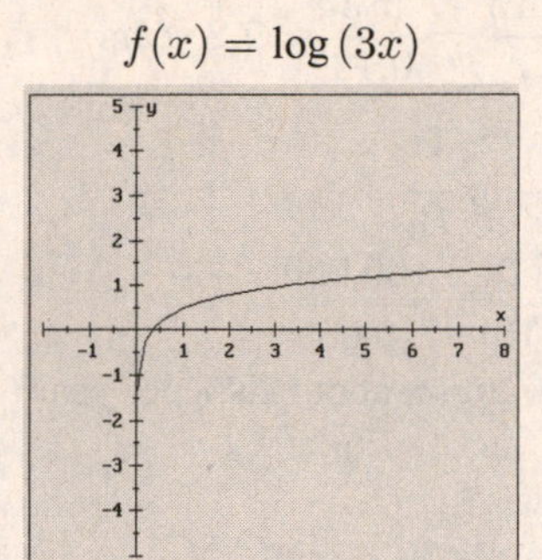

125. $f(x) = \log(-x)$

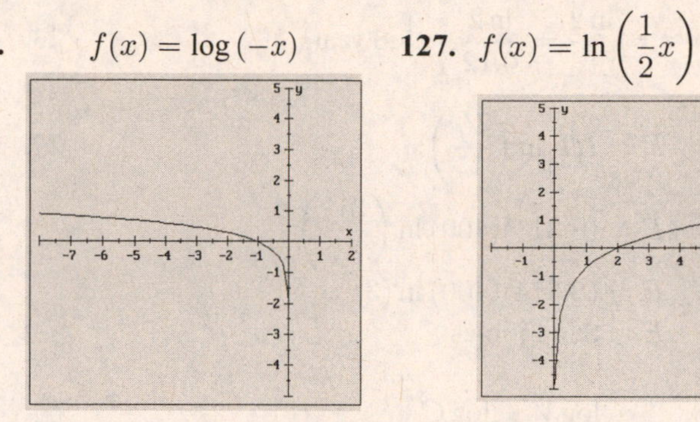

127. $f(x) = \ln\left(\frac{1}{2}x\right)$

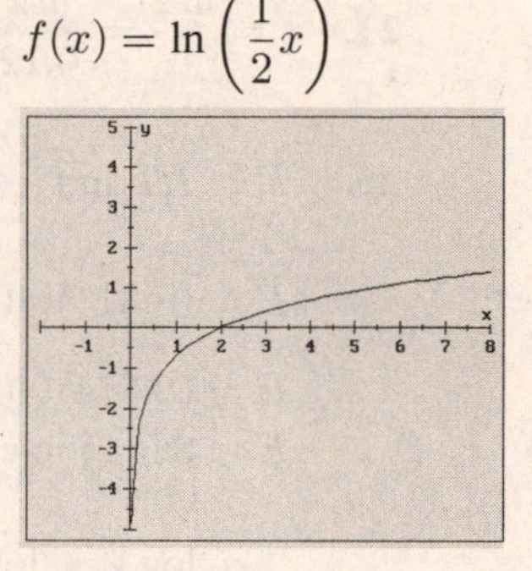

129. $f(x) = \ln(-x)$

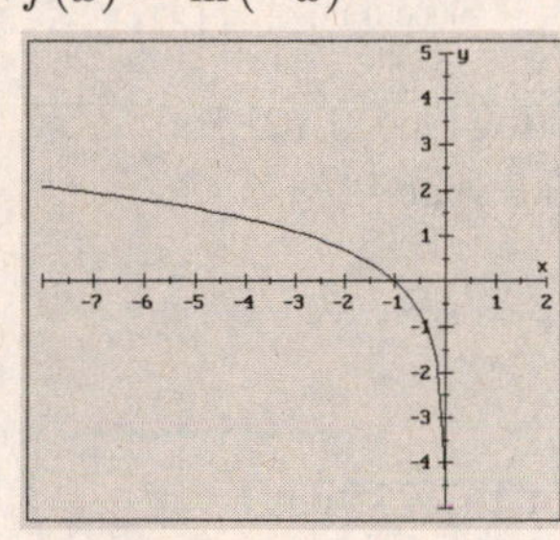

131-139. **Answers may vary.**

141. True.

143. True.

145. False. $\log_2 \frac{1}{1024} = -10$.

147. True.

149. True.

Exercises 5.4 (page 541)

1. $20 \log \frac{E_O}{E_I}$

3. $t = -\frac{1}{k} \ln\left(1 - \frac{C}{M}\right)$

5. $E = RT \ln\left(\frac{V_f}{V_i}\right)$

7. $\text{dB gain} = 20 \log \frac{E_0}{E_I} = 20 \log \frac{17}{0.03} \approx 55 \text{ dB}$

9. $\text{dB gain} = 20 \log \frac{E_0}{E_I} = 20 \log \frac{20}{0.71} \approx 29 \text{ dB}$

11. $\text{dB gain} = 20 \log \frac{E_0}{E_I} = 20 \log \frac{30}{0.1} \approx 49.5 \text{ dB}$

13. $R = \log \frac{A}{P} = \log \frac{5000}{0.2} \approx 4.4$

15. $R = \log \frac{A}{P} = \log \frac{2500}{\frac{1}{4}} = 4$

17. $R = \log \frac{A}{P} = \log \frac{6000}{0.3} \approx 4.3 \Rightarrow$ no damage

19. $t = -\frac{1}{k} \ln\left(1 - \frac{C}{M}\right) = -\frac{1}{0.116} \ln\left(1 - \frac{0.9M}{M}\right) = -\frac{1}{0.116} \ln(1 - 0.9) \approx 19.8$ minutes

21. $t = \frac{\ln 2}{r} = \frac{\ln 2}{0.12} \approx 5.8$ years

23. $t = \frac{\ln 3}{r} = \frac{\ln 3}{0.12} \approx 9.2$ years

25.
$$E = RT\ln\left(\frac{V_f}{V_i}\right)$$
$$E = (8.314)(400)\ln\left(\frac{3V_i}{V_i}\right)$$
$$E = (8.314)(400)\ln(3)$$
$$E \approx 3654 \text{ joules}$$

27.
$$r = \frac{1}{t}\ln\frac{P}{P_0} = \frac{1}{10}\ln\frac{100{,}000}{10{,}000} \approx 0.23 \Rightarrow \text{about 23\% per year}$$

29.
$$n = \frac{\log V - \log C}{\log\left(1 - \frac{2}{N}\right)} = \frac{\log 8000 - \log 37{,}000}{\log\left(1 - \frac{2}{5}\right)} \approx 3 \text{ years old}$$

31.
$$n = \frac{\log\left[\frac{Ar}{P} + 1\right]}{\log(1+r)} = \frac{\log\left[\frac{20000(0.12)}{1000} + 1\right]}{\log(1 + 0.12)} \approx 10.8 \text{ years}$$

33. $V = ER_1 \ln\frac{R_2}{R_1} = (400{,}000)(0.25)\ln\left(\frac{2}{0.25}\right) \approx 208{,}000$ V

35. $f(x) = 12 \Rightarrow \ln x = 12 \Rightarrow x \approx 162755 \Rightarrow$ You would go out 162,755 cm.
$$162{,}755 \text{ cm} = \frac{162{,}755 \text{ cm}}{1} \cdot \frac{1 \text{ in}}{2.54 \text{ cm}} \cdot \frac{1 \text{ ft}}{12 \text{ in}} \cdot \frac{1 \text{ mi}}{5280 \text{ ft}} \approx 1 \text{ mile}$$

37. **Answers may vary.**

39. Set $x = 0$:
$$y = \frac{1}{1 + e^{-2(0)}} = \frac{1}{1 + e^0} = \frac{1}{1+1} = \frac{1}{2};\ y\text{-intercept: } \left(0, \frac{1}{2}\right)$$

Exercises 5.5 (page 554)

1. 0

3. M, N

5. x, y

7. x

9. $\neq$

11. $\log_4 1 = 0$

13. $\log_4 4^7 = 7\log_4 4 = 7$

15. $5^{\log_5 10} = 10$

17. $\log_5 5 = 1$

19-23. **Answers may vary.**

25. $\log_b 2xy = \log_b 2 + \log_b x + \log_b y$

27.
$$\log_b \frac{2x}{y} = \log_b 2x - \log_b y = \log_b 2 + \log_b x - \log_b y$$

EXERCISES 5.5

29. $\log_b x^2y^3 = \log_b x^2 + \log_b y^3$
$= 2\log_b x + 3\log_b y$

31. $\log_b (xy)^{1/3} = \frac{1}{3}\log_b xy$
$= \frac{1}{3}(\log_b x + \log_b y)$

33. $\log_b x\sqrt{z} = \log_b xz^{1/2} = \log_b x + \log_b z^{1/2} = \log_b x + \frac{1}{2}\log_b z$

35. $\log_b \frac{\sqrt[3]{x}}{\sqrt[3]{yz}} = \log_b \sqrt[3]{x} - \log_b \sqrt[3]{yz} = \log_b x^{1/3} - \log_b (yz)^{1/3} = \frac{1}{3}\log_b x - \frac{1}{3}\log_b yz$
$= \frac{1}{3}\log_b x - \frac{1}{3}(\log_b y + \log_b z)$
$= \frac{1}{3}\log_b x - \frac{1}{3}\log_b y - \frac{1}{3}\log_b z$

37. $\ln x^7y^8 = \ln x^7 + \ln y^8$
$= 7\ln x + 8\ln y$

39. $\ln \frac{x}{y^4z} = \ln x - \ln y^4z$
$= \ln x - (\ln y^4 + \ln z)$
$= \ln x - (4\ln y + \ln z)$
$= \ln x - 4\ln y - \ln z$

41. $\log_b (x+1) - \log_b x = \log_b \frac{x+1}{x}$

43. $2\log_b x + \frac{1}{3}\log_b y = \log_b x^2 + \log_b y^{1/3} = \log_b x^2y^{1/3} = \log_b x^2\sqrt[3]{y}$

45. $-3\log_b x - 2\log_b y + \frac{1}{2}\log_b z = \log_b x^{-3} + \log_b y^{-2} + \log_b z^{1/2} = \log_b x^{-3}y^{-2}\sqrt{z} = \log_b \frac{\sqrt{z}}{x^3y^2}$

47. $\log_b\left(\frac{x}{z} + x\right) - \log_b\left(\frac{y}{z} + y\right) = \log_b \frac{\frac{x}{z} + x}{\frac{y}{z} + y} = \log_b \frac{z\left(\frac{x}{z} + x\right)}{z\left(\frac{y}{z} + y\right)} = \log_b \frac{x + xz}{y + yz} = \log_b \frac{x}{y}$

49. $\ln x + \ln (x+5) - \ln 9 = \ln x(x+5) - \ln 9 = \ln \frac{x(x+5)}{9}$

51. $-6\ln x - 2\ln y + \ln z = \ln x^{-6} + \ln y^{-2} + \ln z = \ln x^{-6}y^{-2}z = \ln \frac{z}{x^6y^2}$

53. $\log_b ab = \log_b a + \log_b b = \log_b a + 1$
TRUE

55. $\log_b 0$ is undefined.
FALSE

57. $\log_b (x \cdot y) = \log_b x + \log_b y$, so
$\log_b (x+y) \neq \log_b x + \log_b y$
TRUE (unless $x \cdot y = x + y$)

59. If $\log_a b = c$, then $\log_b a = c$
FALSE

61. $\log_7 7^7 = 7 \Rightarrow 7^7 = 7^7$; TRUE

63. $-\log_b x = \log_b x^{-1} = \log_b \frac{1}{x}$; FALSE

EXERCISES 5.5

65. $\log_b\left(\frac{A}{B}\right) = \log_b A - \log_b B$, so

$$\frac{\log_b(A)}{\log_b(B)} \neq \log_b A - \log_b B$$

FALSE

67. $\log_b \frac{1}{5} = \log_b 5^{-1} = -\log_b 5$

TRUE

69. $\frac{1}{3}\log_b a^3 = \frac{1}{3} \cdot 3\log_b a = \log_b a$

TRUE

71. Let $\log_{1/b} y = c$.

Then $\left(\frac{1}{b}\right)^c = y$.

$$\left(\left(\tfrac{b}{1}\right)^{-1}\right)^c = y$$

$$(b)^{-c} = y \Rightarrow \log_b y = -c.$$

$$\log_{1/b} y + \log_b y = c + (-c) = 0.$$

TRUE

73. $\ln(xy) = \ln x + \ln y$, so

$\ln(xy) \neq (\ln x)(\ln y)$

FALSE

75. $\frac{1}{5}\ln a^5 = \frac{1}{5} \cdot 5\ln a = \ln a$

TRUE

77.
$$\begin{aligned}\log_{10} 28 &= \log_{10}(4 \cdot 7)\\ &= \log_{10} 4 + \log_{10} 7\\ &= 0.6021 + 0.8451 = 1.4472\end{aligned}$$

79.
$$\begin{aligned}\log_{10} 2.25 &= \log_{10}\left(\frac{9}{4}\right)\\ &= \log_{10} 9 - \log_{10} 4\\ &= 0.9542 - 0.6021 = 0.3521\end{aligned}$$

81.
$$\begin{aligned}\log_{10}\left(\frac{63}{4}\right) &= \log_{10} 63 - \log_{10} 4\\ &= \log_{10}(7 \cdot 9) - \log_{10} 4\\ &= \log_{10} 7 + \log_{10} 9 - \log_{10} 4\\ &= 0.8451 + 0.9542 - 0.6021\\ &= 1.1972\end{aligned}$$

83.
$$\begin{aligned}\log_{10} 252 &= \log_{10}(4 \cdot 63)\\ &= \log_{10}(4 \cdot 7 \cdot 9)\\ &= \log_{10} 4 + \log_{10} 7 + \log_{10} 9\\ &= 0.6021 + 0.8451 + 0.9542\\ &= 2.4014\end{aligned}$$

85.
$$\begin{aligned}\log_{10} 112 = \log_{10}(4 \cdot 28) = \log_{10}(4 \cdot 7 \cdot 4) &= \log_{10} 4 + \log_{10} 7 + \log_{10} 4\\ &= 0.6021 + 0.8451 + 0.6021 = 2.0493\end{aligned}$$

87.
$$\begin{aligned}\log_{10}\left(\frac{144}{49}\right) = \log_{10} 144 - \log_{10} 49 &= \log_{10}(4 \cdot 4 \cdot 9) - \log_{10}(7 \cdot 7)\\ &= \log_{10} 4 + \log_{10} 4 + \log_{10} 9 - \log_{10} 7 - \log_{10} 7\\ &= 0.6021 + 0.6021 + 0.9542 - 0.8451 - 0.8451 = 0.4682\end{aligned}$$

89. $\log_3 7 = \frac{\log_{10} 7}{\log_{10} 3} \approx 1.7712$

91. $\log_\pi 3 = \frac{\log_{10} 3}{\log_{10} \pi} \approx 0.9597$

93. $\log_3 8 = \frac{\log_e 8}{\log_e 3} \approx 1.8928$

95. $\log_{\sqrt{2}} \sqrt{5} = \frac{\log_e \sqrt{5}}{\log_e \sqrt{2}} \approx 2.3219$

EXERCISES 5.5

97. $\text{pH} = -\log[\text{H}^+]$
$= -\log(6.3 \times 10^{-8}) \approx 7.20$

99. $\text{pH} = -\log[\text{H}^+]$
$= -\log(1.7 \times 10^{-5}) \approx 4.77$

101.
$$\begin{aligned} \text{pH} &= -\log[\text{H}^+] \\ 2.9 &= -\log[\text{H}^+] \\ -2.9 &= \log[\text{H}^+] \\ 1.26 \times 10^{-3} &\approx [\text{H}^+] \end{aligned} \qquad \begin{aligned} \text{pH} &= -\log[\text{H}^+] \\ 3.3 &= -\log[\text{H}^+] \\ -3.3 &= \log[\text{H}^+] \\ 5.01 \times 10^{-4} &\approx [\text{H}^+] \end{aligned}$$
The hydrogen ion concentration can range from 5.01×10^{-4} to 1.26×10^{-3}.

103. $\text{dB gain} = 10\log\dfrac{P_O}{P_I}$
$= 10\log\dfrac{40}{\frac{1}{2}} \approx 19 \text{ dB}$

105. $L = k\ln I$
$4L = 4k\ln I = k \cdot 4\ln I = k\ln I^4$
The original intensity must be raised to the fourth power.

107. $E = 8300\ln V$
$2E = 2 \cdot 8300\ln V = 8300\ln V^2$
The original volume is squared.

109-111. Answers may vary.

113. $3^{4\log_3 2} + 5^{\frac{1}{2}\log_5 25} = 3^{\log_3 2^4} + 5^{\log_5 25^{1/2}}$
$= 2^4 + 25^{1/2}$
$= 16 + 5 = 21$

115. Let $\log_b M = x$ and $\log_b N = y$.
Then $b^x = M$ and $b^y = N$.
$\dfrac{M}{N} = \dfrac{b^x}{b^y} = b^{x-y}$. So $\log_b \dfrac{M}{N} = x - y$,
or $\log_b \dfrac{M}{N} = \log_b M - \log_b N$.

117. $e^{x\ln a} = e^{\ln a^x} = a^x$

119. $\ln(e^x) = x\ln e = x(1) = x$

121. $\log(0.9) < 0$, so $\ln(\log(0.9))$ is undefined.

123. Answers may vary.

125.
$$\begin{aligned} \log_{500}\frac{w^{100}x^{200}z^{300}}{y^{400}} &= \log_{500}(w^{100}x^{200}z^{300}) - \log_{500}(y^{400}) \\ &= \log_{500}(w^{100}) + \log_{500}(x^{200}) + \log_{500}(z^{300}) - \log_{500}(y^{400}) \\ &= 100\log_{500} w + 200\log_{500} x + 300\log_{500} z - 400\log_{500} y \Rightarrow \textbf{b.} \end{aligned}$$

127. $\log_{200} x\sqrt[100]{y} = \log_{200}(xy^{1/100}) = \log_{200} x + \log_{200}(y^{1/100}) = \log_{200} x + \dfrac{1}{100}\log_{200} y \Rightarrow$ **d.**

129. $\log_{300} x^{100}y^{200} = \log_{300} x^{100} + \log_{300} y^{200} = 100\log_{300} x + 200\log_{300} y \Rightarrow$ **f.**

131. $-\log_{300} x^{100}y^{200} = \log_{300}(x^{100}y^{200})^{-1} = \log_{300}\left(\dfrac{1}{x^{100}} \cdot \dfrac{1}{y^{200}}\right) = \log_{300}\dfrac{1}{x^{100}} + \log_{300}\dfrac{1}{y^{200}} \Rightarrow$ **h.**

Exercises 5.6 (page 569)

1. exponential

3. $A_0 2^{-t/h}$

EXERCISES 5.6

5.
$$\begin{aligned} 2^{3x+2} &= 16^x \\ 2^{3x+2} &= \left(2^4\right)^x \\ 2^{3x+2} &= 2^{4x} \\ 3x+2 &= 4x \\ 2 &= x \end{aligned}$$

7.
$$\begin{aligned} 27^{x+1} &= 3^{2x+1} \\ \left(3^3\right)^{x+1} &= 3^{2x+1} \\ 3^{3x+3} &= 3^{2x+1} \\ 3x+3 &= 2x+1 \\ x &= -2 \end{aligned}$$

9.
$$\begin{aligned} 5^{4x+1} &= 25^{-x-2} \\ 5^{4x+1} &= \left(5^2\right)^{-x-2} \\ 5^{4x+1} &= 5^{-2x-4} \\ 4x+1 &= -2x-4 \\ 6x &= -5 \\ x &= -\frac{5}{6} \end{aligned}$$

11.
$$\begin{aligned} 4^{x-2} &= 8^x \\ \left(2^2\right)^{x-2} &= \left(2^3\right)^x \\ 2^{2x-4} &= 2^{3x} \\ 2x-4 &= 3x \\ -4 &= x \end{aligned}$$

13.
$$\begin{aligned} 81^{2x} &= 27^{2x-5} \\ \left(3^4\right)^{2x} &= \left(3^3\right)^{2x-5} \\ 3^{8x} &= 3^{6x-15} \\ 8x &= 6x-15 \\ 2x &= -15 \\ x &= -\frac{15}{2} \end{aligned}$$

15.
$$\begin{aligned} 2^{x^2-2x} &= 8 \\ 2^{x^2-2x} &= 2^3 \\ x^2-2x &= 3 \\ x^2-2x-3 &= 0 \\ (x-3)(x+1) &= 0 \end{aligned}$$
$x-3=0$ **or** $x+1=0$

$x=3$ $\quad$ $x=-1$

17.
$$\begin{aligned} 36^{x^2} &= 216^{x^2-3} \\ \left(6^2\right)^{x^2} &= \left(6^3\right)^{x^2-3} \\ 6^{2x^2} &= 6^{3x^2-9} \\ 2x^2 &= 3x^2-9 \\ 9 &= x^2 \\ \pm 3 &= x \end{aligned}$$

19.
$$\begin{aligned} 7^{x^2+3x} &= \frac{1}{49} \\ 7^{x^2+3x} &= 7^{-2} \\ x^2+3x &= -2 \\ x^2+3x+2 &= 0 \\ (x+2)(x+1) &= 0 \end{aligned}$$
$x+2=0$ **or** $x+1=0$

$x=-2$ $\quad$ $x=-1$

21.
$$\begin{aligned} e^{-x+6} &= e^x \\ -x+6 &= x \\ 6 &= 2x \\ 3 &= x \end{aligned}$$

23.
$$\begin{aligned} e^{x^2-1} &= e^{24} \\ x^2-1 &= 24 \\ x^2-25 &= 0 \\ (x-5)(x+5) &= 0 \end{aligned}$$
$x-5=0$ **or** $x+5=0$

$x=-5$ $\quad$ $x=5$

25.
$$\begin{aligned} 4^x &= 5 \\ \log 4^x &= \log 5 \\ x\log 4 &= \log 5 \\ x &= \frac{\log 5}{\log 4} \\ x &\approx 1.1610 \end{aligned}$$

EXERCISES 5.6

27.
$$\begin{aligned} 13^{x-1} &= 2 \\ \log 13^{x-1} &= \log 2 \\ (x-1)\log 13 &= \log 2 \\ x\log 13 - \log 13 &= \log 2 \\ x\log 13 &= \log 2 + \log 13 \\ x &= \frac{\log 2 + \log 13}{\log 13} \\ x &\approx 1.2702 \end{aligned}$$

29.
$$\begin{aligned} 2^{x+1} &= 3^x \\ \log 2^{x+1} &= \log 3^x \\ (x+1)\log 2 &= x\log 3 \\ x\log 2 + \log 2 &= x\log 3 \\ x\log 2 - x\log 3 &= -\log 2 \\ x(\log 2 - \log 3) &= -\log 2 \\ x &= \frac{-\log 2}{\log 2 - \log 3} \\ x &\approx 1.7095 \end{aligned}$$

31.
$$\begin{aligned} 2^x &= 3^x \\ \log 2^x &= \log 3^x \\ x\log 2 &= x\log 3 \\ x\log 2 - x\log 3 &= 0 \\ x(\log 2 - \log 3) &= 0 \\ x &= \frac{0}{\log 2 - \log 3} = 0 \end{aligned}$$

33.
$$\begin{aligned} 7^{x^2} &= 10 \\ \log 7^{x^2} &= \log 10 \\ x^2\log 7 &= \log 10 \\ x^2 &= \frac{\log 10}{\log 7} \\ x &= \pm\sqrt{\frac{\log 10}{\log 7}} \\ x &\approx \pm 1.0878 \end{aligned}$$

35.
$$\begin{aligned} 8^{x^2} &= 9^x \\ \log 8^{x^2} &= \log 9^x \\ x^2\log 8 &= x\log 9 \\ x^2\log 8 - x\log 9 &= 0 \\ x(x\log 8 - \log 9) &= 0 \end{aligned}$$
$$\begin{aligned} x = 0 \quad &\textbf{or} \quad x\log 8 - \log 9 = 0 \\ x = 0 \quad &\qquad x\log 8 = \log 9 \\ x = 0 \quad &\qquad x = \frac{\log 9}{\log 8} \\ x = 0 \quad &\qquad x \approx 1.0566 \end{aligned}$$

37.
$$\begin{aligned} e^x &= 10 \\ \ln e^x &= \ln 10 \\ x\ln e &= \ln 10 \\ x &= \ln 10 \end{aligned}$$

39.
$$\begin{aligned} 4e^{2x} &= 24 \\ e^{2x} &= 6 \\ \ln e^{2x} &= \ln 6 \\ 2x\ln e &= \ln 6 \\ 2x &= \ln 6 \\ x &= \frac{\ln 6}{2} = \frac{1}{2}\ln 6 \end{aligned}$$

41.
$$\begin{aligned} 4^{x+2} - 4^x &= 15 \\ 4^x 4^2 - 4^x &= 15 \\ 16\cdot 4^x - 4^x &= 15 \\ 15\cdot 4^x &= 15 \\ 4^x &= 1 \\ x &= 0 \end{aligned}$$

EXERCISES 5.6

43.
$$
\begin{aligned}
2(3^x) &= 6^{2x} \\
\log[2(3^x)] &= \log 6^{2x} \\
\log 2 + \log 3^x &= 2x\log 6 \\
\log 2 + x\log 3 &= 2x\log 6 \\
x\log 3 - 2x\log 6 &= -\log 2 \\
x(\log 3 - 2\log 6) &= -\log 2 \\
x &= \frac{-\log 2}{\log 3 - 2\log 6} \\
x &\approx 0.2789
\end{aligned}
$$

45.
$$
\begin{aligned}
2^{2x} - 10(2^x) + 16 &= 0 \\
y^2 - 10y + 16 &= 0 \\
(y-2)(y-8) &= 0
\end{aligned}
$$
$y - 2 = 0$ **or** $y - 8 = 0$

$y = 2$ $\qquad y = 8$

$2^x = 2$ $\qquad 2^x = 8$

$x = 1$ $\qquad x = 3$

47.
$$
\begin{aligned}
2^{2x+1} - 2^x &= 1 \\
2^{2x}2^1 - 2^x - 1 &= 0 \\
2\left(2^{2x}\right) - 2^x - 1 &= 0 \\
(2(2^x)+1)(2^x-1) &= 0
\end{aligned}
$$
$2(2^x) + 1 = 0$ **or** $2^x - 1 = 0$

$2(2^x) = -1$ $\qquad 2^x = 1$

$2^x = -\frac{1}{2}$ $\qquad x = 0$

impossible

49.
$$
\begin{aligned}
\log x^2 &= 2 \\
x^2 &= 10^2 \\
x^2 &= 100 \\
x &= \pm\sqrt{100} = \pm 10
\end{aligned}
$$

51.
$$
\begin{aligned}
\log\frac{4x+1}{2x+9} &= 0 \\
10^0 &= \frac{4x+1}{2x+9} \\
1 &= \frac{4x+1}{2x+9} \\
2x+9 &= 4x+1 \\
8 &= 2x \\
4 &= x
\end{aligned}
$$

53.
$$
\begin{aligned}
\ln x &= 6 \\
e^6 &= x
\end{aligned}
$$

55.
$$
\begin{aligned}
\ln(2x-7) &= 4 \\
e^4 &= 2x - 7 \\
e^4 + 7 &= 2x \\
\frac{e^4+7}{2} &= x, \text{ or } x = \frac{1}{2}\left(e^4+7\right)
\end{aligned}
$$

57.
$$
\begin{aligned}
\log_2(2x-3) &= \log_2(x+4) \\
2x - 3 &= x + 4 \\
x &= 7
\end{aligned}
$$

59.
$$
\begin{aligned}
\log x + \log(x-48) &= 2 \\
\log x(x-48) &= 2 \\
x(x-48) &= 10^2 \\
x^2 - 48x - 100 &= 0 \\
(x-50)(x+2) &= 0
\end{aligned}
$$
$x - 50 = 0$ **or** $x + 2 = 0$

$x = 50$ $\qquad x = -2$: extraneous

61.
$$
\begin{aligned}
\log x + \log(x-15) &= 2 \\
\log x(x-15) &= 2 \\
x(x-15) &= 10^2 \\
x^2 - 15x - 100 &= 0 \\
(x-20)(x+5) &= 0
\end{aligned}
$$
$x - 20 = 0$ **or** $x + 5 = 0$

$x = 20$ $\qquad x = -5$: extraneous

EXERCISES 5.6

63.
$$\begin{aligned} \log(x+90) &= 3-\log x \\ \log x+\log(x+90) &= 3 \\ \log x(x+90) &= 3 \\ x(x+90) &= 10^3 \\ x^2+90x-1000 &= 0 \\ (x-10)(x+100) &= 0 \end{aligned}$$
$x-10=0$ **or** $x+100=0$

$x=10$ $\quad$ $x=-100$

extraneous

65.
$$\begin{aligned} \log 5000-\log(x-2) &= 3 \\ \log\frac{5000}{x-2} &= 3 \\ \frac{5000}{x-2} &= 10^3 \\ 5000 &= 1000(x-2) \\ 5000 &= 1000x-2000 \\ 7000 &= 1000x \\ 7 &= x \end{aligned}$$

67.
$$\begin{aligned} \log_7 x+\log_7(x-5) &= \log_7 6 \\ \log_7 x(x-5) &= \log_7 6 \\ x(x-5) &= 6 \\ x^2-5x-6 &= 0 \\ (x-6)(x+1) &= 0 \end{aligned}$$
$x-6=0$ **or** $x+1=0$

$x=6$ $\quad$ $x=-1$

extraneous

69.
$$\begin{aligned} \ln 15-\ln(x-2) &= \ln x \\ \ln\frac{15}{x-2} &= \ln x \\ \frac{15}{x-2} &= x \\ 15 &= x(x-2) \\ 0 &= x^2-2x-15 \\ 0 &= (x-5)(x+3) \end{aligned}$$
$x-5=0$ **or** $x+3=0$

$x=5$ $\quad$ $x=-3$

extraneous

71.
$$\begin{aligned} \log_6 8-\log_6 x &= \log_6(x-2) \\ \log_6\frac{8}{x} &= \log_6(x-2) \\ \frac{8}{x} &= x-2 \\ 8 &= x(x-2) \\ 0 &= x^2-2x-8 \\ 0 &= (x-4)(x+2) \end{aligned}$$
$x-4=0$ **or** $x+2=0$

$x=4$ $\quad$ $x=-2$

extraneous

73.
$$\begin{aligned} \log_8(x-1)-\log_8 6 &= \log_8(x-2)-\log_8 x \\ \log_8\tfrac{x-1}{6} &= \log_8\tfrac{x-2}{x} \\ \tfrac{x-1}{6} &= \tfrac{x-2}{x} \\ x(x-1) &= 6(x-2) \\ x^2-x &= 6x-12 \\ x^2-7x+12 &= 0 \\ (x-3)(x-4) &= 0 \end{aligned}$$
$x-3=0$ **or** $x-4=0$

$x=3$ $\quad$ $x=4$

75.
$$\begin{aligned} \log(\log x) &= 1 \\ \log x &= 10^1 \\ \log x &= 10 \\ x &= 10^{10} \end{aligned}$$

77.
$$\begin{aligned} \frac{\log(3x-4)}{\log x} &= 2 \\ \log(3x-4) &= 2\log x \\ \log(3x-4) &= \log x^2 \\ 3x-4 &= x^2 \\ 0 &= x^2-3x+4 \end{aligned}$$
No real solutions

79.
$$\frac{\ln(5x+6)}{2} = \ln x$$
$$\ln(5x+6) = 2\ln x$$
$$\ln(5x+6) = \ln x^2$$
$$5x+6 = x^2$$
$$0 = x^2 - 5x - 6$$
$$0 = (x-6)(x+1)$$
$$x - 6 = 0 \quad \textbf{or} \quad x + 1 = 0$$
$$x = 6 \qquad x = -1$$
extraneous

81.
$$\log_3 x = \log_3\left(\frac{1}{x}\right) + 4$$
$$\log_3 x = \log_3 x^{-1} + 4$$
$$\log_3 x = -\log_3 x + 4$$
$$2\log_3 x = 4$$
$$\log_3 x = 2$$
$$x = 9$$

83.
$$2\log_2 x = 3 + \log_2(x-2)$$
$$\log_2 x^2 - \log_2(x-2) = 3$$
$$\log_2 \tfrac{x^2}{x-2} = 3$$
$$\tfrac{x^2}{x-2} = 2^3$$
$$x^2 = 8(x-2)$$
$$x^2 - 8x + 16 = 0$$
$$(x-4)(x-4) = 0$$
$$x - 4 = 0 \quad \textbf{or} \quad x - 4 = 0$$
$$x = 4 \qquad x = 4$$

85.

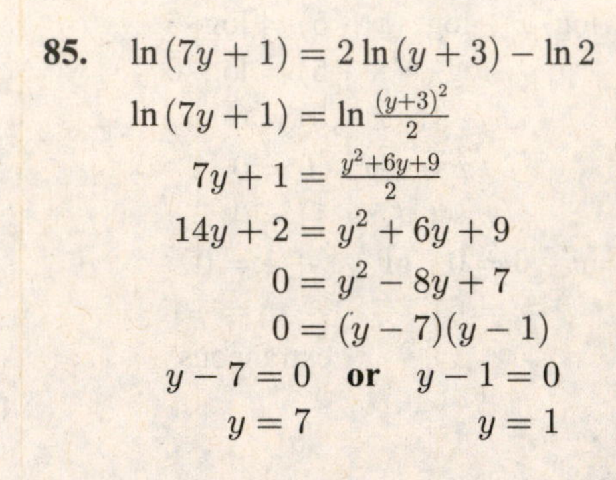

$$\ln(7y+1) = 2\ln(y+3) - \ln 2$$
$$\ln(7y+1) = \ln\tfrac{(y+3)^2}{2}$$
$$7y + 1 = \tfrac{y^2+6y+9}{2}$$
$$14y + 2 = y^2 + 6y + 9$$
$$0 = y^2 - 8y + 7$$
$$0 = (y-7)(y-1)$$
$$y - 7 = 0 \quad \textbf{or} \quad y - 1 = 0$$
$$y = 7 \qquad y = 1$$

87. Graph $y = \log x + \log(x-15)$ and $y = 2$ and find the x-coordinate of the point(s) of intersection: $x = 20$

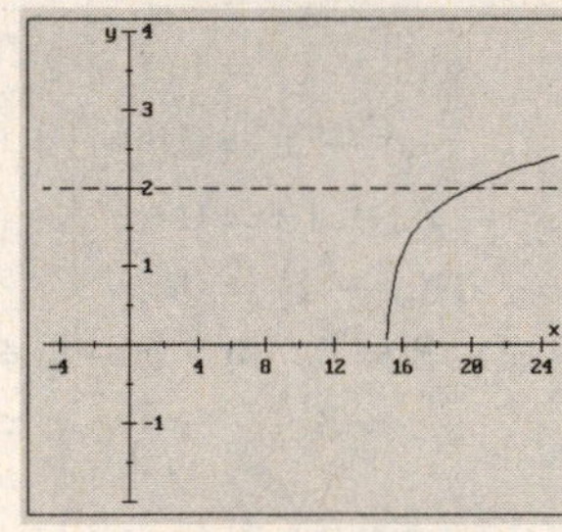

89.

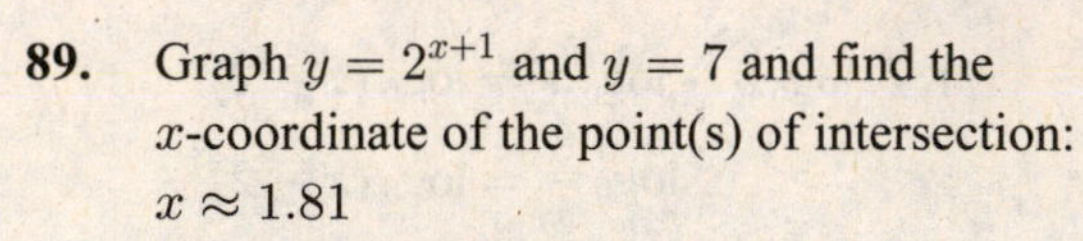

Graph $y = 2^{x+1}$ and $y = 7$ and find the x-coordinate of the point(s) of intersection: $x \approx 1.81$

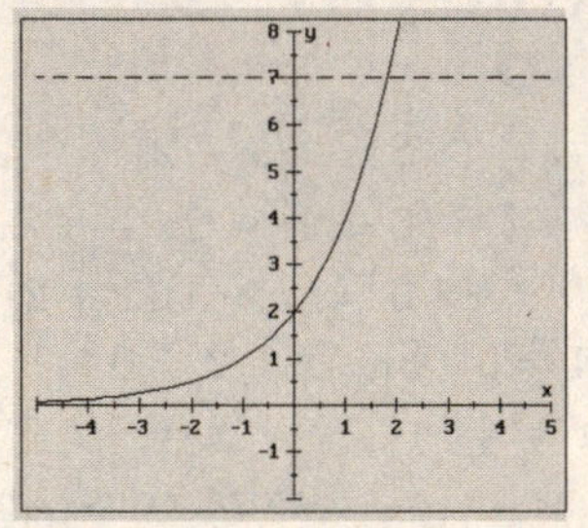

91.
$$A = A_0 2^{-t/h}$$
$$0.75A_0 = A_0 \cdot 2^{-t/(12.4)}$$
$$0.75 = 2^{-t/12.4}$$
$$\log(0.75) = \log\left(2^{-t/12.4}\right)$$
$$\log(0.75) = -\frac{t}{12.4}\log 2$$
$$-\frac{12.4\log(0.75)}{\log 2} = t$$
$$5.1 \text{ years} \approx t$$

93.

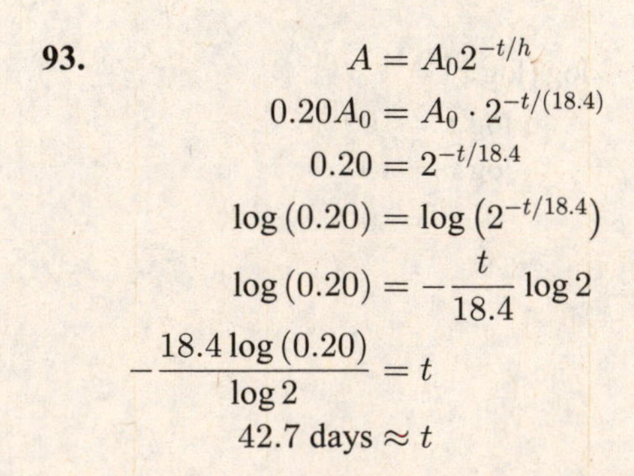

$$A = A_0 2^{-t/h}$$
$$0.20A_0 = A_0 \cdot 2^{-t/(18.4)}$$
$$0.20 = 2^{-t/18.4}$$
$$\log(0.20) = \log\left(2^{-t/18.4}\right)$$
$$\log(0.20) = -\frac{t}{18.4}\log 2$$
$$-\frac{18.4\log(0.20)}{\log 2} = t$$
$$42.7 \text{ days} \approx t$$

EXERCISES 5.6

95.
$$\begin{aligned} A &= A_0 2^{-t/h} \\ 0.70A_0 &= A_0 \cdot 2^{-t/5700} \\ 0.70 &= 2^{-t/5700} \\ \log(0.70) &= \log\left(2^{-t/5700}\right) \\ \log(0.70) &= -\frac{t}{5700}\log 2 \\ -\frac{5700\log(0.70)}{\log 2} &= t \\ 2900 \text{ years} &\approx t \end{aligned}$$

97.
$$\begin{aligned} A &= P\left(1+\frac{r}{k}\right)^{kt} \\ 800 &= 500\left(1+\frac{0.085}{2}\right)^{2t} \\ \frac{8}{5} &= (1.0425)^{2t} \\ \log\left(\frac{8}{5}\right) &= \log(1.0425)^{2t} \\ \log 8 - \log 5 &= 2t\log(1.0425) \\ \frac{\log 8 - \log 5}{2\log(1.0425)} &= t \\ 5.6 \text{ years} &\approx t \end{aligned}$$

99.
$$\begin{aligned} A &= P\left(1+\frac{r}{k}\right)^{kt} \\ 2100 &= 1300\left(1+\frac{0.09}{4}\right)^{4t} \\ \frac{21}{13} &= (1.0225)^{4t} \\ \log\left(\frac{21}{13}\right) &= \log(1.0225)^{4t} \\ \log 21 - \log 13 &= 4t\log(1.0225) \\ \frac{\log 21 - \log 13}{4\log(1.0225)} &= t \\ 5.4 \text{ years} &\approx t \end{aligned}$$

101.
$$\begin{aligned} A &= Pe^{rt} \\ 2P &= Pe^{rt} \\ 2 &= e^{rt} \\ \ln 2 &= \ln e^{rt} \\ \ln 2 &= rt \\ \frac{\ln 2}{r} &= t \\ \frac{0.70}{r} &\approx t \\ \frac{70}{(100 \cdot r)\%} &\approx t \end{aligned}$$

103.
$$\begin{aligned} P &= P_0 a^t \\ 3P_0 &= P_0 a^5 \\ 3 &= a^5 \\ 3^{1/5} &= \left(a^5\right)^{1/5} \\ 3^{1/5} &= a \end{aligned}$$

$$\begin{aligned} P &= P_0 a^t \\ 2P_0 &= P_0\left(3^{1/5}\right)^t \\ 2 &= 3^{t/5} \\ \log 2 &= \log 3^{t/5} \\ \log 2 &= \frac{t}{5}\log 3 \\ \frac{5\log 2}{\log 3} &= t \\ 3.2 \text{ days} &\approx t \end{aligned}$$

105.
$$\begin{aligned} T &= 70 + 110e^{-0.2t} \\ 80 &= 70 + 110e^{-0.2t} \\ 10 &= 110e^{-0.2t} \\ \tfrac{1}{11} &= e^{-0.2t} \\ \ln\left(\tfrac{1}{11}\right) &= \ln e^{-0.2t} \\ \ln\left(\tfrac{1}{11}\right) &= -0.2t \\ \frac{\ln\left(\frac{1}{11}\right)}{-0.2} &= t \\ 12 \approx t &\Rightarrow t \approx 12 \text{ minutes} \end{aligned}$$

107.
$$\begin{aligned} T &= 60 + 40e^{kt} \\ 90 &= 60 + 40e^{k(3)} \\ 30 &= 40e^{3k} \\ 0.75 &= e^{3k} \\ \ln(0.75) &= \ln e^{3k} \\ \ln(0.75) &= 3k \\ \tfrac{\ln(0.75)}{3} &= k \end{aligned}$$

109.
$$\begin{aligned} T &= 300 - 300e^{kt} \\ 100 &= 300 - 300e^{k(5)} \\ -200 &= -300e^{5k} \\ \frac{2}{3} &= e^{5k} \\ \ln\left(\frac{2}{3}\right) &= \ln e^{5k} \\ \ln\left(\frac{2}{3}\right) &= 5k \\ \frac{\ln\left(\frac{2}{3}\right)}{5} &= k \end{aligned}$$

111. **Answers may vary.**

113. **Answers may vary.**

115.
$$\begin{aligned} P &= P_0e^{rt} \\ 2P_0 &= P_0e^{rt} \\ 2 &= e^{rt} \\ \ln 2 &= \ln e^{rt} \\ \ln 2 &= rt \\ \frac{\ln 2}{r} &= t \end{aligned}$$

117. **Answers may vary.**

119.
$$\begin{aligned} \log_2(\log_5(\log_7 x)) &= 2 \\ \log_5(\log_7 x) &= 2^2 = 4 \\ \log_7 x &= 5^4 = 625 \\ x &= 7^{625} \end{aligned}$$

121. True.

123. False. $x^{3/2} = \frac{27}{8}$ is not an exponential equation.

125. True.

127. False. Use the natural logarithm.

129. False. After combining the logarithms, you would end up with $2^5 = \frac{8}{x}$.

Chapter 5 Review (page 572)

1. $5^{\sqrt{2}}5^{\sqrt{2}} = 5^{\sqrt{2}+\sqrt{2}} = 5^{2\sqrt{2}}$

2. $\left(2^{\sqrt{5}}\right)^{\sqrt{2}} = 2^{\sqrt{5}\cdot\sqrt{2}} = 2^{\sqrt{10}}$

3. $f(x) = 3^x$: $(0, 1)$, $(1, 3)$

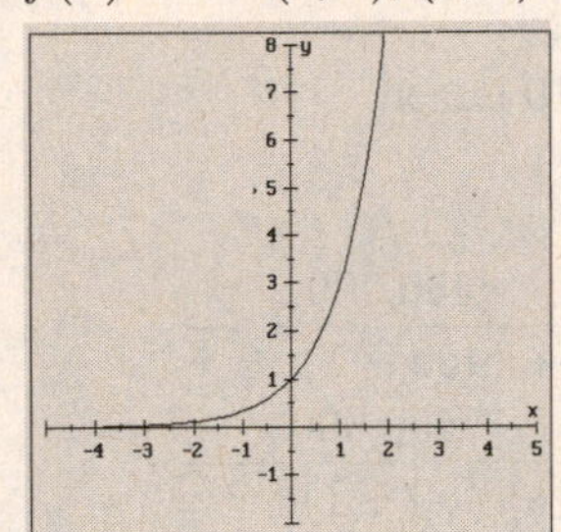

4. $f(x) = \left(\dfrac{1}{3}\right)^x$: $(0, 1)$, $\left(1, \dfrac{1}{3}\right)$

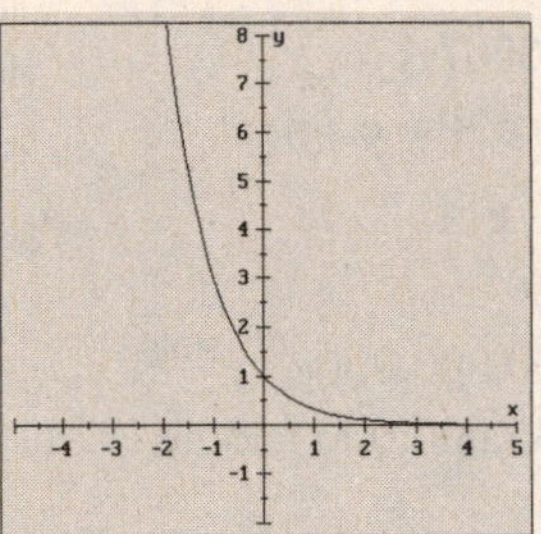

5. $f(x) = 7^x$: goes through $(0, 1)$ and $(1, 7)$
$p = 1, q = 7$

6. $y = b^x$: domain $= (-\infty, \infty)$; range $= (0, \infty)$

7. $f(x) = \left(\dfrac{1}{2}\right)^x - 2$

8. $f(x) = \left(\dfrac{1}{2}\right)^{x+2}$

Shift $y = \left(\dfrac{1}{2}\right)^x$ down 2:

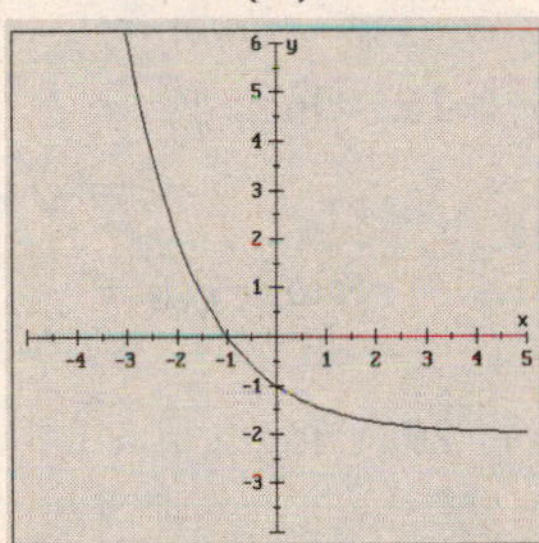

Shift $y = \left(\dfrac{1}{2}\right)^x$ left 2:

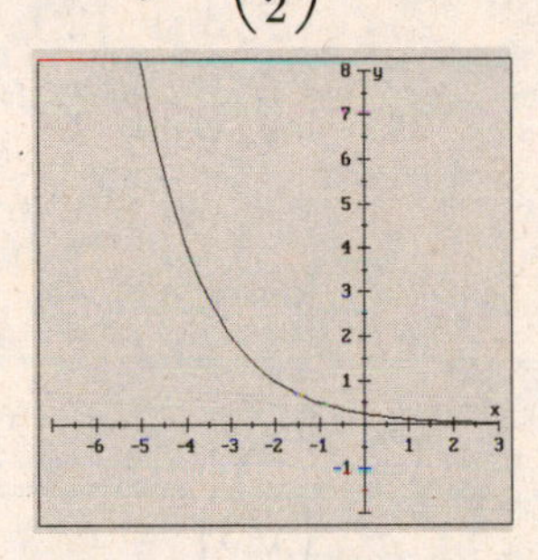

9. $f(x) = -5^x$
Reflect $y = 5^x$ about x-axis.

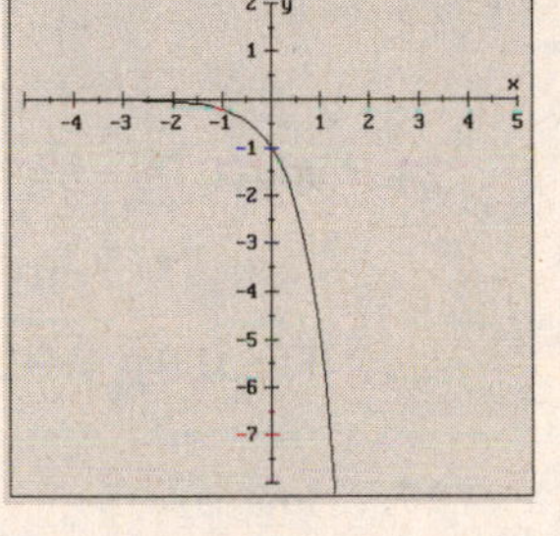

10. $f(x) = -5^x + 4$; Reflect $y = 5^x$ about x. Shift U4.

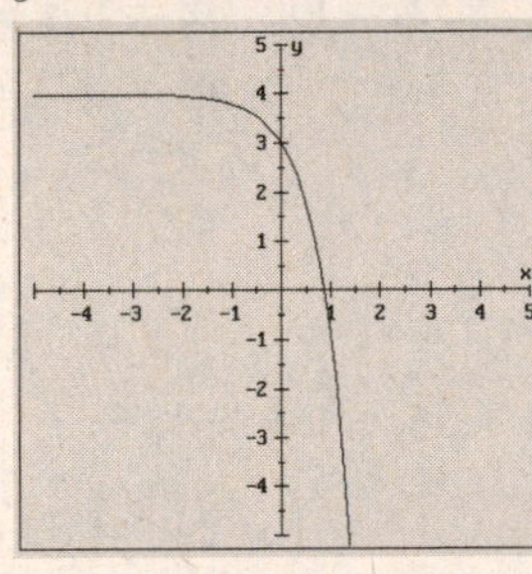

11. $f(x) = e^x + 1$
Shift $y = e^x$ up 1:

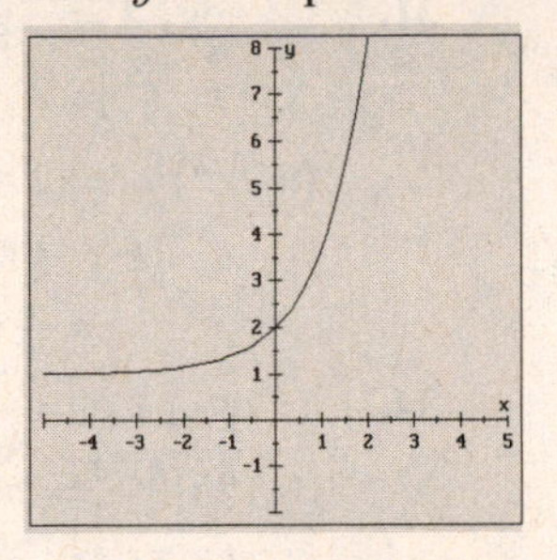

12. $f(x) = e^{x-3}$
Shift $y = e^x$ right 3:

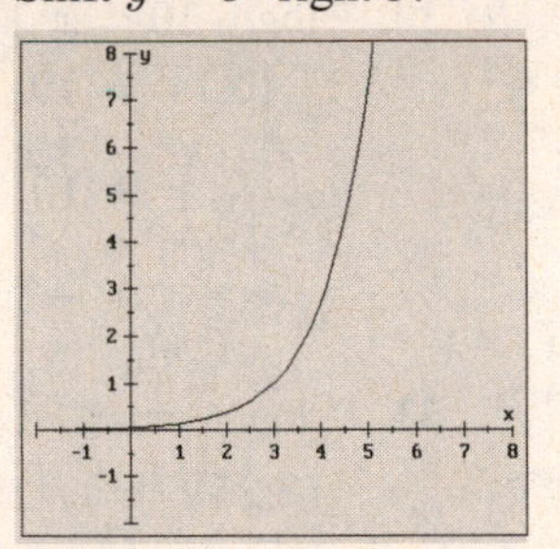

13.
$$\begin{aligned} A &= P\left(1 + \frac{r}{k}\right)^{kt} \\ &= 10{,}500\left(1 + \frac{0.09}{4}\right)^{4(60)} \\ &\approx \$2{,}189{,}703.45 \end{aligned}$$

14.
$$\begin{aligned} A &= Pe^{rt} \\ &= 10{,}500e^{0.09(60)} \\ &\approx 2{,}324{,}767.37 \end{aligned}$$

15. $A = A_0 2^{-t/h}$
$= A_0 \cdot 2^{-20/34.2}$
$\approx 0.6667A_0$
about $\frac{2}{3}$ of the original

16. $I = I_0 k^x$
$= 14(0.7)^{12}$
≈ 0.19 lumen

17. $P = P_0 e^{kt}$
$P = 300{,}000{,}000e^{0.015(50)}$
$P \approx 635{,}000{,}000$ people

18. $P = \dfrac{450{,}000}{1 + (450 - 1)e^{-0.2t}}$
$= \dfrac{450{,}000}{1 + (450 - 1)e^{-0.2(5)}}$
≈ 2708 people

19. domain $= (0, \infty)$; range $= (-\infty, \infty)$

20. domain $= (0, \infty)$; range $= (-\infty, \infty)$

21. $\log_3 9 = ?$
$3^? = 9$
$\log_3 9 = 2$

22. $\log_9 \dfrac{1}{3} = ?$
$9^? = \dfrac{1}{3}$
$\log_9 \dfrac{1}{3} = -\dfrac{1}{2}$

23. $\log_x 1 = ?$
$x^? = 1$
$\log_x 1 = 0$

24. $\log_5 0.04 = ?$
$5^? = 0.04 = \dfrac{1}{25}$
$\log_5 0.04 = -2$

25. $\log_a \sqrt{a} = ?$
$a^? = \sqrt{a}$
$\log_a \sqrt{a} = \dfrac{1}{2}$

26. $\log_a \sqrt[3]{a} = ?$
$a^? = \sqrt[3]{a}$
$\log_a \sqrt[3]{a} = \dfrac{1}{3}$

27. $\log_2 x = 5$
$2^5 = x$
$32 = x$

28. $\log_{\sqrt{3}} x = 4$
$\left(\sqrt{3}\right)^4 = x$
$9 = x$

29. $\log_{\sqrt{2}} x = 6$
$\left(\sqrt{2}\right)^6 = x$
$8 = x$

30. $\log_{0.1} 10 = x$
$(0.1)^x = 10$
$\left(\dfrac{1}{10}\right)^x = 10$
$x = -1$

31. $\log_x 2 = -\frac{1}{3}$
$x^{-1/3} = 2$
$\left(x^{-1/3}\right)^{-3} = 2^{-3}$
$x = \frac{1}{8}$

32. $\log_x 32 = 5$
$x^5 = 32$
$x = 2$

33. $\log_{0.25} x = -1$
$(0.25)^{-1} = x$
$\left(\frac{1}{4}\right)^{-1} = x$
$4 = x$

34. $\log_{0.125} x = -\frac{1}{3}$
$(0.125)^{-1/3} = x$
$\left(\frac{1}{8}\right)^{-1/3} = x$
$2 = x$

35. $\log_{\sqrt{2}} 32 = x$
$\left(\sqrt{2}\right)^x = 32$
$\left(2^{1/2}\right)^x = 2^5$
$\dfrac{1}{2}x = 5$
$x = 10$

36. $\log_{\sqrt{5}} x = -4$
$$\left(\sqrt{5}\right)^{-4} = x$$
$$\frac{1}{25} = x$$

37. $\log_{\sqrt{3}} 9\sqrt{3} = x$
$$\left(\sqrt{3}\right)^{x} = 9\sqrt{3}$$
$$\left(3^{1/2}\right)^{x} = 3^{5/2}$$
$$\frac{1}{2}x = \frac{5}{2}$$
$$x = 5$$

38. $\log_{\sqrt{5}} 5\sqrt{5} = x$
$$\left(\sqrt{5}\right)^{x} = 5\sqrt{5}$$
$$\left(5^{1/2}\right)^{x} = 5^{3/2}$$
$$\frac{1}{2}x = \frac{3}{2}$$
$$x = 3$$

39. $f(x) = \log(x-2)$; Shift $y = \log x$ right 2:

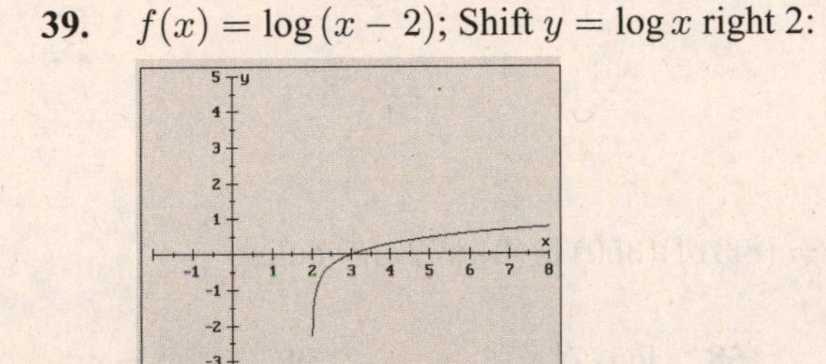

40. $f(x) = 3 + \log x$; Shift $y = \log x$ up 3:

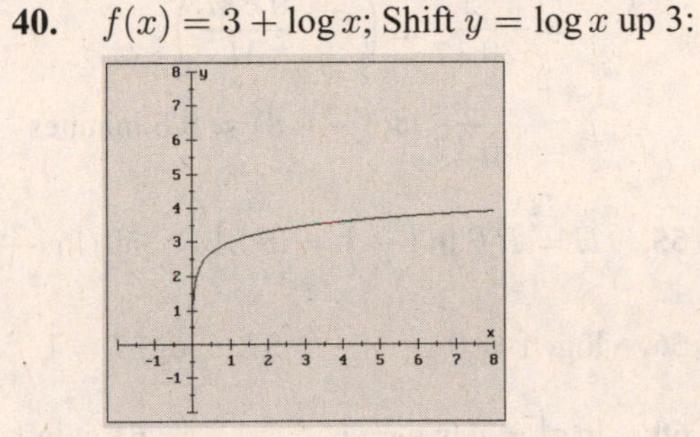

41. $y = 4^x$; $y = \log_4 x$

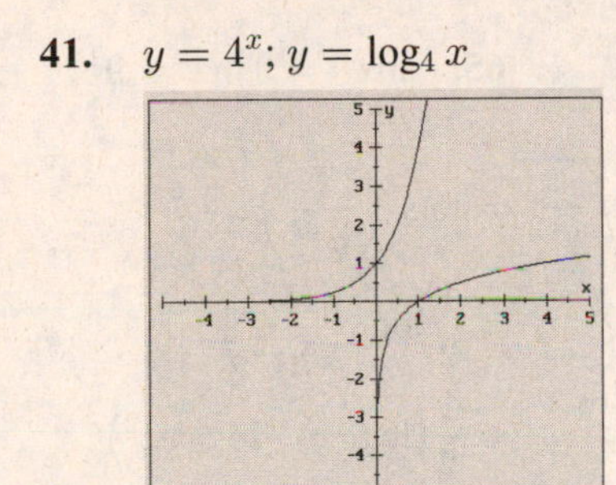

42. $y = \left(\frac{1}{3}\right)^x$; $y = \log_{1/3} x$

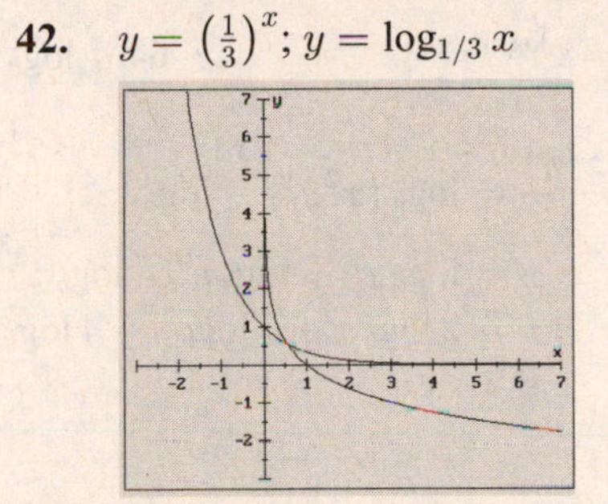

43. $\ln 452 \approx 6.1137$

44. $\ln(\log 7.85) \approx -0.1111$

45. $\ln x = 2.336$
$$x \approx 10.3398$$

46. $\ln x = \log 8.8$
$$x \approx 2.5715$$

47. $y = f(x) = 1 + \ln x$
Shift $y = \ln x$ up 1:

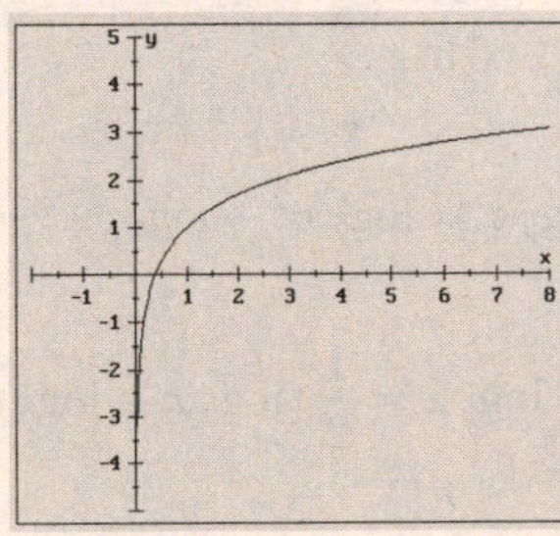

48. $y = f(x) = \ln(x+1)$
Shift $y = \ln x$ left 1:

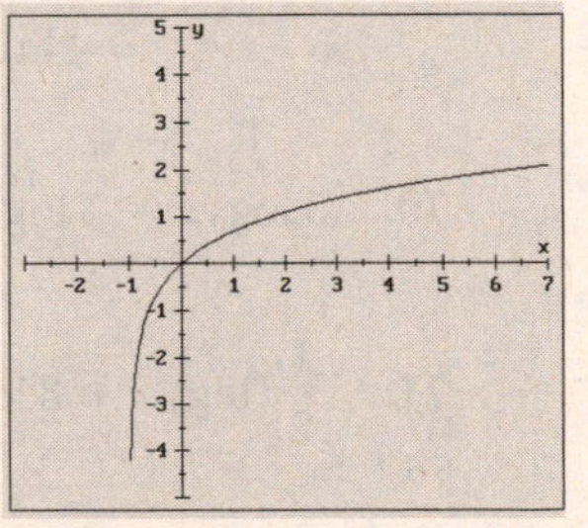

49. $\ln(e^{12}) = 12 \ln e = 12$

50. $e^{\ln 14x} = 14x$

51. $\text{dB gain} = 20 \log \dfrac{E_0}{E_I} = 20 \log \dfrac{18}{0.04} \approx 53 \text{ dB gain}$

52. $R = \log \dfrac{A}{P} = \log \dfrac{7500}{0.3} \approx 4.4$

53. $t = -\dfrac{1}{k} \ln \left(1 - \dfrac{C}{M}\right)$

$t = -\dfrac{1}{0.17} \ln \left(1 - \dfrac{0.8M}{M}\right)$

$t = -\dfrac{1}{0.17} \ln(1 - 0.8) \approx 9.5 \text{ minutes}$

54. $t = \dfrac{\ln 2}{r} = \dfrac{\ln 2}{0.03} \approx 23 \text{ years}$

55. $E = RT \ln \left(\frac{V_f}{V_i}\right) = (8.314)(350) \ln \left(\frac{2V_i}{V_i}\right) = (8.314)(350) \ln (2) \approx 2017 \text{ joules}$

56. $\log_7 1 = 0$

57. $\log_7 7 = 1$

58. $\log_7 7^3 = 3$

59. $7^{\log_7 4} = 4$

60. $\ln e^4 = 4 \ln e = 4$

61. $\ln 1 = 0$

62. $10^{\log_{10} 7} = 7$

63. $e^{\ln 3} = e^{\log_e 3} = 3$

64. $\log_b b^4 = 4 \log_b b = 4$

65. $\ln e^9 = 9 \ln e = 9$

66.
$$\begin{aligned} \log_b \frac{x^2y^3}{z^4} &= \log_b \left(x^2y^3\right) - \log_b \left(z^4\right) \\ &= \log_b x^2 + \log_b y^3 - \log_b z^4 \\ &= 2 \log_b x + 3 \log_b y - 4 \log_b z \end{aligned}$$

67.
$$\begin{aligned} \log_b \sqrt{\frac{x}{yz^2}} &= \log_b \left(\frac{x}{yz^2}\right)^{1/2} \\ &= \frac{1}{2} \log_b \left(\frac{x}{yz^2}\right) \\ &= \frac{1}{2} \left(\log_b x - \log_b yz^2\right) \\ &= \frac{1}{2} (\log_b x - \log_b y - 2 \log_b z) \end{aligned}$$

68.
$$\begin{aligned} \ln \frac{x^4}{y^5z^6} &= \ln x^4 - \ln y^5z^6 \\ &= \ln x^4 - \left(\ln y^5 + \ln z^6\right) \\ &= 4 \ln x - (5 \ln y + 6 \ln z) \\ &= 4 \ln x - 5 \ln y - 6 \ln z \end{aligned}$$

69.
$$\begin{aligned} \ln \sqrt[3]{xyz} &= \ln (xyz)^{1/3} \\ &= \frac{1}{3} \ln xyz \\ &= \frac{1}{3} (\ln x + \ln y + \ln z) \end{aligned}$$

70. $3 \log_b x - 5 \log_b y + 7 \log_b z = \log_b x^3 + \log_b y^{-5} + \log_b z^7 = \log_b \dfrac{x^3z^7}{y^5}$

71.
$$\begin{aligned} \frac{1}{2} (\log_b x + 3 \log_b y) - 7 \log_b z = \frac{1}{2} \left(\log_b x + \log_b y^3\right) + \log_b z^{-7} &= \frac{1}{2} \log_b xy^3 + \log_b z^{-7} \\ &= \log_b \sqrt{xy^3} + \log_b z^{-7} \\ &= \log_b \frac{\sqrt{xy^3}}{z^7} \end{aligned}$$

72. $4\ln x - 5\ln y - 6\ln z = \ln x^4 + \ln y^{-5} + \ln z^{-6} = \ln \dfrac{x^4}{y^5 z^6}$

73. $\dfrac{1}{2}\ln x + 3\ln y - \dfrac{1}{3}\ln z = \ln x^{1/2} + \ln y^3 + \ln z^{-1/3} = \ln \dfrac{y^3\sqrt{x}}{\sqrt[3]{z}}$

74. $\log abc = \log a + \log b + \log c = 0.6 + 0.36 + 2.4 = 3.36$

75. $\log a^2 b = \log a^2 + \log b = 2\log a + \log b = 2(0.6) + 0.36 = 1.56$

76. $\log \dfrac{ac}{b} = \log a + \log c - \log b = 0.6 + 2.4 - 0.36 = 2.64$

77.
$$\log \frac{a^2}{c^3 b^2} = \log a^2 - \log c^3 b^2 = \log a^2 - \log c^3 - \log b^2 = 2\log a - 3\log c - 2\log b$$
$$= 2(0.6) - 3(2.4) - 2(0.36) = -6.72$$

78. $\log_5 17 = \dfrac{\log 17}{\log 5} \approx 1.7604$

79.
$$\begin{aligned} \text{pH} &= -\log[\text{H}^+] \\ 3.1 &= -\log[\text{H}^+] \\ -3.1 &= \log[\text{H}^+] \\ 7.94 \times 10^{-4} &\approx [\text{H}^+] \end{aligned}$$

80. $L = k\ln I$

$k\ln\dfrac{I}{2} = k(\ln I - \ln 2) = k\ln I - k\ln 2 \Rightarrow$ The loudness decreases by $k\ln 2$.

81.
$$\begin{aligned} 81^{x+2} &= 27 \\ (3^4)^{x+2} &= 3^3 \\ 3^{4x+8} &= 3^3 \\ 4x + 8 &= 3 \\ 4x &= -5 \\ x &= -\frac{5}{4} \end{aligned}$$

82.
$$\begin{aligned} 2^{x^2+4x} &= \frac{1}{8} \\ 2^{x^2+4x} &= 2^{-3} \\ x^2 + 4x &= -3 \\ x^2 + 4x + 3 &= 0 \\ (x+1)(x+3) &= 0 \end{aligned}$$
$x + 1 = 0$ **or** $x + 3 = 0$

$x = -1$ $\qquad x = -3$

83.
$$\begin{aligned} e^x &= e^{-6x+14} \\ x &= -6x + 14 \\ 7x &= 14 \\ x &= 2 \end{aligned}$$

84.
$$\begin{aligned} e^{2x^2} &= e^{18} \\ 2x^2 &= 18 \\ x^2 &= 9 \\ x &= \pm 3 \end{aligned}$$

85.
$$\begin{aligned} 3^x &= 7 \\ \log 3^x &= \log 7 \\ x\log 3 &= \log 7 \\ x &= \frac{\log 7}{\log 3} \\ x &\approx 1.7712 \end{aligned}$$

86.
$$\begin{aligned} 2^x &= 3^{x-1} \\ \log 2^x &= \log 3^{x-1} \\ x\log 2 &= (x-1)\log 3 \\ x\log 2 &= x\log 3 - \log 3 \\ \log 3 &= x\log 3 - x\log 2 \\ \log 3 &= x(\log 3 - \log 2) \\ \frac{\log 3}{\log 3 - \log 2} &= x \\ 2.7095 &\approx x \end{aligned}$$

87.
$$\begin{aligned} 2e^x &= 16 \\ e^x &= 8 \\ \ln e^x &= \ln 8 \\ x\ln e &= \ln 8 \\ x &= \ln 8 \approx 2.0794 \end{aligned}$$

88.
$$\begin{aligned} -5e^x &= -35 \\ e^x &= 7 \\ \ln e^x &= \ln 7 \\ x\ln e &= \ln 7 \\ x &= \ln 7 \approx 1.9459 \end{aligned}$$

89.
$$\begin{aligned} \log_7(-7x+2) &= \log_7(3x+32) \\ -7x+2 &= 3x+32 \\ -30 &= 10x \\ -3 &= x \end{aligned}$$

90.
$$\begin{aligned} \ln(x+3) &= \ln(-5x+51) \\ x+3 &= -5x+51 \\ 6x &= 48 \\ x &= 8 \end{aligned}$$

91.
$$\begin{aligned} \log x + \log(29-x) &= 2 \\ \log x(29-x) &= 2 \\ x(29-x) &= 10^2 \\ -x^2+29x-100 &= 0 \\ x^2-29x+100 &= 0 \\ (x-25)(x-4) &= 0 \\ x-25=0 \quad \textbf{or} \quad x-4 &= 0 \\ x=25 \qquad\qquad x &= 4 \end{aligned}$$

92.
$$\begin{aligned} \log_2 x + \log_2(x-2) &= 3 \\ \log_2 x(x-2) &= 3 \\ x(x-2) &= 2^3 \\ x^2-2x-8 &= 0 \\ (x-4)(x+2) &= 0 \\ x-4=0 \quad \textbf{or} \quad x+2 &= 0 \\ x=4 \qquad\qquad x &= -2 \\ &\text{extraneous} \end{aligned}$$

93.
$$\begin{aligned} \log_2(x+2) + \log_2(x-1) &= 2 \\ \log_2(x+2)(x-1) &= 2 \\ (x+2)(x-1) &= 2^2 \\ x^2+x-2 &= 4 \\ x^2+x-6 &= 0 \\ (x-2)(x+3) &= 0 \\ x-2=0 \quad \textbf{or} \quad x+3 &= 0 \\ x=2 \qquad\qquad x &= -3 \\ &\text{extraneous} \end{aligned}$$

94.
$$\begin{aligned} \frac{\log(7x-12)}{\log x} &= 2 \\ \log(7x-12) &= 2\log x \\ \log(7x-12) &= \log x^2 \\ 7x-12 &= x^2 \\ 0 &= x^2-7x+12 \\ 0 &= (x-3)(x-4) \\ x &= 3, x = 4 \end{aligned}$$

95.
$$\begin{aligned} \ln x + \ln(x-5) &= \ln 6 \\ \ln x(x-5) &= \ln 6 \\ x(x-5) &= 6 \\ x^2 - 5x &= 6 \\ x^2 - 5x - 6 &= 0 \\ (x-6)(x+1) &= 0 \end{aligned}$$
$x - 6 = 0$ **or** $x + 1 = 0$

$x = 6$ $\quad$ $x = -1$ extraneous

96.
$$\begin{aligned} \log 3 - \log(x-1) &= -1 \\ \log \frac{3}{x-1} &= -1 \\ \frac{3}{x-1} &= 10^{-1} \\ \frac{3}{x-1} &= \frac{1}{10} \\ 30 &= x - 1 \\ 31 &= x \end{aligned}$$

97.
$$\begin{aligned} e^{x \ln 2} &= 9 \\ \ln e^{x \ln 2} &= \ln 9 \\ x \ln 2 \ln e &= \ln 9 \\ x \ln 2 &= \ln 9 \\ x &= \frac{\ln 9}{\ln 2} \approx 3.1699 \end{aligned}$$

98.
$$\begin{aligned} \ln x &= \ln(x-1) \\ x &= x - 1 \end{aligned}$$
no solution

99.
$$\begin{aligned} \ln x - 3 &= 4 \\ \ln x &= 7 \\ x &= e^7 \approx 1096.6332 \end{aligned}$$

100.
$$\begin{aligned} \ln x &= \ln(x-1) + 1 \\ \ln x - \ln(x-1) &= 1 \\ \ln \frac{x}{x-1} &= 1 \\ \frac{x}{x-1} &= e^1 \\ \frac{x}{x-1} &= \frac{e}{1} \\ x &= e(x-1) \\ x &= ex - e \\ e &= ex - x \\ e &= x(e-1) \\ \frac{e}{e-1} &= x, \text{ or } x \approx 1.5820 \end{aligned}$$

101. Note: $\log_{10} x = \dfrac{\ln x}{\ln 10}$
$$\begin{aligned} \ln x &= \log_{10} x \\ \ln x &= \frac{\ln x}{\ln 10} \\ \ln x \ln 10 &= \ln x \\ \ln x \ln 10 - \ln x &= 0 \\ \ln x(\ln 10 - 1) &= 0 \\ \ln x &= 0 \Rightarrow x = 1 \end{aligned}$$

102.
$$\begin{aligned} A &= A_0 2^{-t/h} \\ \frac{2}{3} A_0 &= A_0 \cdot 2^{-t/5700} \\ \frac{2}{3} &= 2^{-t/5700} \\ \log(2/3) &= \log\left(2^{-t/5700}\right) \\ \log(2/3) &= -\frac{t}{5700} \log 2 \\ -\frac{5700 \log(2/3)}{\log 2} &= t \Rightarrow t \approx 3300 \text{ years} \end{aligned}$$

Chapter 5 Test (page 584)

1. $f(x) = 2^x + 1 \Rightarrow$ Shift $y = 2^x$ up 1.

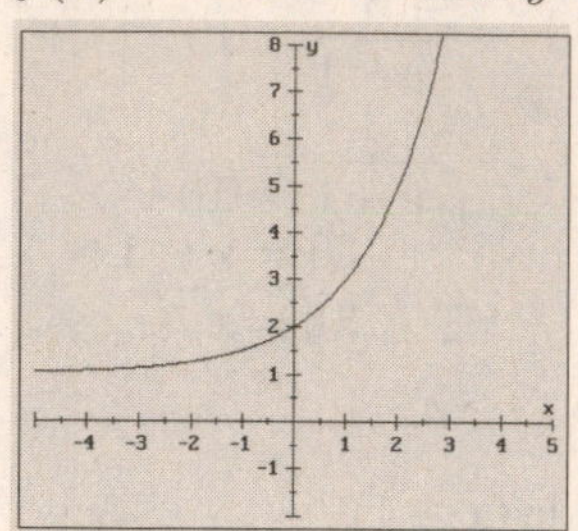

2. $f(x) = e^{x-2} \Rightarrow$ Shift $y = e^x$ right 2.

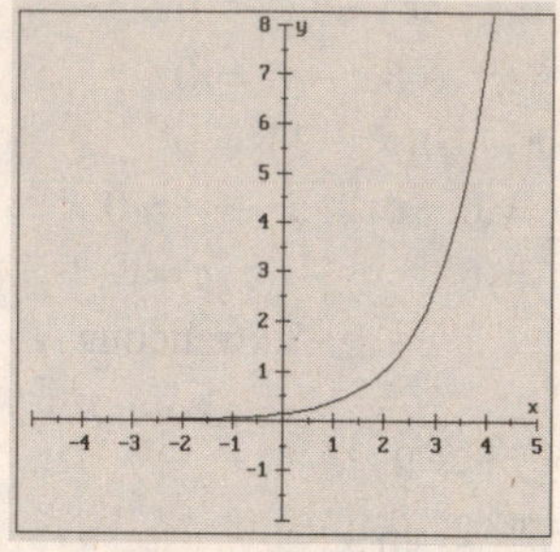

3. $A = 3(2)^{-6} = 3 \cdot \frac{1}{64} = \frac{3}{64}$ gram

4. $A = 1000\left(1 + \frac{0.06}{2}\right)^{2(1)} \approx \$1060.90$

5. $A = 2000e^{0.08(10)} \approx \$4451.08$

6. $\log_7 343 = \log_7 7^3 = 3$

7. $\log_3 \dfrac{1}{27} = \log_3 3^{-3} = -3$

8. $\log_{10} 10^{12} + 10^{\log_{10} 5} = 12 + 5 = 17$

9. $\log_{3/2} \frac{9}{4} = \log_{3/2} \left(\frac{3}{2}\right)^2 = 2$

10. $\log_{2/3} \frac{27}{8} = \log_{2/3} \left(\frac{2}{3}\right)^{-3} = -3$

11. $f(x) = \log(x-1)$; Shift $y = \log x$ right 1.

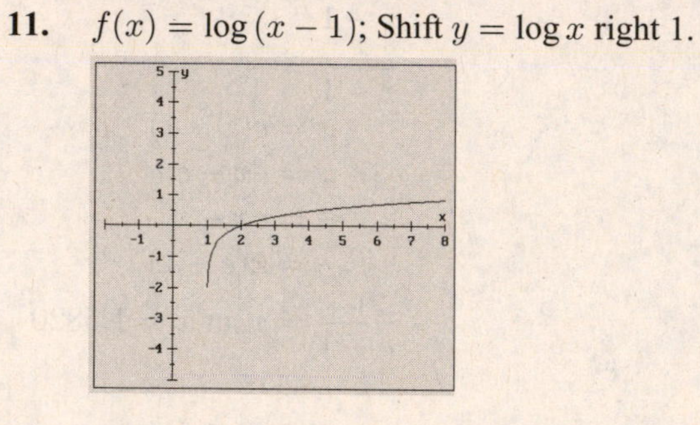

12. $f(x) = 2 + \ln x$; Shift $y = \ln x$ up 2.

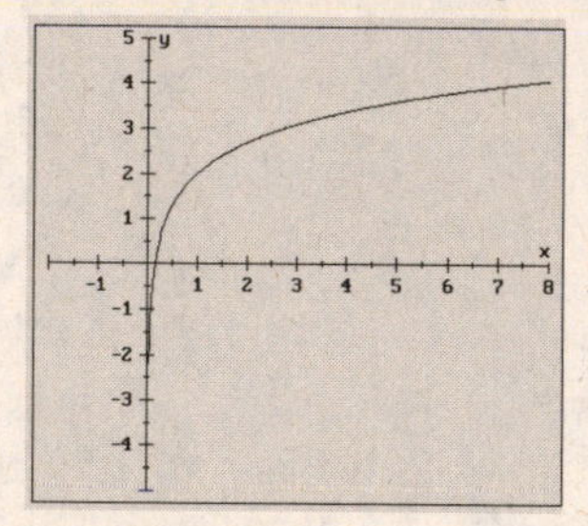

13. $\log a^2bc^3 = \log a^2 + \log b + \log c^3 = 2\log a + \log b + 3\log c$

14. $\ln \sqrt{\dfrac{a}{b^2c}} = \ln \left(\dfrac{a}{b^2c}\right)^{1/2} = \dfrac{1}{2} \ln \dfrac{a}{b^2c} = \dfrac{1}{2}\left(\ln a - \ln b^2 - \ln c\right) = \dfrac{1}{2}(\ln a - 2\ln b - \ln c)$

15. $\dfrac{1}{2}\log(a+2) + \log b - 2\log c = \log\sqrt{a+2} + \log b - \log c^2 = \log \dfrac{b\sqrt{a+2}}{c^2}$

16. $\dfrac{1}{3}(\ln a - 2\ln b) - \ln c = \dfrac{1}{3} \ln \dfrac{a}{b^2} - \ln c = \ln \sqrt[3]{\dfrac{a}{b^2}} - \ln c = \ln \dfrac{\sqrt[3]{\frac{a}{b^2}}}{c}$

17. $\log 24 = \log 8 \cdot 3 = \log 2^3 \cdot 3 = 3\log 2 + \log 3 = 3(0.3010) + 0.4771 = 1.3801$

CHAPTER 5 TEST

18. $\log \frac{8}{3} = \log \frac{2^3}{3} = 3\log 2 - \log 3 = 3(0.3010) - 0.4771 = 0.4259$

19. $\log_7 3 = \frac{\log 3}{\log 7}$ or $\frac{\ln 3}{\ln 7}$

20. $\log_\pi e = \frac{\log e}{\log \pi}$ or $\frac{\ln e}{\ln \pi} = \frac{1}{\ln \pi}$

21. $\log_a ab = \log_a a + \log_a b = 1 + \log_a b$
TRUE

22. $\log \frac{a}{b} = \log a - \log b$
FALSE

23.
$$\begin{aligned} \text{pH} &= -\log\,[\text{H}^+] \\ &= -\log\left(3.7 \times 10^{-7}\right) \\ &\approx 6.4 \end{aligned}$$

24.
$$\begin{aligned} \text{dB gain} &= 20 \log \frac{E_O}{E_I} \\ &= 20 \log \frac{60}{0.3} \\ &\approx 46 \text{ dB gain} \end{aligned}$$

25.
$$\begin{aligned} 3^{x^2-2x} &= 27 \\ 3^{x^2-2x} &= 3^3 \\ x^2 - 2x &= 3 \\ x^2 - 2x - 3 &= 0 \\ (x-3)(x+1) &= 0 \end{aligned}$$
$x - 3 = 0$ **or** $x + 1 = 0$
$x = 3 \qquad x = -1$

26.
$$\begin{aligned} 3^{x-1} &= 100^x \\ \log 3^{x-1} &= \log 100^x \\ (x-1)\log 3 &= x \log 100 \\ x\log 3 - \log 3 &= 2x \\ x\log 3 - 2x &= \log 3 \\ x(\log 3 - 2) &= \log 3 \\ x &= \frac{\log 3}{\log 3 - 2} \\ x &\approx -0.3133 \end{aligned}$$

27.
$$\begin{aligned} 5e^x &= 45 \\ e^x &= 9 \\ \ln e^x &= \ln 9 \\ x \ln e &= \ln 9 \\ x &= \ln 9 \approx 2.1972 \end{aligned}$$

28.
$$\begin{aligned} \ln(5x+2) &= \ln(2x+5) \\ 5x + 2 &= 2x + 5 \\ 3x &= 3 \\ x &= 1 \end{aligned}$$

29.
$$\begin{aligned} \log x + \log(x-9) &= 1 \\ \log x(x-9) &= 1 \\ x(x-9) &= 10^1 \\ x^2 - 9x - 10 &= 0 \\ (x-10)(x+1) &= 0 \end{aligned}$$
$x - 10 = 0$ **or** $x + 1 = 0$
$x = 10 \qquad x = -1$
extraneous

30.
$$\begin{aligned} \log_6 18 - \log_6(x-3) &= \log_6 3 \\ \log_6 \frac{18}{x-3} &= \log_6 3 \\ \frac{18}{x-3} &= 3 \\ 18 &= 3x - 9 \\ 27 &= 3x \\ 9 &= x \end{aligned}$$

Cumulative Review Exercises (page 585)

1. $f(x) = 2(x+5)^2 - 8$

$a = 2 \Rightarrow$ up, vertex: $(-5, -8)$

$$0 = 2(x+5)^2 - 8$$
$$8 = 2(x+5)^2$$
$$4 = (x+5)^2$$
$$\pm 2 = x + 5$$
$$-5 \pm 2 = x$$

$x = -3, x = -7 \Rightarrow (-3, 0), (-7, 0)$

$f(0) = 42 \Rightarrow (0, 42)$

axis of symmetry: $x = -5$

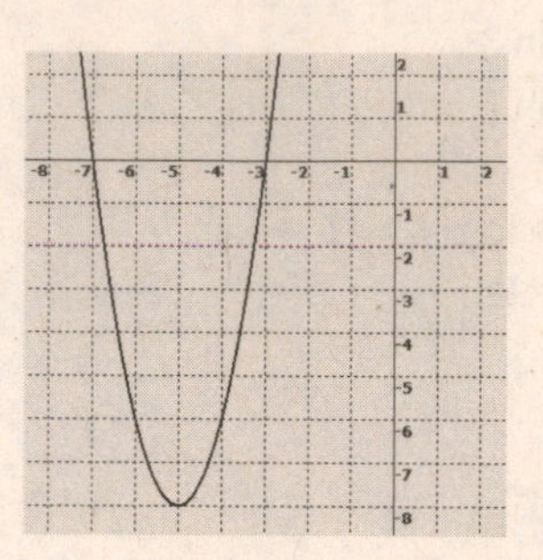

2. $f(x) = -x^2 - 6x - 5;$

$a = -1, b = -6, c = -5$

$$x = -\frac{b}{2a} = -\frac{-6}{2(-1)} = -3$$
$$y = -x^2 - 6x - 5$$
$$= -(-3)^2 - 6(-3) - 5 = 4$$

vertex: $(-3, 4)$, $a = -1 \Rightarrow$ down

$$0 = -x^2 - 6x - 5$$
$$0 = x^2 + 6x + 5$$
$$0 = (x+1)(x+5)$$

$x = -1$ or $x = -5 \Rightarrow (-1, 0), (-5, 0)$

$f(0) = -5 \Rightarrow (0, -5)$

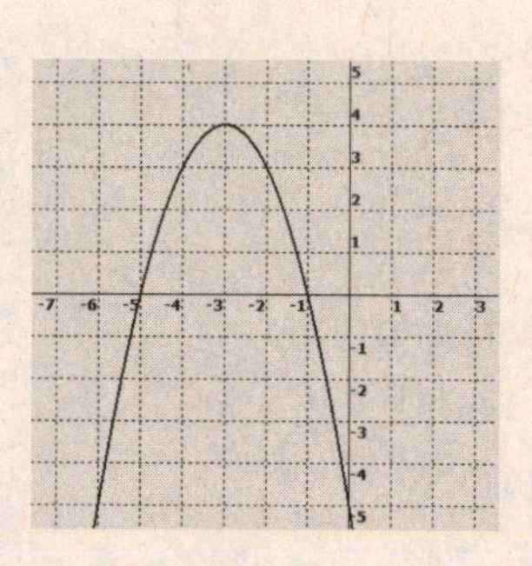

3. $y = f(x) = x^3 + x$

$$f(-x) = (-x)^3 + (-x)$$
$$= -x^3 - x$$
$$= -f(x) \Rightarrow \text{odd}$$

x-int.	y-int.
$x^3 + x = 0$	$y = 0^3 + 0$
$x(x^2 + 1) = 0$	$y = 0$
$x = 0$	$(0, 0)$
$(0, 0)$	

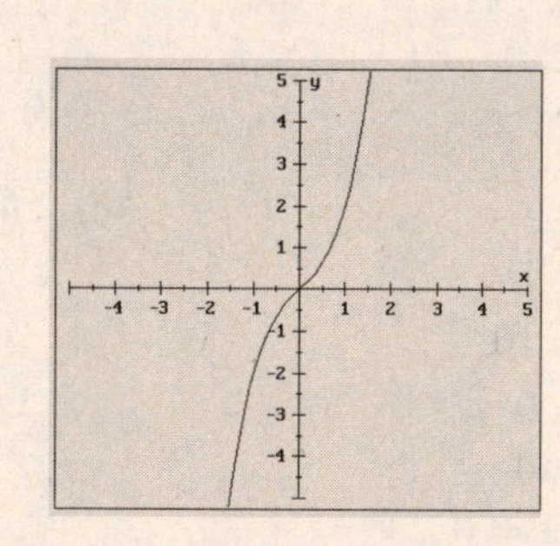

CUMULATIVE REVIEW EXERCISES

4. $f(x) = -x^4 + 2x^2 + 1$

$f(-x) = -(-x)^4 + 2(-x)^2 + 1$

$= -x^4 + 2x^2 + 1 \Rightarrow$ even

x-int.	y-int.
$-x^4 + 2x^2 + 1 = 0$	$f(0) = 1$
$x^4 - 2x^2 - 1 = 0$	$y = 1$
not rational numbers	$(0, 1)$

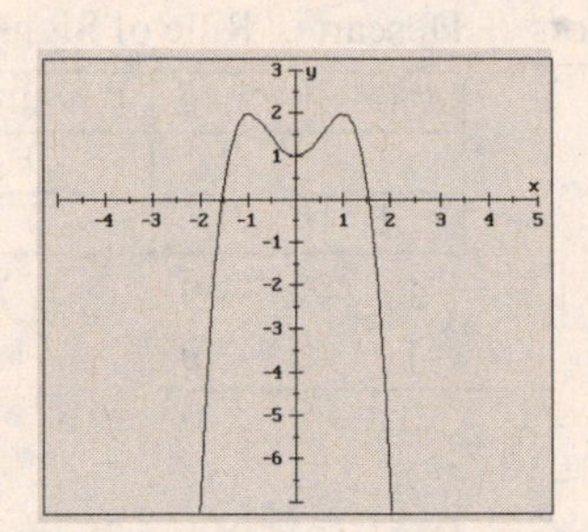

5. $\begin{array}{r|rrrr} 1 & 4 & 0 & 3 & 2 \\ & & 4 & 4 & 7 \\ \hline & 4 & 4 & 7 & 9 \end{array}$ $P(1) = 9$

6. $\begin{array}{r|rrrr} -2 & 4 & 0 & 3 & 2 \\ & & -8 & 16 & -38 \\ \hline & 4 & -8 & 19 & -36 \end{array}$ $P(-2) = -36$

7. $\begin{array}{r|rrrr} \frac{1}{2} & 4 & 0 & 3 & 2 \\ & & 2 & 1 & 2 \\ \hline & 4 & 2 & 4 & 4 \end{array}$ $P(\frac{1}{2}) = 4$

8. $\begin{array}{r|rrrr} i & 4 & 0 & 3 & 2 \\ & & 4i & -4 & -i \\ \hline & 4 & 4i & -1 & 2-i \end{array}$ $P(i) = 2 - i$

9. $\begin{array}{r|rrrr} -1 & 1 & 2 & -1 & -2 \\ & & -1 & -1 & 2 \\ \hline & 1 & 1 & -2 & 0 \end{array}$ factor

10. $\begin{array}{r|rrrr} 2 & 1 & 2 & -1 & -2 \\ & & 2 & 8 & 14 \\ \hline & 1 & 4 & 7 & 12 \end{array}$ not a factor

11. $\begin{array}{r|rrrr} 1 & 1 & 2 & -1 & -2 \\ & & 1 & 3 & 2 \\ \hline & 1 & 3 & 2 & 0 \end{array}$ factor

12. $\begin{array}{r|rrrr} -2 & 1 & 2 & -1 & -2 \\ & & -2 & 0 & 2 \\ \hline & 1 & 0 & -1 & 0 \end{array}$ factor

13. $P(x) = x^{12} - 4x^8 + 2x^4 + 12 = 0$

12 zeros

14. $P(x) = x^{2000} - 1 = 0 \Rightarrow$ 2000 zeros

15. $P(x) = x^4 + 2x^3 - 3x^2 + x + 2$: 2 sign variations $\Rightarrow$ 2 or 0 positive roots

$P(-x) = (-x)^4 + 2(-x)^3 - 3(-x)^2 + (-x) + 2$

$= x^4 - 2x^3 - 3x^2 - x + 2$: 2 sign variations $\Rightarrow$ 2 or 0 negative roots

# pos	# neg	# nonreal
2	2	0
2	0	2
0	2	2
0	0	4

16. $P(x) = x^4 - 3x^3 - 2x^2 - 3x - 5$

1 sign variation $\Rightarrow$ 1 positive root

$P(-x) = (-x)^4 - 3(-x)^3 - 2(-x)^2 - 3(-x) - 5$

$= x^4 + 3x^3 - 2x^2 + 3x - 5$

3 sign variations $\Rightarrow$ 3 or 1 negative roots

# pos	# neg	# nonreal
1	3	0
1	1	2

CUMULATIVE REVIEW EXERCISES

17. Possible rational roots: $\pm 1, \pm 3, \pm 9$

Descartes' Rule of Signs

# pos	# neg	# nonreal
1	2	0
1	0	2

Test $x = -1$:

$$\begin{array}{r|rrrr} -1 & 1 & 1 & -9 & -9 \\ & & -1 & 0 & 9 \\ \hline & 1 & 0 & -9 & 0 \end{array}$$

$$x^3 + x^2 - 9x - 9 = 0$$
$$(x+1)\left(x^2 - 9\right) = 0$$
$$(x+1)(x+3)(x-3) = 0$$

Solution set: $\{-1, -3, 3\}$

18. Possible rational roots: $\pm 1, \pm 2$

Descartes' Rule of Signs

# pos	# neg	# nonreal
2	1	0
0	1	2

Test $x = -1$:

$$\begin{array}{r|rrrr} -1 & 1 & -2 & -1 & 2 \\ & & -1 & 3 & -2 \\ \hline & 1 & -3 & 2 & 0 \end{array}$$

$$x^3 - 2x^2 - x + 2 = 0$$
$$(x+1)\left(x^2 - 3x + 2\right) = 0$$
$$(x+1)(x-1)(x-2) = 0$$

Solution set: $\{-1, 1, 2\}$

19. $y = \dfrac{x}{x-3}$

Vertical: $x = 3$; Horizontal: $y = 1$
Slant: none; x-intercepts: $(0, 0)$
y-intercepts: $(0, 0)$; Symmetry: none

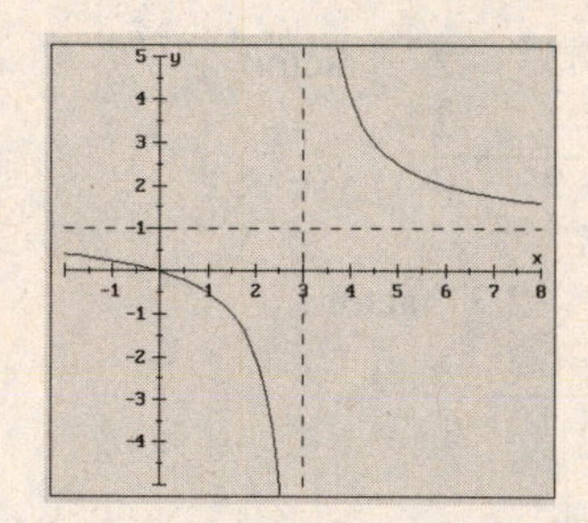

20. $y = \dfrac{x^2 - 1}{x^2 - 9} = \dfrac{(x+1)(x-1)}{(x+3)(x-3)}$

Vertical: $x = -3, x = 3$; Horizontal: $y = 1$
Slant: none; x-intercepts: $(-1, 0), (1, 0)$
y-intercepts: $\left(0, \frac{1}{9}\right)$; Symmetry: y-axis

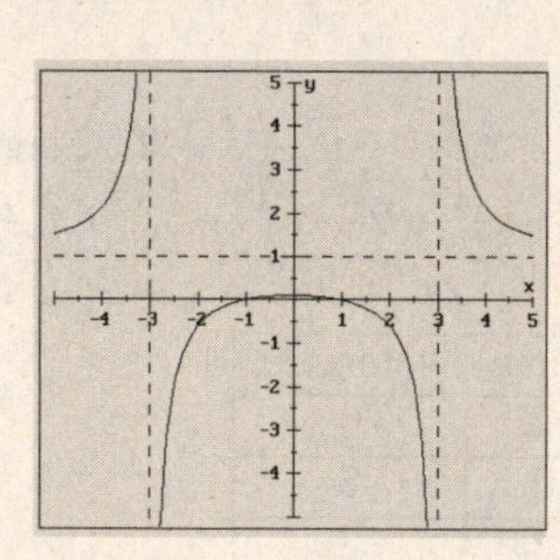

CUMULATIVE REVIEW EXERCISES

21. $f(x) = 3^x - 2$; Shift $y = 3^x$ D2.

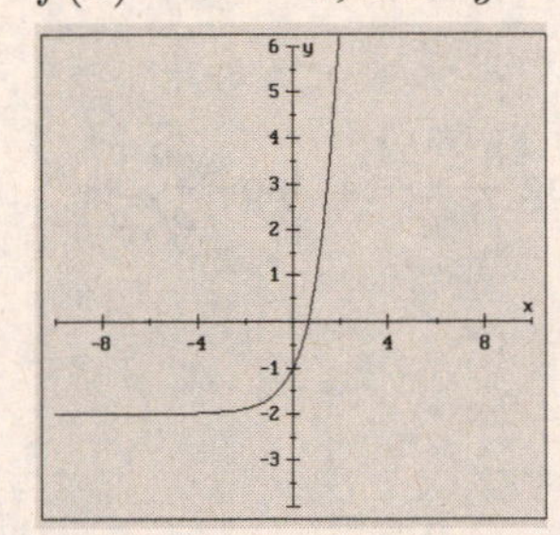

22. $f(x) = 2e^x$; Stretch $y = e^x$ vertically by a factor of 2.

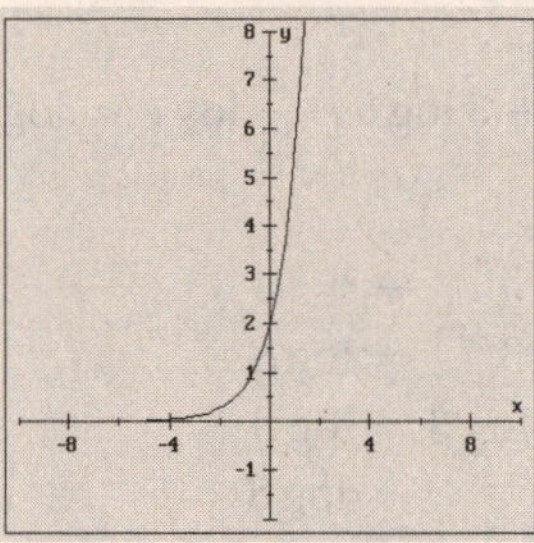

23. $f(x) = \log_3 x$

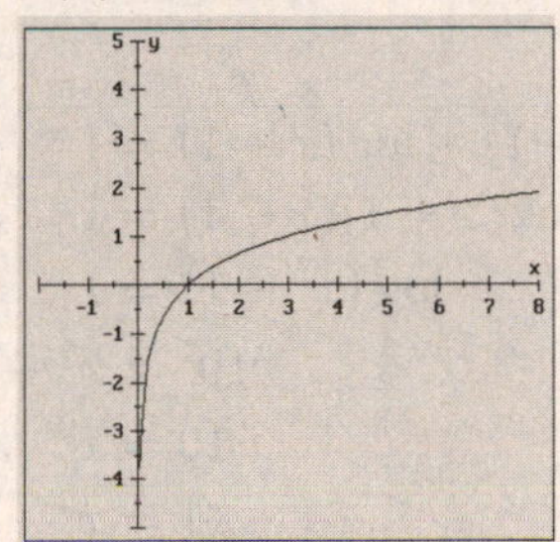

24. $f(x) = \ln(x-2)$; Shift $y = \ln x$ R2.

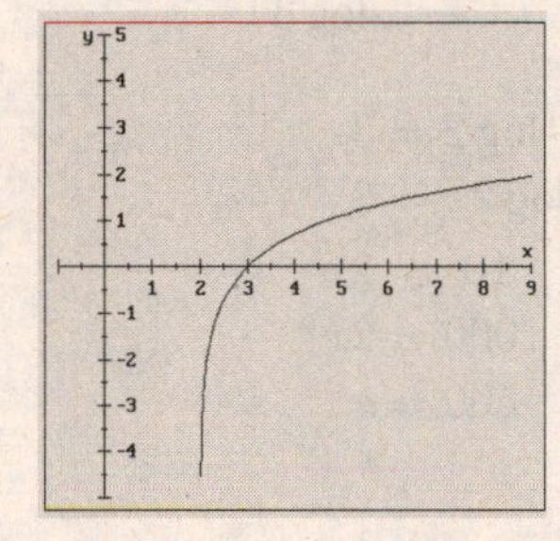

25. $\log_2 64 = 6$ (because $2^6 = 64$)

26. $\log_{1/2} 8 = -3$ $\left(\text{because } \left(\frac{1}{2}\right)^{-3} = 8\right)$

27. $\ln e^3 = 3 \ln e = 3$

28. $2^{\log_2 2} = 2$

29.
$$\log_5 x = -3$$
$$5^{-3} = x$$
$$\frac{1}{125} = x$$

30.
$$\log_x 72 = 2$$
$$x^2 = 72$$
$$x = \sqrt{72} = 6\sqrt{2}$$

31. $\log abc = \log a + \log b + \log c$

32.
$$\begin{aligned}\log \frac{a^2 b}{c} &= \log a^2 b - \log c \\ &= \log a^2 + \log b - \log c \\ &= 2\log a + \log b - \log c\end{aligned}$$

33.
$$\begin{aligned}\log \sqrt{\frac{ab}{c^3}} &= \log \left(\frac{ab}{c^3}\right)^{1/2} \\ &= \tfrac{1}{2} \log \left(\frac{ab}{c^3}\right) \\ &= \tfrac{1}{2}(\log ab - \log c^3) \\ &= \tfrac{1}{2}(\log a + \log b - 3\log c)\end{aligned}$$

34.
$$\begin{aligned}\ln \frac{\sqrt{ab^2}}{c} &= \ln \frac{\left(ab^2\right)^{1/2}}{c} \\ &= \ln \left(ab^2\right)^{1/2} - \ln c \\ &= \tfrac{1}{2} \ln \left(ab^2\right) - \ln c \\ &= \tfrac{1}{2}(\ln a + \ln b^2) - \ln c \\ &= \tfrac{1}{2}(\ln a + 2\ln b) - \ln c \\ &= \tfrac{1}{2} \ln a + \ln b - \ln c\end{aligned}$$

35. $3\ln a - 3\ln b = \ln a^3 - \ln b^3 = \ln \dfrac{a^3}{b^3}$

36. $\dfrac{1}{2}\log a + 3\log b - \dfrac{2}{3}\log c = \log a^{1/2} + \log b^3 - \log c^{2/3} = \log \dfrac{a^{1/2}b^3}{c^{2/3}} = \log \dfrac{\sqrt{a}\, b^3}{\sqrt[3]{c^2}}$

37.
$$\begin{aligned} 3^{x+1} &= 8 \\ \log 3^{x+1} &= \log 8 \\ (x+1)\log 3 &= \log 8 \\ x+1 &= \frac{\log 8}{\log 3} \\ x &= \frac{\log 8}{\log 3} - 1 \end{aligned}$$

38.
$$\begin{aligned} 3^{x-1} &= 3^{2x} \\ x-1 &= 2x \\ -1 &= x \end{aligned}$$

39.
$$\begin{aligned} \log x + \log 2 &= 3 \\ \log 2x &= 3 \\ 10^3 &= 2x \\ 1000 &= 2x \\ 500 &= x \end{aligned}$$

40.
$$\begin{aligned} \log(x+1) + \log(x-1) &= 1 \\ \log(x+1)(x-1) &= 1 \\ \log\left(x^2-1\right) &= 1 \\ 10^1 &= x^2 - 1 \\ 10 &= x^2 - 1 \\ 11 &= x^2 \\ \pm\sqrt{11} &= x \end{aligned}$$

Only the positive answer, $x = \sqrt{11}$, checks.

Exercises 6.1 (page 598)

1. system

3. consistent

5. independent

7. consistent

9. dependent

11. is

13. $\begin{cases} y = -3x + 5 \\ x - 2y = -3 \end{cases}$

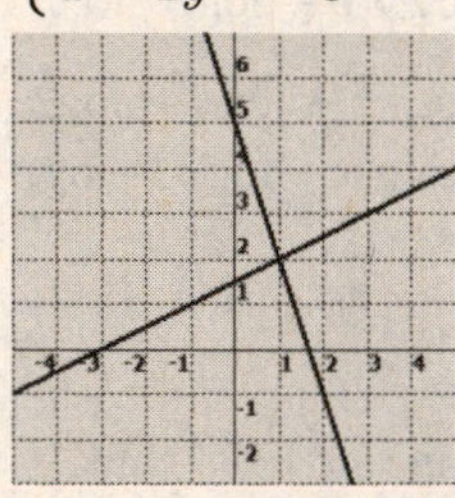

solution: $(1, 2)$

15.

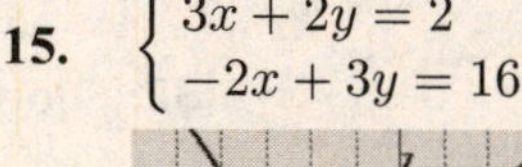

$\begin{cases} 3x + 2y = 2 \\ -2x + 3y = 16 \end{cases}$

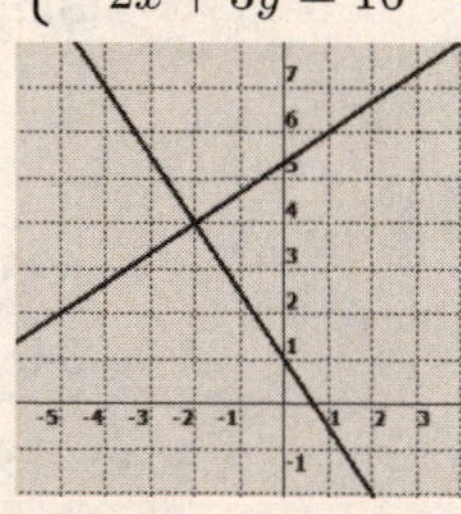

solution: $(-2, 4)$

17. $\begin{cases} y = -x + 5 \\ 3x + 3y = 30 \end{cases}$

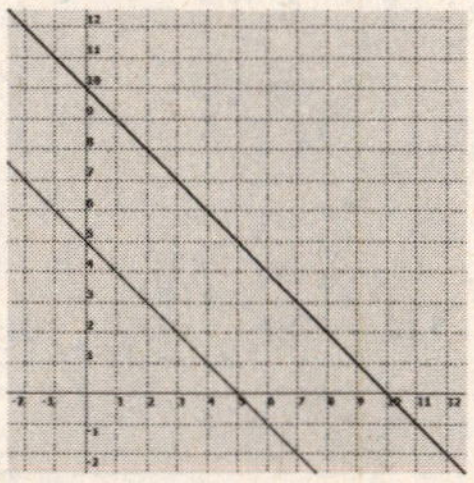

no solution
inconsistent system

EXERCISES 6.1

19. $\begin{cases} y = -x + 6 \\ 5x + 5y = 30 \end{cases}$

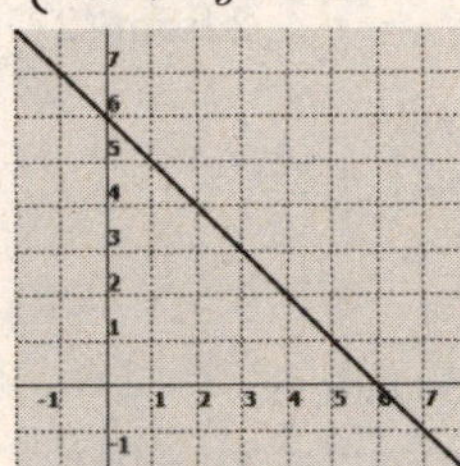

dependent equations
infinitely many solutions

21. $\begin{cases} y = -5.7x + 7.8 \\ y = 37.2 - 19.1x \end{cases}$

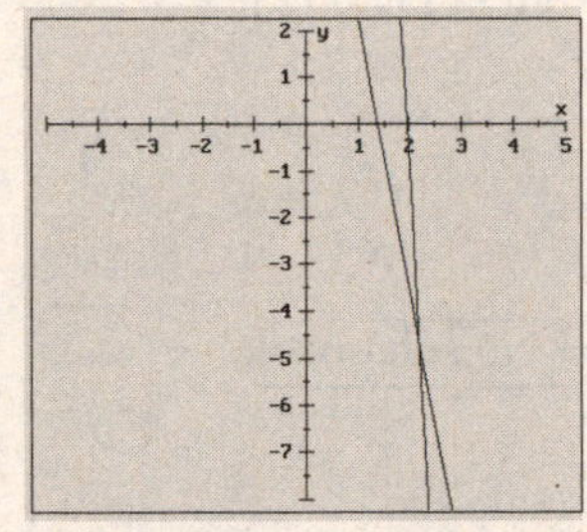

solution: $(2.2, -4.7)$

23. $\begin{cases} y = \dfrac{5.5 - 2.7x}{3.5} \\ 5.3x - 9.2y = 6.0 \end{cases}$

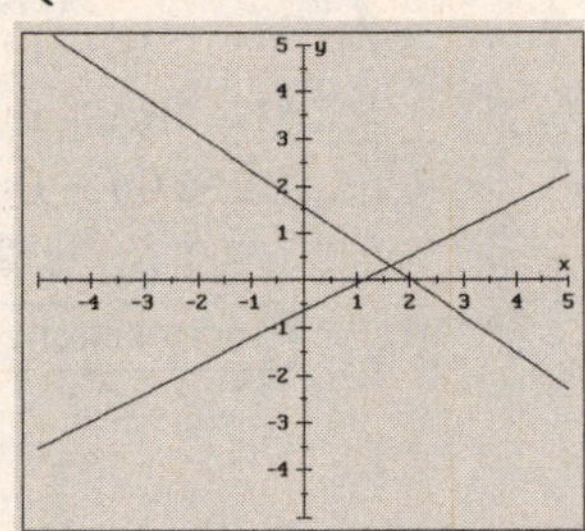

solution: $(1.7, 0.3)$

25. $\begin{cases} (1) \quad y = x - 1 \\ (2) \quad y = 2x \end{cases}$

Substitute $y = x - 1$ from (1) into (2):

$$y = 2x$$
$$x - 1 = 2x$$
$$-1 = x$$

Substitute and solve for y:

$y = 2x$

$y = 2(-1) = -2$

$(-1, -2)$

27. $\begin{cases} (1) \quad 2x + 3y = 0 \\ (2) \quad y = 3x - 11 \end{cases}$

Substitute $y = 3x - 11$ from (2) into (1):

$$2x + 3y = 0$$
$$2x + 3(3x - 11) = 0$$
$$2x + 9x - 33 = 0$$
$$11x = 33$$
$$x = 3$$

Substitute and solve for y:

$y = 3x - 11$

$y = 3(3) - 11 = -2$

$(3, -2)$

29. $\begin{cases} (1) \quad 4x + 3y = 3 \\ (2) \quad 2x - 6y = -1 \end{cases}$

Substitute $y = \dfrac{3 - 4x}{3}$ from (1) into (2):

$$2x - 6y = -1$$
$$2x - 6 \cdot \frac{3 - 4x}{3} = -1$$
$$2x - 2(3 - 4x) = -1$$
$$2x - 6 + 8x = -1$$
$$10x = 5$$
$$x = \tfrac{1}{2}$$

Substitute and solve for y:

$$4x + 3y = 3$$
$$4\left(\tfrac{1}{2}\right) + 3y = 3$$
$$2 + 3y = 3$$
$$3y = 1$$
$$y = \tfrac{1}{3}$$

Solution:

$\left(\tfrac{1}{2}, \tfrac{1}{3}\right)$

EXERCISES 6.1

31. $\begin{cases} (1) & x + 3y = 1 \\ (2) & 2x + 6y = 3 \end{cases}$

Substitute $x = 1 - 3y$ from (1) into (2):

$$\begin{aligned} 2x + 6y &= 3 \\ 2(1 - 3y) + 6y &= 3 \\ 2 - 6y + 6y &= 3 \\ 2 &\neq 3 \end{aligned}$$

$\boxed{\text{Inconsistent system} \Rightarrow \text{No solution}}$

33. $\begin{cases} (1) & y = 3x - 6 \\ (2) & x = \frac{1}{3}y + 2 \end{cases}$

Substitute $x = \frac{1}{3}y + 2$ from (2) into (1):

$$\begin{aligned} y &= 3x - 6 \\ y &= 3\left(\tfrac{1}{3}y + 2\right) - 6 \\ y &= y + 6 - 6 \\ 0 &= 0 \end{aligned}$$

Dependent equations

General solution: $(x, 3x - 6)$

35.

$$\begin{array}{ll} 5x - 3y = 12 \Rightarrow \times\ (-1) & -5x + 3y = -12 \\ \underline{2x - 3y = \ \ 3} & \underline{\ \ \ 2x - 3y = \ \ \ 3} \\ & -3x \qquad = -9 \\ & \ \ \ x \qquad = \ \ \ 3 \end{array}$$

$$\begin{aligned} 2x - 3y &= 3 \\ 2(3) - 3y &= 3 \\ 6 - 3y &= 3 \\ -3y &= -3 \\ y &= 1 \end{aligned}$$

Solution: $(3, 1)$

37.

$$\begin{array}{ll} x - 7y = -11 \Rightarrow \times\ (-8) & -8x + 56y = \ \ 88 \\ \underline{8x + 2y = \ \ 28} & \underline{\ \ \ 8x + \ \ 2y = \ \ 28} \\ & \qquad 58y = 116 \\ & \qquad\ \ \ y = \ \ \ 2 \end{array}$$

$$\begin{aligned} x - 7y &= -11 \\ x - 7(2) &= -11 \\ x - 14 &= -11 \\ x &= 3 \end{aligned}$$

Solution: $(3, 2)$

39.

$$\begin{array}{l} 3(x - y) = y - 9 \Rightarrow 3x - 3y = y - 9 \Rightarrow 3x - 4y = \ \ -9 \Rightarrow \times\ (5) \\ \underline{5(x + y) = \ \ -15} \Rightarrow \underline{5x + 5y = \ \ -15} \Rightarrow \underline{5x + 5y = -15} \Rightarrow \times\ (4) \end{array}$$

$$\begin{array}{ll} 15x - 20y = & -45 \\ \underline{20x + 20y =} & \underline{-60} \\ 35x \qquad = & -105 \\ x \qquad = & -3 \end{array}$$

$$\begin{aligned} 5x + 5y &= -15 \\ 5(-3) + 5y &= -15 \\ -15 + 5y &= -15 \\ 5y &= 0 \\ y &= 0 \end{aligned}$$

Solution: $(-3, 0)$

41.

$$\begin{array}{lll} 2 = \dfrac{1}{x + y} & \Rightarrow 2(x + y) = 1 & \Rightarrow 2x + 2y = 1 \\ \underline{2 = \dfrac{3}{x - y}} & \underline{\Rightarrow 2(x - y) = 3} & \underline{\Rightarrow 2x - 2y = 3} \\ & & \quad 4x \qquad = 4 \\ & & \quad\ \ x \qquad = 1 \end{array}$$

$$\begin{aligned} 2x + 2y &= 1 \\ 2(1) + 2y &= 1 \\ 2 + 2y &= 1 \\ 2y &= -1 \\ y &= -\tfrac{1}{2} \end{aligned}$$

Solution: $\left(1, -\frac{1}{2}\right)$

43.

$$\begin{array}{ll} y + 2x = 5 \Rightarrow 2x + \quad y = \ \ 5 & \\ \underline{0.5y = 2.5 - x \Rightarrow \ \ x + 0.5y = 2.5} \Rightarrow \times\ (-2) & \end{array}$$

$$\begin{array}{l} 2x + y = \ \ \ 5 \\ \underline{-2x - y = -5} \\ \qquad\ 0 = \ \ \ 0 \end{array}$$

$\boxed{\text{Dependent Equations}}$

$$\begin{aligned} 2x + y &= 5 \\ y &= -2x + 5 \end{aligned}$$

$\boxed{\text{Gen. sol.: } (x, -2x + 5)}$

EXERCISES 6.1

45.

$$\begin{array}{l} x + 2(x - y) = 2 \Rightarrow x + 2x - 2y = 2 \Rightarrow \quad 3x - 2y = 2 \\ \underline{3(y - x) - y = 5} \Rightarrow \underline{3y - 3x - y = 5} \Rightarrow \underline{-3x + 2y = 5} \\ \qquad\qquad\qquad\qquad\qquad\qquad\qquad\qquad 0 \neq 7 \end{array}$$

No Solution

Inconsistent system

47.

$$\begin{array}{lll} x + \dfrac{y}{3} = \dfrac{5}{3} \quad \Rightarrow \times\ (3) & 3x + y = 5 \quad \Rightarrow 3x + y = 5 & 3x + y = \ \ 5 \\ \underline{\dfrac{x + y}{3} = 3 - x} \Rightarrow \times\ (3) & \underline{x + y = 9 - 3x} \Rightarrow \underline{4x + y = 9} \Rightarrow \times\ (-1) & \underline{-4x - y = -9} \\ & & -x \quad = -4 \\ & & x \quad = \ \ 4 \end{array}$$

$$\begin{array}{rl} 3x + y = 5 & \text{Solution:} \\ 3(4) + y = 5 & (4, -7) \\ 12 + y = 5 & \\ y = -7 & \end{array}$$

49.

$$\begin{array}{llll} \dfrac{3}{2}x + \dfrac{1}{3}y = 2 \Rightarrow \times(6) & 9x + 2y = 12 \Rightarrow & 9x + 2y = \ \ 12 & 6x + y = 9 \\ \underline{\dfrac{2}{3}x + \dfrac{1}{9}y = 1} \Rightarrow \times(9) & \underline{6x + \ \ y = \ \ 9} \Rightarrow \times(-2) & \underline{-12x - 2y = -18} & 6(2) + y = 9 \\ & & -3x \quad = -6 & 12 + y = 9 \\ & & x \quad = \ \ 2 & y = -3 \end{array}$$

Solution: $(2, -3)$

51.

$$\begin{array}{lll} \dfrac{x - y}{5} + \dfrac{x + y}{2} = 6 \Rightarrow \times(10) & 2(x - y) + 5(x + y) = 60 \Rightarrow & 2x - 2y + 5x + 5y = 60 \\ \underline{\dfrac{x - y}{2} - \dfrac{x + y}{4} = 3} \Rightarrow \times(8) & \underline{4(x - y) - 2(x + y) = 24} \Rightarrow & \underline{4x - 4y - 2x - 2y = 24} \end{array}$$

$$\begin{array}{llll} 7x + 3y = 60 \Rightarrow \times\ (2) & 14x + 6y = 120 & 7x + 3y = 60 & \text{Solution:} \\ \underline{2x - 6y = 24} & \underline{\ \ 2x - 6y = \ \ 24} & 7(9) + 3y = 60 & (9, -1) \\ & 16x \quad = 144 & 63 + 3y = 60 & \\ & x \quad = \ \ 9 & 3y = -3 & \\ & & y = -1 & \end{array}$$

53.

$$\begin{array}{lll} (1)\ \ x + y + z = 3 & \text{Add (1) and (3):} & \text{Add equations (2) and (3):} \\ (2)\ \ 2x + y + z = 4 & (1)\ \ \ x + \ \ y + z = 3 & (2)\ \ 2x + \ \ y + z = 4 \\ (3)\ \ 3x + y - z = 5 & (3)\ \ \underline{3x + \ \ y - z = 5} & (3)\ \ \underline{3x + \ \ y - z = 5} \\ & (4)\ \ 4x + 2y \quad = 8 & (5)\ \ 5x + 2y \quad = 9 \end{array}$$

Solve the system of two equations and two unknowns formed by equations (4) and (5):

$$\begin{array}{lllll} 4x + 2y = 8 \Rightarrow \times(-1) & -4x - 2y = -8 & 4x + 2y = 8 & x + y + z = 3 & \text{Solution:} \\ \underline{5x + 2y = 9} \Rightarrow & \underline{\ \ 5x + 2y = \ \ 9} & 4(1) + 2y = 8 & 1 + 2 + z = 3 & (1, 2, 0) \\ & x \quad = \ \ 1 & 2y = 4 & 3 + z = 3 & \\ & & y = 2 & z = 0 & \end{array}$$

55. (1) $x - y + z = 0$
(2) $x + y + 2z = -1$
(3) $-x - y + z = 0$

Add (1) and (2):

$$\begin{array}{lrcl} (1) & x - y + z &=& 0 \\ (2) & x + y + 2z &=& -1 \\ \hline (4) & 2x \quad + 3z &=& -1 \end{array}$$

Add equations (2) and (3):

$$\begin{array}{lrcl} (2) & x + y + 2z &=& -1 \\ (3) & -x - y + z &=& 0 \\ \hline (5) & 3z &=& -1 \\ & z &=& -\frac{1}{3} \end{array}$$

Solve the system of two equations and two unknowns formed by equations (4) and (5):

$$\begin{aligned} 2x + 3z &= -1 \\ 2x + 3\left(-\tfrac{1}{3}\right) &= -1 \\ 2x &= 0 \\ x &= 0 \end{aligned} \qquad \begin{aligned} x - y + z &= 0 \\ 0 - y + \left(-\tfrac{1}{3}\right) &= 0 \\ -y &= \tfrac{1}{3} \\ y &= -\tfrac{1}{3} \end{aligned}$$

Solution: $\left(0, -\frac{1}{3}, -\frac{1}{3}\right)$

57. (1) $2x + y = 4$
(2) $x - z = 2$
(3) $y + z = 1$

Add (2) and (3):

$$\begin{array}{lrcl} (2) & x \quad - z &=& 2 \\ (3) & y + z &=& 1 \\ \hline (4) & x + y &=& 3 \end{array}$$

Solve the system of two equations and two unknowns formed by equations (1) and (4):

$$\begin{array}{l} 2x + y = 4 \Rightarrow \times(-1) \\ \underline{x + y = 3 \Rightarrow} \end{array} \quad \begin{array}{rcl} -2x - y &=& -4 \\ x + y &=& 3 \\ \hline -x &=& -1 \\ x &=& 1 \end{array}$$

$$\begin{aligned} x + y &= 3 \\ 1 + y &= 3 \\ y &= 2 \end{aligned} \qquad \begin{aligned} y + z &= 1 \\ 2 + z &= 1 \\ z &= -1 \end{aligned}$$

Solution: $(1, 2, -1)$

59. (1) $x + y + z = 6$
(2) $2x + y + 3z = 17$
(3) $x + y + 2z = 11$

Add (1) and $-(2)$:

$$\begin{array}{lrcl} (1) & x + y + z &=& 6 \\ -(2) & -2x - y - 3z &=& -17 \\ \hline (4) & -x \quad - 2z &=& -11 \end{array}$$

Add equations (1) and $-(3)$:

$$\begin{array}{lrcl} (1) & x + y + z &=& 6 \\ -(3) & -x - y - 2z &=& -11 \\ \hline (5) & -z &=& -5 \\ & z &=& 5 \end{array}$$

Solve the system of two equations and two unknowns formed by equations (4) and (5):

$$\begin{aligned} -x - 2z &= -11 \\ -x - 2(5) &= -11 \\ -x &= -1 \\ x &= 1 \end{aligned} \qquad \begin{aligned} x + y + z &= 6 \\ 1 + y + 5 &= 6 \\ y &= 0 \end{aligned}$$

Solution: $(1, 0, 5)$

61. (1) $x + y + z = 3$
(2) $x + z = 2$
(3) $2x + 2y + 2z = 3$

Add $-2 \cdot (1)$ and (3):

$$\begin{array}{lrcl} -2 \cdot (1) & -2x - 2y - 2z &=& -6 \\ (3) & 2x + 2y + 2z &=& 3 \\ \hline (4) & 0 &\neq& -3 \end{array}$$

No solution; inconsistent system

63. (1) $x+2y-z=2$
(2) $2x-y=-1$
(3) $3x+y+z=1$

Add (1) and (3):

$$\begin{array}{ll} (1) & x+2y-z=2 \\ (3) & 3x+\ \ y+z=1 \\ \hline (4) & 4x+3y \qquad\ \ =3 \end{array}$$

Solve the system of two equations and two unknowns formed by equations (2) and (4):

$$\begin{array}{l} 2x-\ \ y=-1 \Rightarrow \times(3) \\ \underline{4x+3y=\ \ 3 \Rightarrow} \end{array} \quad \begin{array}{r} 6x-3y=-3 \\ \underline{4x+3y=\ \ 3} \\ 10x \qquad =\ \ 0 \\ x \qquad =\ \ 0 \end{array} \quad \begin{array}{r} 2x-y=-1 \\ 2(0)-y=-1 \\ -y=-1 \\ y=1 \end{array} \quad \begin{array}{r} 3x+y+z=1 \\ 3(0)+1+z=1 \\ z=0 \end{array}$$

Solution: $(0,1,0)$

65. (1) $3x+4y+2z=4$
(2) $6x-2y+z=4$
(3) $3x-8y-6z=-3$

Add (1) and $2\cdot(2)$:

$$\begin{array}{ll} (1) & 3x+4y+2z=\ \ 4 \\ 2\cdot(2) & 12x-4y+2z=\ \ 8 \\ \hline (4) & 15x \qquad +4z=12 \end{array}$$

Add equations $2\cdot(1)$ and (3):

$$\begin{array}{ll} 2\cdot(1) & 6x+8y+4z=\ \ 8 \\ (3) & 3x-8y-6z=-3 \\ \hline (5) & 9x \qquad -2z=\ \ 5 \end{array}$$

Solve the system of two equations and two unknowns formed by equations (4) and (5):

$$\begin{array}{l} 15x+4z=12 \Rightarrow \\ \underline{\ 9x-2z=\ \ 5 \Rightarrow \times(2)} \end{array} \quad \begin{array}{r} 15x+4z=12 \\ \underline{18x-4z=10} \\ 33x \qquad =22 \\ x \qquad =\frac{2}{3} \end{array} \quad \begin{array}{r} 15x+4z=12 \\ 15\left(\frac{2}{3}\right)+4z=12 \\ 4z=2 \\ z=\frac{1}{2} \end{array} \quad \begin{array}{r} 3x+4y+2z=4 \\ 3\left(\frac{2}{3}\right)+4y+2\left(\frac{1}{2}\right)=4 \\ 4y=1 \\ y=\frac{1}{4} \end{array}$$

Solution: $\left(\frac{2}{3},\frac{1}{4},\frac{1}{2}\right)$

67. (1) $2x-y-z=0$
(2) $x-2y-z=-1$
(3) $x-y-2z=-1$

Add (1) and $-2\cdot(2)$:

$$\begin{array}{ll} (1) & 2x-\ \ y-\ \ z=0 \\ -2\cdot(2) & -2x+4y+2z=2 \\ \hline (4) & 3y+\ \ z=2 \end{array}$$

Add equations (1) and $-2\cdot(3)$:

$$\begin{array}{ll} (1) & 2x-\ \ y-\ \ z=0 \\ (3) & -2x+2y+4z=2 \\ \hline (5) & y+3z=2 \end{array}$$

Solve the system of two equations and two unknowns formed by equations (4) and (5):

$$\begin{array}{l} 3y+\ \ z=2 \Rightarrow \\ \underline{\ y+3z=2 \Rightarrow \times(-3)} \end{array} \quad \begin{array}{r} 3y+\ \ z=\ \ 2 \\ \underline{-3y-9z=-6} \\ -8z=-4 \\ z=\ \ \frac{1}{2} \end{array} \quad \begin{array}{r} y+3z=2 \\ y+3\left(\frac{1}{2}\right)=2 \\ y=\frac{1}{2} \end{array} \quad \begin{array}{r} x-y-2z=-1 \\ x-\frac{1}{2}-2\left(\frac{1}{2}\right)=-1 \\ x=\frac{1}{2} \end{array}$$

Solution: $\left(\frac{1}{2},\frac{1}{2},\frac{1}{2}\right)$

69. (1) $(x+y)+(y+z)+(z+x)=6 \Rightarrow \quad 2x+2y+2z=6$
(2) $(x-y)+(y-z)+(z-x)=0 \Rightarrow \quad 0=0$
(3) $x+y+2z=4 \Rightarrow \quad x+y+2z=4$

Add (1) and $-2\cdot(3)$:

$$\begin{array}{ll} (1) & 2x+2y+2z=\ \ 6 \\ -2\cdot(3) & -2x-2y-4z=-8 \\ \hline (4) & -2z=-2 \\ & z=\ \ 1 \end{array}$$

Since (2) is always true, the equations are dependent. z must equal 1. Then, from (3), $x+y=2$. Let $x=$ any real #. Then $y=2-x$. Solution: $\boxed{(x,2-x,1)}$

71. Let x = cost of a hamburger and let y = cost of the fries. Then $\begin{cases} (1) & 2x+4y=8 \\ (2) & 3x+2y=8 \end{cases}$

$$\begin{array}{ll} 2x+4y=8 & \\ 3x+2y=8 \Rightarrow \times(-2) & \end{array} \quad \begin{array}{rcr} 2x+4y &=& 8 \\ -6x-4y &=& -16 \\ \hline -4x &=& -8 \\ x &=& 2 \end{array} \quad \begin{array}{r} 2x+4y=8 \\ 2(2)+4y=8 \\ 4y=4 \\ y=1 \end{array}$$

A hamburger costs $2, while an order of fries costs $1.

73. Let x = acres of corn and let y = acres of soybeans. Then $\begin{cases} (1) & x+y=350 \\ (2) & x=y+100 \end{cases}$

Substitute $x=y+100$ from (2) into (1):

$$\begin{aligned} x+y&=350 \\ y+100+y&=350 \\ 2y&=250 \\ y&=125 \end{aligned}$$

Substitute and solve for x:

$$\begin{aligned} x&=y+100 \\ &=125+100=225 \end{aligned}$$

The farmer should plant 225 acres of corn and 125 acres of soybeans.

75. Let b = speed in still water. Let c = the speed of the current. Then the following system applies:

$$\begin{array}{ll} 3(b+c)=30 & 3b+3c=30 \Rightarrow \times\ (5) \\ 5(b-c)=30 & 5b-5c=30 \Rightarrow \times\ (3) \end{array} \quad \begin{array}{rcr} 15b+15c &=& 150 \\ 15b-15c &=& 90 \\ \hline 30b &=& 240 \\ b &=& 8 \end{array}$$

The boat has a speed of 8 kilometers per hour in still water.

77. Let x = grams of 9% alloy. Let y = grams of 84% alloy. Then the following system applies: (note: 34% of 60 is $0.34 \times 60 = 20.4$)

$$\begin{array}{ll} x+y=60 \Rightarrow \times\ (-9) \\ 0.09x+0.84y=20.4 \Rightarrow \times\ (100) \end{array} \quad \begin{array}{rcr} -9x-9y &=& -540 \\ 9x+84y &=& 2040 \\ \hline 75y &=& 1500 \\ y &=& 20 \end{array}$$

She must use 40 grams of the 9% and 20 grams of the 84% alloy.

79.

$$\begin{array}{ll} 448x=112y & \Rightarrow 448x-112y=0 \Rightarrow \times(-1) \\ 448(x+1)=192(y-1) & \Rightarrow 448x-192y=-640 \Rightarrow \end{array} \quad \begin{array}{rcr} -448x+112y &=& 0 \\ 448x-192y &=& -640 \\ \hline -80y &=& -640 \\ y &=& 8 \end{array}$$

$$\begin{aligned} 448x&=112y \\ 448x&=112(8) \\ x&=2 \end{aligned}$$

The lever has a length of 10 feet.

81.

$$\begin{aligned} E(x)&=43.53x+742.72 \\ R(x)&=89.95x \end{aligned} \qquad \begin{aligned} E(x)&=R(x) \\ 43.53x+742.72&=89.95x \\ 742.72&=46.42x \\ 16&=x \Rightarrow \end{aligned}$$

Daily production should be 16 pairs.

EXERCISES 6.1

83. Let $x =$ hours at fast food restaurant, $y =$ hours at gas station and $z =$ janitorial hours.

$$\begin{array}{ll} (1) & x = 15 \\ (2) & x + y + z = 30 \\ (3) & 5.7x + 6.3y + 10z = 198.50 \end{array}$$

Substitute $x = 15$ into (2) and (3):

$$\begin{array}{ll} (2) & 15 + y + z = 30 \\ (3) & 5.7(15) + 6.3y + 10z = 198.5 \end{array}$$

Solve the system of two equations and two unknowns formed by equations (2) and (3):

$$\begin{array}{l} y + z = 15 \Rightarrow \times(-10) \\ 6.3y + 10z = 113 \Rightarrow \end{array} \qquad \begin{array}{rcr} -10y - 10z &=& -150 \\ 6.3y + 10z &=& 113 \\ \hline -3.7y &=& -37 \\ y &=& 10 \end{array} \qquad \begin{array}{r} y + z = 15 \\ 10 + z = 15 \\ z = 5 \end{array}$$

He spends 15 hours cooking, 10 hours at the gas station and 5 hours doing janitorial work.

85. Let $x =$ # between 0-14, $y =$ # between 15-49 and $z =$ # 50 or over.

$$\begin{array}{ll} (1) & x + y + z = 3 \\ (2) & x + y = 2.61 \\ (3) & y + z = 1.95 \end{array}$$

Add (1) and $-(2)$:

$$\begin{array}{lrcr} (1) & x + y + z &=& 3 \\ -(2) & -x - y &=& -2.61 \\ \hline & z &=& 0.39 \end{array}$$

Substitute $z = 0.39$ into (3):

$$\begin{aligned} y + z &= 1.95 \\ y + 0.39 &= 1.95 \\ y &= 1.56 \end{aligned}$$

Substitute $y = 1.56$ into (2):

$$\begin{aligned} x + y &= 2.61 \\ x + 1.56 &= 2.61 \\ x &= 1.05 \end{aligned}$$

There are 1.05 million between 0-14, 1.56 million between 15-49 and 0.39 million over 50.

87. Let $x =$ the smallest angle, $y =$ the middle angle and $z =$ the largest angle.

$$\begin{array}{ll} (1) & x + y + z = 180 \\ (2) & z = x + y + 20 \\ (3) & z = 3x + 10 \end{array}$$

Substitute (3) into (1):

$$\begin{aligned} x + y + 3x + 10 &= 180 \\ 4x + y &= 170 \quad (4) \end{aligned}$$

Substitute (3) and (2):

$$\begin{aligned} 3x + 10 &= x + y + 20 \\ 2x - y &= 10 \quad (5) \end{aligned}$$

Solve the system of two equations and two unknowns formed by equations (4) and (5):

$$\begin{array}{rcl} 4x + y &=& 170 \\ 2x - y &=& 10 \\ \hline 6x &=& 180 \\ x &=& 30 \end{array} \qquad \begin{aligned} 4x + y &= 170 \\ 4(30) + y &= 170 \\ 120 + y &= 170 \\ y &= 50 \end{aligned} \qquad \begin{aligned} z &= 3x + 10 \\ z &= 3(30) + 10 \\ z &= 90 + 10 \\ z &= 100 \end{aligned}$$

Solution: The angles have measures of 30°, 50° and 100°.

89-95. Answers may vary.

97. One example:

$$\begin{cases} (1) & x + y = 3 \\ (2) & x - y = -7 \end{cases}$$

99. One example:

$$\begin{cases} (1) & x + y = 3 \\ (2) & x + y = 7 \end{cases}$$

101. False. The lines are parallel and the system has no solutions.

103. False. The solution would consist only of ordered triples that satisfy the system.

105. True.

107. False. Use the substitution or elimination method.

109. True.

Exercises 6.2 (page 612)

1. matrix **3.** coefficient **5.** equation

7. row equivalent **9.** interchanged **11.** adding, multiple

13. $\begin{cases}(1) & x+y=5\\(2) & x-2y=-4\end{cases} \Rightarrow \begin{cases}(1) & x+y=5\\(2) & -3y=-9\end{cases} \Rightarrow \begin{cases}(1) & x+y=5\\(2) & y=3\end{cases}$

$-(1)+(2)\Rightarrow(2)$ $\quad -\frac{1}{3}(2)\Rightarrow(2)$

From (2): $y=3$ From (1): $x+y=5$, $x+3=5$, $x=2$ Solution: $(2,3)$

15. $\begin{cases}(1) & x-y=1\\(2) & 2x-y=8\end{cases} \Rightarrow \begin{cases}(1) & x-y=1\\(2) & y=6\end{cases}$

$-2(1)+(2)\Rightarrow(2)$

From (2): $y=6$ From (1): $x-y=1$, $x-6=1$, $x=7$ Solution: $(7,6)$

17. $\begin{cases}(1) & x+2y-z=2\\(2) & x-3y+2z=1\\(3) & x+y-3z=-6\end{cases} \Rightarrow \begin{cases}(1) & x+2y-z=2\\(2) & -5y+3z=-1\\(3) & -y-2z=-8\end{cases} \Rightarrow \begin{cases}(1) & x+2y-z=2\\(2) & -5y+3z=-1\\(3) & 13z=39\end{cases} \Rightarrow$

$-(1)+(2)\Rightarrow(2)$, $-(1)+(3)\Rightarrow(3)$ $\quad -5(3)+(2)\Rightarrow(3)$

$\begin{cases}(1) & x+2y-z=2\\(2) & y-\frac{3}{5}z=\frac{1}{5}\\(3) & z=3\end{cases}$

$-\frac{1}{5}(2)\Rightarrow(2)$, $\frac{1}{13}(3)\Rightarrow(3)$

From (3): $z=3$

From (2): $y-\dfrac{3}{5}z=\dfrac{1}{5}$, $y-\dfrac{3}{5}(3)=\dfrac{1}{5}$, $y=2$

From (1): $x+2y-z=2$, $x+2(2)-(3)=2$, $x=1$

Solution: $(1,2,3)$

19. $\begin{cases}(1) & x-y-z=-3\\(2) & 5x+y=6\\(3) & y+z=4\end{cases} \Rightarrow \begin{cases}(1) & x-y-z=-3\\(2) & 6y+5z=21\\(3) & y+z=4\end{cases} \Rightarrow \begin{cases}(1) & x-y-z=-3\\(2) & y+\frac{5}{6}z=\frac{7}{2}\\(3) & -z=-3\end{cases} \Rightarrow$

$-5(1)+(2)\Rightarrow(2)$ $\quad \frac{1}{6}(2)\Rightarrow(2)$, $-6(3)+(2)\Rightarrow(3)$

$\begin{cases}(1) & x-y-z=-3\\(2) & y+\frac{5}{6}z=\frac{7}{2}\\(3) & z=3\end{cases} \Rightarrow$

$-(3)\Rightarrow(3)$

From (3): $z=3$

From (2): $y+\frac{5}{6}z=\frac{7}{2}$, $y+\frac{5}{6}(3)=\frac{7}{2}$, $y=1$

From (1): $x-y-z=-3$, $x-(1)-(3)=-3$, $x=1$

Solution: $(1,1,3)$

EXERCISES 6.2

21. row echelon form

23. reduced row echelon form

25. $\left[\begin{array}{cc|c} 2 & 1 & 3 \\ 1 & -3 & 5 \end{array}\right] \Rightarrow \underset{R_1 \Leftrightarrow R_2}{\left[\begin{array}{cc|c} 1 & -3 & 5 \\ 2 & 1 & 3 \end{array}\right]} \Rightarrow \underset{-2R_1 + R_2 \Rightarrow R_2}{\left[\begin{array}{cc|c} 1 & -3 & 5 \\ 0 & 7 & -7 \end{array}\right]} \Rightarrow \underset{\frac{1}{7}R_2 \Rightarrow R_2}{\left[\begin{array}{cc|c} 1 & -3 & 5 \\ 0 & 1 & -1 \end{array}\right]}$

From R_2: $y = -1$ From R_1: $x - 3y = 5$ Solution: $(2, -1)$

$$x - 3(-1) = 5$$
$$x + 3 = 5$$
$$x = 2$$

27. $\left[\begin{array}{cc|c} 1 & -7 & -2 \\ 5 & -2 & -10 \end{array}\right] \Rightarrow \underset{-5R_1 + R_2 \Rightarrow R_2}{\left[\begin{array}{cc|c} 1 & -7 & -2 \\ 0 & 33 & 0 \end{array}\right]} \Rightarrow \underset{\frac{1}{33}R_2 \Rightarrow R_2}{\left[\begin{array}{cc|c} 1 & -7 & -2 \\ 0 & 1 & 0 \end{array}\right]}$

From R_2: $y = 0$ From R_1: $x - 7y = -2$ Solution: $(-2, 0)$

$$x - 7(0) = -2$$
$$x = -2$$

29. $\left[\begin{array}{cc|c} 2 & -1 & 5 \\ 1 & 3 & 6 \end{array}\right] \Rightarrow \underset{R_1 \Leftrightarrow R_2}{\left[\begin{array}{cc|c} 1 & 3 & 6 \\ 2 & -1 & 5 \end{array}\right]} \Rightarrow \underset{-2R_1 + R_2 \Rightarrow R_2}{\left[\begin{array}{cc|c} 1 & 3 & 6 \\ 0 & -7 & -7 \end{array}\right]} \Rightarrow \underset{-\frac{1}{7}R_2 \Rightarrow R_2}{\left[\begin{array}{cc|c} 1 & 3 & 6 \\ 0 & 1 & 1 \end{array}\right]}$

From R_2: $y = 1$ From R_1: $x + 3y = 6$ Solution: $(3, 1)$

$$x + 3(1) = 6$$
$$x = 3$$

31. $\left[\begin{array}{cc|c} 1 & -2 & 3 \\ -2 & 4 & 6 \end{array}\right] \Rightarrow \underset{2R_1 + R_2 \Rightarrow R_2}{\left[\begin{array}{cc|c} 1 & -2 & 3 \\ 0 & 0 & 12 \end{array}\right]} \Rightarrow$ From R_2, $0x + 0y = 12$. This is impossible.

No solution $\Rightarrow$ inconsistent system

33. $\left[\begin{array}{cc|c} 2 & -1 & 7 \\ -1 & \frac{1}{3} & -\frac{7}{3} \end{array}\right] \Rightarrow \underset{R_1 \Leftrightarrow -R_2}{\left[\begin{array}{cc|c} 1 & -\frac{1}{3} & \frac{7}{3} \\ 2 & -1 & 7 \end{array}\right]} \Rightarrow \underset{-2R_1 + R_2 \Rightarrow R_2}{\left[\begin{array}{cc|c} 1 & -\frac{1}{3} & \frac{7}{3} \\ 0 & -\frac{1}{3} & \frac{7}{3} \end{array}\right]} \Rightarrow \underset{-3R_2 \Rightarrow R_2}{\left[\begin{array}{cc|c} 1 & -\frac{1}{3} & \frac{7}{3} \\ 0 & 1 & -7 \end{array}\right]}$

From R_2: $y = -7$ From R_1: $x - \frac{1}{3}y = \frac{7}{3}$ Solution: $(0, -7)$

$$x - \tfrac{1}{3}(-7) = \tfrac{7}{3}$$
$$x + \tfrac{7}{3} = \tfrac{7}{3}$$
$$x = 0$$

35. $\left[\begin{array}{ccc|c} 1 & -1 & 1 & 3 \\ 2 & -1 & 1 & 4 \\ 1 & 2 & -1 & -1 \end{array}\right] \Rightarrow \left[\begin{array}{ccc|c} 1 & -1 & 1 & 3 \\ 0 & 1 & -1 & -2 \\ 0 & 3 & -2 & -4 \end{array}\right] \Rightarrow \left[\begin{array}{ccc|c} 1 & -1 & 1 & 3 \\ 0 & 1 & -1 & -2 \\ 0 & 0 & 1 & 2 \end{array}\right]$

$-2R_1 + R_2 \Rightarrow R_2$, $-R_1 + R_3 \Rightarrow R_3$; $-3R_2 + R_3 \Rightarrow R_3$

From R_3: $z = 2$

From R_2: $y - z = -2$, $y - (2) = -2$, $y = 0$

From R_1: $x - y + z = 3$, $x - (0) + (2) = 3$, $x = 1$

Solution: $(1, 0, 2)$

37. $\left[\begin{array}{ccc|c} 1 & 1 & -1 & -1 \\ 3 & 1 & 0 & 4 \\ 0 & 1 & -2 & -4 \end{array}\right] \Rightarrow \left[\begin{array}{ccc|c} 1 & 1 & -1 & -1 \\ 0 & -2 & 3 & 7 \\ 0 & 1 & -2 & -4 \end{array}\right] \Rightarrow \left[\begin{array}{ccc|c} 1 & 1 & -1 & -1 \\ 0 & 1 & -\frac{3}{2} & -\frac{7}{2} \\ 0 & 0 & -1 & -1 \end{array}\right] \Rightarrow \left[\begin{array}{ccc|c} 1 & 1 & -1 & -1 \\ 0 & 1 & -\frac{3}{2} & -\frac{7}{2} \\ 0 & 0 & 1 & 1 \end{array}\right]$

$-3R_1 + R_2 \Rightarrow R_2$; $-\frac{1}{2}R_2 \Rightarrow R_2$, $R_2 + 2R_3 \Rightarrow R_3$; $-R_3 \Rightarrow R_3$

From (3): $z = 1$

From (2): $y - \dfrac{3}{2}z = -\dfrac{7}{2}$, $y - \dfrac{3}{2}(1) = -\dfrac{7}{2}$, $y = -2$

From (1): $x + y - z = -1$, $x + (-2) - (1) = -1$, $x = 2$

Solution: $(2, -2, 1)$

39. $\left[\begin{array}{ccc|c} 1 & -1 & 1 & 2 \\ 2 & 1 & 1 & 5 \\ 3 & 0 & -4 & -5 \end{array}\right] \Rightarrow \left[\begin{array}{ccc|c} 1 & -1 & 1 & 2 \\ 0 & 3 & -1 & 1 \\ 0 & 3 & -7 & -11 \end{array}\right] \Rightarrow \left[\begin{array}{ccc|c} 1 & -1 & 1 & 2 \\ 0 & 1 & -\frac{1}{3} & \frac{1}{3} \\ 0 & 0 & 6 & 12 \end{array}\right] \Rightarrow$

$-2R_1 + R_2 \Rightarrow R_2$, $-3R_1 + R_3 \Rightarrow R_3$; $\frac{1}{3}R_2 \Rightarrow R_2$, $-R_3 + R_2 \Rightarrow R_3$

$\left[\begin{array}{ccc|c} 1 & -1 & 1 & 2 \\ 0 & 1 & -\frac{1}{3} & \frac{1}{3} \\ 0 & 0 & 1 & 2 \end{array}\right]$

$\frac{1}{6}R_3 \Rightarrow R_3$

From (3): $z = 2$

From (2): $y - \frac{1}{3}z = \frac{1}{3}$, $y - \frac{1}{3}(2) = \frac{1}{3}$, $y = 1$

From (1): $x - y + z = 2$, $x - (1) + (2) = 2$, $x = 1$

Solution: $(1, 1, 2)$

41. $\left[\begin{array}{ccc|c} 1 & 1 & -1 & 5 \\ 1 & 1 & 1 & 2 \\ 3 & 3 & -1 & 12 \end{array}\right] \Rightarrow \left[\begin{array}{ccc|c} 1 & 1 & -1 & 5 \\ 0 & 0 & 2 & -3 \\ 0 & 0 & 2 & -3 \end{array}\right] \Rightarrow \left[\begin{array}{ccc|c} 1 & 1 & -1 & 5 \\ 0 & 0 & 1 & -\frac{3}{2} \\ 0 & 0 & 1 & -\frac{3}{2} \end{array}\right]$

$-R_1 + R_2 \Rightarrow R_2$, $-3R_1 + R_3 \Rightarrow R_3$; $\frac{1}{2}R_2 \Rightarrow R_2$, $\frac{1}{2}R_3 \Rightarrow R_3$

From R_2: $z = -\frac{3}{2}$

From R_1: $x + y - z = 5$, $x + y - \left(-\frac{3}{2}\right) = 5$, $y = -x + \frac{7}{2}$

Solution: $\left(x, -x + \frac{7}{2}, -\frac{3}{2}\right)$

EXERCISES 6.2

43. $\left[\begin{array}{ccc|c} 2 & -1 & 1 & 6 \\ 3 & 1 & -1 & 2 \\ -1 & 3 & -3 & 8 \end{array}\right] \Rightarrow \underset{R_1 \Leftrightarrow R_3}{\left[\begin{array}{ccc|c} -1 & 3 & -3 & 8 \\ 3 & 1 & -1 & 2 \\ 2 & -1 & 1 & 6 \end{array}\right]} \Rightarrow \underset{\substack{3R_1 + R_2 \Rightarrow R_2 \\ 2R_1 + R_3 \Rightarrow R_3}}{\left[\begin{array}{ccc|c} -1 & 3 & -3 & 8 \\ 0 & 10 & -10 & 26 \\ 0 & 5 & -5 & 22 \end{array}\right]} \Rightarrow$

$\underset{-2R_3 + R_2 \Rightarrow R_3}{\left[\begin{array}{ccc|c} -1 & 3 & -3 & 8 \\ 0 & 10 & -10 & 26 \\ 0 & 0 & 0 & -18 \end{array}\right]}$ R_3 indicates $0x + 0y + 0z = -18$. This is impossible.

No solution $\Rightarrow$ inconsistent system

45. $\left[\begin{array}{cc|c} 1 & -2 & 7 \\ 0 & 1 & 3 \end{array}\right] \Rightarrow \underset{2R_2 + R_1 \Rightarrow R_1}{\left[\begin{array}{cc|c} 1 & 0 & 13 \\ 0 & 1 & 3 \end{array}\right]} \Rightarrow$ Solution: $(13, 3)$

47. $\left[\begin{array}{ccc|c} 1 & 2 & -1 & 3 \\ 0 & 1 & 3 & 1 \\ 0 & 0 & 1 & -2 \end{array}\right] \Rightarrow \underset{-2R_2 + R_1 \Rightarrow R_1}{\left[\begin{array}{ccc|c} 1 & 0 & -7 & 1 \\ 0 & 1 & 3 & 1 \\ 0 & 0 & 1 & -2 \end{array}\right]} \Rightarrow \underset{\substack{7R_3 + R_1 \Rightarrow R_1 \\ -3R_3 + R_2 \Rightarrow R_2}}{\left[\begin{array}{ccc|c} 1 & 0 & 0 & -13 \\ 0 & 1 & 0 & 7 \\ 0 & 0 & 1 & -2 \end{array}\right]}$ Solution: $(-13, 7, -2)$

49. $\left[\begin{array}{cc|c} 1 & -1 & 7 \\ 1 & 1 & 13 \end{array}\right] \Rightarrow \underset{-R_1 + R_2 \Rightarrow R_2}{\left[\begin{array}{cc|c} 1 & -1 & 7 \\ 0 & 2 & 6 \end{array}\right]} \Rightarrow \underset{\frac{1}{2}R_2 \Rightarrow R_2}{\left[\begin{array}{cc|c} 1 & -1 & 7 \\ 0 & 1 & 3 \end{array}\right]} \Rightarrow \underset{R_2 + R_1 \Rightarrow R_1}{\left[\begin{array}{cc|c} 1 & 0 & 10 \\ 0 & 1 & 3 \end{array}\right]}$ Solution: $(10, 3)$

51. $\left[\begin{array}{cc|c} 1 & -\frac{1}{2} & 0 \\ 1 & 2 & 0 \end{array}\right] \Rightarrow \underset{-R_1 + R_2 \Rightarrow R_2}{\left[\begin{array}{cc|c} 1 & -\frac{1}{2} & 0 \\ 0 & \frac{5}{2} & 0 \end{array}\right]} \Rightarrow \underset{\frac{2}{5}R_2 \Rightarrow R_2}{\left[\begin{array}{cc|c} 1 & -\frac{1}{2} & 0 \\ 0 & 1 & 0 \end{array}\right]} \Rightarrow \underset{\frac{1}{2}R_2 + R_1 \Rightarrow R_1}{\left[\begin{array}{cc|c} 1 & 0 & 0 \\ 0 & 1 & 0 \end{array}\right]}$ Solution: $(0, 0)$

53. $\left[\begin{array}{ccc|c} 1 & 1 & 2 & 0 \\ 1 & 1 & 1 & 2 \\ 1 & 0 & 1 & 1 \end{array}\right] \Rightarrow \underset{\substack{-R_1 + R_2 \Rightarrow R_2 \\ -R_1 + R_3 \Rightarrow R_3}}{\left[\begin{array}{ccc|c} 1 & 1 & 2 & 0 \\ 0 & 0 & -1 & 2 \\ 0 & -1 & -1 & 1 \end{array}\right]} \Rightarrow \underset{R_2 \Leftrightarrow R_3}{\left[\begin{array}{ccc|c} 1 & 1 & 2 & 0 \\ 0 & -1 & -1 & 1 \\ 0 & 0 & -1 & 2 \end{array}\right]} \Rightarrow \underset{\substack{-R_2 \Rightarrow R_2 \\ -R_3 \Rightarrow R_3}}{\left[\begin{array}{ccc|c} 1 & 1 & 2 & 0 \\ 0 & 1 & 1 & -1 \\ 0 & 0 & 1 & -2 \end{array}\right]}$

$\underset{-R_2 + R_1 \Rightarrow R_1}{\left[\begin{array}{ccc|c} 1 & 0 & 1 & 1 \\ 0 & 1 & 1 & -1 \\ 0 & 0 & 1 & -2 \end{array}\right]} \Rightarrow \underset{\substack{-R_3 + R_1 \Rightarrow R_1 \\ -R_3 + R_2 \Rightarrow R_2}}{\left[\begin{array}{ccc|c} 1 & 0 & 0 & 3 \\ 0 & 1 & 0 & 1 \\ 0 & 0 & 1 & -2 \end{array}\right]}$ Solution: $(3, 1, -2)$

EXERCISES 6.2

55.

$$\left[\begin{array}{ccc|c} 2 & 1 & -2 & 1 \\ -1 & 1 & -3 & 0 \\ 4 & 3 & 0 & 4 \end{array}\right] \Rightarrow \left[\begin{array}{ccc|c} 2 & 1 & -2 & 1 \\ 0 & 3 & -8 & 1 \\ 0 & 1 & 4 & 2 \end{array}\right] \Rightarrow \left[\begin{array}{ccc|c} -6 & 0 & -2 & -2 \\ 0 & 3 & -8 & 1 \\ 0 & 0 & -20 & -5 \end{array}\right] \Rightarrow \left[\begin{array}{ccc|c} -6 & 0 & -2 & -2 \\ 0 & 3 & -8 & 1 \\ 0 & 0 & 1 & \frac{1}{4} \end{array}\right]$$

$R_1 + 2R_2 \Rightarrow R_2$, $-2R_1 + R_3 \Rightarrow R_3$; $-3R_1 + R_2 \Rightarrow R_1$, $-3R_3 + R_2 \Rightarrow R_3$; $-\frac{1}{20}R_3 \Rightarrow R_3$

$$\left[\begin{array}{ccc|c} -6 & 0 & 0 & -\frac{3}{2} \\ 0 & 3 & 0 & 3 \\ 0 & 0 & 1 & \frac{1}{4} \end{array}\right] \Rightarrow \left[\begin{array}{ccc|c} 1 & 0 & 0 & \frac{1}{4} \\ 0 & 1 & 0 & 1 \\ 0 & 0 & 1 & \frac{1}{4} \end{array}\right]$$

$2R_3 + R_1 \Rightarrow R_1$, $8R_3 + R_2 \Rightarrow R_2$; $-\frac{1}{6}R_1 \Rightarrow R_1$, $\frac{1}{3}R_2 \Rightarrow R_2$

Solution: $\left(\frac{1}{4}, 1, \frac{1}{4}\right)$

57.

$$\left[\begin{array}{cccc|c} 2 & -2 & 3 & 1 & 2 \\ 1 & 1 & 1 & 1 & 5 \\ -1 & 2 & -3 & 2 & 2 \\ 1 & 1 & 2 & -1 & 4 \end{array}\right] \Rightarrow \left[\begin{array}{cccc|c} 2 & -2 & 3 & 1 & 2 \\ 0 & -4 & 1 & -1 & -8 \\ 0 & 2 & -3 & 5 & 6 \\ 0 & -4 & -1 & 3 & -6 \end{array}\right] \Rightarrow \left[\begin{array}{cccc|c} -4 & 0 & -5 & -3 & -12 \\ 0 & -4 & 1 & -1 & -8 \\ 0 & 0 & -5 & 9 & 4 \\ 0 & 0 & 2 & -4 & -2 \end{array}\right]$$

$-2R_2 + R_1 \Rightarrow R_2$, $2R_3 + R_1 \Rightarrow R_3$, $-2R_4 + R_1 \Rightarrow R_4$; $-2R_1 + R_2 \Rightarrow R_1$, $2R_3 + R_2 \Rightarrow R_3$, $-R_4 + R_2 \Rightarrow R_4$

$$\left[\begin{array}{cccc|c} -4 & 0 & -5 & -3 & -12 \\ 0 & -4 & 1 & -1 & -8 \\ 0 & 0 & -5 & 9 & 4 \\ 0 & 0 & 1 & -2 & -1 \end{array}\right] \Rightarrow \left[\begin{array}{cccc|c} 4 & 0 & 0 & 12 & 16 \\ 0 & -20 & 0 & 4 & -36 \\ 0 & 0 & -5 & 9 & 4 \\ 0 & 0 & 0 & -1 & -1 \end{array}\right] \Rightarrow \left[\begin{array}{cccc|c} 1 & 0 & 0 & 3 & 4 \\ 0 & -5 & 0 & 1 & -9 \\ 0 & 0 & -5 & 9 & 4 \\ 0 & 0 & 0 & 1 & 1 \end{array}\right]$$

$\frac{1}{2}R_4 \Rightarrow R_4$; $-R_1 + R_3 \Rightarrow R_1$, $5R_2 + R_3 \Rightarrow R_2$, $5R_4 + R_3 \Rightarrow R_4$; $\frac{1}{4}R_1 \Rightarrow R_1$, $\frac{1}{4}R_2 \Rightarrow R_2$, $-R_4 \Rightarrow R_4$

$$\Rightarrow \left[\begin{array}{cccc|c} 1 & 0 & 0 & 0 & 1 \\ 0 & -5 & 0 & 0 & -10 \\ 0 & 0 & -5 & 0 & -5 \\ 0 & 0 & 0 & 1 & 1 \end{array}\right] \Rightarrow \left[\begin{array}{cccc|c} 1 & 0 & 0 & 0 & 1 \\ 0 & 1 & 0 & 0 & 2 \\ 0 & 0 & 1 & 0 & 1 \\ 0 & 0 & 0 & 1 & 1 \end{array}\right]$$

$-3R_4 + R_1 \Rightarrow R_1$, $-R_4 + R_2 \Rightarrow R_2$, $-9R_4 + R_3 \Rightarrow R_3$; $-\frac{1}{5}R_2 \Rightarrow R_2$, $-\frac{1}{5}R_3 \Rightarrow R_3$

Solution: $(1, 2, 1, 1)$

59.

$$\left[\begin{array}{cccc|c} 1 & 1 & 0 & 1 & 4 \\ 1 & 0 & 1 & 1 & 2 \\ 2 & 2 & 1 & 2 & 8 \\ 1 & -1 & 1 & -1 & -2 \end{array}\right] \Rightarrow \left[\begin{array}{cccc|c} 1 & 1 & 0 & 1 & 4 \\ 0 & 1 & -1 & 0 & 2 \\ 0 & 0 & 1 & 0 & 0 \\ 0 & 2 & -1 & 2 & 6 \end{array}\right] \Rightarrow \left[\begin{array}{cccc|c} 1 & 0 & 1 & 1 & 2 \\ 0 & 1 & -1 & 0 & 2 \\ 0 & 0 & 1 & 0 & 0 \\ 0 & 0 & 1 & 2 & 2 \end{array}\right] \Rightarrow$$

$-R_2 + R_1 \Rightarrow R_2$, $-2R_1 + R_3 \Rightarrow R_3$, $-R_4 + R_1 \Rightarrow R_4$; $-R_2 + R_1 \Rightarrow R_1$, $-2R_2 + R_4 \Rightarrow R_4$

continued on next page...

59. continued

$$\left[\begin{array}{cccc|c} 1 & 0 & 0 & 1 & 2 \\ 0 & 1 & 0 & 0 & 2 \\ 0 & 0 & 1 & 0 & 0 \\ 0 & 0 & 0 & 2 & 2 \end{array}\right] \Rightarrow \left[\begin{array}{cccc|c} 1 & 0 & 0 & 0 & 1 \\ 0 & 1 & 0 & 0 & 2 \\ 0 & 0 & 1 & 0 & 0 \\ 0 & 0 & 0 & 1 & 1 \end{array}\right] \text{ Solution: } (1, 2, 0, 1)$$

$-R_3 + R_1 \Rightarrow R_1$, $R_3 + R_2 \Rightarrow R_2$, $-R_3 + R_4 \Rightarrow R_4$; $-\frac{1}{2}R_4 + R_1 \Rightarrow R_1$, $\frac{1}{2}R_4 \Rightarrow R_4$

61.

$$\left[\begin{array}{ccc|c} \frac{1}{3} & \frac{3}{4} & -\frac{2}{3} & -2 \\ 1 & \frac{1}{2} & \frac{1}{3} & 1 \\ \frac{1}{6} & -\frac{1}{8} & -1 & 0 \end{array}\right] \Rightarrow \left[\begin{array}{ccc|c} 1 & \frac{9}{4} & -2 & -6 \\ 6 & 3 & 2 & 6 \\ 4 & -3 & -24 & 0 \end{array}\right] \Rightarrow \left[\begin{array}{ccc|c} 1 & \frac{9}{4} & -2 & -6 \\ 0 & -\frac{21}{2} & 14 & 42 \\ 0 & -12 & -16 & 24 \end{array}\right] \Rightarrow$$

$3R_1 \Rightarrow R_1$, $6R_2 \Rightarrow R_2$, $24R_3 \Rightarrow R_3$; $-6R_1 + R_2 \Rightarrow R_2$, $-4R_1 + R_3 \Rightarrow R_3$

$$\left[\begin{array}{ccc|c} 1 & \frac{9}{4} & -2 & -6 \\ 0 & 1 & -\frac{4}{3} & -4 \\ 0 & 3 & 4 & -6 \end{array}\right] \Rightarrow \left[\begin{array}{ccc|c} 1 & 0 & 1 & 3 \\ 0 & 1 & -\frac{4}{3} & -4 \\ 0 & 0 & 8 & 6 \end{array}\right] \Rightarrow \left[\begin{array}{ccc|c} 1 & 0 & 0 & \frac{9}{4} \\ 0 & 1 & 0 & -3 \\ 0 & 0 & 1 & \frac{3}{4} \end{array}\right] \text{ Solution: } \left(\tfrac{9}{4}, -3, \tfrac{3}{4}\right)$$

$-\frac{2}{21}R_2 \Rightarrow R_2$, $-\frac{1}{4}R_3 \Rightarrow R_3$; $-\frac{9}{4}R_2 + R_1 \Rightarrow R_1$, $-3R_2 + R_3 \Rightarrow R_3$; $-\frac{1}{8}R_3 + R_1 \Rightarrow R_1$, $\frac{1}{6}R_3 + R_2 \Rightarrow R_2$, $\frac{1}{8}R_3 \Rightarrow R_3$

63.

$$\left[\begin{array}{ccc|c} \frac{1}{2} & \frac{1}{4} & -1 & 2 \\ \frac{2}{3} & \frac{1}{4} & \frac{1}{2} & \frac{3}{2} \\ \frac{2}{3} & 0 & 1 & -\frac{1}{3} \end{array}\right] \Rightarrow \left[\begin{array}{ccc|c} 1 & \frac{1}{2} & -2 & 4 \\ 8 & 3 & 6 & 18 \\ 2 & 0 & 3 & -1 \end{array}\right] \Rightarrow \left[\begin{array}{ccc|c} 1 & \frac{1}{2} & -2 & 4 \\ 0 & -1 & 22 & -14 \\ 0 & -1 & 7 & -9 \end{array}\right] \Rightarrow$$

$2R_1 \Rightarrow R_1, 12R_2 \Rightarrow R_2$, $3R_3 \Rightarrow R_3$; $-8R_1 + R_2 \Rightarrow R_2$, $-2R_1 + R_3 \Rightarrow R_3$

$$\left[\begin{array}{ccc|c} 1 & \frac{1}{2} & -2 & 4 \\ 0 & 1 & -22 & 14 \\ 0 & -1 & 7 & -9 \end{array}\right] \Rightarrow \left[\begin{array}{ccc|c} 1 & 0 & 9 & -3 \\ 0 & 1 & -22 & 14 \\ 0 & 0 & -15 & 5 \end{array}\right] \Rightarrow \left[\begin{array}{ccc|c} 1 & 0 & 9 & -3 \\ 0 & 1 & -22 & 14 \\ 0 & 0 & 1 & -\frac{1}{3} \end{array}\right] \Rightarrow$$

$-R_2 \Rightarrow R_2$; $-\frac{1}{2}R_2 + R_1 \Rightarrow R_1$, $R_2 + R_3 \Rightarrow R_3$; $-\frac{1}{15}R_3 \Rightarrow R_3$

$$\left[\begin{array}{ccc|c} 1 & 0 & 0 & 0 \\ 0 & 1 & 0 & \frac{20}{3} \\ 0 & 0 & 1 & -\frac{1}{3} \end{array}\right] \text{ Solution: } \left(0, \tfrac{20}{3}, -\tfrac{1}{3}\right)$$

$-9R_3 + R_1 \Rightarrow R_1$, $22R_3 + R_2 \Rightarrow R_2$

EXERCISES 6.2

65. $\left[\begin{array}{ccc|c} 3 & -6 & 9 & 18 \\ 2 & -4 & 3 & 12 \\ 1 & -2 & 3 & 6 \end{array}\right] \Rightarrow \underset{R_1 \Leftrightarrow R_3}{\left[\begin{array}{ccc|c} 1 & -2 & 3 & 6 \\ 2 & -4 & 3 & 12 \\ 3 & -6 & 9 & 18 \end{array}\right]} \Rightarrow \underset{\substack{-2R_1 + R_2 \Rightarrow R_2 \\ -3R_1 + R_3 \Rightarrow R_3}}{\left[\begin{array}{ccc|c} 1 & -2 & 3 & 6 \\ 0 & 0 & -3 & 0 \\ 0 & 0 & 0 & 0 \end{array}\right]} \Rightarrow \underset{-\frac{1}{3}R_2 \Rightarrow R_2}{\left[\begin{array}{ccc|c} 1 & -2 & 3 & 6 \\ 0 & 0 & 1 & 0 \\ 0 & 0 & 0 & 0 \end{array}\right]}$

From (2): $z = 0$ From (1): $x - 2y + 3z = 6$

$x - 2y + 3(0) = 6$

$2y = -x + 6$

$y = -\frac{1}{2}x - 3$

Solution: $\left(x, -\frac{1}{2}x - 3, 0\right)$

67. $\left[\begin{array}{cc|c} 1 & 1 & -2 \\ 3 & -1 & 6 \\ 2 & 2 & -4 \\ 1 & -1 & 4 \end{array}\right] \Rightarrow \underset{\substack{-3R_1 + R_2 \Rightarrow R_2 \\ -2R_1 + R_3 \Rightarrow R_3 \\ -R_1 + R_4 \Rightarrow R_4}}{\left[\begin{array}{cc|c} 1 & 1 & -2 \\ 0 & -4 & 12 \\ 0 & 0 & 0 \\ 0 & -2 & 6 \end{array}\right]} \Rightarrow \underset{\substack{-\frac{1}{4}R_2 \Rightarrow R_2 \\ R_3 \Leftrightarrow R_4}}{\left[\begin{array}{cc|c} 1 & 1 & -2 \\ 0 & 1 & -3 \\ 0 & -2 & 6 \\ 0 & 0 & 0 \end{array}\right]} \Rightarrow \underset{\substack{-R_2 + R_1 \Rightarrow R_1 \\ 2R_2 + R_3 \Rightarrow R_3}}{\left[\begin{array}{cc|c} 1 & 0 & 1 \\ 0 & 1 & -3 \\ 0 & 0 & 0 \\ 0 & 0 & 0 \end{array}\right]}$

Solution: $(1, -3)$

69. $\left[\begin{array}{ccc|c} 1 & 2 & 1 & 4 \\ 3 & -1 & -1 & 2 \end{array}\right] \Rightarrow \underset{-3R_1 + R_2 \Rightarrow R_2}{\left[\begin{array}{ccc|c} 1 & 2 & 1 & 4 \\ 0 & -7 & -4 & -10 \end{array}\right]} \Rightarrow \underset{-\frac{1}{7}R_2 \Rightarrow R_2}{\left[\begin{array}{ccc|c} 1 & 2 & 1 & 4 \\ 0 & 1 & \frac{4}{7} & \frac{10}{7} \end{array}\right]} \Rightarrow \underset{-2R_2 + R_1 \Rightarrow R_1}{\left[\begin{array}{ccc|c} 1 & 0 & -\frac{1}{7} & \frac{8}{7} \\ 0 & 1 & \frac{4}{7} & \frac{10}{7} \end{array}\right]}$

From R_1: $x - \frac{1}{7}z = \frac{8}{7}$

$x = \frac{8}{7} + \frac{1}{7}z$

From R_2: $y + \frac{4}{7}z = \frac{10}{7}$

$y = \frac{10}{7} - \frac{4}{7}z$

Solution: $x = \frac{8}{7} + \frac{1}{7}z, y = \frac{10}{7} - \frac{4}{7}z$, $z =$ any real number

71. $\left[\begin{array}{cccc|c} 1 & 1 & 0 & 0 & 1 \\ 1 & 0 & 1 & 0 & 0 \\ 0 & 1 & 0 & 1 & 0 \end{array}\right] \Rightarrow \underset{-R_2 + R_1 \Rightarrow R_2}{\left[\begin{array}{cccc|c} 1 & 1 & 0 & 0 & 1 \\ 0 & 1 & -1 & 0 & 1 \\ 0 & 1 & 0 & 1 & 0 \end{array}\right]} \Rightarrow \underset{\substack{-R_2 + R_1 \Rightarrow R_1 \\ -R_2 + R_3 \Rightarrow R_3}}{\left[\begin{array}{cccc|c} 1 & 0 & 1 & 0 & 0 \\ 0 & 1 & -1 & 0 & 1 \\ 0 & 0 & 1 & 1 & -1 \end{array}\right]} \Rightarrow$

$\underset{\substack{-R_3 + R_1 \Rightarrow R_1 \\ R_3 + R_2 \Rightarrow R_2}}{\left[\begin{array}{cccc|c} 1 & 0 & 0 & -1 & 1 \\ 0 & 1 & 0 & 1 & 0 \\ 0 & 0 & 1 & 1 & -1 \end{array}\right]} \Rightarrow$ From R_1: $w - z = 1$, $w = 1 + z$ From R_2: $x + z = 0$, $x = -z$ From R_3: $y + z = -1$, $y = -1 - z$

Solution: $w = 1 + z$, $x = -z$, $y = -1 - z$, $z =$ any real #

73. $\left[\begin{array}{cc|c} 1 & 1 & 3 \\ 2 & 1 & 1 \\ 3 & 2 & 2 \end{array}\right] \Rightarrow \underset{\substack{-2R_1 + R_2 \Rightarrow R_2 \\ -3R_1 + R_3 \Rightarrow R_3}}{\left[\begin{array}{cc|c} 1 & 1 & 3 \\ 0 & -1 & -5 \\ 0 & -1 & -7 \end{array}\right]} \Rightarrow \underset{-R_2 + R_3 \Rightarrow R_3}{\left[\begin{array}{cc|c} 1 & 1 & 3 \\ 0 & -1 & -5 \\ 0 & 0 & -2 \end{array}\right]}$

R_3 indicates that $0x + 0y = -2$. This is impossible. The system is inconsistent. $\Rightarrow$ no solution

EXERCISES 6.2

75. Let $p =$ speed with no wind. Let $w =$ the speed of the wind. Then the following system applies:

$$\begin{cases} p + w = 300 \\ p - w = 220 \end{cases} \quad \left[\begin{array}{cc|c} 1 & 1 & 300 \\ 1 & -1 & 220 \end{array}\right] \Rightarrow \underset{-R_2 + R_1 \Rightarrow R_2}{\left[\begin{array}{cc|c} 1 & 1 & 300 \\ 0 & 2 & 80 \end{array}\right]} \Rightarrow \underset{\frac{1}{2}R_2 \Rightarrow R_2}{\left[\begin{array}{cc|c} 1 & 1 & 300 \\ 0 & 1 & 40 \end{array}\right]} \Rightarrow \underset{-R_2 + R_1 \Rightarrow R_1}{\left[\begin{array}{cc|c} 1 & 0 & 260 \\ 0 & 1 & 40 \end{array}\right]}$$

The plane has a speed of 260 miles per hour with no wind, so it could travel 1300 miles in 5 hours.

77. Let $d =$ width of a dictionary, $a =$ width of an atlas and $t =$ width of a thesaurus.

$$\begin{cases} 3d + 5a + t = 35 \\ 6d + 2t = 35 \\ 2d + 4a + 3t = 35 \end{cases} \quad \left[\begin{array}{ccc|c} 3 & 5 & 1 & 35 \\ 6 & 0 & 2 & 35 \\ 2 & 4 & 3 & 35 \end{array}\right] \Rightarrow \underset{\substack{-2R_1 + R_2 \Rightarrow R_2 \\ -\frac{2}{3}R_1 + R_3 \Rightarrow R_3}}{\left[\begin{array}{ccc|c} 3 & 5 & 1 & 35 \\ 0 & -10 & 0 & -35 \\ 0 & \frac{2}{3} & \frac{7}{3} & \frac{35}{3} \end{array}\right]} \Rightarrow \underset{\substack{-\frac{1}{10}R_2 \Rightarrow R_2 \\ 3R_3 \Rightarrow R_3}}{\left[\begin{array}{ccc|c} 3 & 5 & 1 & 35 \\ 0 & 1 & 0 & 3.5 \\ 0 & 2 & 7 & 35 \end{array}\right]}$$

$$\Rightarrow \underset{\substack{-5R_2 + R_1 \Rightarrow R_1 \\ -2R_2 + R_3 \Rightarrow R_3}}{\left[\begin{array}{ccc|c} 3 & 0 & 1 & 17.5 \\ 0 & 1 & 0 & 3.5 \\ 0 & 0 & 7 & 28 \end{array}\right]} \Rightarrow \underset{\substack{-\frac{1}{7}R_3 + R_1 \Rightarrow R_1 \\ \frac{1}{7}R_3 \Rightarrow R_3}}{\left[\begin{array}{ccc|c} 3 & 0 & 0 & 13.5 \\ 0 & 1 & 0 & 3.5 \\ 0 & 0 & 1 & 4 \end{array}\right]} \Rightarrow \underset{\frac{1}{3}R_1 \Rightarrow R_1}{\left[\begin{array}{ccc|c} 1 & 0 & 0 & 4.5 \\ 0 & 1 & 0 & 3.5 \\ 0 & 0 & 1 & 4 \end{array}\right]}$$

Dictionaries are 4.5 in. wide. Atlases are 3.5 in. wide. Thesauruses are 4 in. wide.

79. Let A, B and C represent the number of ounces of each food.

$$\begin{cases} A + 2B + 2C = 22 \\ A + B + C = 12 \\ 2A + B + 2C = 20 \end{cases} \quad \left[\begin{array}{ccc|c} 1 & 2 & 2 & 22 \\ 1 & 1 & 1 & 12 \\ 2 & 1 & 2 & 20 \end{array}\right] \Rightarrow \underset{\substack{-R_1 + R_2 \Rightarrow R_2 \\ -2R_1 + R_3 \Rightarrow R_3}}{\left[\begin{array}{ccc|c} 1 & 2 & 2 & 22 \\ 0 & -1 & -1 & -10 \\ 0 & -3 & -2 & -24 \end{array}\right]} \Rightarrow$$

$$\underset{\substack{2R_2 + R_1 \Rightarrow R_1 \\ -3R_2 + R_3 \Rightarrow R_3 \\ -R_2 \Rightarrow R_2}}{\left[\begin{array}{ccc|c} 1 & 0 & 0 & 2 \\ 0 & 1 & 1 & 10 \\ 0 & 0 & 1 & 6 \end{array}\right]} \Rightarrow \underset{-R_3 + R_2 \Rightarrow R_2}{\left[\begin{array}{ccc|c} 1 & 0 & 0 & 2 \\ 0 & 1 & 0 & 4 \\ 0 & 0 & 1 & 6 \end{array}\right]}$$

2 ounces of Food A, 4 ounces of Food B, and 6 ounces of Food C should be used.

81-87. Answers may vary.

89.

$$\left[\begin{array}{ccc|c} 1 & 1 & 1 & 14 \\ 2 & 3 & -2 & -7 \\ 1 & -5 & 1 & 8 \end{array}\right] \Rightarrow \underset{\substack{-2R_1 + R_2 \Rightarrow R_2 \\ -R_1 + R_3 \Rightarrow R_3}}{\left[\begin{array}{ccc|c} 1 & 1 & 1 & 14 \\ 0 & 1 & -4 & -35 \\ 0 & -6 & 0 & -6 \end{array}\right]} \Rightarrow \underset{R_2 \Leftrightarrow R_3}{\left[\begin{array}{ccc|c} 1 & 1 & 1 & 14 \\ 0 & -6 & 0 & -6 \\ 0 & 1 & -4 & -35 \end{array}\right]} \Rightarrow$$

$$\underset{\substack{\frac{1}{6}R_2 + R_1 \Rightarrow R_1 \\ \frac{1}{6}R_2 + R_3 \Rightarrow R_3}}{\left[\begin{array}{ccc|c} 1 & 0 & 1 & 13 \\ 0 & -6 & 0 & -6 \\ 0 & 0 & -4 & -36 \end{array}\right]} \Rightarrow \underset{\substack{\frac{1}{4}R_3 + R_1 \Rightarrow R_1 \\ -\frac{1}{6}R_2 \Rightarrow R_2 \\ -\frac{1}{4}R_3 \Rightarrow R_3}}{\left[\begin{array}{ccc|c} 1 & 0 & 0 & 4 \\ 0 & 1 & 0 & 1 \\ 0 & 0 & 1 & 9 \end{array}\right]}$$

$x^2 = 4 \Rightarrow \boxed{x = \pm 2}$

$y^2 = 1 \Rightarrow \boxed{y = \pm 1}$

$z^2 = 9 \Rightarrow \boxed{z = \pm 3}$

EXERCISES 6.2

91. False. It has 7 rows.

93. True.

95. True.

97. False. It has no solution.

99. True.

Exercises 6.3 (page 628)

1. i, j

3. corresponding

5. columns, rows

7. additive identity

9. $x = 2, y = 5$

11. $x + y = 3$
$3 + x = 4 \Rightarrow x = 1$
$5y = 10 \Rightarrow y = 2$

13. $$A + B = \begin{bmatrix} 2 & 1 & -1 \\ -3 & 2 & 5 \end{bmatrix} + \begin{bmatrix} -3 & 1 & 2 \\ -3 & -2 & -5 \end{bmatrix} = \begin{bmatrix} -1 & 2 & 1 \\ -6 & 0 & 0 \end{bmatrix}$$

15. additive inverse of $A = \begin{bmatrix} -5 & 2 & -7 \\ 5 & 0 & -3 \\ 2 & -3 & 5 \end{bmatrix}$

17. $$A - B = \begin{bmatrix} -3 & 2 & -2 \\ -1 & 4 & -5 \end{bmatrix} - \begin{bmatrix} 3 & -3 & -2 \\ -2 & 5 & -5 \end{bmatrix} = \begin{bmatrix} -6 & 5 & 0 \\ 1 & -1 & 0 \end{bmatrix}$$

19. $5A = 5\begin{bmatrix} 3 & -3 \\ 0 & -2 \end{bmatrix} = \begin{bmatrix} 15 & -15 \\ 0 & -10 \end{bmatrix}$

21. $5A = 5\begin{bmatrix} 5 & 15 & -2 \\ -2 & -5 & 1 \end{bmatrix} = \begin{bmatrix} 25 & 75 & -10 \\ -10 & -25 & 5 \end{bmatrix}$

23. $$5A + 3B = 5\begin{bmatrix} 3 & 1 & -2 \\ -4 & 3 & -2 \end{bmatrix} + 3\begin{bmatrix} 1 & -2 & 2 \\ -5 & -5 & 3 \end{bmatrix} = \begin{bmatrix} 15 & 5 & -10 \\ -20 & 15 & -10 \end{bmatrix} + \begin{bmatrix} 3 & -6 & 6 \\ -15 & -15 & 9 \end{bmatrix} = \begin{bmatrix} 18 & -1 & -4 \\ -35 & 0 & -1 \end{bmatrix}$$

25. $X + A = B$
$$X = B - A = \begin{bmatrix} 5 & 2 & -1 \\ 4 & 1 & -2 \end{bmatrix} - \begin{bmatrix} 1 & -3 & 4 \\ 2 & -1 & 2 \end{bmatrix} = \begin{bmatrix} 4 & 5 & -5 \\ 2 & 2 & -4 \end{bmatrix}$$

27. $2B - X = A$
$-X = -2B + A$
$$X = 2B - A = 2\begin{bmatrix} 5 & 2 & -1 \\ 4 & 1 & -2 \end{bmatrix} - \begin{bmatrix} 1 & -3 & 4 \\ 2 & -1 & 2 \end{bmatrix} = \begin{bmatrix} 9 & 7 & -6 \\ 6 & 3 & -6 \end{bmatrix}$$

29. $X + 2A = 3B$
$$X = 3B - 2A = 3\begin{bmatrix} 5 & 2 & -1 \\ 4 & 1 & -2 \end{bmatrix} - 2\begin{bmatrix} 1 & -3 & 4 \\ 2 & -1 & 2 \end{bmatrix} = \begin{bmatrix} 13 & 12 & -11 \\ 8 & 5 & -10 \end{bmatrix}$$

EXERCISES 6.3

31. $2X - 3A = B$

$$2X = 3A + B$$

$$X = \frac{3}{2}A + \frac{1}{2}B = \frac{3}{2}\begin{bmatrix} 1 & -3 & 4 \\ 2 & -1 & 2 \end{bmatrix} + \frac{1}{2}\begin{bmatrix} 5 & 2 & -1 \\ 4 & 1 & -2 \end{bmatrix} = \begin{bmatrix} 4 & -\frac{7}{2} & \frac{11}{2} \\ 5 & -1 & 2 \end{bmatrix}$$

33. $3A + 5B = -3X$

$$X = -A - \frac{5}{3}B = -\begin{bmatrix} 1 & -3 & 4 \\ 2 & -1 & 2 \end{bmatrix} - \frac{5}{3}\begin{bmatrix} 5 & 2 & -1 \\ 4 & 1 & -2 \end{bmatrix} = \begin{bmatrix} -\frac{28}{3} & -\frac{1}{3} & -\frac{7}{3} \\ -\frac{26}{3} & -\frac{2}{3} & \frac{4}{3} \end{bmatrix}$$

35. $$\begin{bmatrix} 2 & 3 \\ 3 & -2 \end{bmatrix}_{2\times 2}\begin{bmatrix} 1 & 2 \\ 0 & -2 \end{bmatrix}_{2\times 2} = \begin{bmatrix} (2)(1)+(3)(0) & (2)(2)+(3)(-2) \\ (3)(1)+(-2)(0) & (3)(2)+(-2)(-2) \end{bmatrix}_{2\times 2} = \begin{bmatrix} 2 & -2 \\ 3 & 10 \end{bmatrix}$$

37. $$\begin{bmatrix} -4 & -2 \\ 21 & 0 \end{bmatrix}_{2\times 2}\begin{bmatrix} -5 & 6 \\ 21 & -1 \end{bmatrix}_{2\times 2} = \begin{bmatrix} (-4)(-5)+(-2)(21) & (-4)(6)+(-2)(-1) \\ (21)(-5)+(0)(21) & (21)(6)+(0)(-1) \end{bmatrix}_{2\times 2}$$

$$= \begin{bmatrix} -22 & -22 \\ -105 & 126 \end{bmatrix}$$

39. $$\begin{bmatrix} 2 & 1 & 3 \\ 1 & 2 & -1 \\ 0 & 1 & 0 \end{bmatrix}_{3\times 3}\begin{bmatrix} 1 & 2 & 3 \\ 2 & -2 & 1 \\ 0 & 0 & 1 \end{bmatrix}_{3\times 3}$$

$$= \begin{bmatrix} (2)(1)+(1)(2)+(3)(0) & (2)(2)+(1)(-2)+(3)(0) & (2)(3)+(1)(1)+(3)(1) \\ (1)(1)+(2)(2)+(-1)(0) & (1)(2)+(2)(-2)+(-1)(0) & (1)(3)+(2)(1)+(-1)(1) \\ (0)(1)+(1)(2)+(0)(0) & (0)(2)+(1)(-2)+(0)(0) & (0)(3)+(1)(1)+(0)(1) \end{bmatrix}_{3\times 3}$$

$$= \begin{bmatrix} 4 & 2 & 10 \\ 5 & -2 & 4 \\ 2 & -2 & 1 \end{bmatrix}$$

41. $$\begin{bmatrix} 1 \\ -2 \\ -3 \end{bmatrix}_{3\times 1}\begin{bmatrix} 4 & -5 & -6 \end{bmatrix}_{1\times 3} = \begin{bmatrix} (1)(4) & (1)(-5) & (1)(-6) \\ (-2)(4) & (-2)(-5) & (-2)(-6) \\ (-3)(4) & (-3)(-5) & (-3)(-6) \end{bmatrix}_{3\times 3} = \begin{bmatrix} 4 & -5 & -6 \\ -8 & 10 & 12 \\ -12 & 15 & 18 \end{bmatrix}$$

43. $$\begin{bmatrix} 1 & 2 & 3 \end{bmatrix}_{1\times 3}\begin{bmatrix} 4 & 5 & 6 \\ 7 & 8 & 9 \end{bmatrix}_{2\times 3}$$

Not possible

45. $$\begin{bmatrix} 2 & 3 & 4 \\ 1 & 2 & 3 \\ -2 & 2 & 2 \end{bmatrix}_{3\times 3}\begin{bmatrix} -1 \\ 2 \\ 3 \end{bmatrix}_{3\times 1} = \begin{bmatrix} (2)(-1)+(3)(2)+(4)(3) \\ (1)(-1)+(2)(2)+(3)(3) \\ (-2)(-1)+(2)(2)+(2)(3) \end{bmatrix}_{3\times 1} = \begin{bmatrix} 16 \\ 12 \\ 12 \end{bmatrix}$$

47. $$\begin{bmatrix} 1 & 2 & 3 \\ 4 & 5 & 6 \\ 7 & 8 & 9 \end{bmatrix}_{3\times 3}\begin{bmatrix} 1 & 2 \\ 3 & 4 \end{bmatrix}_{2\times 2}$$

Not possible

EXERCISES 6.3

49. $AB = \begin{bmatrix} 2.3 & -1.7 & 3.1 \\ -2 & 3.5 & 1 \\ -8 & 4.7 & 9.1 \end{bmatrix} \begin{bmatrix} -2.5 \\ 5.2 \\ -7 \end{bmatrix} = \begin{bmatrix} -36.29 \\ 16.2 \\ -19.26 \end{bmatrix}$

51. $A^2 = \begin{bmatrix} 2.3 & -1.7 & 3.1 \\ -2 & 3.5 & 1 \\ -8 & 4.7 & 9.1 \end{bmatrix} \begin{bmatrix} 2.3 & -1.7 & 3.1 \\ -2 & 3.5 & 1 \\ -8 & 4.7 & 9.1 \end{bmatrix} = \begin{bmatrix} -16.11 & 4.71 & 33.64 \\ -19.6 & 20.35 & 6.4 \\ -100.6 & 72.82 & 62.71 \end{bmatrix}$

53. $A(B+C) = \begin{bmatrix} 2 & 3 \\ 1 & 3 \end{bmatrix} \left(\begin{bmatrix} 2 & 1 & -5 \\ 1 & 1 & 2 \end{bmatrix} + \begin{bmatrix} -2 & -1 & 6 \\ 0 & -1 & -1 \end{bmatrix} \right) = \begin{bmatrix} 2 & 3 \\ 1 & 3 \end{bmatrix} \begin{bmatrix} 0 & 0 & 1 \\ 1 & 0 & 1 \end{bmatrix} = \begin{bmatrix} 3 & 0 & 5 \\ 3 & 0 & 4 \end{bmatrix}$

$$AB + AC = \begin{bmatrix} 2 & 3 \\ 1 & 3 \end{bmatrix} \begin{bmatrix} 2 & 1 & -5 \\ 1 & 1 & 2 \end{bmatrix} + \begin{bmatrix} 2 & 3 \\ 1 & 3 \end{bmatrix} \begin{bmatrix} -2 & -1 & 6 \\ 0 & -1 & -1 \end{bmatrix}$$
$$= \begin{bmatrix} 7 & 5 & -4 \\ 5 & 4 & 1 \end{bmatrix} + \begin{bmatrix} -4 & -5 & 9 \\ -2 & -4 & 3 \end{bmatrix} = \begin{bmatrix} 3 & 0 & 5 \\ 3 & 0 & 4 \end{bmatrix}$$

55. $3(AB) = 3\left(\begin{bmatrix} 2 & 3 \\ 1 & 3 \end{bmatrix} \begin{bmatrix} 2 & 1 & -5 \\ 1 & 1 & 2 \end{bmatrix} \right) = 3\begin{bmatrix} 7 & 5 & -4 \\ 5 & 4 & 1 \end{bmatrix} = \begin{bmatrix} 21 & 15 & -12 \\ 15 & 12 & 3 \end{bmatrix}$

$$(3A)B = \left(3\begin{bmatrix} 2 & 3 \\ 1 & 3 \end{bmatrix} \right) \begin{bmatrix} 2 & 1 & -5 \\ 1 & 1 & 2 \end{bmatrix} = \begin{bmatrix} 6 & 9 \\ 3 & 9 \end{bmatrix} \begin{bmatrix} 2 & 1 & -5 \\ 1 & 1 & 2 \end{bmatrix} = \begin{bmatrix} 21 & 15 & -12 \\ 15 & 12 & 3 \end{bmatrix}$$

57. $A - BC = \begin{bmatrix} 1 & 3 \\ 2 & 5 \end{bmatrix} - \begin{bmatrix} -1 \\ 3 \end{bmatrix} \begin{bmatrix} 3 & 2 \end{bmatrix} = \begin{bmatrix} 1 & 3 \\ 2 & 5 \end{bmatrix} - \begin{bmatrix} -3 & -2 \\ 9 & 6 \end{bmatrix} = \begin{bmatrix} 4 & 5 \\ -7 & -1 \end{bmatrix}$

59. $CB - AB = \begin{bmatrix} 3 & 2 \end{bmatrix} \begin{bmatrix} -1 \\ 3 \end{bmatrix} - \begin{bmatrix} 1 & 3 \\ 2 & 5 \end{bmatrix} \begin{bmatrix} -1 \\ 3 \end{bmatrix} = [3] - \begin{bmatrix} 8 \\ 13 \end{bmatrix} \Rightarrow$ not possible

61. $ABC = \begin{bmatrix} 1 & 3 \\ 2 & 5 \end{bmatrix} \begin{bmatrix} -1 \\ 3 \end{bmatrix} \begin{bmatrix} 3 & 2 \end{bmatrix} = \begin{bmatrix} 8 \\ 13 \end{bmatrix} \begin{bmatrix} 3 & 2 \end{bmatrix} = \begin{bmatrix} 24 & 16 \\ 39 & 26 \end{bmatrix}$

63. $A^2B = \begin{bmatrix} 1 & 3 \\ 2 & 5 \end{bmatrix} \begin{bmatrix} 1 & 3 \\ 2 & 5 \end{bmatrix} \begin{bmatrix} -1 \\ 3 \end{bmatrix} = \begin{bmatrix} 7 & 18 \\ 12 & 31 \end{bmatrix} \begin{bmatrix} -1 \\ 3 \end{bmatrix} = \begin{bmatrix} 47 \\ 81 \end{bmatrix}$

65. $Q = \begin{bmatrix} 200 & 300 & 100 \\ 100 & 200 & 200 \end{bmatrix}, C = \begin{bmatrix} 5 \\ 2 \\ 4 \end{bmatrix}$

$QC = \begin{bmatrix} 200 & 300 & 100 \\ 100 & 200 & 200 \end{bmatrix} \begin{bmatrix} 5 \\ 2 \\ 4 \end{bmatrix} = \begin{bmatrix} 2000 \\ 1700 \end{bmatrix}$ Cost of balls from Supplier 1 / Cost of balls from Supplier 2

EXERCISES 6.3

67. $Q = \begin{bmatrix} 217 & 23 & 319 \\ 347 & 24 & 340 \\ 3 & 97 & 750 \end{bmatrix}, P = \begin{bmatrix} 0.75 \\ 1.00 \\ 1.25 \end{bmatrix}$

$QP = \begin{bmatrix} 217 & 23 & 319 \\ 347 & 24 & 340 \\ 3 & 97 & 750 \end{bmatrix} \begin{bmatrix} 0.75 \\ 1.00 \\ 1.25 \end{bmatrix} = \begin{bmatrix} 584.50 \\ 709.25 \\ 1036.75 \end{bmatrix}$ \$ spent by adult males, \$ spent by adult females, \$ spent by children

69. $A^2 = \begin{bmatrix} 0 & 1 & 1 \\ 1 & 0 & 0 \\ 0 & 1 & 0 \end{bmatrix}^2 = \begin{bmatrix} 0 & 1 & 1 \\ 1 & 0 & 0 \\ 0 & 1 & 0 \end{bmatrix} \begin{bmatrix} 0 & 1 & 1 \\ 1 & 0 & 0 \\ 0 & 1 & 0 \end{bmatrix} = \begin{bmatrix} 1 & 1 & 0 \\ 0 & 1 & 1 \\ 1 & 0 & 0 \end{bmatrix}$

71. $A^2 = \begin{bmatrix} 0 & 2 & 1 & 0 \\ 2 & 0 & 1 & 0 \\ 1 & 1 & 0 & 2 \\ 0 & 0 & 2 & 0 \end{bmatrix}^2 = \begin{bmatrix} 0 & 2 & 1 & 0 \\ 2 & 0 & 1 & 0 \\ 1 & 1 & 0 & 2 \\ 0 & 0 & 2 & 0 \end{bmatrix} \begin{bmatrix} 0 & 2 & 1 & 0 \\ 2 & 0 & 1 & 0 \\ 1 & 1 & 0 & 2 \\ 0 & 0 & 2 & 0 \end{bmatrix} = \begin{bmatrix} 5 & 1 & 2 & 2 \\ 1 & 5 & 2 & 2 \\ 2 & 2 & 6 & 0 \\ 2 & 2 & 0 & 4 \end{bmatrix}$

A^2 represents the number of ways two cities can be linked through one intermediary.

73. **Answers may vary.**

75. **Answers may vary.**

77. Let $A = \begin{bmatrix} 1 & 1 \\ 1 & 1 \end{bmatrix}$ and $B = \begin{bmatrix} 1 & 0 \\ 0 & 0 \end{bmatrix}$.

$(AB)^2 = \left(\begin{bmatrix} 1 & 1 \\ 1 & 1 \end{bmatrix} \begin{bmatrix} 1 & 0 \\ 0 & 0 \end{bmatrix} \right)^2 = \begin{bmatrix} 1 & 0 \\ 1 & 0 \end{bmatrix}^2 = \begin{bmatrix} 1 & 0 \\ 1 & 0 \end{bmatrix} \begin{bmatrix} 1 & 0 \\ 1 & 0 \end{bmatrix} = \begin{bmatrix} 1 & 0 \\ 1 & 0 \end{bmatrix}$

$A^2B^2 = \begin{bmatrix} 1 & 1 \\ 1 & 1 \end{bmatrix}^2 \begin{bmatrix} 1 & 0 \\ 0 & 0 \end{bmatrix}^2 = \begin{bmatrix} 2 & 2 \\ 2 & 2 \end{bmatrix} \begin{bmatrix} 1 & 0 \\ 0 & 0 \end{bmatrix} = \begin{bmatrix} 2 & 0 \\ 2 & 0 \end{bmatrix}$. $(AB)^2 \neq A^2B^2$

79. Let $A = \begin{bmatrix} 1 & 2 \\ 1 & 2 \end{bmatrix}$ and $B = \begin{bmatrix} 2 & 2 \\ -1 & -1 \end{bmatrix}$. $AB = \begin{bmatrix} 1 & 2 \\ 1 & 2 \end{bmatrix} \begin{bmatrix} 2 & 2 \\ -1 & -1 \end{bmatrix} = \begin{bmatrix} 0 & 0 \\ 0 & 0 \end{bmatrix}$

81. False.

$AB = \begin{bmatrix} 2 & 2 \\ 2 & 2 \end{bmatrix} \begin{bmatrix} 5 & 5 \\ 5 & 5 \end{bmatrix} = \begin{bmatrix} 20 & 20 \\ 20 & 20 \end{bmatrix}$.

83. True.

85. False. The number of columns of the 1st matrix must equal the number of rows of the 2nd.

87. True.

89. True.

Exercises 6.4 (page 638)

1. $AB = BA = I$

3. $[I \mid A^{-1}]$

EXERCISES 6.4

5. $\left[\begin{array}{cc|cc} 3 & -4 & 1 & 0 \\ -2 & 3 & 0 & 1 \end{array}\right] \Rightarrow \underset{\frac{1}{3}R_1 \Rightarrow R_1}{\left[\begin{array}{cc|cc} 1 & -\frac{4}{3} & \frac{1}{3} & 0 \\ -2 & 3 & 0 & 1 \end{array}\right]} \Rightarrow \underset{2R_1 + R_2 \Rightarrow R_2}{\left[\begin{array}{cc|cc} 1 & -\frac{4}{3} & \frac{1}{3} & 0 \\ 0 & \frac{1}{3} & \frac{2}{3} & 1 \end{array}\right]} \Rightarrow$

$\underset{4R_2 + R_1 \Rightarrow R_1}{\left[\begin{array}{cc|cc} 1 & 0 & 3 & 4 \\ 0 & \frac{1}{3} & \frac{2}{3} & 1 \end{array}\right]} \Rightarrow \underset{3R_2 \Rightarrow R_2}{\left[\begin{array}{cc|cc} 1 & 0 & 3 & 4 \\ 0 & 1 & 2 & 3 \end{array}\right]} \Rightarrow$ Inverse: $\begin{bmatrix} 3 & 4 \\ 2 & 3 \end{bmatrix}$

7. $\left[\begin{array}{cc|cc} 3 & 7 & 1 & 0 \\ 2 & 5 & 0 & 1 \end{array}\right] \Rightarrow \underset{\frac{1}{3}R_1 \Rightarrow R_1}{\left[\begin{array}{cc|cc} 1 & \frac{7}{3} & \frac{1}{3} & 0 \\ 2 & 5 & 0 & 1 \end{array}\right]} \Rightarrow \underset{-2R_1 + R_2 \Rightarrow R_2}{\left[\begin{array}{cc|cc} 1 & \frac{7}{3} & \frac{1}{3} & 0 \\ 0 & \frac{1}{3} & -\frac{2}{3} & 1 \end{array}\right]} \Rightarrow$

$\underset{-7R_2 + R_1 \Rightarrow R_1}{\left[\begin{array}{cc|cc} 1 & 0 & 5 & -7 \\ 0 & \frac{1}{3} & -\frac{2}{3} & 1 \end{array}\right]} \Rightarrow \underset{3R_2 \Rightarrow R_2}{\left[\begin{array}{cc|cc} 1 & 0 & 5 & -7 \\ 0 & 1 & -2 & 3 \end{array}\right]} \Rightarrow$ Inverse: $\begin{bmatrix} 5 & -7 \\ -2 & 3 \end{bmatrix}$

9. $\left[\begin{array}{ccc|ccc} 1 & 0 & 3 & 1 & 0 & 0 \\ -1 & 1 & 3 & 0 & 1 & 0 \\ -2 & 1 & 1 & 0 & 0 & 1 \end{array}\right] \Rightarrow \underset{\substack{R_1 + R_2 \Rightarrow R_2 \\ 2R_1 + R_3 \Rightarrow R_3}}{\left[\begin{array}{ccc|ccc} 1 & 0 & 3 & 1 & 0 & 0 \\ 0 & 1 & 6 & 1 & 1 & 0 \\ 0 & 1 & 7 & 2 & 0 & 1 \end{array}\right]} \Rightarrow \underset{-R_2 + R_3 \Rightarrow R_3}{\left[\begin{array}{ccc|ccc} 1 & 0 & 3 & 1 & 0 & 0 \\ 0 & 1 & 6 & 1 & 1 & 0 \\ 0 & 0 & 1 & 1 & -1 & 1 \end{array}\right]} \Rightarrow$

$\underset{\substack{-3R_3 + R_1 \Rightarrow R_1 \\ -6R_3 + R_2 \Rightarrow R_2}}{\left[\begin{array}{ccc|ccc} 1 & 0 & 0 & -2 & 3 & -3 \\ 0 & 1 & 0 & -5 & 7 & -6 \\ 0 & 0 & 1 & 1 & -1 & 1 \end{array}\right]} \Rightarrow$ Inverse: $\begin{bmatrix} -2 & 3 & -3 \\ -5 & 7 & -6 \\ 1 & -1 & 1 \end{bmatrix}$

11. $\left[\begin{array}{ccc|ccc} 3 & 2 & 1 & 1 & 0 & 0 \\ 1 & 1 & -1 & 0 & 1 & 0 \\ 4 & 3 & 1 & 0 & 0 & 1 \end{array}\right] \Rightarrow \underset{R_1 \Leftrightarrow R_2}{\left[\begin{array}{ccc|ccc} 1 & 1 & -1 & 0 & 1 & 0 \\ 3 & 2 & 1 & 1 & 0 & 0 \\ 4 & 3 & 1 & 0 & 0 & 1 \end{array}\right]} \Rightarrow \underset{\substack{-3R_1 + R_2 \Rightarrow R_2 \\ -4R_1 + R_3 \Rightarrow R_3}}{\left[\begin{array}{ccc|ccc} 1 & 1 & -1 & 0 & 1 & 0 \\ 0 & -1 & 4 & 1 & -3 & 0 \\ 0 & -1 & 5 & 0 & -4 & 1 \end{array}\right]} \Rightarrow$

$\underset{-R_2 \Rightarrow R_2}{\left[\begin{array}{ccc|ccc} 1 & 1 & -1 & 0 & 1 & 0 \\ 0 & 1 & -4 & -1 & 3 & 0 \\ 0 & -1 & 5 & 0 & -4 & 1 \end{array}\right]} \Rightarrow \underset{\substack{-R_2 + R_1 \Rightarrow R_1 \\ R_2 + R_3 \Rightarrow R_3}}{\left[\begin{array}{ccc|ccc} 1 & 0 & 3 & 1 & -2 & 0 \\ 0 & 1 & -4 & -1 & 3 & 0 \\ 0 & 0 & 1 & -1 & -1 & 1 \end{array}\right]} \Rightarrow$

$\underset{\substack{-3R_3 + R_1 \Rightarrow R_1 \\ 4R_3 + R_2 \Rightarrow R_2}}{\left[\begin{array}{ccc|ccc} 1 & 0 & 0 & 4 & 1 & -3 \\ 0 & 1 & 0 & -5 & -1 & 4 \\ 0 & 0 & 1 & -1 & -1 & 1 \end{array}\right]} \Rightarrow$ Inverse $= \begin{bmatrix} 4 & 1 & -3 \\ -5 & -1 & 4 \\ -1 & -1 & 1 \end{bmatrix}$

EXERCISES 6.4

13. $\left[\begin{array}{ccc|ccc} 1 & 3 & 5 & 1 & 0 & 0 \\ 0 & 1 & 6 & 0 & 1 & 0 \\ 1 & 4 & 11 & 0 & 0 & 1 \end{array}\right] \Rightarrow \underset{-R_1 + R_3 \Rightarrow R_3}{\left[\begin{array}{ccc|ccc} 1 & 3 & 5 & 1 & 0 & 0 \\ 0 & 1 & 6 & 0 & 1 & 0 \\ 0 & 1 & 6 & -1 & 0 & 1 \end{array}\right]} \Rightarrow \underset{-R_2 + R_3 \Rightarrow R_3}{\left[\begin{array}{ccc|ccc} 1 & 3 & 5 & 1 & 0 & 0 \\ 0 & 1 & 6 & 0 & 1 & 0 \\ 0 & 0 & 0 & -1 & -1 & 1 \end{array}\right]}$

Since the original matrix cannot be changed into the identity, there is no inverse matrix.

15. $\left[\begin{array}{ccc|ccc} 1 & 2 & 3 & 1 & 0 & 0 \\ 0 & 1 & 2 & 0 & 1 & 0 \\ 0 & 0 & 1 & 0 & 0 & 1 \end{array}\right] \Rightarrow \underset{-2R_2 + R_1 \Rightarrow R_1}{\left[\begin{array}{ccc|ccc} 1 & 0 & -1 & 1 & -2 & 0 \\ 0 & 1 & 2 & 0 & 1 & 0 \\ 0 & 0 & 1 & 0 & 0 & 1 \end{array}\right]} \Rightarrow \underset{\substack{R_3 + R_1 \Rightarrow R_1 \\ -2R_3 + R_2 \Rightarrow R_2}}{\left[\begin{array}{ccc|ccc} 1 & 0 & 0 & 1 & -2 & 1 \\ 0 & 1 & 0 & 0 & 1 & -2 \\ 0 & 0 & 1 & 0 & 0 & 2 \end{array}\right]} \Rightarrow$

Inverse $= \begin{bmatrix} 1 & -2 & 1 \\ 0 & 1 & -2 \\ 0 & 0 & 1 \end{bmatrix}$

17. $\left[\begin{array}{ccc|ccc} 1 & 6 & 4 & 1 & 0 & 0 \\ 1 & -2 & -5 & 0 & 1 & 0 \\ 2 & 4 & -1 & 0 & 0 & 1 \end{array}\right] \Rightarrow \underset{\substack{-R_1 + R_2 \Rightarrow R_2 \\ -2R_1 + R_3 \Rightarrow R_3}}{\left[\begin{array}{ccc|ccc} 1 & 6 & 4 & 1 & 0 & 0 \\ 0 & -8 & -9 & -1 & 1 & 0 \\ 0 & -8 & -9 & -2 & 0 & 1 \end{array}\right]} \Rightarrow \underset{-R_2 + R_3 \Rightarrow R_3}{\left[\begin{array}{ccc|ccc} 1 & 6 & 4 & 1 & 0 & 0 \\ 0 & -8 & -9 & -1 & 1 & 0 \\ 0 & 0 & 0 & -1 & -1 & 1 \end{array}\right]}$

Since the original matrix cannot be changed into the identity, there is no inverse matrix.

19. Inverse $= \begin{bmatrix} 1 & -2 & 1 & 0 \\ 0 & 1 & -2 & 1 \\ 0 & 0 & 1 & -2 \\ 0 & 0 & 0 & 1 \end{bmatrix}$

21. Inverse $= \begin{bmatrix} 8 & -2 & -6 \\ -5 & 2 & 4 \\ 2 & 0 & -2 \end{bmatrix}$

23. Inverse $= \begin{bmatrix} -2.5 & 5 & 3 & 5.5 \\ 5.5 & -8 & -6 & -9.5 \\ -1 & 3 & 1 & 3 \\ -5.5 & 9 & 6 & 10.5 \end{bmatrix}$

25. $\begin{bmatrix} 3 & -4 \\ -2 & 3 \end{bmatrix}\begin{bmatrix} x \\ y \end{bmatrix} = \begin{bmatrix} 1 \\ 5 \end{bmatrix}$

$\begin{bmatrix} x \\ y \end{bmatrix} = \begin{bmatrix} 3 & -4 \\ -2 & 3 \end{bmatrix}^{-1}\begin{bmatrix} 1 \\ 5 \end{bmatrix}$

$\begin{bmatrix} x \\ y \end{bmatrix} = \begin{bmatrix} 3 & 4 \\ 2 & 3 \end{bmatrix}\begin{bmatrix} 1 \\ 5 \end{bmatrix}$

$\begin{bmatrix} x \\ y \end{bmatrix} = \begin{bmatrix} 23 \\ 17 \end{bmatrix}$

27. $\begin{bmatrix} 3 & -4 \\ -2 & 3 \end{bmatrix}\begin{bmatrix} x \\ y \end{bmatrix} = \begin{bmatrix} 0 \\ 0 \end{bmatrix}$

$\begin{bmatrix} x \\ y \end{bmatrix} = \begin{bmatrix} 3 & -4 \\ -2 & 3 \end{bmatrix}^{-1}\begin{bmatrix} 0 \\ 0 \end{bmatrix}$

$\begin{bmatrix} x \\ y \end{bmatrix} = \begin{bmatrix} 3 & 4 \\ 2 & 3 \end{bmatrix}\begin{bmatrix} 0 \\ 0 \end{bmatrix}$

$\begin{bmatrix} x \\ y \end{bmatrix} = \begin{bmatrix} 0 \\ 0 \end{bmatrix}$

EXERCISES 6.4

29. $$\begin{bmatrix} 2 & 1 & -1 \\ 2 & 2 & -1 \\ -1 & -1 & 1 \end{bmatrix}\begin{bmatrix} x \\ y \\ z \end{bmatrix} = \begin{bmatrix} 2 \\ 4 \\ -1 \end{bmatrix}$$

$$\begin{bmatrix} x \\ y \\ z \end{bmatrix} = \begin{bmatrix} 2 & 1 & -1 \\ 2 & 2 & -1 \\ -1 & -1 & 1 \end{bmatrix}^{-1}\begin{bmatrix} 2 \\ 4 \\ -1 \end{bmatrix}$$

$$\begin{bmatrix} x \\ y \\ z \end{bmatrix} = \begin{bmatrix} 1 & 0 & 1 \\ -1 & 1 & 0 \\ 0 & 1 & 2 \end{bmatrix}\begin{bmatrix} 2 \\ 4 \\ -1 \end{bmatrix} = \begin{bmatrix} 1 \\ 2 \\ 2 \end{bmatrix}$$

31. $$\begin{bmatrix} -2 & 1 & -3 \\ 2 & 3 & 0 \\ 1 & 0 & 1 \end{bmatrix}\begin{bmatrix} x \\ y \\ z \end{bmatrix} = \begin{bmatrix} 2 \\ -3 \\ 5 \end{bmatrix}$$

$$\begin{bmatrix} x \\ y \\ z \end{bmatrix} = \begin{bmatrix} -2 & 1 & -3 \\ 2 & 3 & 0 \\ 1 & 0 & 1 \end{bmatrix}^{-1}\begin{bmatrix} 2 \\ -3 \\ 5 \end{bmatrix}$$

$$\begin{bmatrix} x \\ y \\ z \end{bmatrix} = \begin{bmatrix} 3 & -1 & 9 \\ -2 & 1 & -6 \\ -3 & 1 & -8 \end{bmatrix}\begin{bmatrix} 2 \\ -3 \\ 5 \end{bmatrix} = \begin{bmatrix} 54 \\ -37 \\ -49 \end{bmatrix}$$

33. $$\begin{bmatrix} 5 & 3 \\ -7 & 5 \end{bmatrix}\begin{bmatrix} x \\ y \end{bmatrix} = \begin{bmatrix} 13 \\ -9 \end{bmatrix}$$

$$\begin{bmatrix} x \\ y \end{bmatrix} = \begin{bmatrix} 5 & 3 \\ -7 & 5 \end{bmatrix}^{-1}\begin{bmatrix} 13 \\ -9 \end{bmatrix} = \begin{bmatrix} 2 \\ 1 \end{bmatrix}$$

35. $$\begin{bmatrix} 5 & 2 & 3 \\ 2 & 0 & 5 \\ 3 & 0 & 1 \end{bmatrix}\begin{bmatrix} x \\ y \\ z \end{bmatrix} = \begin{bmatrix} 12 \\ 7 \\ 4 \end{bmatrix}$$

$$\begin{bmatrix} x \\ y \\ z \end{bmatrix} = \begin{bmatrix} 5 & 2 & 3 \\ 2 & 0 & 5 \\ 3 & 0 & 1 \end{bmatrix}^{-1}\begin{bmatrix} 12 \\ 7 \\ 4 \end{bmatrix} = \begin{bmatrix} 1 \\ 2 \\ 1 \end{bmatrix}$$

37. $$\begin{bmatrix} 23 & 27 \\ 21 & 22 \end{bmatrix}\begin{bmatrix} x \\ y \end{bmatrix} = \begin{bmatrix} 127 \\ 108 \end{bmatrix}$$

$$\begin{bmatrix} x \\ y \end{bmatrix} = \begin{bmatrix} 23 & 27 \\ 21 & 22 \end{bmatrix}^{-1}\begin{bmatrix} 127 \\ 108 \end{bmatrix}$$

$$\begin{bmatrix} x \\ y \end{bmatrix} = \begin{bmatrix} 2 \\ 3 \end{bmatrix}$$ $\Rightarrow$ 2 of model A and 3 of model B can be made.

39. $AB = \begin{bmatrix} 17 \\ 43 \end{bmatrix}$

$A^{-1}AB = IB = B$

$B = A^{-1}AB = \begin{bmatrix} 1 & 1 \\ 2 & 3 \end{bmatrix}^{-1} \begin{bmatrix} 17 \\ 43 \end{bmatrix}$

$= \begin{bmatrix} 8 \\ 9 \end{bmatrix} \Rightarrow$ "HI"

41-43. Answers may vary.

45. $A^2 = \begin{bmatrix} -1 & -1 \\ 1 & 1 \end{bmatrix} \begin{bmatrix} -1 & -1 \\ 1 & 1 \end{bmatrix} = \begin{bmatrix} 0 & 0 \\ 0 & 0 \end{bmatrix}$

47. $\begin{bmatrix} 1 & 0 & 0 \\ -2 & -3 & -2 \\ 3 & 6 & 1 \end{bmatrix} \begin{bmatrix} x \\ y \\ z \end{bmatrix} = \begin{bmatrix} 0 \\ 0 \\ 0 \end{bmatrix}$

$\begin{bmatrix} x \\ y \\ z \end{bmatrix} = \begin{bmatrix} 1 & 0 & 0 \\ -2 & -3 & -2 \\ 3 & 6 & 1 \end{bmatrix}^{-1} \begin{bmatrix} 0 \\ 0 \\ 0 \end{bmatrix} = \begin{bmatrix} 0 \\ 0 \\ 0 \end{bmatrix}$

49.

$AB = AC$

$A^{-1}AB = A^{-1}AC$

$IB = IC$

$B = C$

51. $(I - B)(I + B) = I^2 + IB - BI - B^2$

$= I + B - B - B^2$

$= I - B^2$

$= I - \mathbf{0} = I$

Thus, $I - B$ and $I + B$ are inverses.

53. False. Some square matrices have inverses.

55. False. $A^{-1} = \begin{bmatrix} -1 & \frac{1}{2} \\ \frac{3}{4} & -\frac{1}{4} \end{bmatrix}$

57. True.

59. True.

Exercises 6.5 (page 652)

1. $|A|$, $\det A$

3. 0

5. 0

7. $\begin{vmatrix} 2 & 1 \\ -2 & 3 \end{vmatrix} = (2)(3) - (1)(-2)$

$= 6 - (-2) = 8$

9. $\begin{vmatrix} 2 & -3 \\ -3 & 5 \end{vmatrix} = (2)(5) - (-3)(-3)$

$= 10 - 9 = 1$

11. $M_{21} = \begin{vmatrix} -2 & 3 \\ 8 & 9 \end{vmatrix} = (-2)(9) - (3)(8)$

$= -18 - 24 = -42$

13. $M_{33} = \begin{vmatrix} 1 & -2 \\ 4 & 5 \end{vmatrix} = (1)(5) - (-2)(4)$

$= 5 + 8 = 13$

EXERCISES 6.5

15. $C_{21} = -\begin{vmatrix} -2 & 3 \\ 8 & 9 \end{vmatrix} = -[(-2)(9) - (3)(8)]$
$= -[-18 - 24] = 42$

17. $C_{33} = \begin{vmatrix} 1 & -2 \\ 4 & 5 \end{vmatrix} = (1)(5) - (-2)(4)$
$= 5 + 8 = 13$

19. $$\begin{vmatrix} 2 & -3 & 5 \\ -2 & 1 & 3 \\ 1 & 3 & -2 \end{vmatrix} = 2\begin{vmatrix} 1 & 3 \\ 3 & -2 \end{vmatrix} - (-3)\begin{vmatrix} -2 & 3 \\ 1 & -2 \end{vmatrix} + 5\begin{vmatrix} -2 & 1 \\ 1 & 3 \end{vmatrix}$$
$$= 2(-11) + 3(1) + 5(-7) = -22 + 3 - 35 = -54$$

21. $$\begin{vmatrix} 1 & -1 & 2 \\ 2 & 1 & 3 \\ 1 & 1 & -1 \end{vmatrix} = 1\begin{vmatrix} 1 & 3 \\ 1 & -1 \end{vmatrix} - (-1)\begin{vmatrix} 2 & 3 \\ 1 & -1 \end{vmatrix} + 2\begin{vmatrix} 2 & 1 \\ 1 & 1 \end{vmatrix}$$
$$= 1(-4) + 1(-5) + 2(1) = -4 - 5 + 2 = -7$$

23. $$\begin{vmatrix} 2 & 1 & -1 \\ 1 & 3 & 5 \\ 2 & -5 & 3 \end{vmatrix} = 2\begin{vmatrix} 3 & 5 \\ -5 & 3 \end{vmatrix} - 1\begin{vmatrix} 1 & 5 \\ 2 & 3 \end{vmatrix} + (-1)\begin{vmatrix} 1 & 3 \\ 2 & -5 \end{vmatrix}$$
$$= 2(34) - 1(-7) - 1(-11) = 68 + 7 + 11 = 86$$

25. $$\begin{vmatrix} 0 & 1 & -3 \\ -3 & 5 & 2 \\ 2 & -5 & 3 \end{vmatrix} = 0\begin{vmatrix} 5 & 2 \\ -5 & 3 \end{vmatrix} - 1\begin{vmatrix} -3 & 2 \\ 2 & 3 \end{vmatrix} + (-3)\begin{vmatrix} -3 & 5 \\ 2 & -5 \end{vmatrix}$$
$$= 0 - 1(-13) - 3(5) = 0 + 13 - 15 = -2$$

27. $$\begin{vmatrix} 0 & 0 & 1 & 0 \\ -2 & 1 & 0 & 1 \\ 1 & 0 & 1 & 2 \\ 2 & 0 & 1 & 2 \end{vmatrix} = 0(***) - 0(***) + 1\begin{vmatrix} -2 & 1 & 1 \\ 1 & 0 & 2 \\ 2 & 0 & 2 \end{vmatrix} - 0(***)$$
$$= 1\left(-2\begin{vmatrix} 0 & 2 \\ 0 & 2 \end{vmatrix} - 1\begin{vmatrix} 1 & 2 \\ 2 & 2 \end{vmatrix} + 1\begin{vmatrix} 1 & 0 \\ 2 & 0 \end{vmatrix}\right) = -2(0) - 1(-2) + 1(0) = 2$$

29. $$\begin{vmatrix} 10 & 20 & 10 & 30 \\ -2 & 1 & -3 & 1 \\ -1 & 0 & 1 & -2 \\ 2 & -1 & -1 & 3 \end{vmatrix} = \begin{vmatrix} 10 & 20 & 10 & 30 \\ 0 & 0 & -4 & 4 \\ 0 & 2 & 2 & 1 \\ 0 & -5 & -3 & -3 \end{vmatrix} \quad \begin{matrix} R_2 + R_4 \\ \frac{1}{10}R_1 + R_3 \\ -\frac{1}{5}R_1 + R_4 \end{matrix}$$
$$= 10\begin{vmatrix} 0 & -4 & 4 \\ 2 & 2 & 1 \\ -5 & -3 & -3 \end{vmatrix}$$ (expanded along first column of previous matrix)
$$= 10\left(0\begin{vmatrix} 2 & 1 \\ -3 & -3 \end{vmatrix} - (-4)\begin{vmatrix} 2 & 1 \\ -5 & -3 \end{vmatrix} + 4\begin{vmatrix} 2 & 2 \\ -5 & -3 \end{vmatrix}\right)$$
$$= 10[0 + 4(-1) + 4(4)] = 10(12) = 120$$

EXERCISES 6.5

31. R_1 and R_2 have been switched. This multiplies the determinant by -1. TRUE

33. R_1 and R_2 have both been multiplied by -1. This multiplies the determinant by -1 twice. FALSE

35. R_1 and R_2 have been switched. This multiplies the determinant by -1. However, R_3 has been multiplied by -1, which also multiplies the determinant by -1. Thus, the determinant remains equal to 3.

37. R_1 has been added to R_3. This leaves the determinant equal to 3.

39. $x = \dfrac{\begin{vmatrix} 7 & 2 \\ -4 & -3 \end{vmatrix}}{\begin{vmatrix} 3 & 2 \\ 2 & -3 \end{vmatrix}} = \dfrac{-13}{-13} = 1 \quad y = \dfrac{\begin{vmatrix} 3 & 7 \\ 2 & -4 \end{vmatrix}}{\begin{vmatrix} 3 & 2 \\ 2 & -3 \end{vmatrix}} = \dfrac{-26}{-13} = 2$

41. $x = \dfrac{\begin{vmatrix} 3 & -1 \\ 9 & -7 \end{vmatrix}}{\begin{vmatrix} 1 & -1 \\ 3 & -7 \end{vmatrix}} = \dfrac{-12}{-4} = 3 \quad y = \dfrac{\begin{vmatrix} 1 & 3 \\ 3 & 9 \end{vmatrix}}{\begin{vmatrix} 1 & -1 \\ 3 & -7 \end{vmatrix}} = \dfrac{0}{-4} = 0$

43. $x = \dfrac{\begin{vmatrix} 2 & 2 & 1 \\ 2 & -1 & 1 \\ 4 & 1 & 3 \end{vmatrix}}{\begin{vmatrix} 1 & 2 & 1 \\ 1 & -1 & 1 \\ 1 & 1 & 3 \end{vmatrix}} = \dfrac{-6}{-6} = 1 \quad y = \dfrac{\begin{vmatrix} 1 & 2 & 1 \\ 1 & 2 & 1 \\ 1 & 4 & 3 \end{vmatrix}}{\begin{vmatrix} 1 & 2 & 1 \\ 1 & -1 & 1 \\ 1 & 1 & 3 \end{vmatrix}} = \dfrac{0}{-6} = 0 \quad z = \dfrac{\begin{vmatrix} 1 & 2 & 2 \\ 1 & -1 & 2 \\ 1 & 1 & 4 \end{vmatrix}}{\begin{vmatrix} 1 & 2 & 1 \\ 1 & -1 & 1 \\ 1 & 1 & 3 \end{vmatrix}} = \dfrac{-6}{-6} = 1$

45. $x = \dfrac{\begin{vmatrix} 5 & -1 & 1 \\ 10 & -3 & 2 \\ 0 & 3 & 1 \end{vmatrix}}{\begin{vmatrix} 2 & -1 & 1 \\ 3 & -3 & 2 \\ 1 & 3 & 1 \end{vmatrix}} = \dfrac{-5}{-5} = 1 \quad y = \dfrac{\begin{vmatrix} 2 & 5 & 1 \\ 3 & 10 & 2 \\ 1 & 0 & 1 \end{vmatrix}}{\begin{vmatrix} 2 & -1 & 1 \\ 3 & -3 & 2 \\ 1 & 3 & 1 \end{vmatrix}} = \dfrac{5}{-5} = -1 \quad z = \dfrac{\begin{vmatrix} 2 & -1 & 5 \\ 3 & -3 & 10 \\ 1 & 3 & 0 \end{vmatrix}}{\begin{vmatrix} 2 & -1 & 1 \\ 3 & -3 & 2 \\ 1 & 3 & 1 \end{vmatrix}} = \dfrac{-10}{-5} = 2$

47. Rewrite system:

$3x + 2y + 3z = 66$
$2x + 6y - z = 36$
$3x + 1y + 6z = 96$

$x = \dfrac{\begin{vmatrix} 66 & 2 & 3 \\ 36 & 6 & -1 \\ 96 & 1 & 6 \end{vmatrix}}{\begin{vmatrix} 3 & 2 & 3 \\ 2 & 6 & -1 \\ 3 & 1 & 6 \end{vmatrix}} = \dfrac{198}{33} = 6$

continued on next page...

47. continued

$$y=\frac{\begin{vmatrix}3&66&3\\2&36&-1\\3&96&6\end{vmatrix}}{\begin{vmatrix}3&2&3\\2&6&-1\\3&1&6\end{vmatrix}}=\frac{198}{33}=6 \qquad z=\frac{\begin{vmatrix}3&2&66\\2&6&36\\3&1&96\end{vmatrix}}{\begin{vmatrix}3&2&3\\2&6&-1\\3&1&6\end{vmatrix}}=\frac{396}{33}=12$$

49.

$$p=\frac{\begin{vmatrix}0&-1&3&-1\\-1&1&0&-1\\2&0&-1&0\\7&-2&0&3\end{vmatrix}}{\begin{vmatrix}2&-1&3&-1\\1&1&0&-1\\3&0&-1&0\\1&-2&0&3\end{vmatrix}}=\frac{-15}{-18}=\frac{5}{6} \qquad q=\frac{\begin{vmatrix}2&0&3&-1\\1&-1&0&-1\\3&2&-1&0\\1&7&0&3\end{vmatrix}}{\begin{vmatrix}2&-1&3&-1\\1&1&0&-1\\3&0&-1&0\\1&-2&0&3\end{vmatrix}}=\frac{-12}{-18}=\frac{2}{3}$$

$$r=\frac{\begin{vmatrix}2&-1&0&-1\\1&1&-1&-1\\3&0&2&0\\1&-2&7&3\end{vmatrix}}{\begin{vmatrix}2&-1&3&-1\\1&1&0&-1\\3&0&-1&0\\1&-2&0&3\end{vmatrix}}=\frac{-9}{-18}=\frac{1}{2} \qquad s=\frac{\begin{vmatrix}2&-1&3&0\\1&1&0&-1\\3&0&-1&2\\1&-2&0&7\end{vmatrix}}{\begin{vmatrix}2&-1&3&-1\\1&1&0&-1\\3&0&-1&0\\1&-2&0&3\end{vmatrix}}=\frac{-45}{-18}=\frac{5}{2}$$

51.

$$\begin{vmatrix}x&y&1\\0&0&1\\4&6&1\end{vmatrix}=0$$

$$x\begin{vmatrix}0&1\\6&1\end{vmatrix}-y\begin{vmatrix}0&1\\4&1\end{vmatrix}+1\begin{vmatrix}0&0\\4&6\end{vmatrix}=0$$

$$x(-6)-y(-4)+1(0)=0$$

$$-6x+4y=0$$

$$3x-2y=0$$

53.

$$\begin{vmatrix}x&y&1\\-2&3&1\\5&-3&1\end{vmatrix}=0$$

$$x\begin{vmatrix}3&1\\-3&1\end{vmatrix}-y\begin{vmatrix}-2&1\\5&1\end{vmatrix}+1\begin{vmatrix}-2&3\\5&-3\end{vmatrix}=0$$

$$x(6)-y(-7)+1(-9)=0$$

$$6x+7y-9=0$$

$$6x+7y=9$$

55.

$$\pm\frac{1}{2}\begin{vmatrix}0&0&1\\12&0&1\\12&5&1\end{vmatrix}=\pm\frac{1}{2}(60)$$

$= 30$ square units

57.

$$\pm\frac{1}{2}\begin{vmatrix}2&3&1\\10&8&1\\0&20&1\end{vmatrix}=\pm\frac{1}{2}(146)$$

$= 73$ square units

59.

$$\begin{vmatrix}a&b\\c&d\end{vmatrix}=ad-bc$$

$$\begin{vmatrix}b&a\\d&c\end{vmatrix}=bc-ad=-(ad-bc)$$

EXERCISES 6.5

61. $\begin{vmatrix} a & b \\ c & d \end{vmatrix} = ad - bc$

$\begin{vmatrix} a & b+ka \\ c & d+kc \end{vmatrix} = a(d+kc) - (b+ka)c = ad + akc - bc - akc = ad - bc$

63. $\begin{vmatrix} 3 & x \\ 1 & 2 \end{vmatrix} = \begin{vmatrix} 2 & -1 \\ x & -5 \end{vmatrix}$

$6 - x = -10 + x$

$-2x = -16$

$x = 8$

65. $\begin{vmatrix} 3 & x & 1 \\ x & 0 & -2 \\ 4 & 0 & 1 \end{vmatrix} = \begin{vmatrix} 2 & x \\ x & 4 \end{vmatrix}$

$-x\begin{vmatrix} x & -2 \\ 4 & 1 \end{vmatrix} = 8 - x^2$

$-x(x+8) = 8 - x^2$

$-x^2 - 8x = -x^2 + 8$

$x = -1$

67. $\begin{vmatrix} 2.3 & 5.7 & 6.1 \\ 3.4 & 6.2 & 8.3 \\ 5.8 & 8.2 & 9.2 \end{vmatrix} = 21.468$

69. Let $x =$ \$ invested in HiTech, $y =$ \$ invested in SaveTel, and $z =$ \$ invested in OilCo.

$$\begin{cases} x + y + z = 20{,}000 \\ y + z = 3x \\ 0.10x + 0.05y + 0.06z = 0.066(20{,}000) \end{cases} \qquad \begin{aligned} x + y + z &= 20000 \\ -3x + y + z &= 0 \\ 10x + 5y + 6z &= 132000 \end{aligned}$$

$$x = \frac{\begin{vmatrix} 20000 & 1 & 1 \\ 0 & 1 & 1 \\ 132000 & 5 & 6 \end{vmatrix}}{\begin{vmatrix} 1 & 1 & 1 \\ -3 & 1 & 1 \\ 10 & 5 & 6 \end{vmatrix}} = \frac{20000}{4} = 5000, \; y = \frac{\begin{vmatrix} 1 & 20000 & 1 \\ -3 & 0 & 1 \\ 10 & 132000 & 6 \end{vmatrix}}{\begin{vmatrix} 1 & 1 & 1 \\ -3 & 1 & 1 \\ 10 & 5 & 6 \end{vmatrix}} = \frac{32000}{4} = 8000$$

$$z = \frac{\begin{vmatrix} 1 & 1 & 20000 \\ -3 & 1 & 0 \\ 10 & 5 & 132000 \end{vmatrix}}{\begin{vmatrix} 1 & 1 & 1 \\ -3 & 1 & 1 \\ 10 & 5 & 6 \end{vmatrix}} = \frac{28000}{4} = 7000 \Rightarrow$$ He should invest \$5000 in HiTech, \$8000 in SaveTel, and \$7000 in OilCo.

71. **Answers may vary.**

73. **Answers may vary.**

75. $\begin{vmatrix} 1 & 3 & 4 \\ 0 & 5 & 2 \\ 0 & 0 & 2 \end{vmatrix} = 10$

$1 \cdot 5 \cdot 2 = 10$

77. $\begin{vmatrix} 1 & 2 & 4 & 3 \\ 0 & 2 & 2 & 1 \\ 0 & 0 & 3 & 2 \\ 0 & 0 & 0 & 4 \end{vmatrix} = 24$

$1 \cdot 2 \cdot 3 \cdot 4 = 24$

79. $3(1)(1) + 2(-2)(1) + (-1)(2)(3) = -7$

$1(1)(-1) + 3(-2)(3) + 1(2)(2) = -15$

$-7 - (-15) = 8$

81. domain: $n \times n$ matrices

range: all real numbers

83. Yes.

85. False. In general, $|A+B| \neq |A|+|B|$.

87. False. $\begin{vmatrix} 999 & 888 \\ 777 & 666 \end{vmatrix} = -\begin{vmatrix} 777 & 666 \\ 999 & 888 \end{vmatrix}$.

89. True.

Exercises 6.6 (page 662)

1. first-degree, second-degree

3.
$$\frac{3x-1}{x(x-1)} = \frac{A}{x} + \frac{B}{x-1}$$
$$\frac{3x-1}{x(x-1)} = \frac{A(x-1)}{x(x-1)} + \frac{Bx}{x(x-1)}$$
$$\frac{3x-1}{x(x-1)} = \frac{Ax-A+Bx}{x(x-1)}$$
$$\frac{3x-1}{x(x-1)} = \frac{(A+B)x-A}{x(x-1)}$$
$$\begin{cases} A+B = 3 \\ -A = -1 \end{cases} \Rightarrow A=1, B=2$$
$$\frac{3x-1}{x(x-1)} = \frac{1}{x} + \frac{2}{x-1}$$

5.
$$\frac{2x-15}{x(x-3)} = \frac{A}{x} + \frac{B}{x-3}$$
$$\frac{2x-15}{x(x-3)} = \frac{A(x-3)}{x(x-3)} + \frac{Bx}{x(x-3)}$$
$$\frac{2x-15}{x(x-3)} = \frac{Ax-3A+Bx}{x(x-3)}$$
$$\frac{2x-15}{x(x-3)} = \frac{(A+B)x-3A}{x(x-3)}$$
$$\begin{cases} A+B = 2 \\ -3A = -15 \end{cases} \Rightarrow A=5, B=-3$$
$$\frac{2x-15}{x(x-3)} = \frac{5}{x} - \frac{3}{x-3}$$

7.
$$\frac{3x+1}{(x+1)(x-1)} = \frac{A}{x+1} + \frac{B}{x-1}$$
$$\frac{3x+1}{(x+1)(x-1)} = \frac{A(x-1)}{(x+1)(x-1)} + \frac{B(x+1)}{(x+1)(x-1)}$$
$$\frac{3x+1}{(x+1)(x-1)} = \frac{Ax-A+Bx+B}{(x+1)(x-1)}$$
$$\frac{3x+1}{(x+1)(x-1)} = \frac{(A+B)x+(-A+B)}{(x+1)(x-1)}$$
$$\begin{cases} A+B=3 \\ -A+B=1 \end{cases} \Rightarrow A=1, B=2$$
$$\frac{3x+1}{(x+1)(x-1)} = \frac{1}{x+1} + \frac{2}{x-1}$$

9.
$$\frac{-4}{x^2-2x} =$$
$$\frac{-4}{x(x-2)} = \frac{A}{x} + \frac{B}{x-2}$$
$$\frac{-4}{x(x-2)} = \frac{A(x-2)}{x(x-2)} + \frac{Bx}{x(x-2)}$$
$$\frac{-4}{x(x-2)} = \frac{Ax-2A+Bx}{x(x-2)}$$
$$\frac{-4}{x(x-2)} = \frac{(A+B)x-2A}{x(x-2)}$$
$$\frac{-4}{x(x-2)} = \frac{(A+B)x-2A}{x(x-2)}$$
$$\begin{cases} A+B = 0 \\ -2A = -4 \end{cases} \Rightarrow A=2, B=-2$$
$$\frac{4}{x^2-2x} = \frac{2}{x} - \frac{2}{x-2}$$

EXERCISES 6.6

11. $$\frac{-2x+11}{(x+2)(x-3)}=\frac{A}{x+2}+\frac{B}{x-3} \qquad \frac{-2x+11}{x^2-x-6}=\frac{-2x+11}{(x+2)(x-3)}$$

$$\frac{-2x+11}{(x+2)(x-3)}=\frac{A(x-3)}{(x+2)(x-3)}+\frac{B(x+2)}{(x+2)(x-3)}$$

$$\begin{cases} A+\ \ B=-2 \\ -3A+2B=\ \ 11 \end{cases} \Rightarrow A=-3, B=1$$

$$\frac{-2x+11}{(x+2)(x-3)}=\frac{Ax-3A+Bx+2B}{(x+2)(x-3)}$$

$$\frac{-2x+11}{(x+2)(x-3)}=\frac{-3}{x+2}+\frac{1}{x-3}$$

$$\frac{-2x+11}{(x+2)(x-3)}=\frac{(A+B)x+(-3A+2B)}{(x+2)(x-3)}$$

13. $$\frac{3x-23}{(x+3)(x-1)}=\frac{A}{x+3}+\frac{B}{x-1} \qquad \frac{3x-23}{x^2+2x-3}=\frac{3x-23}{(x+3)(x-1)}$$

$$\frac{3x-23}{(x+3)(x-1)}=\frac{A(x-1)}{(x+3)(x-1)}+\frac{B(x+3)}{(x+3)(x-1)}$$

$$\begin{cases} A+\ \ B=\ \ 3 \\ -A+3B=-23 \end{cases} \Rightarrow A=8, B=-5$$

$$\frac{3x-23}{(x+3)(x-1)}=\frac{Ax-A+Bx+3B}{(x+3)(x-1)}$$

$$\frac{3x-23}{(x+3)(x-1)}=\frac{8}{x+3}-\frac{5}{x-1}$$

$$\frac{3x-23}{(x+3)(x-1)}=\frac{(A+B)x+(-A+3B)}{(x+3)(x-1)}$$

15. $$\frac{9x-31}{2x^2-13x+15}=\frac{9x-31}{(2x-3)(x-5)}=\frac{A}{2x-3}+\frac{B}{x-5}$$

$$\frac{9x-31}{(2x-3)(x-5)}=\frac{A(x-5)}{(2x-3)(x-5)}+\frac{B(2x-3)}{(2x-3)(x-5)}$$

$$\frac{9x-31}{(2x-3)(x-5)}=\frac{Ax-5A+2Bx-3B}{(2x-3)(x-5)}$$

$$\frac{9x-31}{(2x-3)(x-5)}=\frac{(A+2B)x+(-5A-3B)}{(2x-3)(x-5)}$$

$$\begin{cases} A+2B=\ \ 9 \\ -5A-3B=-31 \end{cases} \Rightarrow A=5, B=2 \qquad \frac{9x-31}{(2x-3)(x-5)}=\frac{5}{2x-3}+\frac{2}{x-5}$$

17. $$\frac{4x^2+4x-2}{x(x^2-1)}=\frac{4x^2+4x-2}{x(x+1)(x-1)}=\frac{A}{x}+\frac{B}{x+1}+\frac{C}{x-1}$$

$$\frac{4x^2+4x-2}{x(x+1)(x-1)}=\frac{A(x+1)(x-1)}{x(x+1)(x-1)}+\frac{Bx(x-1)}{x(x+1)(x-1)}+\frac{Cx(x+1)}{x(x+1)(x-1)}$$

$$\frac{4x^2+4x-2}{x(x+1)(x-1)}=\frac{Ax^2-A+Bx^2-Bx+Cx^2+Cx}{x(x+1)(x-1)}$$

$$\frac{4x^2+4x-2}{x(x+1)(x-1)}=\frac{(A+B+C)x^2+(-B+C)x+(-A)}{x(x+1)(x-1)}$$

$$\begin{cases} A+B+C=\ \ 4 \\ \quad -B+C=\ \ 4 \\ -A \qquad\quad =-2 \end{cases} \Rightarrow \begin{matrix} A=2 \\ B=-1 \\ C=3 \end{matrix} \qquad \frac{4x^2+4x-2}{x(x+1)(x-1)}=\frac{2}{x}-\frac{1}{x+1}+\frac{3}{x-1}$$

19. $$\begin{aligned}\frac{x^2+x+3}{x(x^2+3)} &= \frac{A}{x}+\frac{Bx+C}{x^2+3}\\ &= \frac{A(x^2+3)}{x(x^2+3)}+\frac{(Bx+C)x}{x(x^2+3)}\\ &= \frac{Ax^2+3A+Bx^2+Cx}{x(x^2+3)}\\ &= \frac{(A+B)x^2+Cx+3A}{x(x^2+3)}\end{aligned}$$

$$\begin{cases} A+B &= 1 \\ C &= 1 \\ 3A &= 3 \end{cases} \Rightarrow \begin{matrix} A=1 \\ B=0 \\ C=1 \end{matrix}$$

$$\frac{x^2+x+3}{x(x^2+3)} = \frac{1}{x}+\frac{1}{x^2+3}$$

21. $$\frac{3x^2+8x+11}{(x+1)(x^2+2x+3)} = \frac{A}{x+1}+\frac{Bx+C}{x^2+2x+3}$$

$$\frac{3x^2+8x+11}{(x+1)(x^2+2x+3)} = \frac{A(x^2+2x+3)}{(x+1)(x^2+2x+3)}+\frac{(Bx+C)(x+1)}{(x+1)(x^2+2x+3)}$$

$$\frac{3x^2+8x+11}{(x+1)(x^2+2x+3)} = \frac{Ax^2+2Ax+3A+Bx^2+Bx+Cx+C}{(x+1)(x^2+2x+3)}$$

$$\frac{3x^2+8x+11}{(x+1)(x^2+2x+3)} = \frac{(A+B)x^2+(2A+B+C)x+(3A+C)}{(x+1)(x^2+2x+3)}$$

$$\begin{cases} A+B &= 3 \\ 2A+B+C &= 8 \\ 3A \quad +C &= 11 \end{cases} \Rightarrow \begin{matrix} A=3 \\ B=0 \\ C=2 \end{matrix} \qquad \frac{3x^2+8x+11}{(x+1)(x^2+2x+3)} = \frac{3}{x+1}+\frac{2}{x^2+2x+3}$$

23. $$\begin{aligned}\frac{5x^2+9x+3}{x(x+1)^2} &= \frac{A}{x}+\frac{B}{x+1}+\frac{C}{(x+1)^2}\\ &= \frac{A(x+1)^2}{x(x+1)^2}+\frac{Bx(x+1)}{x(x+1)^2}+\frac{Cx}{x(x+1)^2}\\ &= \frac{Ax^2+2Ax+A+Bx^2+Bx+Cx}{x(x+1)^2}\\ &= \frac{(A+B)x^2+(2A+B+C)x+A}{x(x+1)^2}\end{aligned}$$

$$\begin{cases} A+B &= 5 \\ 2A+B+C &= 9 \\ A &= 3 \end{cases} \Rightarrow \begin{matrix} A=3 \\ B=2 \\ C=1 \end{matrix} \qquad \frac{5x^2+9x+3}{x(x+1)^2} = \frac{3}{x}+\frac{2}{x+1}+\frac{1}{(x+1)^2}$$

EXERCISES 6.6

25.

$$\begin{aligned}\frac{-2x^2+x-2}{x^2(x-1)} &= \frac{A}{x}+\frac{B}{x^2}+\frac{C}{x-1}\\ &= \frac{Ax(x-1)}{x^2(x-1)}+\frac{B(x-1)}{x^2(x-1)}+\frac{Cx^2}{x^2(x-1)}\\ &= \frac{Ax^2-Ax+Bx-B+Cx^2}{x^2(x-1)}\\ &= \frac{(A+C)x^2+(-A+B)x+(-B)}{x^2(x-1)}\end{aligned}$$

$$\begin{cases} A \quad\quad + C = -2 \\ -A+B \quad\quad = 1 \\ \quad -B \quad\quad = -2 \end{cases} \Rightarrow \begin{matrix} A=1 \\ B=2 \\ C=-3 \end{matrix} \qquad \frac{-2x^2+x-2}{x^2(x-1)} = \frac{1}{x}+\frac{2}{x^2}-\frac{3}{x-1}$$

27.

$$\begin{aligned}\frac{3x^2-13x+18}{x^3-6x^2+9x} = \frac{3x^2-13x+18}{x(x-3)^2} &= \frac{A}{x}+\frac{B}{x-3}+\frac{C}{(x-3)^2}\\ &= \frac{A(x-3)^2}{x(x-3)^2}+\frac{Bx(x-3)}{x(x-3)^2}+\frac{Cx}{x(x-3)^2}\\ &= \frac{Ax^2-6Ax+9A+Bx^2-3Bx+Cx}{x(x-3)^2}\\ &= \frac{(A+B)x^2+(-6A-3B+C)x+9A}{x(x-3)^2}\end{aligned}$$

$$\begin{cases} A+B \quad\quad = 3 \\ -6A-3B+C = -13 \\ 9A \quad\quad = 18 \end{cases} \Rightarrow \begin{matrix} A=2 \\ B=1 \\ C=2 \end{matrix} \qquad \frac{3x^2-13x+18}{x(x-3)^2} = \frac{2}{x}+\frac{1}{x-3}+\frac{2}{(x-3)^2}$$

29.

$$\begin{aligned}\frac{x^2-2x-3}{(x-1)^3} &= \frac{A}{x-1}+\frac{B}{(x-1)^2}+\frac{C}{(x-1)^3}\\ &= \frac{A(x-1)^2}{(x-1)^3}+\frac{B(x-1)}{(x-1)^3}+\frac{C}{(x-1)^3}\\ &= \frac{Ax^2-2Ax+A+Bx-B+C}{(x-1)^3}\\ &= \frac{Ax^2+(-2A+B)x+(A-B+C)}{(x-1)^3}\end{aligned}$$

$$\begin{cases} A \quad\quad = 1 \\ -2A+B \quad\quad = -2 \\ A-B+C = -3 \end{cases} \Rightarrow \begin{matrix} A=1 \\ B=0 \\ C=-4 \end{matrix} \qquad \frac{x^2-2x-3}{(x-1)^3} = \frac{1}{x-1}-\frac{4}{(x-1)^3}$$

31. $$\frac{x^3+4x^2+2x+1}{x^4+x^3+x^2}=$$

$$\begin{aligned}\frac{x^3+4x^2+2x+1}{x^2(x^2+x+1)}&=\frac{A}{x}+\frac{B}{x^2}+\frac{Cx+D}{x^2+x+1}\\&=\frac{Ax(x^2+x+1)}{x^2(x^2+x+1)}+\frac{B(x^2+x+1)}{x^2(x^2+x+1)}+\frac{(Cx+D)x^2}{x^2(x^2+x+1)}\\&=\frac{Ax^3+Ax^2+Ax+Bx^2+Bx+B+Cx^3+Dx^2}{x^2(x^2+x+1)}\\&=\frac{(A+C)x^3+(A+B+D)x^2+(A+B)x+B}{x^2(x^2+x+1)}\end{aligned}$$

$$\begin{cases}A\quad\;\;+C\qquad=1\\A+B\qquad+D=4\\A+B\qquad\quad\;=2\\\quad\;\;B\qquad\quad\;\;=1\end{cases}\Rightarrow\begin{matrix}A=1\\B=1\\C=0\\D=2\end{matrix}\qquad\frac{x^3+4x^2+2x+1}{x^2(x^2+x+1)}=\frac{1}{x}+\frac{1}{x^2}+\frac{2}{x^2+x+1}$$

33. $$\begin{aligned}\frac{4x^3+5x^2+3x+4}{x^2(x^2+1)}&=\frac{A}{x}+\frac{B}{x^2}+\frac{Cx+D}{x^2+1}\\&=\frac{Ax(x^2+1)}{x^2(x^2+1)}+\frac{B(x^2+1)}{x^2(x^2+1)}+\frac{(Cx+D)x^2}{x^2(x^2+1)}\\&=\frac{Ax^3+Ax+Bx^2+B+Cx^3+Dx^2}{x^2(x^2+1)}\\&=\frac{(A+C)x^3+(B+D)x^2+Ax+B}{x^2(x^2+1)}\end{aligned}$$

$$\begin{cases}A\quad+C\qquad=4\\\quad B\qquad+D=5\\A\qquad\qquad\;\;=3\\\quad B\qquad\quad\;\;=4\end{cases}\Rightarrow\begin{matrix}A=3\\B=4\\C=1\\D=1\end{matrix}\qquad\frac{4x^3+5x^2+3x+4}{x^2(x^2+1)}=\frac{3}{x}+\frac{4}{x^2}+\frac{x+1}{x^2+1}$$

35. $$\frac{-x^2-3x-5}{x^3+x^2+2x+2}=\frac{-x^2-3x-5}{x^2(x+1)+2(x+1)}=\frac{-x^2-3x-5}{(x+1)(x^2+2)}$$

$$\begin{aligned}\frac{-x^2-3x-5}{(x+1)(x^2+2)}&=\frac{A}{x+1}+\frac{Bx+C}{x^2+2}\\&=\frac{A(x^2+2)}{(x+1)(x^2+2)}+\frac{(Bx+C)(x+1)}{(x+1)(x^2+2)}\\&=\frac{Ax^2+2A+Bx^2+Bx+Cx+C}{(x+1)(x^2+2)}\\&=\frac{(A+B)x^2+(B+C)x+(2A+C)}{(x+1)(x^2+2)}\end{aligned}$$

$$\begin{cases}A+B\qquad=-1\\\quad\;\;B+C=-3\\2A\qquad+C=-5\end{cases}\Rightarrow\begin{matrix}A=-1\\B=0\\C=-3\end{matrix}\qquad\frac{-x^2-3x-5}{(x+1)(x^2+2)}=\frac{-1}{x+1}-\frac{3}{x^2+2}$$

EXERCISES 6.6

37. $$\frac{x^3+4x^2+3x+6}{(x^2+2)(x^2+x+2)} = \frac{Ax+B}{x^2+2} + \frac{Cx+D}{x^2+x+2}$$

$$= \frac{(Ax+B)(x^2+x+2)}{(x^2+2)(x^2+x+2)} + \frac{(Cx+D)(x^2+2)}{(x^2+2)(x^2+x+2)}$$

$$= \frac{Ax^3+Ax^2+2Ax+Bx^2+Bx+2B+Cx^3+2Cx+Dx^2+2D}{(x^2+2)(x^2+x+2)}$$

$$= \frac{(A+C)x^3+(A+B+D)x^2+(2A+B+2C)x+(2B+2D)}{(x^2+2)(x^2+x+2)}$$

$$\begin{cases} A \quad + C \quad = 1 \\ A + B \quad + D = 4 \\ 2A + B + 2C \quad = 3 \\ 2B \quad + 2D = 6 \end{cases} \Rightarrow \begin{matrix} A=1 \\ B=1 \\ C=0 \\ D=2 \end{matrix} \quad \frac{x^3+4x^2+3x+6}{(x^2+2)(x^2+x+2)} = \frac{x+1}{x^2+2} + \frac{2}{x^2+x+2}$$

39. $$\frac{2x^4+6x^3+20x^2+22x+25}{x(x^2+2x+5)^2}$$

$$= \frac{A}{x} + \frac{Bx+C}{x^2+2x+5} + \frac{Dx+E}{(x^2+2x+5)^2}$$

$$= \frac{A(x^2+2x+5)^2}{x(x^2+2x+5)^2} + \frac{(Bx+C)(x)(x^2+2x+5)}{x(x^2+2x+5)^2} + \frac{(Dx+E)x}{x(x^2+2x+5)^2}$$

$$= \frac{(A+B)x^4+(4A+2B+C)x^3+(14A+5B+2C+D)x^2+(20A+5C+E)x+(25A)}{x(x^2+2x+5)^2}$$

$$\begin{cases} A + B = 2 \\ 4A + 2B + C = 6 \\ 14A + 5B + 2C + D = 20 \\ 20A + 5C + E = 22 \\ 25A = 25 \end{cases} \Rightarrow \begin{matrix} A=1 \\ B=1 \\ C=0 \\ D=1 \\ E=2 \end{matrix}$$

$$\frac{2x^4+6x^3+20x^2+22x+25}{x(x^2+2x+5)^2} = \frac{1}{x} + \frac{x}{x^2+2x+5} + \frac{x+2}{(x^2+2x+5)^2}$$

41. Use long division first: $\frac{x^3}{x^2+3x+2} = x-3+\frac{7x+6}{x^2+3x+2} = x-3+\frac{7x+6}{(x+1)(x+2)}$

$$\frac{7x+6}{(x+1)(x+2)} = \frac{A}{x+1} + \frac{B}{x+2}$$

$$= \frac{A(x+2)}{(x+1)(x+2)} + \frac{B(x+1)}{(x+1)(x+2)}$$

$$= \frac{Ax+2A+Bx+B}{(x+1)(x+2)}$$

$$= \frac{(A+B)x+(2A+B)}{(x+1)(x+2)}$$

$$\begin{cases} A+B=7 \\ 2A+B=6 \end{cases} \Rightarrow A=-1, B=8 \quad x-3+\tfrac{7x+6}{(x+1)(x+2)} = x-3-\tfrac{1}{x+1}+\tfrac{8}{x+2}$$

EXERCISES 6.6

43. Use long division first: $\dfrac{3x^3+3x^2+6x+4}{3x^3+x^2+3x+1}=1+\dfrac{2x^2+3x+3}{3x^3+x^2+3x+1}=1+\dfrac{2x^2+3x+3}{(3x+1)(x^2+1)}$

$$\begin{aligned}\frac{2x^2+3x+3}{(3x+1)(x^2+1)}&=\frac{A}{3x+1}+\frac{Bx+C}{x^2+1}\\&=\frac{A(x^2+1)}{(3x+1)(x^2+1)}+\frac{(Bx+C)(3x+1)}{(3x+1)(x^2+1)}\\&=\frac{Ax^2+A+3Bx^2+Bx+3Cx+C}{(3x+1)(x^2+1)}\\&=\frac{(A+3B)x^2+(B+3C)x+(A+C)}{(3x+1)(x^2+1)}\end{aligned}$$

$$\begin{cases}A+3B&=2\\ \quad B+3C&=3\\ A\quad +C&=3\end{cases}\Rightarrow\begin{matrix}A=2\\B=0\\C=1\end{matrix}\quad 1+\frac{2x^2+3x+3}{(3x+1)(x^2+1)}=1+\frac{2}{3x+1}+\frac{1}{x^2+1}$$

45. Use long division first: $\dfrac{x^3+3x^2+2x+1}{x^3+x^2+x}=1+\dfrac{2x^2+x+1}{x^3+x^2+x}=1+\dfrac{2x^2+x+1}{x(x^2+x+1)}$

$$\begin{aligned}\frac{2x^2+x+1}{x(x^2+x+1)}&=\frac{A}{x}+\frac{Bx+C}{x^2+x+1}\\&=\frac{A(x^2+x+1)}{x(x^2+x+1)}+\frac{(Bx+C)x}{x(x^2+x+1)}\\&=\frac{Ax^2+Ax+A+Bx^2+Cx}{x(x^2+x+1)}\\&=\frac{(A+B)x^2+(A+C)x+(A)}{x(x^2+x+1)}\end{aligned}$$

$$\frac{2x^2+x+1}{x(x^2+x+1)}=\frac{(A+B)x^2+(A+C)x+(A)}{x(x^2+x+1)}$$

$$\begin{cases}A+B&=2\\ A\quad +C&=1\\ A&=1\end{cases}\Rightarrow\begin{matrix}A=1\\B=1\\C=0\end{matrix}\quad 1+\frac{2x^2+x+1}{x(x^2+x+1)}=1+\frac{1}{x}+\frac{x}{x^2+x+1}$$

47. Use long division first: $\dfrac{2x^4+2x^3+3x^2-1}{(x^2-x)(x^2+1)}=2+\dfrac{4x^3+x^2+2x-1}{x(x-1)(x^2+1)}$

$$\begin{aligned}\frac{4x^3+x^2+2x-1}{x(x-1)(x^2+1)}&=\frac{A}{x}+\frac{B}{x-1}+\frac{Cx+D}{x^2+1}\\&=\frac{A(x-1)(x^2+1)}{x(x-1)(x^2+1)}+\frac{Bx(x^2+1)}{x(x-1)(x^2+1)}+\frac{(Cx+D)(x)(x-1)}{x(x-1)(x^2+1)}\\&=\frac{Ax^3-Ax^2+Ax-A+Bx^3+Bx+Cx^3-Cx^2+Dx^2-Dx}{x(x-1)(x^2+1)}\\&=\frac{(A+B+C)x^3+(-A-C+D)x^2+(A+B-D)x+(-A)}{x(x-1)(x^2+1)}\end{aligned}$$

continued on next page...

47. continued

$$\begin{cases} A+B+C = 4 \\ -A - C + D = 1 \\ A+B - D = 2 \\ -A = -1 \end{cases} \Rightarrow \begin{matrix} A=1 \\ B=3 \\ C=0 \\ D=2 \end{matrix} \quad 2+\frac{4x^3+x^2+2x-1}{x(x-1)(x^2+1)} = 2+\frac{1}{x}+\frac{3}{x-1}+\frac{2}{x^2+1}$$

49-53. Answers may vary.

55. $x^3+1=(x+1)(x^2-x+1)$
not prime

57. e

59. d

61. a

Exercises 6.7 (page 672)

1. half-plane, boundary

3. is not

5. $2x+3y<12$

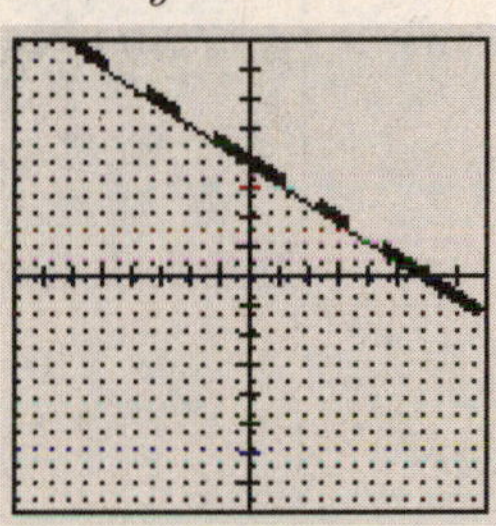

7. $x<3$

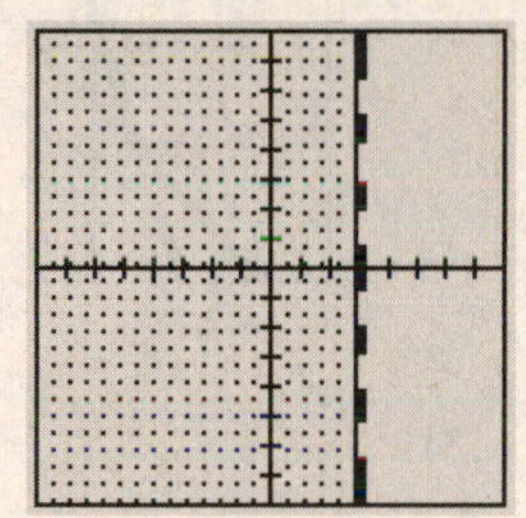

9. $4x-y>4$

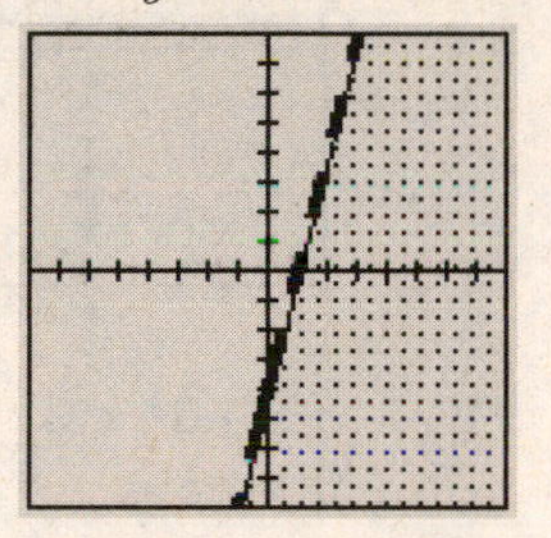

11. $y>2x$

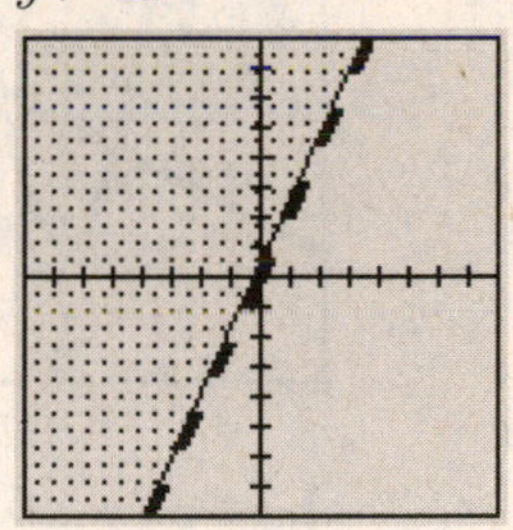

13. $y \leq \frac{1}{2}x+1$

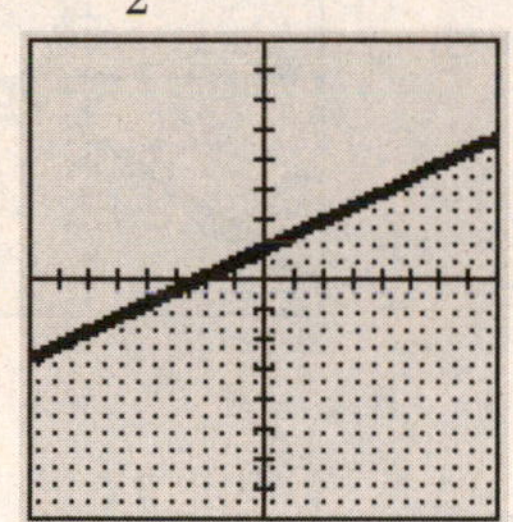

15. $2y \geq 3x-2$

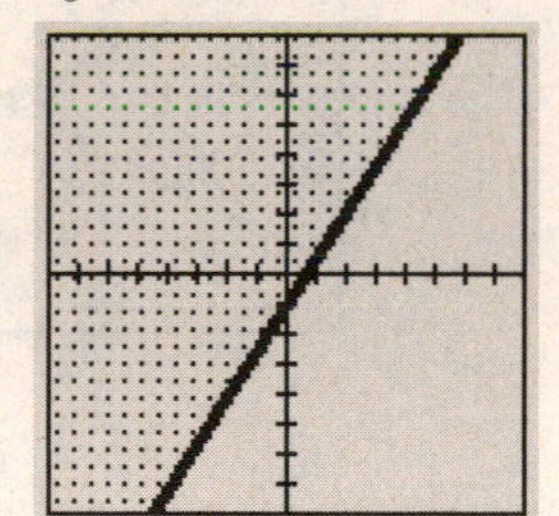

EXERCISES 6.7

17. $y < x^2$

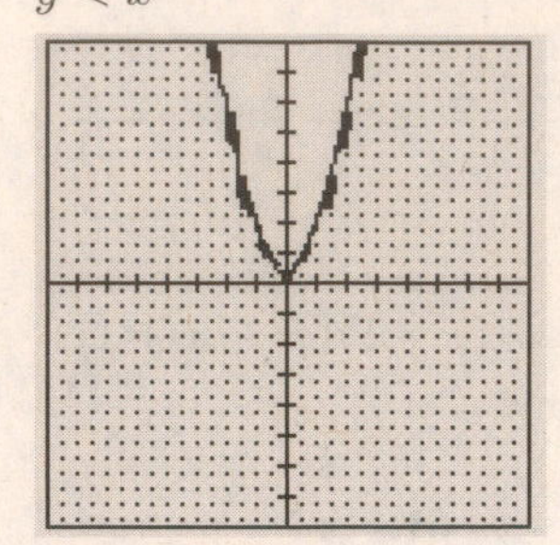

19. $x^2 + y^2 \leq 4$

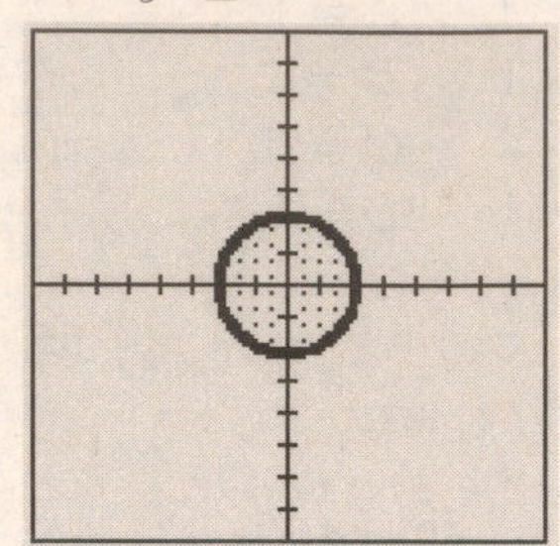

21. $\begin{cases} y < 3 \\ x \geq 2 \end{cases}$

23. $\begin{cases} y \geq 1 \\ x < 2 \end{cases}$

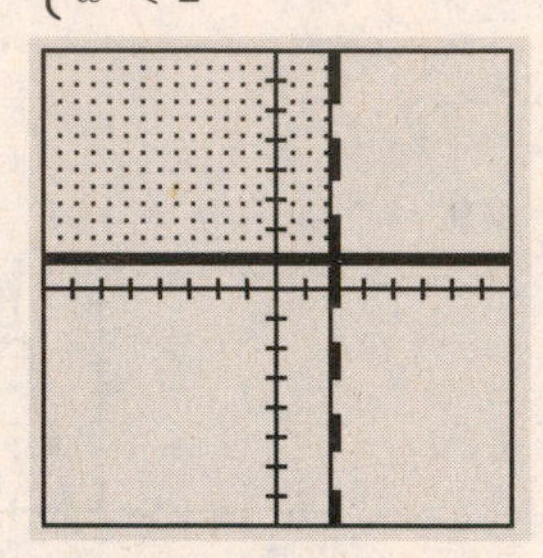

25. $\begin{cases} y \leq x - 2 \\ y \geq 2x + 1 \end{cases}$

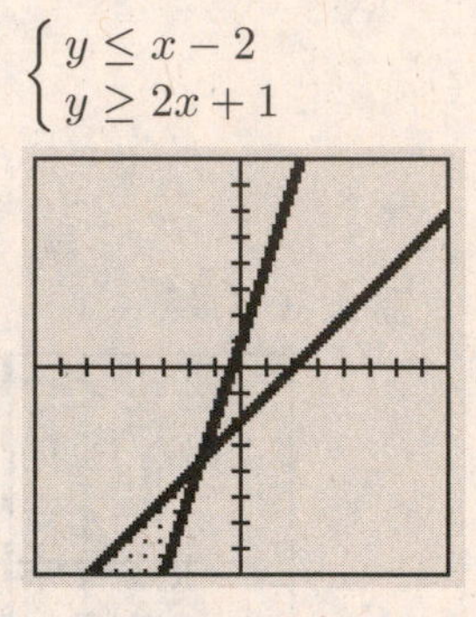

27. $\begin{cases} x + y < 2 \\ x + y \leq 1 \end{cases}$

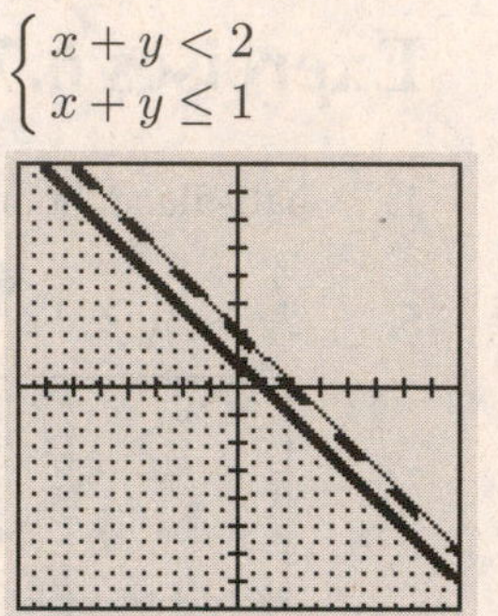

29. $\begin{cases} x + 2y < 3 \\ 2x - 4y < 8 \end{cases}$

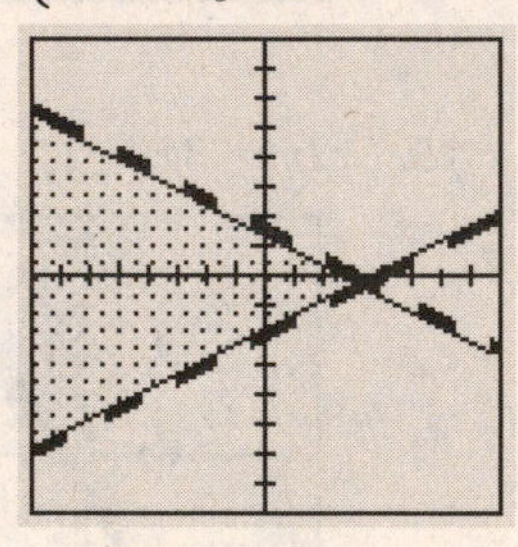

31. $\begin{cases} 2x - 3y \geq 6 \\ 3x + 2y < 6 \end{cases}$

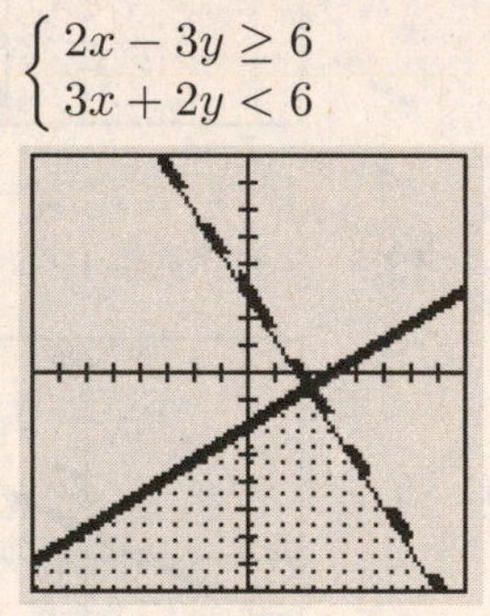

33. $\begin{cases} y \geq x^2 - 4 \\ y \leq \frac{1}{2}x \end{cases}$

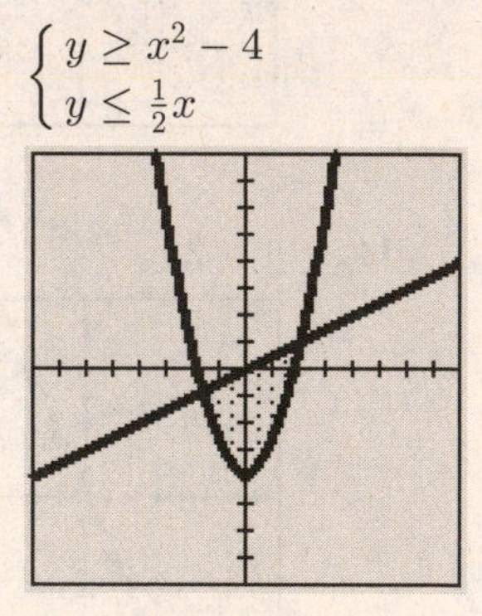

35. $\begin{cases} y \geq x^2 \\ y < 4 - x^2 \end{cases}$

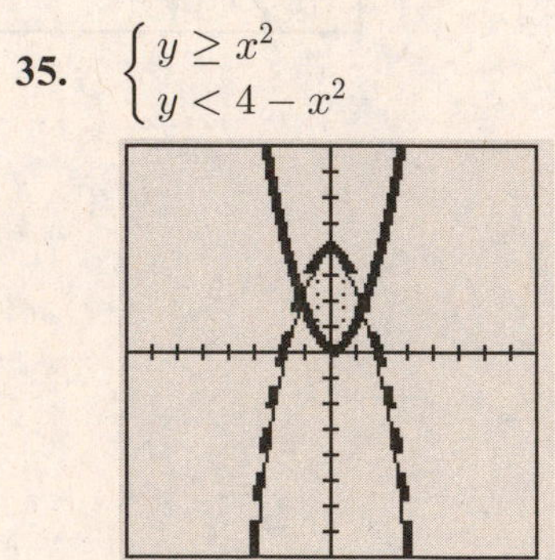

37. $\begin{cases} 2x - y \leq 0 \\ x + 2y \leq 10 \\ y \geq 0 \end{cases}$

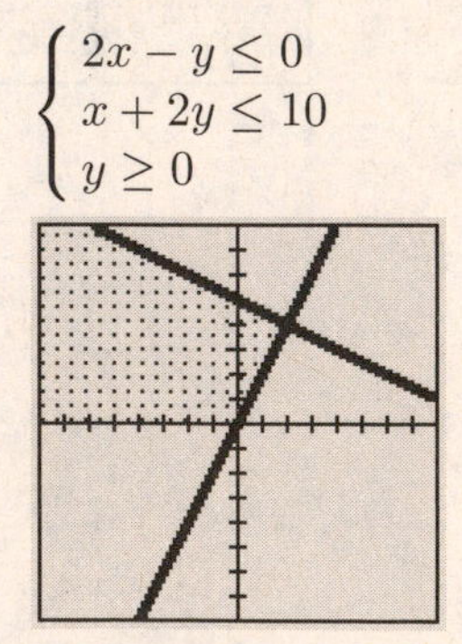

39. $\begin{cases} 3x - 2y \geq 5 \\ 2x + y \geq 8 \\ x \leq 5 \end{cases}$

EXERCISES 6.7

41. $\begin{cases} x+y \le 4 \\ x-y \le 4 \\ x \ge 0, y \ge 0 \end{cases}$

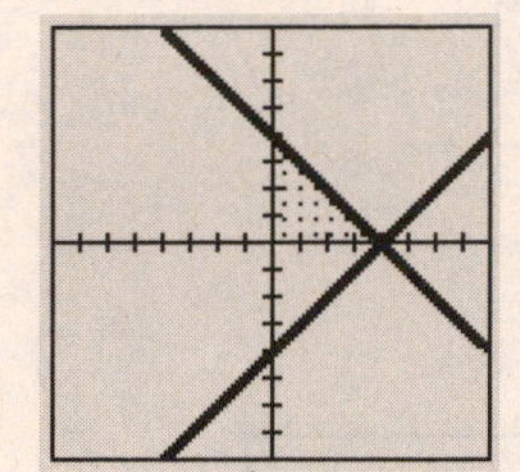

43. $\begin{cases} 3x-2y \le 6 \\ x+2y \le 10 \\ x \ge 0, y \ge 0 \end{cases}$

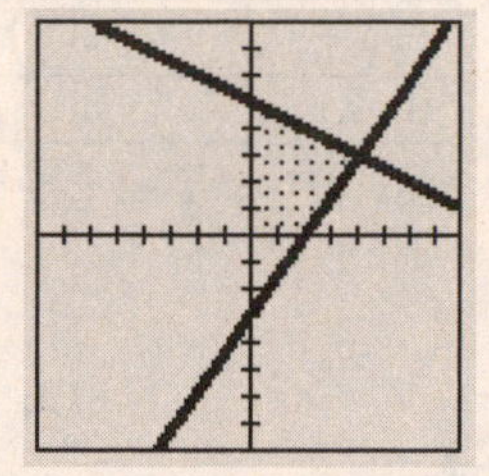

45. $\begin{cases} 6s+4l \le 60 \\ s \ge 0, l \ge 0 \end{cases}$

47. **a.** $\begin{cases} 5x+6y \ge 600 \\ x \ge 0, y \ge 0 \end{cases}$

b.

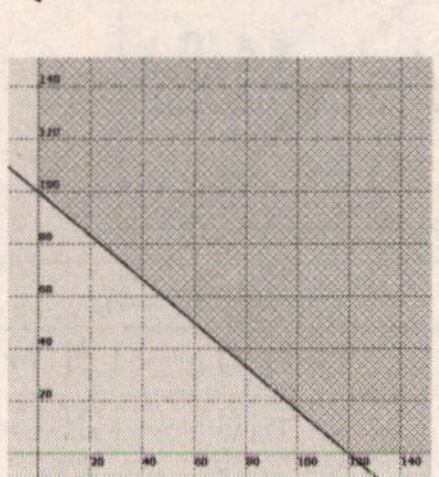

49-53. Answers may vary.

55. False. If $(0,0)$ is on the boundary, it cannot be used as the test point.

57. False. A system of inequalities can have no solution.

59. True.

61. True.

Exercises 6.8 (page 683)

1. constraints

3. objective

5.

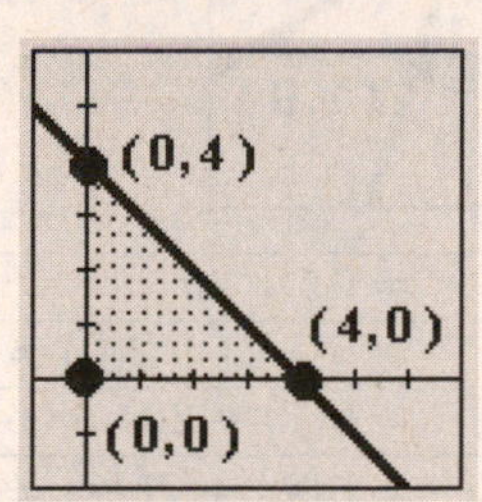

Point	$P = 2x + 3y$
$(0, 0)$	$= 2(0) + 3(0) = 0$
$(0, 4)$	$= 2(0) + 3(4) = 12$
$(4, 0)$	$= 2(4) + 3(0) = 8$

Max: $P = 12$ at $(0, 4)$

7.

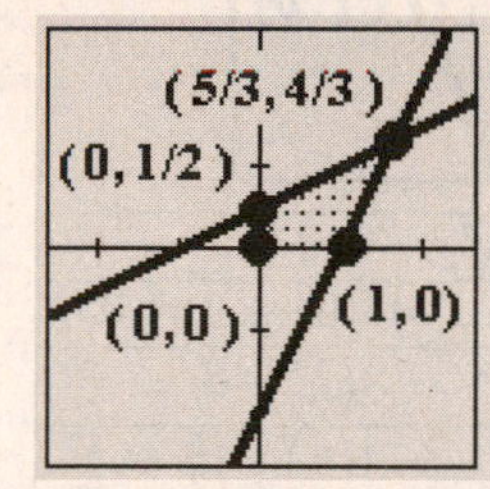

Point	$P = y + \frac{1}{2}x$
$(0, 0)$	$= 0 + \frac{1}{2}(0) = 0$
$\left(0, \frac{1}{2}\right)$	$= \frac{1}{2} + \frac{1}{2}(0) = \frac{1}{2}$
$\left(\frac{5}{3}, \frac{4}{3}\right)$	$= \frac{4}{3} + \frac{1}{2}\left(\frac{5}{3}\right) = \frac{13}{6}$
$(1, 0)$	$= 0 + \frac{1}{2}(1) = \frac{1}{2}$

Max: $P = \frac{13}{6}$ at $\left(\frac{5}{3}, \frac{4}{3}\right)$

EXERCISES 6.8

9.

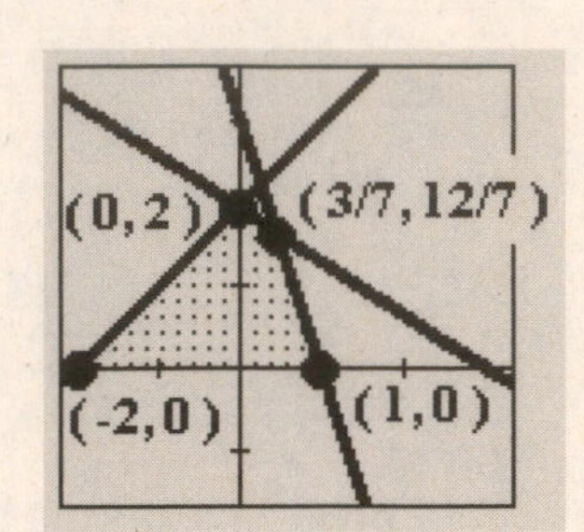

Point	$P = 2x + y$
$(-2, 0)$	$= 2(-2) + 0 = -4$
$(1, 0)$	$= 2(1) + 0 = 2$
$\left(\frac{3}{7}, \frac{12}{7}\right)$	$= 2\left(\frac{3}{7}\right) + \frac{12}{7} = \frac{18}{7}$
$(0, 2)$	$= 2(0) + 2 = 2$

Max: $P = \frac{18}{7}$ at $\left(\frac{3}{7}, \frac{12}{7}\right)$

11.

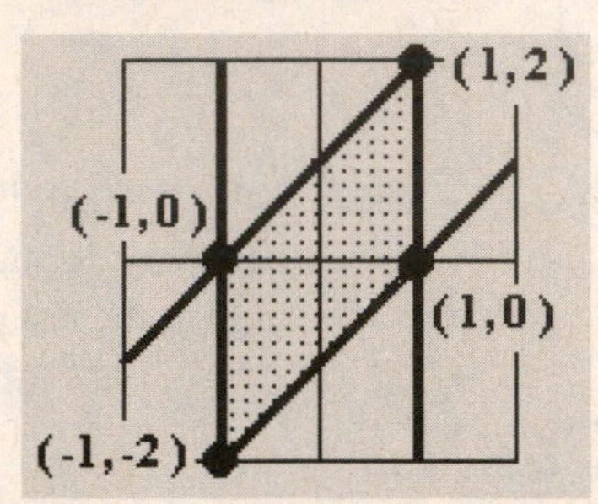

Point	$P = 3x - 2y$
$(1, 0)$	$= 3(1) - 2(0) = 3$
$(1, 2)$	$= 3(1) - 2(2) = -1$
$(-1, 0)$	$= 3(-1) - 2(0) = -3$
$(-1, -2)$	$= 3(-1) - 2(-2) = 1$

Max: $P = 3$ at $(1, 0)$

13.

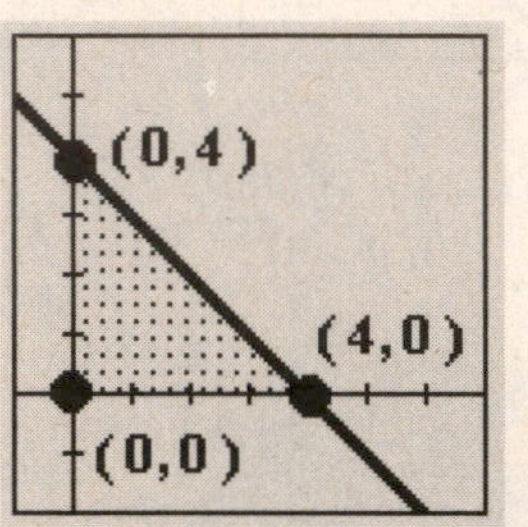

Point	$P = 5x + 12y$
$(0, 0)$	$= 5(0) + 12(0) = 0$
$(0, 4)$	$= 5(0) + 12(4) = 48$
$(4, 0)$	$= 5(4) + 12(0) = 20$

Min: $P = 0$ at $(0, 0)$

15.

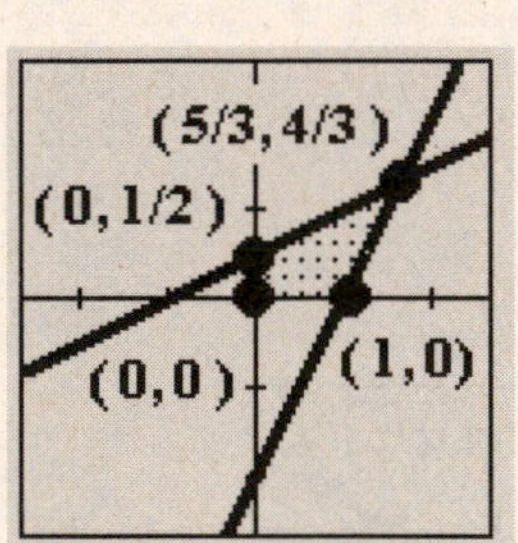

Point	$P = 3y + x$
$(0, 0)$	$= 3(0) + 0 = 0$
$\left(0, \frac{1}{2}\right)$	$= 3\left(\frac{1}{2}\right) + 0 = \frac{3}{2}$
$\left(\frac{5}{3}, \frac{4}{3}\right)$	$= 3\left(\frac{4}{3}\right) + \frac{5}{3} = \frac{17}{3}$
$(1, 0)$	$= 3(0) + 1 = 1$

Min: $P = 0$ at $(0, 0)$

17.

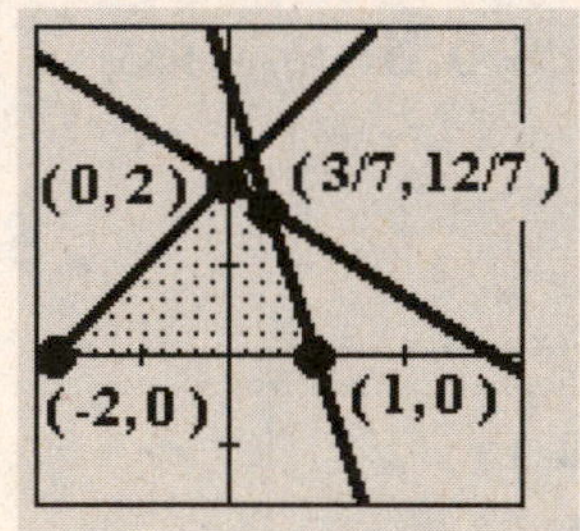

Point	$P = 6x + 2y$
$(-2, 0)$	$= 6(-2) + 2(0) = -12$
$(1, 0)$	$= 6(1) + 2(0) = 6$
$\left(\frac{3}{7}, \frac{12}{7}\right)$	$= 6\left(\frac{3}{7}\right) + 2\left(\frac{12}{7}\right) = 6$
$(0, 2)$	$= 6(0) + 2(2) = 4$

Min: $P = -12$ at $(-2, 0)$

EXERCISES 6.8

19.

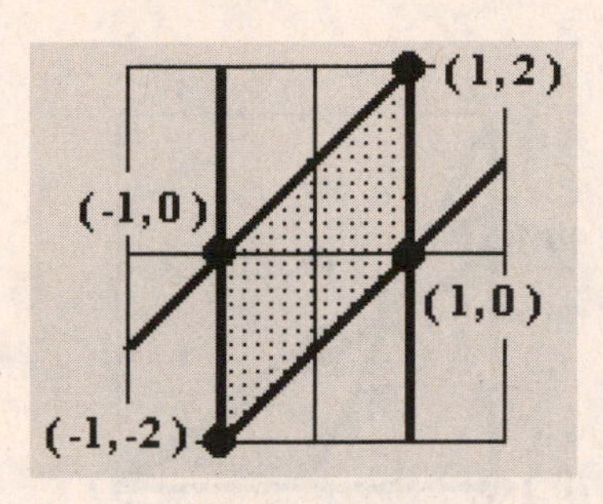

Point	$P = 2x - 2y$
$(1, 0)$	$= 2(1) - 2(0) = 2$
$(1, 2)$	$= 2(1) - 2(2) = -2$
$(-1, 0)$	$= 2(-1) - 2(0) = -2$
$(-1, -2)$	$= 2(-1) - 2(-2) = 2$

Min: $P = -2$ on the edge joining $(1, 2)$ and $(-1, 0)$

21. Let x = # tables and y = # chairs.

Maximize $P = 100x + 80y$

subject to $\begin{cases} 2x + 3y \le 42 \\ 6x + 2y \le 42 \\ x \ge 0, y \ge 0 \end{cases}$

Point	$P = 100x + 80y$
$(7, 0)$	$= 100(7) + 80(0) = 700$
$(3, 12)$	$= 100(3) + 80(12) = 1260$
$(0, 14)$	$= 100(0) + 80(14) = 1120$

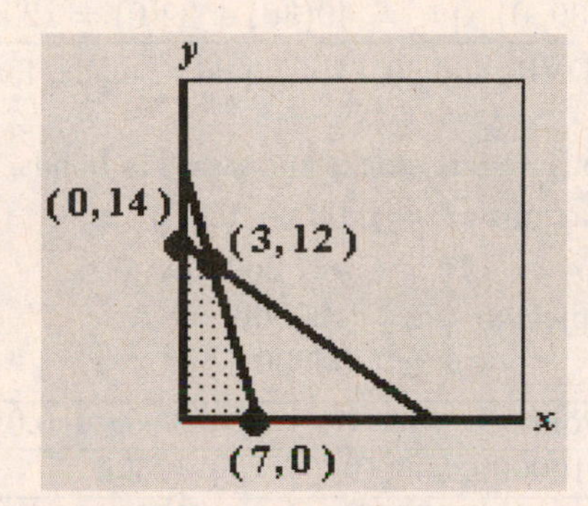

They should make 3 tables and 12 chairs, for a maximum profit of \$1260.

23. Let x = # IBM and y = # Apple.

Maximize $P = 50x + 40y$

subject to $\begin{cases} x + y \le 60 \\ 20 \le x \le 30 \\ 30 \le y \le 50 \end{cases}$

Point	$P = 50x + 40y$
$(20, 30)$	$= 50(20) + 40(30) = 2200$
$(30, 30)$	$= 50(30) + 40(30) = 2700$
$(20, 40)$	$= 50(20) + 40(40) = 2600$

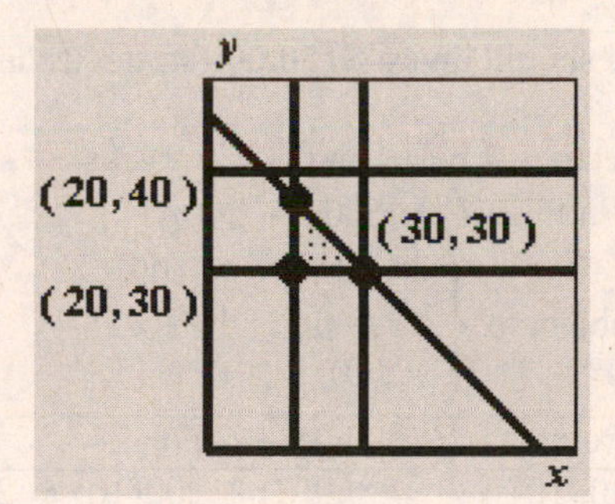

She should stock 30 IBM and 30 Apple computers, for a maximum commission of \$2700.

25. Let $x = \#$ DVRs and $y = \#$ TVs.
Maximize $P = 40x + 32y$

subject to $\begin{cases} 3x + 4y \le 180 \\ 2x + 3y \le 120 \\ 2x + y \le 60 \\ x \ge 0, y \ge 0 \end{cases}$

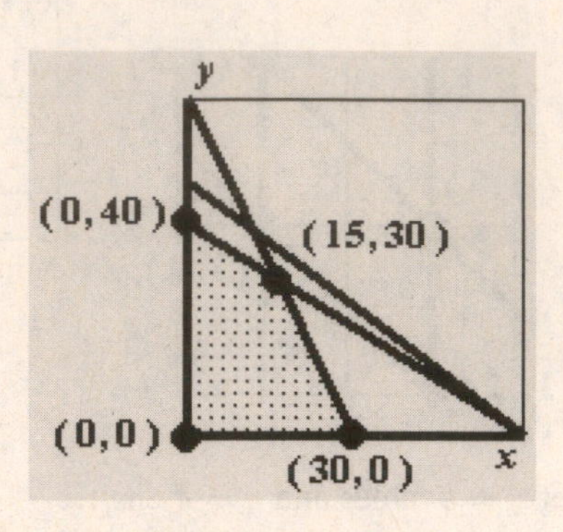

Point	$P = 40x + 32y$
(0, 0)	$= 40(0) + 32(0) = 0$
(0, 40)	$= 40(0) + 32(40) = 1280$
(15, 30)	$= 40(15) + 32(30) = 1560$
(30, 0)	$= 40(30) + 32(0) = 1200$

15 DVRs and 30 TVs should be made, for a maximum profit of \$1560.

27. Let $x = \$$ in stocks and $y = \$$ in bonds.
Maximize $P = 0.09x + 0.07y$

subject to $\begin{cases} x + y \le 200000 \\ x \ge 100000 \\ y \ge 50000 \end{cases}$

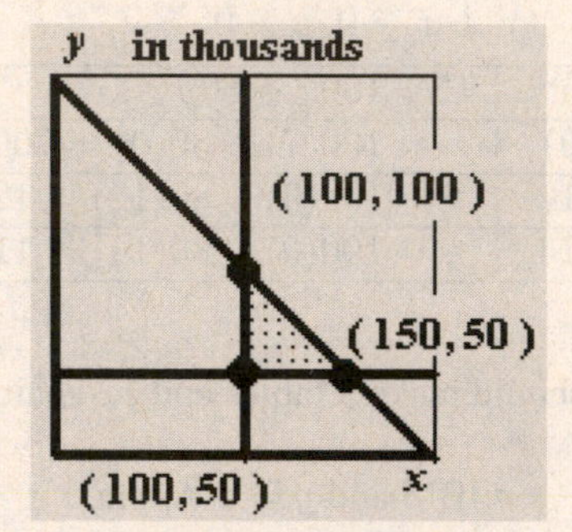

Point	$P = 0.09x + 0.07y$
(100000, 50000)	$= 12500$
(150000, 50000)	$= 17000$
(100000, 100000)	$= 16000$

She should invest \$150,000 in stocks and \$50,000 in bonds, for a maximum return of \$17,000.

29. Let $x = \#$ buses and $y = \#$ trucks.
Minimize $P = 350x + 200y$

subject to $\begin{cases} 40x + 10y \ge 100 \\ 3x + 6y \ge 18 \\ x \ge 0, y \ge 0 \end{cases}$

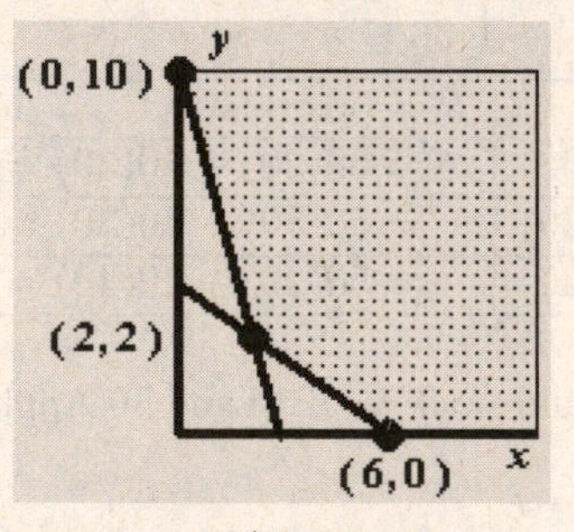

Point	$P = 350x + 200y$
(0, 10)	$= 350(0) + 200(10) = 2000$
(2, 2)	$= 350(2) + 200(2) = 1100$
(6, 0)	$= 350(6) + 200(0) = 2100$

2 buses and 2 trucks should be rented, for a minimum cost of \$1100.

31-35. Answers may vary.

37. False. There must be constraints that describe a bounded region.

39. False. The minimum value can occur at more than one point.

Chapter 6 Review (page 686)

1. $\begin{cases} 2x - y = -1 \\ x + y = 7 \end{cases}$

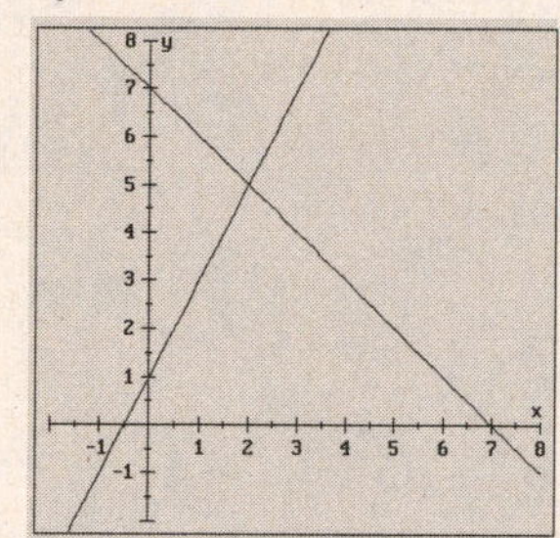

solution: $(2, 5)$

2. $\begin{cases} 5x + 2y = 1 \\ 2x - y = -5 \end{cases}$

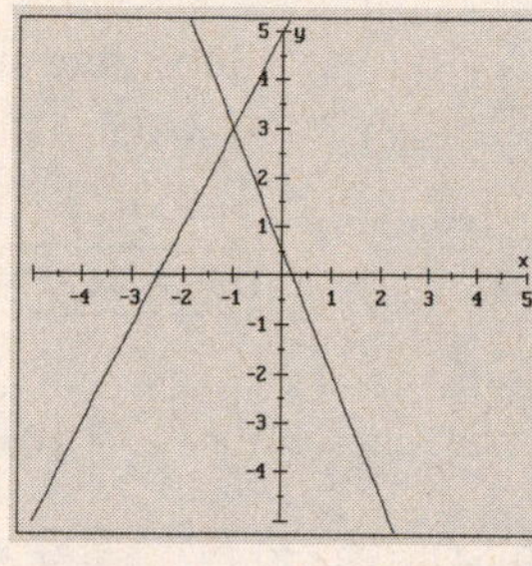

solution: $(-1, 3)$

3. 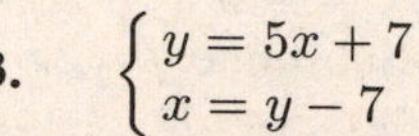
$\begin{cases} y = 5x + 7 \\ x = y - 7 \end{cases}$

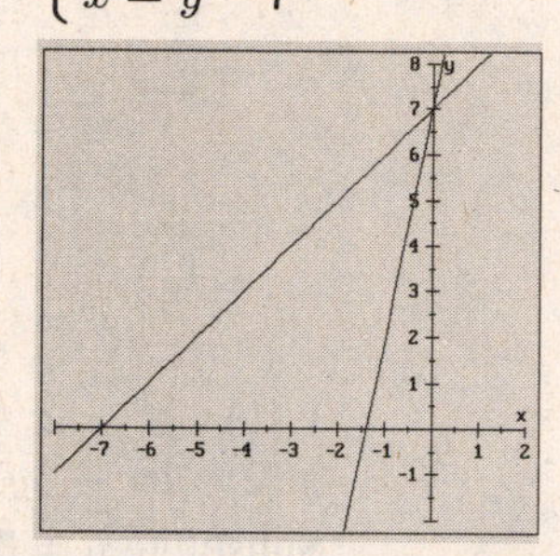

solution: $(0, 7)$

4. $\begin{cases} 3x + 2y = 6 \\ y = -\frac{3}{2}x + 3 \end{cases}$

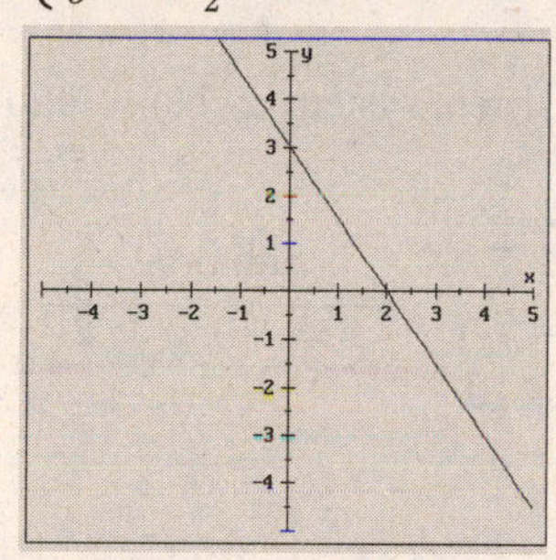

infinitely many solutions
dependent equations

5. 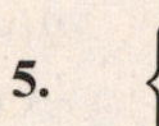
$\begin{cases} 4x - y = 4 \\ y = 4(x - 2) \end{cases}$

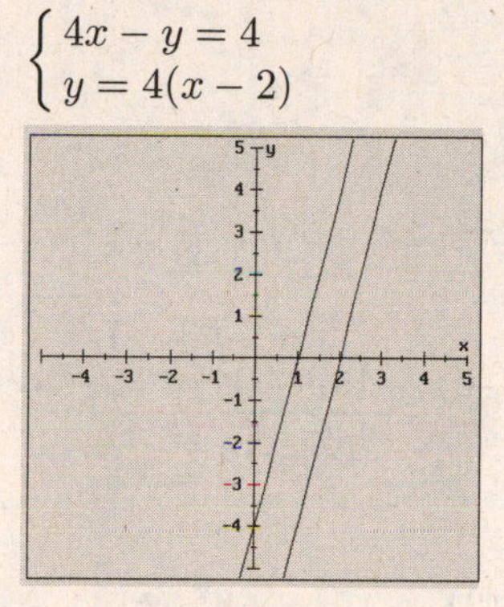

no solutions
inconsistent system

6. $\begin{cases} (1) \quad 2y + x = 0 \\ (2) \quad x = y + 3 \end{cases}$

Substitute $x = y + 3$ into (1):

$$\begin{aligned} 2y + x &= 0 \\ 2y + (y + 3) &= 0 \\ 3y + 3 &= 0 \\ 3y &= -3 \\ y &= -1 \end{aligned}$$

Substitute and solve for x:

$x = y + 3$

$x = -1 + 3 = 2$

$\boxed{x = 2, y = -1}$

7. $\begin{cases} (1) \quad 2x + y = -3 \\ (2) \quad x - y = 3 \end{cases}$

Substitute $x = y + 3$ into (1):

$$\begin{aligned} 2x + y &= -3 \\ 2(y + 3) + y &= -3 \\ 2y + 6 + y &= -3 \\ 3y &= -9 \\ y &= -3 \end{aligned}$$

Substitute and solve for x:

$x = y + 3$

$x = -3 + 3 = 0$

$\boxed{x = 0, y = -3}$

8. $\begin{cases} (1) & \dfrac{x+y}{2} + \dfrac{x-y}{3} = 1 \\ (2) & y = 3x - 2 \end{cases}$

Substitute $y = 3x - 2$ into (1):

$$\frac{x+3x-2}{2} + \frac{x-(3x-2)}{3} = 1$$
$$\frac{4x-2}{2} + \frac{-2x+2}{3} = 1$$
$$3(4x-2) + 2(-2x+2) = 6$$
$$12x - 6 - 4x + 4 = 6$$
$$8x = 8$$
$$x = 1$$

Substitute and solve for y:

$y = 3x - 2$

$y = 3(1) - 2 = 1$

$\boxed{x = 1, y = 1}$

9. $\begin{cases} (1) & y = 3x - 4 \\ (2) & 9x - 3y = 12 \end{cases}$

Substitute $y = 3x - 4$ into (2):

$$9x - 3y = 12$$
$$9x - 3(3x-4) = 12$$
$$9x - 9x + 12 = 12$$
$$0 = 0$$

Dependent equations

General solution: $(x, 3x - 4)$

10. $\begin{cases} (1) & x = -\frac{3}{2}y + 3 \\ (2) & 2x + 3y = 4 \end{cases}$

Substitute $x = -\frac{3}{2}y + 3$ into (2):

$$2x + 3y = 4$$
$$2\left(-\tfrac{3}{2}y + 3\right) + 3y = 4$$
$$-3y + 6 + 3y = 4$$
$$6 \neq 4$$

Inconsistent system $\Rightarrow$ No solution

11.
$$\begin{aligned} x + 5y &= 7 \\ 3x + y &= -7 \Rightarrow \times\ (-5) \end{aligned} \qquad \begin{aligned} x + 5y &= 7 \\ -15x - 5y &= 35 \\ \hline -14x &= 42 \\ x &= -3 \end{aligned} \qquad \begin{aligned} x + 5y &= 7 \\ -3 + 5y &= 7 \\ 5y &= 10 \\ y &= 2 \end{aligned}$$

Solution: $\boxed{x = -3, y = 2}$

12.
$$\begin{aligned} 2x + 3y &= 11 \Rightarrow \times\ (3) \\ 3x - 7y &= -41 \Rightarrow \times\ (-2) \end{aligned} \qquad \begin{aligned} 6x + 9y &= 33 \\ -6x + 14y &= 82 \\ \hline 23y &= 115 \\ y &= 5 \end{aligned} \qquad \begin{aligned} 2x + 3y &= 11 \\ 2x + 3(5) &= 11 \\ 2x &= -4 \\ x &= -2 \end{aligned}$$

Solution: $\boxed{x = -2, y = 5}$

13.
$$\begin{aligned} 2(x+y) - x = 0 \Rightarrow x + 2y &= 0 \Rightarrow \times(-3) \\ 3(x+y) + 2y = 1 \Rightarrow 3x + 5y &= 1 \Rightarrow \end{aligned} \qquad \begin{aligned} -3x - 6y &= 0 \\ 3x + 5y &= 1 \\ \hline -y &= 1 \\ y &= -1 \end{aligned} \qquad \begin{aligned} x + 2y &= 0 \\ x + 2(-1) &= 0 \\ x &= 2 \end{aligned}$$

Solution: $\boxed{\begin{aligned} x &= 2 \\ y &= -1 \end{aligned}}$

14.
$$\begin{aligned} 8x + 12y &= 24 \\ 2x + 3y &= 4 \Rightarrow \times\ (-4) \end{aligned} \qquad \begin{aligned} 8x + 12y &= 24 \\ -8x - 12y &= -16 \\ \hline 0 &\neq 8 \end{aligned}$$

$\Rightarrow$ Inconsistent system: No solution

15.
$$\begin{aligned} 3x - y &= 4 \Rightarrow \times\ (-3) \\ 9x - 3y &= 12 \end{aligned} \qquad \begin{aligned} -9x + 3y &= -12 \\ 9x - 3y &= 12 \\ \hline 0 &= 0 \end{aligned} \qquad \begin{aligned} 3x - y &= 4 \\ 3x - 4 &= y \end{aligned}$$

Dependent equations, General Solution: $(x, 3x - 4)$

16. (1) $3x+2y-z=2$
(2) $x+y-z=0$
(3) $2x+3y-z=1$

Add (1) and $-(2)$:

$$\begin{array}{rl} (1) & 3x+2y-z=2 \\ -(2) & -x-\ \ y+z=0 \\ \hline (4) & 2x+\ \ y \qquad =2 \end{array}$$

Add equations (1) and $-(3)$:

$$\begin{array}{rl} (1) & 3x+2y-z=\ \ 2 \\ -(3) & -2x-3y+z=-1 \\ \hline (5) & x-\ \ y \qquad =\ \ 1 \end{array}$$

Solve the system of two equations and two unknowns formed by equations (4) and (5):

$$\begin{array}{l} 2x+y=2 \\ \underline{x-y=1} \\ 3x \quad =3 \\ x \quad =1 \end{array} \qquad \begin{array}{l} 2x+y=2 \\ 2(1)+y=2 \\ y=0 \end{array} \qquad \begin{array}{l} x+y-z=0 \\ 1+0-z=0 \\ -z=-1 \\ z=1 \end{array}$$

Solution: $\boxed{x=1, y=0, z=1}$

17. (1) $5x-y+z=3$
(2) $3x+y+2z=2$
(3) $x+y=2$

Add (1) and (2):

$$\begin{array}{rl} (1) & 5x-y+\ \ z=3 \\ (2) & 3x+y+2z=2 \\ \hline (4) & 8x \qquad +3z=5 \end{array}$$

Add equations (1) and (3):

$$\begin{array}{rl} (1) & 5x-y+z=3 \\ (3) & x+y \qquad =2 \\ \hline (5) & 6x \qquad +z=5 \end{array}$$

Solve the system of two equations and two unknowns formed by equations (4) and (5):

$$\begin{array}{l} 8x+3z=5 \\ \underline{6x+\ \ z=5} \Rightarrow \times\ (-3) \end{array} \qquad \begin{array}{r} 8x+3z=\ \ 5 \\ \underline{-18x-3z=-15} \\ -10x \quad =-10 \\ x \quad =\ \ 1 \end{array} \qquad \begin{array}{l} 6x+z=5 \\ 6(1)+z=5 \\ z=-1 \end{array} \qquad \begin{array}{l} x+y=2 \\ 1+y=2 \\ y=1 \end{array}$$

Solution: $\boxed{x=1, y=1, z=-1}$

18. (1) $2x-y+z=1$
(2) $x-y+2z=3$
(3) $x-y+z=1$

Add (1) and $-(2)$:

$$\begin{array}{rl} (1) & 2x-y+\ \ z=\ \ 1 \\ -(2) & -x+y-2z=-3 \\ \hline (4) & x \qquad -\ z=-2 \end{array}$$

Add equations (1) and $-(3)$:

$$\begin{array}{rl} (1) & 2x-y+z=\ \ 1 \\ -(3) & -x+y-z=-1 \\ \hline (5) & x \qquad\quad =\ \ 0 \end{array}$$

Solve the system of two equations and two unknowns formed by equations (4) and (5):

$$\begin{array}{l} x-z=-2 \\ 0-z=-2 \\ z=2 \end{array} \qquad \begin{array}{l} x-y+z=1 \\ 0-y+2=1 \\ -y=-1 \\ y=1 \end{array}$$

Solution: $\boxed{x=0, y=1, z=2}$

19. Let x = cost of fake fur and let y = cost of leather. Then $\begin{cases} (1) & 25x+15y=9300 \\ (2) & 10x+30y=12600 \end{cases}$

$$\begin{array}{l} 25x+15y=\ \ 9300 \Rightarrow \times\ (-2) \\ \underline{10x+30y=12600} \end{array} \qquad \begin{array}{r} -50x-30y=-18600 \\ \underline{10x+30y=\ \ 12600} \\ -40x \quad =\ -6000 \\ x \quad =\ \ \ 150 \end{array} \qquad \begin{array}{r} 25x+15y=9300 \\ 25(150)+15y=9300 \\ 15y=5550 \\ y=370 \end{array}$$

The fake fur coats cost \$150 while the leather coats cost \$370. The cost will be \$10,400.

CHAPTER 6 REVIEW

20. Let x = # adult tickets, y = # senior tickets and z = # children tickets.

$$\begin{array}{ll} (1) & x+y+z=1800 \\ (2) & 5x+4y+2.5z=7425 \\ (3) & y+z=900 \end{array}$$

Add $-4(1)$ and (2):

$$\begin{array}{lrcl} -4(1) & -4x-4y-4z & = & -7200 \\ (2) & 5x+4y+2.5z & = & 7425 \\ \hline (4) & x\quad -1.5z & = & 225 \end{array}$$

Add equations (1) and $-(3)$:

$$\begin{array}{lrcl} (1) & x+y+z & = & 1800 \\ -(3) & -y-z & = & -900 \\ \hline (5) & x & = & 900 \end{array}$$

Solve the system of two equations and two unknowns formed by equations (4) and (5):

$$\begin{aligned} x-1.5z&=225 \\ 900-1.5z&=225 \\ -1.5z&=-675 \\ z&=450 \end{aligned} \qquad \begin{aligned} y+z&=900 \\ y+450&=900 \\ y&=450 \end{aligned}$$

There were 900 adult tickets, 450 senior tickets, and 450 children's tickets sold.

21.
$$\left[\begin{array}{cc|c} 2 & 5 & 7 \\ 3 & -1 & 2 \end{array}\right] \Rightarrow \underset{-R_1+R_2 \Rightarrow R_1}{\left[\begin{array}{cc|c} 1 & -6 & -5 \\ 3 & -1 & 2 \end{array}\right]} \Rightarrow \underset{-3R_1+R_2 \Rightarrow R_2}{\left[\begin{array}{cc|c} 1 & -6 & -5 \\ 0 & 17 & 17 \end{array}\right]} \Rightarrow \underset{\frac{1}{17}R_2 \Rightarrow R_2}{\left[\begin{array}{cc|c} 1 & -6 & -5 \\ 0 & 1 & 1 \end{array}\right]} \Rightarrow$$

$$\underset{6R_2+R_1 \Rightarrow R_1}{\left[\begin{array}{cc|c} 1 & 0 & 1 \\ 0 & 1 & 1 \end{array}\right]} \quad \text{Solution: } \boxed{x=1, y=1}$$

22.
$$\left[\begin{array}{cc|c} 3 & -1 & -4 \\ -6 & 2 & 8 \end{array}\right] \Rightarrow \underset{2R_1+R_2 \Rightarrow R_2}{\left[\begin{array}{cc|c} 3 & -1 & -4 \\ 0 & 0 & 0 \end{array}\right]} \Rightarrow \begin{aligned} 3x-y&=-4 \\ 3x+4&=y \end{aligned}$$

Dependent equations

General Solution: $(x, 3x+4)$

23.
$$\left[\begin{array}{ccc|c} 1 & 3 & -1 & 8 \\ 2 & 1 & -2 & 11 \\ 1 & -1 & 5 & -8 \end{array}\right] \Rightarrow \underset{\substack{-2R_1+R_2 \Rightarrow R_2 \\ -R_1+R_3 \Rightarrow R_3}}{\left[\begin{array}{ccc|c} 1 & 3 & -1 & 8 \\ 0 & -5 & 0 & -5 \\ 0 & -4 & 6 & -16 \end{array}\right]} \Rightarrow \underset{-\frac{1}{5}R_2 \Rightarrow R_2}{\left[\begin{array}{ccc|c} 1 & 3 & -1 & 8 \\ 0 & 1 & 0 & 1 \\ 0 & -4 & 6 & -16 \end{array}\right]} \Rightarrow$$

$$\underset{\substack{-3R_2+R_1 \Rightarrow R_1 \\ 4R_2+R_3 \Rightarrow R_3}}{\left[\begin{array}{ccc|c} 1 & 0 & -1 & 5 \\ 0 & 1 & 0 & 1 \\ 0 & 0 & 6 & -12 \end{array}\right]} \Rightarrow \underset{\substack{\frac{1}{6}R_3+R_1 \Rightarrow R_1 \\ \frac{1}{6}R_3 \Rightarrow R_3}}{\left[\begin{array}{ccc|c} 1 & 0 & 0 & 3 \\ 0 & 1 & 0 & 1 \\ 0 & 0 & 1 & -2 \end{array}\right]} \quad \text{Solution: } \boxed{x=3, y=1, z=-2}$$

24.
$$\left[\begin{array}{ccc|c} 1 & 3 & 1 & 3 \\ 2 & -1 & 1 & -11 \\ 3 & 2 & 3 & 2 \end{array}\right] \Rightarrow \underset{\substack{-2R_1+R_2 \Rightarrow R_2 \\ -3R_1+R_3 \Rightarrow R_3}}{\left[\begin{array}{ccc|c} 1 & 3 & 1 & 3 \\ 0 & -7 & -1 & -17 \\ 0 & -7 & 0 & -7 \end{array}\right]} \Rightarrow \underset{R_2 \Leftrightarrow -\frac{1}{7}R_3}{\left[\begin{array}{ccc|c} 1 & 3 & 1 & 3 \\ 0 & 1 & 0 & 1 \\ 0 & -7 & -1 & -17 \end{array}\right]} \Rightarrow$$

$$\underset{\substack{-3R_2+R_1 \Rightarrow R_1 \\ 7R_2+R_3 \Rightarrow R_3}}{\left[\begin{array}{ccc|c} 1 & 0 & 1 & 0 \\ 0 & 1 & 0 & 1 \\ 0 & 0 & -1 & -10 \end{array}\right]} \Rightarrow \underset{\substack{R_3+R_1 \Rightarrow R_1 \\ -R_3 \Rightarrow R_3}}{\left[\begin{array}{ccc|c} 1 & 0 & 0 & -10 \\ 0 & 1 & 0 & 1 \\ 0 & 0 & 1 & 10 \end{array}\right]} \quad \text{Solution: } \boxed{x=-10, y=1, z=10}$$

CHAPTER 6 REVIEW

25. $\left[\begin{array}{ccc|c} 1 & 1 & 1 & 4 \\ 3 & -2 & -2 & -3 \\ 4 & -1 & -1 & 0 \end{array}\right] \Rightarrow \left[\begin{array}{ccc|c} 1 & 1 & 1 & 4 \\ 0 & -5 & -5 & -15 \\ 0 & -5 & -5 & -16 \end{array}\right] \Rightarrow \left[\begin{array}{ccc|c} 1 & 1 & 1 & 4 \\ 0 & -5 & -5 & -15 \\ 0 & 0 & 0 & -1 \end{array}\right]$

$-3R_1 + R_2 \Rightarrow R_2$; $-4R_1 + R_3 \Rightarrow R_3$; $-R_2 + R_3 \Rightarrow R_3$

The last row indicates $0x + 0y + 0z = -1$. This is impossible. $\Rightarrow$ no solution

26. $\left[\begin{array}{cccc|c} 1 & 1 & 1 & -1 & 4 \\ 2 & -1 & 2 & 3 & -8 \\ -1 & 2 & -3 & 1 & 4 \\ 3 & 1 & 2 & -3 & 9 \end{array}\right] \Rightarrow \left[\begin{array}{cccc|c} 1 & 1 & 1 & -1 & 4 \\ 0 & -3 & 0 & 5 & -16 \\ 0 & 3 & -2 & 0 & 8 \\ 0 & -2 & -1 & 0 & -3 \end{array}\right] \Rightarrow \left[\begin{array}{cccc|c} 1 & 1 & 1 & -1 & 4 \\ 0 & 1 & -3 & 0 & 5 \\ 0 & 3 & -2 & 0 & 8 \\ 0 & -3 & 0 & 5 & -16 \end{array}\right]$

$-2R_1 + R_2 \Rightarrow R_2$; $R_1 + R_3 \Rightarrow R_3$; $-3R_1 + R_3 \Rightarrow R_4$; $R_3 + R_4 \Rightarrow R_2$; $R_2 \Rightarrow R_4$

$\left[\begin{array}{cccc|c} 1 & 0 & 4 & -1 & -1 \\ 0 & 1 & -3 & 0 & 5 \\ 0 & 0 & 7 & 0 & -7 \\ 0 & 0 & -9 & 5 & -1 \end{array}\right] \Rightarrow \left[\begin{array}{cccc|c} 1 & 0 & 4 & -1 & -1 \\ 0 & 1 & -3 & 0 & 5 \\ 0 & 0 & 1 & 0 & -1 \\ 0 & 0 & -9 & 5 & -1 \end{array}\right] \Rightarrow \left[\begin{array}{cccc|c} 1 & 0 & 0 & -1 & 3 \\ 0 & 1 & 0 & 0 & 2 \\ 0 & 0 & 1 & 0 & -1 \\ 0 & 0 & 0 & 5 & -10 \end{array}\right]$

$-R_2 + R_1 \Rightarrow R_1$; $-3R_2 + R_3 \Rightarrow R_3$; $3R_2 + R_4 \Rightarrow R_4$; $\frac{1}{7}R_3 \Rightarrow R_3$; $-4R_3 + R_1 \Rightarrow R_1$; $3R_3 + R_2 \Rightarrow R_2$; $9R_3 + R_4 \Rightarrow R_4$

$\Rightarrow \left[\begin{array}{cccc|c} 1 & 0 & 0 & -1 & 3 \\ 0 & 1 & 0 & 0 & 2 \\ 0 & 0 & 1 & 0 & -1 \\ 0 & 0 & 0 & 1 & -2 \end{array}\right] \Rightarrow \left[\begin{array}{cccc|c} 1 & 0 & 0 & 0 & 1 \\ 0 & 1 & 0 & 0 & 2 \\ 0 & 0 & 1 & 0 & -1 \\ 0 & 0 & 0 & 1 & -2 \end{array}\right]$

$\frac{1}{5}R_4 \Rightarrow R_4$; $R_4 + R_1 \Rightarrow R_1$

Solution: $(1, 2, -1, -2)$

27. $-4 = x, x = -4, 0 = x + 4, x + 7 = y \Rightarrow \boxed{x = -4, y = 3}$

28. $\begin{bmatrix} 3 & 2 & 1 \\ 3 & 2 & 1 \end{bmatrix} + \begin{bmatrix} -2 & 1 & 3 \\ 1 & -2 & 1 \end{bmatrix} = \begin{bmatrix} 1 & 3 & 4 \\ 4 & 0 & 2 \end{bmatrix}$

29. $\begin{bmatrix} 2 & 3 & 5 \\ 1 & -2 & 4 \\ 2 & 1 & -2 \end{bmatrix} - \begin{bmatrix} 0 & -2 & 1 \\ 3 & 4 & -2 \\ 6 & -4 & 1 \end{bmatrix} = \begin{bmatrix} 2 & 5 & 4 \\ -2 & -6 & 6 \\ -4 & 5 & -3 \end{bmatrix}$

30. $\begin{bmatrix} 1 & -2 \\ -3 & 1 \end{bmatrix}\begin{bmatrix} 2 & 3 \\ -1 & 2 \end{bmatrix} = \begin{bmatrix} 4 & -1 \\ -7 & -7 \end{bmatrix}$

31. $\begin{bmatrix} -2 & 3 & 5 \\ 1 & -2 & -3 \end{bmatrix}\begin{bmatrix} 2 & 1 \\ -1 & 2 \\ -2 & 3 \end{bmatrix} = \begin{bmatrix} -17 & 19 \\ 10 & -12 \end{bmatrix}$

32. $\begin{bmatrix} 1 & -3 & 2 \end{bmatrix}\begin{bmatrix} 2 \\ 1 \\ 3 \end{bmatrix} = \begin{bmatrix} 5 \end{bmatrix}$

CHAPTER 6 REVIEW

33. $\begin{bmatrix} 1 \\ 2 \\ 1 \\ 5 \end{bmatrix} \begin{bmatrix} 2 & -1 & 1 & 3 \end{bmatrix} = \begin{bmatrix} 2 & -1 & 1 & 3 \\ 4 & -2 & 2 & 6 \\ 2 & -1 & 1 & 3 \\ 10 & -5 & 5 & 15 \end{bmatrix}$

34. $\begin{bmatrix} 1 & -5 & 3 \\ 2 & 1 & -1 \end{bmatrix} \begin{bmatrix} 2 \\ -2 \\ 3 \end{bmatrix} \begin{bmatrix} 1 & -1 \\ -1 & 3 \end{bmatrix} \begin{bmatrix} 1 \\ -2 \end{bmatrix} = \begin{bmatrix} 21 \\ -1 \end{bmatrix} \begin{bmatrix} 1 & -1 \\ -1 & 3 \end{bmatrix} \begin{bmatrix} 1 \\ -2 \end{bmatrix} \Rightarrow$ not possible

35. $\begin{bmatrix} 1 & -3 & 2 \end{bmatrix} \begin{bmatrix} 2 \\ 1 \\ -5 \end{bmatrix} + \begin{bmatrix} 1 & -3 \end{bmatrix} \begin{bmatrix} 2 \\ 5 \end{bmatrix} = [-11] + [-13] = [-24]$

36. $\left(\begin{bmatrix} 1 & -3 \\ 3 & 1 \end{bmatrix} + \begin{bmatrix} -1 & 3 \\ 1 & 1 \end{bmatrix} \right) \begin{bmatrix} 1 \\ -5 \end{bmatrix} = \begin{bmatrix} 0 & 0 \\ 4 & 2 \end{bmatrix} \begin{bmatrix} 1 \\ -5 \end{bmatrix} = \begin{bmatrix} 0 \\ -6 \end{bmatrix}$

37. $X + A = -B$

$$X = -A - B = -\begin{bmatrix} 0 & -2 \\ -3 & 3 \\ -1 & 0 \end{bmatrix} - \begin{bmatrix} 1 & -2 \\ 3 & 9 \\ 5 & 1 \end{bmatrix} = \begin{bmatrix} -1 & 4 \\ 0 & -12 \\ -4 & -1 \end{bmatrix}$$

38. $4X - A = B$

$$4X = A + B$$

$$X = \frac{1}{4}A + \frac{1}{4}B = \frac{1}{4}\begin{bmatrix} 0 & -2 \\ -3 & 3 \\ -1 & 0 \end{bmatrix} + \frac{1}{4}\begin{bmatrix} 1 & -2 \\ 3 & 9 \\ 5 & 1 \end{bmatrix} = \begin{bmatrix} \frac{1}{4} & -1 \\ 0 & 3 \\ 1 & \frac{1}{4} \end{bmatrix}$$

39. $\left[\begin{array}{cc|cc} 2 & 3 & 1 & 0 \\ 3 & 5 & 0 & 1 \end{array}\right] \Rightarrow \underset{\frac{1}{2}R_1 \Rightarrow R_1}{\left[\begin{array}{cc|cc} 1 & \frac{3}{2} & \frac{1}{2} & 0 \\ 3 & 5 & 0 & 1 \end{array}\right]} \Rightarrow \underset{-3R_1 + R_2 \Rightarrow R_2}{\left[\begin{array}{cc|cc} 1 & \frac{3}{2} & \frac{1}{2} & 0 \\ 0 & \frac{1}{2} & -\frac{3}{2} & 1 \end{array}\right]} \Rightarrow$

$\underset{-3R_2 + R_1 \Rightarrow R_1}{\left[\begin{array}{cc|cc} 1 & 0 & 5 & -3 \\ 0 & \frac{1}{2} & -\frac{3}{2} & 1 \end{array}\right]} \Rightarrow \underset{2R_2 \Rightarrow R_2}{\left[\begin{array}{cc|cc} 1 & 0 & 5 & -3 \\ 0 & 1 & -3 & 2 \end{array}\right]} \Rightarrow$ Inverse: $\begin{bmatrix} 5 & -3 \\ -3 & 2 \end{bmatrix}$

40. $\left[\begin{array}{cc|cc} 2 & -1 & 1 & 0 \\ -6 & 4 & 0 & 1 \end{array}\right] \Rightarrow \underset{\frac{1}{2}R_1 \Rightarrow R_1}{\left[\begin{array}{cc|cc} 1 & -\frac{1}{2} & \frac{1}{2} & 0 \\ -6 & 4 & 0 & 1 \end{array}\right]} \Rightarrow \underset{6R_1 + R_2 \Rightarrow R_2}{\left[\begin{array}{cc|cc} 1 & -\frac{1}{2} & \frac{1}{2} & 0 \\ 0 & 1 & 3 & 1 \end{array}\right]} \Rightarrow \underset{\frac{1}{2}R_2 + R_1 \Rightarrow R_1}{\left[\begin{array}{cc|cc} 1 & 0 & 2 & \frac{1}{2} \\ 0 & 1 & 3 & 1 \end{array}\right]}$

Inverse: $\begin{bmatrix} 2 & \frac{1}{2} \\ 3 & 1 \end{bmatrix}$

41. $\left[\begin{array}{cc|cc} -6 & 4 & 1 & 0 \\ -3 & 2 & 0 & 1 \end{array}\right] \Rightarrow \underset{-2R_2 + R_1 \Rightarrow R_2}{\left[\begin{array}{cc|cc} -6 & 4 & 1 & 0 \\ 0 & 0 & 1 & -2 \end{array}\right]} \Rightarrow$ No inverse exists.

42. $\left[\begin{array}{ccc|ccc} 1 & 0 & 0 & 1 & 0 & 0 \\ 2 & 0 & -2 & 0 & 1 & 0 \\ 1 & 2 & 2 & 0 & 0 & 1 \end{array}\right] \Rightarrow \underset{R_2 \Leftrightarrow R_3}{\left[\begin{array}{ccc|ccc} 1 & 0 & 0 & 1 & 0 & 0 \\ 1 & 2 & 2 & 0 & 0 & 1 \\ 2 & 0 & -2 & 0 & 1 & 0 \end{array}\right]} \Rightarrow \underset{\substack{-R_1 + R_2 \Rightarrow R_2 \\ -2R_1 + R_3 \Rightarrow R_3}}{\left[\begin{array}{ccc|ccc} 1 & 0 & 0 & 1 & 0 & 0 \\ 0 & 2 & 2 & -1 & 0 & 1 \\ 0 & 0 & -2 & -2 & 1 & 0 \end{array}\right]} \Rightarrow$

$\underset{R_2 + R_3 \Rightarrow R_2}{\left[\begin{array}{ccc|ccc} 1 & 0 & 0 & 1 & 0 & 0 \\ 0 & 2 & 0 & -3 & 1 & 1 \\ 0 & 0 & -2 & -2 & 1 & 0 \end{array}\right]} \Rightarrow \underset{\substack{\frac{1}{2}R_2 \Rightarrow R_2 \\ -\frac{1}{2}R_3 \Rightarrow R_3}}{\left[\begin{array}{ccc|ccc} 1 & 0 & 0 & 1 & 0 & 0 \\ 0 & 1 & 0 & -\frac{3}{2} & \frac{1}{2} & \frac{1}{2} \\ 0 & 0 & 1 & 1 & -\frac{1}{2} & 0 \end{array}\right]}$: Inverse $= \begin{bmatrix} 1 & 0 & 0 \\ -\frac{3}{2} & \frac{1}{2} & \frac{1}{2} \\ 1 & -\frac{1}{2} & 0 \end{bmatrix}$

43. $\left[\begin{array}{ccc|ccc} 1 & 0 & 8 & 1 & 0 & 0 \\ 3 & 7 & 6 & 0 & 1 & 0 \\ 1 & 2 & 3 & 0 & 0 & 1 \end{array}\right] \Rightarrow \underset{\substack{-3R_1 + R_2 \Rightarrow R_2 \\ -R_1 + R_3 \Rightarrow R_3}}{\left[\begin{array}{ccc|ccc} 1 & 0 & 8 & 1 & 0 & 0 \\ 0 & 7 & -18 & -3 & 1 & 0 \\ 0 & 2 & -5 & -1 & 0 & 1 \end{array}\right]} \Rightarrow \underset{\frac{1}{7}R_2 \Rightarrow R_2}{\left[\begin{array}{ccc|ccc} 1 & 0 & 8 & 1 & 0 & 0 \\ 0 & 1 & -\frac{18}{7} & -\frac{3}{7} & \frac{1}{7} & 0 \\ 0 & 2 & -5 & -1 & 0 & 1 \end{array}\right]} \Rightarrow$

$\underset{-2R_2 + R_3 \Rightarrow R_3}{\left[\begin{array}{ccc|ccc} 1 & 0 & 8 & 1 & 0 & 0 \\ 0 & 1 & -\frac{18}{7} & -\frac{3}{7} & \frac{1}{7} & 0 \\ 0 & 0 & \frac{1}{7} & -\frac{1}{7} & -\frac{2}{7} & 1 \end{array}\right]} \Rightarrow \underset{\substack{-56R_3 + R_1 \Rightarrow R_1 \\ 18R_3 + R_2 \Rightarrow R_2 \\ 7R_3 \Rightarrow R_3}}{\left[\begin{array}{ccc|ccc} 1 & 0 & 0 & 9 & 16 & -56 \\ 0 & 1 & 0 & -3 & -5 & 18 \\ 0 & 0 & 1 & -1 & -2 & 7 \end{array}\right]} \Rightarrow$ Inverse: $\begin{bmatrix} 9 & 16 & -56 \\ -3 & -5 & 18 \\ -1 & -2 & 7 \end{bmatrix}$

44. $\left[\begin{array}{ccc|ccc} 4 & 4 & 1 & 1 & 0 & 0 \\ 1 & 1 & 1 & 0 & 1 & 0 \\ -1 & -1 & 0 & 0 & 0 & 1 \end{array}\right] \Rightarrow \underset{R_1 \Leftrightarrow R_2}{\left[\begin{array}{ccc|ccc} 1 & 1 & 1 & 0 & 1 & 0 \\ 4 & 4 & 1 & 1 & 0 & 0 \\ -1 & -1 & 0 & 0 & 0 & 1 \end{array}\right]} \Rightarrow$

$\underset{\substack{-4R_1 + R_2 \Rightarrow R_2 \\ R_1 + R_3 \Rightarrow R_3}}{\left[\begin{array}{ccc|ccc} 1 & 1 & 1 & 0 & 1 & 0 \\ 0 & 0 & -3 & 1 & -4 & 0 \\ 0 & 0 & 1 & 0 & 1 & 1 \end{array}\right]} \Rightarrow \underset{3R_3 + R_2 \Rightarrow R_2}{\left[\begin{array}{ccc|ccc} 1 & 1 & 1 & 0 & 1 & 0 \\ 0 & 0 & 0 & 1 & -1 & 3 \\ 0 & 0 & 1 & 0 & 1 & 1 \end{array}\right]}$: No inverse exists.

45. $\begin{bmatrix} 3 & -1 \\ 1 & 2 \end{bmatrix}\begin{bmatrix} x \\ y \end{bmatrix} = \begin{bmatrix} 8 \\ 5 \end{bmatrix}$

$$\begin{bmatrix} x \\ y \end{bmatrix} = \begin{bmatrix} 3 & -1 \\ 1 & 2 \end{bmatrix}^{-1}\begin{bmatrix} 8 \\ 5 \end{bmatrix} = \begin{bmatrix} \frac{2}{7} & \frac{1}{7} \\ -\frac{1}{7} & \frac{3}{7} \end{bmatrix}\begin{bmatrix} 8 \\ 5 \end{bmatrix} = \begin{bmatrix} 3 \\ 1 \end{bmatrix}$$

46. $\begin{bmatrix} 4 & -1 & 2 \\ 1 & 1 & 2 \\ 1 & 0 & 1 \end{bmatrix} \begin{bmatrix} x \\ y \\ z \end{bmatrix} = \begin{bmatrix} 0 \\ 1 \\ 0 \end{bmatrix}$

$$\begin{bmatrix} x \\ y \\ z \end{bmatrix} = \begin{bmatrix} 4 & -1 & 2 \\ 1 & 1 & 2 \\ 1 & 0 & 1 \end{bmatrix}^{-1} \begin{bmatrix} 0 \\ 1 \\ 0 \end{bmatrix} = \begin{bmatrix} 1 & 1 & -4 \\ 1 & 2 & -6 \\ -1 & -1 & 5 \end{bmatrix} \begin{bmatrix} 0 \\ 1 \\ 0 \end{bmatrix} = \begin{bmatrix} 1 \\ 2 \\ -1 \end{bmatrix}$$

47. $\begin{bmatrix} 1 & 3 & 1 & 3 \\ 1 & 4 & 1 & 3 \\ 0 & 1 & 1 & 0 \\ 1 & 2 & -1 & 2 \end{bmatrix} \begin{bmatrix} w \\ x \\ y \\ z \end{bmatrix} = \begin{bmatrix} 1 \\ 2 \\ 1 \\ 1 \end{bmatrix}$

$$\begin{bmatrix} w \\ x \\ y \\ z \end{bmatrix} = \begin{bmatrix} 1 & 3 & 1 & 3 \\ 1 & 4 & 1 & 3 \\ 0 & 1 & 1 & 0 \\ 1 & 2 & -1 & 2 \end{bmatrix}^{-1} \begin{bmatrix} 1 \\ 2 \\ 1 \\ 1 \end{bmatrix} = \begin{bmatrix} 3 & -5 & 5 & 3 \\ -1 & 1 & 0 & 0 \\ 1 & -1 & 1 & 0 \\ 0 & 1 & -2 & -1 \end{bmatrix} \begin{bmatrix} 1 \\ 2 \\ 1 \\ 1 \end{bmatrix} = \begin{bmatrix} 1 \\ 1 \\ 0 \\ -1 \end{bmatrix}$$

48. $\begin{vmatrix} 3 & -2 \\ 1 & -3 \end{vmatrix} = (3)(-3) - (-2)(1) = -9 + 2 = -7$

49. $\begin{vmatrix} 1 & -2 & 3 \\ 2 & -1 & 3 \\ 1 & -1 & 0 \end{vmatrix} = 1\begin{vmatrix} -1 & 3 \\ -1 & 0 \end{vmatrix} - (-2)\begin{vmatrix} 2 & 3 \\ 1 & 0 \end{vmatrix} + 3\begin{vmatrix} 2 & -1 \\ 1 & -1 \end{vmatrix}$

$$= 1(3) + 2(-3) + 3(-1) = 3 - 6 - 3 = -6$$

50. $\begin{vmatrix} 1 & 3 & -1 \\ 1 & 2 & 1 \\ 1 & 0 & 2 \end{vmatrix} = 1\begin{vmatrix} 2 & 1 \\ 0 & 2 \end{vmatrix} - 3\begin{vmatrix} 1 & 1 \\ 1 & 2 \end{vmatrix} + (-1)\begin{vmatrix} 1 & 2 \\ 1 & 0 \end{vmatrix}$

$$= 1(4) - 3(1) - 1(-2) = 4 - 3 + 2 = 3$$

51. Expand along 3rd row... $\begin{vmatrix} 1 & 2 & 3 & 4 \\ -1 & 3 & -3 & 2 \\ 0 & 0 & 0 & -1 \\ 3 & 3 & 4 & 3 \end{vmatrix} = 0(*) - 0(*) + 0(*) - (-1)\begin{vmatrix} 1 & 2 & 3 \\ -1 & 3 & -3 \\ 3 & 3 & 4 \end{vmatrix}$

$$= 1\left(1\begin{vmatrix} 3 & -3 \\ 3 & 4 \end{vmatrix} - 2\begin{vmatrix} -1 & -3 \\ 3 & 4 \end{vmatrix} + 3\begin{vmatrix} -1 & 3 \\ 3 & 3 \end{vmatrix}\right)$$

$$= 1(21) - 2(5) + 3(-12) = -25$$

52. $x = \dfrac{\begin{vmatrix} -5 & 3 \\ -4 & 1 \end{vmatrix}}{\begin{vmatrix} 1 & 3 \\ -2 & 1 \end{vmatrix}} = \dfrac{7}{7} = 1 \quad y = \dfrac{\begin{vmatrix} 1 & -5 \\ -2 & -4 \end{vmatrix}}{\begin{vmatrix} 1 & 3 \\ -2 & 1 \end{vmatrix}} = \dfrac{-14}{7} = -2$

53. $x = \dfrac{\begin{vmatrix} -1 & -1 & 1 \\ -4 & -1 & 3 \\ -1 & -3 & 1 \end{vmatrix}}{\begin{vmatrix} 1 & -1 & 1 \\ 2 & -1 & 3 \\ 1 & -3 & 1 \end{vmatrix}} = \dfrac{2}{2} = 1$ $\quad y = \dfrac{\begin{vmatrix} 1 & -1 & 1 \\ 2 & -4 & 3 \\ 1 & -1 & 1 \end{vmatrix}}{\begin{vmatrix} 1 & -1 & 1 \\ 2 & -1 & 3 \\ 1 & -3 & 1 \end{vmatrix}} = \dfrac{0}{2} = 0$ $\quad z = \dfrac{\begin{vmatrix} 1 & -1 & -1 \\ 2 & -1 & -4 \\ 1 & -3 & -1 \end{vmatrix}}{\begin{vmatrix} 1 & -1 & 1 \\ 2 & -1 & 3 \\ 1 & -3 & 1 \end{vmatrix}} = \dfrac{-4}{2} = -2$

54. $x = \dfrac{\begin{vmatrix} 7 & -3 & 1 \\ -9 & 1 & -3 \\ 3 & 1 & 1 \end{vmatrix}}{\begin{vmatrix} 1 & -3 & 1 \\ 1 & 1 & -3 \\ 1 & 1 & 1 \end{vmatrix}} = \dfrac{16}{16} = 1$ $\quad y = \dfrac{\begin{vmatrix} 1 & 7 & 1 \\ 1 & -9 & -3 \\ 1 & 3 & 1 \end{vmatrix}}{\begin{vmatrix} 1 & -3 & 1 \\ 1 & 1 & -3 \\ 1 & 1 & 1 \end{vmatrix}} = \dfrac{-16}{16} = -1$

$z = \dfrac{\begin{vmatrix} 1 & -3 & 7 \\ 1 & 1 & -9 \\ 1 & 1 & 3 \end{vmatrix}}{\begin{vmatrix} 1 & -3 & 1 \\ 1 & 1 & -3 \\ 1 & 1 & 1 \end{vmatrix}} = \dfrac{48}{16} = 3$

55. $w = \dfrac{\begin{vmatrix} 4 & 1 & -1 & 1 \\ 4 & 1 & 0 & 1 \\ 0 & 1 & 2 & 1 \\ 2 & 0 & 1 & 1 \end{vmatrix}}{\begin{vmatrix} 1 & 1 & -1 & 1 \\ 2 & 1 & 0 & 1 \\ 0 & 1 & 2 & 1 \\ 1 & 0 & 1 & 1 \end{vmatrix}} = \dfrac{-4}{-4} = 1$ $\quad x = \dfrac{\begin{vmatrix} 1 & 4 & -1 & 1 \\ 2 & 4 & 0 & 1 \\ 0 & 0 & 2 & 1 \\ 1 & 2 & 1 & 1 \end{vmatrix}}{\begin{vmatrix} 1 & 1 & -1 & 1 \\ 2 & 1 & 0 & 1 \\ 0 & 1 & 2 & 1 \\ 1 & 0 & 1 & 1 \end{vmatrix}} = \dfrac{0}{-4} = 0$

$y = \dfrac{\begin{vmatrix} 1 & 1 & 4 & 1 \\ 2 & 1 & 4 & 1 \\ 0 & 1 & 0 & 1 \\ 1 & 0 & 2 & 1 \end{vmatrix}}{\begin{vmatrix} 1 & 1 & -1 & 1 \\ 2 & 1 & 0 & 1 \\ 0 & 1 & 2 & 1 \\ 1 & 0 & 1 & 1 \end{vmatrix}} = \dfrac{4}{-4} = -1$ $\quad z = \dfrac{\begin{vmatrix} 1 & 1 & -1 & 4 \\ 2 & 1 & 0 & 4 \\ 0 & 1 & 2 & 0 \\ 1 & 0 & 1 & 2 \end{vmatrix}}{\begin{vmatrix} 1 & 1 & -1 & 1 \\ 2 & 1 & 0 & 1 \\ 0 & 1 & 2 & 1 \\ 1 & 0 & 1 & 1 \end{vmatrix}} = \dfrac{-8}{-4} = 2$

56. $\begin{vmatrix} 3a & 3b & 3c \\ d & e & f \\ g & h & i \end{vmatrix} = 3\begin{vmatrix} a & b & c \\ d & e & f \\ g & h & i \end{vmatrix} = 21$

57. $\begin{vmatrix} a & b & c \\ d+g & e+h & f+i \\ g & h & i \end{vmatrix} = \begin{vmatrix} a & b & c \\ d & e & f \\ g & h & i \end{vmatrix} = 7$

58. $$\frac{7x+3}{x^2+x}=\frac{7x+3}{x(x+1)}=\frac{A}{x}+\frac{B}{x+1}=\frac{A(x+1)}{x(x+1)}+\frac{Bx}{x(x+1)}=\frac{Ax+A+Bx}{x(x+1)}=\frac{(A+B)x+A}{x(x+1)}$$

$$\begin{cases} A+B=7 \\ A \qquad = 3 \end{cases} \Rightarrow \begin{matrix} A=3 \\ B=4 \end{matrix} \quad \frac{7x+3}{x(x+1)}=\frac{3}{x}+\frac{4}{x+1}$$

59. $$\frac{4x^3+3x+x^2+2}{x^4+x^2}=\frac{4x^3+x^2+3x+2}{x^2(x^2+1)}=\frac{A}{x}+\frac{B}{x^2}+\frac{Cx+D}{x^2+1}=\frac{Ax(x^2+1)}{x^2(x^2+1)}+\frac{B(x^2+1)}{x^2(x^2+1)}+\frac{(Cx+D)x^2}{x^2(x^2+1)}=\frac{Ax^3+Ax+Bx^2+B+Cx^3+Dx^2}{x^2(x^2+1)}=\frac{(A+C)x^3+(B+D)x^2+Ax+B}{x^2(x^2+1)}$$

$$\begin{cases} A \quad +C \qquad = 4 \\ \quad B \qquad +D = 1 \\ A \qquad\qquad = 3 \\ \quad B \qquad\quad = 2 \end{cases} \Rightarrow \begin{matrix} A=3 \\ B=2 \\ C=1 \\ D=-1 \end{matrix} \quad \frac{4x^3+x^2+3x+2}{x^2(x^2+1)}=\frac{3}{x}+\frac{2}{x^2}+\frac{x-1}{x^2+1}$$

60. $$\frac{x^2+5}{x^3+x^2+5x}=\frac{x^2+5}{x(x^2+x+5)}=\frac{A}{x}+\frac{Bx+C}{x^2+x+5}=\frac{A(x^2+x+5)}{x(x^2+x+5)}+\frac{(Bx+C)x}{x(x^2+x+5)}=\frac{Ax^2+Ax+5A+Bx^2+Cx}{x(x^2+x+5)}=\frac{(A+B)x^2+(A+C)x+(5A)}{x(x^2+x+5)}$$

$$\begin{cases} A+B \qquad = 1 \\ A \qquad +C = 0 \\ 5A \qquad\quad = 5 \end{cases} \Rightarrow \begin{matrix} A=1 \\ B=0 \\ C=-1 \end{matrix} \quad \frac{x^2+5}{x(x^2+x+5)}=\frac{1}{x}-\frac{1}{x^2+x+5}$$

61.
$$\frac{x^2+1}{(x+1)^3} = \frac{A}{x+1} + \frac{B}{(x+1)^2} + \frac{C}{(x+1)^3}$$
$$= \frac{A(x+1)^2}{(x+1)^3} + \frac{B(x+1)}{(x+1)^3} + \frac{C}{(x+1)^3}$$
$$= \frac{Ax^2+2Ax+A+Bx+B+C}{(x+1)^3}$$
$$= \frac{Ax^2+(2A+B)x+(A+B+C)}{(x+1)^3}$$

$$\begin{cases} A & = 1 \\ 2A+B & = 0 \\ A+B+C & = 1 \end{cases} \Rightarrow \begin{matrix} A = 1 \\ B = -2 \\ C = 2 \end{matrix} \qquad \frac{x^2+1}{(x+1)^3} = \frac{1}{x+1} - \frac{2}{(x+1)^2} + \frac{2}{(x+1)^3}$$

62. $y \geq -2x - 1$

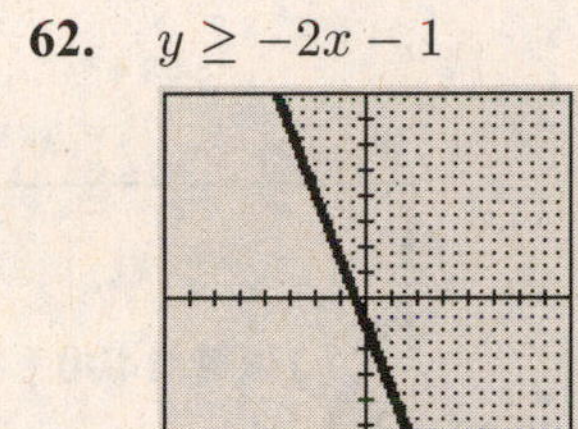
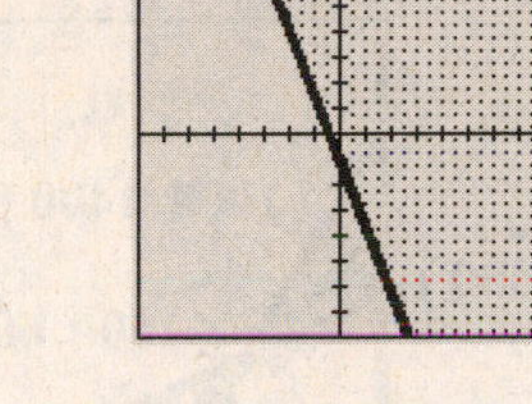

63. $x^2 + y^2 > 4$

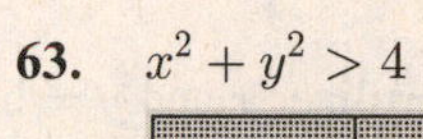
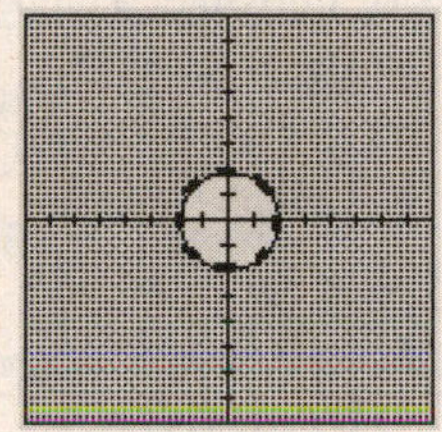

64. $\begin{cases} 3x + 2y \leq 6 \\ x - y > 3 \end{cases}$

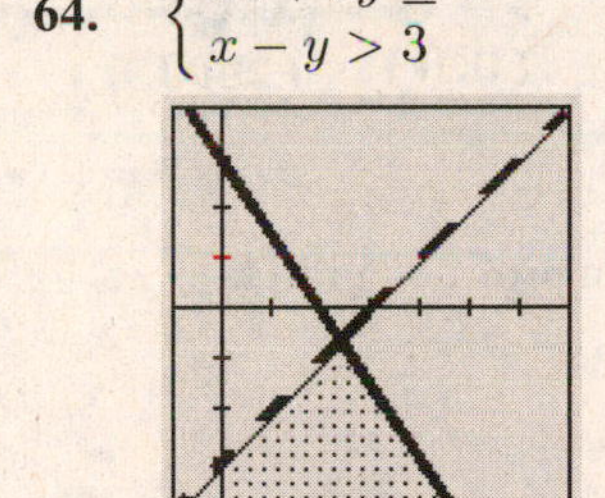

65. $\begin{cases} y \leq x^2 + 1 \\ y \geq x^2 - 1 \end{cases}$

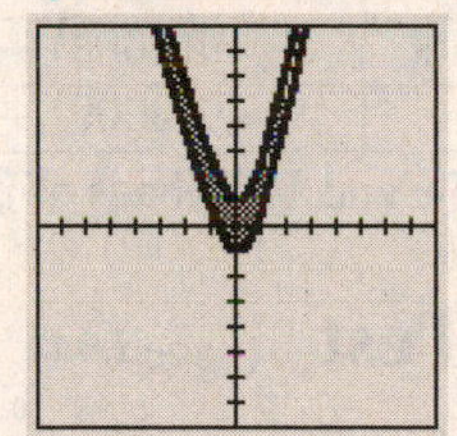

66.

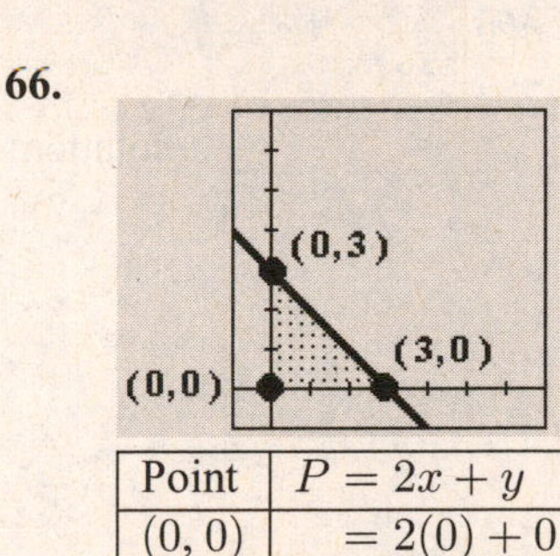

Point	$P = 2x + y$
$(0, 0)$	$= 2(0) + 0 = 0$
$(0, 3)$	$= 2(0) + 3 = 3$
$(3, 0)$	$= 2(3) + 0 = 6$

Max: $P = 6$ at $(3, 0)$

67.

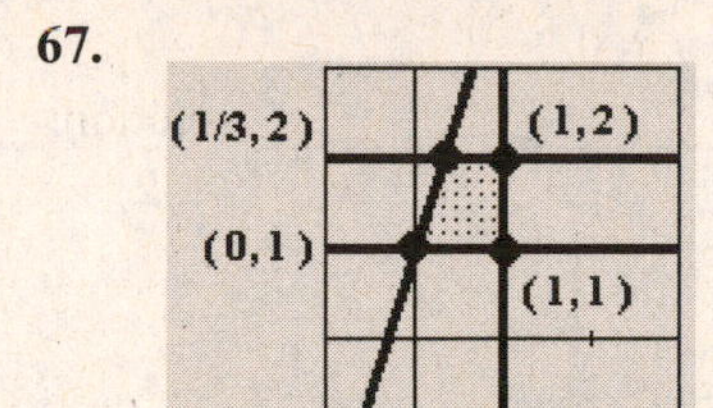

Point	$P = 3x - y$
$(0, 1)$	$= 3(0) - 1 = -1$
$(1, 1)$	$= 3(1) - 1 = 2$
$(1, 2)$	$= 3(1) - 2 = 1$
$(\frac{1}{3}, 2)$	$= 3(\frac{1}{3}) - 2 = -1$

Max: $P = 2$ at $(1, 1)$

68.

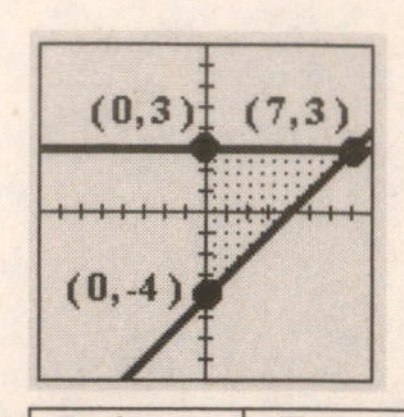

Point	$P = 2x - 3y$
$(0, 3)$	$= 2(0) - 3(3) = -9$
$(7, 3)$	$= 2(7) - 3(3) = 5$
$(0, -4)$	$= 2(0) - 3(-4) = 12$

Max: $P = 12$ at $(0, -4)$

69.

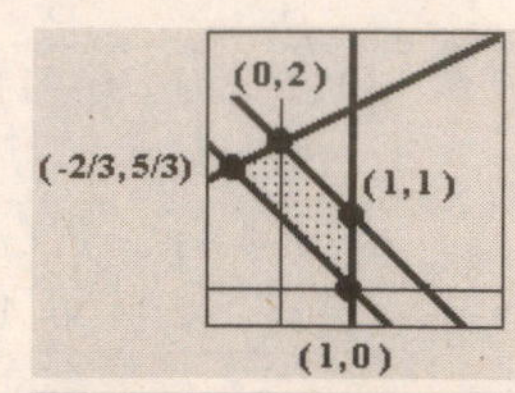

Point	$P = y - 2x$
$(0, 2)$	$= 2 - 2(0) = 2$
$(1, 1)$	$= 1 - 2(1) = -1$
$(1, 0)$	$= 0 - 2(1) = -2$
$\left(-\frac{2}{3}, \frac{5}{3}\right)$	$= \frac{5}{3} - 2\left(-\frac{2}{3}\right) = 3$

Max: $P = 3$ at $\left(-\frac{2}{3}, \frac{5}{3}\right)$

70. Let $x =$ bags of Fertilizer x and $y =$ bags of Fertilizer y.
Maximize $P = 6x + 5y$

$$\text{subject to } \begin{cases} 6x + 10y \leq 20000 \\ 8x + 6y \leq 16400 \\ 6x + 4y \leq 12000 \\ x \geq 0, y \geq 0 \end{cases}$$

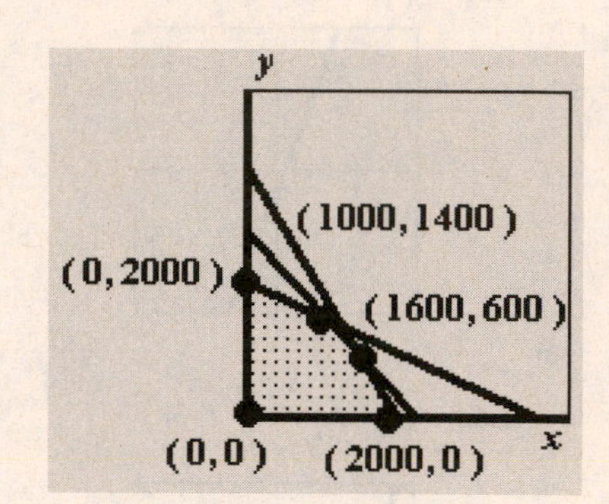

Point	$P = 6x + 5y$
$(0, 0)$	$= 0$
$(0, 2000)$	$= 10000$
$(1000, 1400)$	$= 13000$
$(1600, 600)$	$= 6(1600) + 5(600) = 12600$
$(2000, 0)$	$= 6(2000) + 5(0) = 12000$

1000 bags of x and 1400 bags of y should be made, for a maximum profit of $13,000.

Chapter 6 Test (page 694)

1. $\begin{cases} x - 3y = -5 \\ 2x - y = 0 \end{cases}$

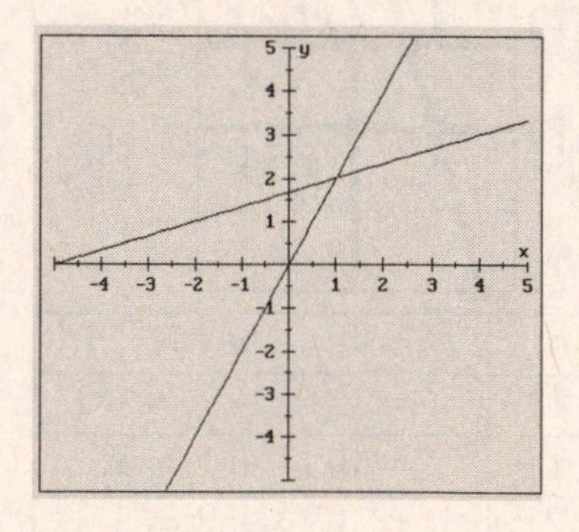

solution: $(1, 2)$

2. $\begin{cases} x = 2y + 5 \\ y = 2x - 4 \end{cases}$

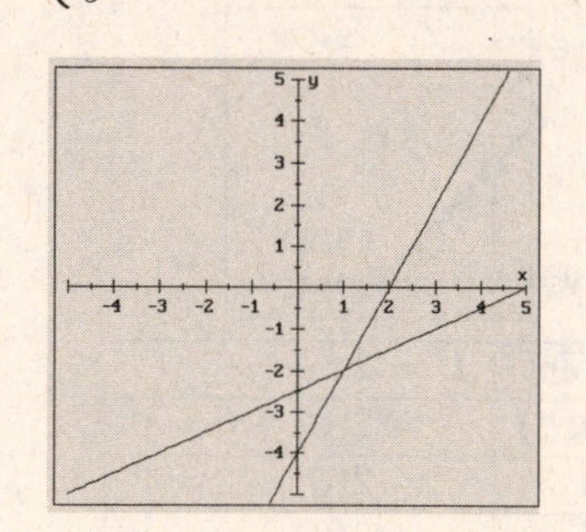

solution: $(1, -2)$

3.
$$\begin{aligned} 3x + y &= 0 \Rightarrow \times (5) \\ 2x - 5y &= 17 \end{aligned} \qquad \begin{aligned} 15x + 5y &= 0 \\ 2x - 5y &= 17 \\ \hline 17x &= 17 \\ x &= 1 \end{aligned} \qquad \begin{aligned} 3x + y &= 0 \\ 3(1) + y &= 0 \\ y &= -3 \end{aligned}$$

Solution: $\boxed{x = 1, y = -3}$

4.
$$\begin{aligned} \frac{x+y}{2} + x &= 7 \Rightarrow 3x + y = 14 \Rightarrow 3x + y = 14 \\ \frac{x-y}{2} - y &= -6 \Rightarrow x - 3y = -12 \Rightarrow -3x + 9y = 36 \\ \hline 10y &= 50 \\ y &= 5 \end{aligned} \qquad \begin{aligned} x - 3y &= -12 \\ x - 3(5) &= -12 \\ x &= 3 \end{aligned}$$

Solution: $\boxed{x = 3, \; y = 5}$

5. Let $x =$ liters of 20% solution and $y =$ liters of 45% solution. The following system applies:

$$\begin{aligned} x + y &= 10 \Rightarrow \times (-2) \\ 0.2x + 0.45y &= 3 \Rightarrow \times (10) \end{aligned} \qquad \begin{aligned} -2x - 2y &= -20 \\ 2x + 4.5y &= 30 \\ \hline 2.5y &= 10 \\ y &= 4 \end{aligned}$$

She must use 4 liters of the 45% and 6 liters of the 20% solution.

6. Let $x =$ # from Ace, $y =$ # from Hi-Fi and $z =$ # from CD World.

$$\begin{aligned} (1) \quad x + y + z &= 175 \\ (2) \quad -x - y + z &= 25 \\ (3) \quad 170x + 165y + 160z &= 28500 \end{aligned}$$

Add (1) and (2):

$$\begin{aligned} (1) \quad x + y + z &= 175 \\ (2) \quad -x - y + z &= 25 \\ \hline (4) \quad 2z &= 200 \\ z &= 100 \end{aligned}$$

Add equations 170(2) and (3):

$$\begin{aligned} 170(2) \quad -170x - 170y + 170z &= 4250 \\ (3) \quad 170x + 165y + 160z &= 28500 \\ \hline (5) \quad -5y + 330z &= 32750 \end{aligned}$$

Solve the system of two equations and two unknowns formed by equations (4) and (5):

$$\begin{aligned} -5y + 330z &= 32750 \\ -5y + 330(100) &= 32750 \\ -5y &= -250 \\ y &= 50 \end{aligned} \qquad \begin{aligned} x + y + z &= 175 \\ x + 50 + 100 &= 175 \\ x &= 25 \end{aligned}$$

Ace buys 25 units per month. Hi-Fi buys 50 units per month. CD World buys 100 units per month.

7.
$$\left[\begin{array}{cc|c} 3 & -2 & 4 \\ 2 & 3 & 7 \end{array}\right] \Rightarrow \underset{-R_2 + R_1 \Rightarrow R_1}{\left[\begin{array}{cc|c} 1 & -5 & -3 \\ 2 & 3 & 7 \end{array}\right]} \Rightarrow \underset{-2R_1 + R_2 \Rightarrow R_2}{\left[\begin{array}{cc|c} 1 & -5 & -3 \\ 0 & 13 & 13 \end{array}\right]} \Rightarrow \underset{\frac{1}{13}R_2 \Rightarrow R_2}{\left[\begin{array}{cc|c} 1 & -5 & -3 \\ 0 & 1 & 1 \end{array}\right]}$$

From R_2: $y = 1$ From R_1:
$$\begin{aligned} x - 5y &= -3 \\ x - 5(1) &= -3 \\ x &= 2 \end{aligned}$$

Solution: $\boxed{x = 2, y = 1}$

CHAPTER 6 TEST

8. $\left[\begin{array}{ccc|c} 1 & 3 & -1 & 6 \\ 2 & -1 & -2 & -2 \\ 1 & 2 & 1 & 6 \end{array}\right] \Rightarrow \left[\begin{array}{ccc|c} 1 & 3 & -1 & 6 \\ 0 & -7 & 0 & -14 \\ 0 & 1 & -2 & 0 \end{array}\right] \Rightarrow \left[\begin{array}{ccc|c} 1 & 3 & -1 & 6 \\ 0 & 1 & 0 & 2 \\ 0 & 0 & -14 & -14 \end{array}\right] \Rightarrow$

$-2R_1 + R_2 \Rightarrow R_2$, $-R_3 + R_1 \Rightarrow R_3$; $-\frac{1}{7}R_2 \Rightarrow R_2$, $7R_3 + R_2 \Rightarrow R_3$

$\left[\begin{array}{ccc|c} 1 & 3 & -1 & 6 \\ 0 & 1 & 0 & 2 \\ 0 & 0 & 1 & 1 \end{array}\right]$ $-\frac{1}{14}R_3 \Rightarrow R_3$

From (3): $z = 1$
From (2): $y = 2$
From (1): $x + 3y - z = 6$
$x + 3(2) - (1) = 6$
$x = 1$

Solution: $\boxed{x = 1, y = 2, z = 1}$

9. $\left[\begin{array}{ccc|c} 1 & 2 & 3 & -5 \\ 3 & 1 & -2 & 7 \\ 0 & 1 & -1 & 2 \end{array}\right] \Rightarrow \left[\begin{array}{ccc|c} 1 & 2 & 3 & -5 \\ 0 & -5 & -11 & 22 \\ 0 & 1 & -1 & 2 \end{array}\right] \Rightarrow \left[\begin{array}{ccc|c} 1 & 2 & 3 & -5 \\ 0 & 1 & -1 & 2 \\ 0 & -5 & -11 & 22 \end{array}\right] \Rightarrow$

$-3R_1 + R_2 \Rightarrow R_2$; $R_2 \Leftrightarrow R_3$

$\left[\begin{array}{ccc|c} 1 & 0 & 5 & -9 \\ 0 & 1 & -1 & 2 \\ 0 & 0 & -16 & 32 \end{array}\right] \Rightarrow \left[\begin{array}{ccc|c} 1 & 0 & 5 & -9 \\ 0 & 1 & -1 & 2 \\ 0 & 0 & 1 & -2 \end{array}\right] \Rightarrow \left[\begin{array}{ccc|c} 1 & 0 & 0 & 1 \\ 0 & 1 & 0 & 0 \\ 0 & 0 & 1 & -2 \end{array}\right]$ Solution: $\boxed{\begin{array}{l} x = 1 \\ y = 0 \\ z = -2 \end{array}}$

$-2R_2 + R_1 \Rightarrow R_1$, $5R_2 + R_3 \Rightarrow R_3$; $-\frac{1}{16}R_3 \Rightarrow R_3$; $-5R_3 + R_1 \Rightarrow R_1$, $R_2 + R_3 \Rightarrow R_2$

10. $\left[\begin{array}{ccc|c} 1 & 2 & 1 & 0 \\ 3 & -2 & -2 & 7 \\ 4 & 0 & -1 & 7 \end{array}\right] \Rightarrow \left[\begin{array}{ccc|c} 1 & 2 & 1 & 0 \\ 0 & -8 & -5 & 7 \\ 0 & -8 & -5 & 7 \end{array}\right] \Rightarrow \left[\begin{array}{ccc|c} 1 & 2 & 1 & 0 \\ 0 & 1 & \frac{5}{8} & -\frac{7}{8} \\ 0 & 0 & 0 & 0 \end{array}\right] \Rightarrow$

$-3R_1 + R_2 \Rightarrow R_2$, $-4R_1 + R_3 \Rightarrow R_3$; $-R_2 + R_3 \Rightarrow R_3$, $-\frac{1}{8}R_2 \Rightarrow R_2$

$\left[\begin{array}{ccc|c} 1 & 0 & -\frac{1}{4} & \frac{7}{4} \\ 0 & 1 & \frac{5}{8} & -\frac{7}{8} \\ 0 & 0 & 0 & 0 \end{array}\right] \Rightarrow$ Solution: $\boxed{\begin{array}{l} x = \frac{7}{4} + \frac{1}{4}z \\ y = -\frac{7}{8} - \frac{5}{8}z \\ z = \text{any real number} \end{array}}$

$-2R_2 + R_1 \Rightarrow R_1$

Note: This answer is equivalent to the answer provided in the textbook.

11. $3\begin{bmatrix} 2 & -3 & 5 \\ 0 & 3 & -1 \end{bmatrix} - 5\begin{bmatrix} -2 & 1 & -1 \\ 0 & 3 & 2 \end{bmatrix} = \begin{bmatrix} 6 & -9 & 15 \\ 0 & 9 & -3 \end{bmatrix} - \begin{bmatrix} -10 & 5 & -5 \\ 0 & 15 & 10 \end{bmatrix}$

$= \begin{bmatrix} 16 & -14 & 20 \\ 0 & -6 & -13 \end{bmatrix}$

12. $\begin{bmatrix} 1 & 2 & 3 \end{bmatrix}\begin{bmatrix} 2 & -2 \\ -2 & 2 \\ 1 & 0 \end{bmatrix}\begin{bmatrix} 3 \\ -2 \end{bmatrix} = \begin{bmatrix} 1 & 2 \end{bmatrix}\begin{bmatrix} 3 \\ -2 \end{bmatrix} = \begin{bmatrix} -1 \end{bmatrix}$

CHAPTER 6 TEST

13. $\left[\begin{array}{cc|cc} 5 & 19 & 1 & 0 \\ 2 & 7 & 0 & 1 \end{array}\right] \Rightarrow \underset{\frac{1}{5}R_1 \Rightarrow R_1}{\left[\begin{array}{cc|cc} 1 & \frac{19}{5} & \frac{1}{5} & 0 \\ 2 & 7 & 0 & 1 \end{array}\right]} \Rightarrow \underset{-2R_1 + R_2 \Rightarrow R_2}{\left[\begin{array}{cc|cc} 1 & \frac{19}{5} & \frac{1}{5} & 0 \\ 0 & -\frac{3}{5} & -\frac{2}{5} & 1 \end{array}\right]} \Rightarrow$

$\underset{-\frac{5}{3}R_2 \Rightarrow R_2}{\left[\begin{array}{cc|cc} 1 & \frac{19}{5} & \frac{1}{5} & 0 \\ 0 & 1 & \frac{2}{3} & -\frac{5}{3} \end{array}\right]} \Rightarrow \underset{-\frac{19}{5}R_2 + R_1 \Rightarrow R_1}{\left[\begin{array}{cc|cc} 1 & 0 & -\frac{7}{3} & \frac{19}{3} \\ 0 & 1 & \frac{2}{3} & -\frac{5}{3} \end{array}\right]} \Rightarrow$ Inverse: $\begin{bmatrix} -\frac{7}{3} & \frac{19}{3} \\ \frac{2}{3} & -\frac{5}{3} \end{bmatrix}$

14. $\left[\begin{array}{ccc|ccc} -1 & 3 & -2 & 1 & 0 & 0 \\ 4 & 1 & 4 & 0 & 1 & 0 \\ 0 & 3 & -1 & 0 & 0 & 1 \end{array}\right] \Rightarrow \underset{\substack{4R_1 + R_2 \Rightarrow R_2 \\ -R_1 \Rightarrow R_1}}{\left[\begin{array}{ccc|ccc} 1 & -3 & 2 & -1 & 0 & 0 \\ 0 & 13 & -4 & 4 & 1 & 0 \\ 0 & 3 & -1 & 0 & 0 & 1 \end{array}\right]} \Rightarrow$

$\underset{-4R_3 + R_2 \Rightarrow R_2}{\left[\begin{array}{ccc|ccc} 1 & -3 & 2 & -1 & 0 & 0 \\ 0 & 1 & 0 & 4 & 1 & -4 \\ 0 & 3 & -1 & 0 & 0 & 1 \end{array}\right]} \Rightarrow \underset{\substack{3R_2 + R_1 \Rightarrow R_1 \\ -3R_2 + R_3 \Rightarrow R_3}}{\left[\begin{array}{ccc|ccc} 1 & 0 & 2 & 11 & 3 & -12 \\ 0 & 1 & 0 & 4 & 1 & -4 \\ 0 & 0 & -1 & -12 & -3 & 13 \end{array}\right]} \Rightarrow$

$\underset{\substack{2R_3 + R_1 \Rightarrow R_1 \\ -R_3 \Rightarrow R_3}}{\left[\begin{array}{ccc|ccc} 1 & 0 & 0 & -13 & -3 & 14 \\ 0 & 1 & 0 & 4 & 1 & -4 \\ 0 & 0 & 1 & 12 & 3 & -13 \end{array}\right]}$ Inverse: $\begin{bmatrix} -13 & -3 & 14 \\ 4 & 1 & -4 \\ 12 & 3 & -13 \end{bmatrix}$

15. $\begin{bmatrix} 5 & 19 \\ 2 & 7 \end{bmatrix}\begin{bmatrix} x \\ y \end{bmatrix} = \begin{bmatrix} 3 \\ 2 \end{bmatrix}$

$$\begin{bmatrix} x \\ y \end{bmatrix} = \begin{bmatrix} 5 & 19 \\ 2 & 7 \end{bmatrix}^{-1}\begin{bmatrix} 3 \\ 2 \end{bmatrix} = \begin{bmatrix} -\frac{7}{3} & \frac{19}{3} \\ \frac{2}{3} & -\frac{5}{3} \end{bmatrix}\begin{bmatrix} 3 \\ 2 \end{bmatrix} = \begin{bmatrix} \frac{17}{3} \\ -\frac{4}{3} \end{bmatrix}$$

16. $\begin{bmatrix} -1 & 3 & -2 \\ 4 & 1 & 4 \\ 0 & 3 & -1 \end{bmatrix}\begin{bmatrix} x \\ y \\ z \end{bmatrix} = \begin{bmatrix} 1 \\ 3 \\ -1 \end{bmatrix}$

$$\begin{bmatrix} x \\ y \\ z \end{bmatrix} = \begin{bmatrix} -1 & 3 & -2 \\ 4 & 1 & 4 \\ 0 & 3 & -1 \end{bmatrix}^{-1}\begin{bmatrix} 1 \\ 3 \\ -1 \end{bmatrix}$$

$$\begin{bmatrix} x \\ y \\ z \end{bmatrix} = \begin{bmatrix} -13 & -3 & 14 \\ 4 & 1 & -4 \\ 12 & 3 & -13 \end{bmatrix}\begin{bmatrix} 1 \\ 3 \\ -1 \end{bmatrix} = \begin{bmatrix} -36 \\ 11 \\ 34 \end{bmatrix}$$

17. $\begin{vmatrix} 3 & -5 \\ -3 & 1 \end{vmatrix} = (3)(1) - (-5)(-3) = 3 - 15 = -12$

18. $$\begin{vmatrix} 3 & 5 & -1 \\ -2 & 3 & -2 \\ 1 & 5 & -3 \end{vmatrix} = 3\begin{vmatrix} 3 & -2 \\ 5 & -3 \end{vmatrix} - 5\begin{vmatrix} -2 & -2 \\ 1 & -3 \end{vmatrix} + (-1)\begin{vmatrix} -2 & 3 \\ 1 & 5 \end{vmatrix}$$
$$= 3(1) - 5(8) - 1(-13) = 3 - 40 + 13 = -24$$

19. $$y = \frac{\begin{vmatrix} 3 & 3 \\ -3 & 2 \end{vmatrix}}{\begin{vmatrix} 3 & -5 \\ -3 & 1 \end{vmatrix}} = \frac{15}{-12} = -\frac{5}{4}$$

20. $$y = \frac{\begin{vmatrix} 3 & 2 & -1 \\ -2 & 1 & -2 \\ 1 & 0 & -3 \end{vmatrix}}{\begin{vmatrix} 3 & 5 & -1 \\ -2 & 3 & -2 \\ 1 & 5 & -3 \end{vmatrix}} = \frac{-24}{-24} = 1$$

21. $$\frac{5x}{2x^2 - x - 3} = \frac{5x}{(2x-3)(x+1)} = \frac{A}{2x-3} + \frac{B}{x+1}$$
$$= \frac{A(x+1)}{(2x-3)(x+1)} + \frac{B(2x-3)}{(2x-3)(x+1)}$$
$$= \frac{Ax + A + 2Bx - 3B}{(2x-3)(x+1)}$$
$$= \frac{(A+2B)x + (A-3B)}{(2x-3)(x+1)}$$

$$\begin{cases} A + 2B = 5 \\ A - 3B = 0 \end{cases} \Rightarrow \begin{matrix} A = 3 \\ B = 1 \end{matrix} \quad \frac{5x}{(2x-3)(x+1)} = \frac{3}{2x-3} + \frac{1}{x+1}$$

22. $$\frac{3x^2 + x + 2}{x^3 + 2x} = \frac{3x^2 + x + 2}{x(x^2+2)} = \frac{A}{x} + \frac{Bx + C}{x^2 + 2}$$
$$= \frac{A(x^2+2)}{x(x^2+2)} + \frac{(Bx+C)x}{x(x^2+2)}$$
$$= \frac{Ax^2 + 2A + Bx^2 + Cx}{x(x^2+2)}$$
$$= \frac{(A+B)x^2 + Cx + 2A}{x(x^2+2)}$$

$$\begin{cases} A + B \quad = 3 \\ \qquad C = 1 \\ 2A \qquad = 2 \end{cases} \Rightarrow \begin{matrix} A = 1 \\ B = 2 \\ C = 1 \end{matrix} \quad \frac{3x^2 + x + 2}{x(x^2+2)} = \frac{1}{x} + \frac{2x+1}{x^2+2}$$

CHAPTER 6 TEST

23. $\begin{cases} x - 3y \geq 3 \\ x + 3y \leq 3 \end{cases}$

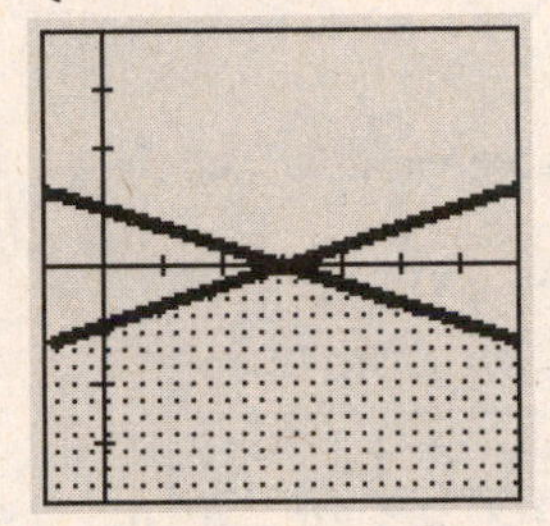

24. $\begin{cases} 3x + 4y \leq 12 \\ 3x + 4y \geq 6 \\ x \geq 0, y \geq 0 \end{cases}$

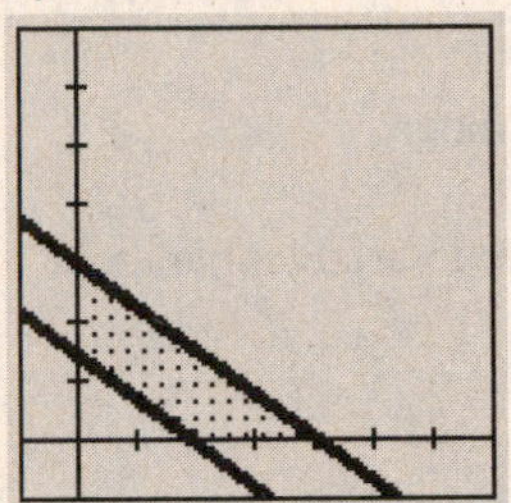

25.

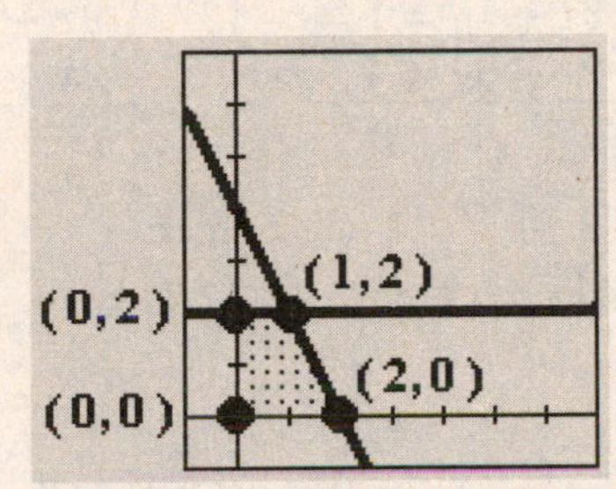

Point	$P = 3x + 2y$
$(0, 2)$	$= 3(0) + 2(2) = 4$
$(1, 2)$	$= 3(1) + 2(2) = 7$
$(2, 0)$	$= 3(2) + 2(0) = 6$
$(0, 0)$	$= 3(0) + 2(0) = 0$

Max: $P = 7$ at $(1, 2)$

26.

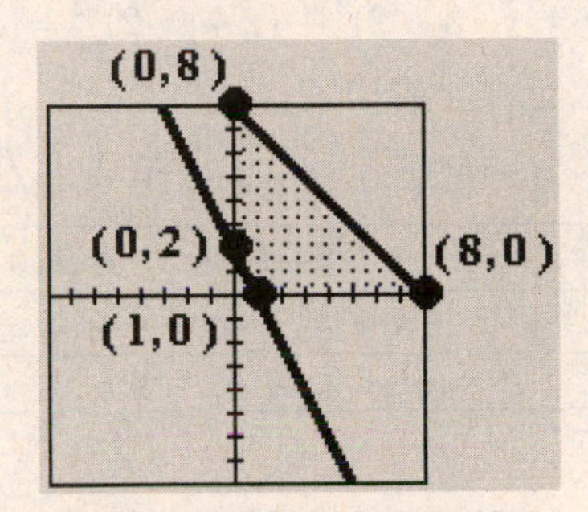

Point	$P = y - x$
$(0, 2)$	$= 2 - 0 = 2$
$(0, 8)$	$= 8 - 0 = 8$
$(8, 0)$	$= 0 - 8 = -8$
$(1, 0)$	$= 0 - 1 = -1$

Min: $P = -8$ at $(8, 0)$

Exercises 7.1 (page 710)

1. $(2, -5), 3$

3. $(0, 0), \sqrt{5}$

5. to the left

7. downward

9. directrix, focus

11. Two squared variables: circle

13. One squared variable: parabola

15.
$$(x-h)^2 + (y-k)^2 = r^2$$
$$(x-0)^2 + (y-0)^2 = 7^2$$
$$\boxed{x^2 + y^2 = 49}$$
$$\boxed{x^2 + y^2 - 49 = 0}$$

17.
$$r = \sqrt{(3-2)^2 + (2-(-2))^2} = \sqrt{17}$$
$$(x-h)^2 + (y-k)^2 = r^2$$
$$(x-2)^2 + (y-(-2))^2 = \left(\sqrt{17}\right)^2$$
$$\boxed{(x-2)^2 + (y+2)^2 = 17}$$
$$x^2 - 4x + 4 + y^2 + 4y + 4 = 17$$
$$\boxed{x^2 + y^2 - 4x + 4y - 9 = 0}$$

19.
$$\begin{array}{ll} 3x + y = 1 \Rightarrow \times (3) & 9x + 3y = 3 \\ \underline{-2x - 3y = 4} & \underline{-2x - 3y = 4} \\ & 7x = 7 \\ & x = 1 \end{array}$$
$$\begin{array}{ll} 3x + y = 1 & \text{Center:} \\ 3(1) + y = 1 & (1, -2) \\ y = -2 & \end{array}$$
$$(x-h)^2 + (y-k)^2 = r^2$$
$$(x-1)^2 + (y-(-2))^2 = 6^2$$
$$\boxed{(x-1)^2 + (y+2)^2 = 36}$$
$$x^2 - 2x + 1 + y^2 + 4y + 4 = 36$$
$$\boxed{x^2 + y^2 - 2x + 4y - 31 = 0}$$

21.
$$x^2 + y^2 = 4$$
$$(x-0)^2 + (y-0)^2 = 2^2$$
$$C(0, 0), r = 2$$

23.
$$3x^2 + 3y^2 - 12x - 6y = 12$$
$$x^2 - 4x + y^2 - 2y = 4$$
$$x^2 - 4x + 4 + y^2 - 2y + 1 = 4 + 4 + 1$$
$$(x-2)^2 + (y-1)^2 = 3^2$$
$$C(2, 1), r = 3$$

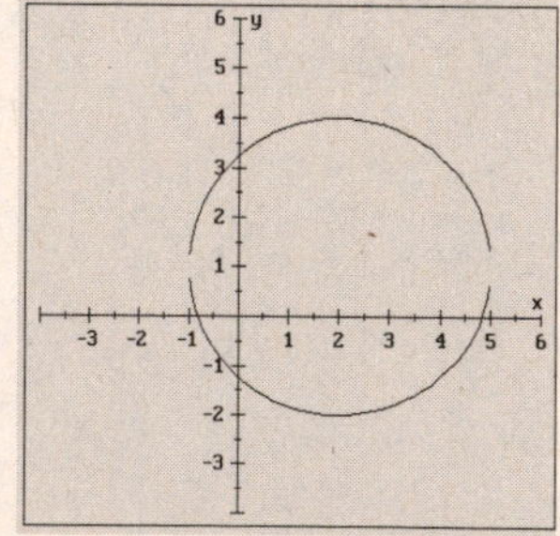

EXERCISES 7.1

25.
$$x^2 = 12y$$
$$(x-0)^2 = 4\cdot 3(y-0)$$
$p = 3$, opens up
V $(0,0)$, $F(0,3)$, D: $y=-3$

27.
$$(y-3)^2 = 20x$$
$$(y-3)^2 = 4\cdot 5(x-0)$$
$p = 5$, opens right
V $(0,3)$, $F(5,3)$, D: $x=-5$

29.
$$(x+2)^2 = -24(y-1)$$
$$(x-(-2))^2 = 4\cdot(-6)(y-1)$$
$p=-6$, opens down
V $(-2,1)$, $F(-2,-5)$, D: $y=7$

31. Vertical (up), $p=3$
$$(x-h)^2 = 4p(y-k)$$
$$(x-0)^2 = 4(3)(y-0)$$
$$x^2 = 12y$$

33. Horizontal (left), $p=-3$
$$(y-k)^2 = 4p(x-h)$$
$$(y-0)^2 = 4(-3)(x-0)$$
$$y^2 = -12x$$

35. Vertical (down), $p=-3$
$$(x-h)^2 = 4p(y-k)$$
$$(x-3)^2 = 4(-3)(y-5)$$
$$(x-3)^2 = -12(y-5)$$

37. Vertical (down), $p=-7$
$$(x-h)^2 = 4p(y-k)$$
$$(x-3)^2 = 4(-7)(y-5)$$
$$(x-3)^2 = -28(y-5)$$

39. Vertical (down), $p=-1$
$$(x-h)^2 = 4p(y-k)$$
$$(x-0)^2 = 4(-1)(y-2)$$
$$x^2 = -4(y-2)$$

41. Horizontal (right), $p=2$
$$(y-k)^2 = 4p(x-h)$$
$$(y-(-5))^2 = 4(2)(x-1)$$
$$(y+5)^2 = 8(x-1)$$

43.
$$(x-2)^2 = 4p(y-2)$$
$$(0-2)^2 = 4p(0-2)$$
$$4 = -8p$$
$$-\frac{1}{2} = p$$
$$-2 = 4p$$
$$(x-2)^2 = -2(y-2)$$

OR
$$(y-2)^2 = 4p(x-2)$$
$$(0-2)^2 = 4p(0-2)$$
$$4 = -8p$$
$$-\frac{1}{2} = p$$
$$-2 = 4p$$
$$(y-2)^2 = -2(x-2)$$

45.
$$(x-(-4))^2 = 4p(y-6)$$
$$(0+4)^2 = 4p(3-6)$$
$$16 = -12p$$
$$-\frac{4}{3} = p$$
$$-\frac{16}{3} = 4p$$
$$(x+4)^2 = -\tfrac{16}{3}(y-6)$$

OR
$$(y-6)^2 = 4p(x-(-4))$$
$$(3-6)^2 = 4p(0+4)$$
$$9 = 16p$$
$$\frac{9}{16} = p$$
$$\frac{9}{4} = 4p$$
$$(y-6)^2 = \tfrac{9}{4}(x+4)$$

47. $(x-6)^2 = 4p(y-8)$ **OR** $(y-8)^2 = 4p(x-6)$

$$\begin{aligned}(x-6)^2 &= 4p(y-8) \\ (5-6)^2 &= 4p(10-8) \\ 1 &= 8p \\ \frac{1}{8} &= p \\ \frac{1}{2} &= 4p \\ (x-6)^2 &= \tfrac{1}{2}(y-8)\end{aligned} \qquad \textbf{OR} \qquad \begin{aligned}(y-8)^2 &= 4p(x-6) \\ (10-8)^2 &= 4p(5-6) \\ 4 &= -4p \\ -1 &= p \\ -4 &= 4p \\ (y-8)^2 &= -4(x-6)\end{aligned}$$

Check to see which equation is satisfied by $(5, 6)$ as well. Answer: $(y-8)^2 = -4(x-6)$

49.

$$\begin{aligned}(x-3)^2 &= 4p(y-1) \\ (4-3)^2 &= 4p(3-1) \\ 1 &= 8p \\ \frac{1}{8} &= p \\ \frac{1}{2} &= 4p \\ (x-3)^2 &= \tfrac{1}{2}(y-1)\end{aligned} \qquad \textbf{OR} \qquad \begin{aligned}(y-1)^2 &= 4p(x-3) \\ (3-1)^2 &= 4p(4-3) \\ 4 &= 4p \\ (y-1)^2 &= 4(x-3)\end{aligned}$$

Check to see which equation is satisfied by $(2, 3)$ as well. Answer: $(x-3)^2 = \frac{1}{2}(y-1)$

51.

$$\begin{aligned}y &= x^2+4x+5 \\ y-5 &= x^2+4x \\ y-5+4 &= x^2+4x+4 \\ y-1 &= (x+2)^2\end{aligned}$$

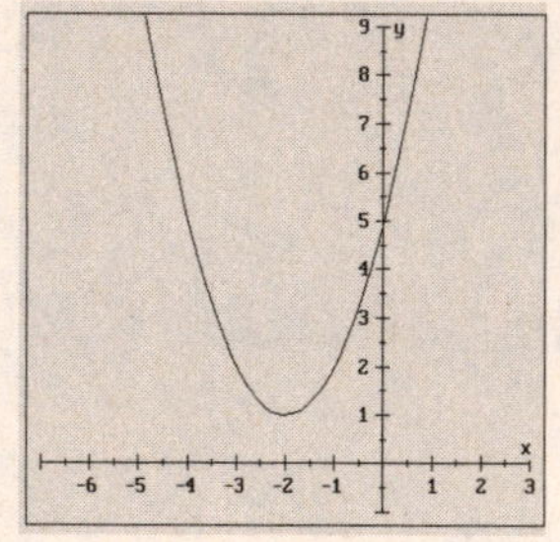

53.

$$\begin{aligned}y^2+4x-6y &= -1 \\ y^2-6y &= -4x-1 \\ y^2-6y+9 &= -4x-1+9 \\ (y-3)^2 &= -4x+8 \\ (y-3)^2 &= -4(x-2)\end{aligned}$$

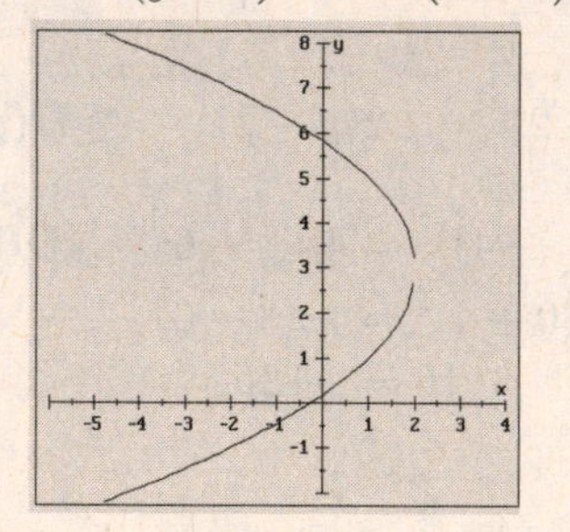

EXERCISES 7.1

55.
$$\begin{aligned} y^2 - 4y &= 4x - 8 \\ y^2 - 4y + 4 &= 4x - 8 + 4 \\ (y-2)^2 &= 4x - 4 \\ (y-2)^2 &= 4(x-1) \end{aligned}$$

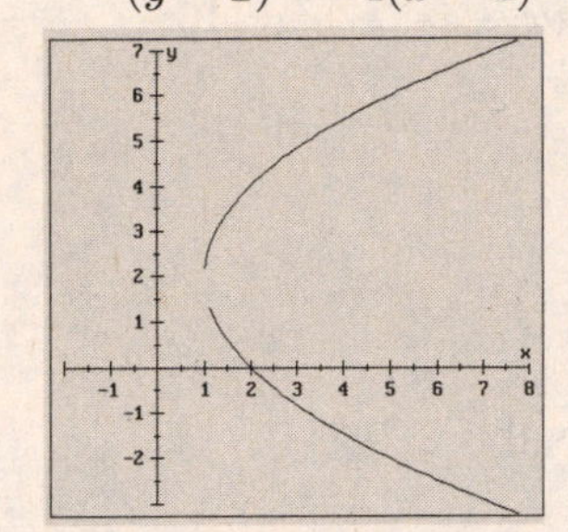

57.
$$\begin{aligned} y^2 - 4y &= -8x + 20 \\ y^2 - 4y + 4 &= -8x + 20 + 4 \\ (y-2)^2 &= -8x + 24 \\ (y-2)^2 &= -8(x-3) \end{aligned}$$

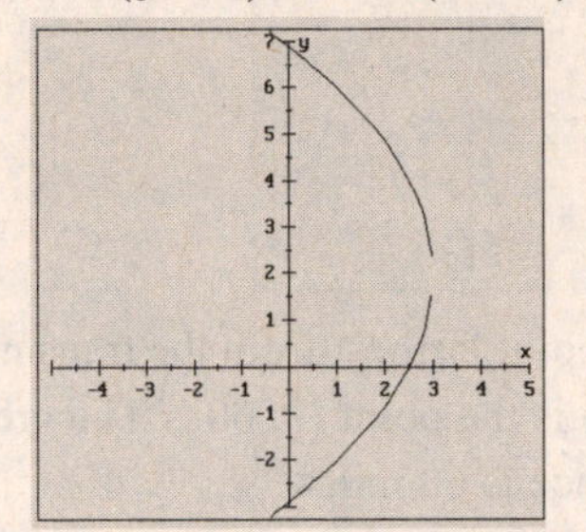

59.
$$\begin{aligned} x^2 - 6y + 22 &= -4x \\ x^2 + 4x &= 6y - 22 \\ x^2 + 4x + 4 &= 6y - 22 + 4 \\ (x+2)^2 &= 6y - 18 \\ (x+2)^2 &= 6(y-3) \end{aligned}$$

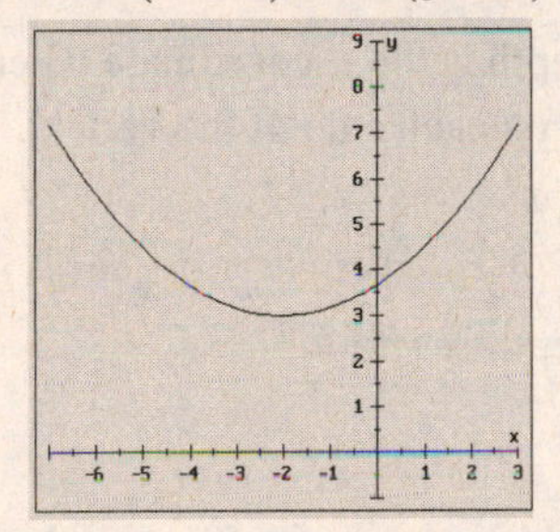

61.
$$\begin{aligned} 4x^2 - 4x + 32y &= 47 \\ x^2 - x &= -8y + \tfrac{47}{4} \\ x^2 - x + \tfrac{1}{4} &= -8y + \tfrac{47}{4} + \tfrac{1}{4} \\ \left(x - \tfrac{1}{2}\right)^2 &= -8y + 12 \\ \left(x - \tfrac{1}{2}\right)^2 &= -8\left(y - \tfrac{3}{2}\right) \end{aligned}$$

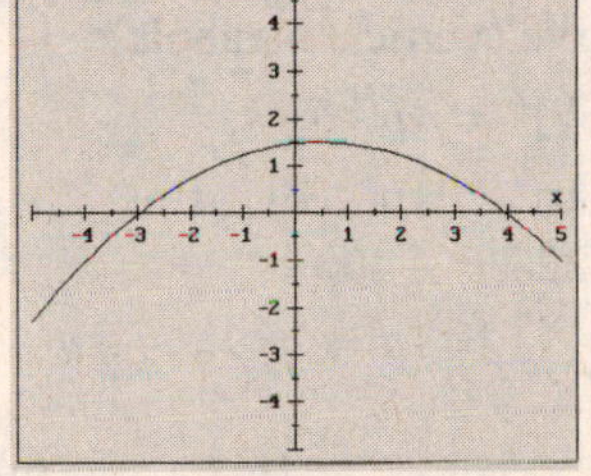

63.
$$\begin{aligned} y^2 &= 4x - 12 \\ y &= \pm\sqrt{4x - 12} \end{aligned}$$

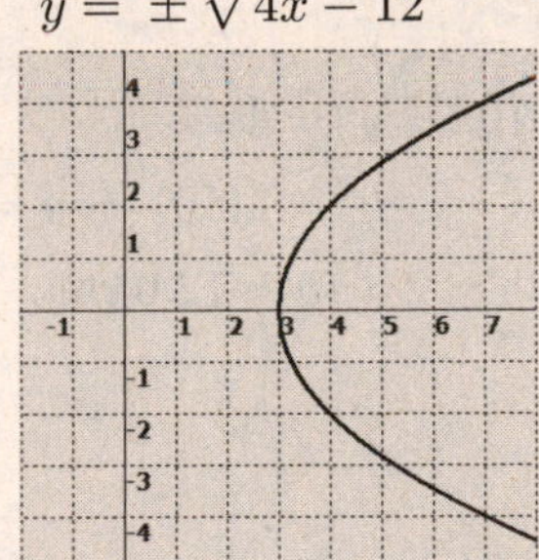

65. Check the coordinates:
$$\begin{aligned} x^2 + y^2 &= 50^2 + 70^2 \\ &= 2500 + 4900 = 7400 \end{aligned}$$
$7400 < 8100$, so the city can receive.

67. Graph both circles: $\begin{cases} x^2 + y^2 = 1600 \\ x^2 + (y - 35)^2 = 625 \end{cases}$

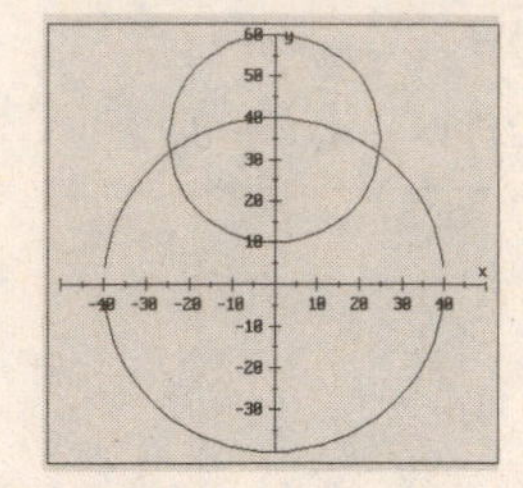

The point farthest from the transmitter $(0, 0)$ is the point $(0, 60)$. The greatest distance is 60 miles.

69. $C(4, 0), r = 4$
$$(x - h)^2 + (y - k)^2 = r^2$$
$$(x - 4)^2 + (y - 0)^2 = 4^2$$
$$(x - 4)^2 + y^2 = 16$$

71. $x^2 + y^2 = 16$: $C(0, 0), r = 4$
Small gear: $C(7, 0), r = 3$
$$(x - h)^2 + (y - k)^2 = r^2$$
$$(x - 7)^2 + (y - 0)^2 = 3^2$$
$$(x - 7)^2 + y^2 = 9$$

73. Find the distance to the focus:
$4p = 8 \Rightarrow p = 2$
It will be hottest 2 feet from the vertex.

75. The vertex is $(0, 0)$, while $(15, -10)$ is on the curve (vertical parabola):
$$(x - h)^2 = 4p(y - k)$$
$$(15 - 0)^2 = 4p(-10 - 0)$$
$$225 = -40p$$
$$-\tfrac{45}{2} = 4p \Rightarrow x^2 = -\tfrac{45}{2}y$$

77. The depth is the y-coordinate when $x = 4$. [4 feet on each side of the vertex]
$$y = \tfrac{1}{16}x^2$$
$$y = \tfrac{1}{16}(4)^2 = 1$$
The depth is 1 foot.

79.
$$s = -16t^2 + 80\sqrt{3}t$$
$$-\tfrac{1}{16}s = t^2 - 5\sqrt{3}t$$
$$-\tfrac{1}{16}s + \tfrac{75}{4} = t^2 - 5\sqrt{3}t + \tfrac{75}{4}$$
$$-\tfrac{1}{16}(s - 300) = \left(t - \tfrac{5\sqrt{3}}{2}\right)^2$$
The maximum height is 300 feet.

81. Place the vertex at $(0, 0)$, with the focus at $(1, 0) \Rightarrow p = 1, 4p = 4$.
$$(y - k)^2 = 4p(x - h)$$
$$y^2 = 4x$$
Let $x = 10$: $\quad y^2 = 4x$
$$y^2 = 4(10) \Rightarrow y = \pm\sqrt{40}$$
The width $= 2\sqrt{40} \approx 12.6$ cm.

83. The vertex is $(0, 0)$, while $(315, -630)$ is on the curve (vertical parabola):
$$(x - h)^2 = 4p(y - k)$$
$$(315 - 0)^2 = 4p(-630 - 0)$$
$$99{,}225 = -2520p$$
$$-\tfrac{19845}{504} = p \Rightarrow 4p = -\tfrac{19845}{126}$$

Let $y = -430$:
$$x^2 = -\tfrac{19845}{126}y$$
$$x^2 = -\tfrac{19845}{126}(-430)$$
$$x = \pm\sqrt{-\tfrac{19845}{126}(-430)}$$
$$x \approx \pm 260$$
The width is about 520 feet.

85. **Answers may vary.**

87.
$$(y-2)^2 = 8(x-1)$$
$$y^2 - 4y + 4 = 8x - 8$$
$$y^2 - 8x - 4y + 12 = 0$$
$$0x^2 + 0xy + y^2 - 8x - 4y + 12 = 0$$

89.

$(x-h)^2 + (y-k)^2 = r^2$ $(0-h)^2 + (8-k)^2 = r^2$ $h^2 + 64 - 16k + k^2 = r^2$ $h^2 + k^2 - r^2 = 16k - 64$	$(x-h)^2 + (y-k)^2 = r^2$ $(5-h)^2 + (3-k)^2 = r^2$ $25 - 10h + h^2 + 9 - 6k + k^2 = r^2$ $h^2 + k^2 - r^2 = 10h + 6k - 34$
$(x-h)^2 + (y-k)^2 = r^2$ $(4-h)^2 + (6-k)^2 = r^2$ $16 - 8h + h^2 + 36 - 12k + k^2 = r^2$ $h^2 + k^2 - r^2 = 8h + 12k - 52$	

$$\begin{cases} 16k - 64 = 10h + 6k - 34 & \Rightarrow & 10k - 10h = 30 \\ 16k - 64 = 8h + 12k - 52 & \Rightarrow & 4k - 8h = 12 \end{cases} \Rightarrow k = 3,\, h = 0$$

Substitute into one of the above equations to get $r = 5$. Circle: $x^2 + (y-3)^2 = 25$

91.

$$y = ax^2 + bx + c \qquad y = ax^2 + bx + c \qquad y = ax^2 + bx + c$$
$$8 = a(1)^2 + b(1) + c \qquad -1 = a(-2)^2 + b(-2) + c \qquad 15 = a(2)^2 + b(2) + c$$
$$8 = a + b + c \qquad -1 = 4a - 2b + c \qquad 15 = 4a + 2b + c$$

$$\begin{cases} a + b + c = 8 \\ 4a - 2b + c = -1 \\ 4a + 2b + c = 15 \end{cases} \Rightarrow a = 1,\, b = 4,\, c = 3 \Rightarrow y = x^2 + 4x + 3$$

93. The stone hits the ground when $s = 0$:
$$0 = -16t^2 + 128t$$
$$0 = -16t(t-8)$$
It hits the ground after 8 seconds.
Find s when $t = x$:
$$s = -16x^2 + 128x$$

Find s when $t = 8 - x$:
$$s = -16(8-x)^2 + 128(8-x)$$
$$= -16(64 - 16x + x^2) + 1024 - 128x$$
$$= -1024 + 256x - 16x^2 + 1024 - 128x$$
$$= -16x^2 + 128x$$

95. b

97. c

99. b

101. d

Exercises 7.2 (page 728)

1. sum, constant

3. vertices

5. $(a, 0), (-a, 0)$

EXERCISES 7.2

7. $2a = 26 \Rightarrow$ String: 26 inches long
$2b = 10 \Rightarrow b = 5$
$b^2 = a^2 - c^2$
$5^2 = 13^2 - c^2 \Rightarrow c = 12$
Thumbtacks: $2c = 24$ inches apart

9. Both variables squared with equal coefficients: circle

11. One variable squared: parabola

13. Both variables squared with unequal coefficients: ellipse

15. $a = 4, b = 3$; horizontal
$C(0,0)$
$$\frac{x^2}{16} + \frac{y^2}{9} = 1$$

17. $c = 3, a = 5$; horizontal
$$b^2 = a^2 - c^2$$
$$= 25 - 9 = 16$$
$$\frac{x^2}{25} + \frac{y^2}{16} = 1$$

19. $c = 1, b = \frac{4}{3}$; vertical
$$a^2 = b^2 + c^2$$
$$= \frac{16}{9} + 1 = \frac{25}{9}$$
$$\frac{x^2}{16/9} + \frac{y^2}{25/9} = 1$$
$$\frac{9x^2}{16} + \frac{9y^2}{25} = 1$$

21. $c = 3, a = 4$; vertical
$$b^2 = a^2 - c^2$$
$$= 16 - 9 = 7$$
$$\frac{x^2}{7} + \frac{y^2}{16} = 1$$

23. vertical
$$\frac{(x-3)^2}{4} + \frac{(y-4)^2}{9} = 1$$

25. horizontal
$$\frac{(x-3)^2}{9} + \frac{(y-4)^2}{4} = 1$$

27. Center: $(3, 4)$, $b = 4$, $c = 5$, horizontal
$$a^2 = b^2 + c^2 = 16 + 25 = 41$$
$$\frac{(x-3)^2}{41} + \frac{(y-4)^2}{16} = 1$$

29. Center: $(0, 4)$, $c = 4$, $a = 6$, horizontal
$$b^2 = a^2 - c^2 = 36 - 16 = 20$$
$$\frac{x^2}{36} + \frac{(y-4)^2}{20} = 1$$

31. Center: $(0, 0)$, $c = 6$, $a = 10$, horizontal
$$b^2 = a^2 - c^2 = 100 - 36 = 64$$
$$\frac{x^2}{100} + \frac{y^2}{64} = 1$$

33. $$\frac{x^2}{25} + \frac{y^2}{9} = 1$$
Center: $(0, 0)$, $a = 5$, $b = 3$, horizontal

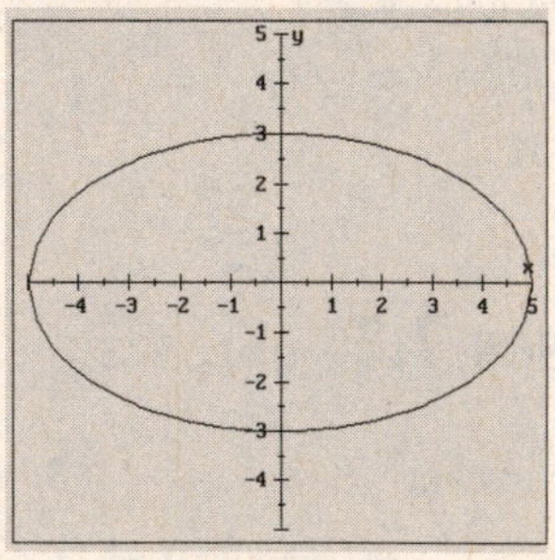

EXERCISES 7.2

35. $\dfrac{x^2}{25} + \dfrac{y^2}{49} = 1$

Center: $(0, 0)$, $a = 7$, $b = 5$, vertical

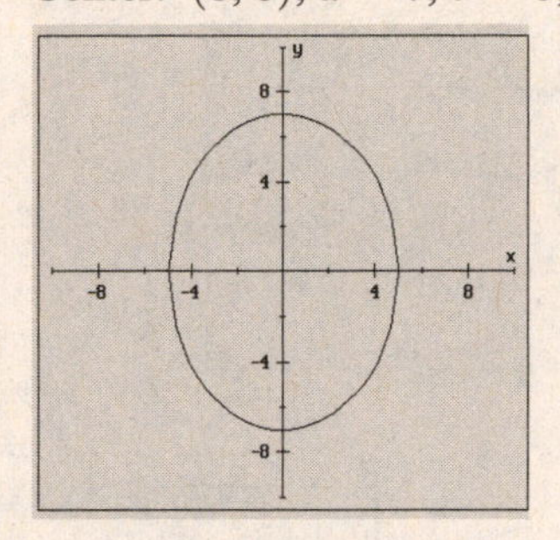

37. $\dfrac{x^2}{16} + \dfrac{(y+2)^2}{36} = 1$

Center: $(0, -2)$, $a = 6$, $b = 4$, vertical

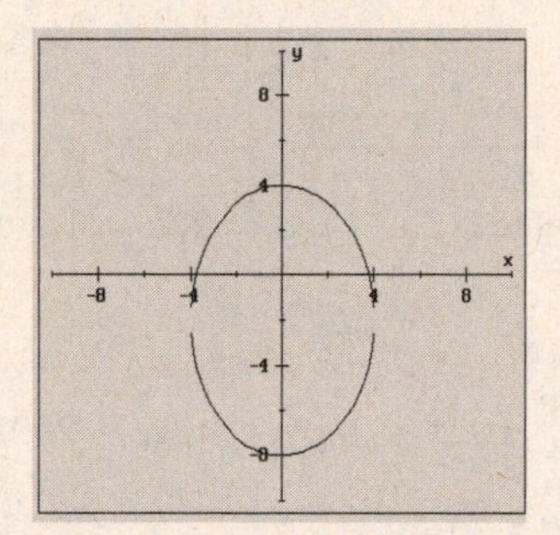

39. $\dfrac{(x-4)^2}{49} + \dfrac{(y-2)^2}{9} = 1$

Center: $(4, 2)$, $a = 7$, $b = 3$, horizontal

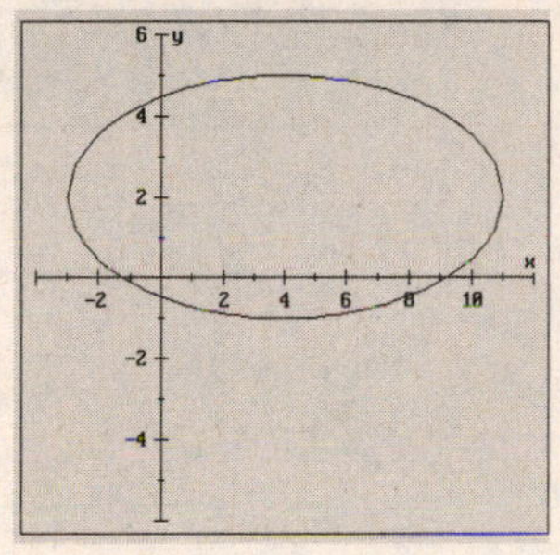

41.

$$4x^2 + y^2 - 2y = 15$$
$$4x^2 + y^2 - 2y + 1 = 15 + 1$$
$$4x^2 + (y-1)^2 = 16$$
$$\frac{4x^2}{16} + \frac{(y-1)^2}{16} = \frac{16}{16}$$
$$\frac{x^2}{4} + \frac{(y-1)^2}{16} = 1$$

43.

$$9x^2 + 4y^2 + 18x + 16y - 11 = 0$$
$$9(x^2 + 2x) + 4(y^2 + 4y) = 11$$
$$9(x^2 + 2x + 1) + 4(y^2 + 4y + 4) = 11 + 9 + 16$$
$$9(x+1)^2 + 4(y+2)^2 = 36$$
$$\frac{9(x+1)^2}{36} + \frac{4(y+2)^2}{36} = \frac{36}{36}$$
$$\frac{(x+1)^2}{4} + \frac{(y+2)^2}{9} = 1$$

EXERCISES 7.2

45.

$$x^2 + 4y^2 - 4x + 8y + 4 = 0$$
$$x^2 - 4x + 4(y^2 + 2y) = -4$$
$$x^2 - 4x + 4 + 4(y^2 + 2y + 1) = -4 + 4 + 4$$
$$(x-2)^2 + 4(y+1)^2 = 4$$
$$\frac{(x-2)^2}{4} + \frac{(y+1)^2}{1} = 1$$

Center: $(2, -1)$, $a = 2$, $b = 1$, horizontal

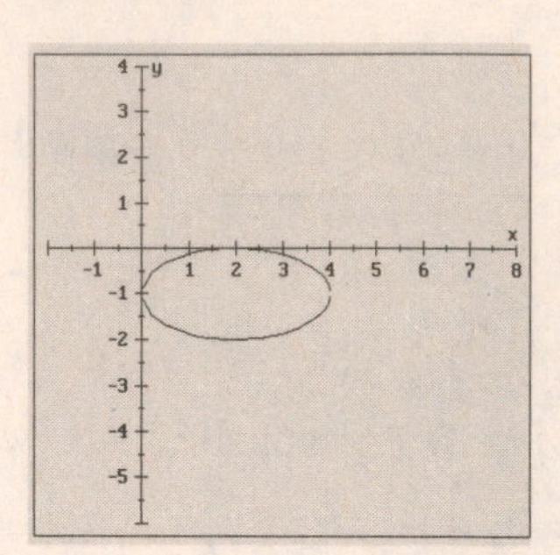

47.

$$16x^2 + 25y^2 - 160x - 200y + 400 = 0$$
$$16(x^2 - 10x) + 25(y^2 - 8y) = -400$$
$$16(x^2 - 10x + 25) + 25(y^2 - 8y + 16) = -400 + 400 + 400$$
$$16(x-5)^2 + 25(y-4)^2 = 400$$
$$\frac{(x-5)^2}{25} + \frac{(y-4)^2}{16} = 1$$

Center: $(5, 4)$, $a = 5$, $b = 4$, horizontal

49.

$$\frac{x^2}{4} + \frac{y^2}{36} = 1$$
$$9x^2 + 4y^2 = 36$$
$$4y^2 = 36 - 9x^2$$
$$y^2 = \frac{36 - 9x^2}{4}$$
$$y = \pm\sqrt{\frac{36 - 9x^2}{4}}$$

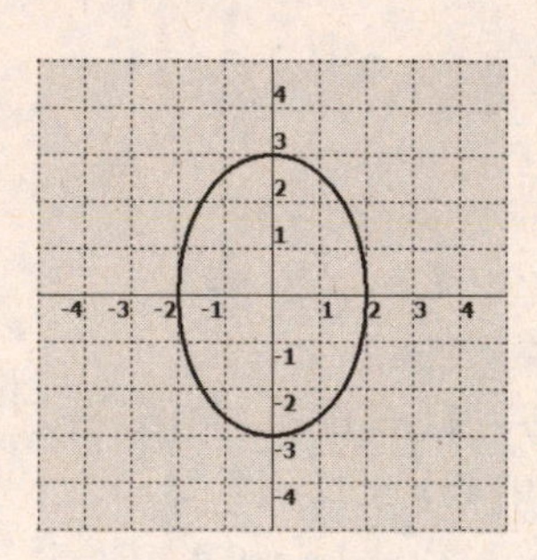

51. $C(0,0)$, $a = 30$, $b = 20$, horizontal

$$\frac{(x-0)^2}{30^2} + \frac{(y-0)^2}{20^2} = 1$$
$$\frac{x^2}{900} + \frac{y^2}{400} = 1$$

53. $a = 50, b = 30$

$$\frac{x^2}{2500} + \frac{y^2}{900} = 1$$

$$c^2 = a^2 - b^2$$
$$= 2500 - 900$$
$$= 1600$$
$$c = 40$$

$$\frac{40^2}{2500} + \frac{y^2}{900} = 1$$
$$\frac{y^2}{900} = \frac{900}{2500}$$
$$\frac{y}{30} = \pm\frac{30}{50}$$
$$y = \pm 18$$

The focal width is 36 meters.

55. $a = 24, b = 12$

$$\frac{x^2}{144} + \frac{y^2}{576} = 1$$
$$\frac{x^2}{144} + \frac{(-12)^2}{576} = 1$$
$$\frac{x^2}{144} = \frac{432}{576}$$
$$x \approx \pm 10.4$$

The width is about 20.8 inches.

57. The farthest distance $= a + c$:

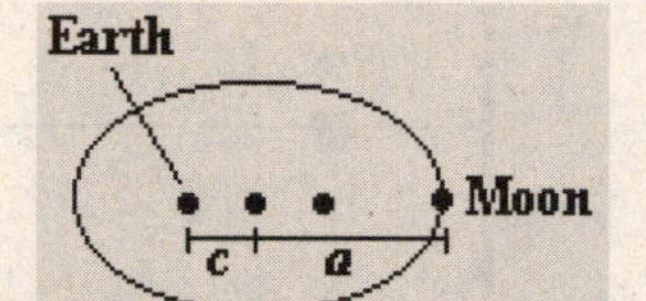

$$a = \frac{378000}{2} = 189000$$
$$\frac{c}{a} = \frac{11}{200}$$
$$c = \frac{11a}{200} = 10395$$

distance $= a + c =$ 199,395 miles

59. Answers may vary.

61. Answers may vary.

63. The equation of the ellipse is $\frac{x^2}{a^2} + \frac{y^2}{b^2} = 1$.

P is a point on the ellipse, so solve for y^2:

$$\frac{x^2}{a^2} + \frac{y^2}{b^2} = 1$$
$$\frac{y^2}{b^2} = 1 - \frac{x^2}{a^2}$$
$$y^2 = b^2\left(1 - \frac{x^2}{a^2}\right)$$

$$\begin{aligned} FP &= \sqrt{(c-x)^2 + (0-y)^2} \\ &= \sqrt{(c-x)^2 + y^2} \\ &= \sqrt{(c-x)^2 + b^2\left(1 - \frac{x^2}{a^2}\right)} \\ &= \sqrt{(c-x)^2 + (a^2 - c^2)\left(1 - \frac{x^2}{a^2}\right)} \\ &= \sqrt{c^2 - 2cx + x^2 + a^2 - x^2 - c^2 + \frac{c^2}{a^2}x^2} \\ &= \sqrt{a^2 - 2cx + \frac{c^2}{a^2}x^2} \\ &= \sqrt{\left(a - \frac{c}{a}x\right)^2} = a - \frac{c}{a}x \end{aligned}$$

65. By the result of **#63**, the distance between a point $P(x, y)$ on the ellipse and a focus $F(c, 0)$ is $D = a - \frac{c}{a}x$. Since P is a point on the ellipse, x must take on values from $-a$ to a. To make D as small as possible, x must be positive and as large as possible. This occurs when $x = a$. If $x = a$, then point P is actually at point V.

67. Consider the following diagram:

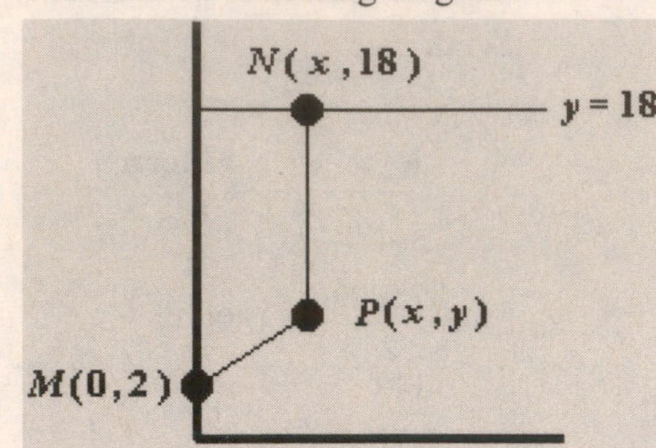

$$PM = \tfrac{1}{3}PN$$
$$\sqrt{(x-0)^2+(y-2)^2} = \tfrac{1}{3}\sqrt{(x-x)^2+(y-18)^2}$$
$$x^2+(y-2)^2 = \tfrac{1}{9}\left[0+(y-18)^2\right]$$
$$x^2+y^2-4y+4 = \tfrac{1}{9}(y-18)^2$$
$$9x^2+9y^2-36y+36 = y^2-36y+324$$
$$9x^2+8y^2 = 288$$
$$\frac{x^2}{32}+\frac{y^2}{36} = 1$$

69.
$$\frac{(x-h)^2}{a^2}+\frac{(y-k)^2}{b^2} = 1$$
$$a^2b^2\left[\frac{(x-h)^2}{a^2}+\frac{(y-k)^2}{b^2}\right] = a^2b^2(1)$$
$$b^2(x-h)^2+a^2(y-k)^2 = a^2b^2$$
$$b^2\left(x^2-2hx+h^2\right)+a^2\left(y^2-2ky+k^2\right) = a^2b^2$$
$$b^2x^2-2b^2hx+b^2h^2+a^2y^2-2a^2ky+a^2k^2-a^2b^2 = 0$$
$$\boldsymbol{b^2}x^2+\mathbf{0}xy+\boldsymbol{a^2}y^2+\boldsymbol{(-2b^2h)}x+\boldsymbol{(-2a^2k)}y+\boldsymbol{(b^2h^2+a^2y^2-a^2b^2)} = 0$$

71. d **73.** a **75.** b **77.** c

Exercises 7.3 (page 744)

1. difference, constant

3. $(a, 0), (-a, 0)$

5. transverse axis

7. Both variables squared with equal coefficients and same sign: circle

9. One variable squared: parabola

11. Both variables squared with unequal coefficients and same sign: ellipse

13. Both variables squared with opposite signs: hyperbola

15. $a = 5, c = 7$; horizontal
$$b^2 = c^2 - a^2$$
$$= 49 - 25 = 24$$
$$\frac{x^2}{25} - \frac{y^2}{24} = 1$$

17. $a = 2, b = 3$; horizontal
$$\frac{(x-2)^2}{4} - \frac{(y-4)^2}{9} = 1$$

EXERCISES 7.3

19. $a = 3$; vertical

$$\frac{(y-3)^2}{9} - \frac{(x-5)^2}{b^2} = 1$$

$$\frac{(8-3)^2}{9} - \frac{(1-5)^2}{b^2} = 1$$

$$\frac{25}{9} - \frac{16}{b^2} = 1$$

$$-\frac{16}{b^2} = -\frac{16}{9}$$

$$b^2 = 9 \Rightarrow \frac{(y-3)^2}{9} - \frac{(x-5)^2}{9} = 1$$

21. Center: $(0, 0)$, $a = 3$, $c = 5$; vertical

$$b^2 = c^2 - a^2$$
$$= 25 - 9 = 16$$
$$\frac{y^2}{9} - \frac{x^2}{16} = 1$$

23. $c = 6$, $a = 2$; horizontal

$$b^2 = c^2 - a^2$$
$$= 36 - 4 = 32$$
$$\frac{(x-1)^2}{4} - \frac{(y-4)^2}{32} = 1$$

25.

$$\frac{x^2}{a^2} - \frac{y^2}{b^2} = 1$$
$$\frac{4^2}{a^2} - \frac{2^2}{b^2} = 1$$
$$\frac{16}{a^2} - \frac{4}{b^2} = 1$$
$$\frac{16}{a^2} = 1 + \frac{4}{b^2}$$
$$\frac{64}{a^2} = 4 + \frac{16}{b^2}$$

$$\frac{x^2}{a^2} - \frac{y^2}{b^2} = 1$$
$$\frac{8^2}{a^2} - \frac{(-6)^2}{b^2} = 1$$
$$\frac{64}{a^2} - \frac{36}{b^2} = 1$$

$$\frac{64}{a^2} - \frac{36}{b^2} = 1$$
$$4 + \frac{16}{b^2} - \frac{36}{b^2} = 1$$
$$3 = \frac{20}{b^2}$$
$$b^2 = \frac{20}{3}$$
$$a^2 = 10$$

$$\frac{x^2}{10} - \frac{3y^2}{20} = 1$$

27.

$$4(x-1)^2 - 9(y+2)^2 = 36$$
$$\frac{4(x-1)^2}{36} - \frac{9(y+2)^2}{36} = \frac{36}{36}$$
$$\frac{(x-1)^2}{9} - \frac{(y+2)^2}{4} = 1$$

$a = 3, b = 2$

Area $= (2a)(2b) = (6)(4) = 24$ sq. units

29.

$$\begin{aligned}
x^2 + 6x - y^2 + 2y &= -11 \\
x^2 + 6x - \left(y^2 - 2y\right) &= -11 \\
x^2 + 6x + 9 - \left(y^2 - 2y + 1\right) &= -11 + 9 - 1 \\
(x+3)^2 - (y-1)^2 &= -3 \\
\frac{(x+3)^2}{-3} - \frac{(y-1)^2}{-3} &= 1 \\
\frac{(y-1)^2}{3} - \frac{(x+3)^2}{3} &= 1 \Rightarrow a = \sqrt{3}, b = \sqrt{3};
\end{aligned}$$

$\text{Area} = (2a)(2b) = \left(2\sqrt{3}\right)\left(2\sqrt{3}\right) = 12$ sq. units

31.

$$\begin{aligned}
(2a)(2b) &= 36 \\
4(2b) &= 36 \\
b &= \tfrac{9}{2}
\end{aligned}$$

$$\frac{(x+2)^2}{4} - \frac{4(y+4)^2}{81} = 1$$

OR

$$\frac{(y+4)^2}{4} - \frac{4(x+2)^2}{81} = 1$$

33. Center: $(0,0)$, $a = 6$, $b = \frac{5}{4}$

$$\frac{x^2}{6^2} - \frac{y^2}{\left(\frac{5}{4}\right)^2} = 1$$

$$\frac{x^2}{36} - \frac{16y^2}{25} = 1$$

35. $\dfrac{x^2}{9} - \dfrac{y^2}{4} = 1$

Center: $(0,0)$, $a = 3$, $b = 2$, horizontal

37.

$$\begin{aligned}
4x^2 - 3y^2 &= 36 \\
\frac{4x^2}{36} - \frac{3y^2}{36} &= \frac{36}{36} \\
\frac{x^2}{9} - \frac{y^2}{12} &= 1
\end{aligned}$$

Center: $(0,0)$, $a = 3$, $b = 2\sqrt{3}$, horizontal

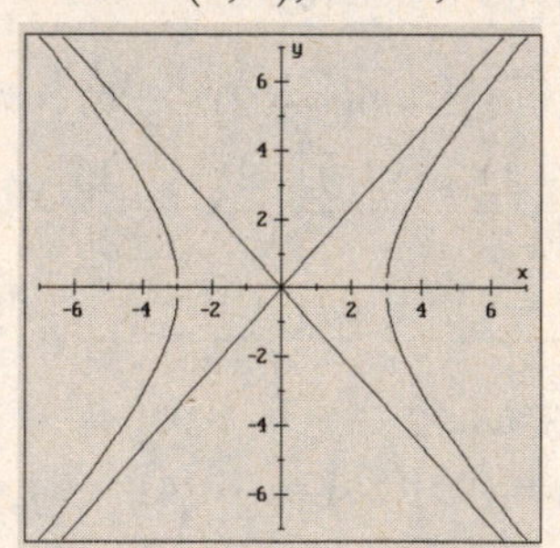

EXERCISES 7.3

39. $y^2 - x^2 = 1$

$$\frac{y^2}{1} - \frac{x^2}{1} = 1$$

Center: $(0, 0)$, $a = 1$, $b = 1$, vertical

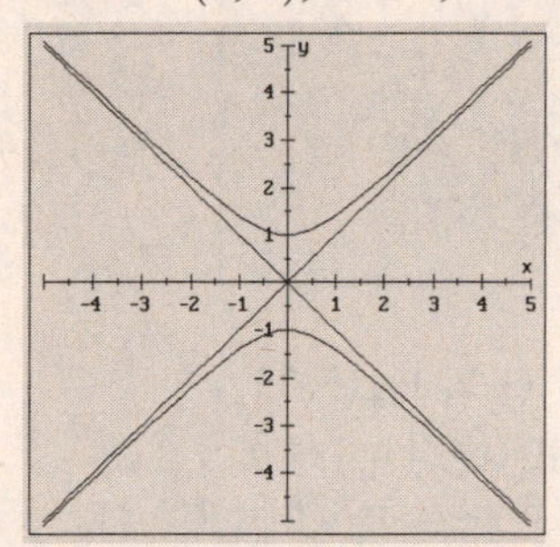

41. $\dfrac{(x+2)^2}{9} - \dfrac{y^2}{4} = 1$

Center: $(-2, 0)$, $a = 3$, $b = 2$, horizontal

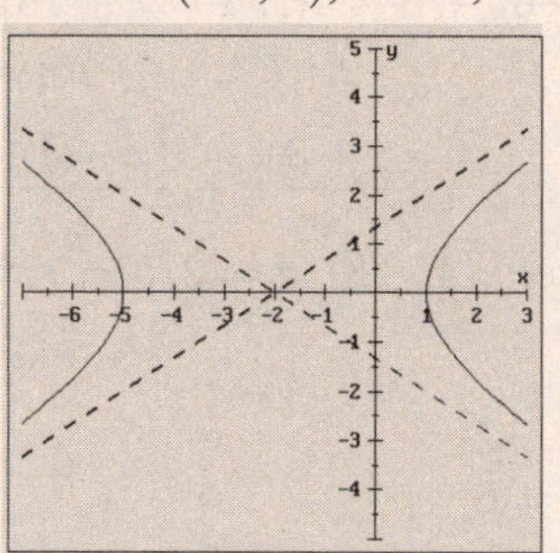

43. $4(y-2)^2 - 9(x+1)^2 = 36$

$$\frac{4(y-2)^2}{36} - \frac{9(x+1)^2}{36} = \frac{36}{36}$$

$$\frac{(y-2)^2}{9} - \frac{(x+1)^2}{4} = 1$$

Center: $(-1, 2)$, $a = 3$, $b = 2$, vertical

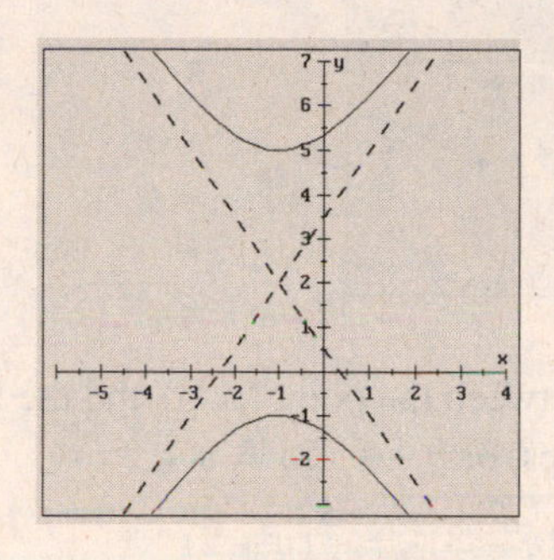

45.

$$4x^2 - 2y^2 + 8x - 8y = 8$$

$$4(x^2 + 2x) - 2(y^2 + 4y) = 8$$

$$4(x^2 + 2x + 1) - 2(y^2 + 4y + 4) = 8 + 4 - 8$$

$$4(x+1)^2 - 2(y+2)^2 = 4$$

$$\frac{4(x+1)^2}{4} - \frac{2(y+2)^2}{4} = \frac{4}{4}$$

$$\frac{(x+1)^2}{1} - \frac{(y+2)^2}{2} = 1$$

Center: $(-1, -2)$, $a = 1$, $b = \sqrt{2}$, horizontal

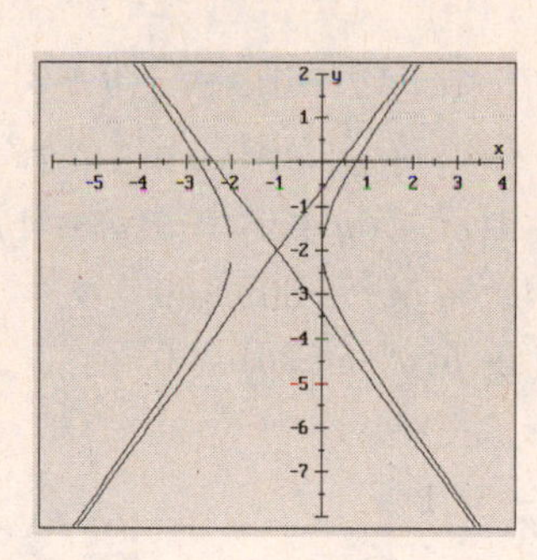

47.

$$\begin{aligned} y^2 - 4x^2 + 6y + 32x &= 59 \\ y^2 + 6y - 4(x^2 - 8x) &= 59 \\ y^2 + 6y + 9 - 4(x^2 - 8x + 16) &= 59 + 9 - 64 \\ (y+3)^2 - 4(x-4)^2 &= 4 \\ \frac{(y+3)^2}{4} - \frac{(x-4)^2}{1} &= 1 \end{aligned}$$

Center: $(4, -3)$, $a = 2$, $b = 1$, vertical

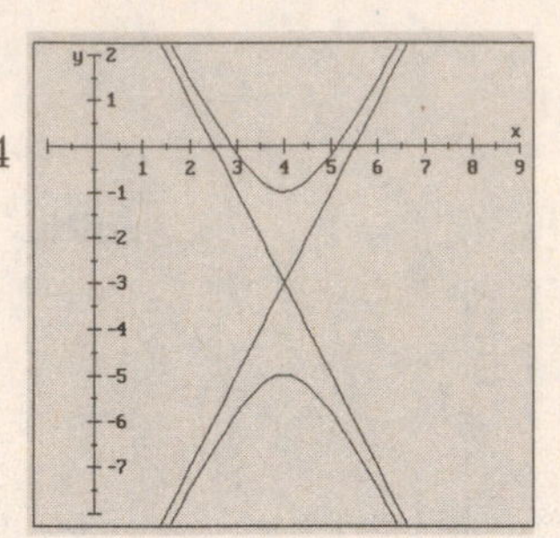

49. $-xy = 6$

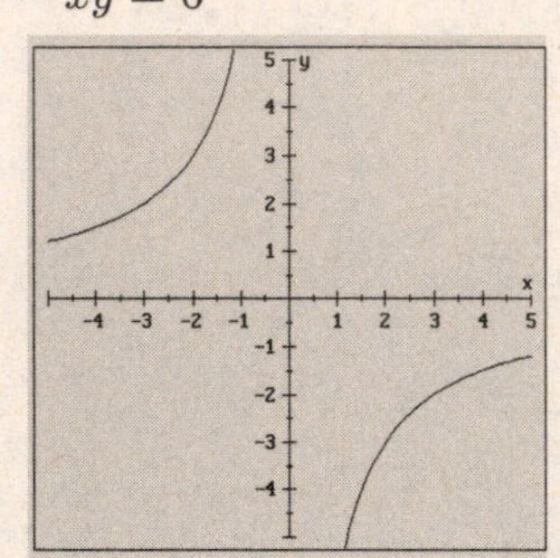

51. Foci: $(-2, 1)$, $(8, 1)$

Center: $(3, 1)$, $c = 5$

$2a = 6 \Rightarrow a = 3, b = 4$

$$\frac{(x-3)^2}{9} - \frac{(y-1)^2}{16} = 1$$

53. The distance between the point (x, y) and the line $y = -2$ is the difference between the y-coordinates, or $y - (-2) = y + 2$.

$$\begin{aligned} \sqrt{(x-0)^2 + (y-3)^2} &= \frac{3}{2}(y+2) \\ x^2 + (y-3)^2 &= \frac{9}{4}(y+2)^2 \\ 4x^2 + 4(y-3)^2 &= 9(y+2)^2 \\ 4x^2 + 4(y^2 - 6y + 9) &= 9(y^2 + 4y + 4) \\ 4x^2 - 5y^2 - 60y &= 0 \\ 4x^2 - 5(y^2 + 12y) &= 0 \end{aligned}$$

$$\begin{aligned} 4x^2 - 5(y^2 + 12y) &= 0 \\ 4x^2 - 5(y^2 + 12y + 36) &= 0 - 5(36) \\ 4x^2 - 5(y+6)^2 &= -180 \\ \frac{4x^2}{-180} - \frac{5(y+6)^2}{-180} &= \frac{-180}{-180} \\ \frac{(y+6)^2}{36} - \frac{x^2}{45} &= 1 \end{aligned}$$

55.

$$\begin{aligned} x^2 - \frac{y^2}{4} &= 1 \\ -\frac{y^2}{4} &= -x^2 + 1 \\ y^2 &= 4x^2 - 4 \\ y &= \pm\sqrt{4x^2 - 4} \end{aligned}$$

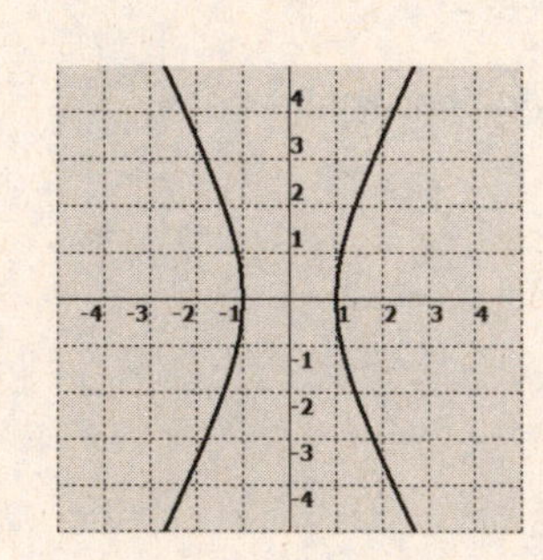

57. $xy = k$
$(12)(2) = k$
$24 = k$

59. $9y^2 - x^2 = 81$
$\frac{9y^2}{81} - \frac{x^2}{81} = \frac{81}{81}$
$\frac{y^2}{9} - \frac{x^2}{81} = 1$
$a = 3 \Rightarrow 3$ units

61. $2a = 24$: $a = 12$
$c = 13$: $b = 5$
$\frac{x^2}{144} - \frac{y^2}{25} = 1$

63. $2a = 12$: $a = 6$
$c = 10$: $b = 8$
$\frac{x^2}{36} - \frac{y^2}{64} = 1$

65-71. Answers may vary.

73. b

75. c

77. b

79. a

Exercises 7.4 (page 754)

1. graphs

3. $\begin{cases} 8x^2 + 32y^2 = 256 \\ x = 2y \end{cases}$

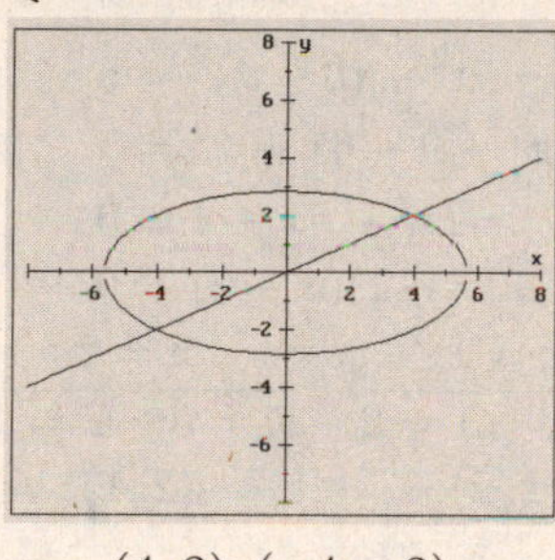

$(4, 2), (-4, -2)$

5. $\begin{cases} x^2 + y^2 = 90 \\ y = x^2 \end{cases}$

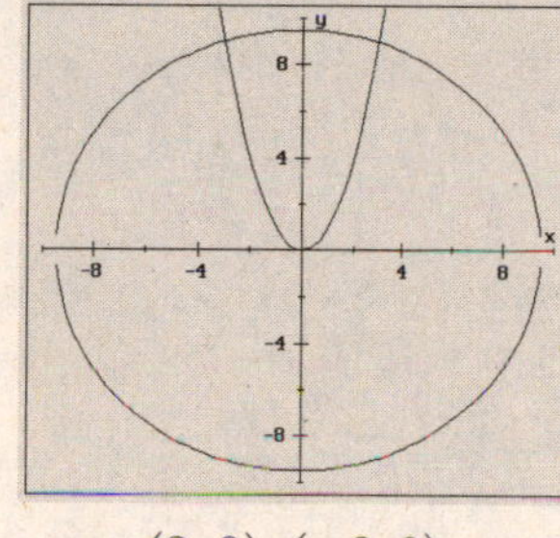

$(3, 9), (-3, 9)$

7. $\begin{cases} x^2 + y^2 = 25 \\ 12x^2 + 64y^2 = 768 \end{cases}$

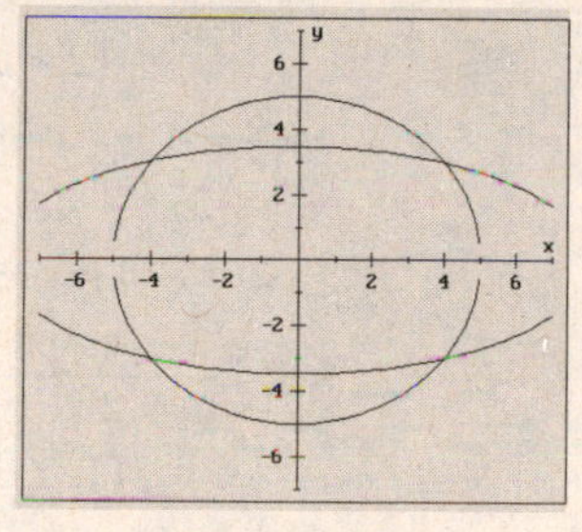

$(-4, 3), (4, 3)$
$(-4, -3), (4, -3)$

9. $\begin{cases} x^2 - 13 = -y^2 \\ y = 2x - 4 \end{cases}$

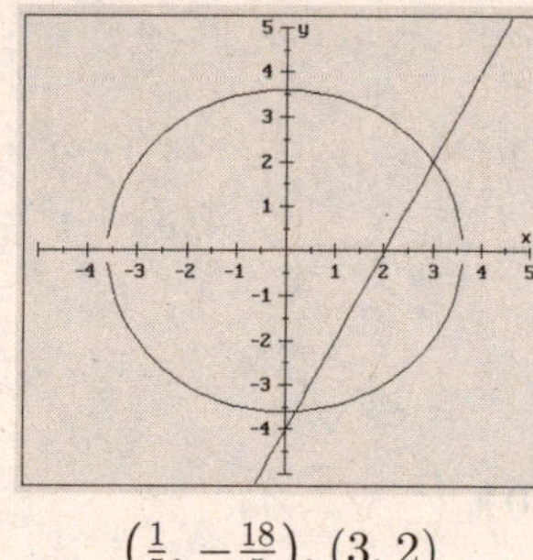

$\left(\frac{1}{5}, -\frac{18}{5}\right), (3, 2)$

11. $\begin{cases} x^2 - 6x - y = -5 \\ x^2 - 6x + y = -5 \end{cases}$

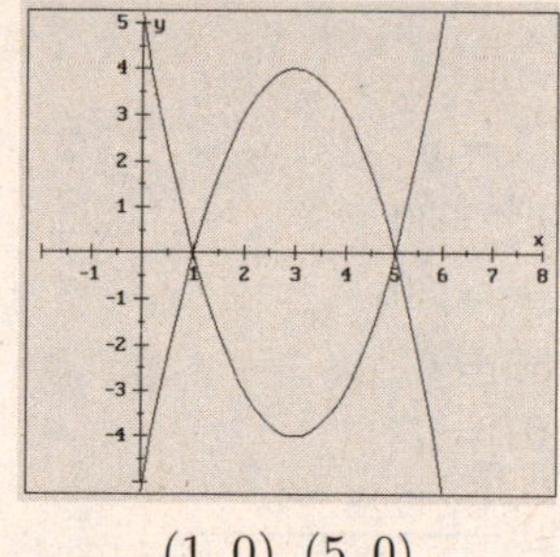

$(1, 0), (5, 0)$

13. $\begin{cases} y = x + 1 \\ y = x^2 + x \end{cases}$

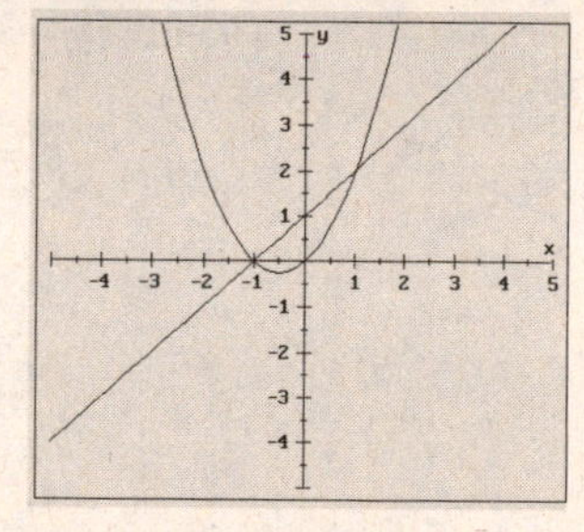

$(1, 2), (-1, 0)$

EXERCISES 7.4

15. $\begin{cases} 6x^2 + 9y^2 = 10 \Rightarrow y = \pm\sqrt{\frac{10-6x^2}{9}} \\ 3y - 2x = 0 \quad\Rightarrow y = \frac{2}{3}x \end{cases}$

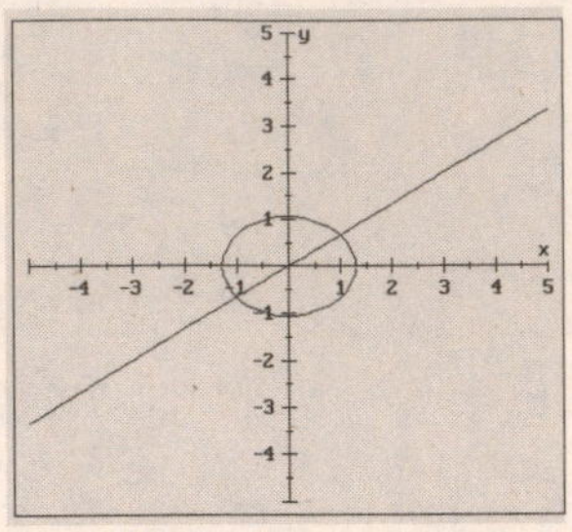

(1, 0.67), (−1, −0.67)

17. $5x + 3y = 15 \Rightarrow y = \dfrac{15-5x}{3}$

$$25x^2 + 9y^2 = 225$$
$$25x^2 + 9\left(\frac{15-5x}{3}\right)^2 = 225$$
$$25x^2 + (15-5x)^2 = 225$$
$$25x^2 + 225 - 150x + 25x^2 = 225$$
$$50x^2 - 150x = 0$$
$$50x(x-3) = 0$$

$x = 0$: $y = \dfrac{15-5(0)}{3} = 5$, (0, 5)

$x = 3$: $y = \dfrac{15-5(3)}{3} = 0$, (3, 0)

19. $x + y = 2 \Rightarrow y = 2 - x$

$$x^2 + y^2 = 2$$
$$x^2 + (2-x)^2 = 2$$
$$x^2 + 4 - 4x + x^2 = 2$$
$$2x^2 - 4x + 2 = 0$$
$$2(x-1)(x-1) = 0$$

$x = 1$: $y = 2 - 1 = 1$, (1, 1)

21. $x + y = 3 \Rightarrow y = 3 - x$

$$x^2 + y^2 = 5$$
$$x^2 + (3-x)^2 = 5$$
$$x^2 + 9 - 6x + x^2 = 5$$
$$2x^2 - 6x + 4 = 0$$
$$2(x-1)(x-2) = 0$$

$x = 1$: $y = 3 - 1 = 2$, (1, 2)

$x = 2$: $y = 3 - 2 = 1$, (2, 1)

23. $y = x^2 - 1 \Rightarrow x^2 = y + 1$

$$x^2 + y^2 = 13$$
$$y + 1 + y^2 = 13$$
$$y^2 + y - 12 = 0$$
$$(y-3)(y+4) = 0$$

$y = 3$: $3 + 1 = x^2 \Rightarrow x = \pm 2$, (2, 3), (−2, 3)

$y = -4$: $-4 + 1 = x^2$, no real solutions

25. $y = x^2 \Rightarrow x^2 = y$

$$x^2 + y^2 = 30$$
$$y + y^2 = 30$$
$$y^2 + y - 30 = 0$$
$$(y-5)(y+6) = 0$$

$y = 5$: $5 = x^2 \Rightarrow x = \pm\sqrt{5}$, $\left(\sqrt{5}, 5\right), \left(-\sqrt{5}, 5\right)$

$y = -6$: $-6 = x^2$, no real solutions

27.
$$\begin{aligned} 2x^2 + y^2 &= 6 \\ x^2 - y^2 &= 3 \\ \hline 3x^2 &= 9 \\ x^2 &= 3 \\ x &= \pm\sqrt{3} \end{aligned}$$

$$2x^2 + y^2 = 6$$
$$2(3) + y^2 = 6$$
$$y^2 = 0$$
$$y = 0$$

$\left(\sqrt{3}, 0\right), \left(-\sqrt{3}, 0\right)$

EXERCISES 7.4

29.
$$\begin{array}{rcl} x^2+y^2 &=& 20 \\ x^2-y^2 &=& -12 \\ \hline 2x^2 &=& 8 \\ x^2 &=& 4 \\ x &=& \pm 2 \end{array} \qquad \begin{array}{r} x^2+y^2=20 \\ 4+y^2=20 \\ y^2=16 \\ y=\pm 4 \end{array}$$
$(2, 4), (-2, 4), (2, -4), (-2, -4)$

31.
$$\begin{array}{r} y=x^2-10 \Rightarrow x^2=y+10 \\ y^2=40-x^2 \\ y^2=40-y-10 \\ y^2+y-30=0 \\ (y-5)(y+6)=0 \end{array} \qquad \begin{array}{c} y=5 \\ \hline x^2=5+10 \Rightarrow x=\pm\sqrt{15} \\ \left(\sqrt{15},5\right), \left(-\sqrt{15},5\right) \end{array} \qquad \begin{array}{c} y=-6 \\ \hline x^2=-6+10 \Rightarrow x=\pm 2 \\ (2,-6), (-2,-6) \end{array}$$

33.
$$\begin{array}{r} y=x^2-4 \Rightarrow x^2=y+4 \\ x^2-y^2=-16 \\ y+4-y^2=-16 \\ y^2-y-20=0 \\ (y-5)(y+4)=0 \end{array} \qquad \begin{array}{c} y=5 \\ \hline x^2=5+4 \Rightarrow x=\pm 3 \\ (3,5), (-3,5) \end{array} \qquad \begin{array}{c} y=-4 \\ \hline x^2=-4+4 \Rightarrow x=0 \\ (0,-4) \end{array}$$

35.
$$\begin{array}{rcl} x^2-y^2 &=& -5 \\ 3x^2+2y^2 &=& 30 \\ \hline \\ 2x^2-2y^2 &=& -10 \\ 3x^2+2y^2 &=& 30 \\ \hline 5x^2 &=& 20 \\ x^2 &=& 4 \\ x &=& \pm 2 \end{array} \qquad \begin{array}{r} 3x^2+2y^2=30 \\ 3(4)+2y^2=30 \\ 2y^2=18 \\ y^2=9 \\ y=\pm 3 \end{array}$$
$(2, 3), (-2, 3), (2, -3), (-2, -3)$

37.
$$\begin{array}{r} \frac{1}{x}+\frac{2}{y}=1 \\ \frac{2}{x}-\frac{1}{y}=\frac{1}{3} \\ \hline \end{array} \qquad \begin{array}{rcl} \frac{1}{x}+\frac{2}{y} &=& 1 \\ \frac{4}{x}-\frac{2}{y} &=& \frac{2}{3} \\ \hline \frac{5}{x} &=& \frac{5}{3} \\ x &=& 3 \end{array} \qquad \begin{array}{r} \frac{1}{x}+\frac{2}{y}=1 \\ \frac{1}{3}+\frac{2}{y}=1 \\ \frac{2}{y}=\frac{2}{3} \\ y=3 \end{array}$$
$(3, 3)$

39.
$$\begin{array}{r} 3y^2=xy \\ 3y^2-xy=0 \\ y(3y-x)=0 \\ y=0 \text{ or } x=3y \end{array} \qquad \begin{array}{r} y=0 \\ \hline 2x^2+xy-84=0 \\ 2x^2+x(0)-84=0 \\ 2x^2=84 \\ x^2=42 \\ x=\pm\sqrt{42} \\ \left(\sqrt{42},0\right), \left(-\sqrt{42},0\right) \end{array} \qquad \begin{array}{r} x=3y \\ \hline 2x^2+xy-84=0 \\ 2(3y)^2+(3y)y-84=0 \\ 18y^2+3y^2-84=0 \\ 21y^2=84 \\ y^2=4 \Rightarrow y=\pm 2 \\ (6,2), (-6,-2) \end{array}$$

EXERCISES 7.4

41.

$$xy = \tfrac{1}{6} \Rightarrow y = \tfrac{1}{6x}$$
$$y + x = 5xy$$
$$\frac{1}{6x} + x = \frac{5x}{6x}$$
$$1 + 6x^2 = 5x$$
$$6x^2 - 5x + 1 = 0$$
$$(2x-1)(3x-1) = 0$$

$x = \frac{1}{2}$	$x = \frac{1}{3}$
$y = \dfrac{1}{6(1/2)} = \dfrac{1}{3}$	$y = \dfrac{1}{6(1/3)} = \dfrac{1}{2}$
$\left(\frac{1}{2}, \frac{1}{3}\right)$	$\left(\frac{1}{3}, \frac{1}{2}\right)$

43. Let $x =$ width and $y =$ length.

$$\begin{cases} xy = 63 \\ 2x + 2y = 32 \end{cases}$$
$$xy = 63 \Rightarrow y = \tfrac{63}{x}$$

$$2x + 2y = 32$$
$$2x + 2\left(\frac{63}{x}\right) = 32$$
$$2x^2 + 126 = 32x$$
$$2x^2 - 32x + 126 = 0$$
$$2(x-9)(x-7) = 0$$

$x = 9$	$x = 7$
$y = \frac{63}{9} = 7$	$y = \frac{63}{7} = 9$

The dimensions are 9 cm by 7 cm.

45. Let $x =$ width and $y =$ length.

$$\begin{cases} xy = 8000 \\ 2x + y = 260 \end{cases}$$
$$xy = 8000 \Rightarrow y = \tfrac{8000}{x}$$

$$2x + y = 260$$
$$2x + \left(\frac{8000}{x}\right) = 260$$
$$2x^2 + 8000 = 260x$$
$$2x^2 - 260x + 8000 = 0$$
$$2(x-50)(x-80) = 0$$

$x = 50$	$x = 80$
$y = \frac{8000}{50} = 160$	$y = \frac{8000}{80} = 100$

The dimensions are 50 ft by 160 ft or 80 ft by 100 ft.

47. Let $x =$ Carol's principal.

	I	P	r
Carol	67.50	x	$\frac{67.50}{x}$
John	94.50	$x + 150$	$\frac{94.50}{x+150}$

[John's rate] = [Carol's rate] + 0.015

$$\frac{94.50}{x+150} = \frac{67.50}{x} + 0.015$$
$$94.5x = 67.5(x+150) + 0.015x(x+150)$$
$$0.015x^2 - 24.75x + 10{,}125 = 0$$
$$x^2 - 1650x + 675{,}000 = 0$$
$$(x-750)(x-900) = 0$$

$x - 750 = 0$ **or** $x - 900 = 0$

$x = 750$ $\quad$ $x = 900 \Rightarrow$

Carol invested either $750 at 9% or she invested $900 at 7.5% interest.

49.

$$\begin{cases} y = \frac{1}{10}x \\ y = -\frac{1}{300}x^2 + \frac{1}{5}x \end{cases}$$
$$\tfrac{1}{10}x = -\tfrac{1}{300}x^2 + \tfrac{1}{5}x$$
$$30x = -x^2 + 60x$$
$$x^2 - 30x = 0$$
$$x(x-30) = 0$$

$x = 0$ **or** $x - 30 = 0$

$x = 30$

$x = 30 \Rightarrow y = \frac{1}{10}(30) = 3$ $\quad$ $\boxed{(30, 3)}$

51. $\begin{cases} y = x^2 \\ x + y = 2 \end{cases}$

$$\begin{aligned} x + y &= 2 \\ x + x^2 &= 2 \\ x^2 + x - 2 &= 0 \\ (x+2)(x-1) &= 0 \end{aligned}$$

$x + 2 = 0$ **or** $x - 1 = 0$

$x = -2$ $\quad$ $x = 1$

$x = -2 \Rightarrow y = (-2)^2 = 4$

$x = 1 \Rightarrow y = 1^2 = 1$

There are potential collision points at $(-2, 4)$ and $(1, 1)$.

53. $\begin{cases} x = 2y \\ (x-120)^2 + y^2 = 100^2 \end{cases}$

The x-coordinate of the point where the line crosses the circle closest to the origin has the approximate coordinates $(20.5, 10.25)$. The distance to the origin is about 23 miles.

55. **Answers may vary.**

57. **Answers may vary.**

59. False. The substitution and elimination methods are more precise.

61. False. Use the elimination method.

63. True.

65. False. The difference is 4.

Chapter 7 Review (page 757)

1.
$$\begin{aligned} (x-h)^2 + (y-k)^2 &= r^2 \\ (x-0)^2 + (y-0)^2 &= 4^2 \\ x^2 + y^2 &= 16 \end{aligned}$$

2. $r = \sqrt{(6-0)^2 + (8-0)^2} = 10$
$$\begin{aligned} (x-h)^2 + (y-k)^2 &= r^2 \\ (x-0)^2 + (y-0)^2 &= 10^2 \\ x^2 + y^2 &= 100 \end{aligned}$$

3.
$$\begin{aligned} (x-h)^2 + (y-k)^2 &= r^2 \\ (x-3)^2 + (y-(-2))^2 &= 5^2 \\ (x-3)^2 + (y+2)^2 &= 25 \end{aligned}$$

4. $r = \sqrt{(-2-1)^2 + (4-0)^2} = 5$
$$\begin{aligned} (x-h)^2 + (y-k)^2 &= r^2 \\ (x-(-2))^2 + (y-4)^2 &= 5^2 \\ (x+2)^2 + (y-4)^2 &= 25 \end{aligned}$$

5. $C\left(\frac{-2+12}{2}, \frac{4+16}{2}\right) = C(5, 10)$

$r = \sqrt{(12-5)^2 + (16-10)^2} = \sqrt{85}$

$(x-h)^2 + (y-k)^2 = r^2$

$(x-5)^2 + (y-10)^2 = \left(\sqrt{85}\right)^2$

$(x-5)^2 + (y-10)^2 = 85$

6. $C\left(\frac{-3+7}{2}, \frac{-6+10}{2}\right) = C(2, 2)$

$r = \sqrt{(7-2)^2 + (10-2)^2} = \sqrt{89}$

$(x-h)^2 + (y-k)^2 = r^2$

$(x-2)^2 + (y-2)^2 = \left(\sqrt{89}\right)^2$

$(x-2)^2 + (y-2)^2 = 89$

7.

$$x^2 + y^2 - 6x + 4y = 3$$
$$x^2 - 6x + y^2 + 4y = 3$$
$$x^2 - 6x + 9 + y^2 + 4y + 4 = 3 + 9 + 4$$
$$(x-3)^2 + (y+2)^2 = 16$$

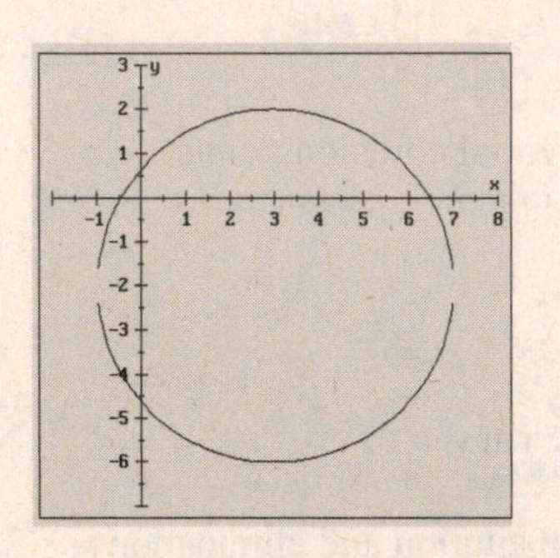

8.

$$x^2 + 4x + y^2 - 10y = -13$$
$$x^2 + 4x + 4 + y^2 - 10y + 25 = -13 + 4 + 25$$
$$(x+2)^2 + (y-5)^2 = 16$$

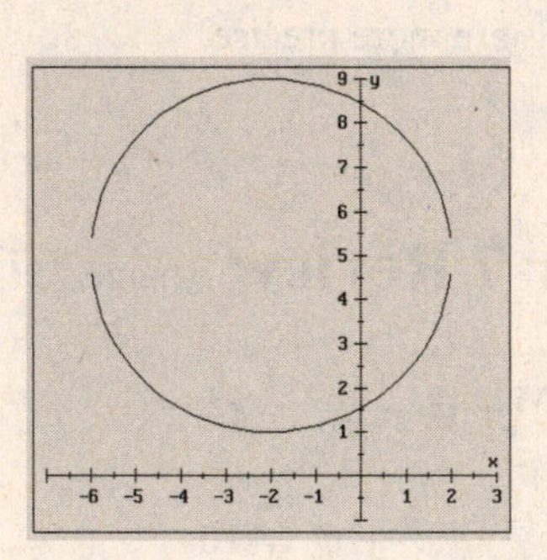

9. Horizontal

$$(y-0)^2 = 4p(x-0)$$
$$(4-0)^2 = 4p(-8-0)$$
$$16 = -32p$$
$$-2 = 4p$$
$$y^2 = -2x$$

10. Vertical

$$(x-0)^2 = 4p(y-0)$$
$$(-8-0)^2 = 4p(4-0)$$
$$64 = 16p$$
$$16 = 4p$$
$$x^2 = 16y$$

11. Vertical

$$(x+2)^2 = 4p(y-3)$$
$$(-4+2)^2 = 4p(-8-3)$$
$$4 = 4p(-11)$$
$$-\frac{4}{11} = 4p$$
$$(x+2)^2 = -\frac{4}{11}(y-3)$$

12.
$$\begin{aligned} x^2 - 4y - 2x + 9 &= 0 \\ x^2 - 2x &= 4y - 9 \\ x^2 - 2x + 1 &= 4y - 9 + 1 \\ (x-1)^2 &= 4(y-2) \end{aligned}$$

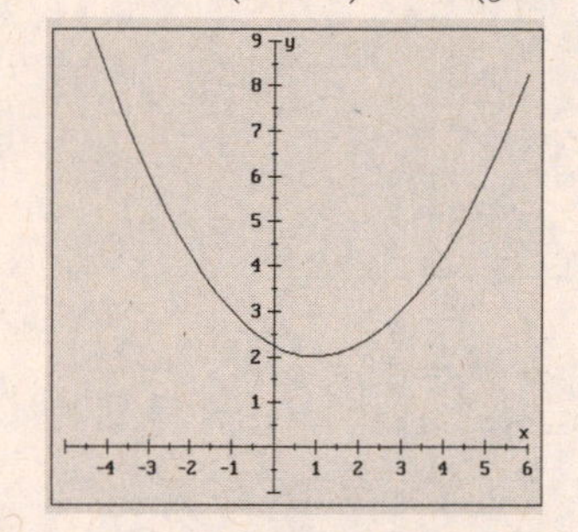

13.
$$\begin{aligned} y^2 - 6y &= 4x - 13 \\ y^2 - 6y + 9 &= 4x - 13 + 9 \\ (y-3)^2 &= 4(x-1) \end{aligned}$$

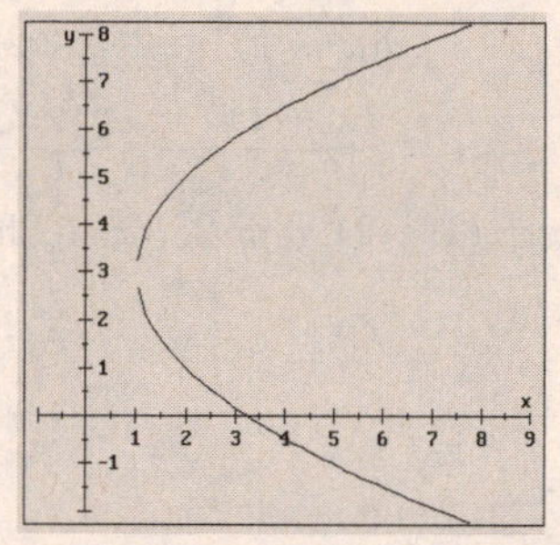

14. $a = 6, b = 4$, horizontal

$$\frac{x^2}{36} + \frac{y^2}{16} = 1$$

15. $a = 5, b = 2$, vertical

$$\frac{x^2}{4} + \frac{y^2}{25} = 1$$

16. $a = 4, b = 3$, horizontal $\Rightarrow \dfrac{(x+2)^2}{16} + \dfrac{(y-3)^2}{9} = 1$

17.
$$\begin{aligned} 4x^2 + y^2 - 16x + 2y &= -13 \\ 4(x^2 - 4x) + y^2 + 2y &= -13 \\ 4(x^2 - 4x + 4) + y^2 + 2y + 1 &= -13 + 16 + 1 \\ 4(x-2)^2 + (y+1)^2 &= 4 \\ \frac{(x-2)^2}{1} + \frac{(y+1)^2}{4} &= 1 \end{aligned}$$

Center: $(2, -1)$, $a = 2$, $b = 1$, vertical

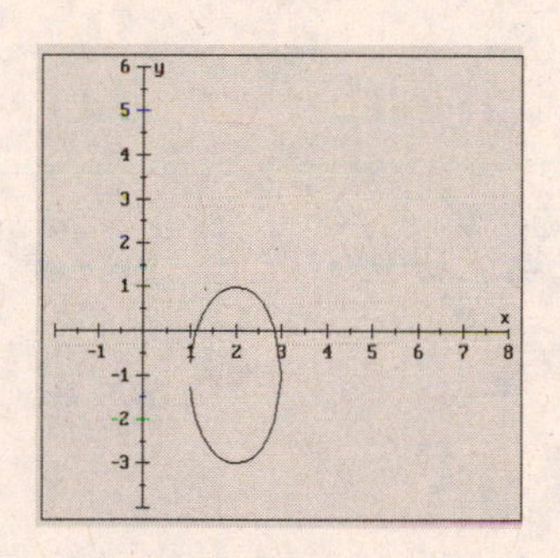

18. $a = 2, c = 4$; horizontal

$b^2 = c^2 - a^2 = 16 - 4 = 12$

$$\frac{x^2}{4} - \frac{y^2}{12} = 1$$

19. $a = 3, c = 5$; vertical

$b^2 = c^2 - a^2 = 25 - 9 = 16$

$$\frac{y^2}{9} - \frac{x^2}{16} = 1$$

20. $C(0, 3)$, $a = 3, c = 5$; horizontal

$b^2 = c^2 - a^2 = 25 - 9 = 16$

$$\frac{x^2}{9} - \frac{(y-3)^2}{16} = 1$$

21. $C(3, 0)$, $a = 3, c = 5$; vertical

$b^2 = c^2 - a^2 = 25 - 9 = 16$

$$\frac{y^2}{9} - \frac{(x-3)^2}{16} = 1$$

22. $y = \pm \dfrac{b}{a}x \Rightarrow y = \pm \dfrac{4}{5}x$

23.

$$9x^2 - 4y^2 - 16y - 18x = 43$$
$$9(x^2 - 2x) - 4(y^2 + 4y) = 43$$
$$9(x^2 - 2x + 1) - 4(y^2 + 4y + 4) = 43 + 9 - 16$$
$$9(x-1)^2 - 4(y+2)^2 = 36$$
$$\frac{(x-1)^2}{4} - \frac{(y+2)^2}{9} = 1$$

Center: $(1, -2)$, $a = 2$, $b = 3$, horizontal

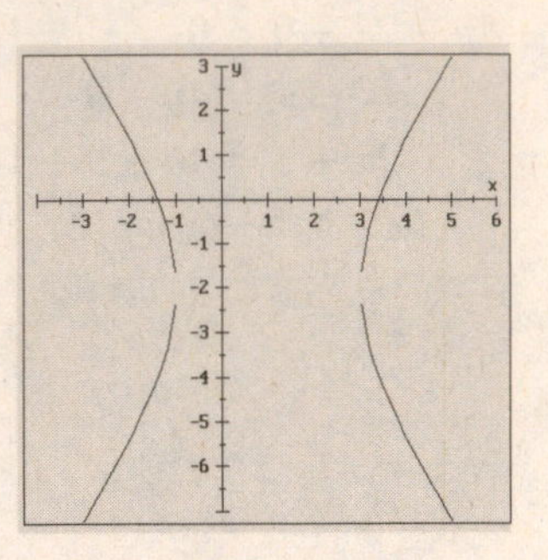

24. $4xy = 1$

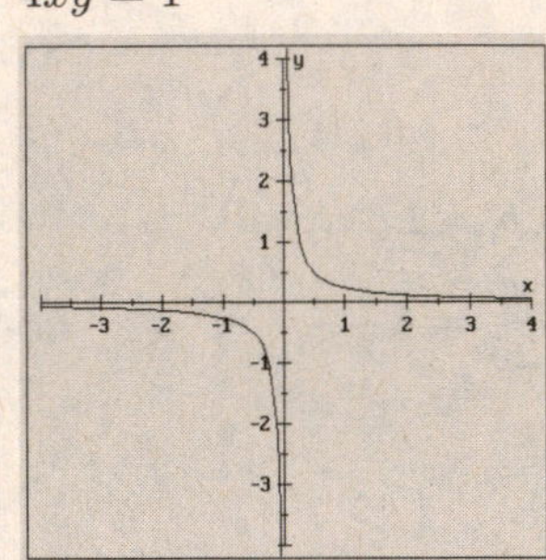

25. $\begin{cases} x^2 + y^2 = 16 \\ y = x + 4 \end{cases}$

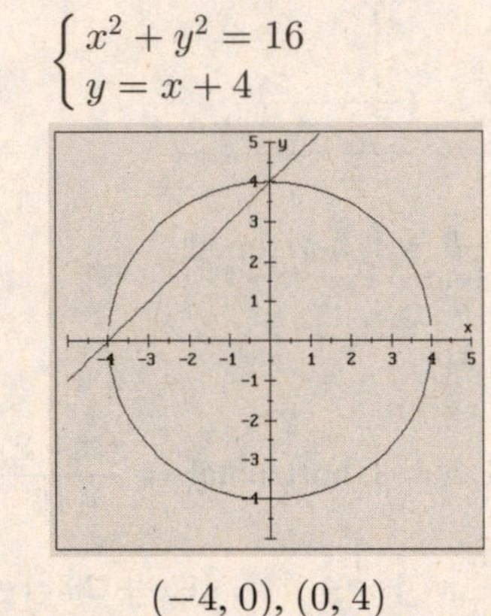

$(-4, 0), (0, 4)$

26.

$\begin{cases} 3x^2 + y^2 = 52 \\ x^2 - y^2 = 12 \end{cases}$

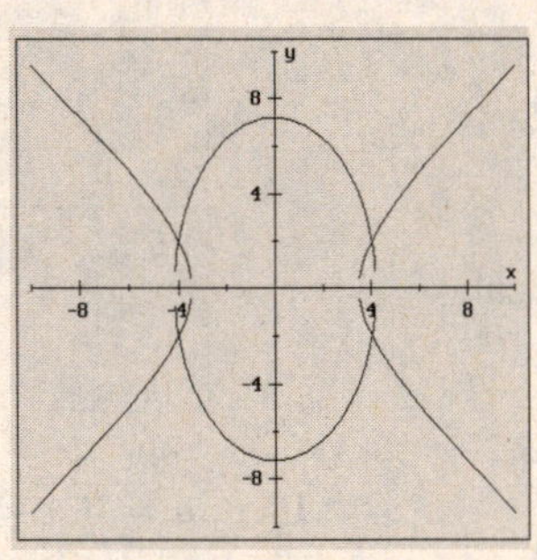

$(-4, 2), (4, 2), (-4, -2), (4, -2)$

27. $\begin{cases} \dfrac{x^2}{16} + \dfrac{y^2}{12} = 1 \\ x^2 - \dfrac{y^2}{3} = 1 \end{cases}$

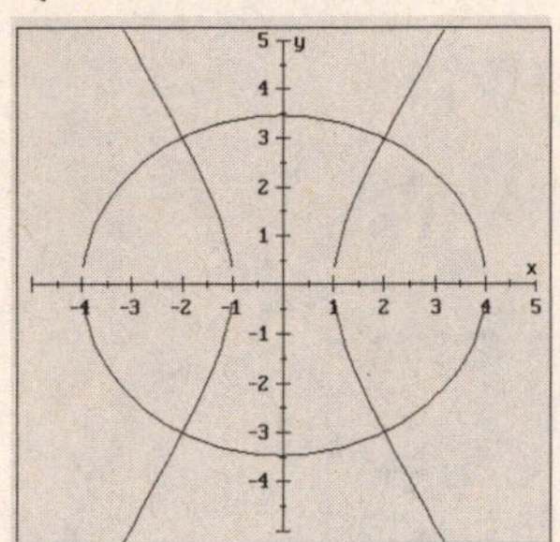

$(-2, 3), (2, 3), (-2, -3), (2, -3)$

28.
$$\begin{aligned} 3x^2 + y^2 &= 52 \\ x^2 - y^2 &= 12 \\ \hline 4x^2 &= 64 \\ x^2 &= 16 \\ x &= \pm 4 \end{aligned} \qquad \begin{aligned} 3x^2 + y^2 &= 52 \\ 3(16) + y^2 &= 52 \\ y^2 &= 4 \\ y &= \pm 2 \end{aligned}$$

$(4, 2), (-4, 2), (4, -2), (-4, -2)$

29.
$$-\sqrt{3}y + 4\sqrt{3} = 3x \Rightarrow x = \frac{-\sqrt{3}y + 4\sqrt{3}}{3}$$
$$x^2 + y^2 = 16$$
$$\left(\frac{-\sqrt{3}y + 4\sqrt{3}}{3}\right)^2 + y^2 = 16$$
$$3y^2 - 24y + 48 + 9y^2 = 144$$
$$12y^2 - 24y - 96 = 0$$
$$12(y - 4)(y + 2) = 0$$

$y = 4$
$$x = \frac{-\sqrt{3}(4) + 4\sqrt{3}}{3} = 0 \Rightarrow (0, 4)$$

$y = -2$
$$x = \frac{-\sqrt{3}(-2) + 4\sqrt{3}}{3} = 2\sqrt{3} \Rightarrow \left(2\sqrt{3}, -2\right)$$

30.
$$\begin{aligned} \tfrac{x^2}{16} + \tfrac{y^2}{12} &= 1 \\ x^2 - \tfrac{y^2}{3} &= 1 \\ \hline 3x^2 + 4y^2 &= 48 \\ 3x^2 - y^2 &= 3 \\ \hline 5y^2 &= 45 \end{aligned} \qquad \begin{aligned} 5y^2 &= 45 \\ y^2 &= 9 \\ y &= \pm 3 \end{aligned} \qquad \begin{aligned} 3x^2 - y^2 &= 3 \\ 3x^2 - 9 &= 3 \\ 3x^2 &= 12 \\ x^2 &= 4 \\ x &= \pm 2 \Rightarrow (2, 3), (-2, 3), (2, -3), (-2, -3) \end{aligned}$$

Chapter 7 Test (page 766)

1.
$$(x - h)^2 + (y - k)^2 = r^2$$
$$(x - 2)^2 + (y - 3)^2 = 3^2$$
$$(x - 2)^2 + (y - 3)^2 = 9$$

2.
$$C\left(\frac{-2 + 6}{2}, \frac{-2 + 8}{2}\right) = C(2, 3)$$
$$r = \sqrt{(6 - 2)^2 + (8 - 3)^2} = \sqrt{41}$$
$$(x - h)^2 + (y - k)^2 = r^2$$
$$(x - 2)^2 + (y - 3)^2 = \left(\sqrt{41}\right)^2$$
$$(x - 2)^2 + (y - 3)^2 = 41$$

CHAPTER 7 TEST

3. $r = \sqrt{(7-2)^2 + (7-(-5))^2} = 13$

$$(x-h)^2 + (y-k)^2 = r^2$$
$$(x-2)^2 + (y-(-5))^2 = 13^2$$
$$(x-2)^2 + (y+5)^2 = 169$$

4.
$$x^2 + y^2 - 4x + 6y + 4 = 0$$
$$x^2 - 4x + y^2 + 6y = -4$$
$$x^2 - 4x + 4 + y^2 + 6y + 9 = -4 + 4 + 9$$
$$(x-2)^2 + (y+3)^2 = 9$$
$$C(2, -3),\ r = 3$$

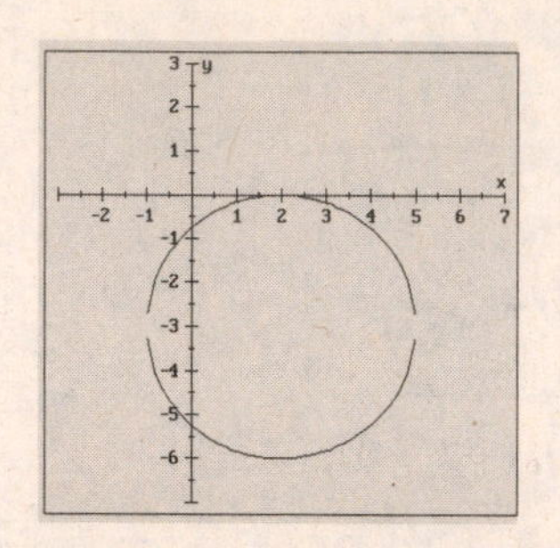

5. Vertical (up), $p = 4$

$$(x-h)^2 = 4p(y-k)$$
$$(x-3)^2 = 4(4)(y-2)$$
$$(x-3)^2 = 16(y-2)$$

6. Horizontal

$$(y+6)^2 = 4p(x-4)$$
$$(-4+6)^2 = 4p(3-4)$$
$$4 = -4p$$
$$-4 = 4p$$
$$(y+6)^2 = -4(x-4)$$

7.
$$(x-2)^2 = 4p(y+3) \quad \textbf{OR} \quad (y+3)^2 = 4p(x-2)$$
$$(0-2)^2 = 4p(0+3) \qquad (0+3)^2 = 4p(0-2)$$
$$4 = 4p(3) \qquad 9 = 4p(-2)$$
$$\frac{4}{3} = 4p \qquad -\frac{9}{2} = 4p$$
$$(x-2)^2 = \tfrac{4}{3}(y+3) \qquad (y+3)^2 = -\tfrac{9}{2}(x-2)$$

8.

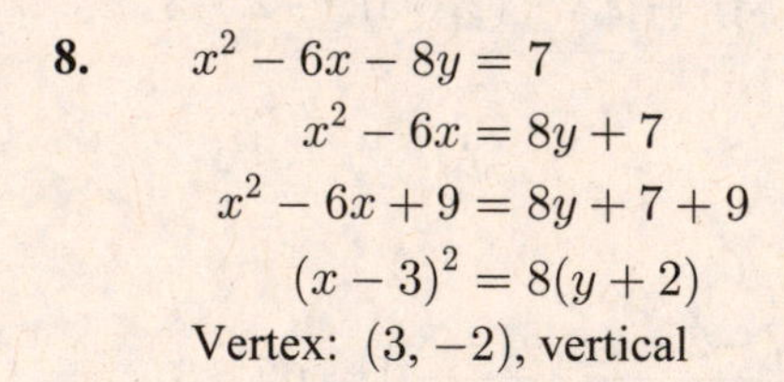

$$x^2 - 6x - 8y = 7$$
$$x^2 - 6x = 8y + 7$$
$$x^2 - 6x + 9 = 8y + 7 + 9$$
$$(x-3)^2 = 8(y+2)$$

Vertex: $(3, -2)$, vertical

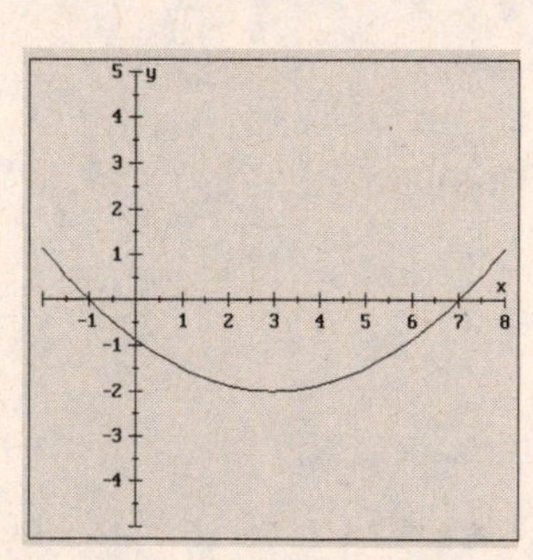

CHAPTER 7 TEST

9. $a = 10, c = 6$, horizontal
$$b^2 = a^2 - c^2$$
$$= 100 - 36 = 64$$
$$\frac{x^2}{100} + \frac{y^2}{64} = 1$$

10. $b = 12, c = 5$, horizontal
$$a^2 = b^2 + c^2$$
$$= 144 + 25 = 169$$
$$\frac{x^2}{169} + \frac{y^2}{144} = 1$$

11. $a = 6, b = 2$, vertical
$$\frac{(x-2)^2}{4} + \frac{(y-3)^2}{36} = 1$$

12.
$$9x^2 + 4y^2 - 18x - 16y - 11 = 0$$
$$9(x^2 - 2x) + 4(y^2 - 4y) = 11$$
$$9(x^2 - 2x + 1) + 4(y^2 - 4y + 4) = 11 + 9 + 16$$
$$9(x-1)^2 + 4(y-2)^2 = 36$$
$$\frac{(x-1)^2}{4} + \frac{(y-2)^2}{9} = 1$$
Center: $(1, 2)$, $a = 3$, $b = 2$, vertical

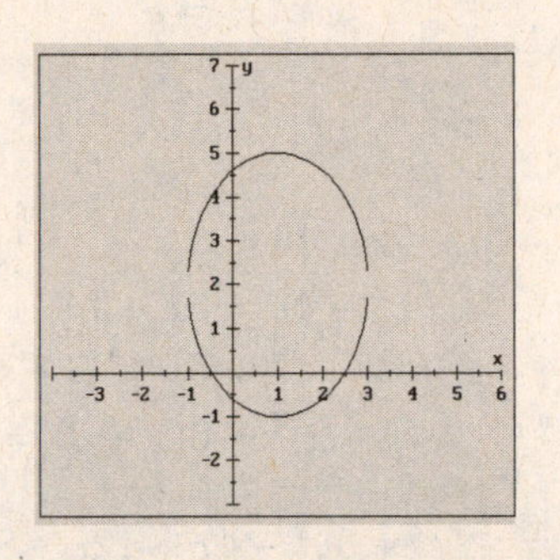

13. $a = 5, c = 13$; horizontal
$$b^2 = c^2 - a^2$$
$$= 169 - 25 = 144$$
$$\frac{x^2}{25} - \frac{y^2}{144} = 1$$

14. $C(0, 0)$, $a = 6$, $c = \frac{13}{2}$
horizontal
$$b^2 = c^2 - a^2$$
$$= \tfrac{169}{4} - 36 = \tfrac{25}{4}$$
$$\frac{x^2}{36} - \frac{y^2}{\frac{25}{4}} = 1$$

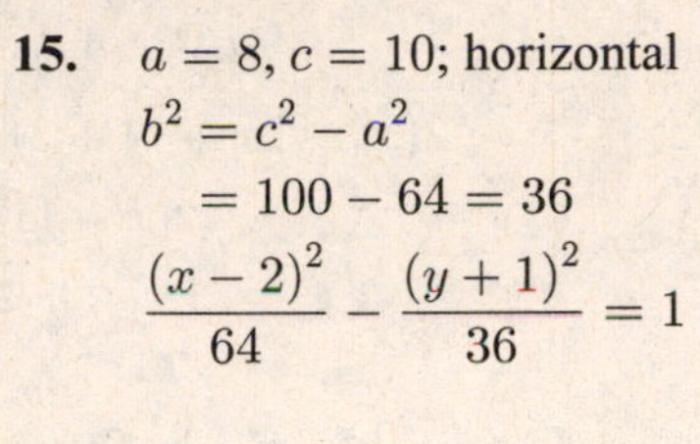

15. $a = 8, c = 10$; horizontal
$$b^2 = c^2 - a^2$$
$$= 100 - 64 = 36$$
$$\frac{(x-2)^2}{64} - \frac{(y+1)^2}{36} = 1$$

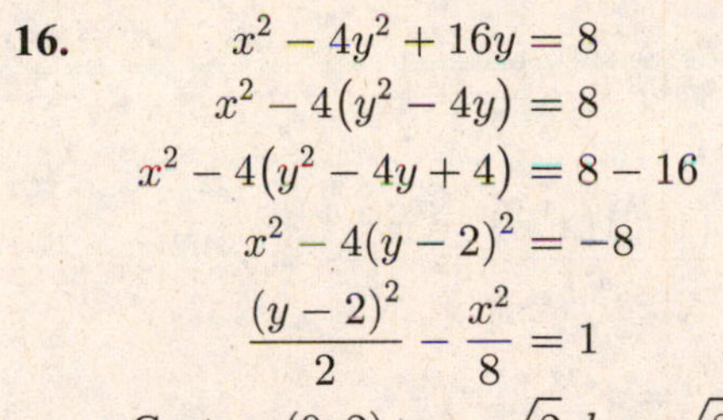

16.
$$x^2 - 4y^2 + 16y = 8$$
$$x^2 - 4(y^2 - 4y) = 8$$
$$x^2 - 4(y^2 - 4y + 4) = 8 - 16$$
$$x^2 - 4(y-2)^2 = -8$$
$$\frac{(y-2)^2}{2} - \frac{x^2}{8} = 1$$
Center: $(0, 2)$, $a = \sqrt{2}$, $b = \sqrt{8}$, vertical

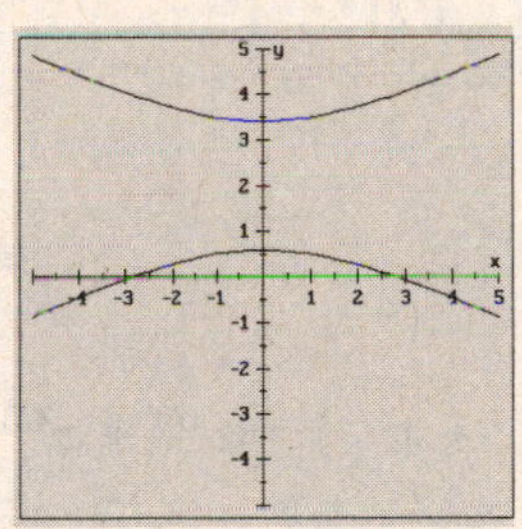

17.
$$y = x^2 - 3 \Rightarrow x^2 = y + 3$$
$$x^2 + y^2 = 23$$
$$y + 3 + y^2 = 23$$
$$y^2 + y - 20 = 0$$
$$(y-4)(y+5) = 0$$

$y = 4$
$$4 + 3 = x^2 \Rightarrow x = \pm\sqrt{7}$$
$$\left(\sqrt{7}, 4\right), \left(-\sqrt{7}, 4\right)$$

$y = -5$
$$-5 + 3 = x^2$$
no real solutions

18.

$$\begin{aligned} x^2+y^2=27 \Rightarrow x^2 &= 27-y^2 \\ 2x^2-3y^2 &= 9 \\ 2(27-y^2)-3y^2 &= 9 \\ 54-2y^2-3y^2 &= 9 \\ 45 &= 5y^2 \\ 9 &= y^2 \end{aligned}$$

$$\frac{y=3}{x^2=27-3^2 \Rightarrow x=\pm 3\sqrt{2}}$$
$$\left(3\sqrt{2},3\right), \left(-3\sqrt{2},3\right)$$

$$\frac{y=-3}{x^2=27-(-3)^2 \Rightarrow x=\pm 3\sqrt{2}}$$
$$\left(3\sqrt{2},-3\right), \left(-3\sqrt{2},-3\right)$$

19.

$$\begin{aligned} y^2-4y-6x-14 &= 0 \\ y^2-4y &= 6x+14 \\ y^2-4x+4 &= 6x+14+4 \\ (y-2)^2 &= 6(x+3) \Rightarrow \text{Parabola} \end{aligned}$$

20.

$$\begin{aligned} 2x^2+3y^2-4x+12y+8 &= 0 \\ 2(x^2-2x)+3(y^2+4y) &= -8 \\ 2(x^2-2x+1)+3(y^2+4y+4) &= -8+2+12 \\ 2(x-1)^2+3(y+2)^2 &= 6 \\ \frac{(x-1)^2}{3}+\frac{(y+2)^2}{2} &= 1 \Rightarrow \text{ellipse} \end{aligned}$$

Cumulative Review Exercises (page 767)

1. $64^{2/3}=\left(64^{1/3}\right)^2=4^2=16$

2. $8^{-1/3}=\frac{1}{8^{1/3}}=\frac{1}{2}$

3. $\frac{y^{2/3}y^{5/3}}{y^{1/3}}=\frac{y^{7/3}}{y^{1/3}}=y^{6/3}=y^2$

4. $\frac{\left(x^{5/3}\right)\left(x^{1/2}\right)}{x^{3/4}}=\frac{x^{13/6}}{x^{3/4}}=x^{34/24}=x^{17/12}$

5. $\left(x^{2/3}-x^{1/3}\right)\left(x^{2/3}+x^{1/3}\right)=x^{4/3}+x^{3/3}-x^{3/3}-x^{2/3}=x^{4/3}-x^{2/3}$

6. $\left(x^{-1/2}+x^{1/2}\right)^2=\left(x^{-1/2}+x^{1/2}\right)\left(x^{-1/2}+x^{1/2}\right)=x^{-2/2}+x^0+x^0+x^{2/2}=\frac{1}{x}+2+x$

7. $\sqrt[3]{-27x^3}=\sqrt[3]{(-3x)^3}=-3x$

8. $\sqrt{48t^3}=\sqrt{16t^2}\sqrt{3t}=4t\sqrt{3t}$

9. $\sqrt[3]{\frac{128x^4}{2x}}=\sqrt[3]{64x^3}=4x$

10. $\sqrt{x^2+6x+9}=\sqrt{(x+3)^2}=x+3$

11. $\sqrt{50}-\sqrt{8}+\sqrt{32}=5\sqrt{2}-2\sqrt{2}+4\sqrt{2}=7\sqrt{2}$

12. $-3\sqrt[4]{32}-2\sqrt[4]{162}+5\sqrt[4]{48}=-3\cdot 2\sqrt[4]{2}-2\cdot 3\sqrt[4]{2}+5\cdot 2\sqrt[4]{3}=-12\sqrt[4]{2}+10\sqrt[4]{3}$

13. $3\sqrt{2}\left(2\sqrt{3}-4\sqrt{12}\right)=6\sqrt{6}-12\sqrt{24}=6\sqrt{6}-12\cdot 2\sqrt{6}=-18\sqrt{6}$

14. $\dfrac{5}{\sqrt[3]{x}}=\dfrac{5\sqrt[3]{x^2}}{\sqrt[3]{x}\sqrt[3]{x^2}}=\dfrac{5\sqrt[3]{x^2}}{x}$

15. $\dfrac{\sqrt{x}+2}{\sqrt{x}-1}=\dfrac{(\sqrt{x}+2)(\sqrt{x}+1)}{(\sqrt{x}-1)(\sqrt{x}+1)}$
$=\dfrac{x+3\sqrt{x}+2}{x-1}$

16. $\sqrt[6]{x^3y^3}=(x^3y^3)^{1/6}=x^{3/6}y^{3/6}=x^{1/2}y^{1/2}=(xy)^{1/2}=\sqrt{xy}$

17.
$$\begin{aligned}
5\sqrt{x+2}&=x+8\\
\left(5\sqrt{x+2}\right)^2&=(x+8)^2\\
25(x+2)&=x^2+16x+64\\
25x+50&=x^2+16x+64\\
0&=x^2-9x+14\\
0&=(x-2)(x-7)\\
x=2 \text{ or } x&=7 \text{ (both check)}
\end{aligned}$$

18.
$$\begin{aligned}
\sqrt{x}+\sqrt{x+2}&=2\\
\sqrt{x+2}&=2-\sqrt{x}\\
\left(\sqrt{x+2}\right)^2&=\left(2-\sqrt{x}\right)^2\\
x+2&=4-4\sqrt{x}+x\\
4\sqrt{x}&=2\\
\left(4\sqrt{x}\right)^2&=2^2\\
16x&=4\\
x&=\frac{1}{4}
\end{aligned}$$

19.
$$\begin{aligned}
2x^2+x-3&=0\\
x^2+\frac{1}{2}x&=\frac{3}{2}\\
x^2+\frac{1}{2}x+\frac{1}{16}&=\frac{24}{16}+\frac{1}{16}\\
\left(x+\tfrac{1}{4}\right)^2&=\tfrac{25}{16}\\
x+\tfrac{1}{4}&=\pm\tfrac{5}{4}\\
x&=-\tfrac{1}{4}\pm\tfrac{5}{4}\\
x=1 \text{ or } x&=-\tfrac{3}{2}
\end{aligned}$$

20. $3x^2+4x-1=0\Rightarrow a=3, b=4, c=-1$
$$\begin{aligned}
x&=\frac{-b\pm\sqrt{b^2-4ac}}{2a}\\
&=\frac{-4\pm\sqrt{4^2-4(3)(-1)}}{2(3)}\\
&=\frac{-4\pm\sqrt{16+12}}{6}\\
&=\frac{-4\pm\sqrt{28}}{6}=\frac{-2\pm\sqrt{7}}{3}
\end{aligned}$$

21. $(3+5i)+(4-3i)=3+5i+4-3i=7+2i$

22. $(7-4i)-(12+3i)=7-4i-12-3i=-5-7i$

23. $(2-3i)(2+3i)=4+6i-6i-9i^2=4-9(-1)=4+9=13+0i$

24. $(3+i)(3-3i)=9-9i+3i-3i^2=9-6i-3(-1)=9-6i+3=12-6i$

25. $(3-2i)-(4+i)^2=3-2i-\left(16+8i+i^2\right)=3-2i-(15+8i)=3-2i-15-8i$
$=-12-10i$

26. $\frac{5}{3-i} = \frac{5(3+i)}{(3-i)(3+i)} = \frac{5(3+i)}{9-i^2} = \frac{5(3+i)}{10} = \frac{3+i}{2} = \frac{3}{2} + \frac{1}{2}i$

27. $|3+2i| = \sqrt{3^2+2^2} = \sqrt{13}$

28. $|5-6i| = \sqrt{5^2+(-6)^2} = \sqrt{61}$

29. $2x^2 + 4x = k \Rightarrow 2x^2 + 4x - k = 0$

$a = 2, b = 4, c = -k$: Set $b^2 - 4ac = 0$.

$$\begin{aligned} b^2 - 4ac &= 0 \\ 4^2 - 4(2)(-k) &= 0 \\ 16 + 8k &= 0 \\ k &= -2 \end{aligned}$$

30. $y = \frac{1}{2}x^2 - x + 1$: $a = \frac{1}{2}, b = -1, c = 1$

$$x = -\frac{b}{2a} = -\frac{-1}{2\left(\frac{1}{2}\right)} = 1$$

$$y = \frac{1}{2}(1)^2 - 1 + 1 = \frac{1}{2}$$

31.

$$\begin{aligned} x^2 - x - 6 &> 0 \\ (x+2)(x-3) &> 0 \end{aligned}$$

factors = 0: $x = -2, x = 3$

intervals: $(-\infty, -2), (-2, 3), (3, \infty)$

interval	test number	value of x^2-x-6
$(-\infty, -2)$	-3	$+6$
$(-2, 3)$	0	-6
$(3, \infty)$	4	$+6$

Solution: $(-\infty, -2) \cup (3, \infty)$

32.

$$\begin{aligned} x^2 - x - 6 &\le 0 \\ (x+2)(x-3) &\le 0 \end{aligned}$$

factors = 0: $x = -2, x = 3$

intervals: $(-\infty, -2), (-2, 3), (3, \infty)$

interval	test number	value of x^2-x-6
$(-\infty, -2)$	-3	$+6$
$(-2, 3)$	0	-6
$(3, \infty)$	4	$+6$

Solution: $[-2, 3]$

33. $f(-1) = 3(-1)^2 + 2 = 3 + 2 = 5$

34.

$$\begin{aligned} (g \circ f)(2) = g(f(2)) &= g\left(3(2)^2 + 2\right) \\ &= g(14) \\ &= 2(14) - 1 = 27 \end{aligned}$$

35.

$$\begin{aligned} (f \circ g)(x) &= f(g(x)) \\ &= f(2x-1) \\ &= 3(2x-1)^2 + 2 \\ &= 3\left(4x^2 - 4x + 1\right) + 2 \\ &= 12x^2 - 12x + 3 + 2 \\ &= 12x^2 - 12x + 5 \end{aligned}$$

36.

$$\begin{aligned} (g \circ f)(x) &= g(f(x)) \\ &= g\left(3x^2 + 2\right) \\ &= 2\left(3x^2 + 2\right) - 1 \\ &= 6x^2 + 4 - 1 \\ &= 6x^2 + 3 \end{aligned}$$

37. $y = \log_2 x \Rightarrow 2^y = x$

38. $3^b = a \Rightarrow \log_3 a = b$

39. $\log_x 25 = 2 \Rightarrow x^2 = 25 \Rightarrow x = 5$

40. $\log_5 125 = x \Rightarrow 5^x = 125 \Rightarrow x = 3$

41. $\log_3 x = -3 \Rightarrow 3^{-3} = x \Rightarrow x = \frac{1}{27}$

42. $\log_5 x = 0 \Rightarrow 5^0 = x \Rightarrow x = 1$

43. $y = \log_2 x$; inverse: $y = 2^x$

44. $\log_{10} 10^x = x$, so $y = x$.

45. $\log 98 = \log(14 \cdot 7) = \log 14 + \log 7 = 1.1461 + 0.8451 = 1.9912$

CUMULATIVE REVIEW EXERCISES

46. $\log 2 = \log \frac{14}{7} = \log 14 - \log 7 = 1.1461 - 0.8451 = 0.3010$

47. $\log 49 = \log 7^2 = 2 \log 7 = 2(0.8451) = 1.6902$

48. $\log \frac{7}{5} = \log \frac{14}{10} = \log 14 - \log 10 = 1.14641 - 1 = 0.1461$

49.
$$\begin{aligned} 2^{x+2} &= 3^x \\ \log 2^{x+2} &= \log 3^x \\ (x+2)\log 2 &= x \log 3 \\ x \log 2 + 2 \log 2 &= x \log 3 \\ 2 \log 2 &= x \log 3 - x \log 2 \\ 2 \log 2 &= x(\log 3 - \log 2) \\ \frac{2 \log 2}{\log 3 - \log 2} &= x \end{aligned}$$

50.
$$\begin{aligned} 2 \log 5 + \log x - \log 4 &= 2 \\ \log 5^2 + \log x - \log 4 &= 2 \\ \log \frac{25x}{4} &= 2 \\ 10^2 &= \frac{25x}{4} \\ 400 &= 25x \\ 16 &= x \end{aligned}$$

51.
$$\begin{aligned} A &= A_0\left(1 + \frac{r}{k}\right)^{kt} \\ &= 9000\left(1 + \frac{-0.12}{1}\right)^{1(9)} \\ &\approx \$2848.31 \end{aligned}$$

52. $\log_6 8 = \frac{\log 8}{\log 6} \approx 1.16056$

53. $\begin{cases} 2x + y = 5 \\ x - 2y = 0 \end{cases}$

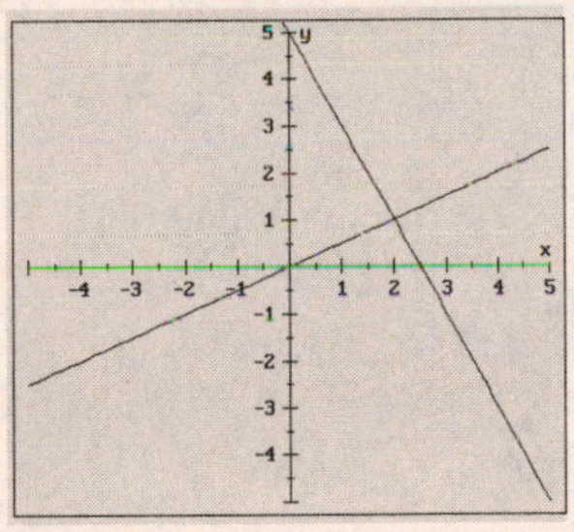

solution: $(2, 1)$

54. $\begin{cases} (1) & 3x + y = 4 \\ (2) & 2x - 3y = -1 \end{cases}$

Substitute $y = -3x + 4$ from (1) into (2):
$$\begin{aligned} 2x - 3y &= -1 \\ 2x - 3(-3x + 4) &= -1 \\ 2x + 9x - 12 &= -1 \\ 11x &= 11 \\ x &= 1 \end{aligned}$$

Substitute and solve for y:

$y = -3x + 4 = -3(1) + 4 = 1$

$\boxed{x = 1, y = 1}$

55.
$$\begin{aligned} x + 2y &= -2 \\ 2x - y &= 6 \Rightarrow \times 2 \end{aligned} \qquad \begin{aligned} x + 2y &= -2 \\ 4x - 2y &= 12 \\ \hline 5x &= 10 \\ x &= 2 \end{aligned} \qquad \begin{aligned} 2x - y &= 6 \\ 2(2) - y &= 6 \\ -y &= 2 \\ y &= -2 \end{aligned}$$

Solution: $\boxed{x = 2, y = -2}$

CUMULATIVE REVIEW EXERCISES

56. $\begin{array}{l} \dfrac{x}{10} + \dfrac{y}{5} = \dfrac{1}{2} \Rightarrow \times 10 \\ \dfrac{x}{2} - \dfrac{y}{5} = \dfrac{13}{10} \Rightarrow \times 10 \end{array}$ $\begin{array}{rl} x + 2y = & 5 \\ 5x - 2y = & 13 \\ \hline 6x \quad = & 18 \\ x \quad = & 3 \end{array}$ $\begin{array}{r} x + 2y = 5 \\ 3 + 2y = 5 \\ 2y = 2 \\ y = 1 \end{array}$ Solution: $\boxed{x = 3,\ y = 1}$

57. $\begin{vmatrix} 3 & -2 \\ 1 & -1 \end{vmatrix} = 3(-1) - (-2)1$
$= -3 + 2 = -1$

58. $y = \dfrac{\begin{vmatrix} 4 & -1 \\ 3 & -7 \end{vmatrix}}{\begin{vmatrix} 4 & -3 \\ 3 & 4 \end{vmatrix}} = \dfrac{-25}{25} = -1$

59. $$\begin{bmatrix} x \\ y \\ z \end{bmatrix} = \begin{bmatrix} 1 & 1 & 1 \\ 2 & -1 & -1 \\ 1 & -2 & 1 \end{bmatrix}^{-1} \begin{bmatrix} 1 \\ -4 \\ 4 \end{bmatrix} = \begin{bmatrix} \frac{1}{3} & \frac{1}{3} & 0 \\ \frac{1}{3} & 0 & -\frac{1}{3} \\ \frac{1}{3} & -\frac{1}{3} & \frac{1}{3} \end{bmatrix} \begin{bmatrix} 1 \\ -4 \\ 4 \end{bmatrix} = \begin{bmatrix} -1 \\ -1 \\ 3 \end{bmatrix}$$

60. $$\begin{bmatrix} x \\ y \\ z \end{bmatrix} = \begin{bmatrix} 1 & 2 & 3 \\ 3 & 2 & 1 \\ 2 & 3 & 1 \end{bmatrix}^{-1} \begin{bmatrix} 6 \\ 6 \\ 6 \end{bmatrix} = \begin{bmatrix} -\frac{1}{12} & \frac{7}{12} & -\frac{1}{3} \\ -\frac{1}{12} & -\frac{5}{12} & \frac{2}{3} \\ \frac{5}{12} & \frac{1}{12} & -\frac{1}{3} \end{bmatrix} \begin{bmatrix} 6 \\ 6 \\ 6 \end{bmatrix} = \begin{bmatrix} 1 \\ 1 \\ 1 \end{bmatrix}$$

61. $(y-3)^2 = 8(x+3)$
Vertex: $(-3, 3)$

62. 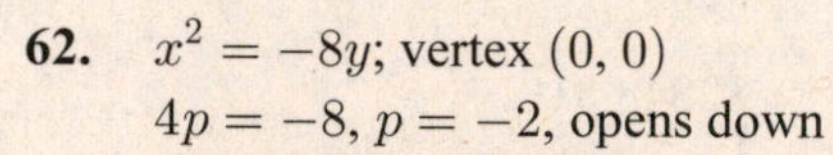
$x^2 = -8y$; vertex $(0, 0)$
$4p = -8$, $p = -2$, opens down

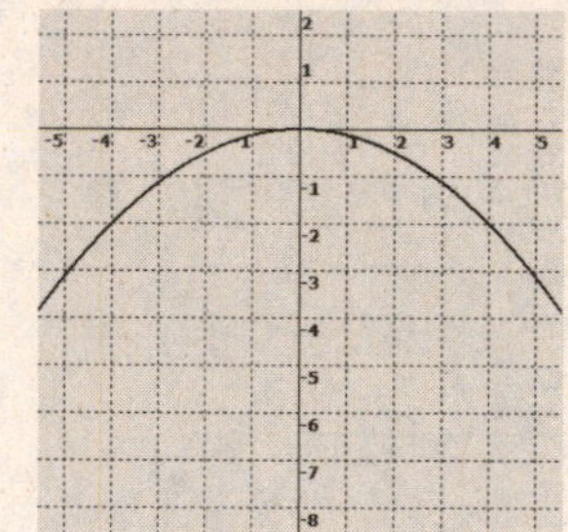

63. $\dfrac{(x-1)^2}{9} + \dfrac{(y-3)^2}{25} = 1$
Center: $(1, 3)$, $a = 5$, $b = 3$, vertical

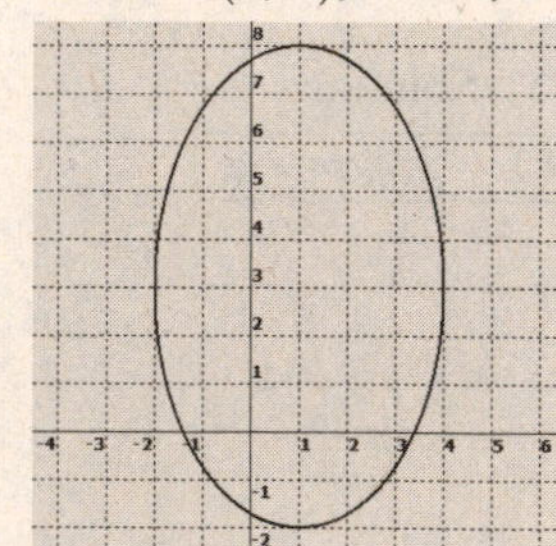

64. $\dfrac{(y-3)^2}{9} - \dfrac{(x-2)^2}{16} = 1$
Center: $(2, 3)$, $a = 3$, $b = 4$, vertical

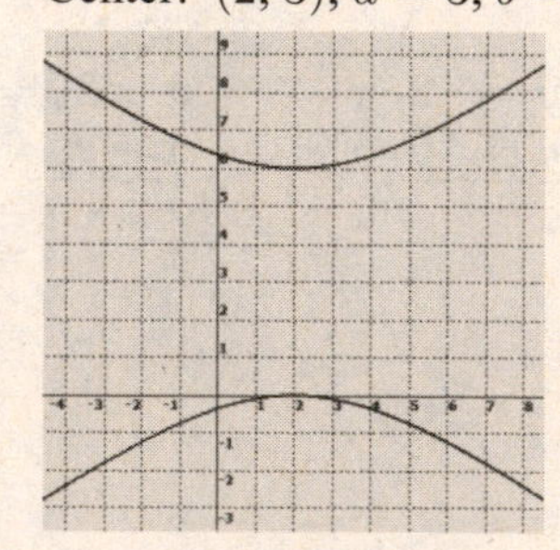

Exercises 8.1 (page 775)

1. power

3. first

5. $7 \cdot 6 \cdot 5 \cdot 4 \cdot 3 \cdot 2 \cdot 1$

7. $(n-1)!$

9. $5! = 5 \cdot 4 \cdot 3 \cdot 2 \cdot 1 = 120$

11. $3! \cdot 6! = 6 \cdot 720 = 4320$

13. $6! + 6! = 720 + 720 = 1440$

15. $\dfrac{9!}{12!} = \dfrac{9!}{12 \cdot 11 \cdot 10 \cdot 9!} = \dfrac{1}{1320}$

17. $\dfrac{5! \cdot 7!}{9!} = \dfrac{5! \cdot 7!}{9 \cdot 8 \cdot 7!} = \dfrac{120}{72} = \dfrac{5}{3}$

19. $\dfrac{18!}{6!(18-6)!} = \dfrac{18!}{6! \cdot 12!} = \dfrac{18 \cdot 17 \cdot 16 \cdot 15 \cdot 14 \cdot 13 \cdot 12!}{6! \cdot 12!} = \dfrac{13{,}366{,}080}{720} = 18{,}564$

21. Row 5 of Pascal's triangle: 1 5 10 10 5 1
$(a+b)^5 = a^5 + 5a^4b + 10a^3b^2 + 10a^2b^3 + 5ab^4 + b^5$

23. Row 3 of Pascal's triangle: 1 3 3 1
$(x-y)^3 = x^3 + 3x^2(-y) + 3x(-y)^2 + (-y)^3 = x^3 - 3x^2y + 3xy^2 - y^3$

25. $(a+b)^3 = a^3 + \dfrac{3!}{1!2!}a^2b + \dfrac{3!}{2!1!}ab^2 + b^3 = a^3 + 3a^2b + 3ab^2 + b^3$

27. $(a-b)^5 = a^5 + \dfrac{5!}{1!4!}a^4(-b) + \dfrac{5!}{2!3!}a^3(-b)^2 + \dfrac{5!}{3!2!}a^2(-b)^3 + \dfrac{5!}{4!1!}a(-b)^4 + (-b)^5$
$= a^5 - 5a^4b + 10a^3b^2 - 10a^2b^3 + 5ab^4 - b^5$

29. $(2x+y)^3 = (2x)^3 + \dfrac{3!}{1!2!}(2x)^2y + \dfrac{3!}{2!1!}(2x)y^2 + y^3 = 8x^3 + 12x^2y + 6xy^2 + y^3$

31. $(x-2y)^3 = x^3 + \dfrac{3!}{1!2!}x^2(-2y) + \dfrac{3!}{2!1!}x(-2y)^2 + (-2y)^3 = x^3 - 6x^2y + 12xy^2 - 8y^3$

33. $(2x+3y)^4 = (2x)^4 + \dfrac{4!}{1!3!}(2x)^3(3y) + \dfrac{4!}{2!2!}(2x)^2(3y)^2 + \dfrac{4!}{3!1!}(2x)(3y)^3 + (3y)^4$
$= 16x^4 + 96x^3y + 216x^2y^2 + 216xy^3 + 81y^4$

35. $(x-2y)^4 = x^4 + \dfrac{4!}{1!3!}x^3(-2y) + \dfrac{4!}{2!2!}x^2(-2y)^2 + \dfrac{4!}{3!1!}x(-2y)^3 + (-2y)^4$
$= x^4 - 8x^3y + 24x^2y^2 - 32xy^3 + 16y^4$

37. $(x-3y)^5 = x^5 + \dfrac{5!}{1!4!}x^4(-3y) + \dfrac{5!}{2!3!}x^3(-3y)^2 + \dfrac{5!}{3!2!}x^2(-3y)^3 + \dfrac{5!}{4!1!}x(-3y)^4 + (-3y)^5$
$= x^5 - 15x^4y + 90x^3y^2 - 270x^2y^3 + 405xy^4 - 243y^5$

39. $\left(\frac{x}{2}+y\right)^4 = \left(\frac{x}{2}\right)^4 + \frac{4!}{1!3!}\left(\frac{x}{2}\right)^3 y + \frac{4!}{2!2!}\left(\frac{x}{2}\right)^2 y^2 + \frac{4!}{3!1!}\left(\frac{x}{2}\right) y^3 + y^4$
$= \frac{1}{16}x^4 + \frac{1}{2}x^3y + \frac{3}{2}x^2y^2 + 2xy^3 + y^4$

41. The 3rd term will involve b^2.
$\frac{4!}{2!2!}a^2b^2 = 6a^2b^2$

43. The 5th term will involve b^4.
$\frac{7!}{3!4!}a^3b^4 = 35a^3b^4$

45. The 6th term will involve $(-b)^5$.
$\frac{5!}{0!5!}a^0(-b)^5 = -b^5$

47. The 5th term will involve b^4.
$\frac{17!}{13!4!}a^{13}b^4 = 2380a^{13}b^4$

49. The 2nd term will involve $\left(-\sqrt{2}\right)^1$.
$\frac{4!}{3!1!}a^3\left(-\sqrt{2}\right)^1 = -4\sqrt{2}\,a^3$

51. The 5th term will involve $\left(\sqrt{3}b\right)^4$.
$\frac{9!}{5!4!}a^5\left(\sqrt{3}b\right)^4 = 1134a^5b^4$

53. The 3rd term will involve y^2.
$\frac{4!}{2!2!}\left(\frac{x}{2}\right)^2 y^2 = \frac{3}{2}x^2y^2$

55. The 10th term will involve $\left(-\frac{s}{2}\right)^9$.
$\frac{11!}{2!9!}\left(\frac{r}{2}\right)^2\left(-\frac{s}{2}\right)^9 = -\frac{55}{2048}r^2s^9$

57. The 4th term will involve b^3.
$\frac{n!}{(n-3)!3!}a^{n-3}b^3$

59. The rth term will involve b^{r-1}.
$\frac{n!}{(n-r+1)!(r-1)!}a^{n-r+1}b^{r-1}$

61-67. Answers may vary.

69. $\frac{n!}{0!(n-0)!} = \frac{n!}{0!n!} = \frac{n!}{1 \cdot n!} = 1$

71. False. $0! = 1$

73. True.

75. False. The exponents on b increase by 1 in each successive term.

77. False. The number of terms is 445.

79. False. The constant terms is -252.

Exercises 8.2 (page 786)

1. domain

3. series

5. infinite

7. Summation notation

9. 6

11. $5c$

13. $f(1) = 5(1)(1-1) = 0$ $\quad f(2) = 5(2)(2-1) = 10$ $\quad f(3) = 5(3)(3-1) = 30$
$f(4) = 5(4)(4-1) = 60$ $\quad f(5) = 5(5)(5-1) = 100$ $\quad f(6) = 5(6)(6-1) = 150$

EXERCISES 8.2

15. 1, 6, 11, 16, ... Add 5 to get the next term. The next term is 21.

17. $a, a+d, a+2d, a+3d, \ldots$ Add d to get the next term. The next term is $a+4d$.

19. 1, 3, 6, 10, ... The difference between terms increases by 1 each time. The next term is $10+5=15$.

21.
$$\begin{aligned} a_1 &= 9(1)-1=8 \\ a_2 &= 9(2)-1=17 \\ a_3 &= 9(3)-1=26 \\ a_4 &= 9(4)-1=35 \\ a_5 &= 9(5)-1=44 \\ a_{30} &= 9(30)-1=269 \end{aligned}$$

23.
$$\begin{aligned} a_1 &= -(1)^2+5=4 \\ a_2 &= -(2)^2+5=1 \\ a_3 &= -(3)^2+5=-4 \\ a_4 &= -(4)^2+5=-11 \\ a_5 &= -(5)^2+5=-20 \\ a_{20} &= -(20)^2+5=-395 \end{aligned}$$

25.
$$\begin{aligned} a_1 &= (1)^3+6=7 \\ a_2 &= (2)^3+6=14 \\ a_3 &= (3)^3+6=33 \\ a_4 &= (4)^3+6=70 \\ a_5 &= (5)^3+6=131 \\ a_{10} &= (10)^3+6=1006 \end{aligned}$$

27.
$$\begin{aligned} a_1 &= \frac{(1)-1}{(1)}=0 \\ a_2 &= \frac{(2)-1}{(2)}=\frac{1}{2} \\ a_3 &= \frac{(3)-1}{(3)}=\frac{2}{3} \\ a_4 &= \frac{(4)-1}{(4)}=\frac{3}{4} \\ a_5 &= \frac{(5)-1}{(5)}=\frac{4}{5} \\ a_{30} &= \frac{(30)-1}{(30)}=\frac{29}{30} \end{aligned}$$

29.
$$\begin{aligned} a_1 &= \frac{(-1)^1}{4^1}=-\frac{1}{4} \\ a_2 &= \frac{(-1)^2}{4^2}=\frac{1}{16} \\ a_3 &= \frac{(-1)^3}{4^3}=-\frac{1}{64} \\ a_4 &= \frac{(-1)^4}{4^4}=\frac{1}{256} \\ a_5 &= \frac{(-1)^5}{4^5}=-\frac{1}{1024} \\ a_8 &= \frac{(-1)^8}{4^8}=\frac{1}{65{,}536} \end{aligned}$$

31.
$$\begin{aligned} a_1 &= \frac{(-1)^{1+1}}{1^2}=\frac{1}{1}=1 \\ a_2 &= \frac{(-1)^{2+1}}{2^2}=-\frac{1}{4} \\ a_3 &= \frac{(-1)^{3+1}}{3^2}=\frac{1}{9} \\ a_4 &= \frac{(-1)^{4+1}}{4^2}=-\frac{1}{16} \\ a_5 &= \frac{(-1)^{5+1}}{5^2}=\frac{1}{25} \\ a_{16} &= \frac{(-1)^{16+1}}{16^2}=-\frac{1}{256} \end{aligned}$$

33. $1+2+3+4+5=15$

35. $3+3+3+3+3=15$

37. $2\left(\frac{1}{3}\right)^1+2\left(\frac{1}{3}\right)^2+2\left(\frac{1}{3}\right)^3+2\left(\frac{1}{3}\right)^4+2\left(\frac{1}{3}\right)^5=\frac{2}{3}+\frac{2}{9}+\frac{2}{27}+\frac{2}{81}+\frac{2}{243}=\frac{242}{243}$

39. $[3(1)-2]+[3(2)-2]+[3(3)-2]+[3(4)-2]+[3(5)-2]=1+4+7+10+13=35$

41.
$$\begin{aligned} a_1 &= 3 \\ a_2 &= 2a_1+1=2(3)+1=7 \\ a_3 &= 2a_2+1=2(7)+1=15 \\ a_4 &= 2a_3+1=2(15)+1=31 \end{aligned}$$

43.
$$\begin{aligned} a_1 &= -4 \\ a_2 &= \tfrac{a_1}{2}=\tfrac{-4}{2}=-2 \\ a_3 &= \tfrac{a_2}{2}=\tfrac{-2}{2}=-1 \\ a_4 &= \tfrac{a_3}{2}=\tfrac{-1}{2}=-\tfrac{1}{2} \end{aligned}$$

45.
$$\begin{aligned} a_1 &= k \\ a_2 &= a_1^2=k^2 \\ a_3 &= a_2^2=\left(k^2\right)^2=k^4 \\ a_4 &= a_3^2=\left(k^4\right)^2=k^8 \end{aligned}$$

47.
$$\begin{aligned} a_1 &= 8 \\ a_2 &= \tfrac{2a_1}{k}=\tfrac{2(8)}{k}=\tfrac{16}{k} \\ a_3 &= \tfrac{2a_2}{k}=\tfrac{2\left(\frac{16}{k}\right)}{k}=\tfrac{32}{k^2} \\ a_4 &= \tfrac{2a_3}{k}=\tfrac{2\left(\frac{32}{k^2}\right)}{k}=\tfrac{64}{k^3} \end{aligned}$$

EXERCISES 8.2

49. alternating

51. not alternating

53. $\sum_{k=1}^{5} 2k = 2\sum_{k=1}^{5} k = 2(1+2+3+4+5) = 2(15) = 30$

55. $\sum_{k=3}^{4} (-2k^2) = -2\sum_{k=3}^{4} k^2 = -2(3^2+4^2) = -2(25) = -50$

57. $\sum_{k=1}^{5} (3k-1) = 3\sum_{k=1}^{5} k - \sum_{k=1}^{5} 1 = 3(1+2+3+4+5) - 5(1) = 3(15) - 5 = 40$

59. $\sum_{k=1}^{1000} \frac{1}{2} = 1000\left(\frac{1}{2}\right) = 500$

61. $\sum_{k=3}^{4} \frac{1}{k} = \frac{1}{3} + \frac{1}{4} = \frac{4}{12} + \frac{3}{12} = \frac{7}{12}$

63.
$$\begin{aligned}\sum_{k=1}^{4} (4k+1)^2 - \sum_{k=1}^{4} (4k-1)^2 &= \sum_{k=1}^{4} \left(16k^2+8k+1\right) - \sum_{k=1}^{4} \left(16k^2-8k+1\right) \\ &= \sum_{k=1}^{4} 16k = 16\sum_{k=1}^{4} k = 16(1+2+3+4) = 16(10) = 160\end{aligned}$$

65.
$$\begin{aligned}\sum_{k=6}^{8} (5k-1)^2 + \sum_{k=6}^{8} (10k-1) &= \sum_{k=6}^{8} \left(25k^2-10k+1\right) + \sum_{k=6}^{8} (10k-1) \\ &= \sum_{k=6}^{8} \left(25k^2\right) = 25\sum_{k=6}^{8} k^2 = 25\left(6^2+7^2+8^2\right) = 3725\end{aligned}$$

67-71. Answers may vary.

73. False. The next term is -216.

75. True.

77. False. $\sum_{k=2}^{1000} 5 = 999(5) = 4995$.

79. False. $\sum_{k=1}^{999} 9k^{99} = 9\sum_{k=1}^{999} k^{99}$.

Exercises 8.3 (page 792)

1. $(n-1)d$

3. infinite

5. $a_n = a + (n-1)d$

7. Arithmetic means

9. 1, 3, 5, 7, 9, 11

EXERCISES 8.3

11.
$$\begin{aligned} a_n &= a + (n-1)d \\ a_3 &= 5 + (3-1)d \\ 2 &= 5 + 2d \\ -3 &= 2d \\ -\frac{3}{2} &= d \Rightarrow 5, \frac{7}{2}, 2, \frac{1}{2}, -1, -\frac{5}{2} \end{aligned}$$

13.
$$\begin{aligned} a_n &= a + (n-1)d \\ a_7 &= a + (7-1)\frac{5}{2} \\ 24 &= a + 6\left(\frac{5}{2}\right) \\ 24 &= a + 15 \\ 9 &= a \Rightarrow 9, \frac{23}{2}, 14, \frac{33}{2}, 19, \frac{43}{2} \end{aligned}$$

15.
$$\begin{aligned} a_n &= a + (n-1)d \\ a_{40} &= 6 + (40-1)8 \\ &= 6 + 39(8) = 6 + 312 = 318 \end{aligned}$$

17.
$$\begin{aligned} a_n &= a + (n-1)d \\ a_6 &= -2 + (6-1)d \\ 28 &= -2 + 5d \\ 30 &= 5d \Rightarrow d = 6 \end{aligned}$$

19.
$$\begin{aligned} a_n &= a + (n-1)d \\ a_{55} &= -8 + (55-1)7 \\ &= -8 + 54(7) = -8 + 378 = 370 \end{aligned}$$

21. 5th term: $14 = a + 4d$ 2nd term: $5 = a + d$

Solve the system: $\begin{cases} a + 4d = 14 \\ a + d = 5 \end{cases}$

$a = 2$, $d = 3$

$a_{15} = 2 + 14(3) = 44$

23.
$$\begin{aligned} a &= 10, a_5 = 20 \\ 20 &= 10 + 4d \\ 10 &= 4d \\ \frac{5}{2} &= d \Rightarrow 10, \boxed{\frac{25}{2}, 15, \frac{35}{2}}, 20 \end{aligned}$$

25.
$$\begin{aligned} a &= -7, a_6 = \frac{2}{3} \\ \frac{2}{3} &= -7 + 5d \\ \frac{23}{3} &= 5d \\ \frac{23}{15} &= d \\ &-7, \boxed{-\frac{82}{15}, -\frac{59}{15}, -\frac{12}{5}, -\frac{13}{15}}, \frac{2}{3} \end{aligned}$$

27. $a = 5, d = 2$

$a_{15} = a + (n-1)d = 5 + 14(2) = 33$

$S_{15} = \dfrac{n(a + a_{15})}{2} = \dfrac{15(5 + 33)}{2} = 285$

29. $a = \dfrac{27}{2}, d = \dfrac{3}{2}$

$a_{20} = a + (n-1)d = \dfrac{27}{2} + 19\left(\dfrac{3}{2}\right) = 42$

$S_{20} = \dfrac{n(a + a_{20})}{2} = \dfrac{20\left(\frac{27}{2} + 42\right)}{2} = 555$

31. $d = \dfrac{1}{2}, a_{25} = 10$

$10 = a + 24\left(\dfrac{1}{2}\right)$

$10 = a + 12 \Rightarrow a = -2$

$a_{30} = a + (n-1)d = -2 + 29\left(\dfrac{1}{2}\right) = \dfrac{25}{2}$

$S_{30} = \dfrac{n(a + a_{30})}{2} = \dfrac{30\left(-2 + \frac{25}{2}\right)}{2} = 157\dfrac{1}{2}$

EXERCISES 8.3

33. $a = 1, d = 1, n = a_{200} = 200; S_{200} = \dfrac{n(a + a_{200})}{2} = \dfrac{200(1 + 200)}{2} = 20{,}100$

35. $a = 180, d = 180, n = 8 - 2 = 6$
$a_6 = a + (n-1)d = 180 + 5(180)$
$= 1080°$

$a = 180, d = 180, n = 12 - 2 = 10$
$a_{10} = a + (n-1)d = 180 + 9(180)$
$= 1800°$

37. $a = 5500, d = -105, n = 49$
Note: $n = 49$ occurs at the beginning of the 49th month, right after the 48th payment has been made
49th term $= a + (n-1)d$
$= 5500 + 48(-105)$
$= \$460$

39. $a = 237{,}500; d = 150{,}000;$
$a_{10} = a + (n-1)d$
$= 237{,}500 + 9(150{,}000)$
$= \$1{,}587{,}500$

41. $a_3 = a + (n-1)d$
$= 16 + 2(32) = 80$ feet

43. $a = 1, d = 1, n = 20, a_{20} = 20$
$S = \dfrac{n(a + a_{20})}{2} = \dfrac{20(1 + 20)}{2} = 210$ logs

45-49. Answers may vary.

51. True.

53. True.

55. True.

57. True.

59. False. There are $S_n = \dfrac{n(a + a_n)}{2} = \dfrac{20(5 + 24)}{2} = 290.$

Exercises 8.4 (page 801)

1. r^{n-1}

3. ar^{n-1}

5. infinite

7. Geometric means

9. 10, 20, 40, 80

11. $-2, -6, -18, -54$

13. $3, 3\sqrt{2}, 6, 6\sqrt{2}$

15. $a_4 = ar^{4-1}$
$54 = 2r^3$
$27 = r^3$
$3 = r \Rightarrow 2, 6, 18, 54$

17. $a = \frac{1}{4}, r = 4;\ a_6 = ar^{6-1} = \left(\frac{1}{4}\right)4^5 = 256$

EXERCISES 8.4

19.
$$a_2 = ar^1 \quad 6 = ar^1 \quad a_3 = ar^2 \quad -18 = ar^2$$
$$ar^2 = -18 \quad ar \cdot r = -18 \quad 6r = -18 \quad r = -3$$
$$r = -3: \; a = \frac{6}{r} = \frac{6}{-3} = -2$$
$$a_5 = ar^4 = -2(-3)^4 = -162$$

21.
$$a_5 = ar^4$$
$$20 = 10r^4$$
$$2 = r^4$$
$\sqrt[4]{2} = r$ (problem specifies positive)

10, $\boxed{10\sqrt[4]{2}, 10\sqrt{2}, 10\sqrt[4]{8}}$, 20

23.
$$a_6 = ar^5$$
$$2048 = 2r^5$$
$$1024 = r^5$$
$$4 = r$$
2, $\boxed{8, 32, 128, 512}$, 2048

25. $a = 4, r = 2, n = 5$
$$S_5 = \frac{a - ar^n}{1 - r} = \frac{4 - 4(2)^5}{1 - 2} = \frac{-124}{-1} = 124$$

27. $a = 2, r = -3, n = 10$
$$S_{10} = \frac{a - ar^n}{1 - r} = \frac{2 - 2(-3)^{10}}{1 - (-3)} = \frac{-118{,}096}{4} = -29{,}524$$

29. $a = 3, r = \frac{3}{2}, n = 6$
$$S_6 = \frac{a - ar^n}{1 - r} = \frac{3 - 3\left(\frac{3}{2}\right)^6}{1 - \frac{3}{2}} = \frac{-\frac{1995}{64}}{-\frac{1}{2}} = \frac{1995}{32}$$

31. $a = 6, r = \frac{2}{3}$
$$S_\infty = \frac{a}{1 - r} = \frac{6}{1 - \frac{2}{3}} = \frac{6}{\frac{1}{3}} = 18$$

33. $a = 12, r = -\frac{1}{2}$
$$S_\infty = \frac{a}{1 - r} = \frac{12}{1 - \left(-\frac{1}{2}\right)} = \frac{12}{\frac{3}{2}} = 8$$

35. $a = \frac{5}{10} = \frac{1}{2}, r = \frac{1}{10}$
$$S_\infty = \frac{a}{1 - r} = \frac{\frac{1}{2}}{1 - \frac{1}{10}} = \frac{\frac{1}{2}}{\frac{9}{10}} = \frac{5}{9}$$

37. $a = \frac{25}{100} = \frac{1}{4}, r = \frac{1}{100}$
$$S_\infty = \frac{a}{1 - r} = \frac{\frac{1}{4}}{1 - \frac{1}{100}} = \frac{\frac{1}{4}}{\frac{99}{100}} = \frac{25}{99}$$

39. $a = 623, r = 1.10$
$$a_9 = ar^8 = 623(1.1)^8$$
≈ 1335 students

professors $= \frac{1335}{60} \approx 22.25$

23 professors will be needed.

EXERCISES 8.4

41. $a = 100, r = 0.6, a_5 = 100(0.6)^4$
$= 12.96$ ft

Down: $a = 100, r = 0.6$

$$S_\infty = \frac{a}{1-r} = \frac{100}{1-0.6} = 250$$

Up: $a = 60, r = 0.6$

$$S_\infty = \frac{a}{1-r} = \frac{60}{1-0.6} = 150$$

Total vertical distance: 400 ft

43. $a = 10, r = 0.95$

$$a_{14} = ar^{13} = 10(0.95)^{13} \approx 5.13 \text{ meters}$$

45. $a = 1000, r = 1 + \dfrac{0.0675}{365}, n = 365$

$$ar^{365} = 1000\left(1 + \frac{0.0675}{365}\right)^{365} \approx \$1069.82$$

The interest will be $69.82.

47. $a = c, r = 0.80, n = 5$

$ar^5 = c(0.80)^5 = 0.32768c$, or about $0.33c$

49. $a = 5 \times 10^9, r = 2$

$n = (3020 - 2000)/30 = 34$

$ar^{34} = 5 \times 10^9(2)^{34} \approx 8.6 \times 10^{19}$

51. $a = 50{,}000;\ r = 1.06, n = 22$

$$ar^{22} = 50{,}000(1.06)^{22} \approx \$180{,}176.87$$

53. $a = 1000, r = 1 + \dfrac{0.07}{4}, n = 40$

$$ar^{40} = 1000\left(1 + \frac{0.07}{4}\right)^{40} \approx \$2001.60$$

55. $a = 1000, r = 1 + \dfrac{0.07}{365}, n = 3650$

$$ar^{3650} = 1000\left(1 + \frac{0.07}{365}\right)^{3650} \approx \$2013.62$$

57. $a = 2000, r = 1 + \dfrac{0.11}{4}, n = 180$

$$ar^{180} = 2000\left(1 + \frac{0.11}{4}\right)^{180} \approx \$264{,}094.58$$

59. $a = 1000, r = 0.8$

$$S_\infty = \frac{a}{1-r} = \frac{1000}{1-0.8} = 5000$$

61. $a = 1, r = 2, n = 64$

$$S = \frac{a - ar^n}{1-r} = \frac{1 - 1(2)^{64}}{1-2} \approx 1.8447 \times 10^{19} \text{ grains}$$

63-67. Answers may vary.

69. no

71. True.

73. False. The nth term is ar^{n-1}.

75. False. The common ratio must be between -1 and 1.

77. True.

Exercises 8.5 (page 808)

1. two

3. $n = k + 1$

5.

$n = 1$

$$5(1) \stackrel{?}{=} \frac{5(1)(1+1)}{2}$$
$$5 = 5$$

$n = 2$

$$5 + 5(2) \stackrel{?}{=} \frac{5(2)(2+1)}{2}$$
$$15 \stackrel{?}{=} \frac{10(3)}{2}$$
$$15 = 15$$

$n = 3$

$$5 + 10 + 5(3) \stackrel{?}{=} \frac{5(3)(3+1)}{2}$$
$$30 \stackrel{?}{=} \frac{15(4)}{2}$$
$$30 = 30$$

$n = 4$

$$5 + 10 + 15 + 5(4) \stackrel{?}{=} \frac{5(4)(4+1)}{2}$$
$$50 \stackrel{?}{=} \frac{20(5)}{2}$$
$$50 = 50$$

7.

$n = 1$

$$3(1) + 4 \stackrel{?}{=} \frac{1(3(1)+11)}{2}$$
$$7 \stackrel{?}{=} \frac{1(14)}{2}$$
$$7 = 7$$

$n = 2$

$$7 + 3(2) + 4 \stackrel{?}{=} \frac{2(3(2)+11)}{2}$$
$$17 \stackrel{?}{=} \frac{2(17)}{2}$$
$$17 = 17$$

$n = 3$

$$7 + 10 + 3(3) + 4 \stackrel{?}{=} \frac{3(3(3)+11)}{2}$$
$$30 \stackrel{?}{=} \frac{3(20)}{2}$$
$$30 = 30$$

$n = 4$

$$7 + 10 + 13 + 3(4) + 4 \stackrel{?}{=} \frac{4(3(4)+11)}{2}$$
$$46 \stackrel{?}{=} \frac{4(23)}{2}$$
$$46 = 46$$

9.

Check $n = 1$: $2 \stackrel{?}{=} 1(1+1)$ True for $n = 1$ $2 = 2$	
Assume for $n = k$:	$2 + 4 + 6 + \cdots + 2k = k(k+1)$
Show for $n = k + 1$:	$\boxed{2 + 4 + 6 + \cdots + 2k} + 2(k+1) = \boxed{k(k+1)} + 2(k+1)$ $2 + 4 + 6 + \cdots + 2(k+1) = k^2 + k + 2k + 2$ $2 + 4 + 6 + \cdots + 2(k+1) = k^2 + 3k + 2$ $2 + 4 + 6 + \cdots + 2(k+1) = (k+1)(k+2)$
Since this is what results when $n = k + 1$ is in the formula, we have shown that the formula works for $n = k + 1$ if it works for $n = k$.	

EXERCISES 8.5

11.

Check $n = 1$: $4(1) - 1 \stackrel{?}{=} 1(2(1) + 1)$ True for $n = 1$

$3 = 3$

Assume for $n = k$ and show for $n = k + 1$:

$$3 + 7 + 11 + \cdots + (4k - 1) = k(2k + 1)$$

$$\boxed{3 + 7 + 11 + \cdots + (4k - 1)} + 4(k + 1) - 1 = \boxed{k(2k + 1)} + 4(k + 1) - 1$$

$$3 + 7 + 11 + \cdots + [4(k + 1) - 1] = 2k^2 + k + 4k + 4 - 1$$

$$3 + 7 + 11 + \cdots + [4(k + 1) - 1] = 2k^2 + 5k + 3$$

$$3 + 7 + 11 + \cdots + [4(k + 1) - 1] = (k + 1)(2k + 3)$$

$$3 + 7 + 11 + \cdots + [4(k + 1) - 1] = (k + 1)(2(k + 1) + 1)$$

Since this is what results when $n = k + 1$ is in the formula, we have shown that the formula works for $n = k + 1$ if it works for $n = k$.

13.

Check $n = 1$: $14 - 4(1) \stackrel{?}{=} 12(1) - 2(1)^2$ True for $n = 1$

$10 = 10$

Assume for $n = k$ and show for $n = k + 1$:

$$10 + 6 + 2 + \cdots + (14 - 4k) = 12k - 2k^2$$

$$\boxed{10 + 6 + 2 + \cdots + (14 - 4k)} + 14 - 4(k + 1) = \boxed{12k - 2k^2} + 14 - 4(k + 1)$$

$$10 + 6 + 2 + \cdots + (14 - 4(k + 1)) = 12k - 2k^2 + 14 - 4k - 4$$

$$10 + 6 + 2 + \cdots + (14 - 4(k + 1)) = 12k + 12 - 2k^2 - 4k - 2$$

$$10 + 6 + 2 + \cdots + (14 - 4(k + 1)) = 12(k + 1) - 2(k^2 + 2k + 1)$$

$$10 + 6 + 2 + \cdots + (14 - 4(k + 1)) = 12(k + 1) - 2(k + 1)^2$$

Since this is what results when $n = k + 1$ is in the formula, we have shown that the formula works for $n = k + 1$ if it works for $n = k$.

EXERCISES 8.5

15. Check $n=1$: $3(1)-1 \stackrel{?}{=} \frac{1(3(1)+1)}{2}$ True for $n=1$

$2=2$

Assume for $n=k$ and show for $n=k+1$:

$$2+5+8+\cdots+(3k-1)=\frac{k(3k+1)}{2}$$

$$\boxed{2+5+8+\cdots+(3k-1)}+3(k+1)-1=\boxed{\frac{k(3k+1)}{2}}+3(k+1)-1$$

$$2+5+8+\cdots+(3(k+1)-1)=\frac{k(3k+1)}{2}+\frac{2\cdot(3(k+1)-1)}{2}$$

$$2+5+8+\cdots+(3(k+1)-1)=\frac{3k^2+k+6k+6-2}{2}$$

$$2+5+8+\cdots+(3(k+1)-1)=\frac{3k^2+7k+4}{2}$$

$$2+5+8+\cdots+(3(k+1)-1)=\frac{(k+1)(3k+4)}{2}$$

$$2+5+8+\cdots+(3(k+1)-1)=\frac{(k+1)(3(k+1)+1)}{2}$$

Since this is what results when $n=k+1$ is in the formula, we have shown that the formula works for $n=k+1$ if it works for $n=k$.

17. Check $n=1$: $1^2 \stackrel{?}{=} \frac{1(1+1)(2(1)+1)}{6}$ True for $n=1$

$1=1$

Assume for $n=k$ and show for $n=k+1$:

$$1^2+2^2+3^2+\cdots+k^2=\frac{k(k+1)(2k+1)}{6}$$

$$\boxed{1^2+2^2+3^2+\cdots+k^2}+(k+1)^2=\boxed{\frac{k(k+1)(2k+1)}{6}}+(k+1)^2$$

$$1^2+2^2+3^2+\cdots+(k+1)^2=\frac{(2k^2+k)(k+1)}{6}+\frac{6(k+1)(k+1)}{6}$$

$$1^2+2^2+3^2+\cdots+(k+1)^2=\frac{(2k^2+k+6(k+1))(k+1)}{6}$$

$$1^2+2^2+3^2+\cdots+(k+1)^2=\frac{(2k^2+7k+6)(k+1)}{6}$$

$$1^2+2^2+3^2+\cdots+(k+1)^2=\frac{(2k+3)(k+2)(k+1)}{6}$$

$$1^2+2^2+3^2+\cdots+(k+1)^2=\frac{(k+1)(k+2)(2k+3)}{6}$$

Since this is what results when $n=k+1$ is in the formula, we have shown that the formula works for $n=k+1$ if it works for $n=k$.

EXERCISES 8.5

19. Check $n = 1$: $\frac{5}{3}(1) - \frac{4}{3} \stackrel{?}{=} 1\left(\frac{5}{6}(1) - \frac{1}{2}\right)$ True for $n = 1$

$\frac{1}{3} = \frac{1}{3}$

Assume for $n = k$ and show for $n = k + 1$:

$$\frac{1}{3} + 2 + \frac{11}{3} + \cdots + \left(\frac{5}{3}k - \frac{4}{3}\right) = k\left(\frac{5}{6}k - \frac{1}{2}\right)$$

$$\boxed{\frac{1}{3} + 2 + \frac{11}{3} + \cdots + \left(\frac{5}{3}k - \frac{4}{3}\right)} + \left(\frac{5}{3}(k+1) - \frac{4}{3}\right) = \boxed{k\left(\frac{5}{6}k - \frac{1}{2}\right)} + \left(\frac{5}{3}(k+1) - \frac{4}{3}\right)$$

$$\frac{1}{3} + 2 + \frac{11}{3} + \cdots + \left(\frac{5}{3}(k+1) - \frac{4}{3}\right) = \frac{5}{6}k^2 - \frac{1}{2}k + \frac{5}{3}k + \frac{5}{3} - \frac{4}{3}$$

$$\frac{1}{3} + 2 + \frac{11}{3} + \cdots + \left(\frac{5}{3}(k+1) - \frac{4}{3}\right) = \frac{5}{6}k^2 + \frac{7}{6}k + \frac{1}{3}$$

$$\frac{1}{3} + 2 + \frac{11}{3} + \cdots + \left(\frac{5}{3}(k+1) - \frac{4}{3}\right) = (k+1)\left(\frac{5}{6}k + \frac{1}{3}\right)$$

$$\frac{1}{3} + 2 + \frac{11}{3} + \cdots + \left(\frac{5}{3}(k+1) - \frac{4}{3}\right) = (k+1)\left(\frac{5}{6}k + \frac{5}{6} - \frac{1}{2}\right)$$

$$\frac{1}{3} + 2 + \frac{11}{3} + \cdots + \left(\frac{5}{3}(k+1) - \frac{4}{3}\right) = (k+1)\left(\frac{5}{6}(k+1) - \frac{1}{2}\right)$$

Since this is what results when $n = k + 1$ is in the formula, we have shown that the formula works for $n = k + 1$ if it works for $n = k$.

21. Check $n = 1$: $\left(\frac{1}{2}\right)^1 \stackrel{?}{=} 1 - \left(\frac{1}{2}\right)^1$ True for $n = 1$

$\frac{1}{2} = \frac{1}{2}$

Assume for $n = k$ and show for $n = k + 1$:

$$\frac{1}{2} + \frac{1}{4} + \frac{1}{8} + \cdots + \left(\frac{1}{2}\right)^k = 1 - \left(\frac{1}{2}\right)^k$$

$$\boxed{\frac{1}{2} + \frac{1}{4} + \frac{1}{8} + \cdots + \left(\frac{1}{2}\right)^k} + \left(\frac{1}{2}\right)^{k+1} = \boxed{1 - \left(\frac{1}{2}\right)^k} + \left(\frac{1}{2}\right)^{k+1}$$

$$\frac{1}{2} + \frac{1}{4} + \frac{1}{8} + \cdots + \left(\frac{1}{2}\right)^{k+1} = 1 - 2\left(\frac{1}{2}\right)\left(\frac{1}{2}\right)^k + \left(\frac{1}{2}\right)^{k+1}$$

$$\frac{1}{2} + \frac{1}{4} + \frac{1}{8} + \cdots + \left(\frac{1}{2}\right)^{k+1} = 1 - 2\left(\frac{1}{2}\right)^{k+1} + \left(\frac{1}{2}\right)^{k+1}$$

$$\frac{1}{2} + \frac{1}{4} + \frac{1}{8} + \cdots + \left(\frac{1}{2}\right)^{k+1} = 1 - \left(\frac{1}{2}\right)^{k+1}$$

Since this is what results when $n = k + 1$ is in the formula, we have shown that the formula works for $n = k + 1$ if it works for $n = k$.

EXERCISES 8.5

23.

Check $n = 1$: $2^{1-1} \stackrel{?}{=} 2^1 - 1$ True for $n = 1$ $1 = 1$
Assume for $n = k$ and show for $n = k + 1$: $2^0 + 2^1 + 2^2 + \cdots + 2^{k-1} = 2^k - 1$ $\boxed{2^0 + 2^1 + 2^2 + \cdots + 2^{k-1}} + 2^k = \boxed{2^k - 1} + 2^k$ $2^0 + 2^1 + 2^2 + \cdots + 2^k = 2 \cdot 2^k - 1$ $2^0 + 2^1 + 2^2 + \cdots + 2^k = 2^{k+1} - 1$
Since this is what results when $n = k + 1$ is in the formula, we have shown that the formula works for $n = k + 1$ if it works for $n = k$.

25.

Check $n = 1$: $x - y$ is a factor of $x^1 - y^1$. True for $n = 1$
Assume for $n = k$ and show for $n = k + 1$: Thus, we assume that $x^k - y^k = (x - y)(\text{SOMETHING})$. $x^{k+1} - y^{k+1} = x^{k+1} - xy^k + xy^k - y^{k+1}$ $= x(x^k - y^k) + y^k(x - y)$ $= x(x - y)(\text{SOMETHING}) + y^k(x - y)$ $= (x - y)[x(\text{SOMETHING}) + y^k]$
We have shown that $x - y$ is a factor of $x^{k+1} - y^{k+1}$ if it is a factor of $x^k - y^k$.

27. The formula is true for $n = 3$, since a triangle has $180° = (3 - 2) \cdot 180°$. Next, assume that a polygon with k sides has an angle sum of $(k - 2) \cdot 180°$. Take a polygon with $k + 1$ sides. Consider two adjacent sides with a common endpoint. Connect the endpoints which are NOT common to both sides. The figure is now a polygon with k sides with a triangle adjacent to it.

$$\boxed{\text{Sum of angles of } (k+1)\text{-sided polygon}} = \boxed{\text{Sum of angles of } k\text{-sided polygon}} + \boxed{\text{Sum of angles of triangle}}$$

$$= (k - 2) \cdot 180° + 180° = (k - 1) \cdot 180° = [(k + 1) - 2] \cdot 180°$$

Thus, the formula works for $n = k + 1$ if it works for $n = k$.

29. Assume for $n = k$ and show for $n = k + 1$:

$$1 + 2 + 3 + \cdots + k = \tfrac{k}{2}(k + 1) + 1$$
$$\boxed{1 + 2 + 3 + \cdots + k} + (k + 1) = \boxed{\tfrac{k}{2}(k + 1) + 1} + (k + 1)$$
$$1 + 2 + 3 + \cdots + (k + 1) = \tfrac{1}{2}k(k + 1) + (k + 1) + 1$$
$$1 + 2 + 3 + \cdots + (k + 1) = (k + 1)\left(\tfrac{1}{2}k + 1\right)$$
$$1 + 2 + 3 + \cdots + (k + 1) = \tfrac{1}{2}(k + 1)(k + 2)$$
$$1 + 2 + 3 + \cdots + (k + 1) = \tfrac{(k+1)}{2}(k + 2)$$

The formula works for $n = k + 1$ if it works for $n = k$. However, the formula does not work for $n = 1$. Thus, the formula does not work for all natural numbers.

EXERCISES 8.5

31.

Check $n = 1$: $7^1 - 1 = 7 - 1 = 6$ True for $n = 1$ Thus, $7^1 - 1$ is divisible by 6.
Assume for $n = k$ and show for $n = k + 1$: $7^k - 1$ is divisible by 6, so $7^k - 1 = 6 \cdot x$, where x is some natural number. Then $7^{k+1} - 1 = 7^k \cdot 7 - 1 = (6x + 1) \cdot 7 - 1 = 42x + 6 = 6(7x + 1)$ Thus, $7^{k+1} - 1$ is divisible by 6.
We have shown that $7^{k+1} - 1$ is divisible by 6 if $7^k - 1$ is divisible by 6.

33.

Check $n = 1$: $1 + r^1 \stackrel{?}{=} \dfrac{1 - r^2}{1 - r}$ True for $n = 1$

$$1 + r \stackrel{?}{=} \frac{(1 + r)(1 - r)}{1 - r}$$

$$1 + r = 1 + r$$

Assume for $n = k$ and show for $n = k + 1$:

$$1 + r + r^2 + r^3 + \cdots + r^k = \frac{1 - r^{k+1}}{1 - r}$$

$$\boxed{1 + r + r^2 + r^3 + \cdots + r^k} + r^{k+1} = \boxed{\frac{1 - r^{k+1}}{1 - r}} + r^{k+1}$$

$$1 + r + r^2 + r^3 + \cdots + r^{k+1} = \frac{1 - r^{k+1} + r^{k+1}(1 - r)}{1 - r}$$

$$1 + r + r^2 + r^3 + \cdots + r^{k+1} = \frac{1 - r^{k+1} + r^{k+1} - r^{k+2}}{1 - r}$$

$$1 + r + r^2 + r^3 + \cdots + r^{k+1} = \frac{1 - r^{(k+1)+1}}{1 - r}$$

Since this is what results when $n = k + 1$ is in the formula, we have shown that the formula works for $n = k + 1$ if it works for $n = k$.

35. **Answers may vary.**

37.

Check $n = 1$: $a^m a^1 = a^m \cdot a = a^{m+1}$, by definition True for $n = 1$
Assume for $n = k$ and show for $n = k + 1$: $a^m a^k = a^{m+k}$ $a^m a^k a = a^{m+k} a$ $a^m a^{k+1} = a^{m+k+1}$ $a^m a^{k+1} = a^{m+(k+1)}$
We have shown that the formula works for $n = k + 1$ if it works for $n = k$.

39. **a.** 1 **b.** 3 **c.** 7 **d.** 15

41. False. It is used for natural numbers.

43. False. Assume it is true for $n = k$.

Exercises 8.6 (page 817)

1. $6 \cdot 4 = 24$ **3.** $\dfrac{n!}{(n-r)!}$ **5.** 1 **7.** $\dfrac{n!}{r!(n-r)!}$

9. 1 **11.** $\dfrac{n!}{a! \cdot b! \cdots}$

13. $P(7,4) = \dfrac{7!}{(7-4)!} = \dfrac{7!}{3!} = \dfrac{7 \cdot 6 \cdot 5 \cdot 4 \cdot 3!}{3!} = 7 \cdot 6 \cdot 5 \cdot 4 = 840$

15. $C(7,4) = \dfrac{7!}{4!(7-4)!} = \dfrac{7!}{4!3!} = \dfrac{7 \cdot 6 \cdot 5 \cdot 4!}{4!3!} = \dfrac{7 \cdot 6 \cdot 5}{3 \cdot 2 \cdot 1} = 35$

17. $P(5,5) = \dfrac{5!}{(5-5)!} = \dfrac{5!}{0!} = \dfrac{5!}{1} = 5 \cdot 4 \cdot 3 \cdot 2 \cdot 1 = 120$

19. $\dbinom{5}{4} = \dfrac{5!}{4!(5-4)!} = \dfrac{5 \cdot 4!}{4!1!} = \dfrac{5}{1} = 5$ **21.** $\dbinom{5}{0} = \dfrac{5!}{0!(5-0)!} = \dfrac{5!}{0!5!} = \dfrac{1}{1} = 1$

23. $P(5,4) \cdot C(5,3) = \dfrac{5!}{(5-4)!} \cdot \dfrac{5!}{3!(5-3)!} = \dfrac{5!}{1!} \cdot \dfrac{5!}{3!2!} = 120 \cdot 10 = 1200$

25. $\dbinom{5}{3}\dbinom{4}{3}\dbinom{3}{3} = \dfrac{5!}{3!(5-3)!} \cdot \dfrac{4!}{3!(4-3)!} \cdot \dfrac{3!}{3!(3-3)!} = 10 \cdot 4 \cdot 1 = 40$

27. $\dbinom{68}{66} = \dfrac{68!}{66!(68-66)!} = \dfrac{68 \cdot 67 \cdot 66!}{66!2!} = \dfrac{68 \cdot 67}{2 \cdot 1} = 2278$

29. $8 \cdot 6 \cdot 3 = 144$ **31.** $8 \cdot 10 \cdot 10 \cdot 10 \cdot 10 \cdot 10 \cdot 10 = 8{,}000{,}000$

33. Consider the e and the r to be a block that cannot be divided, say x. Then the problem becomes finding the number of ways to rearrange the letters in the word $numbx$. This can be done in 5!, or 120 ways. For each of these possibilities, the e and the r could be reversed, doubling the number of possibilities. The answer is $2 \cdot 120$, or 240 ways.

35. The word must appear as $\square\, LU \,\square\square$, where one of the Fs must appear in each box. This can be done in 3!, or 6 ways.

37. $8! = 40{,}320$

39. The line will look like this:

$$\frac{5}{\text{M}}\ \frac{4}{\text{M}}\ \frac{3}{\text{M}}\ \frac{2}{\text{M}}\ \frac{1}{\text{M}}\ \frac{5}{\text{W}}\ \frac{4}{\text{W}}\ \frac{3}{\text{W}}\ \frac{2}{\text{W}}\ \frac{1}{\text{W}}$$

Then there are $5 \cdot 4 \cdot 3 \cdot 2 \cdot 1 \cdot 5 \cdot 4 \cdot 3 \cdot 2 \cdot 1 = 14{,}400$ arrangements.

EXERCISES 8.6

41. $P(30, 3) = \frac{30!}{27!} = 24{,}360$

43. $(8 - 1)! = 7! = 5040$

45. Consider the two people who must sit together as a single person, so that there are 5 "people" who must be arranged in a circle. This can be done in $(5 - 1)! = 4! = 24$ ways. However, the two people who have been seated next to each other could be switched, so that the number of arrangements is doubled. There are $2 \cdot 24 = 48$ possible arrangements.

47. Consider Ella and Eli as a single person and Jaydén and Jackson as a single person, so that there are 5 "people" who must be arranged in a circle. This can be done in $(5 - 1)! = 4! = 24$ ways. However, each group of 2 people could be switched, so that the number of arrangements will equal $2 \cdot 2 \cdot 24$, or 96.

49. $\binom{10}{4} = \frac{10!}{4!6!} = 210$

51. $4! = 24$

53. $7! = 5040$

55. $\frac{6!}{3!2!1!} = 60$

57. $25 \cdot 24 \cdot 9 \cdot 9 \cdot 8 \cdot 7 = 2{,}721{,}600$

59. $\binom{6}{3}\binom{8}{3} = 20 \cdot 56 = 1120$

61. $\binom{12}{4}\binom{10}{3} = 495 \cdot 120 = 59{,}400$

63. $\underset{\text{H}}{\underline{17}} \cdot \underset{\text{W}}{\underline{16}} = 272$

65. $\binom{8}{2} = 28$

67. $\binom{10}{5} = 252$

69. $\binom{30}{5} = 142{,}506$

71. $\binom{12}{10} = 66$

73. For each topping, you have two choices, select or do not select:

$\underset{T_1}{\underline{2}} \quad \underset{T_2}{\underline{2}} \quad \underset{T_3}{\underline{2}} \quad \underset{T_4}{\underline{2}} \quad \underset{T_5}{\underline{2}} \quad \underset{T_6}{\underline{2}} \quad \underset{T_7}{\underline{2}} \quad \underset{T_8}{\underline{2}}$

$2 \cdot 2 \cdot 2 \cdot 2 \cdot 2 \cdot 2 \cdot 2 \cdot 2 = 256$

75. 9th row of triangle: 1 8 28 56 70 56 28 8 1; 4th number in row: $\boxed{56}$

77-79. Answers may vary.

81. $C(n, n) = \frac{n!}{n!(n - n)!} = \frac{n!}{n!0!} = \frac{n!}{n!} = 1$

83. $\binom{n}{n-r} = \frac{n!}{(n - r)!(n - (n - r))!} = \frac{n!}{(n - r)!r!} = \frac{n!}{r!(n - r)!} = \binom{n}{r}$

85. Answers may vary.

87. False. Permutations are used when order matters.

89. True.

91. True.

93. True.

Exercises 8.7 (page 824)

1. experiment

3. $\dfrac{n(E)}{n(S)}$

5. $\{(1, H), (2, H), (3, H), (4, H), (5, H), (6, H), (1, T), (2, T), (3, T), (4, T), (5, T), (6, T)\}$

7. $\{A, B, C, D, E, F, G, H, I, J, K, L, M, N, O, P, Q, R, S, T, U, V, W, X, Y, Z\}$

9. $\dfrac{1}{6}$

11. $\dfrac{4}{6} = \dfrac{2}{3}$

13. $\dfrac{19}{42}$

15. $\dfrac{13}{42}$

17. $\dfrac{3}{8}$

19. $\dfrac{0}{8} = 0$

21. rolls of 4: $\{(1, 3), (2, 2), (3, 1)\}$
Probability $= \dfrac{3}{36} = \dfrac{1}{12}$

23. $\dfrac{\text{\# aces}}{\text{\# cards}} \cdot \dfrac{\text{\# aces}}{\text{\# cards}} = \dfrac{4}{52} \cdot \dfrac{4}{52} = \dfrac{1}{169}$

25. $\dfrac{\text{\# red}}{\text{\# eggs}} = \dfrac{5}{12}$

27. $\dfrac{\text{\# ways to get 13 cards of the same suit}}{\text{\# ways to get 13 cards from the deck of 52}} = \dfrac{4 \cdot \binom{13}{13}}{\binom{52}{13}} = \dfrac{4}{6.350136 \times 10^{11}} \approx 6.3 \times 10^{-12}$

29. impossible $\Rightarrow 0$

31. $\dfrac{\text{\# face cards}}{\text{\# cards in deck}} = \dfrac{12}{52} = \dfrac{3}{13}$

33. $\dfrac{\text{\# ways to get 5 orange}}{\text{\# ways to get 5 cubes}} = \dfrac{\binom{5}{5}}{\binom{6}{5}} = \dfrac{1}{6}$

35. rolls of 11: $(1, 4, 6), (1, 5, 5), (1, 6, 4), (2, 3, 6), (2, 4, 5), (2, 5, 4), (2, 6, 3), (3, 2, 6), (3, 3, 5)$
$(3, 4, 4), (3, 5, 3), (3, 6, 2), (4, 1, 6), (4, 2, 5), (4, 3, 4), (4, 4, 3), (4, 5, 2), (4, 6, 1), (5, 1, 5),$
$(5, 2, 4), (5, 3, 3), (5, 4, 2), (5, 5, 1), (6, 1, 4), (6, 2, 3), (6, 3, 2), (6, 4, 1)$
Probability $= \dfrac{27}{216} = \dfrac{1}{8}$

37. $\dfrac{\binom{5}{3}}{2^5} = \dfrac{10}{32} = \dfrac{5}{16}$

39. $SSSS, SSSF, SSFS, SSFF, SFSS, SFSF, SFFS, SFFF,$
$FSSS, FSSF, FSFS, FSFF, FFSS, FFSF, FFFS, FFFF$

41. $\dfrac{4}{16} = \dfrac{1}{4}$

43. $\dfrac{4}{16} = \dfrac{1}{4}$

45. $\dfrac{16}{16} = 1$

47. $\dfrac{32}{119}$

49. $\dfrac{\binom{8}{4}}{\binom{10}{4}} = \dfrac{70}{210} = \dfrac{1}{3}$

51. $P(A \cap B) = P(A) \cdot P(B|A)$
$= 0.3(0.6) = 0.18$

53. $P(A \cap B) = P(A) \cdot P(B|A)$
$= 0.2(0.7) = 0.14$

55. $P(A \cap B) = P(A) \cdot P(B|A)$
$0.25 = 0.75P(B|A)$
$0.33 \approx P(B|A)$
$33\% \approx P(B|A)$

57-59. Answers may vary.

61. no

63. False. Probabilities are between 0 and 1.

65. False. An impossible event has probability 0.

67. True.

69. False the probability is $\frac{1}{2} \cdot \frac{1}{2} \cdot \frac{1}{2} \cdot \frac{1}{2} = \frac{1}{16}$.

Chapter 8 Review (page 827)

1. $6! = 6 \cdot 5 \cdot 4 \cdot 3 \cdot 2 \cdot 1 = 720$

2. $7! \cdot 0! \cdot 1! \cdot 3! = 5040 \cdot 1 \cdot 1 \cdot 6 = 30{,}240$

3. $\frac{8!}{7!} = \frac{8 \cdot 7!}{7!} = 8$

4. $\frac{5! \cdot 7! \cdot 8!}{6! \cdot 9!} = \frac{5! \cdot 7 \cdot 6! \cdot 8!}{6! \cdot 9 \cdot 8!} = \frac{5! \cdot 7}{9} = \frac{280}{3}$

5. $(x+y)^3 = x^3 + \frac{3!}{1!2!}x^2y + \frac{3!}{2!1!}xy^2 + y^3 = x^3 + 3x^2y + 3xy^2 + y^3$

6. $(p+q)^4 = p^4 + \frac{4!}{1!3!}p^3q + \frac{4!}{2!2!}p^2q^2 + \frac{4!}{3!1!}pq^3 + q^4 = p^4 + 4p^3q + 6p^2q^2 + 4pq^3 + q^4$

7. $(a-b)^5 = a^5 + \frac{5!}{1!4!}a^4(-b) + \frac{5!}{2!3!}a^3(-b)^2 + \frac{5!}{3!2!}a^2(-b)^3 + \frac{5!}{4!1!}a(-b)^4 + (-b)^5$
$= a^5 - 5a^4b + 10a^3b^2 - 10a^2b^3 + 5ab^4 - b^5$

8. $(2a-b)^3 = (2a)^3 + \frac{3!}{1!2!}(2a)^2(-b) + \frac{3!}{2!1!}(2a)(-b)^2 + (-b)^3 = 8a^3 - 12a^2b + 6ab^2 - b^3$

9. The 4th term will involve b^3.
$\frac{8!}{5!3!}a^5b^3 = 56a^5b^3$

10. The 3rd term will involve $(-y)^2$.
$\frac{5!}{3!2!}(2x)^3(-y)^2 = 80x^3y^2$

11. The 7th term will involve $(-y)^6$.
$\frac{9!}{3!6!}x^3(-y)^6 = 84x^3y^6$

12. The 4th term will involve 7^3.
$\frac{6!}{3!3!}(4x)^3 7^3 = 439{,}040x^3$

13. $4^3 - 1 = 63$

14. $\frac{4^2+2}{2} = \frac{18}{2} = 9$

CHAPTER 8 REVIEW

15. $a_1 = 2(1)^2 - 2 = 0;\ a_2 = 2(2)^2 - 2 = 6$

$a_3 = 2(3)^2 - 2 = 16;\ a_4 = 2(4)^2 - 2 = 30$

$a_5 = 2(5)^2 - 2 = 48$

$a_{10} = 2(10)^2 - 2 = 198$

16. $a_1 = -2$

$a_2 = 2a_1^2 = 2(-2)^2 = 8$

$a_3 = 2a_2^2 = 2(8)^2 = 128$

$a_4 = 2a_3^2 = 2(128)^2 = 32{,}768$

17. $$\sum_{k=1}^{4} 3k^2 = 3\sum_{k=1}^{4} k^2 = 3(1^2 + 2^2 + 3^2 + 4^2) = 3(30) = 90$$

18. $$\sum_{k=1}^{10} 6 = 10(6) = 60$$

19. $$\sum_{k=5}^{8}(k^3 + 3k^2) = \sum_{k=5}^{8} k^3 + 3\sum_{k=5}^{8} k^2 = (5^3 + 6^3 + 7^3 + 8^3) + 3(5^2 + 6^2 + 7^2 + 8^2) = 1718$$

20. $$\sum_{k=1}^{30}\left(\frac{3}{2}k - 12\right) - \frac{3}{2}\sum_{k=1}^{30} k = \frac{3}{2}\sum_{k=1}^{30} k - \sum_{k=1}^{30} 12 - \frac{3}{2}\sum_{k=1}^{30} k = -\sum_{k=1}^{30} 12 = -360$$

21. $a = 5, d = 4$

$a_{29} = a + (n-1)d$

$= 5 + (29 - 1)4 = 117$

22. $a = 8, d = 7$

$a_{40} = a + (n-1)d$

$= 8 + (40 - 1)7 = 281$

23. $a = 6, d = -7$

$a_{15} = a + (n-1)d$

$= 6 + (15 - 1)(-7) = -92$

24. $a = \frac{1}{2}, d = -2$

$a_{35} = a + (n-1)d$

$= \frac{1}{2} + (35 - 1)(-2) = -\frac{135}{2}$

25. $a = 2, a_5 = 8$

$8 = 2 + 4d$

$6 = 4d$

$\frac{3}{2} = d \Rightarrow 2, \boxed{\frac{7}{2}, 5, \frac{13}{2}}, 8$

26. $a = 10, a_7 = 100$

$100 = 10 + 6d$

$90 = 6d$

$15 = d \Rightarrow 10, \boxed{25, 40, 55, 70, 85}, 100$

27. $a = 5, d = 4$

$a_{40} = a + (n-1)d = 5 + 39(4) = 161$

$S_{40} = \dfrac{n(a + a_{40})}{2} = \dfrac{40(5 + 161)}{2} = 3320$

28. $a = 8, d = 7$

$a_{40} = a + (n-1)d = 8 + 39(7) = 281$

$S_{40} = \dfrac{n(a + a_{40})}{2} = \dfrac{40(8 + 281)}{2} = 5780$

29. $a = 6, d = -7$

$$a_{40} = a + (n-1)d = 6 + 39(-7) = -267;\ S_{40} = \frac{n(a + a_{40})}{2} = \frac{40(6 - 267)}{2} = -5220$$

30. $a = \frac{1}{2}, d = -2$

$$a_{40} = a + (n-1)d = \frac{1}{2} + 39(-2) = -\frac{155}{2};\ S_{40} = \frac{n(a + a_{40})}{2} = \frac{40\left(\frac{1}{2} - \frac{155}{2}\right)}{2} = -1540$$

31. $a = 81, r = \frac{1}{3}$

$a_{11} = ar^{n-1} = 81\left(\frac{1}{3}\right)^{10} = \frac{1}{729}$

32. $a = 2, r = 3$

$a_9 = ar^{n-1} = 2(3)^8 = 13{,}122$

33. $a = 9, r = \frac{1}{2}$

$a_{15} = ar^{n-1} = 9\left(\frac{1}{2}\right)^{14} = \frac{9}{16{,}384}$

34. $a = 8, r = -\frac{1}{5}$

$a_7 = ar^{n-1} = 8\left(-\frac{1}{5}\right)^6 = \frac{8}{15{,}625}$

35.

$$a_5 = ar^4$$
$$8 = 2r^4$$
$$4 = r^4$$
$$\pm\sqrt[4]{4} = r \Rightarrow r = \pm\sqrt{2}$$

Use $r = +\sqrt{2}$:

$2, \boxed{2\sqrt{2}, 4, 4\sqrt{2}}, 8$

36.

$$a_6 = ar^5$$
$$64 = -2r^5$$
$$-32 = r^5$$
$$-2 = r$$
$$-2, \boxed{4, -8, 16, -32}, 64$$

37.

$$a_3 = ar^2$$
$$64 = 4r^2$$

$16 = r^2 \Rightarrow r = 4$ (problem states positive)

$\Rightarrow 4, \boxed{16}, 64$

38. $a = 81, r = \frac{1}{3}, n = 8$

$$S_8 = \frac{a - ar^n}{1 - r} = \frac{81 - 81\left(\frac{1}{3}\right)^8}{1 - \frac{1}{3}} = \frac{81 - \frac{1}{81}}{\frac{2}{3}} = \frac{3280}{27}$$

39. $a = 2, r = 3, n = 8$

$$S_8 = \frac{a - ar^n}{1 - r} = \frac{2 - 2(3)^8}{1 - 3} = \frac{-13{,}120}{-2} = 6560$$

40. $a = 9, r = \frac{1}{2}, n = 8$

$$S_8 = \frac{a - ar^n}{1 - r} = \frac{9 - 9\left(\frac{1}{2}\right)^8}{1 - \frac{1}{2}} = \frac{\frac{2295}{256}}{\frac{1}{2}} = \frac{2295}{128}$$

41. $a = 8, r = -\frac{1}{5}, n = 8$

$$S_8 = \frac{a - ar^n}{1 - r} = \frac{8 - 8\left(-\frac{1}{5}\right)^8}{1 - \left(-\frac{1}{5}\right)} = \frac{\frac{3{,}124{,}992}{390{,}625}}{\frac{6}{5}} = \frac{520{,}832}{78{,}125}$$

42. $a = \frac{1}{3}, r = 3, n = 8$

$$S_8 = \frac{a - ar^n}{1 - r} = \frac{\frac{1}{3} - \frac{1}{3}(3)^8}{1 - 3} = \frac{-\frac{6560}{3}}{-2} = \frac{3280}{3}$$

43.

$$a_7 = ar^6$$
$$= 2\sqrt{2}\left(\sqrt{2}\right)^6$$
$$= 16\sqrt{2}$$

44. $a = \frac{1}{3}, r = \frac{1}{2}$

$$S_\infty = \frac{a}{1 - r} = \frac{\frac{1}{3}}{1 - \frac{1}{2}} = \frac{\frac{1}{3}}{\frac{1}{2}} = \frac{2}{3}$$

45. $a = \frac{1}{5}, r = -\frac{2}{3}$

$$S_\infty = \frac{a}{1 - r} = \frac{\frac{1}{5}}{1 - \left(-\frac{2}{3}\right)} = \frac{\frac{1}{5}}{\frac{5}{3}} = \frac{3}{25}$$

46. $a = 1, r = \frac{3}{2} > 1 \Rightarrow$ no sum

47. $a = \frac{1}{2}, r = \frac{1}{2}$

$$S_\infty = \frac{a}{1-r} = \frac{\frac{1}{2}}{1-\frac{1}{2}} = \frac{\frac{1}{2}}{\frac{1}{2}} = 1$$

48. $a = \frac{3}{10}, r = \frac{1}{10}$

$$S_\infty = \frac{a}{1-r} = \frac{\frac{3}{10}}{1-\frac{1}{10}} = \frac{\frac{3}{10}}{\frac{9}{10}} = \frac{1}{3}$$

49. $a = \frac{9}{10}, r = \frac{1}{10}$

$$S_\infty = \frac{a}{1-r} = \frac{\frac{9}{10}}{1-\frac{1}{10}} = \frac{\frac{9}{10}}{\frac{9}{10}} = 1$$

50. $a = \frac{17}{100}, r = \frac{1}{100}$

$$S_\infty = \frac{a}{1-r} = \frac{\frac{17}{100}}{1-\frac{1}{100}} = \frac{\frac{17}{100}}{\frac{99}{100}} = \frac{17}{99}$$

51. $a = \frac{45}{100}, r = \frac{1}{100}$

$$S_\infty = \frac{a}{1-r} = \frac{\frac{45}{100}}{1-\frac{1}{100}} = \frac{\frac{45}{100}}{\frac{99}{100}} = \frac{5}{11}$$

52. $a = 3000, r = 1 + \frac{0.0775}{365}, n = 2190$

$$ar^{365} = 3000\left(1 + \frac{0.0775}{365}\right)^{2190} \approx \$4775.81$$

53. $a = 4000, r = 1.05, n = 10$

$ar^{10} = 4000(1.05)^{10} \approx 6516$ in 10 years

$ar^{-5} = 4000(1.05)^{-5} \approx 3134$ 5 years ago

54. $a = 10{,}000;\ r = 0.90;\ ar^{10} = 10{,}000(0.90)^{10} \approx \$3486.78$ in 10 years

55.

$n = 1$

$$1^3 \stackrel{?}{=} \frac{1^2(1+1)^2}{4}$$
$$1 = 1$$

$n = 2$

$$1^3 + 2^3 \stackrel{?}{=} \frac{2^2(2+1)^2}{4}$$
$$9 \stackrel{?}{=} \frac{4(9)}{4}$$
$$9 = 9$$

$n = 3$

$$1^3 + 2^3 + 3^3 \stackrel{?}{=} \frac{3^2(3+1)^2}{4}$$
$$36 \stackrel{?}{=} \frac{9(16)}{4}$$
$$36 = 36$$

$n = 4$

$$1^3 + 2^3 + 3^3 + 4^3 \stackrel{?}{=} \frac{4^2(4+1)^2}{4}$$
$$100 \stackrel{?}{=} \frac{16(25)}{4}$$
$$100 = 100$$

56. Check $n = 1$: $1^3 = \dfrac{1^2(1+1)^2}{4}$ True for $n = 1$

$1 = 1$

Assume for $n = k$ and show for $n = k + 1$:

$$1^3 + 2^3 + 3^3 + \cdots + k^3 = \frac{k^2(k+1)^2}{4}$$

$$\boxed{1^3 + 2^3 + 3^3 + \cdots + k^3} + (k+1)^3 = \boxed{\frac{k^2(k+1)^2}{4}} + (k+1)^3$$

$$1^3 + 2^3 + 3^3 + \cdots + (k+1)^3 = \frac{k^2(k+1)^2 + 4(k+1)^3}{4}$$

$$1^2 + 2^2 + 3^2 + \cdots + (k+1)^2 = \frac{(k+1)^2[k^2 + 4(k+1)]}{4}$$

$$1^2 + 2^2 + 3^2 + \cdots + (k+1)^2 = \frac{(k+1)^2(k+2)^2}{4}$$

Since this is what results when $n = k + 1$ is in the formula, we have shown that the formula works for $n = k + 1$ if it works for $n = k$.

57. $P(8,5) = \dfrac{8!}{(8-5)!} = \dfrac{8!}{3!} = 8 \cdot 7 \cdot 6 \cdot 5 \cdot 4 = 6720$

58. $C(7,4) = \dfrac{7!}{4!(7-4)!} = \dfrac{7!}{4!3!} = \dfrac{7 \cdot 6 \cdot 5 \cdot 4!}{4! \cdot 3 \cdot 2 \cdot 1} = 35$

59. $0! \cdot 1! = 1 \cdot 1 = 1$

60. $P(10,2) \cdot C(10,2) = \dfrac{10!}{8!} \cdot \dfrac{10!}{2!8!} = 90 \cdot 45 = 4050$

61. $P(8,6) \cdot C(8,6) = \dfrac{8!}{2!} \cdot \dfrac{8!}{6!2!} = 20{,}160 \cdot 28 = 564{,}480$

62. $C(8,5) \cdot C(6,2) = \dfrac{8!}{5!3!} \cdot \dfrac{6!}{2!4!} = 56 \cdot 15 = 840$

63. $C(7,5) \cdot P(4,0) = \dfrac{7!}{5!2!} \cdot \dfrac{4!}{4!} = 21 \cdot 1 = 21$

64. $C(12,10) \cdot C(11,0) = \dfrac{12!}{10!2!} \cdot \dfrac{11!}{0!11!} = 66 \cdot 1 = 66$

65. $\dfrac{P(8,5)}{C(8,5)} = \dfrac{6720}{56} = 120$

66. $\dfrac{C(8,5)}{C(13,5)} = \dfrac{56}{1287}$

67. $\dfrac{C(6,3)}{C(10,3)} = \dfrac{20}{120} = \dfrac{1}{6}$

68. $\dfrac{C(13,5)}{C(52,5)} = \dfrac{1287}{2{,}598{,}960} = \dfrac{33}{66{,}640}$

69. Consider each set of two people who must sit together as a single person, so that there are 8 "people" who must be arranged in a circle. This can be done in $(8-1)! = 7! = 5040$ ways. However, each pair seated next to each other could be switched, so that the number of arrangements is multiplied by 4. There are $4 \cdot 5040 = 20{,}160$ possible arrangements.

70. $\dfrac{9!}{2!2!} = 90{,}720$

71.

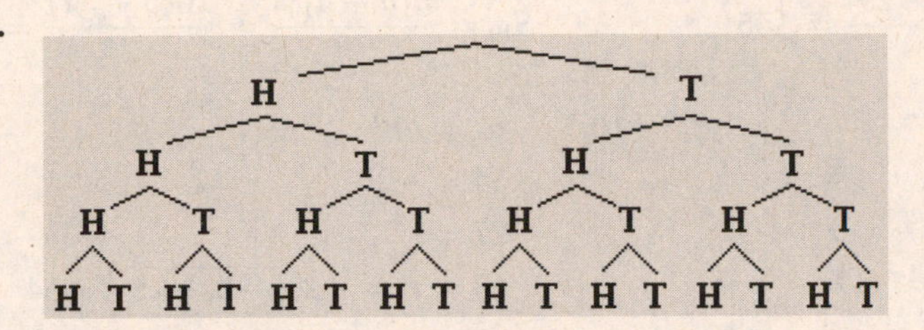

72. $\dbinom{4}{3}\dbinom{4}{2} = 4 \cdot 6 = 24$

73. $\dfrac{\binom{4}{3}\binom{4}{2}}{\binom{52}{5}} = \dfrac{24}{2{,}598{,}960} = \dfrac{1}{108{,}290}$

74. $1 - \dfrac{1}{108{,}290} = \dfrac{108{,}289}{108{,}290}$

75. $\dfrac{\binom{4}{4}\binom{4}{4}\binom{4}{4}\binom{4}{1}}{\binom{52}{13}} \approx \dfrac{4}{6.35 \times 10^{11}} \approx 6.3 \times 10^{-12}$

76. $\dfrac{\binom{8}{3}\binom{6}{2}}{\binom{14}{5}} = \dfrac{840}{2002} = \dfrac{60}{143}$

77. $\dfrac{13 + 13}{52} = \dfrac{1}{2}$

78. $\dfrac{26 + 4 - 2}{52} = \dfrac{7}{13}$

79. $\dfrac{1}{\binom{52}{5}} = \dfrac{1}{2{,}598{,}960}$

80. $\dfrac{4\binom{13}{5}}{\binom{52}{5}} = \dfrac{5148}{2{,}598{,}960} = \dfrac{33}{16{,}660}$

Chapter 8 Test (page 835)

1. $3! \cdot 0! \cdot 4! \cdot 1! = 6 \cdot 1 \cdot 24 \cdot 1 = 144$

2. $\dfrac{2! \cdot 4! \cdot 6! \cdot 8!}{3! \cdot 5! \cdot 7!} = 2! \cdot \dfrac{4!}{3!} \cdot \dfrac{6!}{5!} \cdot \dfrac{8!}{7!}$
$= 2 \cdot 4 \cdot 6 \cdot 8 = 384$

3. The 2nd term will involve $(2y)^1$.
$\dfrac{5!}{4!1!}x^4(2y)^1 = 10x^4y$

4. The 7th term will involve $(-b)^6$.
$\dfrac{8!}{2!6!}(2a)^2(-b)^6 = 112a^2b^6$

5. $\displaystyle\sum_{k=1}^{3}(4k+1) = 4\sum_{k=1}^{3}k + \sum_{k=1}^{3}1 = 4(1+2+3) + 3(1) = 24 + 3 = 27$

6. $\displaystyle\sum_{k=2}^{4}(3k-21) = 3\sum_{k=2}^{4}k - \sum_{k=2}^{4}21 = 3(2+3+4) - 3(21) = 27 - 63 = -36$

7. $a = 2, d = 3$

$a_{10} = a + (n-1)d = 2 + 9(3) = 29$

$S_{10} = \frac{n(a + a_{10})}{2} = \frac{10(2+29)}{2} = 155$

8. $a = 5, d = -4$

$a_{10} = a + (n-1)d = 5 + 9(-4) = -31$

$S_{10} = \frac{n(a + a_{10})}{2} = \frac{10(5-31)}{2} = -130$

9. $a = 4, a_5 = 24$

$24 = 4 + 4d$

$20 = 4d$

$5 = d \Rightarrow 4, \boxed{9, 14, 19}, 24$

10. $a_4 = ar^3$

$-54 = -2r^3$

$27 = r^3$

$3 = r \Rightarrow -2, \boxed{-6, -18}, -54$

11. $a = \frac{1}{4}, r = 2, n = 10$

$$S_{10} = \frac{a - ar^n}{1-r} = \frac{\frac{1}{4} - \frac{1}{4}(2)^{10}}{1-2} = \frac{-\frac{1023}{4}}{-1} = \frac{1023}{4} = 255.75$$

12. $a = 6, r = \frac{1}{3}, n = 10$

$$S_{10} = \frac{a - ar^n}{1-r} = \frac{6 - 6\left(\frac{1}{3}\right)^{10}}{1 - \frac{1}{3}} = \frac{\frac{354,288}{59,049}}{\frac{2}{3}} = \frac{177,144}{19,683} \approx 9$$

13. $a = C;\ r = 0.75$

$ar^3 = C(0.75)^3 \approx \$0.42C$ in 3 years

14. $a = C;\ r = 1.10$

$ar^4 = C(1.10)^4 \approx \$1.46C$ in 4 years

15. Check $n = 1$: $3 = \frac{1}{2}(1)(1+5)$ True for $n = 1$

$3 = 3$

Assume for $n = k$ and show for $n = k + 1$:

$$3 + 4 + 5 + \cdots + (k+2) = \frac{1}{2}k(k+5)$$

$$\boxed{3 + 4 + 5 + \cdots + (k+2)} + ((k+1)+2) = \boxed{\frac{1}{2}k(k+5)} + ((k+1)+2)$$

$$3 + 4 + 5 + \cdots + ((k+1)+2) = \frac{1}{2}k^2 + \frac{7}{2}k + 3$$

$$3 + 4 + 5 + \cdots + ((k+1)+2) = \frac{1}{2}\left(k^2 + 7k + 6\right)$$

$$3 + 4 + 5 + \cdots + ((k+1)+2) = \frac{1}{2}(k+1)(k+6) = \frac{1}{2}(k+1)((k+1)+5)$$

Since this is what results when $n = k + 1$ is in the formula, we have shown that the formula works for $n = k + 1$ if it works for $n = k$.

16. $8 \cdot 10 \cdot 10 \cdot 10 \cdot 10 \cdot 10 = 800,000$

17. $P(7, 2) = \frac{7!}{5!} = 42$

18. $P(4, 4) = \frac{4!}{0!} = 24$

19. $C(8, 2) = \frac{8!}{2!6!} = 28$

20. $C(12, 0) = \frac{12!}{0!12!} = 1$

CHAPTER 8 TEST

21. $4!4! = 576$

22. $(6-1)! = 5! = 120$

23. $\dfrac{5!}{2!} = 60$

24. $\{(H,H,H), (H,H,T), (H,T,H), (H,T,T), (T,H,H), (T,H,T), (T,T,H), (T,T,T)\}$

25. $\dfrac{1}{6}$

26. $\dfrac{4+4}{52} = \dfrac{2}{13}$

27. $\dfrac{\binom{13}{5}}{\binom{52}{5}} = \dfrac{33}{66{,}640}$

28. rolls of 9: $\{(3,6), (4,5), (5,4), (6,3)\}$
Probability $= \dfrac{4}{36} = \dfrac{1}{9}$

29. $\dfrac{\#\text{ blue}}{\#\text{ all}} \cdot \dfrac{\#\text{ blue}}{\#\text{ all}} = \dfrac{30}{50} \cdot \dfrac{29}{49} = \dfrac{87}{245}$

30. $\dfrac{\binom{18}{4}}{\binom{20}{4}} = \dfrac{3060}{4845} = \dfrac{12}{19}$